T0180602

Advances in Intelligent Systems and Computing

Volume 1090

The series "Advances in Intelligent Systems and Computing" contains publications on theory, applications, and design methods of Intelligent Systems and Intelligent Computing. Virtually all disciplines such as engineering, natural sciences, computer and information science, ICT, economics, business, e-commerce, environment, healthcare, life science are covered. The list of topics spans all the areas of modern intelligent systems and computing such as: computational intelligence, soft computing including neural networks, fuzzy systems, evolutionary computing and the fusion of these paradigms, social intelligence, ambient intelligence, computational neuroscience, artificial life, virtual worlds and society, cognitive science and systems, Perception and Vision, DNA and immune based systems, self-organizing and adaptive systems, e-Learning and teaching, human-centered and human-centric computing, recommender systems, intelligent control, robotics and mechatronics including human-machine teaming, knowledge-based paradigms, learning paradigms, machine ethics, intelligent data analysis, knowledge management, intelligent agents, intelligent decision making and support, intelligent network security, trust management, interactive entertainment, Web intelligence and multimedia.

The publications within "Advances in Intelligent Systems and Computing" are primarily proceedings of important conferences, symposia and congresses. They cover significant recent developments in the field, both of a foundational and applicable character. An important characteristic feature of the series is the short publication time and world-wide distribution. This permits a rapid and broad dissemination of research results.

** **Indexing: The books of this series are submitted to ISI Proceedings, EI-Compendex, DBLP, SCOPUS, Google Scholar and Springerlink** **

More information about this series at http://www.springer.com/series/11156

K. Srujan Raju · A. Govardhan ·
B. Padmaja Rani · R. Sridevi ·
M. Ramakrishna Murty

Editors

Proceedings of the Third International Conference on Computational Intelligence and Informatics

ICCII 2018

 Springer

Editors
K. Srujan Raju
CMR Technical Campus
Hyderabad, Telangana, India

B. Padmaja Rani
Department of Computer Science
and Engineering
JNTUH College of Engineering
Hyderabad (Autonomous)
Hyderabad, Telangana, India

M. Ramakrishna Murty
Department of Computer Science
and Engineering
Anil Neerukonda Institute
of Technology & Sciences
Visakhapatnam, Andhra Pradesh, India

A. Govardhan
Department of Computer Science
and Engineering
Jawaharlal Nehru Technological
University Hyderabad
Hyderabad, Telangana, India

R. Sridevi
Department of Computer Science
and Engineering
JNTUH College of Engineering
Hyderabad (Autonomous)
Hyderabad, Telangana, India

ISSN 2194-5357   ISSN 2194-5365   (electronic)
Advances in Intelligent Systems and Computing
ISBN 978-981-15-1479-1   ISBN 978-981-15-1480-7   (eBook)
https://doi.org/10.1007/978-981-15-1480-7

This Springer imprint is published by the registered company Springer Nature Singapore Pte Ltd.
The registered company address is: 152 Beach Road, #21-01/04 Gateway East, Singapore 189721, Singapore

Preface

The Third International Conference on Computational Intelligence and Informatics (ICCII-2018) was hosted by the Department of Computer Science and Engineering, JNTUHCEH, Hyderabad, in association with CSI during December 27–29, 2018. It provided a great platform for researchers from across the world to report, deliberate, and review the latest progress in the cutting-edge research pertaining to computational intelligence and its applications to various engineering fields.

The response to ICCII-2018 was overwhelming with a good number of submissions from different areas relating to computational intelligence and its applications in main tracks. After a rigorous peer review process with the help of program committee members and external reviewers, 79 papers were accepted for publication in this volume of AISC series of Springer.

For the benefit of authors and delegates of the conference, a preconference workshop was held on recent trends in computer science on security, big data and Internet of things. Dr. Kannan Srinathan, IIIT Hyderabad, Dr. Sobhan Babu .Ch, IIT Hyderabad, Prof. A. Govardhan, Rector, JNTUH, Prof. L. Pratap Reddy, JNTUHCEH, have delivered lectures that were very informative.

On the first day of the conference, Sreeganesh Thottempudi, Heidelberg University, and Dr. Hari Eppanapally, Advisory Board Member, Rutgers University, have delivered keynote speeches on December 28, 2018. On the second day of the conference, Dr. Anthony Lolas, School of Information Systems, Walden University, and Dr. Raghu Korrapati, Commissioner for Higher Education (Former), USA, have delivered keynote speeches on December 28, 2018.

We take this opportunity to thank all the speakers and session chairs for their excellent support in making ICCII-2018 a grand success. The quality of a refereed volume depends mainly on the expertise and dedication of the reviewers. We are indebted to the program committee members and external reviewers who not only produced excellent reviews but also reviewed in the stipulated period of time without any delay. We would also like to thank CSI, Hyderabad, for coming forward to support us in organizing this mega-convention. We thank special session chairs Dr. Tanupriya Choudhury and Dr. Praveen of Amity University, Noida.

We express our heartfelt thanks to our Chief Patrons, Prof. A. Venugopal Reddy, Vice Chancellor, JNTUHCEH, Prof. A. Govardhan, Rector, JNTUH, Prof. N. Yadaiah, Registrar, JNTUHCEH, faculty and administrative staff of JNTUHCEH for their continuous support during the course of the convention.

We would also like to thank the authors and participants of this convention, who have considered the convention above all hardships. Finally, we would like to thank all the volunteers who spent tireless efforts in meeting the deadlines and arranging every detail to make sure that the convention ran smoothly. All the efforts are worth and would please us all, if the readers of this proceedings and participants of this convention found the papers and event inspiring and enjoyable. Our sincere thanks to the press, print, and electronic media for their excellent coverage of this convention.

Hyderabad, India	K. Srujan Raju
Hyderabad, India	A. Govardhan
Hyderabad, India	B. Padmaja Rani
Hyderabad, India	R. Sridevi
Visakhapatnam, India	M. Ramakrishna Murty

Contents

Storing Live Sensor Data to the Platforms of Internet of Things (IoT) Using Arduino and Associated Microchips 1
S. Pradeep and Yogesh Kumar Sharma

Comparative Study of Performance of Tabulation and Partition Method for Minimization of DFA 17
Anusha Kolan, K. S. S. Sreevani and H. Jayasree

Click-Through Rate Prediction Using Decision Tree 29
Anusha Kolan, Dasika Moukthika, K. S. S. Sreevani and H. Jayasree

A New Framework Approach to Predict the Wine Dataset Using Cluster Algorithm 39
D. Bhagyalaxmi, R. Manjula and V. Ramanababu

Study and Ranking of Vulnerabilities in the Indian Mobile Banking Applications Using Static Analysis and Bayes Classification 49
Srinadh Swamy Majeti, Faheem Habib, B. Janet and N. P. Dhavale

Analysis of Object Identification Using Quadrature Bank Filter in Wavelet Transforms 65
C. Berin Jones and G. G. Bremiga

Experimental Analysis of the Changes in Speech while Normal Speaking, Walking, Running, and Eating 85
Sakil Ansari, Sanjeev K. Mittal and V. Kamakshi Prasad

UIP—A Smart Web Application to Manage Network Environments 97
T. P. Ezhilarasi, G. Dilip, T. P. Latchoumi and K. Balamurugan

Energy Distribution in a Smart Grid with Load Weight and Time Zone 109
Boddu Rama Devi and K. Srujan Raju

Matrix Approach to Perform Dependent Failure Analysis in Compliance with Functional Safety Standards 123
Gadila Prashanth Reddy and Rangaiah Leburu

A Cloud-Based Privacy-Preserving e-Healthcare System Using Particle Swarm Optimization 133
M. Swathi and K. C. Sreedhar

Improved Approach to Extract Knowledge from Unstructured Data Using Applied Natural Language Processing Techniques 145
U. Mahender, M. Kumara Swamy, Hafeezuddin Shaik and Sheo Kumar

Word Sense Disambiguation in Telugu Language Using Knowledge-Based Approach 153
Neeraja Koppula, B. Padmaja Rani and Koppula Srinivas Rao

A Study of Digital Banking: Security Issues and Challenges 163
B. Vishnuvardhan, B. Manjula and R. Lakshman Naik

Encoding Approach for Intrusion Detection Using PCA and KNN Classifier 187
Nerella Sameera and M. Shashi

Materializing Block Chain Technology to Maintain Digital Ledger of Land Records 201
B. V. Satish Babu and K. Suresh Babu

Review of Optimization Methods of Medical Image Segmentation 213
Thuzar Khin, K. Srujan Raju, G. R. Sinha, Kyi Kyi Khaing and Tin Mar Kyi

Automatic Temperature Control System Using Arduino 219
Kyi Kyi Khaing, K. Srujan Raju, G. R. Sinha and Wit Yee Swe

Performance Analysis of Fingerprint Recognition Using Machine Learning Algorithms 227
Akshay Velapure and Rajendra Talware

A Survey on Digital Forensics Phases, Tools and Challenges 237
Sheena Mohammmed and R. Sridevi

A Brief Survey on Blockchain Technology 249
Vemula Harish and R. Sridevi

Energy-Efficient based Multi-path Routing with Static Sink using Fuzzy Logic Controller in Wireless Sensor Networks 259
Subba Reddy Chavva and Ravi Sankar Sangam

MST Parser for Telugu Language 271
G. Nagaraju, N. Mangathayaru and B. Padmaja Rani

Design of Low-Power Vedic Multipliers Using Pipelining Technology 281
Ansiya Eshack and S. Krishnakumar

Needleman–Wunsch Algorithm Using Multi-threading Approach 289
Sai Reetika Perumalla and Hemalatha Eedi

Emotion-Based Extraction, Classification and Prediction of the Audio Data 301
Anusha Potluri, Ravi Guguloth and Chaitanya Muppala

Web Service Classification and Prediction Using Rule-Based Approach with Recommendations for Quality Improvements 311
M. Swami Das, A. Govardhan and D. Vijaya Lakshmi

Location-Based Alert System Using Twitter Analytics 325
C. S. Lifna and M. Vijayalakshmi

Diagnosis of Diabetes Using Clinical Decision Support System 337
N. Managathayaru, B. Mathura Bai, G. Sunil, G. Hanisha Durga,
C. Anjani Varma, V. Sai Sarath and J. Sai Sandeep

Analysis of Fuel Consumption Characteristics: Insights from the Indian Human Development Survey Using Machine Learning Techniques 349
K. Shyam Sundar, Sangita Khare, Deepa Gupta and Amalendu Jyotishi

Efficient Predictions on Asymmetrical Financial Data Using Ensemble Random Forests 361
Chaitanya Muppala, Sujatha Dandu and Anusha Potluri

Segmentation of Soft Tissues from MRI Brain Images Using Optimized KPCM Clustering Via Level Set Formulation 373
Kama Ramudu and Tummala Ranga Babu

Quicksort Algorithm—An Empirical Study 387
Gampa Rahul, Polamuri Sandeep and Y. L. Malathi Latha

Design an Improved Linked Clustering Algorithm for Spatial Data Mining 403
K. Lakshmaiah, S. Murali Krishna and B. Eswara Reddy

An Efficient Symmetric Key-Based Lightweight Fully Homomorphic Encryption Scheme 417
V. Biksham and D. Vasumathi

MLID: Machine Learning-Based Intrusion Detection from Network Transactions of MEMS Integrated Diversified IoT 427
Ravinder Korani and P. Chandra Sekhar Reddy

A Survey on Deceptive Phishing Attacks in Social Networking Environments 443
Mohammed Mahmood Ali, Mohd S. Qaseem and Md Ateeq Ur Rahman

Implementation of Spatial Images Using Rough Set-Based Classification Techniques 453
D. N. Vasundhara and M. Seetha

Multi-objective Optimization of Composing Tasks from Distributed Workflows in Cloud Computing Networks 467
V. Murali Mohan and K. V. V. Satyanarayana

IP Traceback Through Modified Probabilistic Packet Marking Algorithm Using Record Route 481
Y. Bhavani, V. Janaki and R. Sridevi

Necessity of Fourth Factor Authentication with Multiple Variations as Enhanced User Authentication Technique 491
K. Sharmila and V. Janaki

Performance Analysis of Feature Extraction Methods Based on Genetic Expression for Clustering Video Dataset 501
D. Manju, M. Seetha and P. Sammulal

Identification of Cryptographic Algorithms Using Clustering Techniques 513
Vikas Tiwari, K. V. Pradeepthi and Ashutosh Saxena

A Framework for Evaluating the Quality of Academic Websites 523
Sairam Vakkalanka, Reddi Prasadu, V. V. S. Sasank and A. Surekha

A Survey on Analysis of User Behavior on Digital Market by Mining Clickstream Data 535
Praveen Kumar Padigela and R. Suguna

Optimal Resource Allocation in OFDMA-LTE System to Mitigate Interference Using GA rule-based Mostly HBCCS Technique 547
Kethavath Narender and C. Puttamadappa

Review of Techniques for Automatic Text Summarization 557
B. Shiva Prakash, K. V. Sanjeev, Ramesh Prakash, K. Chandrasekaran, M. V. Rathnamma and V. Venkata Ramana

Data Mining Task Optimization with Soft Computing Approach 567
Lokesh Gagnani and Kalpesh Wandra

Dog Breed Classification Using Transfer Learning 579
Rishabh Jain, Arjeeta Singh, Rishabh Jain and Praveen Kumar

A Framework for Dynamic Access Control System for Cloud Federations Using Blockchain 591
Shaik Raza Sikander and R. Sridevi

ESADSA: Enhanced Self-adaptive Dynamic Software Architecture 601
Sridhar Gummalla, G. Venkateswara Rao and G. V. Swamy

Detection of Parkinson's Disease Through Speech Signatures 619
Jinu James, Shrinidhi Kulkarni, Neenu George, Sneha Parsewar,
Revati Shriram and Mrugali Bhat

Rough Set-Based Classification of Audio Data 627
T. Prathima, A. Govardhan and Y. Ramadevi

Workload Assessment Based on Physiological Parameters 639
Tejaswini Dendage, Vaidehi Deoskar, Pooja Kulkarni, Revati Shriram
and Mrugali Bhat

Realistic Handwriting Generation Using Recurrent Neural Networks and Long Short-Term Networks 651
Suraj Bodapati, Sneha Reddy and Sugamya Katta

Decentralized Framework for Record-Keeping System in Government Using Hyperledger Fabric 663
S. Devidas, N. Rukma Rekha and Y. V. Subba Rao

ECC-Based Secure Group Communication in Energy-Efficient Unequal Clustered WSN (EEUC-ECC) 673
G. Raja Vikram, Addepalli V. N. Krishna and K Shahu Chatrapati

Performance Comparison of Random Forest Classifier and Convolution Neural Network in Predicting Heart Diseases 683
R. P. Ram Kumar and Sanjeeva Polepaka

Food Consumption Monitoring and Tracking in Household Using Smart Container 693
Y. Bevish Jinila, V. Rajalakshmi, L. Mary Gladence and V. Maria Anu

Accuracy Comparison of Classification Techniques for Mouse Dynamics-Based Biometric CaRP 701
Sushama Kulkarni and Hanmant Fadewar

Implementation of Children Activity Tracking System Based on Internet of Things 713
M. Naga Sravani and Samit Kumar Ghosh

Data Mining: Min–Max Normalization Based Data Perturbation Technique for Privacy Preservation 723
Ajmeera Kiran and D. Vasumathi

Role and Impact of Wearables in IoT Healthcare 735
Kamalpreet Singh, Keshav Kaushik, Ahatsham and Vivek Shahare

A Survey of IDS Techniques in MANETs Using Machine-Learning 743
Mohammed Shabaz Hussain and Khaleel Ur Rahman Khan

**Hand Gesture Recognition Based on Saliency and Foveation Features
Using Convolutional Neural Network** 753
Earnest Paul Ijjina

**Human Fall Detection Using Temporal Templates and Convolutional
Neural Networks** 763
Earnest Paul Ijjina

**Action Recognition Using Motion History Information
and Convolutional Neural Networks** 773
Earnest Paul Ijjina

A Review of Various Mechanisms for Botnets Detection 781
Rishikesh Sharma and Abha Thakral

**Malicious URL Classification Using Machine Learning Algorithms
and Comparative Analysis** 791
Anshuman Sharma and Abha Thakral

Optimization in Wireless Sensor Network Using Soft Computing 801
Shruti Gupta, Ajay Rana and Vineet Kansal

**Analyzing Effect of Political Situations on Sensex-Nifty-Rupee—A
Study Based on Election Results** 811
N. Deepika and P. Victer Paul

**Study and Analysis of Modified Mean Shift Method and Kalman Filter
for Moving Object Detection and Tracking** 821
S. Pallavi, K. Ramya Laxmi, N. Ramya and Rohit Raja

**Image Registration and Rectification Using Background Subtraction
Method for Information Security to Justify Cloning Mechanism
Using High-End Computing Techniques** 829
Raj Kumar Patra, Rohit Raja, Tilendra Shishir Sinha
and Md Rashid Mahmood

**Analyzing Heterogeneous Satellite Images for Detecting Flood
Affected Area of Kerala** 839
R. Jeberson Retna Raj and Senduru Srinivasulu

**Action Recognition in Sports Videos Using Stacked Auto Encoder
and HOG3D Features** 849
Earnest Paul Ijjina

An Analytical Study on Gesture Recognition Technology 857
Poorvika Singh Negi and Praveen Kumar

**Artificial Intelligence Approach to Legal Reasoning Evolving 3D
Morphology and Behavior by Computational Artificial Intelligence** 869
Deepakshi Gupta and Shilpi Sharma

Data Mining in Healthcare and Predicting Obesity 877
Anant Joshi, Tanupriya Choudhury, A. Sai Sabitha and K. Srujan Raju

**An Android-Based Mobile Application to Help Alzheimer's
Patients** 889
Sarita, Saurabh Mukherjee and Tanupriya Choudhury

Author Index 905

About the Editors

Dr. K. Srujan Raju is the Professor and Head, Department of CSE, CMR Technical Campus, Hyderabad, India. He has published several papers in refereed international conferences and peer-reviewed journals, and also, he was in the editorial board of Springer AISC, LAIS, LNNS, and other reputed journals. In addition to this, he has served as reviewer for many indexed national and international journals. Prof. Raju is awarded with Significant Contributor, Active Young Member Awards by CSI, authored 4 textbooks, and filed 7 patents and 1 copyright so far.

Dr. A. Govardhan is presently a Professor of CSE, Rector, and Executive Council Member of JNTUH. He served and held several academic and administrative positions. He is the recipient of 31 international and national awards for outstanding services, achievements, contributions, meritorious services, outstanding performance, and remarkable role in the fields of education and service to the nation. He is a Chairman and Member of Boards of Studies of various universities.

Dr. B. Padmaja Rani is working as a Professor in CSE, JNTUHCEH, Hyderabad. She is having 15+ years of professional and 10+ years of research experience. Her areas of interest include Information Retrieval, Natural Language Processing, Information Security, Data Mining, Big Data, and Cloud Computing. Around 50 journals are other credits and she guided five Ph.D. students. She served as Head of the Department and conducted the 1st and 2nd International Conference on Computational Intelligence and Informatics(ICCII-2016, ICCII-2017) organized by the Department of CSE, JNTUHCEH, as one of the conveners.

Dr. R. Sridevi is a Professor in CSE and presently serves as Head of the Department, JNTUHCEH. She served as Addl. Controller of Exams, JNTU, for 4 Years, having 18 years of teaching experience. Her interest includes Network

Security & Cryptography, Steganography, Computer Networks, and Data Structures. She has published about 40 papers at national and international journals and conferences and got Best Paper Award. She is a Life Member of ISTE and IETE.

Dr. M. Ramakrishna Murty is working as an Associate Professor at ANITS, Visakhapatnam. His areas of interest include Data Mining, Databases, Big DataSoft Computing/Computational Intelligence, Machine Learning, Mobile Computing, Computer Networks, Artificial Intelligence, Object-Oriented Programming, and Programming languages. He is a Life Member of ISTE, CSI, & IE. He has 10 international journals, 10 international conferences, 1 national journal, and 3 national conference publications on his credit.

Storing Live Sensor Data to the Platforms of Internet of Things (IoT) Using Arduino and Associated Microchips

S. Pradeep and Yogesh Kumar Sharma

Abstract Internet of Things (IoT) is one of the prominent domains of research in the segment of wireless technologies whereby the smart objects are connected with each other using wireless signals. Despite the enormous areas of research in this stream, the storage and evaluation of live sensor data are widely required so that further analytics on the sensor data can be done. In this research manuscript, the usage of Arduino is presented that is one of the key open-source hardware platforms for programming with Internet of Things (IoT). In open-source hardware platforms of Arduino and NodeMCU, the direct connectivity with the real clouds can be done and is presented in this manuscript. There exist number of open platforms of IoT for logging of data; this manuscript presents the use of ThingSpeak and Carriots so that analytics patterns of sensor data can be implemented effectively. This paper presents the assorted aspects of open-source hardware of Arduino that can be used for different sensor-based applications.

Keywords Arduino open-source hardware · NodeMCU · Internet of Things (IoT) · Storing live sensor data to clouds

1 Introduction

The concept of open-source software is very famous and widely adopted throughout the world but the domain of open-source hardware (OSH) [1] is still under research. Open-source hardware or sometimes called as free and open-source hardware (FOSH) refers to the hardware (firmware, chips, motherboard, sensors) [2] that is available for programming and customization without any issues of proprietary license and distribution. The devices used in electronics and communication engineering are commonly used as open-source hardware in which chips, sensors, and boards are programmed on-demand as per the requirements. In case of

S. Pradeep (✉) · Y. K. Sharma
Department of CSE, JJT University, Jhunjhunu, Rajasthan, India
e-mail: pradeep.sunkari87@gmail.com

© Springer Nature Singapore Pte Ltd. 2020
K. S. Raju et al. (eds.), *Proceedings of the Third International Conference on Computational Intelligence and Informatics*, Advances in Intelligent Systems and Computing 1090, https://doi.org/10.1007/978-981-15-1480-7_1

Fig. 1 Logo of open-source
hardware by open-source
hardware association
(OSHWA)

free and open-source hardware, there is complete flexibility to program the behavior
of motherboard or chipboard as per the requirements [3] (Fig. 1).

2 Types of Open-Source Hardware

There exist different types of open-source hardware which are associated with the
development and programming of electronic boards so that specific and on-demand
applications can be deployed in the integrated circuits on motherboard.

Following are the implementation domains of open-source hardware

- **Computers**: In desktop or laptop computing devices, huge research is going on
  in the customization of motherboard with real-time sensor devices so that live
  signals can be fetched in computers. These live signals may be related to
  satellite data, global positioning system (GPS), temperature, humidity, environmental factors, ultrasonic signals, and many others.
- **Electronics**: It is the key domain whereby the electronics boards are customized
  using programming languages and scripts.
- **Mechatronics**: Merger of mechanical and electronics engineering.

3 Open-Source Hardware in Global Market

Nowadays, the open-source hardware is in use for multiple domains so that the
custom hardware can be programmed with higher degree of performance and
accuracy using programming scripts of C, C++, Python, Java, and many others.

Following are the open-source hardware platforms under assorted domains of
computing, renewable energy, automation, bio-informatics, biotechnology, robotics, telephony, communication, camera, radio, and many others.

Audio Electronics: Monome 40 h, Neuros Digital Audio Computer, Arduinome,
MIDIbox
Telephony: Openmoko, OpenBTS, OsmoBTS, Project Ara, PiPhone and
ZeroPhone, Telecom Infra Project, IP04 IP-PBX

Video electronics: Milkymist One, Neuros OSD
Amateur radio: Homebrew D-STAR Radio
Networking: NetFPGA
Wireless networking: OpenPicus, Sun SPOT, USRP, PowWow Power Optimized Hardware and Software FrameWork for Wireless Motes, Twibright RONJA, SatNOGS
Cameras: AXIOM, Elphel
Computer systems: Arduino, NodeMCU, Chumby, Libre Computer Project, Netduino, CUBIT, Turris Omnia, Parallax Propeller, UDOO, SparkFun Electronics, Parallella, Turris Omnia
Gaming: PocketSprite
Peripherals: Nitrokey
Robotics: ArduCopter, Orb swarm, ICub, OpenRAVE, IOIO, multiplo, Tinkerforge, e-puck mobile robot, Spykee, OpenROV, RobotCub, Thymio
Environmental: Open Source Ecology, Air Quality Egg
Renewable energy: Wind turbines
Lighting and LED: LED Throwies
Architecture and Design: Opendesk, WikiHouse, OpenStructures
Engine control units: SECU-3
Electric vehicle chargers: OpenEVSE
3D printers and scanners: LulzBot, RepRap project
Other hardware: OpenBuilds V-Slot, Multimachine, Open Source Ecology's Global Village Construction Set (GVCS), Lasersaur
Medical devices: Open Prosthetics Project
Scientific hardware: OpenBCI-EEG amplifier, Open-Source Lab
Satellite: UPSat
Open-Source Vehicles: Rally Fighte, Riversimple Urban Car, C, mm, n, LifeTrac tractor, OSCav, SGT01 from Wikispeed, Google Community Vehicle, eCorolla, Luka EV, OSVehicle, OSVehicle Tabby, OpenXC, Wikispeed, OScar.

4 Arduino as an Open-Source Hardware and Electronics Platform for Real-Time Applications

Arduino [1, 4] is available in many variants depending upon the type of application where it is required to be launched. In each of the following categories, there are different boards available which can be programmed for specific application (Fig. 2).

The classes of Arduino boards for programming and hardware customization include the following:

- beginners;
- enhanced and elevated features;

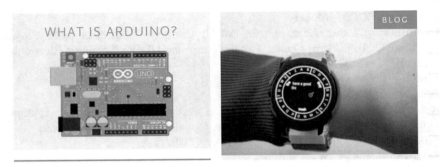

Fig. 2 Arduino board with its use in wearable devices. *Source* Official portal of Arduino

- Internet of Things (IoT);
- education;
- wearable devices.

Arduino UNO [5, 6] is one of the widely used microcontroller boards for the beginners. This board can be easily programmed using Arduino IDE available on the URL: https://www.arduino.cc/en/Main/Software. The Arduino IDE Software is available for multiple platforms including Windows, Mac OSX, and Linux. C and C ++ are the base programming languages which are used to program and customize the behavior of Arduino microcontroller board. Arduino IDE can be downloaded without any cost along with the live updates of the software.

5 Preparing Arduino for Internet of Things (IoT) Applications

The connection of Arduino with live platforms of Internet of Things (IoT) can be created with the use of Wi-Fi or similar network-based chips. ESP8266 is very popular device for communication with Internet that can be easily interfaced with Arduino microcontroller board [7]. ESP8266 is the Wi-Fi chip that is having very low cost and can be directly connected with Arduino board so that its communication with real cloud can be done (Fig. 3).

Fig. 3 ESP8266 microchip
with Wi-Fi connectivity

6 NodeMCU: Open-Source IoT-Based Hardware Platform with in-Built Wi-Fi

To communicate and transfer the data to real cloud, there is another chip titled NodeMCU that can directly communicate with live servers [8, 9]. NodeMCU integrates the firmware with Wi-Fi connectivity in-built so that direct interaction with network can be done without physical wiring using jumper wires. It is Arduino like board that is used for interfacing with real clouds so that the live signals or data can be logged or stored for predictive analysis (Fig. 4).

Fig. 4 NodeMCU firmware for Internet of Things (IoT)

7 Storing Sensor Data To IoT Platforms

A number of IoT platforms are available which can be used for the storage of real-time streaming sensor data from Arduino and similar boards. Once the circuit is created using Arduino, the signals can be transmitted and stored directly on IoT platforms. These open-source IoT platforms can be used for storage, processing, prediction, and visualization of data recorded from the sensors on Arduino board.

Following are few prominent IoT platforms which can be used for device management and visualization of data:

- ThingSpeak: thingspeak.com;
- Carriots: carriots.com;
- KAA: kaaproject.org;
- ThingsBoard: thingsboard.io;
- MainFlux: mainflux.com;
- Thinger: thinger.io;
- DeviceHive: devicehive.com;
  and many others.

8 Transferring Data to ThingSpeak

ThingSpeak is one of the widely used open-source IoT platforms for collection, storage, and visualization of real-time sensor data fetched from Arduino, Raspberry Pi, BeagleBone Black, and similar boards. By this implementation, the live streaming data on cloud can be preserved using channels in ThingSpeak [10, 11].

To work with ThingSpeak, a free account can be created. After successful sign-up on ThingSpeak, a channel is created on the dashboard and a "Write Key" is generated to be used in the code of Arduino.

The key features of ThingSpeak include the following:

- configuration of devices for sending data to cloud-based environment;
- aggregation of sensor data from different devices and motherboards connected with sensors;
- visualization of real-time sensor data including historical records;
- preprocessing and evaluation of sensor data for predictive analysis and knowledge discovery (Fig. 5).

Following is the source code that can be customized and executed on Arduino IDE so that real-time data can be fetched on board and then transmission to ThingSpeak can be done. For simplicity, the execution of code is done for NodeMCU hardware which is having in-built Wi-Fi for sending the data to IoT platform (Fig. 6).

Fig. 5 ThingSpeak: open-source platform for Internet of Things (IoT)

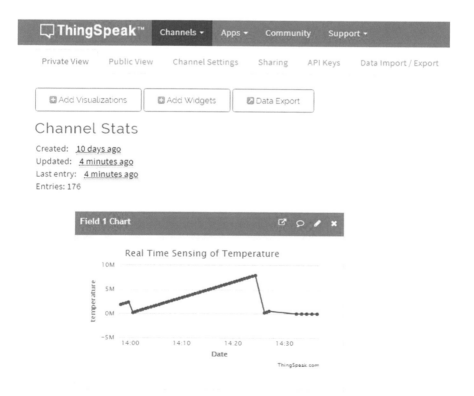

Fig. 6 Visualization of recorded data on ThingSpeak

```
#include <ESP8266WiFi.h>
const char* mynetwork    = "<Wi-Fi Network Name>";
const char* mypassword = "<Wi-Fi Network Mypassword>";
const char* thingspeakhost = "api.thingspeak.com";
char tsaddress[] = "api.thingspeak.com";
String ThingSpeakAPI = "<ThingSpeak Write Key>";
const int UpdateInterval = 30 * 1000;
long ConnectionTime = 0;
boolean connectedlast = false;
int fcounter = 0;
float value = 100;
WiFiClient client;
void setup()
{
  Serial.begin(9600);
  startEthernet();
}
void loop()
{
  String RecordedAnalogInput0 = String(value++, DEC);
  if (client.available())
  {
    char c = client.read();
  }
  if (!client.connected() && connectedlast)
  {
    client.stop();
  }
  if(!client.connected() && (millis() - ConnectionTime > UpdateInterval))
  {
    updateThingSpeak("field1="+RecordedAnalogInput0);
  }
  if (fcounter > 3 ) {startEthernet();}
  connectedlast = client.connected();
}
```

```
void updateThingSpeak(String tsData)
{
  if (client.connect(tsaddress, 80))
  {
    client.print("POST /update HTTP/1.1\n");
    client.print("Thingspeakhost: api.thingspeak.com\n");
    client.print("Connection: close\n");
    client.print("X-THINGSPEAKAPIKEY: "+ThingSpeakAPI+"\n");
    client.print("Content-Type: application/x-www-form-urlencoded\n");
    client.print("Content-Length: ");
    client.print(tsData.length());
    client.print("\n\n");
    client.print(tsData);
    ConnectionTime = millis();
    if (client.connected())
    {
      Serial.println("Connecting to ThingSpeak...");
      Serial.println();
      fcounter = 0;
    }
    else
    {
      fcounter++;
      Serial.println("Connection to ThingSpeak failed ("+String(fcounter, DEC)+")");
      Serial.println();
    }
  }
  else
  {
    fcounter++;
    Serial.println("Connection to ThingSpeak Failed ("+String(fcounter, DEC)+")");
    Serial.println();
    ConnectionTime = millis();
  }
}
```

```
void startEthernet()
{
  fcounter = 0;
  client.stop();
  Serial.println("Connecting");
  WiFi.begin(mynetwork, mypassword);
  while (WiFi.status() != WL_CONNECTED) {
    delay(500); }
}
```

9   Transferring Data to Carriots

Carriots [12, 13] by Altair is another open-source IoT platform for logging and processing of sensor data on cloud-based environment. The use cases of Carriots include the implementation in smart retail, smart cities, smart energy, smart oil, smart agriculture, smart buildings, smart banking, and many others (Fig. 7).

The platform of Carriots can be accessed after free registration on its Web-based environment at carriots.com. After log-into Carriots, a secured Control Panel is provided having options for creation and mapping of devices and streams. There are options to integrate the streaming with e-mail and SMS for live notifications (Fig. 8).

Fig. 7 The IoT platform of Carriots

Following is the source code for Arduino IDE which transmits data to Carriots.

```
#include "ESP8266WiFi.h"
const char* WiFiNetworkName     = "<WiFi Network Name>";
const char* NetworkPassword = "<WiFi NetworkPassword>";
const char* ServerName = "api.carriots.com";
const String CARRIOTSAPI = "<API Key of Carriots>";
const String DEVICE = "defaultDevice@kumargaurav.kumargaurav";
WiFiClient client;
int readval = 0;
void setup() {
 Serial.begin(9600);
 delay(1000);
 Serial.println();
 Serial.print("Connecting to Network");
 Serial.println(WiFiNetworkName);
 WiFi.begin(WiFiNetworkName, NetworkPassword);
 while (WiFi.status() != WL_CONNECTED) {
  delay(1000);
 }
 Serial.println("Connected");
}
void sendStream()
{
 if (client.connect(ServerName, 80)) {
  Serial.println(F("connected"));
  String json = "{\"protocol\":\"v2\",\"device\":\"" + DEVICE +
"\",\"at\":\"now\",\"data\":{\"RecordedData\":\"" + val + "\"}}";
  client.println("POST /streams HTTP/1.1");
  client.println("Host: api.carriots.com");
  client.println("Accept: application/json");
  client.println("User-Agent: Arduino-Carriots");
  client.println("Content-Type: application/json");
  client.print("carriots.CarriotsAPI: ");
  client.println(CARRIOTSAPI);
  client.print("Content-Length: ");
```

```
int thisLength = json.length();
client.println(thisLength);
client.println("Connection: close");
client.println();
client.println(json);
}
else {    Serial.println(F("Not Connected"));  }
}
void loop() {
readval = analogRead(A0);
Serial.println(val);
Serial.println(F("Transmit Data"));
sendStream();
delay(1000);
}
```

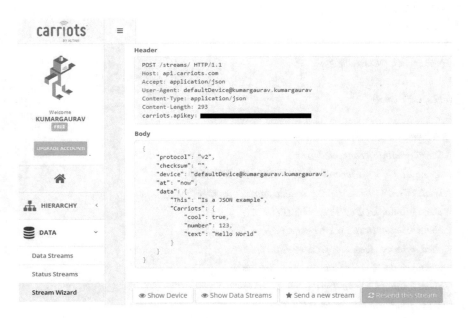

Fig. 8 Generation of API key from the dashboard of Carriots

After executing the code in Arduino IDE, the real-time data will be transmitted to the server of Carriots for recording. There are multiple file formats in which data can be stored for further evaluation.

10 Contribution and Methodology

Following are the key research gaps which are addressed here with the proposed usage of open-source hardware and sensor applications

1. No approach or algorithm is available that can identify and recognize the person drowned or drowning in the water.
2. Manual process is always followed by the swimming experts to save the drowning person.
3. Huge human efforts are required to identify the dead body in the underwater region.

The live sensors-based applications can be integrated with the real-world applications so that the identification of hidden objects can be done using Arduino and related open-source hardware. This work presents the use of sensor-based devices for the development and implementation of a novel and effectual approach for the identification of drowning person with the location in the underwater region whether it is a lake, pond, or particular sea beach. In traditional way, the drown person is searched by the swimming experts to identify the dead body, whereas this research work is having the usage of Underwater Wireless Sensor Networks (UWSNs) with the deep evaluation of the person drown in the water region (Fig. 9).

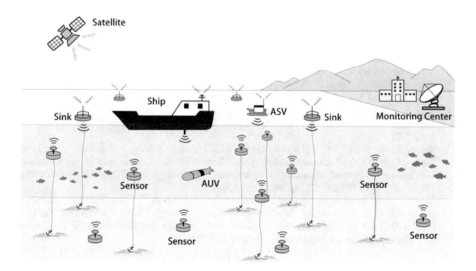

Fig. 9 Sensor integration for identification of hidden objects

The development of UWSN can be done using Arduino or Raspberry Pi for the identification of objects in underwater.

11 Conclusion

Today's world is trying to connect each and everything using network-based devices so that there is higher connectivity for global communication in all the directions and global positions. This type of communication in which every object is connected with each other is covered under Internet of Things (IoT). To work in this domain, there is need to work on real sensor-based devices with the customizable hardware and chips. By this way, a well-connected and programmed environment can be created in which secured and effective communication can take place. There are many open-source hardware available including Arduino, Raspberry Pi, and many others which can be programmed for multiple real-life applications to achieve the low-cost implementation of Internet of Things (IoT).

References

1. Pearce, J.M. 2012. Building Research Equipment with Free, Open-Source Hardware. *Science* 337 (6100): 1303–1304.
2. Sa, I., and P. Corke. 2002. System Identification, Estimation and Control for a Cost Effective Open-Source Quadcopter. In *2012 IEEE International Conference on Robotics and Automation (ICRA)*, May 14, 2202–2209.
3. Marchesini, J., S.W. Smith, O. Wild, J. Stabiner, and A. Barsamian. 2004. Open-Source Applications of TCPA Hardware. In *20th Annual Computer Security Applications Conference*, Dec 6, 294–303, IEEE.
4. Fisher, D.K., and P.J. Gould. 2012. Open-Source Hardware is a Low-Cost Alternative for Scientific Instrumentation and Research. *Modern Instrumentation* 1 (02): 8.
5. Bahrudin, M.S., R.A. Kassim, and N. Buniyamin. 2013. Development of Fire Alarm System Using Raspberry Pi and Arduino Uno. In *2013 International Conference on Electrical, Electronics and System Engineering (ICEECE)*, Dec 4, 43–48, IEEE.
6. Piyare, R. 2013. Internet of Things: Ubiquitous Home Control and Monitoring System Using Android Based Smart Phone. *International Journal of Internet of Things* 2 (1): 5–11.
7. Kodali, R.K., and S. Soratkal. (2016). MQTT Based Home Automation System Using ESP8266. In *2016 IEEE Region Humanitarian Technology Conference (R10-HTC)*, Dec 21, 1–5, IEEE.
8. Ramakrishna Murty, M., J.V.R. Murthy, P.V.G.D. Prasad Reddy, and Suresh C. Sapathy. 2012. A Survey of Cross-Domain Text Categorization Techniques. In *International Conference on Recent Advances in Information Technology RAIT*-2012, ISM-Dhanabad, 978-1-4577-0697-4/12 IEEE Xplorer Proceedings.
9. Shkurti, L., X. Bajrami, E. Canhasi, B. Limani, S. Krrabaj, and A. Hulaj. 2017. Development of Ambient Environmental Monitoring System Through Wireless Sensor Network (WSN) Using NodeMCU and "WSN monitoring". In *2017 6th Mediterranean Conference on InEmbedded Computing (MECO)*, Jun 11 1–5, IEEE.

10. Kodali, R.K., and S. Mandal. 2016. IoT Based Weather station. In *2016 International Conference on Control, Instrumentation, Communication and Computational Technologies (ICCICCT)* Dec 16, 680–683.
11. Pasha, S. 2016. ThingSpeak Based Sensing and Monitoring System for IoT with Matlab Analysis. *International Journal of New Technology and Research (IJNTR)* 2 (6): 19–23.
12. Maureira, M.A., D. Oldenhof, and L. Teernstra. 2011. ThingSpeak—an API and Web Service for the Internet of Things. Retrieved7/11/15World WideWeb, http://www.Mediatechnology. leiden.edu/images/uploads/docs/wt2014_thingspeak.pdf.
13. Botta, A., W. De Donato, V. Persico, and A. Pescapé. 2016. Integration of Cloud Computing and Internet of Things: A Survey. *Future Generation Computer Systems* 1 (56): 684–700.

Comparative Study of Performance of Tabulation and Partition Method for Minimization of DFA

Anusha Kolan, K. S. S. Sreevani and H. Jayasree

Abstract Automata is a machine that executes a set of functions in compliance with a seeded set of instructions. An automata is said to be finite automata if at any point of time, the machine can be in exactly one state among a finite set of states. Finite automata are categorized into two types: DFA and NFA. Deterministic finite automata is a finite and restricted state machine, where for every inserted symbol, there exists one and only one unique and idiosyncratic transition from a given particular state. Non-deterministic finite automata are a finite and restricted state machine where for every inserted symbol or character, there exist zero or one or more transitions on a given input symbol. Minimization of DFA involves reducing the number of states in a DFA and reaches the solution in lesser number of states. Since DFA is a rudimentary element of any computation machine, minimizing the number of states can reduce the computational time and increase the efficiency. Two algorithms for minimization tabulation method and partition method are conferred in this paper. To evaluate the performance, both the methods were implemented in C and Java languages. The execution times of both algorithms are compared to evaluate the performances of the algorithms in these two languages. The initial and final transition diagrams are graphically generated using Graphviz software.

Keywords Automata · DFA · NFA · Partition method · Tabulation method · Minimization

A. Kolan · K. S. S. Sreevani (✉) · H. Jayasree
Department of CSE, MVSR Engineering College, Nadergul, Telangana, India
e-mail: kvaniammalu.14@gmail.com

A. Kolan
e-mail: aaskanu.reddy@gmail.com

H. Jayasree
e-mail: jayasree_cse@mvsrec.edu.in

© Springer Nature Singapore Pte Ltd. 2020
K. S. Raju et al. (eds.), *Proceedings of the Third International Conference on Computational Intelligence and Informatics*, Advances in Intelligent Systems and Computing 1090, https://doi.org/10.1007/978-981-15-1480-7_2

1 Introduction

The action of metamorphosing the given deterministic finite automata (DFA) into another DFA (i.e., its equivalent ones) that has a least count of states comparatively referred to as DFA minimization. Literally, the two DFAs mentioned above are said to be equivalent if they are able to concede the same regular language. For every regular language considered, there is to exist a unique minimal automata that accepts the language, which is a DFA having a minimum figure for the number of states. There are two types of states that have to be either removed or merged in the existing DFA without influencing or infecting the language it accepts for minimization. (I) Unreachable states are the states that are not accessible from the initial state of DFA, when traversed for any given input string. (II)Non-distinguishable states are the states that are difficult to be discerned from each other when traversed for any given input string. The minimization of automata involves diminishing the number of states in a DFA and reaches the final state with fewer transitions. Pattern matching, for example, has been a good application of the minimal DFA [1], where a least cost for computations is being guaranteed. This able to attain such a minimal automata has become an elementary issue in the utilization and enactment of finite automata tools in substructures such as processing of text, analyzation of images, linguistic computer science. There are multiple algorithms available for the minimization of DFA. Two of these existing algorithms for minimization method, tabulation method [2], and partition method [3] are presented and analyzed in this paper. To evaluate the methods, the implementation is carried out in Java and C languages. The time complexity of the algorithm to minimize the automata and the time complexity of the entire application to convert input and output into transition diagrams and for the algorithm to minimize the automata as well as computed and analyzed in Java and C languages. The initial and final transition diagrams are graphically represented. Graphviz [4] software is used for the graphical representation. Intermediate dot language code is generated with Java and C languages. The dot language code is given as input to Graphviz to generate graphical output. The transition diagrams that generated give a clear vision of the minimization. The time complexities of both algorithms assist us to evaluate the performances of the algorithms in these two languages.

2 Methods

2.1 Tabulation Method

To find states that are equivalent, we make our best efforts to find pairs of states that are distinguishable. If p is an accepting state for an input symbol w, and q is a non-accepting state for the same input symbol w, then the pair $\{p, q\}$ is distinguishable. Tabulation method is the standard Myhill–Nerode theorem [5, 6].

Fig. 1 Input transition
diagram

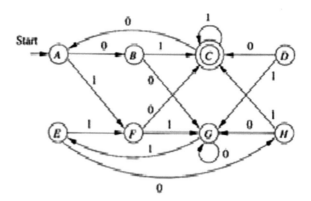

Let us consider an example [3]:

The filling of the table is as follows:

Considering the DFA in Fig. 1, since C is the only accepting state, we put x's in each pair that involves C. We know that some distinguishable pairs, we can discover others. For instance, since {C, H} is distinguishable and states E and F go to H and C respectively on input 0, we know that {E, F} is also a distinguishable pair. Similarly, {A, G} and {E, F} pairs are found as distinguishable. Therefore, the remaining states, i.e., {A, E}, {B, H}, {D, F} are equivalent pairs. Thus, the table-filling algorithm stops with the table as shown in Fig. 2. Thus, the equivalent pairs are joined and states are minimized from 7 to 5 states. The minimized DFA is as shown in Fig. 3. However, if we apply table-filling method after eliminating the inaccessible states, he efficiency increases.

Fig. 2 Table constructed
using tabulation method

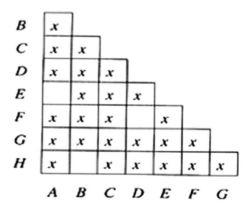

	A	B	C	D	E	F	G
B	x						
C	x	x					
D	x	x	x				
E		x	x	x			
F	x	x	x		x		
G	x	x	x	x	x	x	
H	x		x	x	x	x	x

Fig. 3 Output transition diagram obtained after minimization

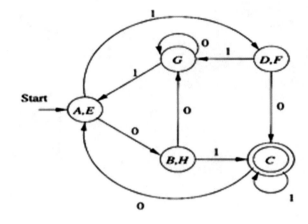

2.2 Partition Method

Partition method is also referred as equivalence method [7]. In this method, we start by removing the unreachable states from the given DFA in Fig. 1. In the example, considered D is an unreachable and hence is removed. However, the minimized DFA when unreachable states are not removed is as shown in Fig. 3.

Step 1: We then draw the transition table for rest states after removing the unreachable states. Now divide these remaining states into table into two partitions as:

1. One partition that contains non-final states, i.e., states {A, B, E, F, G}
2. Other partition that contains final states, i.e., here only C state being 2003 the final state. {C}

Zero—Equivalent: [A B E F G H] [C] 1 2

Step 2: For every state in each of the partitions and for each input symbol represent the partition number of the state it moves into as shown below.

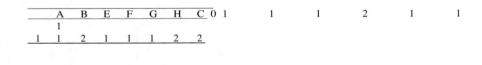

A	B	E	F	G	H	C	0	1		1		1		2		1		1
1																		
1	1	2	1	1	1	2	2											

[Table 1]

Step 3: Identify the states that are moving on to the same partition for given input symbols. For the example from the above table, it is clear that the states A, E, and G move on to states that are in the same partition number '1' on input symbol 0 as well as on input symbol 1. Hence, {A, E, G} is represented as a new partition. Similar is the case with B and H, where both move to states in partition '1' on input symbol 0 and partition '2' on input symbol 1. Hence, {B, H} is now represented as new partition. C is not included in the set of {B, H} as C is a final state, and final sate is represented as a separate partition as per Step 1. State F has a unique combination and hence is kept in a separate partition. The states that are in a partition are referred as equivalent pairs. The process is repeated to verify whether the states in a partition continue to be in the same partition or not.

One—Equivalent:[A E G] [B H] [F] [C] 1 2 3 4
Repeat Step 2 and Step 3, until no new partitions (sets) are formed.

	A	B	E	F	G	H	C
0	2	1	2	4	1	1	1
1	3	4	4	3	1	1	4	4

[Table 2]

We are then left with state G. Form the set {A, E, G}, G is now separated and is kept in a new partition.

Two-equivalent: [A E] [B H] [F] [G] [C] 1 2 3 4 5

	A	B	E	F	G	H	C
0	2	3	2	5	3	3	1
1	4	5	4	3	1	5	5

[Table 3]

Three-equivalent: [A E] [B H] [F] [G] [C]
The process is stopped when $n + 1$ equivalent is equal to n equivalent sets. Thus, the minimized states are {A, E}, {B, H}, {F}, {G}, {C}.

3 Psuedocode

Tabulation Method

Algorithm Tabulation

{

//Generate the initial table

int compute()//compute the table and find all the indistinguishable states

{

//where v0 ,v0 are the states where r1 goes on 0 and 1

//where h0,h1 are the states where c1 goes on 0 and 1 p=find(h0,v0);

q=find(h1,v1);

if(p==1||q==1)//atleast one cell marked then mark

{

return 1;

}

else return

0;

}

dotfilein();//generate the dot code for initial transition diagram and store in an external file

dotfileout(); //generate the dot code for minimized transition diagram and store in an external file

display(); //display the final computed results on the console

}

Partition Method

Algorithm Partition

{

//get all the inputs

//generate and store the dot code for initial transition diagram

//store all the non final states into an array and final states into second

//combine all the similar pairs in each of the tables individually for

i:=0 to n do

{

//to check if the transitions on one and zero of elements i and j are in similar states

if(c[zero[i]]==c[zero[j]] && c[one[i]]==c[one[j]])

{

c[i]=c[j];//states i is assigned same column value as j

break;

}}

//store the minimized states of both the tables in an array

// generate and store the dot code for minimized transition diagram

//display the computed results to console

}

4 Output Screens

4.1 Tabulation Method

The output—minimized DFA after the execution of tabulation method source code for the input as shown in Figs. 4 and 5.

4.2 Partition Method

The output—minimized DFA after the execution of partition method source code for the input as shown in Figs. 6 and 7.

5 Results

Both the algorithms are implemented in C and Java programming languages to compare the performance. Graphviz software is used for the graphical representation. Evaluation is carried out separately to check (1) the performance of the entire

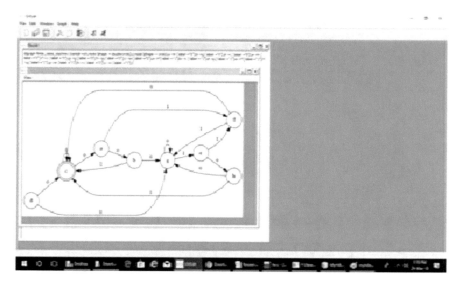

Fig. 4 Graphical representation of input diagram

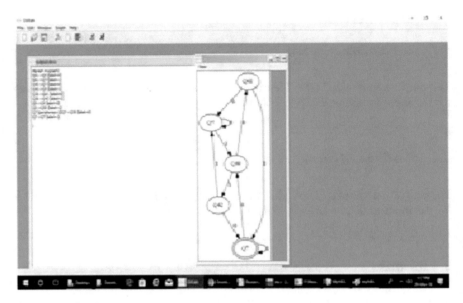

Fig. 5 Graphical representation of output diagram

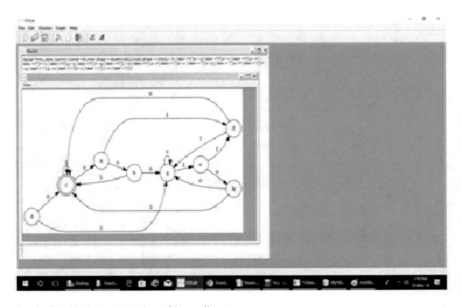

Fig. 6 Graphical representation of input diagram

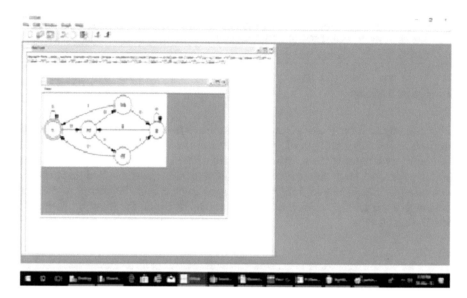

Fig. 7 Graphical representation of output diagram

Table 1 Performance evaluation of the execution of algorithms

Algorithm	Language used	Time complexity
Tabulation method	Java	$O(N^2)$
Tabulation method	C	$O(N^2)$
Partition method	Java	$O(N^3)$
Partition method	C	$O(N^3)$

Table 2 Performance evaluation of minimized output DFA generation

Algorithm	Language used	Time complexity
Tabulation method	Java	$O(N^3)$
Tabulation method	C	$O(N^3)$
Partition method	Java	$O(N^4)$
Partition method	C	$O(N^4)$

application: input-process (algorithm)–output (representing input–output in input–output transition diagrams) and (2) the performance of the algorithm (code segment that determines minimized DFA). The results of evaluation are presented in Tables 1 and 2.

For integrity of performance evaluation and benchmarking the optimization algorithms [8], we included-parsing—the reading (and parsing) of the automata is counted for a run of either of the algorithms. However, caching has not been used as

Hopcroft's partition refinement is not used and works better when caching is avoided and might influence the comparison.

6 Conclusions

DFA is cardinal for any computational machine to obtain a conclusive result based on finite input states and set of transitions on these states. Minimizing of DFA is profitable, as it reduces the computational time, thereby ameliorating the efficiency. In this paper, we have compared the efficiency of tabulation method and partition method algorithms, in two different languages Java and C. We came to the conclusion that the tabulation method is more productive than the partition method in both Java and C.

7 Literature Study

1. Introduction to Automata Theory, Languages, and Computation (3rd Edition) by John E. Hopcroft, Rajeev Motwani and Jeffrey D. Ullman
2. Introduction to the Theory of Computation by Michael Sipser
3. An Introduction to Formal Languages and Automata by Peter Linz 4. Introduction to Automata Theory, Languages, and Computation, 3e by Pearson Education India
4. Theory of Computer Science: Automata, Languages and Computation by Prentice Hall India Learning Private Limited
5. Making Simple Automata by Robert Race.

References

1. Srikanth Reddy, K., Lokesh Kumar Panwar, B.K. Panigrahi, and Rajesh Kumar. 2017. A New Binary Variant of Sine–Cosine Algorithm: Development and Application to Solve Profit-Based Unit Commitment Problem. *Arabian Journal for Science and Engineering.*
2. Motwani, Rajeev, and J.D. Ullman. 1979. *Introduction to Automata, Languages and Computation.*
3. Bjorklund, J., and L. Cleophas. 2009. *A Taxonomy of Minimisation Algorithms for Deterministic Tree Automata.*
4. Sipser, M. 1996. *Introduction to the Theory of Computation.* Michael Sipser Surhone, L.M., M. T. Tennoe, and S.F. Henssonow. 2015. graphviz.
5. Kozen, D.C. 1997. *Automata and Computability.*

6. Liedig, J. 2003. *Journal of Automata, Languages and Combinatorics.* www.jalc.de.
7. Linz, P. 1990. *Formal Languages and Automata.* Race, R. 2014. *Making Simple Automata.*
8. van der Veen, E. 2007. *The Practical Performance of Automata Minimization Algorithms.*

Click-Through Rate Prediction Using Decision Tree

Anusha Kolan, Dasika Moukthika, K. S. S. Sreevani and H. Jayasree

Abstract Advertising click-through prediction is a cardinal machine learning problem in online advertising. Click-through prediction anticipates the prospect of an ad to be viewed by the user. It necessitates binary classification to stratify the ad into either categorical or numerical. The forecast is constructed on ensuing attributes: publisher, user and ad content statistics. The machine learning algorithms when furnished by pertinent data sets about users, advertisers and display platforms can be employed to foretell the tendency of a visitant on an online platform, to click on the ad presented [1]. The quantification of such proficiency is the click-through rate (CTR), which is the ratio of clicks on a specific ad to its total number of views [2]. A higher CTR indicates the pronounced proclivity of an ad and its triumph in the advertising industry. The prediction is rooted to the elementary questionnaire like, how does the mélange of ages influence a client's choice towards a certain ad? What is the impact of people's financial status and net worth in regulating the selections of ads? How does the plethora of people's métiers and limited time factors govern the prediction? and so others. Decision tree classifier is the foundation of our paper, a tree-based algorithm which is a highly espoused algorithm in machine learning [3]. A recursive algorithm, which splits the nodes persistently into subsets by hinging

A. Kolan (✉) · D. Moukthika · K. S. S. Sreevani · H. Jayasree
Department of CSE, MVSR Engineering College, Nadergul, Telangana, India
e-mail: aaskanu.reddy@gmail.com

D. Moukthika
e-mail: moukthidasika@gmail.com

K. S. S. Sreevani
e-mail: kvaniammalu.14@gmail.com

H. Jayasree
e-mail: jayasree_cse@mvsrec.edu.in

© Springer Nature Singapore Pte Ltd. 2020
K. S. Raju et al. (eds.), *Proceedings of the Third International Conference on Computational Intelligence and Informatics*, Advances in Intelligent Systems and Computing 1090, https://doi.org/10.1007/978-981-15-1480-7_3

29

on the antecedent observations. Decision tree classifier is a sequential diagram portraying all the expedient decisions and analogous outcomes. The implementation, coding and development of the algorithm are carried out in Python. The results witness improvement in the prediction by using decision tree algorithm.

Keywords Click-through rate · Prediction · Decision tree · Algorithm · Machine learning

1 Introduction

Internet today is the hub of every industry dealing with huge money. It has occupied fair proportion of the society. Advertisements published online are disseminated within seconds. Nevertheless, the context of posting the ad dominates its marketability than any other facet. Incompatible sites for an ad dwindle the plausibility of users sighting the advertisement. The problem of incompatibility between websites and advertisements is abridged with this algorithm. We are using decision tree algorithm in computing the click-through rate and thereby proliferating the ad by devising the necessary actions. Decision tree is a recursion of greedy searches for calculating the best splitting point at each step based on the training data set. These points further provoke the algorithm in cleaving the data into groups. The central premeditation of the project is to increase the contingency of an ad to be viewed by a user. When the appropriate ad is presented to the germane user, probability of its preference will be increased, following which the advertisement can be bestowed with increasing popularity among the clients. This approach furnishes wide purview to massive companies in selling their goods, creating and presenting the ads related to these goods to customers. Online click-through prediction is lucid and understandable compared to the other manual process [4]. By using decision tree algorithm in pronouncing the click-through rate, the firms can present their ads to apt users and it saves lot of time.

2 Related Work

Click-through prediction of the advertisements can be done by using different methods besides decision tree. Here are some of the methods percolated by the authors:

1. In the journal Click-Through Rate Prediction for Contextual Advertisement Using Linear Regression which is referenced in [5], Muhammad Junaid and Syed Abbas Ali used linear regression method for prediction. In their approach, categorical data is ignored and all continuous-based features like bid and keywords are used to show the direct relation with the outcome CTR [6] using a graphical view. They used contextual ad serving based on highest bid algorithm which picks the ad based on the context of the publisher plus the maximum bidder within that contextually selected pool of advertisements. Thus, the data collected through this is used in the next sections to perform linear regression. But this method suffers decrease in efficiency. 2. The paper A New Approach for Mobile Advertising Click-Through Rate Estimation Based on Deep Belief Nets was published by Jie-Hao Chen. Like in the reference [7], the paper proposes a mobile advertising CTR estimation fusion model based on deep belief nets (DBNs). Their model detects deep level features in place of simple original features and then puts it into a logistic regression model to predict the CTR. Their predicting model contains three steps:

(1) The pre-treatment of original sample feature data [5].
(2) Taking the DBN to build valid features from original ones in a deep level.
(3) Taking the newly built vectors as the input of the logistic regression (LR) model and predicting the CTR.

3 Methods

3.1 Decision Tree Classifier

Among the different classification algorithms such as SVM and linear classifiers used in machine learning, the tree-based learning algorithms are quite common and efficient in data science field. They empower predictive models with high accuracy, stability and ease with which interpretation can be done. Few of the common examples include as follows: decision trees, random forest and boosted trees [8]. Out of these, we have taken the decision tree classifier into consideration for the various advantages it offers in predicting the click-through rate. Decision tree can be constructed relatively fast when compared to other classifications. It can handle both categorical and numerical data and can obtain similar or even better accuracy in the results. In addition to that, the tree is also useful in constructing SQL statements that can be used to access databases very efficiently. One can represent decision alternatives and possible outcomes schematically which helps in comprehending decision flow and outcomes. Moreover, decision trees are capable of handling data sets with errors or missing values. Keeping in mind, the above advantages we have applied the decision tree classifier to predict the click-through rate. The decision tree classifier operates in the form of a decision tree. It maps observations to class assignment that are symbolised leaf nodes through a series of tests represented as internal nodes based on feature values and corresponding conditions [9]. A decision tree is a tree-like graph, a sequential diagram depicting all of the possible decision alternatives and the corresponding outputs. From the root of a tree, every internal node represents what a decision is made based on each branch of a node which represents how a choice may lead to the next nodes, and finally, each terminal node, the leaf, represents an outcome yielded.

3.2   Example Illustrating Decision Tree Classifier

To understand the importance and functioning of a decision tree algorithm, let us take an example of a shoe selling company that is interested in marketing its brand at various websites to promote its sales. Here, we need to consider a fact that not all websites would draw equal and good viewer attention that would increase the brand popularity. So to post the company ads at the right websites, we predict the types of people who would be interested in purchasing a shoe. For this, we follow the basic three steps as shown below:

1. Divide the viewers into multiple categories based on demographics like age, gender, his/her interest in shopping and income levels.
2. Collect the analytics data for the websites on which the company wants to post its advertisement. This data includes the number of viewers of the website, viewers' age range, viewers' place of website access and then give their CTR values.
3. Predict the probability of a person belonging to a particular category in purchasing the shoe. This is done by constructing a decision tree shown in Fig. 1 of features considered and corresponding CTR values taken from past data, traversing through the tree based on a viewer's matching features criteria and finally arriving at the click-through rate value.

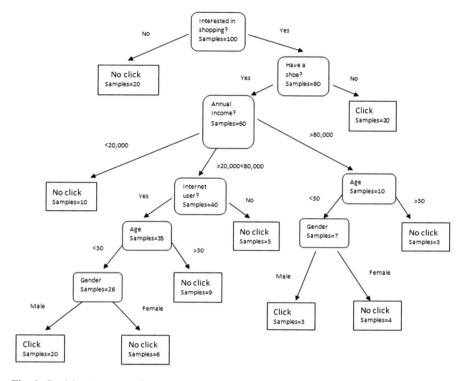

Fig. 1 Decision tree example

4 Availability of Data and Materials

The proposed system requires windows OS, PyCharm text editor for Python scripts to run Python files. Data sets for training and testing the algorithm are taken from kaggle website. The data sets are avazu-ctr-prediction/data.

5 Results

Figure 2 shows the output of the algorithm for test data set of 10000 samples, considering the minimum samples to split as 10. The max_depth parameter is considered to be none, inferring that the tree needs to be split till all the nodes are pure. The investigation on three sets of data consists of first 5000, 7500 and 10000 samples, ranging the min_samples_split to split among 10, 50,100 and 200. The results turned out to give a conclusion that a minimum value of min_samples_split will be giving more clicks prediction (Table 1).

Fig. 2 Result of the decision tree algorithm for 10000 samples

	No. of samples	Clicks predicted	CTR
Table 1 Data analysis for minimum samples to split of 10, considered for each node in forming the tree	5000	232	0.0464
	7500	419	0.0558
	10000	578	0.0578

The results encompass details of clicks predicted and CTR values along with the samples considered for all the four cases of min_samples_split. The min_samples_split gives the number of values considered for a split of a node at each stage of forming the decision tree. The formation of the tree and thereby the prediction of clicks for the ad are greatly affected by this factor. Our research primarily focuses on this factor and finding an optimum value for this factor in order to increase the accuracy for the prediction. Click-through rate is used to measure effectively the ratio of the number of clicks on a particular to the total number of samples. In the first table [Table 1], we have contemplated 10 samples per split to efficiently calculate the CTR. Initially we considered 5000 samples, and then out of them, 232 samples are clicked which gives us CTR to be 232/5000 equal to 0.0464. Then we increased the number of samples to 7500. Out of 7500 samples, 419 clicks are conjectured. CTR is determined as 419/7500 equal to 0.0558. Later we have considered 10000 samples. Out of 10000 samples, the number of clicks obtained is 578. Hence, we calculated CTR as 578/10000 equal to 0.0578 which makes us to consummate that by increasing the number of samples, clicks predicted are also heightened. In Table 2, we increased the min_sample_split value to 50. Then we espied the observations as follows. We initiated the procedure by considering 5000 samples then we got 217 clicks. We got CTR as 217/5000 equal to 0.0434. By taking 7500 samples, we got 356 clicks and CTR is obtained as 354/7500 equal to 0.0474. By increasing number of samples to 10000, clicks predicted are 504. CTR obtained is 504/10000 equal to 0.0504 [Table 2]. In Table 3, we considered the

Table 2 Data analysis for minimum samples to split of 50, considered for each node in forming the tree

No. of samples	Clicks predicted	CTR
5000	217	0.0434
7500	356	0.0474
10000	504	0.0504

Table 3 Data analysis for minimum samples to split of 100, considered for each node in forming the tree

No. of samples	Clicks predicted	CTR
5000	232	0.0464
7500	325	0.0433
10000	391	0.0391

Table 4 Data analysis for minimum samples to split of 200, considered for each node in forming the tree

No. of samples	Clicks predicted	CTR
5000	106	0.0212
7500	253	0.0337
10000	443	0.043

value of min samples to be split as 100 and again repeated the procedure by considering 5000, 75000, 10000 samples. We got 232, 325, 391 number of clicks, respectively. Then CTR for 232 clicks is calculated as 232/5000 equal to 0.0464, and for 325 clicks, CTR obtained is 325/7500 equal to 0.0474. For 391 number of clicks, CTR is 391/10000 equal to 0.0.391. From this result, we can infer that when we have increased the min_sample_split, CTR value is decreased (Table 4).

Figure 3 shows the graphical analysis for the three samples of data, against their respective CTR outputs obtained. It is evident from the graph that CTR value predicted is maximum for minimum value of min_samples_split for all three

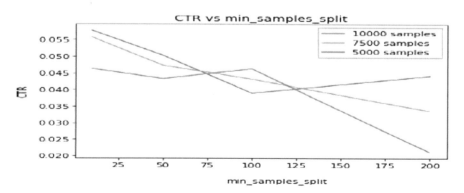

Fig. 3 Graphical analysis of CTR values against min_samples_split values for 10000, 7500 and 5000 samples

samples of data. All the sets of data predict maximum CTR values at minimum value of 10, of min_samples_split and minimum CTR value, for higher values of min_samples_split.

6 Conclusions

Advertising click-through rate prediction is helpful to dwindle the comprehensive cost invested in ads by multitudinous companies. This is evident through the principle of commensuration of the ad with the user; the algorithm helps in presenting the suitable ad to the appropriate user thereby decreasing the undesired projection of ads. Online display advertising is an industry that heavily relies on the ability of machine learning models to predict the ad targeting effectiveness, i.e. how likely the audience belonging to a specific demographic category will be interested in this product. This requisite is addressed with CTR prediction. Decision tree algorithm is proven to be efficacious today. Our research concludes that having a minimum number of samples for a split enhances the prediction capacity of the algorithm, compared to higher values of min_samples_split. Thereby enclosing, decision tree classifier with minimum value of min_samples_split is more effective in predicting the click-through rate, in machine learning.

References

1. https://www.ncbi.nlm.nih.gov/pmc/articles/PMC5676483/.
2. Effendi, Muhammad Junaid, and Syed Abbas Ali. 2007. *Click Through Rate Prediction for Contextual Advertisment Using Linear Regression*. Cornell University Library, 1701.08744. https://arxiv.org/ftp/arxiv/papers/1701/1701.08744.pdf.
3. Bhatnagar, V., R. Majhi, and P.R. Jena. 2017. Comparative Performance Evaluation of Clustering Algorithms for Grouping Manufacturing Firms. *Arabian Journal for Science and Engineering*.
4. https://www.researchgate.net/publication/315778757_Predicting_Click_Through_Rate_of_ new_Ads_Considering_CTRs_as_Manual_Ratings.
5. Chen, Jie-Hao, Zi-Qian Zhao, Ji-Yun Shi, and Chong Zhao. 2017. A New Approach for Mobile Advertising Click-Through Rate Estimation Based on Deep Belief Nets. In *Computational Intelligence Neuroscience*, vol. 2017, PMC.
6. Ramakrishna Murty, M., J.V.R. Murthy, and P.V.G.D. Prasad Reddy. 2011. Text Document Classification Based on a Least Square Support Vector Machines with Singular Value Decomposition. *International Journal of Computer Application (IJCA)* 27 (7): 21–26 (Impact factor 0.821).
7. https://arxiv.org/ftp/arxiv/papers/1701/1701.08744.pdf.

8. Lannani, Xuemin Zhang. 2010. Study of Data Mining Algorithm Based on Decision Tree. In *Browse International Conference*, IEEE explorer.
9. Navya Sri, M., M. Ramakrishna Murty, et al. 2017. Robust Features for Emotion Recognition from Speech by using Gaussian Mixture Model Classification. In *International Conference and Published Proceeding in SIST Series*, vol. 2, 437–444. Berlin: Springer (August-2017).

A New Framework Approach to Predict the Wine Dataset Using Cluster Algorithm

D. Bhagyalaxmi, R. Manjula and V. Ramanababu

Abstract Examination, plan, and usage of programming frameworks for online administrations are a repetitive and testing. Amazon programming gives item proposals, Yahoo! powerfully suggests web pages, afflux makes proposals for films, and Google makes ads on the Internet. Things are suggested in view of the inclinations, needs, attributes, and conditions of clients. The wine informational index has been used in examine for quite a while, and still, it stays as the benchmark informational collection. Nature of wines is hard to characterize as there are numerous variables that impact the apparent quality. This paper exhibits a basic survey of research slants on wine quality and client-driven closeness measures also. A novel client-driven likeness measure in item bunching is proposed to assess the prominent wine informational index named red wine dataset. The test results got in this work can give preferable proposals to item purchasers over the current frameworks. The proposed approach is skilled to gather the red wine dataset into requested gatherings of favored wine variations and can judge the wine quality in view of these client inclination gatherings.

Keywords Computational wine wheel · SMOTE · KNN · PCA

1 Introduction

Wine datasets of the essential characteristics (smell, taste, visual), conservation characteristics (region, climate, site), managing practices (viticulture exercise) and physicochemical elements (pH, acid, etc.) [1] are the factors of attention in assessing the excellence of wine. Data mining procedures in foreseeing wine excellence are in progress, with some auspicious results in the province. Sensory and physicochemical tests are central in wine authorization. It is the standard

D. Bhagyalaxmi (✉) · R. Manjula · V. Ramanababu
Faculty of CSE Department, KU College of Engineering and Technology, KU Campus, Warangal, Telangana, India
e-mail: bhagyalaxmi519@gmail.com

© Springer Nature Singapore Pte Ltd. 2020
K. S. Raju et al. (eds.), *Proceedings of the Third International Conference on Computational Intelligence and Informatics*, Advances in Intelligent Systems and Computing 1090, https://doi.org/10.1007/978-981-15-1480-7_4

practice in physicochemical research center tests, to portray wine by assurance of thickness, liquor, or pH esteems, yet tangible tests depend fundamentally on human specialists. Wine characterization is a troublesome assignment as taste is the slightest comprehended of the human detects. The connection between the physicochemical and tangible examination is mind-boggling to get it. In the nourishment business, notwithstanding the sustenance quality research, machine learning strategies have additionally been connected in order of wine quality. Machine learning strategies give the best approach to manufacture models from information of known class names to anticipate the nature of a wine. In days of yore, wine was considered as an extravagance thing. Today, it is mainstream and delighted in by a wide assortment of individuals. Proficient wine audits offer experiences on wines accessible in vast amount in every year. An efficient way is expected to use those expansive number of surveys to profit wine purchasers, merchants, and producers. No two people judge the wine alike even they taste the wine all the while having the capacity to share and distinguish all similar traits. Experience helps a considerable measure and obstructs the tester. So surveying the nature of wine depending just on the tester understands, and detecting is a major procedure.

2 Related Work

The undertaking of any prescribed framework is to give the client's data about finding the favored things from the simple extensive arrangement of things. The inclination of a client is chosen by voting/rating reaction of the customer to the recommended framework [2, 3].

These days the data on the planet is creating more than a large number of times quicker than the real information process limit. Communitarian separating is an announcement of-the-workmanship procedure utilized for productive information preparing. Two fundamental issues of shared sifting are versatility and quality. Existing community-oriented separating calculations will have the capacity to see lakhs of neighboring items, yet the genuine requests of present-day electronic frameworks are in many lakhs.

Many of the current calculations are experiencing execution issues. Another vital point is that there is a need to enhance the nature of the proposals for the clients. Adaptability and quality for the most part negate each other. A decent harmony between these two manufacturing plants is required. The connection between items is most vital than the connection between clients. Suggestions for the clients are assessed by discovering items that resemble different items as per client likes.

Uncommon indexing systems are required for effective preparing weighted inclination-based inquiry execution. These ordering systems are valuable for getting direct capacity inclination-based improved question results. Direct capacity streamlining inquiries are called inclination determination questions claiming such questions recover tuples expanding a straight capacity characterized over the property estimations of the tuples in a connection. Direct inclination work likewise needs to change for acquiring better question results.

Authors proposed a brand-new data science area named wine informatics. In order to automatically retrieve wines' flavors and characteristics from reviews, which are stored in the human language format, authors proposed a novel "computational wine wheel" to extract keywords. Two completely different public-available datasets are produced based on the new technique in their paper. The hierarchical clustering algorithm is applied to the primary dataset and got purposeful clustering results.

Authors proposed a data analysis approach to classify wine into different quality categories. A dataset of white wines of 4898 records was used in the analysis. As the dataset was imbalanced with about 93% of the observations are from one category with respect to the occurrence of events in it, Synthetic Minority Over-sampling Technique (SMOTE) [4] was applied to oversample the minority class. These categories include high quality, normal quality, and poor quality. Three classification techniques used in this work include decision tree, adaptive boosting, and random forest. Among the techniques, random forest produced to produce the desired results with the minimal error.

Creators proposed an enhanced KNN technique with weights, which impressively enhances the execution of KNN strategy [5]. The creators utilized a sort of preprocessing on prepare information. They presented another esteem named validity to prepare tests which cause more data about the circumstance of preparing information tests in the property space. This new esteem takes into accounts the estimation of dependability and power of any prepare tests in regard to with its neighbors. KNN with connected weights utilizes legitimacy as the increase factor respects more strong grouping as opposed to straightforward KNN strategy, productively.

The wine dataset is the aftereffect of a substance investigation of wines developed in a similar area in Italy that got from three distinct cultivars. The examination decided the amounts of 13 properties found in every one of the three sorts of wines. This informational collection has been used with numerous others for contrasting different classifiers. With regard to arrangement, this is a very much postured issue with well-behaved class structures.

Information mining strategies to characterize the nature of wines utilizing a bigger physicochemical informational collection were utilized in later works. Cortez and his partners [6] constructed models utilizing bolster vector machine, numerous relapse, and neural systems. A dataset with countless is considered (vinho Verde tests from the Minho district of Portugal). A computational technique was produced that performs synchronous variable and model choice. Bolster vector machine accomplished attractive outcomes, "outflanking the numerous relapse and neural system techniques."

Creators upheld unsupervised neural system (NN) in view of adaptive resonance theory (ART1) [7] as another option to measurable classifier in order to segregate among the 178 examples of wine having 13 quantities of traits. The dimensionality of the element factors was diminished to five by important part investigation (PCA).

Out of 13, the initial two quantities of main segments caught more than 55.4% of the change of the wine dataset. Nonhierarchical K-implies grouping calculation was utilized to pick the classes accessible among the examples of wine.

Creators utilized analytical hierarchy process (AHP) arrangement calculation. This calculation gives the best approach to suggest wine based on the parts of the wine. Wine determination based on its traits is an alternate approach. The machine learning techniques utilized here aided in finding the part exactness of wine qualities. The analytical hierarchy process (AHP) is utilized for orchestrating and analyzing convoluted issues by scientific figuring. AHP characterizes various ascribe issue to exhort a specific product to a person. The procedure of AHP is basically used to figure weights. The contributions for AHP are relative inclinations and properties. The creators have taken red wine dataset, and weights were assigned to them in light of AHP [8].

Creators attempted to enhance the learning speed by part the bunch tree into sub-groups and by utilizing investigation and misuse stages and totals also. From client-driven sensor information, Friend book finds ways of life of clients and measures the comparability of ways of life between clients. It prescribes companions to clients if their ways of life have high closeness.

Creators proposed an enhanced community separating calculation that joins k-implies calculation with CHARM calculation [5]. This crossover approach enhanced the expectation nature of suggestion framework.

Creators proposed a calculation that consolidates client-based approach, thing-based, and Bhattacharyya approach. The principle preferred standpoint of this half breed approach is its capacity to discover more dependable things for suggestion. Synergistic sifting innovation works by making a database of inclinations for items by their clients [9]. Community-oriented innovation is getting to be well known in the most recent research zones, for example, e-business, banking, space, share market, etc. Creators demonstrate the day by day life of clients as life records. Utilizing these records, ways of life are removed by utilizing the latent Dirichlet allocation calculation.

3   The Progression in the Field of Wine Excellence Research Is Threefold

3.1   Dataset

The greater part of the exploration about wine has been doing in view of two well-known informational indexes named wine and wine quality which are freely accessible on UCI machine learning stage. Two datasets are accessible of which one dataset is in red wine and has 1599 totally extraordinary assortments, and hence, the option is on white wine and has 4898 assortments. All wines are created in a specific territory of Portugal. Information on the 12 most unique characteristics of wines is collected, one of which reinforces quality, clear learning, [2] and, as a result, the wine's aromatic properties of wines, with thickness, acidity, alcohol content and so on. Every concoction property of wines is consistent factor. Quality is an ordinal variable with the conceivable positioning from 1 (most noticeably bad) to 10 (best). Each type of wine is tasted by three free testers, and in this way, the last rank allocated is that the middle rank given by the testers. Learning in regard to surveys given by clients is furthermore out there.

3.2 Objectives

The basic target of the majority of the analysts is to evaluate/foresee wine quality in light of the wine attributes and client commentators and utilize the outcomes to suggest winemakers, advertisers, and clients too.

3.3 Methodology

Most of the strategies and instruments are planned in light of machine learning and information mining procedures. Arrangement, bunching, and affiliation manage mining are the regular methodologies watched. Wine traits and client surveys are the principle wellsprings of information. These datasets contain genuine esteemed characteristics. The ongoing works in the field represent the reception of cutting edge machine learning systems. To keep up the consistency, the first qualities are standardized in a few papers. A few scientists changed over the standardized qualities to double qualities to get the simplicity of calculations. Methods like guideline segment examination are utilized to diminish the dimensionality of the datasets. K-implies bunching and its altered variations are utilized for grouping. Bayesian systems, neural systems [10, 11], and other characterization procedures are embraced. A couple of papers address the weighted measures and crossover profound learning systems to enhance the ease of use of the outcomes.

4 Proposed Work

(a) To present novel inquiry approaches and efficient calculations for their execution.
(b) To present novel client-driven similarity measures in item grouping.
(c) To apply the proposed grouping to assess the popular wine dataset

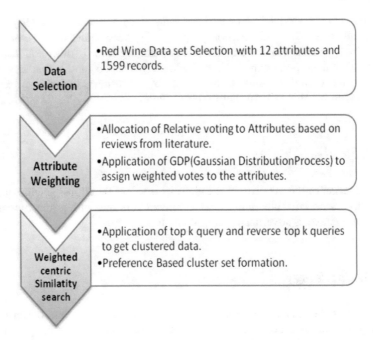

Data Selection
•Red Wine Data set Selection with 12 attributes and 1599 records.

Attribute Weighting
•Allocation of Relative voting to Attributes based on reviews from literature.
•Application of GDP(Gaussian DistributionProcess) to assign weighted votes to the attributes.

Weighted centric Similatity search
•Application of top k query and reverse top k queries to get clustered data.
•Preference Based cluster set formation.

Fig. 1 Process flow

    i. To test whether the wine quality is well supported by its concoction properties.
   ii. To evaluate wine quality in a novel way.

(d) To contrast the client-driven measurements and traditional measurements.
(e) To break down the outcomes and present the discoveries (Fig. 1).

The item or administration suggestion needs a basic assessment of item includes and the client inclinations for the item [1]. A cross-breed calculation proposed in this work involves the accompanying subtasks:

(a) Top-k and switch top-k inquiries ways to deal with rank the items in light of the client inclinations.
(b) Incorporating the weighted positions in similitude figuring, to group the items.
(c) Nearest neighbor likeness look utilizing Jaccard coefficient and altered Jaccard coefficient.

4.1 Rationality/Explanation of the Proposed Work

The proposed procedure bunches the wine dataset records into need-based groups. The bunched information utilizing grouping structures a model to allot the test information records with a prescribed voting mark [1, 12]. The majority of the past research on wine information restricted to ordinary grouping and arrangement approaches relying upon the tester detecting information, while the proposed novel half-and-half approach can suggest the client a superior wine blend without relying upon the tester detecting information.

5 Algorithm and Description

The clustering algorithm using Jacquard constant similarity amount, weighted query method with top-k, and reverse top-k approaches for clustering red wine dataset is as follows.

5.1 Algorithm

1. Read the dataset of 'Mn' tuples into the array data structure.
2. Produce rating/voting/preferences details of' a number of Weighted users for all the items using " Attribute Group" method.
3. Prepare top-k query results for all the items for all the' m' number of user rating/voting/preferences
4. Compute reverse top-k queries for all the items found in control go to the step-3.
5. Sm = get Reverse Set Items Count.
6. T threshold = get Threshold Value
7. m count(minimum count) = Sm * threshold
8. Read the dataset of 'Mn' tuples into the array data structure.
9. Produce rating/voting/preferences details of' a number of Weighted users for all the items using " Attribute Group" method.
10. Prepare top-k query results for all the items for all the' m' number of user rating/voting/preferences
11. Compute reverse top-k queries for all the items found in control go to the step-3.

12. Sm = get Reverse Set Items Count.
13. T threshold = get Threshold Value
14. m count (minimum count) = Sm * threshold
15. While(Sm>mcount) do

    I. {

    II. SCluster(StartCluster) = first cluster of the present list of items

    III. For cluster im = 2 to last in the current list compute similarity measure, Sim (SCluster, im) and store the result.

    IV. Combine all the groups whose comparison measure value > than the specified threshold value into one cluster.

    V. PCount (presentCount) = number of groups combined in control go to the step11.

    VI. Sm = Sm − pCountvalue 'Mn' is stored efficiently in

    VII. the memory. Each User specifies rating/voting/performance details of items. Traditional

    VIII. }

The assessment of results in various contexts.

From Fig. 2, it was experimental that the top-k query value manipulating the number of clusters well for large values of k and as the number of evaluations increased the number of clusters is greater.

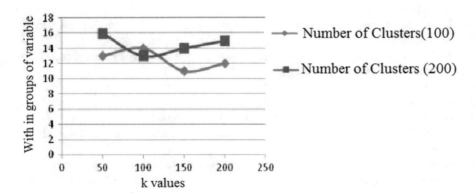

Fig. 2 Number of clusters for 100 and 200 evaluations with variable *k* value

6   Conclusion

A client-driven likeness structure is presented in which the comparability of items is evaluated by client inclinations. A mainstream dataset named "red wine quality "is considered in this work to evaluate the nature of wine by gathering the individual items into bunches and after that review the gatherings in view of inclinations. The user-centric approach gave very unique and intriguing outcomes than the customary methodologies have that do not consider the inclinations that the clients have communicated. It is watched that the inquiry composes presented in this work require higher execution times as the quantity of clients and the inclinations expanded. This may prompt an adaptability issue of the proposed system. This can be smoothened by presenting R-tree like information structures for seeking and ordering reason that can streamline the execution time of the proposed system. The motivation behind building up this sort of a framework is to help and exhort wine clients for better choice and winemakers for giving a superior quality.

References

1. Hu, Gongzhu, Tan Xi, and Faraz Mohammed. 2013. *Classification of Wine Quality with Imbalanced Data*. March 2016 IEEE. https://doi.org/10.1109/icit.2016.7475021.
2. Haydar, Charif, Anne Boyer. 2017. *A New Statistical Density Clustering Algorithm based on Mutual Vote and Subjective Logic Applied to Recommender Systems*. UMAP'17, July 9–12, 2017, Bratislava, Slovakia.
3. Song, Linqi, Cem Tekin, Mihaela van der Schaar. 2014. *Clustering Based Online Learning in Recommender Systems: A Bandit Approach*, May 2014, ICASSP2014. https://doi.org/10.1109/icassp.2014.6854459.
4. Chen, Bernard, Christopher Rhodes, and Aaron Crawford. 2014. Wine Informatics: Applying Data Mining on Wine Sensory Reviews Processed by the Computational Wine Wheel. In *2014 IEEE International Conference on Data Mining Workshop*. https://doi.org/10.1109/icdmw.2014.149.
5. Parvin, Hamid, Hoseinali Alizadeh, and Behrouz Minati. 2010. A Modification on K-Nearest Neighbor Classifier. GJCST 10 (14) (Ver.1.0) November 2010.
6. Nagarnaik, Paritosh, and A. Thomas. 2015. Survey on Recommendation System Methods. In *IEEE Sponsored 2nd International Conference on Electronics and Communication System (ICECS)*, Feb 27, 2015. https://doi.org/10.1109/ecs.2015.7124857.
7. Kavuri, N.C., and Madhusree Kundu. 2011. ART1 Network: Application in Wine Classification. *International Journal of Chemical Engineering and Applications* 2 (3).
8. Thakkar, Kunal, et al. 2016. AHP and Machine Learning Techniques for Wine Recommendation. *International Journal of Computer Science and Information Technologies (IJCSIT)* 7 (5): 2349–2352.
9. Latorre, Maria J., Carmen Garcia-Jares, Bernard Medina, and Carlos Herrero. 1994. Pattern Recognition Analysis Applied to Classification of Wines from Galicia (Northwestern Spain) with the Certified Brand of Origin. *Journal of Agricultural and Food Chemistry* 42 (7): 1451–1455.
10. Cortez, Paulo, Antonio Cerdeira, Fernando Almeida, Telmo Matos, and José Reis. 2009. *Modeling Wine Preferences by Data Mining from Physicochemical Properties*. https://doi.org/10.1016/j.dss.2009.05.016.

11. Panda, Mohit Ranjan, Shubham Dutta, and Saroj Pradhan. 2017. Hybridizing Invasive Weed Optimization with Firefly Algorithm for Multi-Robot Motion Planning. *Arabian Journal for Science and Engineering*.
12. Lahari, K., M. Ramakrishna Murty. 2015. Partition Based Clustering Using Genetic Algorithms and Teaching Learning Based Optimization: Performance Analysis. In *International Conference and Published the Proceedings in AISC*, vol. 2, 191–200. Berlin: Springer. https://doi.org/10.1007/978-3-319-13731-5_22.

Study and Ranking of Vulnerabilities in the Indian Mobile Banking Applications Using Static Analysis and Bayes Classification

Srinadh Swamy Majeti, Faheem Habib, B. Janet and N. P. Dhavale

Abstract Banking has stepped into the world with high-tech makeover by making the services as digitalized by means of mobile applications. Due to this digitalization, customer satisfaction and ease of use improved, especially in the case of retail banking. At the same time, there is a chance of getting our data compromised due to vulnerabilities in the mobile banking applications. These vulnerabilities exposed to threats may lead to security risk and finally cause damage to our assets. The quest to identify vulnerabilities in the mobile applications is now an emerging research area. Because, in previous days, hackers did damage to our assets for their fame but now, they are trying for espionage action and for getting the financial gain. We analyzed mobile applications of reputed banks in India. The main focus of this work is twofold. First, static code analysis (SCA) tools are used in this work to identify the vulnerabilities. But SCA tools are infeasible because of raising unexploitable vulnerabilities. Second, to partially solve this issue, we used machine learning classification algorithm for calculating the occurrence rate of the vulnerability in the mobile applications. We are alerting the banks by assigning rank to each vulnerability in the application based on the impact caused by that vulnerability by coupling the occurrence rate with severity score calculated by using common vulnerability scoring system (CVSS) score.

Keywords Vulnerabilities · Mobile applications · Mobile banking · Static analysis · Bayes classification

S. S. Majeti (✉) · B. Janet
Department of Computer Applications, National Institute of Technology Tiruchirappalli, Tiruchirappalli, India
e-mail: srinathswamy.majety@gmail.com

S. S. Majeti · N. P. Dhavale
Centre for Mobile Banking, IDRBT, Hyderabad, India

F. Habib
Information Systems, IIIT Bhubaneswar, Bhubaneswar, India

© Springer Nature Singapore Pte Ltd. 2020
K. S. Raju et al. (eds.), *Proceedings of the Third International Conference on Computational Intelligence and Informatics*, Advances in Intelligent Systems and Computing 1090, https://doi.org/10.1007/978-981-15-1480-7_5

1 Introduction

Organizations are moving from "digital-first" to "mobile-first" strategy because of substantial growth in the popularity and usage of mobile devices and also outstanding technological advancements happened during the last decade. This advancement made users to complete their tasks faster with less effort. Simply, it enriches our digital lives. Banking, financial services, and insurance (BFSI) [1] improved their market by entering into the mobile world by providing better services to their customers and yields good results. During the last decade, banking sector has witnessed huge development [2]. Due to technology advancement, banking sector changed from paper-based banking to paperless digitized mobile banking [3]. Customers of the banks are easily habituated using banking services by mobile devices because of ease of use, accessibility, and increased security. Especially, in the case of retail banking, the usage of mobile banking by the customers is growing at an astonishing rate in India. India ranks number 4 in mobile banking penetration across the world (Figs. 1 and 2).

By observing the above statistics, it clearly shows banking customers in India are more tech-savvy and are demanding faster, secured transactions in less time. At the same time, consumers are also worried about security concerns. Because of this demand, we tested mobile banking applications whether banks are following tough security measures for protecting the assets of the customers or not.

With the technology growing faster, malicious activities by cyberattackers have also intensified. According to Trustware Global Security Report [6], 95% of tested applications are having at least one vulnerability and 60% of mobile malware targets financial information on devices. So, cyberattackers are concentrating on financial apps to get financial gain [7]. Mobile applications have vulnerabilities exposed to threats that may lead to security risk and may finally cause damage to our assets.

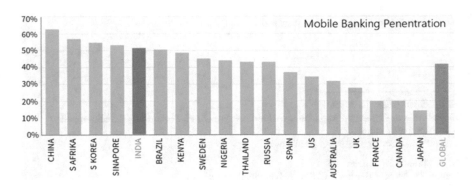

Fig. 1 Mobile banking penetration across the world (*Source* UBS evidence lab) [4]

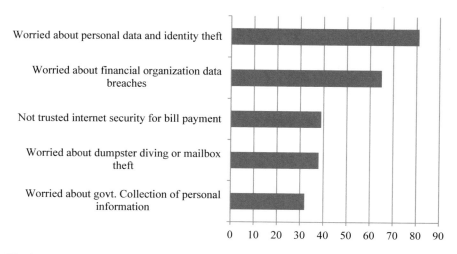

Fig. 2 Statistics about mobile banking features despite security concerns (*Source* Fiserv) [5]

Our approach identified the vulnerabilities in the application and prepared the vulnerability rank report by assigning ranks to each vulnerability based on occurrence rate and CVSS score. Based on this vulnerability rank report, we are alerting the banks that their mobile applications have vulnerabilities and hackers may use these vulnerabilities as an opportunity to gain access and made the system insecure.

The remainder of this document is as follows. Section 2 describes the overview of the proposed model, and Sect. 3 describes results we found after testing the banking apps.

2 Overview of the Proposed Approach

We are alerting the banks by following the approach shown in Fig. 3 that finally produces the ranking report of the vulnerabilities in the mobile application.

We selected the four top reputed banks in India (Bank A, Bank B, Bank C, and Bank D) and downloaded 37 mobile applications of those banks. We extracted the apk files of those banks for testing. After extraction, we get different artifacts at different abstraction levels. Manifest file is the high-level configuration file which stores the permissions. Native codes are represented at lower level of abstraction mainly used for performance requirements in the Android apps. Intents are stored in Java source code. We parsed all the files by using appropriate parsers in the pre-processing stage.

Vulnerability detection rules are formed by considering the reports of different sources like Open Web Application Security Project (OWASP) and Common Vulnerability Exposure (CVE). By using these rules, initially we did manual inspection

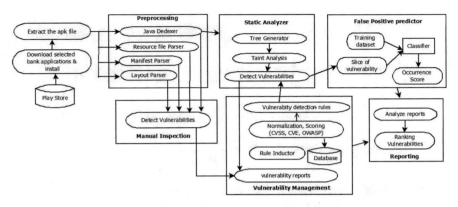

Fig. 3 Proposed approach

and detected the vulnerabilities. To find more vulnerabilities in the mobile applications, we used SCA tools. Rising false positives is the main drawback of static analysis. We tried for reducing the false positives by calculating the occurrence rate of each vulnerability. We used Naive Bayes classification algorithm for calculating the occurrence rate. We calculated CVSS for each vulnerability and normalized both CVSS score and occurrence rate and correlated by calculating the geometric mean.

2.1   Static Analysis

Static analysis is the process which analyzes the application without executing it. Static analysis observes the structure of the program and representation of the code. We analyzed the program data structure and its sensitivity. Control-flow graph (CFG) uses graphical notation that shows all paths must be visited in a program at the time of execution. Call graph (CG) is one type of CFG which gives the relationship between subroutines. Inter-procedural control flow graph (ICFG) is a combination of CFG and CG [8]. For detecting the vulnerabilities, we analyzed the sensitivity with respect to flow, context, and path. After analyzing the Java code, Java bytecode, and Dalvik bytecode and manifested layout and native codes. Static analyzers are the automated tools used for static analysis. It takes input as Java code. From this source code, it constructs abstract syntax tree and performs the taint analysis. From that, it will detect the vulnerabilities based on the vulnerability rules defined in the vulnerability management.

2.2 Naive Bayes Classification

To reduce the unexploitable vulnerabilities raised in the static analysis, we used Naive Bayes classifier. Naive Bayes classifiers are probabilistic classifiers based on Bayes theorem [9]. For estimating the vulnerability occurrence, we used Gaussian distribution. We trained the classifier by supplying the training data set. First, we separate the vulnerabilities based on the type as class and calculated the mean and variance of each class. Let class C_i and mean of the values y in C_i be μ_i and variance be σ_i^2. Let the tested vulnerability be k. Then probability of k in C_i, $p(y = k|C_i)$ calculated by using the below formula 1:

$$p(y = k|C_i) = \left(1 \left/ \sqrt{2 \prod \sigma_i^2}\right.\right) e^{(k-\mu_k)^2} \left/ 2\sigma_i^2\right. \tag{1}$$

By using the above formula, we calculated the probability of occurrence of each vulnerability in each bank.

2.3 CVSS and CVE

"Common vulnerability scoring system gives the characteristics of vulnerability and gives a standardized numerical score reflecting its severity" [10]. For assessing and prioritizing the vulnerabilities, CVSS represented severity of the vulnerabilities as either low, medium, high, or critical. Common vulnerability exposure (CVE) [11] provides a unique number to each vulnerability for public reference. We detected the vulnerabilities in the bank applications, and by using CVE and CVSS, we calculated the severity score of that vulnerability. For example, file unsafe check vulnerability found in one of the Bank 4 apps. Unique identification assigned by CVE is CVE-2017-17672. Its CVSS score is 9.8, and its severity is high. In this way, we identified vulnerabilities in all 37 apps and calculated the severity score of each vulnerability (Fig. 4).

2.4 OWASP

Open Web Application Security Project organization mainly concentrates on improving the software security. It provides web and mobile app security guidelines to individuals, corporations, groups, companies, etc. [12]. OWASP announced top 10 vulnerabilities in the mobile applications. They are

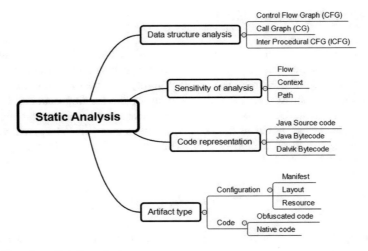

Fig. 4 Overview of static analysis

1. Improper platform usage,
2. Insecure data storage,
3. Insecure communication,
4. Insecure authentication,
5. Insufficient cryptography,
6. Insecure authorization,
7. Poor code quality,
8. Code tampering,
9. Reverse engineering,
10. Extraneous functionality.

3 Results

We tested the mobile applications with OWASP guidelines and identified the issues which may lead to vulnerabilities. In our experiments, we identified weak random number generation, lack of certificate pinning, information leakage, crash in Java code, broken cryptography, client-side SQL injection, and constant IV are major issues for occurring vulnerabilities in the apps. Table 1 shows issues found in various Bank A apps, and Table 2 shows issues which may lead to OWASP top 10 vulnerabilities in Bank A apps.

We tested 37 mobile banking applications. In these applications, we found 24 major vulnerabilities. Each vulnerability is numbered as unique number in this paper for easier representation. For example, file unsafe delete check vulnerability numbered as Vul-01. Table 3 lists the occurrence score and severity scores of all

Table 1 Issues found in Bank A apps

S. no.	Name of the issue	No. of issues found			
		Bank A1 app	Bank A2 app	Bank A3 app	Bank A4 app
1	Weak random number generator	1	2	0	3
2	Information leakage	1	2	0	0
3	Lack of certificate pinning	1	2	0	3
4	Broken cryptographic hash functions	1	0	2	0
5	crash in java code	1	0	0	0
6	Client side SQL injection	0	2	0	0
7	Constant IV	0	5	0	6
8	Lack of Cryptographic MAC	0	0	2	0

Table 2 Issues may lead to vulnerabilities in Bank A apps

S. no.	OWASP top 10	Issues may lead to vulnerability			
		Bank A1 app	Bank A2 app	Bank A3 app	Bank A4 app
1	Improper platform usage				
2	Insecure data storage	✓	✓		
3	Insecure communication	✓	✓		✓
4	Insecure authentication				
5	Insufficient cryptography	✓	✓	✓	✓
6	Insecure authorization				
7	Poor code quality	✓	✓		
8	Code tampering				
9	Reverse engineering				
10	Extraneous functionality				

vulnerabilities identified in the bank apps, and Table 4 shows the correlated scores of all vulnerabilities in all apps. Figures 5, 6, 7, 8 and 9 show the correlated scores of vulnerabilities graphically.

Table 5 shows the rank wise vulnerabilities found in bank applications.

Table 3 Occurrence score and severity scores for vulnerabilities found in all apps

Vul. no	Name of the Vulnerability found	Bank A Apps		Bank B Apps		Bank C Apps		Bank D Apps	
		Occurrence rate	Severity score	Occurrence rate	Severity score	Occurrence rate	Severity score	Occurrence rate	Severity score
Vul-01	File unsafe Delete Check	0.06096	9.8	0.0838	9.8	0.07457	9.8	0.09309	9.8
Vul-02	Using Services/Improper Export of Android Application Services	0.01463	9.8	0.02811	9.8	0.03185	9.8	0.06838	9.3
Vul-03	Improper Export of Android Application Components - Content	0.01463	9.8	0.01267	9.8	0.0104	9.8	0.01497	9.8
Vul-04	Missing usage of native(C, C++) code	0.06263	9.3	0.0624	9.3	0.0104	9.8	0.00939	9.8
Vul-05	Android PackageInfo Signature Verification/Android Fake ID Vulnerability	0.00234	9.8	0.00876	9.8	0.00745	9.8	0.00939	9.8
Vul-06	Using Activities/Improper Export of Android Application Activities	0.00026	9.8	0.00876	9.8	0.05155	9.3	0.00016	9.8
Vul-07	WebView addJavascriptInterface Remote Code Execution	0.03583	9.3	0.02562	9.3	0.03887	9.3	0.0193	9.3
Vul-08	Usage of Root/SuperUser Permission	0.08919	7.8	0.06777	7.8	0.05858	7.8	0.08694	7.8
Vul-09	Usage of Installer verification code	0.05053	7.8	0.07306	7.5	0.05795	7.8	0.07884	7.5
Vul-10	Protection of text fields from copying the text and paste outside your app	0.1001	7.2	0.04118	7.8	0.04369	7.8	0.06875	7.5
Vul-11	SSL Implementation Check - SSL Certificate Verification	0.06415	7.5	0.03183	7.8	0.06989	7.5	0.03334	7.8
Vul-12	Usage of Adb Backup	0.02788	7.8	0.08971	7.2	0.09552	7.2	0.08922	7.2
Vul-13	Use of base64 Encoded Strings	0.02455	7.8	0.05717	7.5	0.01698	7.8	0.01839	7.8
Vul-14	Executing "root" or System Privilege Check	0.00792	7.5	0.01505	7.8	0.0285	7.5	0.0116	7.8
Vul-15	Fragment Vulnerability Check	0.0005	7.5	0.01978	7.5	0.01549	7.5	0.00111	7.5

(continued)

Table 3 (continued)

Vul. no	Name of the Vulnerability found	Bank A Apps		Bank B Apps		Bank C Apps		Bank D Apps	
		Occurrence rate	Severity score	Occurrence rate	Severity score	Occurrence rate	Severity score	Occurrence rate	Severity score
Vul-16	SQLite Journal Information Disclosure Vulnerability	0.03111	6.5	0.01121	6.5	0.01785	6.5	0.00939	6.5
Vul-17	Receiving Broadcasts/Intent Spoofing/Improper Verification of Intent by Broadcast Receiver	0.01463	6.5	0.00876	6.5	0.0104	6.5	0.0083	6.5
Vul-18	Protection of app screens by blurring when the app is running in background	0.09869	5.5	0.09255	5.5	0.09527	5.5	0.09732	5.5
Vul-19	Certificate Pinning	0.01463	5.9	0.00876	5.9	0.0104	5.9	0.00939	5.9
Vul-20	Protection of capturing screenshots & sharing screens outside your app	0.09869	5	0.09255	5	0.09527	5	0.09732	5
Vul-21	MODE_WORLD_READABLE or MODE_WORLD_WRITEABLE Vulnerability check	0	5.5	0.01341	5.5	0.00037	5.5	0.01512	5.5
Vul-22	Outputting Logs to log cat/Logging Sensitive information	0.10039	4.4	0.08147	4.4	0.08711	4.4	0.08835	4.4
Vul-23	Emulator Detection Check	0.04524	4.4	0.03424	4.4	0.03573	4.4	0.0428	4.4
Vul-24	Unencrypted Credentials in Databases (sqlite db) Vulnerability check	0.04052	3	0.0314	3	0.03591	3	0.02913	3

Table 4 Correlated score for vulnerabilities in all apps

Vul. no	Name of the Vulnerability	Correlated Score			
		Bank A	Bank B	Bank C	Bank D
Vul-01	File unsafe Delete Check	24.441931	28.657285	27.033054	30.204006
Vul-02	Using Services/Improper Export of Android Application Services	11.973888	16.59753	17.667201	25.217732
Vul-03	Improper Export of Android Application Components - Content	11.973888	11.14298	10.095544	12.112225
Vul-04	Missing usage of native(C, C++) code	24.134187	24.089832	10.095544	9.5928098
Vul-05	Android PackageInfo Signature Verification/Android Fake ID Vulnerability	4.7887368	9.2654196	8.5445889	9.5928098
Vul-06	Using Activities/Improper Export of Android Application Activities	1.5962456	9.2654196	21.895547	1.2521981
Vul-07	WebView addJavascriptInterface Remote Code Execution	18.254287	15.435867	19.012917	13.397388
Vul-08	Usage of Root/SuperUser Permission	26.375784	22.991433	21.375781	26.040968
Vul-09	Usage of Installer verification code	19.852808	23.408332	21.260527	24.316661
Vul-10	Protection of text fields from copying the text and paste outside your app	26.846229	17.922165	18.460282	22.707378
Vul-11	SSL Implementation Check - SSL Certificate Verification	21.934562	15.756713	22.894868	16.126128
Vul-12	Usage of Adb Backup	14.746661	25.414799	26.224874	25.345295
Vul-13	Use of base64 Encoded Strings	13.837991	20.706883	11.508432	11.976727
Vul-14	Executing "root" or System Privilege Check	7.7071395	10.834667	14.620192	9.5120976
Vul-15	Fragment Vulnerability Check	1.9364917	12.179902	10.778451	2.8853076
Vul-16	SQLite Journal Information Disclosure Vulnerability	14.220232	8.5360998	10.77149	7.81249
Vul-17	Receiving Broadcasts/Intent Spoofing/Improper Verification of Intent by Broadcast Receiver	9.7516665	7.5458598	8.2219219	7.3450664
Vul-18	Protection of app screens by blurring when the app is running in background	23.297961	22.561582	22.890719	23.135687
Vul-19	Certificate Pinning	9.2906943	7.1891585	7.8332624	7.4431848
Vul-20	Protection of capturing screenshots & sharing screens outside your app	22.213734	21.511625	21.825444	22.059012

(continued)

Table 4 (continued)

Vul. no	Name of the Vulnerability	Correlated Score			
		Bank A	Bank B	Bank C	Bank D
Vul-21	MODE_WORLD_READABLE or MODE_WORLD_WRITEABLE Vulnerability check	0	8.5880731	1.4265343	9.1192105
Vul-22	Outputting Logs to logCat/Logging Sensitive information	21.017041	18.933251	19.57764	19.716491
Vul-23	Emulator Detection Check	14.108721	12.274201	12.538421	13.722973
Vul-24	Unencrypted Credentials in Databases (sqlite db) Vulnerability check	11.025425	9.7056685	10.379306	9.3482619

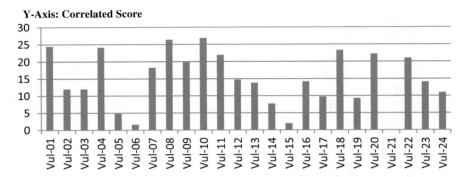

Fig. 5 Graph showing correlated score for vulnerabilities in Bank A apps

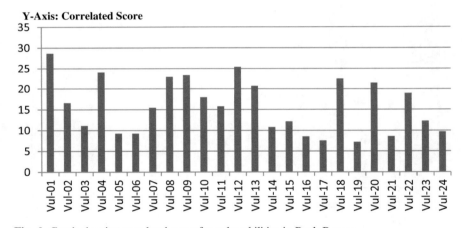

Fig. 6 Graph showing correlated score for vulnerabilities in Bank B apps

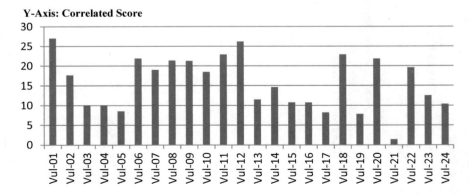

Fig. 7 Graph showing correlated score for vulnerabilities in Bank C apps

Y-Axis: Correlated Score

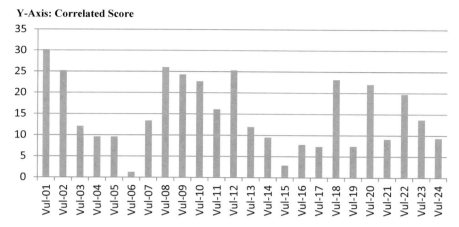

Fig. 8 Graph showing correlated score for vulnerabilities in Bank D apps

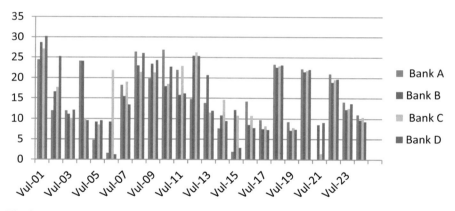

Fig. 9 Graph showing consolidated correlated score (**Y-Axis**) for vulnerabilities in all bank apps

4 Conclusion

From reputed top banks in India, we selected four banks for testing and tested 37 mobile applications of these banks. We did static analysis for identifying vulnerabilities in the mobile banking applications. To reduce false positives, we used Naive Bayes classifier for determining the occurrence rate. For calculating the severity score, we used CVSS and CVE. Occurrence rate and severity score are correlated for calculating the impact caused by that vulnerability, and then we ranked them. We identified issues that lead to OWASP top 10 vulnerabilities. By using these results, we give alerts to banks that their mobile applications have vulnerabilities causes which may lead to severe threat in future.

Table 5 Ranking the vulnerabilities based on correlated score

Rank	Bank A apps	Bank B apps	Bank C apps	Bank D apps
1	Vul-10	Vul-01	Vul-01	Vul-01
2	Vul-08	Vul-12	Vul-12	Vul-08
3	Vul-01	Vul-04	Vul-11	Vul-12
4	Vul-04	Vul-09	Vul-18	Vul-02
5	Vul-18	Vul-08	Vul-06	Vul-09
6	Vul-20	Vul-18	Vul-20	Vul-18
7	Vul-11	Vul-20	Vul-08	Vul-10
8	Vul-22	Vul-13	Vul-09	Vul-20
9	Vul-09	Vul-22	Vul-22	Vul-22
10	Vul-07	Vul-10	Vul-07	Vul-11
11	Vul-12	Vul-02	Vul-10	Vul-23
12	Vul-16	Vul-11	Vul-02	Vul-07
13	Vul-23	Vul-07	Vul-14	Vul-03
14	Vul-13	Vul-23	Vul-23	Vul-13
15	Vul-02	Vul-15	Vul-13	Vul-04
16	Vul-03	Vul-03	Vul-15	Vul-05
17	Vul-24	Vul-14	Vul-16	Vul-14
18	Vul-17	Vul-24	Vul-24	Vul-24
19	Vul-19	Vul-05	Vul-03	Vul-21
20	Vul-14	Vul-06	Vul-04	Vul-16
21	Vul-05	Vul-21	Vul-05	Vul-19
22	Vul-15	Vul-16	Vul-17	Vul-17
23	Vul-06	Vul-17	Vul-19	Vul-15
24	Vul-21	Vul-19	Vul-21	Vul-06

References

1. https://en.wikipedia.org/wiki/BFSI.
2. He, Wu, et al. 2015. *Understanding Mobile Banking Applications' Security risks through Blog Mining and the Workflow Technology.*
3. Maiya, Rajashekara. 2017. How to be a Truly Digital Bank. *Journal of Digital Banking* 1 (4): 338–348.
4. Mobile Banking 2015. *Global Trends and their Impact on Banks Produced in Collaboration with and Using Primary Survey Data Supplied by UBS Evidence Lab.*
5. Ramakrishna Murty, M., J.V.R. Murthy, and P.V.G.D. Prasad Reddy. 2011. Text document Classification Based on a Least Square Support Vector Machines with Singular Value Decomposition. *International Journal of Computer Application (IJCA)* 27 (7): 21–26.
6. https://thefinancialbrand.com/74044/mobile-banking-features-digital-security.
7. https://www.infopoint-security.de/media/Trustwave_2018-GSR_20180329_Interactive.pdf.
8. Nath, Hiran V., and Babu M. Mehtre. 2014. Static Malware Analysis Using Machine Learning Methods. In *International Conference on Security in Computer Networks and Distributed Systems.* Berlin: Springer.

9. Sadeghi, Alireza. 2017. *Efficient Permission-Aware Analysis of Android Apps Dissertation.* Diss. University of California, IRVINE.
10. Koc, Ugur, et al. 2017. Learning a Classifier for False Positive Error Reports Emitted by Static Code Analysis Tools. In *Proceedings of the 1st ACM SIGPLAN International Workshop on Machine Learning and Programming Languages.* ACM.
11. Bhatnagar, V., R. Majhi, and P.R. Jena. 2017. Comparative Performance Evaluation of Clustering Algorithms for Grouping Manufacturing Firms. *Arabian Journal for Science and Engineering*, August 2017.
12. https://www.first.org/cvss/.

Analysis of Object Identification Using Quadrature Bank Filter in Wavelet Transforms

C. Berin Jones and G. G. Bremiga

Abstract Nowadays guided missiles are widely used in military airbase. Weaponry advancements are developing in the last few decades. It has become an aggressive threat from the enemies in the battlefield. By using the effective electronic countermeasure (ECM) toward the targeted missile, it helps to miss the approaching target. For the missile tracking, parameters like phase, frequency, and intensity of the signal play a vital role to guide the signal toward the approaching target. In this paper, analysis of approaching missile has been done through quadrature bank filter (QBF), and it identifies the parameters using wavelet transform. It helps to improve the precision of the input signal by compressing the image through QBF. In addition, the simulation result has been evaluated, and the effectiveness of parameters in each stage of decomposition, filtration, and reconstruction has been done through wavelet transform.

Keywords Electronic countermeasure (ECM) · Object identification · Missile tracking · Quadrature bank filter (QBF) · Wavelet transform

1 Introduction

Recently, more advanced technologies have been used in battlefield to achieve mass destruction. To protect the battle space in modern war, ECM is used to identify the target and terminate the target before it reaches the targeted place. ECM generates the signals that are radiated to compete with the true signals for disrupting the

C. Berin Jones (✉)
Department of Computer Science Engineering,
Bhoj Reddy Engineering College for Women, Hyderabad 500059, India
e-mail: jonesberin@gmail.com; cberinjones@gmail.com

G. G. Bremiga
Department of Electronics and Communication Engineering,
Bhoj Reddy Engineering College for Women, Hyderabad 500059, India
e-mail: dr.bremiga.gg@gmail.com

© Springer Nature Singapore Pte Ltd. 2020
K. S. Raju et al. (eds.), *Proceedings of the Third International Conference on Computational Intelligence and Informatics*, Advances in Intelligent Systems and Computing 1090, https://doi.org/10.1007/978-981-15-1480-7_6

opponent's device and hide the true target position. When the identified target is given as an input image, most of them are corrupted by noisy signal at the time of target identification and transmission. Here, for better performance in terms of time and frequency domain, wavelet transform has been selected.

Basically, de-noising performance follows two steps. Obtained signal m(t) is combined along with the noisy signal $n(t)$ while transmission process.

$$\text{i.e., } x(t) = m(t) + n(t). \tag{1}$$

Noisy signal can be in any form. It can be speckle noise, Gaussian noise, salt and pepper noise. By applying the wavelet transform, $w(t)$ is obtained.

$$\text{i.e., } x(t) \rightarrow \text{Wavelet Transform} \rightarrow w'(t). \tag{2}$$

Then by applying filters like median, mean, Wiener, and Laplacian filters, noise signal can be removed. Finally, inverse discrete wavelet transform is performed to reconstruct the original signal $w'(t)$. Here, input image undergoes wavelet transform to filter the noise from input target image. Also, QBF is used to de-noise the corrupted noisy image, enhance the quality of image, and recover the targeting object. Furthermore, inverse wavelet transform is applied to obtain the image of the target [1]. Recently, many researches have focused to create an optimized filter to filter the noise and solve the noisy problem by weighted noisy square error to achieve better performance analysis after the reconstruction of the identified object. Rest of the paper focuses to improve the recognition speed and reduce the error rate. Related works are reviewed. Wavelet transform-based decomposition and reconstruction and QBF are covered. Finally, experimental result and performance are discussed, and the paper is concluded.

2   Related Work

[2] analyzed and explained the image compression method using wavelet transform and stated that by compact filter and quantizer, the quality of the obtained image is improved. [3] proposed an effective iterative algorithm to recover the noise from image in discrete wavelet transform. [4] explained the wavelet transforms, filter banks, and its various application in the field of image and video processing. [5] presented a complex wavelet transform for analysis and filtering the signals and explained image extension over multiple dimensions and summarized the application range and advantages like de-noising, analysis and synthesis, and classification. [6] derived image de-noising using Gaussian scale mixture and wavelet coefficient model and estimated the mean square error. [7] performed wavelet-based de-noising method and Gaussian smoothing method for de-noising the image, and the experimental results show a reduction in error. [8] proposed an algorithm which minimizes the image noise by combining threshold and linear filter method along

with wavelet transform which improved the efficiency of the performance. [9] proposed a new fuzzy-based image de-noising technique which handles different abnormalities of image by general functions. It optimized the effectiveness of better image identification and reconstruction of image by reducing the noise hidden in image. [10] performed de-noising of image using luminance and segmentation of image parts through wavelet domain and achieved the effective result. [11] performed the restoration process through varying the noise present in the image through Gaussian scale mixture. [12] proposed wavelet-based image de-noising technique. In general, wavelet algorithms are used as a tool for signal processing, which comprises of de-noising and image compression. Additionally, the technique of multi-wavelets is proposed here to modify the coefficient of scalar wavelets, which is computationally faster and shows better results. [13] given a solution for noise elimination in fingerprint image using median filter. Here with the median filter enhancement of the gray-level image showed preferable noise suppression with re-establishment of actual information than the binary images. With an insecure channel, [14] established a highly secured transmission of fingerprints using an effective fingerprint encryption technique. It used fractional random wavelet transform (FrRnWT), which in turn inherits the properties of both the fractional random transform and the wavelet transform. [15] explained various noise removal algorithms and filtering techniques. Also, it highlighted the BM3D filters, an adaptive filter as the best filter to de-noise the salt and pepper noisy images. [16] proposed an efficient de-noising technique to remove the salt and pepper noise using cubic B-spline method.

These conventional techniques concentrate mainly on noise removal techniques, evaluating coefficient values, image decomposition and reconstruction, etc., but none combines all these approaches simultaneously. So to reduce this research gap, this paper presents an integrated approach to address the complexity of radar targeting objects.

3 Implementation in Wavelet Processing

In this paper, we deal with 2D discrete wavelet transform and QBF which ultimately proposed to identify and classify the targeting object. Once the target is identified, it goes through the normalization process for the enhancement of the obtained image. After that, it performs next processing stage of feature extraction by means of wavelet decomposition. The important principle exhibited by wavelet transform is the decomposing process of the functional data components into different frequency components. Every component obtained after decomposition performs independent analysis by considering the resolution parameter. Then QBF is used to filter the noise of the identified object. In the next stage, wavelet reconstruction process is handled which helps to combine the achieve image along with the database and identify the distance and position of the target and classify the identified object. Here, basic scaling and mother wavelet are the functions considered to perform the task.

$$\phi_{i,k}(t) = 2_j/2\phi\left(2_{jt-k}\right). \tag{3a}$$

$$\Psi_{j,k}(t) = 2_j/2_\Psi\left(2_{jt-k}\right). \tag{3b}$$

$$f(t) = \sum_{k=-\infty}^{\infty} a_{j0}(k)\phi_{j0k(t)} + \sum_{j=-jo}^{\infty}\sum_{k=-\infty}^{\infty} d(j,k)\Psi_{j,k}(t). \tag{4}$$

3.1 Radar Object Identification and Classification

The purpose of radar object identification relies on assigning input vectors to different types. When the radar identifies the target, it helps to identify the position of the object. Normally, the targeted missile image consists of certain noise. Wavelet transform helps to model and deals with approximate coefficient of images. Usually in real-time image, noises corrupt the image. Wavelet transform is an excellent method to solve this problem. Major advantage of this transform is that it consumes less energy and performs faster. This wavelet transform gives an idea about sequencing the signal by means of decomposition property. Figure 1 shows flow graph of object identification approach.

3.2 Quadrature Bank Filter

Generally, quadrature bank filter starts to work by selecting the discrete signal. Then the signal is split into a number of sub-band signals. Each sub-band signal goes through the filter $H(z)$ using decimation and interpolation process. After completion, final step of filtration process takes place. Then all the process get added, and the final original signal is obtained at $x'(n)$ which is shown in Fig. 2. In such filter, variation in reconstructed signal from original exists because of the distortion in amplitude and phases, aliasing process, etc. Usually, QBF comprises low-pass and high-pass filter circuits; thus, overlapping of two-response signal occurs and that further results in aliasing effect. To overcome this impact, filters are designed with the assumed considerations of coefficient values.

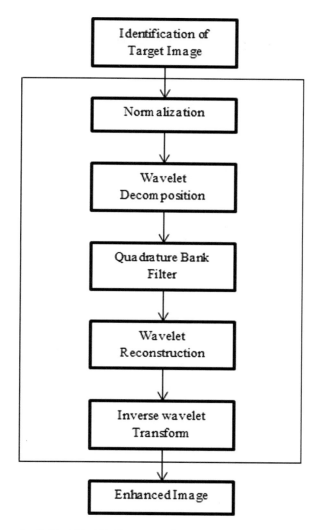

Fig. 1 Flow graph of object identification approach

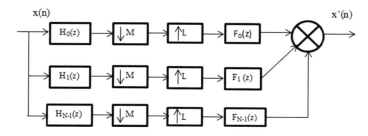

Fig. 2 Structure of quadrature bank filter

Analysis filter

Quadrature mirror filter bank is designed in linear phase which is given as,

$$H_1(z) = H_0(-z). \tag{5}$$

$$h_1(n) = (-1)^n h_0(n). \tag{6}$$

where $H_0(z)$ and $H_1(z)$ are high-pass filter and low-pass filter.

For the decimation and interpolation process, output is defined by

$$y_D(n) = x(m^* n). \tag{7}$$

Synthesis filter

$$F_0(z) = H_1(-z) = H_0(z). \tag{8}$$

$$F_1(z) = -H_1(z) = -H_0(-z). \tag{9}$$

By combining input and output, Z-transform $X'(z)$ is then given by

$$\begin{aligned} X'(z) &= 1/2[H_0(z)F_0(z) + H_1(z)F_1(z)]X(z) \\ &\quad + 1/2[H_0(-z)F_0(z) + H_1(-z)F_1(z)F_1(z)]X(-z). \end{aligned} \tag{10}$$

Then the final output completely depends on

$$\begin{aligned} X'(z) &= 1/2[H_0(z)F_0(z) + H_1(z)F_1(z)]X(z)] \\ &= 1/2\big[H_0^2(z) - H_0^2(-z)X(z). \end{aligned} \tag{11}$$

[17, 18] Filter bank (two channels) is categorized into three major types. They are biorthogonal filter, quadrature bank filter, and orthogonal filter. [19] These filter banks are held to design and perform the operation of filtration process.

3.3   Wavelet Decomposition and Reconstruction

This section describes the representation of the wavelet decomposition and reconstruction. For simplicity, transfer function representation of filters used in QBF is assumed as follows:

$$H(z) = \sum_{n=0}^{L-1} h(z)z^{-n}. \tag{12}$$

$$F(z) = \sum_{n=0}^{L-1} f(z)z^{-n}. \tag{13}$$

Practically, wavelets are analyzed accurately by performing multi-stage decomposition. Certain multi-stage decomposition algorithms are introduced for providing clear understanding of the problems. Input images are transformed into matrix elements with N rows and M columns, respectively. Each decomposition level filters the horizontal data, and further, the details and approximation obtained from the filtrations are subjected to column filtration. Finally, output images are obtained with approximation, vertical detail, horizontal detail, and the diagonal detail. Also, the sub-images are further decomposed in similar manner. The horizontal and vertical edges exhibited by the input image are evaluated using the horizontal and vertical detail coefficients. Both the edge information are combined by considering zero approximation coefficients.

Stage 1 computation is performed by

$$
\begin{aligned}
d1(0) = &\, h(0)x(0) + h(1)x(-1) + h(2)x(-2) \\
&+ h(3)x(-3) + h(4)x(-4) + h(5)x(-5).
\end{aligned}
\tag{14}
$$

$$
\begin{aligned}
d1(2) = &\, h(0)x(2) + h(1)x(1) + h(2)x(0) \\
&+ h(3)x(-1) + h(4)x(-2) + h(5)x(-3).
\end{aligned}
\tag{15}
$$

$$
\begin{aligned}
d1(4) = &\, h(0)x(4) + h(1)x(3) + h(2)x(2) \\
&+ h(3)x(1) + h(4)x(0) + h(5)x(-1).
\end{aligned}
\tag{16}
$$

$$
\begin{aligned}
d1(6) = &\, h(0)x(6) + h(1)x(5) + h(2)x(4) \\
&+ h(3)x(3) + h(4)x(2) + h(5)x(1).
\end{aligned}
\tag{17}
$$

$$
\begin{aligned}
xi1(0) = &\, f(0)x(0) + f(1)x(-1) + f(2)x(-2) \\
&+ f(3)x(-3) + f(4)x(-4) + f(5)x(-5).
\end{aligned}
\tag{18}
$$

$$
\begin{aligned}
xil(2) = &\, f(0)x(2) + f(1)x(1) + f(2)x(0) \\
&+ f(3)x(-1) + f(4)x(-2) + f(5)x(-3).
\end{aligned}
\tag{19}
$$

$$
\begin{aligned}
xil(4) = &\, f(0)x(4) + f(1)x(3) + f(2)x(2) \\
&+ f(3)x(1) + f(4)x(0) + f(5)x(-1).
\end{aligned}
\tag{20}
$$

$$
\begin{aligned}
xil(6) = &\, f(0)x(6) + f(1)x(5) + f(2)x(4) \\
&+ f(3)x(3) + f(4)x(2) + f(5)x(1).
\end{aligned}
\tag{21}
$$

Stage 2 computation is performed by

$$
\begin{aligned}
d2(0) = &\, h(0)xi1(0) + h(1)xi1(-2) + h(2)xi1(-4) \\
&+ h(3)xi1(-6) + h(4)xi1(-8) + h(5)xi1(-10).
\end{aligned}
\tag{22}
$$

$$d2(4) = h(0)xi1(4) + h(1)xi1(2) + h(2)xi1(0) \\ + h(3)xi1(-2) + h(4)xi1(-4) + h(5)xi1(-6). \tag{23}$$

$$xi2(0) = g(0)xi1(0) + g(1)xi1(-2) + g(2)xi1(-4) \\ + g(3)xi1(-6) + g(4)xi1(-8) + g(5)xi1(-10). \tag{24}$$

$$Xi2(4) = g(0)xil(4) + g(1)xi1(2) + g(2)xi1(0) \\ + g(3)xil(-2) + g(4)xi1(-4) + g(5)xi1(-6). \tag{25}$$

Wavelet computation process is performed along with QBF. Similar operations are performed to reconstruct the image. The results of decomposition process through QBF are shown in Figs. 4 and 5. Coefficient value of the wavelet has been adjusted for identifying the parameter value. Horizontal and vertical detail coefficient, diagonal detail coefficient, and approximation coefficient values are identified. By using different function, input values can be changed to convolve and cause different effects in the resolution of the image.

While performing image reconstruction process, the wavelet transforms are applied on the columns and the rows, respectively. As the decomposition stages get completed, periodical symmetric extensions are accomplished on the sub-bands. After this, up-sampling of the band arrays by factor 2 takes place. Further, zero padding of coefficients taking place subsequently follows the filtering process by LPF and HPF. Now the reconstructed images are obtained by applying the wavelet transform to the row elements. Thereby, the obtained reconstructed images seem similar to that of the original images, but the use of QBF filter improves the quality.

4 Scheme of Embedding and Extraction

Embedding and extraction algorithm is applied for de-noising received approaching target. It is divided into two stages to decrease the computation complexity, and hence, it improves the processing speed.

The first stage is embedding algorithm.

1. Pre-processing: Computation of the captured image takes place. It converts RGB into grayscale. Then the computation of wavelet coefficient is achieved from the captured signal.
2. DWT decomposition: Originally captured image is computed to decompose the signal.
3. QBF filter: Coefficient value of each pixel is computed through QBF filter, where decimation and interpolation process has to be identified.

The second stage is extraction algorithm.

Feature extraction maps the input signal into a vectorized components so that input space complexities are further reduced. The feature vector components are of smaller dimensions, and it represents an important property of transformation. Practically, noises are common in every device, and it is necessary to extract the useful data from the noise signals. To achieve this, coefficient values are required to compute and are compared with the threshold limit.

1. DWT reconstruction: Reconstruction process involves the computation of coefficient values.
2. IDWT: Finally, reconstructed image is applied through inverse discrete wavelet transform to identify the position of the target.

If the threshold limit is found to be greater than that of the wavelet coefficients, then it indicates the distorted signal and lower threshold limit retains the signal from the noises.

5　Computational Analysis

To identify the approaching missile, various performance and quality metrics are required to be considered to identify the position of the target precisely. For that normalized absolute error (NAE), image fidelity (IF), normalized cross-correlation (NCC), mean structural similarity index (MSSIM), and peak signal-to-noise ratio (PSNR) are measured.

To find out the value of NCC, values of IF and NAE are essentially needed which is expressed as,

$$\text{NAE} = \sum_{i=1}^{N} \sum_{j=1}^{M} \left(C_{i,j} - C'_{i,j} \right) \Big/ \sum_{i=1}^{N} \sum_{i=1}^{M} \left(C_{i,j} \right). \tag{26}$$

$$\text{IF} = 1 - \left[\sum_{i=1}^{N} \sum_{j=1}^{M} \left(C_{i,j} - C'_{i,j} \right) \Big/ \sum_{i=1}^{N} \sum_{j=1}^{M} \left(C_{i,j} \right)^2 \right] \tag{27}$$

where

N and M　　represent the size of the image,
$C_{(i,j)}$　　represents captured image, and
$C'_{i,j}$　　represents reconstructed image.

Calculation of NCC gives the value of difference between captured image and reconstructed image. It suggests the value between two images from −1 to +1. If the images show same value, it results in 1. If the images show different value, it would be −1. If two images are uncorrelated, then it shows 0. The value of normalized cross-correlation is measured from,

$$\text{NCC} = \sum_{i=1}^{M} \sum_{j=1}^{N} \left[C_{i,j} - \mu_s \right] \left[C'_{i,j} - \mu'_s \right] \Big/ \sqrt{\sum_{i=1}^{M} \sum_{j=1}^{N} \left[C_{i,j} - \mu_c \right]^2} \sqrt{\sum_{i=1}^{M} \sum_{j=1}^{N} \left[C'_{i,j} - \mu'_c \right]^2}$$

(28)

To calculate the MSSIM, it compares all the possible patterns and identifies the structure, contrast, and luminance of the object. Then it is measured from x and y window of common size image $N * M$.

$$\text{MSSIM}(X, Y) = 1/M \sum_{i=1}^{M} \text{SSIM}(x_j, y_j)$$

(29)

where

X and Y are captured image and reconstructed image.
x_j and y_j are image windows.

SSIM can be measured from

$$\text{SSIM}(x, y) = \left(2\mu_x \mu_y + C_1 \right) \left(2\sigma_{xy} + C_2 \right) \Big/ \left(\mu_x^2 + \mu_y^2 + C_1 \right) \left(\sigma_x^2 + \sigma_y^2 + C_2 \right).$$

(30)

where

σ_x represents the average value of x,
σ_y represents the average value of y,
σ_{x2} represents x variance,
σ_{y2} represents y variance,
σ_{xy} represents variance of x and y,
$C_1 = (K_1 L)^2$ and $C_2 = (K_2 L)^2$ are the variables used to stabilize the division,

L = dynamic range of the pixel values,
$K_1 = 0.01$, and $K_2 = 0.03$.

PSNR improves the image quality. It is a common parameter to measure the image quality.

$$\text{PSNR} = 20 \log_{10}(255/\text{RMSE})$$

(31)

where root-mean-square error (RMSE) is given as,

$$\text{RMSE} = \sqrt{[1/MN]} \sum_{i=1}^{N} \sum_{j=1}^{M} \left(C_{i,j} - C'_{i,j} \right)^2$$

(32)

6   Result and Discussion

Here, the visual impact of the targeting object is identified, and the RGB image is converted into the grayscale image for the normalization of the image to identify the position of the target by comparing the object with the database. The normalization process of captured image is shown in Fig. 3. The quality of the image is considered in each stage. Wavelet coefficient values are achieved using wavelet transform.

Normalization is also referred as stretching of images based on contrast and histogram. An important aim of dynamic range expansion is to provide the images or signals within a range suitable for sensing thus normalization term arises. It also further motivates to provide consistent output related to dynamic ranges of data or image signals thereby avoiding distractions. The use of image normalization is to make the image features common to humans or machine visions so that contrast of an image is improved to obtain better feature extraction. Linear normalization stretches the image intensity values, and the histogram intensity is also stretched to offer enhanced contrast. Discrete wavelet transform performs the first- and second-order decomposition processes. Wavelet coefficient value of each pixel is computed along with QBF. Figures 4 and 5 show the output representation of the wavelet first- and second-order decomposition with the help of the QBF, and it also shows individual histogram representation of the decomposition process.

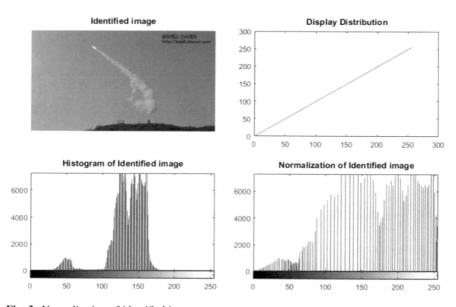

Fig. 3 Normalization of identified image

wavelet First order decomposition

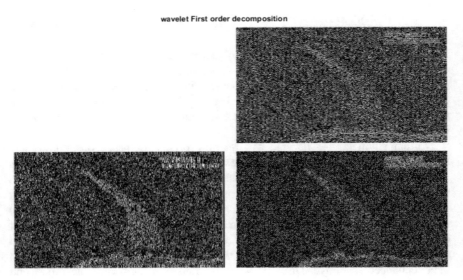

Fig. 4 Wavelet first-order decomposition

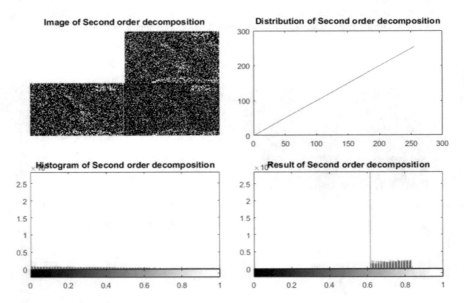

Fig. 5 Wavelet second-order decomposition

In second-order decomposition process, the distribution of signal is said to be linear. The signal remains constant in case of histogram. Thus, the output of decomposition shows a high peak value at time period of 0.62 and then variations in values results. After the processing of first-order and second-order decomposition, reconstruction process of image takes place. Decimation and interpolation process is performed in analysis bank and synthesis bank of QBF which are shown in Figs. 6 and 7.

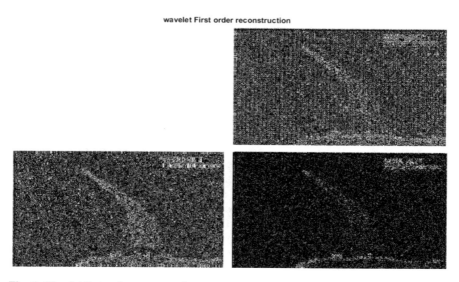

Fig. 6 Wavelet first-order reconstruction

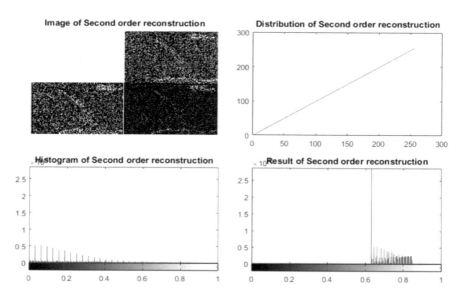

Fig. 7 Wavelet second-order reconstruction

Similar to decomposition, reconstruction of first and second order also possess similar results. But the only difference is the histogram. Here, the amplitude of the signals decreases as the time period increases and finally reaches 0 value at time period 0.5. Finally, the obtained reconstructed image is shown in Fig. 8. PSNR, MSSIM, and NCC parameters are obtained from the reconstructed image. The histogram and the output results show a dependent relationship between the amplitude and its time period due to its signification variations.

Tables 1 and 2 show the quality and scaling factors of PSNR, MSSIM, and NCC, respectively. Quality factor is measured from various ranges of values from 35 to 100. Similarly, scaling factor is measured from −10 to +10, respectively. The PSNR, MSSIM, and NCC increase as the quality and scalar factors tend to increase. While comparing PSNR, MSSIM, and NCC, greater values are obtained.

Figure 9 demonstrates the PSNR quality factor and scaling factor of the approaching target image. DWT along with QBF method decomposes and reconstructs each sub-bands. Mainly, it concerns about the precision and confidential position of the target. Both the PSNR quality and scalar factor vary linearly.

Figure 10 demonstrates the MSSIM quality factor and scaling factor of the approaching target image. Here, the major concern about the luminance, contrast, and structure of the target is adjusted, and values have been measured. In this case, the quality factor tends to increase linearly whereas in scaling factor, MSSIM tends to increase for positive scaling factor but for negative values, decreased values of MSSIM are obtained.

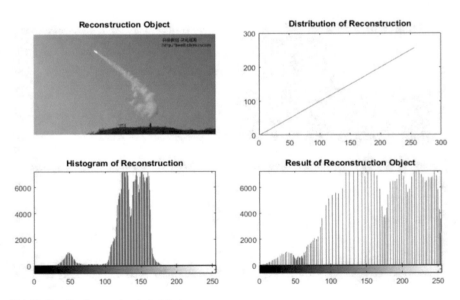

Fig. 8 Result of reconstructed object

Table 1 PSNR, MSSIM, and NCC quality factor value of identified missile

Quality factor	PSNR	MSSIM	NCC
35	39.5596	0.9539	0.997617
40	39.9741	0.9568	0.997927
45	40.3057	0.9597	0.998031
50	40.6373	0.9618	0.998031
55	40.886	0.9634	0.998238
60	41.2176	0.9663	0.998342
65	41.6321	0.9684	0.998549
70	42.0466	0.9705	0.998864
75	42.4611	0.973	0.998964
80	49.2985	0.9754	0.999067
85	43.7047	0.9788	0.999275
90	44.6166	0.9821	0.999482
95	46.3575	0.9866	0.999585
100	50.3368	0.9882	0.999585

Table 2 PSNR, MSSIM, and NCC scaling factor value of identified missile

Scaling factor	PSNR	MSSIM	NCC
−10	21.9378	0.43	0.9266
−9	22.8497	0.46	0.9382
−8	23.9896	0.49	0.949
−7	24.9016	0.54	0.9589
−6	26.2694	0.59	0.968
−5	27.8653	0.65	0.9755
−4	29.8031	0.72	0.9821
−3	32.1969	0.8	0.9875
−2	35.3886	0.88	0.9908
−1	40.0622	0.96	0.9933
1	40.1762	0.96	0.9933
2	35.3886	0.88	0.9912
3	32.1969	0.8	0.9875
4	29.8031	0.72	0.9821
5	27.8653	0.66	0.9759
6	26.2694	0.6	0.9676
7	25.0155	0.55	0.9593
8	23.9896	0.5	0.9494
9	22.8497	0.46	0.9386
10	21.9378	0.43	0.927

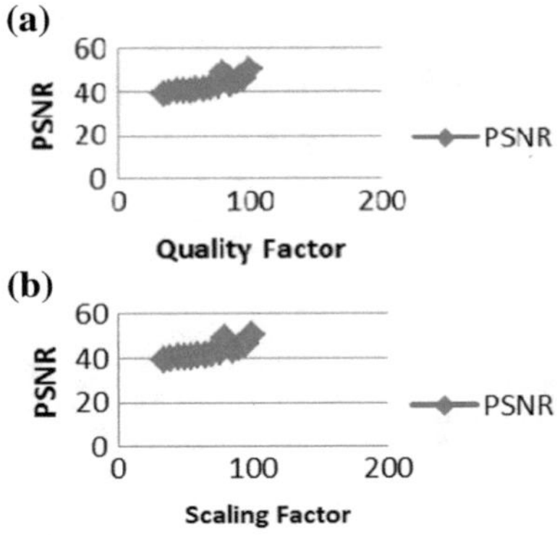

C. Berin Jones and G. G. Bremiga

Fig. 9 PSNR **a** quality factor
and **b** scaling factor

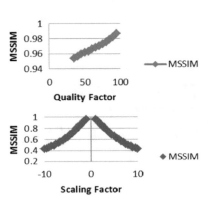

Fig. 10 MSSIM **a** quality
factor and **b** scaling factor

Figure 11 demonstrates the NCC quality factor and scaling factor of the approaching target image. This suggests the similarity between reconstructed image and identified target image. After performing the discrete wavelet transform, QBF is used to identify the target by removing the noises in the image. All the computational values of PSNR, MSSIM, and NCC are calculated and tabulated in Tables 1 and 2. After performing the computational analysis, the obtained results are compared with the performance of previous best performance approaches like clust method, PSNR obtained as 34.64 [10], vector GSM and neighbourhood vector GSM [20], PSNR obtained as 35.29 and 37.05, and ESURE method [21] obtained as 38.15. The proposed method shows optimal values of PSNR 50.33, MSSIM 0.98, and NCC 0.99, which are greater when compared with conventional and biorthogonal wavelets [22]. Decomposition and reconstruction of the process take less time and less computational complexity. While comparing with existing method, this concept improves the precision of the input signal by compressing the image through QBF which helps to identify the target and hack the signal position of the missile.

Fig. 11 NCC **a** quality factor
and **b** scaling factor

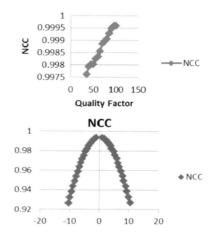

7 Conclusion

In this paper, radar target identification is performed without any knowledge of the position and location of the target. Here, the wavelet transform, decomposition, and reconstruction have been done along with QBF. This method takes minimum time to execute and provides better performance with less computational complexity. Also, it is verified using the parameter PSNR, MSSIM, and NCC, in which parameter shows better-proposed results when compared with the traditional methods. With some advanced sophisticated technique, we can further improve the strategies of identifying the object. Extension of this work can be performed in different directions. This algorithm can also be used to analyze the images that are affected by salt and pepper noise, speckle noise, and other noise models. DWT offers flexible operations by appropriate wavelet selection that improves PSNR and produce visually clear images. Further, the noise can be minimized without actually losing the main data. A thorough research can be done in choosing a mother wavelet suitable for de-noising a particular image.

References

1. Chen, C.K., and J.H. Lee. 1992. Design of quadrature mirror filters with linear phase in the frequency domain. *IEEE Transactions on Circuits and Systems II: Analog and Digital Signal Processing* 39: 593–605.
2. Lewis, A.S., and G. Knowles. 1992. Image Compression Using the 2-D Wavelet Transform. *IEEE Transactions on Image Processing* 1.
3. Enis Cetin, A., and Rashid Ansari. 1994. Signal Recovery from Wavelet Transform Maxima. *IEEE Transactions on Signal Processing* 42.

 4. Strang, G., and T.Q. Nguyen. 1996. *Wavelets and Filter Banks*, 103–108. Cambridge MA: Welleseley-Cambridge Press.
 5. Kingsbury, N.G. 2001. Complex Wavelets for Shift Invariant Analysis and Filtering of Signals. *Journal of Applied and Computational Harmonic Analysis* 10: 234–253.
 6. Portilla, J., V. Strela, M.J. Wainwright, and E.P. Simoncelli. 2003. Image Denoising Using Gaussian Scale Mixtures in the Wavelet Domain. *IEEE Transactions on Image Processing* 12: 1338–1351.
 7. Wink, A.M., and J.B.T.M. Roerdink. 2004. Denoising Functional MR Images: A Comparison of Wavelet Denoising and Gaussian Smoothing. *IEEE Transactions on Medical Imaging* 23: 374–387.
 8. Lee, Yong-Hwan, Sang-Burm Rhee. 2005. Wavelet-Based Image Denoising with Optimal Filter. *Journal of Information Processing Systems*, 32–35.
 9. Schulte, S., B. Huysmans, A. Pižurica, E. Kerre, and W. Philips. 2006. A New Fuzzy-Based Wavelet Shrinkage Image Denoising Technique. In: *Advanced Concepts for Intelligent Vision Systems*.
10. Jovanov, L., A. Pižurica, and W. Philips. 2007. Wavelet Based Joint Denoising of Depth and Luminance Images. In *3D TV Conference*, Kos Island, Greece.
11. Guerrero-Colon, J.A., L. Mancera, and J. Portilla. 2008. Image Restoration Using Space-Variant Gaussian Scale Mixtures in Overcomplete Pyramids. *IEEE Transactions on Image Processing* 17: 27–41.
12. Himabindu, G,, M. Ramakrishna Murty et al. 2018. Extraction of Texture Features and Classification of Renal Masses from Kidney Images. *International Journal of Engineering &Technology* 7 (2.33): 1057–1063.
13. Ruikar, Sachin D., and Dharmpal D Doye. 2011. Wavelet Based Image Denoising Technique. *International Journal of Advanced Computer Science and Applications* 2.
14. Chandra, E., and K. Kanagalakshmi. 2011. Noise Suppression Scheme Using Median Filter in Gray and Binary Images. *International Journal of Computer Applications* 26: 49–57.
15. Bhatnagar, G., Q.M.J. Wu, and B. Raman. 2012. A New Fractional Random Wavelet Transform for Fingerprint Security. *IEEE Transactions on Systems, Man, and Cybernetics Part A: Systems and Humans* 42: 262–275.
16. Verma, Rohit, and Jahid Ali. 2013. A Comparative Study of Various Types of Image Noise and Efficient Noise Removal Techniques. *International Journal of Advanced Research in Computer Science and Software Engineering* 3.
17. Ibrahem, Hani M. 2014. An Efficient Switching Filter Based on Cubic B-Spline for Removal of Salt-and-Pepper Noise. *International Journal of Image, Graphics and Signal Processing* 5: 45–52.
18. Vaidyanathan, P.P. 1993. *Multirate Systems and Filter Banks*. Englewood Cliffs, NJ: Prentice Hall.
19. Bregovic, R., and T. Saramaki. 2003. A General-Purpose Optimization Approach for Designing Two-Channel Fir Filterbanks. *IEEE Transactions on Signal Processing* 51: 1783–1791.
20. Bregovic, R., and T. Saramaki. 1999 Two-Channel FIR Filterbanks—A Tutorial Review and New Results. In *Proceedings of the 2nd International Workshop on Transforms Filter Banks*, vol. 4, 507–558.

21. De Backer, S., A. Pižurica, B. Huysmans, W. Philips, and P. Scheunders. 2008. Denoising of Multicomponent Images Using Wavelet Least-Squares Estimators. *Image Vision Computing* 26: 1038–1051.
22. Benazza-Benyahia, A., and J. Pesquet. 2005. Building Robust Wavelet Estimators for Multicomponent Images Using Stein's Principle. *IEEE Transactions on Image Processing* 14: 1814–1830.

Experimental Analysis of the Changes in Speech while Normal Speaking, Walking, Running, and Eating

Sakil Ansari, Sanjeev K. Mittal and V. Kamakshi Prasad

Abstract Speech communication is a mundane activity. People talk while ambling, running, biking, eating, and so on. This paper analyzes the variability in speech parameters while a person speaking when walking, eating, or running with the normal speaking. In this paper, we get information about the analysis of speech signals while normally speaking and speaking along with walking, eating, or running. The analysis of speech signal, according to minimum pitch, maximum pitch, mean energy intensity and mean F1 (first formant), has been carried out and also experimented in Praat, and then the problem is implemented in MATLAB. This paper meets the variation in speech parameters in different scenarios. The clustering is performed, and the paper presents an experimental comparison of two different features of speech.

Keywords Formants · K means algorithm · MFCC · Spectrogram

1 Introduction

The voiced speech of a typical adult male and a typical adult female have a fundamental frequency from 85 to 180 Hz and from 165 to 255 Hz, respectively. The Maximum range of human hearing includes sound frequencies from about 15 to about 18,000 waves, and the general range of hearing for young people is 20 Hz to 20 kHz. Most of the literature on combined activities which are walking, running,

S. Ansari (✉) · V. Kamakshi Prasad
Department of Computer Science and Engineering,
Jawaharlal Nehru Technological University, Hyderabad, India
e-mail: sakilansari4@gmail.com

V. Kamakshi Prasad
e-mail: kamakshiprasad@jntuh.ac.in

S. K. Mittal
Department of Electrical Engineering, Indian Institute of Science, Bangalore, India
e-mail: rsr.skm@gmail.com

© Springer Nature Singapore Pte Ltd. 2020
K. S. Raju et al. (eds.), *Proceedings of the Third International Conference on Computational Intelligence and Informatics*, Advances in Intelligent Systems and Computing 1090, https://doi.org/10.1007/978-981-15-1480-7_7

eating is concerned with the reduced attention caused by the dual process and with the speed of cognitive processing. For example, talking on a mobile phone changes the way we talk [1], and mathematical operations are slower when people additionally walk on a treadmill than when they just focus on the operations [2]. So, here we investigate how speech production changes when people do different activities in terms of speech parameters like pitch, mean energy intensity, and first formant, and then classification is performed by using K means algorithm. A spectrogram is the spectrum of frequencies of sound that varies with time or some other variables [3].

Speech production was recorded while speaking alone, speaking and walking, speaking and eating, speaking after running. For these instances, we have to analyze the change in speech parameters like minimum pitch, maximum pitch, and intensity. The main challenge in speech recognition systems is tolerance to speaker differences and variations in context and environment [4]. One more challenge is speakers may adopt different strategies when talking while walking, running, and eating [5]. Each speaker may not give pause, and at the same time, it depends on speaker to speaker; one can speak slowly or fast with different pause intervals.

2 Literature survey

In the entitled paper, automatic classification of eating conditions from speech using acoustic feature selection and a set of hierarchical support vector machine classifiers [6], the task of classification of eating conditions from the information embedded in the speech of a speaker while eating has been addressed. In this work, they recognize whether the subject is eating or not and if the subject is eating, what the type of food, the subject eats while speaking? They consider six types of foods—apple, banana, Haribo, biscuit, crisp, and nectarine.

In the literature, there are a few attempts for the automatic classification of eating conditions from speech [7, 8]. Hantke et al. [8] have proposed an automatic classification by using low-level acoustic features as well as high-level features related to intelligibility from an automatic speech recognizer. An audio–visual dataset is used for this classification. On the other hand, Schuller et al [7] have proposed a set of acoustic-only features with multi-class support vector machine (SVM) classifier for the eating condition classification task, which is used as the baseline scheme for the work in this paper. In this paper [9], they study the effect of biking on respiratory, speech parameters, and breathing, and speech production was recorded in eleven subjects while speaking alone and while speaking and biking with different rates. In this work [10], a novel method for the improvement of speech and audio signal classification using spectral flux (SF) pattern recognition for the MPEG Unified Speech and Audio Coding (USAC) standard is presented. This paper [11] talks about a method which deals with the problem of sports classification through audio analysis.

3 Proposed method

The methodology presented in this paper analyzes the speech parameters in terms of pitch, mean energy intensity, and first formant. This analysis is reaffirmed in Praat. The current state-of-the-art accomplishes clustering using two different features, and finally, the accuracy of the system is evaluated. Figure 1 is an architectural diagram of the proposed system.

3.1 Speech parameters analysis

3.1.1 Dataset

The dataset was recorded using Audacity. In Audacity, we recorded the speech for the given text material about TK solver plus for about 60 seconds. The dataset was recorded speech at 32,000 Hz as sample frequency rate, and the rest of the settings were default for all the four conditions which are normal speaking and with walking, running, eating. Four categories of recording were done: normal speaking, speaking while walking, speaking while eating, and speaking after running. For normal speaking conditions, external microphones were connected to laptop in a closed room is used. For walking and speaking condition, a handy recorder is used to record. For eating condition, we used a soft food to eat like idli or curd rice to avoid crispy or crunchy noise. Totally, we recorded for 30 speakers for all four mentioned conditions. Out of 30 participants, 20 were male participants and 10 were female participants. After the completion of data collection, all participants were offered chocolates.

We had conducted the experiments with prior approval with the project committee. For more details, further communications could be established with the corresponding and the first author as Mr. Sakil Ansari: sakilansari4@gmail.com

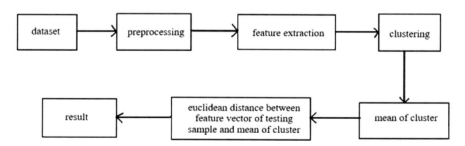

Fig. 1 Architectural diagram of the proposed system

3.1.2 Preprocessing

An individual required to present with four sorts of recordings (normal speaking, walking, eating, and running).The desired recording for a participant was of 60 s of each type. But sometimes the recording may be more than 60 seconds so each recording is framed of 60 s by using Praat.

3.1.3 Analysis

We analyze the speech parameters of all four conditions(Normal speaking, speaking while walking, speaking after running, and speaking while eating) in terms of pitch, mean energy intensity, and first formant. Pitch value relies on the activities of the participants. An individual's pitch conceives in the vocal cords/folds, and the rate at which the vocal folds vibrate is the frequency of the pitch. Pitch detector algorithm is granted to show the variations in pitch values in different activities of the participants and also tested on Praat. The current methodology reveals the pitch values in different cases, and we evaluate the most and the least pitch values.

Intensity varies on the person's activities. In different scenarios like normal speaking, speaking while walking, speaking while eating, and speaking after running, formant also alters. Energy distribution is varied while an individual is speaking in different scenarios. This is measured by the intensity of speech signal. The system describes the utmost intensity and the least intensity. It is also experimented in Praat.

3.2 Evaluation and analysis in MATLAB

3.2.1 Dataset

The dataset is explained in the above section.

3.2.2 Preprocessing

This section is explained in the above section.

3.2.3 Feature extraction

The approach deploys with two features to classify the speech signal whether the speech belongs to normal speaking, walking and speaking, running and speaking, and eating and speaking. MFCC and normalized energy are the features used. The methodology compares the result in terms of accuracy.

MFCC:

To calculate MFCC, following steps are carried out [12, 13]:

- Frame the signal into short frames. 25 ms is taken.
- For each frame, calculate the periodogram estimate of the power spectrum.

$$s_i(k) = \sum_{i=1}^{N} s_i(n)h(n)e^{-j2\pi kn/N} \, 1 \leq k \leq K \tag{1}$$

where $s_i(k)$ is DFT, $s_i(n)$ is speech frame, $h(n)$ is N sample long analysis window, K is the length of DFT

$$p_i(k) = \frac{1}{N}|s_i(k)|^2 \tag{2}$$

where $p_i(k)$ is the periodogram estimate of power spectrum.

- Apply the mel filter bank to the power spectra, sum the energy in each filter.
- Take the logarithm of all filter bank energies.
- Take the DCT of the log filter bank energies.
- Keep DCT coefficients 1–13, discard the rest.

Normalized energy:

Following steps are followed for the calculation of normalized energy:

Audio files:

We take our input sample from preprocessing part.

Total Number of Frames:

From audio files, we find the total number of frames by taking sample frequency (fs) = 320,00 Hz, frame duration (fd) = 25 ms, frameshift (fr) = 10 ms. Now finally, we find the total number of frames by using this data. This frame contains both silence frame and non-silence frames. To get better accuracy, we remove non-silence frames.

To remove silence frames:

The energy of each frame is determined by using

$$\text{energy}(i) = \sum_{i=1}^{n} x_i^2 \tag{3}$$

where $i = 1, 2, 3,..., n$.

Find the mean of energy, i.e., mean (energy). We take threshold of 1% of mean to remove silence frames from input files. If energy of the frame greater than this threshold, then we keep the frames; otherwise, we discard the frame. Now we have only non-silence frame.

Normalized energy of the frame:

$$\text{energy}(i) = \sum_{i=1}^{n} x_i^2 = E \tag{4}$$

where E is the total energy.

Finding root of E

$$\text{Normalizedenergy}(i) = \frac{|x_i|^2}{\sqrt{E}} \tag{5}$$

Now the first 60 *coefficients of normalized energy are arranged* in descending order, and we take first 13 coefficients as features for the proposed system.

Clustering:

In our dataset, we have four types of recordings: normal speaking, walking and speaking, running and speaking, and eating and speaking. So we need four clusters. In our designed system, we apply K means algorithm [14, 15] on the features.

K means algorithm:

K means algorithm(K,D)

k = number of clusters (k = 4), D = Dataset

1. Initial centroids (cluster centers)
2. repeat
3. for each datapoint x D do
4. compute the distance from x to each centroid
5. assign x to the closest centroid
6. endfor
7. re-compute the centroids using the current cluster memberships
8. until the stopping criterion is met.

Finding mean of each cluster:

Now from the above step, we get four clusters. We find the mean of each cluster. So finally, we have four means (each cluster has one mean) (Fig 2).

Finding Euclidean distance between feature vector of testing sample and mean of all 4 clusters:

Suppose $r1$ is the value between feature vector of testing sample and mean of cluster1 (mean1), $r2$ is the value between feature vector of testing sample and mean of cluster2 (mean2) and similarly $r3$ and $r4$. Let four clusters be $c1$, $c2$, $c3$, and $c4$ and their respective means are mean1, mean2, mean3, and mean4.

The minimum Euclidean distance shows that the sample data belongs to that cluster, and we know that the sample data is known so we can also say that cluster belongs to the same class either normal speaking, walking, running, and eating.

Fig. 2 Euclidean distance
between feature vector and
mean of clusters

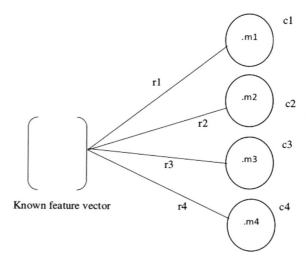

4 Experimental results

In this section, we benchmark the result of the system to evaluate the effectiveness of the proposed methodology. We first present speech parameters evaluation in different scenarios in order to show how the speech parameters differ in different cases. Then, we conduct experiments on two different features of speech and then give evaluation results.

4.1 Result for speech parameters

Table 1 manifests the alternations of speech parameters of participants in performing different activities. We benchmark from the experiments that pitch is the most in normal speaking while the least in walking and speaking, mean energy intensity is the maximum while walking and speaking and the least in normal speaking and the mean F1 is the most in eating and speaking and the least in walking and speaking.

Table 1 Speech parameter analysis

Speech parameters	Result
Pitch	walking < running < eating < normal
Mean energy intensity	normal < eating < running < walking
mean F1	walking < running < normal < eating

4.2 Result by using MFCC

As We experimented our result with two features, i.e., MFCC features and normalized energy as features. In our designed system, we have taken 13 MFCCs coefficients. Table 2 shows the result of the project of male participants. The numerical values in the table notify the number of participants classified. Table 3 shows the result of female participants, and Table 4 shows the result of both male and female participants.

Here, Model1 = normal speaking (N), Model2 = walking and speaking (W), Model3 = running and speaking (R), Model4 = Eating and speaking (E).

As we have taken MFCC features, the system shows about 74% accuracy. For male participants, the system shows about 71 % accuracy, and for female participants, also the system shows 80 % accuracy. For both participants, the system shows 74 % accuracy. So to improve the accuracy of the system, we prefer another feature called as normalized energy.

Table 2 Male participants (using MFCC)

Categories	model1	model2	model3	model4
Normal	15	2	2	1
Walking	3	14	2	1
Eating	2	3	13	2
Running	1	2	1	15

Table 3 Female participants (using MFCC)

Categories	model1	model2	model3	model4
Normal	8	1	1	0
Walking	1	9	0	0
Eating	1	1	8	0
Running	1	1	1	7

Table 4 Both male and female participants (using MFCC)

Categories	model1	model2	model3	model4
Normal	24	2	3	1
Walking	3	23	2	2
Eating	4	4	20	2
Running	2	2	4	22

Table 5 Male participants (normalized energy)

Categories	model1	model2	model3	model4
Normal	16	2	1	1
Walking	2	15	2	1
Eating	2	2	15	1
Running	2	1	1	16

Table 6 Female participants (normalized energy)

Categories	model1	model2	model3	model4
Normal	9	1	0	0
Walking	1	8	1	0
Eating	2	1	7	0
Running	0	0	1	9

Table 7 Both male and female participants (normalized energy)

Categories	model1	model2	model3	model4
Normal	26	2	1	1
Walking	2	25	2	1
Eating	2	3	23	2
Running	3	1	2	24

4.3 Result by using normalized energy

We divide the whole dataset into four clusters, namely Model1 = normal speaking (N), Model2 = walking and speaking (W), Model3 = running and speaking (R), Model4 = eating and speaking (E).

Here also, we tested on male participants, female participants, and finally on both participants. Table 5 shows the result of male participants, Table 6 shows the result of female participants, and Table 7 shows the result of both male and female participants. The system shows 77% accuracy for male participants and 82.25% for female participants, and overall the system shows 81% accuracy.

From the above comparison, we can say that normalized energy is giving better accuracy than MFCC features in our case.

5 Conclusion and future work

We have presented results from an experimental comparison of different features of speech. We have conducted our experiments to show the variations in speech parameters in different scenarios and reaffirmed the experiments in Praat. The designed system is tested on two types of features: MFCC and normalized energy. The system confirms that normalized energy gives better accuracy than MFCC features in our case.

To improve the accuracy of the system, we should take more dataset and the dataset should be collected carefully, the recording should be mono. Also to improve the accuracy of the system, we need to search other features like plp and by observing the spectrogram. Future work involves the implementation of the problem using DBSCAN algorithm by which the shape of the cluster can be figured out.

Acknowledgement The authors would like to express their gratitude for all participants for their precious time that they have spent for data recording.

References

1. Lamberg, E.M., and L.M. Muratori. 2012. Cell phones change the way we walk. *Gait and Posture* 35: 688–690.
2. Al-Yahyaa, E., H. Dawesa, L. Smith, A. Dennisa, K. Howells, and J. Cockburn. 2011. Cognitive Motor Interference While Walking: A Systematic Review and Meta-Analysis. *Neuroscience and Biobehavioral Reviews* 35: 715–728.
3. Reichel, U.D. 2012. PermA and Balloon: Tools for String Alignment and Text Processing. In *Proceedings of Interspeech*, Portland, Oregon, paper no. 346.
4. Schuller, B., S. Steidl, A. Batliner, F. Burkhardt, L. Devillers, C. MuLler, and S. Narayanan. 2013. Paralinguistics in Speech and Language State-of-the-Art and the Challenge. *Computer Speech Language* 27 (1): 4–39.
5. Schuller, B., S. Steidl, A. Batliner, S. Hantke, F. Honig, J.R. Orozco Arroyave, E. Noth, Y. Zhang, and F. Weninger. 2015. The Interspeech 2015 Computational Paralinguistics Challenge: Nativeness, Parkinson's Eating Condition. In *Proceedings INTERSPEECH*.
6. Prasad, A. and P.K. Ghosh. 2015. Automatic Classification of Eating Conditions from Speech Using Acoustic Feature Selection and a Set of Hierarchical Support Vector Machine Classifiers. In *Sixteenth Annual Conference of the International Speech Communication Association*.
7. Schuller, B., S. Steidl, A. Batliner, S. Hantke, F. Honig, J.R. Orozco-Arroyave, E. Noth, Y. Zhang, and F. Weninger. 2015. The Interspeech 2015 Computational Paralinguistics Challenge: Nativeness, Parkinson's and Eating Condition. In *Proceedings INTERSPEECH*.
8. Hantke, S., F. Weninger, R. Kurle, A. Batliner, and B. Schuller. 2015. I Hear You Eat and Speak: Automatic Recognition of Eating Condition and Food Type, ms. to appear.
9. Fuchs, S., U.D. Reichel, and A. Rochet-Capellan. 2015. Changes in Speech and Breathing Rate While Speaking and Biking.
10. Lee, Sangkil, Jieun Kim, and Insung Lee. Speech/Audio Signal Classification Using Spectral Flux Pattern Recognition. 978-0-7695-4856-2/12 26.00 © 2012 IEEE https://doi.org/10.1109/sips.2012.36. Bayram, S., I. Avcibas, B. Sankur, and N. Memon. 2006. Image Manipulation Detection. *Journal of Electronic Imaging* 15 (4): 041102-1–041102-17.
11. Lu, Li, Qingwei Zhao, Yonghong Yan and Kun Liu. 2010. A Study on Sports Video Classification Based on Audio Analysis and Speech Recognition. 978-1-4244-5858-5/10/26.00 ©2010 IEEE.
12. https://www.stanford.edu/.
13. http://practicalcryptography.com/miscellaneous/machine-learning/guide-mel-frequency-cepstral-coefficients-mfccs/#implementation-steps.

14. Wagstaff, Kiri, Seth Rogers. 2001. Constrained K-means Clustering with Background Knowledge. In *Proceedings of the Eighteenth International Conference on Machine Learning*, 577–584.
15. Xiong, Z., R. Radhakrishnan, A. Divakaran, and T.S. Huang. 2005. Highlights Extraction from Sports Video Based on an Audio-Visual Marker Detection Framework. In *IEEE International Conference on Multimedia and Expo, ICME 2005*, 4.

UIP—A Smart Web Application to Manage Network Environments

T. P. Ezhilarasi, G. Dilip, T. P. Latchoumi and K. Balamurugan

Abstract UIP—Ultra-Info Portal is an advanced application for accessing the Web-related application that permits users to handle tasks using browser network environments. UIP application was developed by using ASP.NET and.NET framework. It gains attention through it several e-content packages which a great way is to arrived solution. Ultra-Info Portal is common in every big organization to provide services to customers. In today's world, consumers expect easy access to information about organizations, and meanwhile, Internet and Intranet are becoming common to all organizations. Organizations like banks and other service provider organizations; they compete for each other to provide services to the customers and to attract them. To provide easy services, they are searching for new and popular technology. To satisfy their wants Ultra-Info Portal will become a good tool to use. They can easily interact with organizations with this user-friendly tool.

Keywords Web service · Ultra-Info portal · Database · Web portal · Networks · Web applications

1 Introduction

UIP is an excellent Web application that permits users to handle tasks using the browser in Intranet and Internet environment. This innovative Web portal application can be used in Internet group activities to share messages among several

T. P. Ezhilarasi · G. Dilip
Department of Computer Science and Engineering,
Saveetha School of Engineering, Chennai, India

T. P. Latchoumi
Department of Computer Science and Engineering,
VFSTR (Deemed to be University), Guntur, India

K. Balamurugan (✉)
Department of Mechanical Engineering, VFSTR (Deemed to be University),
Guntur, India
e-mail: kbalan2000@gmail.com

© Springer Nature Singapore Pte Ltd. 2020
K. S. Raju et al. (eds.), *Proceedings of the Third International Conference on Computational Intelligence and Informatics*, Advances in Intelligent Systems and Computing 1090, https://doi.org/10.1007/978-981-15-1480-7_8

people within an association, club and society, and any non-profit organizations [1]. A more specialized function in UIP is the combination of several online functional tools as a package. The UIP portal released with most popular e-content modules, such as news and discussion event list, classifieds links which results in more beneficiary aspects in the user point of view [2, 3]. UIP serves as the best tool in sharing information in a specified user group.

Ultra-Info Portal is common in every big organization to provide services to customers. In today's world, consumers expect easy access to information about organizations, and meanwhile, Internet and Intranet are becoming common to all organizations [4]. Organizations like banks and other service provider organizations compete for each other to provide services to the customers and to attract them [5, 6]. They can easily interact with organizations with this user-friendly tool. It satisfies all requirements of both users and organizations.

The UIP portal operates on the central database on the server where the data can be stored and accessed through database languages such as MY SQL, SQL Server, and Microsoft Access. It can be installed virtually on any Web server, whether internal within the organization or external, hosted by a Web hosting company. Certain functional options can be configured using the administrative module [7].

The flow of this paper is defined as follows: Sect. 2 gives the background of Ultra-Info Portal and its role in Web server. Section 3 gives the existing system and issues derived from that. In Sect. 4, we describe the proposed work with the help of use case diagrams. Section 5 gives the implementation of test cases. In Sect. 6, screenshots of our work have been shown. Finally, in Sect. 6, a conclusion is made.

2   Background

Ultra-Info Portal is common in every big organization to provide services to customers. Organizations like banks and other service provider organizations compete for each other to provide services to the customers and to attract them. To provide easy services, they are searching for new and popular technology. To satisfy their wants, "Ultra-Info Portal" will become a good tool to use [8, 9]. The UIP portal can be operated on any commercial database languages, i.e., a portal can be supported by any databases. It can be framed and worked virtually on any commercial Web server, maybe as an internal Web browser or as an external Web browser which always maintained by an hosting (Web) company, and in order to the activation of some administrative modules in UIP, the permission from the Web hosting company's administrator is required [10].To add more features to the UIP, any related information of the user can be altered depends upon the user's wish who has been enrolled in the UIP [11].

The UIP portal is facilitated with a search engine API to enable users to search Intranet content as well as Internet content, and also Extranet content can be restricted from the user during a search. In addition to the searching feature, UIP Web portals offer other important services like stock management, mailing services, news management [12]. This application can also be combined and worked with any type of information related to any working environment of organizations.

Many types of Web portals exist: They are personal tender, cultural stock, government Web site, domain-specific, search corporate Web portals. Nowadays, the portal plays a vital role in determining and accessing Web functions in a very ease of manner with the defined security level. The portal authentication is familiar, and single use of Web access through portal plays a vital role [13].

3 Existing System

The present system and functions are operated manually. There is no defined specific operation to be operated automatically. The existing portal requires stimulation for each action which is executed by the portal in charge or the portal administrator. Either it will be operated by the portal administrator or it will seek portal administrator permission for each course of action [14]. This has created a great hurdle in the user point as well as the portal in charge present system is manual. In this current system of portal, each action should be recognized and validated by the portal administrator, and grant permission for the authenticated user to carry out their own set of operations.

Issues in the existing system are time consumption, error occurrence, manpower consumption, insufficient data security, late reporting time, and less accuracy. The above-listed drawbacks are the occurrence in the existing system of the portal. Even in modern computerization, the portal user feels inconvenient in accessing these portals.

4 Proposed System

In the present work, the proposed UIP application which is a Web-based application is developed using.NET compact framework and ASP.NET. For the present time, this automated portal serves as an excellent portal in serving user's operation and be useful in serving the administrator's rights. Advantages are recognition of user access, grant of access to e-mail and messages to group members, exporting group information to any database format, stylish, advanced tool operating options, compatible with standard browsers.

The class diagram for the proposed UIP portal is portrayed in Fig. 1; it represents what are the common methods which are used in all classes. It describes the system structure by using classes, attributes, methods, and relationship among all objects. Activity diagram of the proposed system is shown in Fig. 2 and describes the overall flow of the control. Figure 3 represents the sequence diagram which gives an interaction between the users by using lifelines. In Fig. 4, it summarizes the overall architecture of the proposed system.

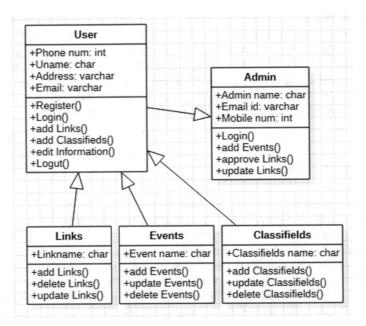

Fig. 1 Class diagram

5 Implementation

5.1 Modules

The UIP is furnished with two major modules:

1. Admin
2. Visitor.

Admin Module: In this, administrator maintains a club, association, and corporation, company or organization and maintains online community, events, discussions, articles, and forums.

Visitor Module: As an initial stage, the user who needs to access the portal can register with the portal with the personal details. The user collects all information like community details and every events detail. He can post his queries and get replies for the same.

Fig. 2 Activity diagram

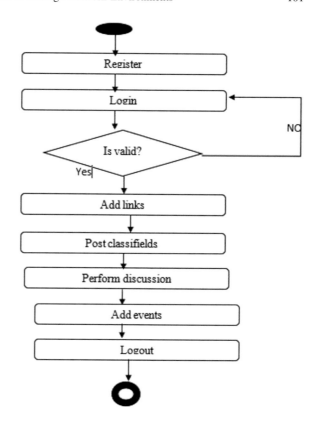

5.2 Test Cases

To verify log-in details, user details, and the e-mail address. Test cases are performed for success as well as for failure results. All the test cases are shown in Tables 1, 2, 3, 4 and 5.

Test cases: 01
Object: Validation test cases for Ultra-Information Portal through credit limit.
Test case id: Ultra-Information _Portal _ 01
Test case name: Validation test for Ultra-Information Portal through credit limit for username and password.
Test procedure: Look and feel for the test cases with Ultra-Information Portal through credit limit.
Test Case: 02
Object: Verifying and validating Ultra-Information Portal through credit limit.
Test Case Id: Ultra-Information _Portal _ 02
Test Case Name: Validation test cases for user details

T. P. Ezhilarasi et al.

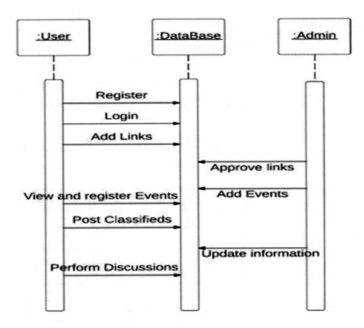

Fig. 3 Sequence diagram

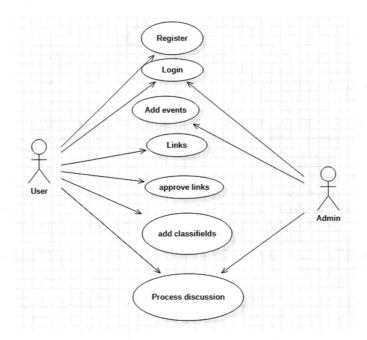

Fig. 4 Use case diagram

Table 1 Data matrix test case 1

Input attribute	Equivalence class partitioning		Boundary value analysis	
	Valid	Invalid	Minimum	Maximum
User name	[A–Z] & [a–z]	[0–9] & Special characters	3	30
Password	[0–9] & [a–z] [A–Z] & special characters	–	3	10

Table 2 Data matrix test case 2

Input	Equivalence class partitioning		Boundary value analysis	
	Valid	Invalid	Minimum	Maximum
First and last name	[a–z] & [A–Z]	[0–9]	–	–
Pwd (password)	[a–z] & [0–9] [A–Z] & Special characters	–	4	10
e-mail	[a–z] & [A–Z] [0–9] & Special characters	–	10	20
Phone no	[0–9]	Alphabets & Special characters	10	10
Address	[a–z] & [A–Z] [0–9] & Special characters	–	–	–
Pin code	[0–9]	[a–z] & [A–Z] characters	6	6

Table 3 Data matrix test case 3

Input	Equivalence class partitioning		Boundary value analysis	
	Valid	Invalid	Minimum	Maximum
E-mail Id	abcd@gmail.com	123@sdf	5 characters	50 characters

Test Procedure: Look and feel of user Id and password is tested with the positive behavior of an application.

Test Case: 03

Object: Verifying and validating Ultra-Information Portal through credit limit.

Test Case Id: Ultra-Information _Portal _ 03

Test Case Name: Validation test cases for the e-mail address of the user

Test Procedure: Look and feel the application is tested with the negative behavior of the application.

Test cases: 04

Object: Verifying and validating Ultra-Information Portal through credit limit.

Test Case Id: Ultra-Information _Portal _ 04

Test Case Name: Success test cases for Ultra-Information Portal

Table 4 Data matrix test case 4

S. no.	Description	Input required	Expected value	Actual value	Comparison of results
1	Enter user name	Vaish	V	V	S
2	Enter password	908	V	V	S
3	Enter E-mail	Anuu@gmail.com	V	V	S
4	Enter address	#2-11-202,subedari	V	V	S
5	Enter phone no	9898989898	V	V	S
6	Enter Pin code	506002	V	V	S

V—Valid S—Success

Table 5 Data matrix test case 5

S.No	Description	Input required	Expected value	Actual value	Comparison of results
1	Enter user name	123sa4	V	IV	F
2	Enter password	123	V	IV	F
3	Enter E-mail	Tejagmail	V	IV	F
4	Enter address	2-11-202, subedari	V	IV	F
5	Enter phone no	984889	V	IV	F
6	Enter Pin code	506	V	IV	F

V—Valid IV—In Valid F—Failure

Test Procedure: Look and feel Ultra-Information Portal through credit limit the application tested with the Positive behavior of an application.
Test case: 05
Object: Verifying and validating Ultra-Information Portal through credit limit.
Test Case Id: Ultra-Information _Portal _ 05.
Test Case Name: Failure test cases for Ultra-Information Portal
Test Procedure: Look and feel of Ultra-Information Portal through credit limit the application tested with the positive behavior of an application.

The home page of the UIP portal is shown in Fig. 5. It contains the portal description, events added, deleted and updated as well as links all the events notification. For utilizing this UIP portal, the user needs to register by filling all the information on the registration (Join us) page is shown in Fig. 6.

The administrator has responsibility for the approval of the event link is shown in Fig. 7. Once the user added the link in the portal, admin needs to approve the link. Figure 8 shows the discussion page (forum) that is where the user and admin can share their ideas, online conversation between them and put users can put their feedbacks on portal.

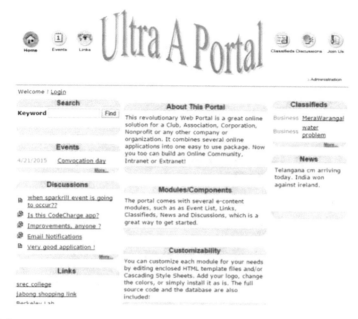

Fig. 5 Index page

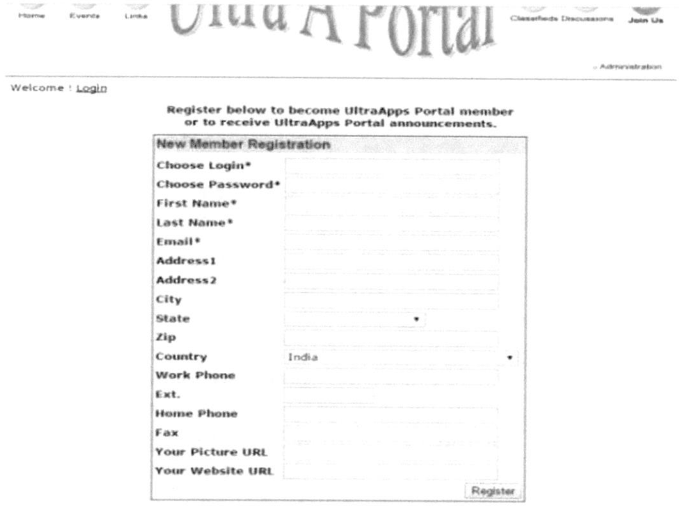

Fig. 6 Join us page

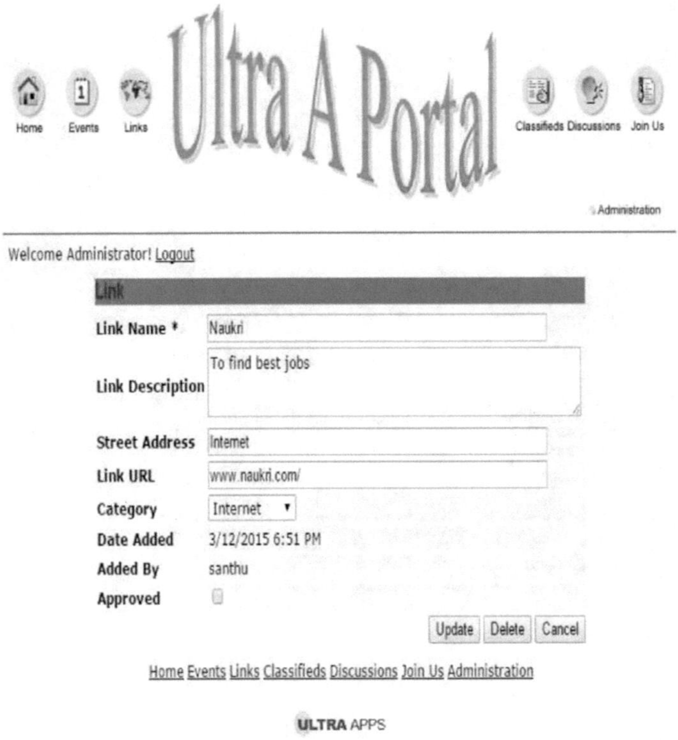

Fig. 7 Admin link approval page

6 Conclusion

Ultra-Info Portal is common in every big organization to provide services to customers. In today's world, consumers expect easy access to information about organizations, and meanwhile, Internet and Intranet are becoming common to all organizations. Organizations like banks and other service provider organizations, they compete for each other to provide services to the customers and to attract them. To provide easy services, they are searching for new and popular technology. To satisfy their wants, Ultra-Info Portal will become a good tool to use.

Finally, we like to conclude that we put all our efforts throughout the development of our project and tried to fulfill most of the requirements of the user. The UIP serves as the best portal in transferring information from the user point of view to the portal and vice versa, accessing user functionalities, and allowing to maintain information in all sorts of view except the restricted contents from the

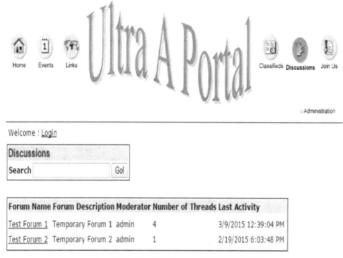

Fig. 8 Discussion page

Internet via search engine. Since the technology we have taken is full on-demand and easy to learn, we can have many applications being developed using these (.NET, MySQL, etc.,) as the base.

References

1. Bouras, C., A. Konidaris, and I. Misedakis. 2004. A Webpage Fragmentation Technique for Personalized Browsing. In *Proceedings of the 2004 ACM Symposium on Applied Computing*, 1146–1147. New York: ACM.
2. Bouras, C., and A. Konidaris. 2001. Web Components: A Concept for Improving Personalization and Reducing User Perceived Latency on the World Wide Web. In *Proceedings of the 2nd International Conference on Internet*, vol. 2, 238–244, USA.
3. Rathi, R., and D.P. Acharjya. 2017. A Framework for Prediction Using Rough Set and Real Coded Genetic Algorithm. *Arabian Journal for Science and Engineering*.
4. Zhao, L., I. Matsuo, F. Salehi, Y. Zhou, and W.J. Lee. 2018. Development of a Real-Time Web-Based Power Monitoring System for the Substation of Petrochemical Facilities. In *2018 IEEE/IAS 54th Industrial and Commercial Power Systems Technical Conference (I&CPS)*, 1–6, Niagara Falls, Canada.
5. Sofyan, Z. 2017. Development of RENAKSI Web-based Application to Support Open Government Anti-corruption Action Plan in Banda Aceh Municipality. In *IEEE International Conference on Electrical Engineering and Informatics (ICELTICs 2017)*, 201–204, Banda Aceh, Indonesia.

6. Newman, C., Z. Agioutantis, and N. Schaefer. 2018. Development of a Web-Platform for Mining Applications. *International Journal of Mining Science and Technology* 28: 95–99.
7. Jin, X., B. Cui, D. Li, Z. Cheng, and C. Yin. 2018. An Improved Payload-Based Anomaly Detector for Web Applications. *Journal of Network and Computer Applications* 106: 111–116.
8. Tunc, H., T. Taddese, P. Volgyesi, J. Sallai, P. Valdastri, and A. Ledeczi. 2016. Web-Based Integrated Development Environment for Event-Driven Applications. *Southeast Con*, 1–8. Norfolk, VA.
9. Amontamavut, P., and E. Hayakawa. 2016. ROS Extension of Blue-Sky Web based Development Environment for IoT. In *IEEE Fifth ICT International Student Project Conference (ICT-ISPC)*, 1–4. Mahidol University, Thailand.
10. Buttler, D., L. Liu, and C.V. Pu. 2001. A Fully Automated Object Extraction System for the World Wide Web. In *Proceedings 21st International Conference on Distributed Computing Systems*, 361–370. Mesa, AZ, USA.
11. Chang, C.H., S.C. Lui, and Y.C. Wu. 2001. Applying Pattern Mining to Web Information Extraction. In *5th Pacific Asia Conference on Knowledge Discovery and Data Mining*, 4–16. Hong Kong.
12. Chang, C.H., S.C. Lui. 2001. IEPAD: Information Extraction Based on Pattern Discovery. In *10th International Conference on World Wide Web*, 595–609. Hong Kong.
13. Bouras, C., A. Konidaris, and I. Misedakis. 2004. Web Page Fragmentation for Personalized Portal Construction. In *International Conference on Information Technology: Coding and Computing (ITCC'04)*, vol. 1, 332–336. Nevada, USA.
14. Lopez, A.J., I.L. Zorrozua, E.S. Ruiz, M.R. Artacho, and M.C. Gil. 2016. Design and Development of a Responsive Web Application Based on Scaffolding Learning. In *IEEE Global Engineering Education Conference (EDUCON)*, 308–313. Abu Dhabi.

Energy Distribution in a Smart Grid with Load Weight and Time Zone

Boddu Rama Devi and K. Srujan Raju

Abstract Energy generation and distribution according to the customer demand are major challenging tasks in the smart grid. In this paper, allocation and distribution of energy in the grid network based on load weight and time are proposed. Energy allocated to different levels in the grid based on load weight during high demand. Finally, based on the time zone and demand, the energy is distributed to the different categories of customers. Various test cases for high and low demand in peak and nonpeak hours with dynamic load are considered for investigation. The requests from three different types of load, i.e., shiftable, non-shiftable and shiftable with noncontinuous slots are considered. The proposed techniques give better flexibility in energy distribution to the customers by maintaining high fairness index. An adaptive shifting intelligent control is applied further to enhance the performance.

Keywords Energy distribution · Fairness index · Load weight · Smart grid

1 Introduction

The power consumption is increasing day by day due to modern electrical appliances, and energy allocation becomes a major challenge at the grid during peak load. It requires more sophisticated techniques at grid to handle energy distribution and pricing [1–12].

The primary and secondary energy sources like solar, wind, etc., can be used for energy generation at the smart grid [13, 14]. It is limited by the available resources,

B. Rama Devi (✉)
Department of Electronics and Communication Engineering, Kakatiya Institute
of Technology and Sciences, Warangal, Telangana 506015, India
e-mail: ramadevikitsw@gmail.com

K. Srujan Raju
Department of CSE, CMR Technical Campus, Medchal, Hyderabad, Telangana, India
e-mail: ksrujanraju@gmail.com

© Springer Nature Singapore Pte Ltd. 2020
K. S. Raju et al. (eds.), *Proceedings of the Third International Conference on Computational Intelligence and Informatics*, Advances in Intelligent Systems and Computing 1090, https://doi.org/10.1007/978-981-15-1480-7_9

109

economy, and the number of projects in a region. To create awareness and meet the requirements, we should use secondary energy sources at the customer level.

Few techniques energy distribution techniques in smart grid based on users, profit, and average slots per user, Modified Round Robin methods (MRRM) were compared with conventional Priority-based and Round Robin methods in [15]. Among all MRRM results high fairness [15]. Further, in smart grid, a dynamic weight-based energy allocation was proposed to optimize different energy slots for customer and maintain high fairness index [16]. In this paper, three different categories of loads at grid such as shiftable and non-shiftable loads with continuous slots, shiftable load with noncontinuous slots are considered with the following objectives:

- To evaluate load request of the customers
- To optimize the energy allocation using load weight and time zone with high fairness
- Adopting adaptive shifting intelligent control during energy distribution.

The work is organized as follows: Sect. 2 describes a load request in a distributed smart grid; Sect. 3 explains load weighted-based energy allocation; and Sect. 4 describes distribution using time zone. Simulation results are explained in Sect. 5. Conclusions about the work are described in Sect. 6.

2   Evaluation of Load Request

The grid can be modeled as a Distributive Smart Grid Tree Network (DSGTN). The DSGTN has I primary substations (PS), J secondary substations (SS), and K customers as shown in Fig. 1. The energy demand request of the customers and the available energy at grid vary dynamically with time. A timing window T, with different time slots $t_i = \{1, 2, \ldots, T\}$, is considered for calculating the energy request, allocation, and total energy distribution. The total price charged for the consumed energy by the load of a customer is calculated for the entire timing window T.

The customer uses different appliances $A = \{1, 2, \ldots, A\}$, each fall under the categories of non-shiftable and shiftable load. Each appliance a requires L_a continuous slots or noncontinuous slots where a represents the customer appliance number from $[1, A]$. Let $L_a^{t_i}(i, j, k)$ represents the load request from an appliance a of customer k connected to (i, j) where i is the primary substation, and j is the secondary substation. Total request from the customer k is $\sum_{a=1}^{A} L_a^{t_i}(i, j, k)$. Let $L_{a\_new}^{t_i}(i, j, k)$ is a new load request and $N_{a\_new}^{t_i}(i, j, k)$ is a new slots required to complete the task. The load continued from the previous slot after the allocation to this slot t_i for continuous task or just shifted from the previous slot to this slot t_i due to non-availability of the energy slot is $L_{aS}^{t_i}(i, j, k)$ and slots required is $N_{aS}^{t_i}(i, j, k)$.

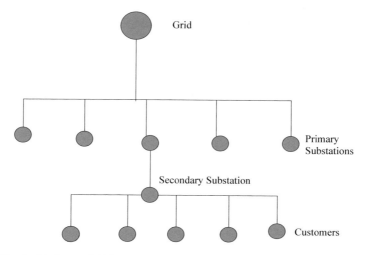

Fig. 1 Distributive Smart Grid Tree Network (DSGTN) model with three levels

Then, the load request from the appliance in the time slot t_i can be expressed as

$$L_a^{t_i}(i,j,k) = L_{a\_new}^{t_i}(i,j,k) + L_{aS}^{t_i}(i,j,k) \tag{1}$$

The number of continuous-time slots required for an appliance a to complete a task can be expressed as

$$N_a^{t_i}(i,j,k) = N_{a\_new}^{t_i}(i,j,k) + N_{aS}^{t_i}(i,j,k) \tag{2}$$

Depending on these categories, the load of the appliance is categorized into four types. They are: (i) C_0: with no load (NL) (ii) C_1: non-shiftable load (NSL) with continuous slots (iii) C_2: shiftable with continuous slots (SLCS), and (iv) C_3: shiftable with noncontinuous slots (SLNCS) and is given by

$$C_a^{t_i}(i,j,k) = \begin{cases} C_0 \text{ if } L_a^{t_i}(i,j,k) \text{ is NL} \\ C_1 \text{ if } L_a^{t_i}(i,j,k) \text{ is NSL} \\ C_2 \text{ if } L_a^{t_i}(i,j,k) \text{ is SLCS} \\ C_3 \text{ if } L_a^{t_i}(i,j,k) \text{ is SLNCS} \end{cases} \tag{3}$$

Let us assume that each customer can generate energy from secondary sources which can be utilized by them. The customer requests for the excess load energy required for them. The total load request of k customers can be expressed as

$$\bar{L}^{t_i}(i,j,k) = \sum_{a=1}^{A} L_a^{t_i}(i,j,k) - E_S^{t_i}(i,j,k) \text{ (kWh)} \tag{4}$$

where $E_S^{t_i}(i,j,k)$ = energy generated by secondary source at the customer node.

The energy generated from secondary sources at the customer is small, smaller amount of energy remained if it considered as per (4), and most of the secondary energy is wasted. So to count this, total secondary energy generated at this level is to be added at j and can be written as

$$L^{t_i}(i,j,k) = \sum_{a=1}^{A} L_a^{t_i}(i,j,k) \tag{5}$$

Total load request (TLR) from K customers at secondary substation j is given by

$$\bar{L}^{t_i}(i,j) = \sum_{k=1}^{K} L^{t_i}(i,j,k) \tag{6}$$

The actual request is sent to the next level after considering the sum of all the secondary energies from customers connected to j and is given by

$$L^{t_i}(i,j) = \sum_{k=1}^{K} L^{t_i}(i,j,k) - E_S^{t_i}(i,j) \tag{7}$$

$$E_S^{t_i}(i,j) = \sum_{k=1}^{K} E_S^{t_i}(i,j,k) \tag{8}$$

The TLR at i from J SSs is given by

$$L^{t_i}(i) = \sum_{j=1}^{J} L^{t_i}(i,j) \tag{9}$$

The TLR at G from I PSs can be expressed as

$$L_G^{t_i} = \sum_{i=1}^{I} L^{t_i}(i) = \sum_{i=1}^{I} \sum_{j=1}^{J} L^{t_i}(i,j) = \sum_{i=1}^{I} \sum_{j=1}^{J} \sum_{k=1}^{K} L^{t_i}(i,j,k) \tag{10}$$

3 Load Weighted-Based Energy Allocation

Load request and demand vary based on the customer requirement which is given by

$$\text{Demand} = \begin{cases} \text{High } L_G^{t_i} > L_{GH}^{t_i} \\ \text{Low } L_G^{t_i} \leq L_{GH}^{t_i} \end{cases} \tag{11}$$

where $L_{GH}^{t_i}$ is energy at grid.

The demand is low, if the load request is less than the available energy at the grid. Then requested energy can be allotted to the customer. When demand is high, the energy allocation to i and j is done by the proposed load weighted-based energy allocation (LWBEA) method. The LWBEA Algorithm is explained in Algorithm I. During high demand, energy allotted to PSs from G is given by

$$L_A^{t_i}(i) = w(i)L_{GH}^{t_i} = \frac{L^{t_i}(i)}{L_G^{t_i}} L_{GH}^{t_i} \tag{12}$$

when demand is high, energy allotted to j from i can be expressed as

$$L_A^{t_i}(i,j) = w(i,j)L_{GH}^{t_i} = \frac{L^{t_i}(i,j)}{L^{t_i}(i)} L_A^{t_i}(i) \tag{13}$$

where $w(i), w(i,j)$ indicates the load based weights of i and j, respectively.

During high demand, actual energy allotted for distribution at j is given by

$$\bar{L}_A^{t_i}(i,j) = L^{t_i}(i,j) + E_S^{t_i}(i,j) \tag{14}$$

4 Energy Distribution Based on the Time Zone with Load Category (EDTZLC)

The day is divided into four time zones depending on the expected demand $\hat{L}_G^{t_i}$ at t_i. They are: (i) Day peak: Expected demand $\hat{L}_G^{t_i}$ is peaking during day time 7.00–11.00 a.m. and time slot $t_i = 0-4$; (ii) day nonpeak: Expected demand $\hat{L}_G^{t_i}$ is lower during day time 11.00 a.m.–5.00 p.m. and time slot $t_i = 4-10$; (iii) night peak: Expected demand is peaking during night time 5.00–9.00 p.m. and time slot $t_i = 10-14$; and (iv) night nonpeak: $\hat{L}_G^{t_i}$ is low during the night time and in the early morning 9.00 p.m.–7.00 a.m. and time slot $t_i = 14-24$.

In this section, the proposed EDTZLC is given in Algorithm II. When demand is low, i.e. $L_G^{t_i} \leq L_{GH}^{t_i}$, the requested energy is distributed to the customers. When demand is high, the slots are allotted to primary substation, secondary substation using LWBEA method is distributed to the customer using EDTZLC Algorithm. Two cases are considered during high demand. They are: (i) day/night non-peak when $(4 \leq t_i < 10) \,||\, (14 \leq t_i < 24)$ and (ii) day/night peak when $(0 \leq t_i < 4) \,||\, (10 \leq t_i < 14)$.

During day/night peak, the energy is first distributed to non-shiftable loads, shiftable load with noncontinuous slots, and then to the shiftable load with continuous slots. If energy is not available, non-shiftable and shiftable load with noncontinuous slots are shifted to next nonpeak slot $t_i + 1$. The loadings with continuous slots are shifted to next nonpeak slots to avoid over the rush. Due to the availability of energy, if the load of shiftable with continuous slots is granted, its category is changed to C_1 as it requires continuous slots after allocation as mentioned in Algorithm II.

ALGORITHM I. Load Weighted-based Energy Allocation (LWBEA) Method

Description
1.   **if** $L_G^{t_i} \leq L_{GH}^{t_i}$ **then** - - Demand is Low, allot requested energy
2.   **for** $i = 1 \, to \, I$ **do**
3. $L_A^{t_i}(i) = L^{t_i}(i)$
4.   **for** $j = 1 \, to \, J$ **do**
5. $L_A^{t_i}(i, j) = L^{t_i}(i, j)$
6. $\overline{L}_A^{t_i}(i, j) = L^{t_i}(i, j) + E_S^{t_i}(i, j)$
7.   **end for**
8.   **end for**
9.   **else** - - Demand is High, use LWBEA method
10.   { **for** $i = 1 \, to \, I$ **do**
11. $L_A^{t_i}(i) = \dfrac{L^{t_i}(i)}{L_G^{t_i}} L_{GH}^{t_i}$
12.   **for** $j = 1 \, to \, J$ **do**
13. $L_A^{t_i}(i, j) = \dfrac{L^{t_i}(i, j)}{L^{t_i}(i)} L_A^{t_i}(i)$
14. $\overline{L}_A^{t_i}(i, j) = L^{t_i}(i, j) + E_S^{t_i}(i, j)$
15.   **end for**
16.   **end for** }
17. **end if**

ALGORITHM II. EDTZLC Algorithm

Description
1.   **if** $L_G^{t_i} \leq L_{GH}^{t_i}$ **then** - - Demand is Low, allocate all
2.   **if** $L_G^{t_i} \leq L_{GH}^{t_i}$ **then** - - Demand is Low, allocate all
3.   **for** $i = 1$ *to* I **do**
4.   **for** $j = 1$ *to* J **do**
5.   **for** $k = 1$ *to* K **do**
6. $L_A^{t_i}(i,j,k) = L^{t_i}(i,j,k)$
7.   **end for**
8.   **end for**
9.   **end for**
10. **elsif** $(0 \leq t_i < 4) \,\|\, (10 \leq t_i < 14)$ **then** - - Peak , use LWBEA
11. {**for** $i = 1$ *to* I **do**
12.   **for** $j = 1$ *to* J **do** { $L_{Dist}^{t_i}(i,j) = L_A^{t_i}(i,j)$ }
13. **for** $k = 1$ *to* K **do**
14. **for** $c = 0,1,3$ **do**
15. **for** $a = 1$ *to* A **do**
16. **if** $C_a^{t_i}(i,j,k) = C_c$ **then**
17. { **if** $E_S^{t_i}(i,j,k) \geq L_a^{t_i}(i,j,k)$ **then** $L_{a,dist}^{t_i}(i,j,k) \leftarrow L_a^{t_i}(i,j,k)$
$E_S^{t_i}(i,j,k) = E_S^{t_i}(i,j,k) - L_a^{t_i}(i,j,k)$;
{**if** $(N_a^{t_i} - 1) \neq 0$ **then** $L_{aS}^{t_i+1}(i,j,k) \leftarrow L_a^{t_i}(i,j,k); N_{aS}^{t_i+1} = N_a^{t_i} - 1$
else $L_{aS}^{t_i+1}(i,j,k) \leftarrow 0; N_{aS}^{t_i+1} = 0$ }
18. **elsif** $L_{Dist}^{t_i}(i,j) > 0$ **then** $L_{a,dist}^{t_i}(i,j,k) \leftarrow L_a^{t_i}(i,j,k)$; $L_{Dist}^{t_i}(i,j) = L_{Dist}^{t_i}(i,j) - L_{a,dist}^{t_i}(i,j,k)$;
{**if** $(N_a^{t_i} - 1) \neq 0$ **then** $L_{aS}^{t_i+1}(i,j,k) \leftarrow L_a^{t_i}(i,j,k); N_{aS}^{t_i+1} = N_a^{t_i} - 1$
else $L_{aS}^{t_i+1}(i,j,k) \leftarrow 0; N_{aS}^{t_i+1} = 0$ }
19. **else** $L_{a,dist}^{t_i}(i,j,k) \leftarrow Not \, alloted$;
Shifted to $t_i + 1 \Rightarrow L_{aS}^{t_i+1}(i,j,k) \leftarrow L_a^{t_i}(i,j,k); N_{aS}^{t_i+1} = N_a^{t_i}$;
20. **end if**}
21. **end if**
22. **end for**
23. **end for**
24. **end for**
25. **end for**
26. **end for**}
27. **for** $i = 1$ *to* I **do**
28. **for** $j = 1$ *to* J **do**
29. **for** $k = 1$ *to* K **do**
30. **for** $a = 1$ *to* A **do**
31. **if** $C_a^{t_i}(i,j,k) = C_2$ **then**

32. {**if** $E_S^{t_i}(i,j,k) \geq L_a^{t_i}(i,j,k)$ **then** $L_{a,dist}^{t_i}(i,j,k) \leftarrow L_a^{t_i}(i,j,k)$; $E_S^{t_i}(i,j,k) = E_S^{t_i}(i,j,k) - L_a^{t_i}(i,j,k)$;

    $C_a^{t_i+1}(i,j,k) = C_1$;

      {**if** $(N_a^{t_i} - 1) \neq 0$ **then** $L_{aS}^{t_i+1}(i,j,k) \leftarrow L_a^{t_i}(i,j,k)$; $N_{aS}^{t_i+1} = N_a^{t_i} - 1$

    else $L_{aS}^{t_i+1}(i,j,k) \leftarrow 0$; $N_{aS}^{t_i+1} = 0$ }

33. **elsif** $L_{Dist}^{t_i}(i,j) > 0$ **then** $L_{a,dist}^{t_i}(i,j,k) \leftarrow L_a^{t_i}(i,j,k)$;

    $L_{Dist}^{t_i}(i,j) = L_{Dist}^{t_i}(i,j) - L_{a,dist}^{t_i}(i,j,k)$; $C_a^{t_i+1}(i,j,k) = C_1$;

    {**if** $(N_a^{t_i} - 1) \neq 0$ **then** $L_{aS}^{t_i+1}(i,j,k) \leftarrow L_a^{t_i}(i,j,k)$; $N_{aS}^{t_i+1} = N_a^{t_i} - 1$

    else $L_{aS}^{t_i+1}(i,j,k) \leftarrow 0$; $N_{aS}^{t_i+1} = 0$ }

34. **else** $L_{a,dist}^{t_i}(i,j,k) \leftarrow$ *Not alloted*; -- *Shifted to* $t_i = 4 \| 14$

  {**if** $(0 \leq t_i < 4)$ **then** $L_{aS}^{t_i=4}(i,j,k) \leftarrow L_a^{t_i}(i,j,k)$; $N_{aS}^{t_i=4} = N_a^{t_i}$;

    else $L_{aS}^{t_i=14}(i,j,k) \leftarrow L_a^{t_i}(i,j,k)$; $N_{aS}^{t_i=14} = N_a^{t_i}$; }

35. **end if**}
36. **end if**
37. **end for**
38. **end for**
39. **end for**
40. **end for**}
41. {**for** $i = 1$ *to* I **do** -- non peak with high demand
42.   **for** $j = 1$ *to* J **do** { $L_{Dist}^{t_i}(i,j) = L_A^{t_i}(i,j)$ }
43.   **for** $k = 1$ *to* K **do**
44.   **for** $c = 0,1,3$ **do**
45.   **for** $a = 1$ *to* A **do**
46.   **if** $C_a^{t_i}(i,j,k) = C_c$ **then**

    { **if** $L_{Dist}^{t_i}(i,j) > 0$ **then** $L_{a,dist}^{t_i}(i,j,k) \leftarrow L_a^{t_i}(i,j,k)$; $L_{Dist}^{t_i}(i,j) = L_{Dist}^{t_i}(i,j) - L_{a,dist}^{t_i}(i,j,k)$;

  {**if** $(N_a^{t_i} - 1) \neq 0$ **then** $L_{aS}^{t_i+1}(i,j,k) \leftarrow L_a^{t_i}(i,j,k)$; $N_{aS}^{t_i+1} = N_a^{t_i} - 1$

    else $L_{aS}^{t_i+1}(i,j,k) \leftarrow 0$; $N_{aS}^{t_i+1} = 0$ }

47. **elsif** $E_S^{t_i}(i,j,k) \geq L_a^{t_i}(i,j,k)$ **the** $L_{a,dist}^{t_i}(i,j,k) \leftarrow L_a^{t_i}(i,j,k)$ $E_S^{t_i}(i,j,k) = E_S^{t_i}(i,j,k) - L_a^{t_i}(i,j,k)$;

  {**if** $(N_a^{t_i} - 1) \neq 0$ **then** $L_{aS}^{t_i+1}(i,j,k) \leftarrow L_a^{t_i}(i,j,k)$; $N_{aS}^{t_i+1} = N_a^{t_i} - 1$

    else $L_{aS}^{t_i+1}(i,j,k) \leftarrow 0$; $N_{aS}^{t_i+1} = 0$ }

48. **else** $L_{a,distribute}^{t_i}(i,j,k) \leftarrow$ *Not alloted*;

49. *Shifted to* $t_i + 1$, $L_{aS}^{t_i+1}(i,j,k) \leftarrow L_a^{t_i}(i,j,k)$; $N_{aS}^{t_i+1} = N_a^{t_i}$;

50. **end if**}
51. **end if**
52. **end for**
53. **end for**
54. **end for**
55. **end for**
56. **end for**}

57. **for** $i = 1$ *to* I **do**
58.   **for** $j = 1$ *to* J **do**
59.   **for** $k = 1$ *to* K **do**
60.   **for** $a = 1$ *to* A **do**
61.   **if** $C_a^{t_i}(i,j,k) = C_2$ **then**
62.   { **if** $L_{Distribute}^{t_i}(i,j) > 0$ **then** $L_{a,dist}^{t_i}(i,j,k) \leftarrow L_a^{t_i}(i,j,k)$;
63. $L_{Dist}^{t_i}(i,j) = L_{Dist}^{t_i}(i,j) - L_{a,dist}^{t_i}(i,j,k);\ \ C_a^{t_i+1}(i,j,k) = C_1$
64.   **elsif** $E_S^{t_i}(i,j,k) \geq L_a^{t_i}(i,j,k)$ **then** $L_{a,dist}^{t_i}(i,j,k) \leftarrow L_a^{t_i}(i,j,k)$

                 $E_S^{t_i}(i,j,k) = E_S^{t_i}(i,j,k) - L_a^{t_i}(i,j,k)$
65.   **else** $L_{a,distribute}^{t_i}(i,j,k) \leftarrow$ *Not alloted*;
66.   *Shifted to* $t_i + 1 \Rightarrow L_{aS}^{t_i+1}(i,j,k) \leftarrow L_a^{t_i}(i,j,k); N_a^{t_i+1} = N_a^{t_i}$;
67.   **end if}**
68.   **end if**
69.   **end for**
70.   **end for**
71.   **end for**
72. **end for}**
73. **end if**

As shown in Fig. 2, the demand is very high at $t_i = 0$–2 and only 70 slots are allotted out of the 79 requested slot at the secondary substation $J = 1$ connected to the primary substation $I = 1$. Among 70 allotted slots, 39 slots are first distributed to C_1 load, 18 slots to C_3 load and 13 slots out of 22 requested slots to C_2 load.

Non-shiftable loads are handled in that slot only. In this approach, the loads of C_1 and C_3 get an allocation of their maximum requested slots. The allotted slots of the C_2 are now continuing in the next slot under the category C_1 (second table, Fig. 2) as they require continuous slots to accomplish the work.

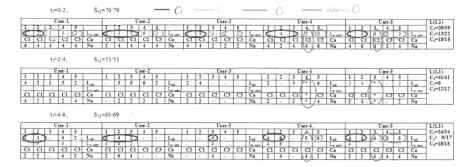

Fig. 2 Shifting C_2 slots from $t_i = 0$–2 to nonpeak time zone $t_i = 4$–6 to reduce further huge demand in peak time slot $t_i = 2$–4

Using EDTZLC Algorithm, the unallocated load C_2 in the time slot $t_i = 0–2$ (represented in blue color) is shifted to $t_i = 4–6$ instead of next time slot $t_i = 2–4$, which further reduces the huge demand and unavailable allotted slot in the peak demand slot $t_i = 2–4$ as shown in Fig. 2. Due to the shifting of the C_2 load into nonpeak demand slot, the demand in nonpeak slot increases temporarily, but the customer gets allocation during next nonpeak slot comfortably. This technique is very much helpful to handle the peak load time zones very effectively.

One of the major drawbacks is, if the demand request in the peak time slot $t_i = 2–4$ is not high and if the requested energy is low then few slots are left unused (not allotted). It is not a major problem, and in next slots using this scheduling the customers get allocation as the demand is low in nonpeak slots. But it becomes very crucial in an urban city with huge demand. This drawback can be overcome by EDTZLC Algorithm with adaptive shifting intelligent control (ASIC). The unallocated slots are shifted to the next time slot and consequent slots if the demand in that slot is low. If the demand is high in the next time slot, and consequent slots, these shifted slots will not get a chance of allocation of slots due to unavailability of energy. The demand in next consequent slots is checked, and the slots are shifted to $t_i + 1$ or $t_i + 2$ or $t_i + 3$, …. slots. Block-based shifting is used for effective functioning.

In EDTZLC Algorithm with ASIC, the unallocated C_2 slots are shifted to the next slots, to use few left slots available more effectively. But due to this, in every time slot, the demand is high due to shifting. It can be adopted when demand is so much high all time and to use each slot effectively.

5   Simulation Results

A DSGTN with $I = J = 3$ is considered for implementation. Each secondary substation with five different users and each with five different appliances of different categories is considered. The test case is implemented in the MATLAB [17]. The customers with C_1, C_2, and C_3 category load appliances of 1, 2, 3, 4, 5, and 6 energy slots with duration of 2, 4, 6, and 8 slots are considered. 1kWh is considered as a unit energy slot. The total load request and the sum of energy generated by the secondary sources from all the customers are calculated at the secondary substation. It avoids the wastage at the individual level and helps the cumulative energy saving during energy allocation in the later stage. The remaining requested load energy is sent to the primary substation and then to the grid. Depending on the available energy on the grid, the energy is allotted PSs and SSs use LWBEA method.

Later, the energy is distributed to the various categories of load based on the EDTZLC Algorithm as shown in Table 1. The demand in time slot $t_i = 2–4$ is actually high, but using EDTZLC Algorithm, the demand is reduced to low, due to the shifting of C_2 load to nonpeak slot. Hence, the demand at $t_i = 4-6$ is low, and it now became high due to the shifting of C_2 slots. Later, the unallocated slots are allotted during the next nonpeak slot $t_i = 6–8$ which has low demand (not shown due to space constraints). For Table 1, it is observed that, the demand at $t_i = 2$–4 is reduced and very few slots are remained due to low demand.

To use the energy slots more effectively during huge demand regions, adaptive shifting intelligent control (ASIC) can be used to handle the distribution properly. The simulation results for the test case using EDTZLC and ASIC are given in Table 2. Here, slots are shifted to next slots $t_i = 2$–4 as demand is low before shifting. After shifting, demand increased to high and few left slots are used effectively.

The fairness index [15] of a node is calculated by

$$FI = \frac{\left(\sum_{q=1}^{Q} X_q\right)^2}{Q \sum_{q=1}^{N} X_q^2}$$ (14)

where X_q represents slots allotted to the child nodes and Q is the number of child nodes. The fairness at primary and secondary stations is equal to 0.99.

Table 1 Energy allocation and distribution with LWBEA and EDTZLC algorithms

Time slot t_i	0–2	2–4	4–6
$L_{GH}^{t_i}/L_G^{t_i}$	400/469	300/291	350/393
Demand High/ low	High	Low	High
$L_A^{t_i}(1)/L^{t_i}(1)$	121/143	92/92	123/138
$L_A^{t_i}(2)/L^{t_i}(2)$	140/164	110/110	102/115
$L_A^{t_i}(3)/L^{t_i}(3)$	139/162	89/89	125/140
$L_A^{t_i}(1,1)/L^{t_i}(1,1)$	70/79 ($C_1 = 39$, $C_3 = 18$, $C_2 = 13/22$)	53/53 ($C_1 = 41$, $C_3 = 12$, $C_2 = 0$)	63/69 ($C_1 = 34$, $C_3 = 18$, $C_2 = 8/17$)
$L_A^{t_i}(1,2)/L^{t_i}(1,2)$	48/55 ($C_1 = 31$, $C_3 = 13$, $C_2 = 4/11$)	42/42 ($C_1 = 32$, $C_3 = 10$, $C_2 = 0$)	51/55 ($C_1 = 18$, $C_3 = 20$, $C_2 = 11/17$)
$L_A^{t_i}(1,3)/L^{t_i}(1,3)$	57/63($C_1 = 28$, $C_3 = 21$, $C_2 = 8/14$)	51/51 ($C_1 = 33$, $C_3 = 18$, $C_2 = 0$)	58/63 ($C_1 = 33$, $C_3 = 16$, $C_2 = 10/14$)
$L_A^{t_i}(2,1)/L^{t_i}(2,1)$	64/72 ($C_1 = 35$, $C_3 = 16$, $C_2 = 13/21$)	60/60 ($C_1 = 46$, $C_3 = 14$, $C_2 = 0$)	60/66 ($C_1 = 30$, $C_3 = 8$, $C_2 = 22/24$)
$L_A^{t_i}(2,2)/L^{t_i}(2,2)$	59/67($C_1 = 36$, $C_3 = 13$, $C_2 = 10/18$)	41/41 ($C_1 = 34$, $C_3 = 7$, $C_2 = 0$)	46/49 ($C_1 = 26$, $C_3 = 7$, $C_2 = 14/16$)
$L_A^{t_i}(2,3)/L^{t_i}(2,3)$	65/74($C_1 = 28$, $C_3 = 30$, $C_2 = 7/16$)	56/56 ($C_1 = 30$, $C_3 = 26$, $C_2 = 0$)	44/48 ($C_1 = 13$, $C_3 = 8$, $C_2 = 23/27$)
$L_A^{t_i}(3,1)/L^{t_i}(3,1)$	68/75 ($C_1 = 37$, $C_3 = 22$, $C_2 = 9/16$)	41/41 ($C_1 = 35$, $C_3 = 6$, $C_2 = 0$)	54/59 ($C_1 = 28$, $C_3 = 20$, $C_2 = 7/11$)
$L_A^{t_i}(3,2)/L^{t_i}(3,2)$	63/71 ($C_1 = 30$, $C_3 = 24$, $C_2 = 9/17$)	49/49 ($C_1 = 33$, $C_3 = 16$, $C_2 = 0$)	66/72($C_1 = 26$, $C_3 = 26$, $C_2 = 12/20$)
$L_A^{t_i}(3,3)/L^{t_i}(3,3)$	65/72 ($C_1 = 32$, $C_3 = 22$, $C_2 = 10/18$)	46/46 ($C_1 = 30$, $C_3 = 16$, $C_2 = 0$)	57/61($C_1 = 25$, $C_3 = 16$, $C_2 = 16/20$)

Table 2 Energy slots allocation with LWBEA and EDTZLC method with ASIC

Time slot t_i	0–2	2–4	4–6
$L_{GH}^{t_i}/L_G^{t_i}$	400/469	300/361	350/365
Demand High/ low	High	High	High
$L_A^{t_i}(1)/L^{t_i}(1)$	121/143	95/114	129/135
$L_A^{t_i}(2)/L^{t_i}(2)$	140/164	112/135	93/97
$L_A^{t_i}(3)/L^{t_i}(3)$	139/162	93/112	128/133
$L_A^{t_i}(1,1)/L^{t_i}(1,1)$	70/79(C_1 = 39, C_3 = 18, C_2 = 13/22)	54/62(C_1 = 41, C_3 = 12, C_2 = 0/9)	67/69(C_1 = 34, C_3 = 18, C_2 = 13/17)
$L_A^{t_i}(1,2)/L^{t_i}(1,2)$	48/55(C_1 = 31, C_3 = 13, C_2 = 4/11)	44/49(C_1 = 32, C_3 = 10, C_2 = 2/7)	53/55(C_1 = 18, C_3 = 20, C_2 = 17/17)
$L_A^{t_i}(1,3)/L^{t_i}(1,3)$	57/63(C_1 = 28, C_3 = 21, C_2 = 8/14)	51/57(C_1 = 33, C_3 = 18, C_2 = 0/6)	58/60(C_1 = 33, C_3 = 16, C_2 = 7/11)
$L_A^{t_i}(2,1)/L^{t_i}(2,1)$	64/72(C_1 = 35, C_3 = 16, C_2 = 13/21)	59/68(C_1 = 46, C_3 = 14, C_2 = 0/8)	52/54(C_1 = 30, C_3 = 8, C_2 = 16/16)
$L_A^{t_i}(2,2)/L^{t_i}(2,2)$	59/67(C_1 = 36, C_3 = 13, C_2 = 10/18)	43/49(C_1 = 34, C_3 = 7, C_2 = 4/8)	42/43(C_1 = 26, C_3 = 7, C_2 = 8/10)
$L_A^{t_i}(2,3)/L^{t_i}(2,3)$	65/74(C_1 = 28, C_3 = 30, C_2 = 7/16)	57/65(C_1 = 30, C_3 = 26, C_2 = 0/9)	47/48(C_1 = 13, C_3 = 8, C_2 = 22/27)
$L_A^{t_i}(3,1)/L^{t_i}(3,1)$	68/75(C_1 = 37, C_3 = 22, C_2 = 9/16)	43/48(C_1 = 35, C_3 = 6, C_2 = 3/7)	53/55(C_1 = 28, C_3 = 20, C_2 = 4/7)
$L_A^{t_i}(3,2)/L^{t_i}(3,2)$	63/71(C_1 = 30, C_3 = 24, C_2 = 9/17)	50/5(C_1 = 33, C_3 = 16, C_2 = 0/8)	67/69(C_1 = 26, C_3 = 26, C_2 = 15/17)
$L_A^{t_i}(3,3)/L^{t_i}(3,3)$	65/72(C_1 = 32, C_3 = 22, C_2 = 10/18)	47/54(C_1 = 30, C_3 = 16, C_2 = 2/8)	60/61(C_1 = 25, C_3 = 16, C_2 = 20/20)

6   Conclusion

In this paper, energy slots allocation at primary and secondary stations based on the load weight by considering the secondary energy storage of the customers is proposed. The load request from various appliances of three different types shiftable, non-shiftable, shiftable with continuous slots at the customer is considered. The energy distribution to the different categories of the load during peak and nonpeak hours is handled smoothly by the proposed Energy Distribution based on the Time Zone with Load Category Algorithm, and if required, adaptive shifting intelligent control can be used to smoothen the distribution for heavy demand urban scenarios.

The proposed algorithm with adaptive shifting intelligent control helps the handling of the loads very efficiently and reduces the huge load at peak hour and long queues also. The customer can accomplish the work in nonpeak hour by shifting it to nonpeak time slot without any pressure.

References

1. Yamamoto, S., T. Tazoe, H. Onda, H. Takeshita, S. Okamoto, and N. Yamanaka. 2013. Distributed Demand Scheduling Method to Reduce Energy Cost in Smart Grid. In *IEEE Region 10 Humanitarian Technology Conference (R10-HTC)*, 148–1153.
2. Chao, H., and P. Hsiung. 2016. A Fair Energy Resource Allocation Strategy for Micro Grid. *Microprocessors and Microsystems* 42, 235–244.
3. Nunna, K., and S. Dolla. 2011. Demand Response in Smart Micro Grids. In *IEEE PES Innovative Smart Grid Technologies—India (ISGT India)*, 131–136.
4. Rama chandran, B., S. Srivastava, C. Edrington, and D. Cartes. 2011. An Intelligent Auction Scheme for Smart Grid Market Using a Hybrid Immune Algorithm. *IEEE Transactions on Industrial Electronics* 58 (10): 4603–4612.
5. Nunna, H.K., and S. Dolla. 2012. Demand Response in Smart Distribution System with Multiple Microgrids. *IEEE Transactions on Smart Grid* 3 (4): 1641–1649.
6. Nunna, H.K., and S. Dolla. 2013. Energy Management in Micro Grids Using Demand Response and Distributed Storage a Multi-Agent Approach. *IEEE Transactions on Power Delivery* 28 (2): 939–947.
7. Nunna, H.K., and S. Dolla. 2013. Intelligent Demand Side Management in Smart-Micro Grid. In *IEEE International Workshop on Intelligent Energy Systems (IWIES)*, 125–130.
8. Nunna, H.K., and S. Dolla. 2014. Responsive End-User-Based Demand Side Management in Multi Micro Grid Environment. *IEEE Transactions on Industrial Informatics* 10 (2): 1262–1272.
9. Yingjie, Z., M. Nicholas, Q. Xiangying, and W. Chen. 2014. The Fair Distribution of Power to Electric Vehicles: An Alternative to Pricing. In *5th IEEE International Conference on Smart Grid Communications*.
10. Ardakanian, O., C. Rosenberg, and S. Keshav. 2013. Distributed Control of Electric Vehicle Charging, e-energy 13. In *Fourth International Conference on Future Energy Systems*, 101–112.
11. Pilloni, V., A. Floris, A. Meloni, and L. Atzor. 2016. Smart Home Energy Management Including Renewable Sources: A QoE-Driven Approach. *IEEE Transactions on Industrial Informatics*.
12. Hom Chaudhuri, B., and M. Kumar. 2011. Market Based Allocation of Power in Smart Grid. In *American Control Conference*.
13. Neely, M.J., A.S. Tehrani, and A.G. Dimakis. (2010) Efficient Algorithms for Renewable Energy Allocation to Delay Tolerant Consumers. In *IEEE International Conference on Smart Grid Communications*.
14. Smart Grids and Renewable—A Guide for Effective Deployment. 2013.
15. Rama Devi, Boddu, Manjubala Bisi, and Rashmi Ranjan Rout. 2017. Fairness Index of Efficient Energy Allocation Schemes in a Tree Based Smart Grid. *Pakistan Journal of Biotechnology* 14 (2): 120–127.
16. Rama Devi, Boddu. Dynamic Weight Based Energy Slots Allocation and Pricing with Load in a Distributive Smart Grid. *Journal of Advance Research in Dynamical & Control Systems* 11: 419–433.
17. MATLAB. https://www.mathworks.com/products/matlab.html. Dated: 24.11.2018.

Matrix Approach to Perform Dependent Failure Analysis in Compliance with Functional Safety Standards

Gadila Prashanth Reddy and Rangaiah Leburu

Abstract The main goal of Functional Safety is to implement accident avoidance, employee safety and machine safety systems. Emerging technologies like Artificial Intelligence (AI), Advanced Driver Assistance System (ADAS), autonomous driving and autonomous industries needs very high computing devices (like SoC) with functional safety implemented. This puts System on Chip (SoC) in critical path of functional safety due to its complex design and low process safety time. Every System on Chip Intellectual property (SoC IP) used in these market segments should be carefully analyzed and comply with automotive standard ISO26262 and industrial standard IEC61508, respectively. FMEDA is one such kind of analysis to identify design failure modes and its detection ability. Another analysis is dependent failure analysis (DFA) which explains SOC IP's freedom from interference with respect to common cause failures or inactive/disable IPs associated with its design. Multiple authors proposed various ways to carry FMEDA analysis and automation. But best practices for dependent failure analysis and automation are still lacking. In this paper, we proposed a novel approach to perform dependent failure analysis using matrix method in compliance with industry standards. This novel method helps analyst to visualize interaction of IPs and enhances the quality of analysis. A new matrix method is used to execute DFA which gives systematic and attributable analysis.

Keywords DFA · FMEDA · ISO · IEC · SoC IP

G. Prashanth Reddy · R. Leburu (✉)
RajaRajeswari College of Engineering, Bangalore, India
e-mail: rleburu@gmail.com

G. Prashanth Reddy
e-mail: prashanth.trr@gmail.com

G. Prashanth Reddy
VTU, Belagavi, India

© Springer Nature Singapore Pte Ltd. 2020
K. S. Raju et al. (eds.), *Proceedings of the Third International Conference on Computational Intelligence and Informatics*, Advances in Intelligent Systems and Computing 1090, https://doi.org/10.1007/978-981-15-1480-7_10

1 Introduction

With the help of hardware and software, SoCs are heavily used in both automotive and heavy industries. Applications running on this SoCs can be classified into safety related (SR) and non-safety related (NSR). A critical SR application could be breaking system, speed control, steering and obstacle avoidance, and NSR application could be infotainment, air conditioning system, etc. Providing a best user experience with high performance it is equally important for these devices to function accurately by complying with industry standards. We have two major standards International Organization for Standardization 26262 (ISO 26262) for automotive [1] and International Electrotechnical Commission 61508 (IEC 61508) for industrial applications [2], where IEC 61508 is considered as mother of all standards.

ISO/IEC calls out for industry standards to meet the reliability and safety goals. ISO 26262 provides Automotive Safety Integrity Levels (ASIL) A, B, C, D (where A is for minimum safety and D is for maximum safety), and IEC 61508 provides industry standards in Safety Integrity Level (SIL) 1, 2, 3, 4 (4 is maximum safety and 1 is minimum safety). These standards improvements and safety goals for Electronic and Electrical (EE) systems.

The terminologies frequently used in this paper are faults, failure modes, effect analysis and safety mechanisms [3]. Faults could be permanent faults like stuck at 0, stuck at 1, stuck at unknown values or transient faults due to alpha particle hitting a bit node of Static Random-Access Memory (SRAM) cell and potentially flipping a bit. Failure can be caused by above faults and failure modes can be further classified as random hardware or systematic failure as shown in Fig. 1. Random hardware failures are physically initiated at a node in a device and systematic failures are non-physical but due to the design process or development process.

Once a fault occurs in EE system there is a need to understand the effect caused by it which is called as an effect analysis. Safety mechanisms talk about identifying or correcting the error in EE systems that few examples are parity or Cyclic

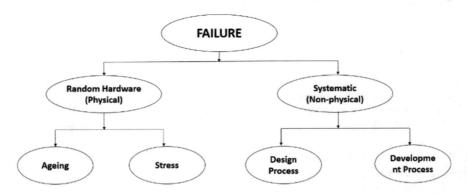

Fig. 1 Failure classification by cause of failure

Redundancy Check (CRC), correcting data by Error Correcting Code (ECC), redundant transmission of data or loosely/hardly coupled lock steps of a processor or a software correcting mechanisms. A detailed analysis on how to implement different safety mechanisms is summarized in [4] and few modern techniques like advance ECC are proposed in [5].

The failure modes effect and diagnostic analysis (FMEDA) talk about fault propagation through design [6], fault effect, possible diagnostics and safety mechanism to handle the fault. This is done by either tool or manual method. To carry out such analysis we need leaf level schematics and functionality of the design. The output of this analysis that is understanding fault and how much percentage of faults can be detected by safety mechanisms will help us to understand diagnostic coverage which in turn defines the SIL levels of the EE system.

Another key analysis is dependent failure analysis [2]. It helps to identify common cause failures with IP blocks and freedom from interference (hardware/software/firmware) with respect to the other digital blocks within IP. Any interference of NSR or disabled digital logic could potentially violate a safety goal.

Further DFA consists of two parts:

1. Validate Freedom from Interference (FFI) between elements.
2. Validate independence within elements.

The dependent failure can be categorized in one of the two types: functional dependency, classic dependency. Functional dependency is where two or more unit's functionality depends on the correct functionality of one single unit. Classic dependency is generally defined as the common cause failure/common mode failure. It may exist where two or more functional units are same and are functioning under the identical conditions.

2 Literature Review on Fault Analysis

The goal of this section is to gather the current industrial practices in fault analysis and to convey background knowledge about the proposed work [6]. This paper analyzes the fault propagation inside the product. Controlling fault propagation increases the accuracy with efficiency, [7] describes a novel methodology of reversing the build-up process helps to recognize each system failure and composition of sources for failure at different levels and [8] helps to understand failure rate assessment like mean time to failure (MTTF), hardware categories and diagnostic coverage level and proposed few analytical technique to calculate failure rate in compliance with ISO 19014, ISO 25119, IEC 61508 [3]. This paper presents a system related failure anamnesis approach. This helps to find main reason for electronic component failure. Different features for designing ADAS functionality and Malfunction behaviors are proposed in [9]. In complement to AUTOSAR, EAST-ADL (architecture description language) intended for automotive embedded system [10]. Paper introduces integration of EAST-ADL with HiP-HOPS method. This is used for automated temporal fault tree analysis. This kind of error model evaluates failure mode, their transformation to/from other states and their propagation within specific systems. Today's safety engineers and automotive industry must understand the intention and content of the standard [11]. This makes them to know about requirements for design of the vehicles. General workflow follows the life cycle of the automobile. This is adopted in safety engineering. The paper [12] presents classical failure analysis case of the SoC for information processing and control.

3   DFA & Proposed Work

In this paper, we presented a novel approach to execute dependent failure analysis using matrix method. In this method, we will plot all leaf level hardware elements and common cause failures/influencers in a matrix format [7]. Each intersection point represents a dependent failure and needs further deep-dive analysis.

Figure 2 helps in understand process flow of proposed work. The main inputs to DFA matrix tool or matrix manual method are list of subblocks in Intellectual Property (IP) and list of all influencers. Typically these two inputs can be obtained by FMEDA analysis. The tool will extract these data and plot all the sub IP blocks in rows and plot all common cause failures/influencers in columns. User has to analyze functional or classical dependency at each node and mark. This might need a thorough understanding of design implementation. Analyst has to further deep dive on each intersection and provide possible safety mechanisms overcome dependent failures.

Applying this process to Serial AT Attachment IP (SATA IP) for simplicity and better understanding of the proposed work, SATA is a CPU bus interface that connects host bus to storage devices such as HDD and SSD. SATA has faster data transfer through higher signaling rates and more efficient transfer via I/O queuing protocol. Figures 3 and 4 explain SATA transmitter and receiver SW protocol and HW implementation.

Let us consider a typical SoC architecture where SATA and PCI are shared to a common bus as shown in Fig. 5, SATA 0 and 1 are two parallel ports shared on bus2 and connected to bus1 with PCI. As an assumption let us consider SATA0 will be used for running safety-related application and SATA1 is used for non-safety-related application and further assumption is PCI is disabled.

A functional safe SATA0 operation is desired however influencers SATA1 which is non-safety critical could stall bus2 and does not allow a SATA0 to run safety-critical application, also when SATA0 wants to interface with other IPs via bus1 due to in advert activation of PCI could stall bus1. Analyzing such scenario is dependent failure analysis and proposed matrix method will help us to understand it better. As shown in Fig. 6, user has to identify and mark all the intersection points and further analyze those interferences. Each crossover point represents a possible

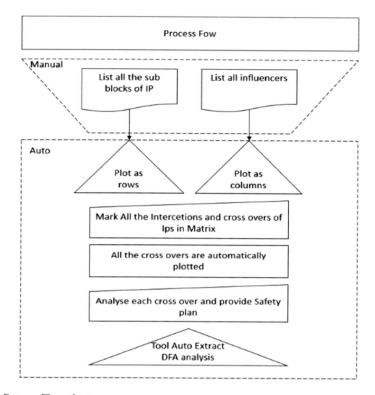

Fig. 2 Process Flow chart

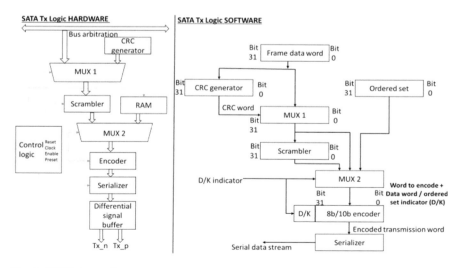

Fig. 3 SATA transmitter

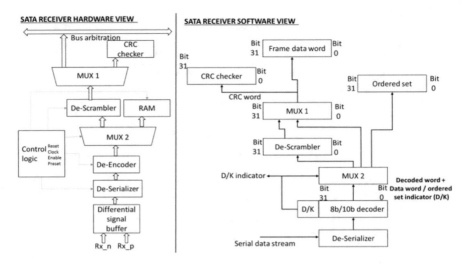

Fig. 4 SATA receiver

Fig. 5 SATA connectivity to other peripherals

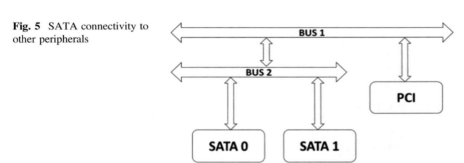

dependent failure also called as depended failure initiators (DFI). In addition to design connectivity, there could be a common DFIs like clock, power and unintentional activation of debug ports and sleep states. Each crossover point defeats safety goal if not address properly, for example, if voltage goes beyond lowest limit, it could defeat both SATA0 safety application and associated safety mechanism. A conventional method to do this is start from functionality of each subblock or element and identifies its failure modes and respective safety mechanisms. This creates complexity in the analysis, for example if a certain IP has 50 subblocks with unique functionality and needs to analyze for 10 DFI, it would take 500 unique elements to analyze. However in Matrix approach, you can directly start with crossover points and reduce the complexity.

Fig. 6 SATA DFA _Matrix
Method

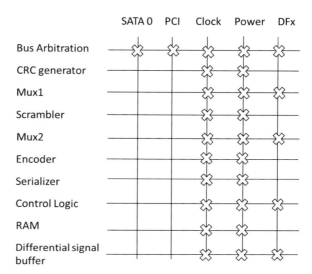

4 Automated Matrix Method

A simple Visual Basic for Application (VBA)-based Graphical User Interface (GUI) can be created to develop automated matrix method, inputs are can be either pre-defined format or from FMEDA analysis with subblock and common cause failure information. Auto-populate the matrix for ease of analysis as shown in Fig. 6. Provide manual button to identify the subblocks and influencers which could potentially result in dependent failures. Auto-populate all the intersection points in GUI and provide a text box for rationale or possible safety mechanisms associated with the dependent failure. Finally, auto-populate complete analysis and store a file for future reference.

5 Advantages & Result of Matrix Method

Reduced complexity considers a manual analysis of DFA where there are 10 subelements and 10 influencers. Executing this in a manual way of tabular columns could result in 100 cells which need to analyze in parallel, on the other hand, matrix method could help in analyzing it in a simple graphical user interface form on a single page. Hence, if a number of subelements and influencers in a digital circuit increase, analyzing DFA manually would result in very high complexity. The experimental results below are based on post-automation and automation cost is not included below. Figure 7 shows the comparison of manual versus autocomplexity when number of elements increases.

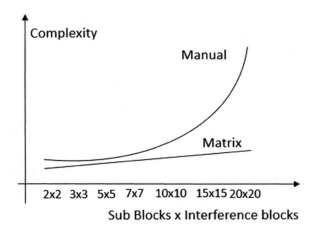

Fig. 7 Manual versus autocomplexity comparison

High accuracy with less executing time, for a moderately complex IP it may take close to 3 to 4 weeks of manual effort to execute DFA however with a manual matrix method it can be significantly brought down by a factor of 70% and by fully automating matrix DFA executing time can be further brought down by 85% with a better accuracy compared (Fig. 8).

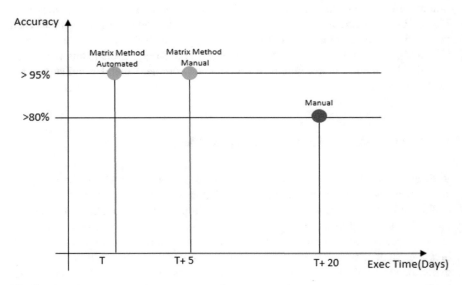

Fig. 8 Manual versus Auto Accuracy comparison

6 Conclusion

Every SoC IP used in automotive industry should be analyzed and comply with industry standards. DFA is one of the critical analyses that need to be done to achieve higher SIL levels. The analyst should carefully analyze possible interference across the platform and address potential dependent failures. If this analysis is neglected or omitted, dependent failures could cause an unsafe environment and could harm human safety and machine safety. This paper suggests an efficient way of doing DFA analysis with high accuracy.

References

1. ISO 26260 Automotive Standards.
2. IEC 16508 Industry Standards.
3. Jacob, P. 2015. Failure Analysis and Reliability on System Level. *Microelectronics Reliability* 55: 2154–2158. (ELSEVIER).
4. Shibahara, Shinichi. Functional Safety SoC for Autonomous Driving. In *2018 IEEE Custom Integrated Circuits Conference (CICC)*.
5. Shaheen, H., G. Boschi, G. Haruthyunyan, and Y. Zorian. 2017. Advanced ECC Solution for Automotive SoCs. In *2017 IEEE 23rd International Symposium on On-Line Testing and Robust System Design (IOLTS)*.
6. Bagalini, E., J. Sini, M. Sonza Reorda, M. Violante, H. Klimesch, and P. Sarson. 2017. An Automatic Approach to Perform the Verification of Hardware Designs According to the ISO26262 Functional Safety Standard. In *2017 18th IEEE Latin American Test Symposium (LATS)*.
7. Herrin, Stephanie A. 1981. Member IEEE, Maintainability Applications Using the Matrix FMEA Technique. *IEEE Transaction on Reliability* R-30 (3).
8. De Rosa, Francesco, Raffaello Cesonib, Stefano Gentac, and Paolo Maggioreb. 2017. Failure Rate Evaluation Method for HW Architecture Derived from Functional Safety Standards (ISO 19014, ISO 25119, IEC 61508). *Reliability Engineering and System Safety* 165: 124–133. (ELSEVIER).
9. Stolte, Torben, Rene S. Hosse, Uwe Becker, and Markus Maurer. 2016. On Functional Safety of Vehicle Actuation Systems in the Context of Automated Driving. In *IFAC (International Federation of Automatic Control)* Hosting by Elsevier Ltd.
10. Chen, DeJiu, Nidhal Mahmud, Martin Walker, Lei Feng, Henrik Lonn, and Yiannis Papadopoulos. 2013. Systems Modeling with EAST-ADL for Fault Tree Analysis through HiP-HOPS. In *4th IFAC Workshop on Dependable Control of Discrete Systems The International Federation of Automatic Control* Sept 4–6, University of York, York, UK (ELSEVIER).
11. Kafka, Peter. 2012. The Automotive Standard ISO 26262, The Innovative Driver for Enhanced Safety Assessment & Technology for Motor Cars. In *2012 International Symposium on Safety Science and Technology*, Procedia Engineering, vol. 45, 2–10 (ELSEVIER).
12. Chen, Yuan, Hui Chen, Xiaowen Zhang, and Ping Lai. 2012. Failure Localization and Mechanism Analysis in System-on-Chip (SOC) using Advanced Failure Analysis Techniques. In *2012 International Conference on Electronic Packaging Technology & High Density Packaging*, 1348, 2012 IEEE.

A Cloud-Based Privacy-Preserving e-Healthcare System Using Particle Swarm Optimization

M. Swathi and K. C. Sreedhar

Abstract Internet has become an integral element of our daily lives owing to its increasing usage. During this model, users will share their information and collaborate with others simply through social communities. The *e*-healthcare community service significantly resolves the issues of individual patients who are remotely situated, have embarrassing medical conditions, or have caretaker responsibilities which will prohibit them from getting satisfactory face-to-face medical and emotional support. Participation in online social collaborations may not be easy due to cultural and language barriers. This paper proposes a privacy-preserving collaborative *e*-healthcare system that connects and integrates patients or caretakers into different groups. This system enables patients or caretakers to chat with other patients with similar problems, understand their feelings, and share many issues of their own. But during this process, private and sensitive information cannot be disclosed to anyone at any point of time. The recommended model uses a special technique, particle swarm optimization to cluster e-profiles based on their similarities. Finally, clustered profiles are encrypted using distributed hashing technique to persevere patients' personal information. The results of proposed framework are compared with well-known privacy-preserving clustering algorithms by using popular similarity measures.

Keywords e-profile · Health care · Particle swarm optimization · Symptoms · Disease · Cluster

M. Swathi
Department of Information Technology, Sri Indu College of Engineering
and Technology, Hyderabad, India
e-mail: swathi.madiraju46@gmail.com

K. C. Sreedhar (✉)
Department of Computer Science and Engineering, Sreenidhi Institute
of Science and Technology, Hyderabad, India
e-mail: simplykakarla@gmail.com

© Springer Nature Singapore Pte Ltd. 2020
K. S. Raju et al. (eds.), *Proceedings of the Third International Conference
on Computational Intelligence and Informatics*, Advances in Intelligent
Systems and Computing 1090, https://doi.org/10.1007/978-981-15-1480-7_11

1 Introduction

In spite of functional and profitable scope in a cloud platform, many clients are hesitating to migrate to the cloud platform due to privacy and security concerns. The *e*-healthcare system significantly impacts the disease diagnostics system and the quality of treatments Sreedhar et al. [1, 2]. The best-known partitional clustering algorithm is the *k*-means algorithm and its variants. In addition to the *k*-means algorithm, Upmanyu et al. [2], several algorithms like SVD, KNN are used for healthcare profile clustering. In this study, a healthcare clustering algorithm based on particle swarm optimization (PSO) is projected. The novelty of this paper shows the best results of hybrid PSO compared to that of *k*-means algorithm.

2 Related Works

Most healthcare applications use cloud computing software such as Microsoft HealthVault, Sreedhar [1]. But, they rely on data security and privacy measures to provide patient health information through the personal health record and electronic health record (EHR) format. Paper [3] discussed clustering using map-reduce framework. Service providers can have direct access to all stored data and can sell these data sets without appropriate permission. Sreedhar et al. [4] propose a multiview similarity approach to implement a cloud-based healthcare system. Cui et al. [5] implement applying the technique of particle swarm optimization to document clustering. This idea is now applied to healthcare system, and moreover, it preserves the privacy by applying the technique addressed in Li et al. [6].

3 Prerequisites

3.1 Patient's e-Profile Representation

In most of the clustering algorithms, the dataset to be clustered is depicted as a collection of vectors $Y = \{y_1, y_2, y_3, \ldots, y_n\}$ wherever the vector y_i corresponds to one patient and is named the feature vector. Each vector y_i is of the form $y_i = \{h_1, h_2, \ldots, h_m\}$ where h_1, h_2, and so on are degree of severity of m diseases whose weights are given with vector $W = \{w_1, w_2, \ldots, w_m\}$. The degree of severity of a disease in a profile can be taken as the value between 0 and 1 or 1–5.

3.2 The Similarity Metric

The similarity between two profiles p_1 and p_2 can be measured using Minkowski distances [5], given by:

$$D_n(p_k, p_l) = \sqrt[n]{\left(\sum_{i=1}^{d_m} |p_{i,k} - p_{i,l}|^n\right)}$$

For $n = 2$, we acquire the Euclidean distance. But during this paper, we use weighted Euclidean distance. The weighted Euclidean distance between two profiles A and B is given as follows:

$$D(A, B) = \sqrt{\sum_i w_i (A_i - B_i)^2}$$

The other normally used similarity measure in healthcare clustering is the cosine similarity metric. It is given as

$$\cos(p_k, p_l) = \frac{p_k^t p_l}{|p_k||p_l|}$$

where $p_k^t p_l$ denotes the dot-product of the two profile vectors. $|.|$ denotes the length of the vector. The formula for weighted cosine similarity measure is given by

$$\cos(p_k, p_l) = \frac{\sum_{i=1}^{d_m} w_i * p_{k,i} * p_{l,i}}{\sqrt{\left(\sum_{i=1}^{d_m} w_i * p_{k,i}^2\right)} * \sqrt{\left(\sum_{i=1}^{d_m} w_i * p_{l,i}^2\right)}}$$

Weighted Cosine measure is used in this paper.

4 Patient e-Profile and Group Creation

Algorithm: Group Creation

Initiation: Each patient must have his or her EMR connected to the e-profile
The entire system allows all patients who have valid e-profiles
All patients R in the system
For
each patient $r \in R$; M_r is the expected patient group size of r
while not all patients in R are in the group
do
for each $r \in R$ who has not joined any other patient group in the system
do

r chooses to be the host of the current patient group with probability
$Prob(O_v) = M_v/|R|$
if r is host then r invites its known people with illness to join its group
end if
end for
for each do
Let K_r be the set of r's known people who are the hosts of other groups;
If $K_r \neq \emptyset$ then
r randomly chooses $v \in K_r$ and joins the group of v;
else
Let L_r be the set of r's known people who already joined other groups;
if $L_r \neq \emptyset$ then
r randomly chooses $v \in L_r$ and joins the group of v;
end if
end if
end for
for each patient group Z do
Let r be the host of Z
if $|R_Z| >= M_r$ then
Z is said to be formed;
end if
end for
end while

5   Proposed *e*-Healthcare Model Using Particle Swarm Optimization(PSO)

In PSO algorithm, the winged creatures in a rush emblematically speak to particles. These particles are spoken to as straightforward specialists flying through an issue space. When a particle moves to another location, a distinctive issue solution is generated. This solution is assessed by a wellness formula that gives a quantitative estimation of the solution's viability. As every particle moves to another situation along each measurement of the issue space, the velocity and direction of every particle will be updated. In combination, the ith particle's close to home experience, P_i, and the worldwide best experience P_g impact the development of every particle through an issue space. Keeping in mind the end goal to investigate the pursuit space widely, the estimations of rand1 and rand2 are changed before uniting around

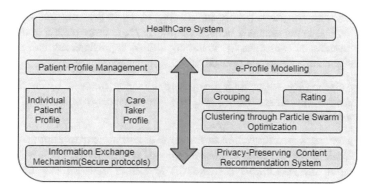

Fig. 1 Overall *e*-healthcare system model

the ideal solution. The estimations of $c1$ and $c2$ control the weight adjust of P_i and P_g to choose the particle's next development velocity. At every iteration, every particle's new location is registered by summing the particle's new velocity V_i with its old location X_i. Figure 1 shows overall *e*-healthcare model.

Given a multidimensional issue space, the *i*th particle changes its velocity and location as indicated by the accompanying Eqs. (1) and (2).

$$V_i = w * V_i + c1 * \text{rand1} * (P_i - X_i) + c2 * \text{rand2} * (P_g - X_i) \tag{1}$$

$$X_i = X_i + V_i \tag{2}$$

where w means the dormancy weight factor; P_i is the location of the particle that encounters the best wellness value; P_g is the location of the particles that experience a global best wellness esteem; $c1$ and $c2$ are constants and are known as speeding up coefficients; and rand1 and rand2 are arbitrary esteems in the range of (0,1).

5.1 The Basic PSO Clustering Algorithm

The wellness (g) of the clustering procedure is ascertained by computing the average distance of e-profiles to the cluster centroid.

$$g = \frac{\sum_{i=1}^{N_c} \left\{ \frac{\sum_{j=1}^{c_i} d(O_i, D_{ij})}{c_i} \right\}}{N_c} \tag{3}$$

where D_{ij} denotes the jth e-profile in ith cluster; O_i is the centroid vector of ith cluster; $d(O_i, D_{ij})$ is the distance between e-profile D_{ij} and the cluster centroid O_i; c_i represents number of e-profiles in cluster C_i; and N_c represents number of clusters. The PSO algorithm can be composed as:

(1) *As an underlying step, each particle arbitrarily picks k distinctive profile vectors from the dataset as the underlying cluster centroid vectors.*
(2) *For every particle:*

   (a) *Assign each profile vector in the dataset to the nearest centroid vector.*
   (b) *Calculate the wellness esteem in view of condition (3).*
   (c) *Update the conditions (1) and (2) to create new velocity and position bringing about new conceivable solutions.*

(3) *Repeat step (2) until one of the underneath end conditions is fulfilled.*

   (a) *The most extreme number of cycles is surpassed or*
   (b) *The average change in centroid vectors is not exactly a predefined esteem.*

5.2   The Hybrid PSO Clustering

The algorithm for hybrid PSO can be composed as:

(1) *Execute the PSO algorithm with initially chosen set of profiles as initial centroids with specified number of particles and cycles.*
(2) *Make the output of step (1) as initial seed for k-means algorithm.*
(3) *Run the k-means algorithm for specified number of cycles.*

5.3   Distributed Privacy Preservation for Clustered Profiles

In our proposed model, the distributed min-hashing technique is implemented for any communication between profile members, Li et al. [6].

Algorithm: Privacy_Preservation (P,C$_H$,P$_{Prob}$)
Initially: SCode →ø
          P→Set of Patients in the system
          C$_H$→ Cluster of e-profiles
          P$_{Prob}$ → Pre_defined probability of C_H
          R$_{Prob}$ →Random probability
The server randomly chooses user p∈P while not all profiles in C$_H$ are in SCode
          if p obtains SCode for the first time
               then
          p arbitrarily generates $0 \leq R_{Prob} \leq 1$

          if $R_{Prob} \leq P_{Prob}$ then

               p arbitrarily generates GID(p)
               C$_H$(p)={C$_H$-C$_H$(SCode)};
               SCode.Assign (GID(p),C$_H$(p));
     end if
     end if
if   p   connects to other related profile p' from the same cluster arbitrarily,
          p' ∈R(p)
Then
 p need to send SCode to p' for arbitral check
          p=p';
while   SCode is sent to admin do
p sends SCode to admin with the possibility of P_{Prob} for unsigned communication
if SCode is not sent to admin
Then
p runs one more arbitral check
end if
end while

6   Experimental Results and Analysis

We have implemented the algorithm in MATLAB (R2016b). We have assumed zero network latency between two clients and between client and server. The developed prototype model is executed on a detailed evaluation of the Symptom-X dataset [7]. In the simulated environment, around 5,000 e-profiles are created by combining both patient and caretaker profiles. The interactions are shown as various interesting posts, comments, and recommendations on a daily basis. Figure 2 indicates that k-means algorithm converges quickly whereas PSO algorithm takes more than 100 iterations to converge and finally hybrid PSO gives fruitful clustering result for the case of $k = 5$ using dataset 4 and Euclidean measure. All the datasets were synthetically generated using MATLAB (Tables 1, 2).

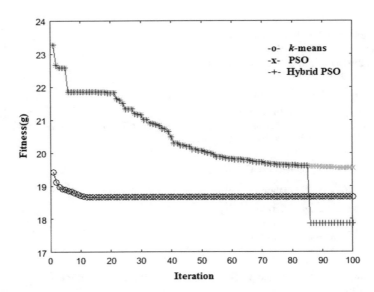

Fig. 2 The convergence behaviors of different clustering algorithms for $k = 5$ using Euclidean measure and dataset 4(k-means, PSO, and Hybrid PSO algorithms)

Table 1 Fitness values for Euclidean measure

Dataset	k	List of schemes			Dataset	K	List of schemes		
		k-means	PSO	Hybrid PSO			k-means	PSO	Hybrid PSO
Dataset 1	5	18.642	19.458	18.698	Dataset 3	5	18.613	19.441	18.606
	10	17.971	19.118	16.619		10	17.984	19.124	16.417
	15	17.615	19.202	17.665		15	17.54	19.173	17.619
	20	17.339	18.904	17.394		20	17.314	19.234	17.33
	25	17.123	19.043	16.753		25	17.073	18.868	17.124
	30	16.914	18.704	17.018		30	16.872	18.861	16.899
	35	16.725	18.608	16.561		35	16.718	18.693	16.76
	40	16.605	18.636	16.739		40	16.582	18.499	16.232
	45	16.479	18.27	16.563		45	16.437	18.401	16.488
	50	16.356	18.151	16.489		50	16.347	18.079	16.029
Dataset 2	5	18.670	19.415	18.673	Dataset 4	5	18.66	19.532	17.86
	10	18.000	19.487	18.032		10	17.977	19.286	18.029
	15	17.641	19.062	17.654		15	17.609	19.005	15.758
	20	17.339	19.007	17.368		20	17.298	19.088	17.33
	25	17.107	18.731	16.876		25	17.096	18.86	17.12
	30	16.925	18.731	16.953		30	16.911	18.627	16.507
	35	16.739	18.644	16.785		35	16.72	18.601	16.762
	40	16.609	18.261	16.260		40	16.599	18.61	16.639
	45	16.451	18.392	16.488		45	16.475	18.525	16.017
	50	16.356	18.340	16.390		50	16.36	18.563	16.378

Table 2 Fitness values for Cosine measure

Dataset	k	List of schemes			Dataset	k	List of schemes		
		k-means	PSO	Hybrid PSO			k-means	PSO	Hybrid PSO
Dataset 1	5	0.10117	0.11049	0.097	Dataset 3	5	0.1009	0.1058	0.0978
	10	0.0945	0.1023	0.0904		10	0.0944	0.1023	0.0904
	15	0.091	0.1011	0.0877		15	0.0888	0.1029	0.0899
	20	0.0868	0.0988	0.0853		20	0.0864	0.1003	0.0855
	25	0.0844	0.0980	0.0829		25	0.0848	0.0987	0.0839
	30	0.0830	0.0972	0.0823		30	0.0821	0.0955	0.0818
	35	0.0809	0.0959	0.0798		35	0.0807	0.0965	0.0793
	40	0.0795	0.0973	0.0787		40	0.0799	0.0948	0.0784
	45	0.0786	0.0939	0.0775		45	0.0781	0.0942	0.0772
	50	0.0770	0.0948	0.0762		50	0.0770	0.0926	0.0763
Dataset 2	5	0.1005	0.1054	0.0971	Dataset 4	5	0.1034	0.1101	0.1000
	10	0.0945	0.1003	0.0887		10	0.0942	0.1013	0.0900
	15	0.0901	0.1003	0.0870		15	0.0908	0.0984	0.0852
	20	0.0866	0.0975	0.0850		20	0.0862	0.1008	0.0848
	25	0.0842	0.0966	0.0833		25	0.0846	0.0979	0.0841
	30	0.0821	0.0975	0.0811		30	0.0820	0.0975	0.0820
	35	0.0808	0.0965	0.0804		35	0.0806	0.0965	0.0800
	40	0.0794	0.0959	0.0789		40	0.0787	0.0970	0.0787
	45	0.0777	0.0971	0.0775		45	0.0779	0.0948	0.0775
	50	0.0776	0.0940	0.0765		50	0.0772	0.0929	0.0748

6.1 Fitness of Clustering

See Fig. 2.

6.2 Latency

The algorithm is tested using data from Fudan BBS's doctor subcommunity dataset, a popular online social community among Chinese universities. The numbers of profiles, posts, and reads may vary for each cluster. The entire evaluation is repeated 100 times by using the same datasets. The latency values of k-means and PSO algorithm for various values of k are as follows. Figure 3 indicates that as value of k increases, the value of latency also increases and PSO algorithm gives always best latency compared to k-means algorithm (Table 3).

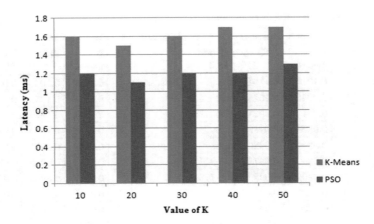

Fig. 3 The latency of *k*-means and PSO

Table 3 Latency values of *k*-means and PSO

	Latency for different value of *k (Seconds)*				
Scheme	10	20	30	40	50
k-means	0.0016	0.0015	0.0016	0.0017	0.0017
PSO	0.0012	0.0011	0.0012	0.0012	0.0013

7 Conclusion

This paper projects an adaptive *e*-healthcare system by using particle swarm optimization clustering scheme. The experimental results demonstrate that the PSO achieves compact results for large number of iterations and hybrid PSO achieves best fitness value for small number of iterations compared to PSO. The novelty in this paper is that hybrid PSO technique gives best results compared to that of *k*-means algorithm.

References

1. Sreedhar, K.C., M.N. Faruk, and B. Venkateswarlu. 2017. A Genetic TDS and BUG With Pseudo-Identifier for Privacy Preservation Over Incremental Data Sets. *Journal of Intelligent and Fuzzy Systems* 32 (4): 2863–2873.
2. Upmanyu, M., A.M. Namboodiri, K. Srinathan, and C.V. Jawahar, Efficient Privacy Preserving k-means Clustering. In: *PAISI'10 Proceedings of the 2010 Pacific Asia conference on Intelligence and Security Informatics*, 154–166.
3. Satish, M. and M. Ramakrishna Murt. 2015. Clustering with Mapreduce using Hadoop Framework. *International Journal on Recent and Innovation Trends in Computing and Communication*. 3(1): 409–413, ISSN 2321-8169.

4. Sreedhar, K.C., and N. Suresh Kumar 2018. An Optimal Cloud-Based *e*-healthcare System using *k*- Centroid MVS Clustering Scheme. *Journal of Intelligent and Fuzzy Systems* 34: 1595–1607.
5. Cui, Xiaohui, Thomas E. Potok, and Paul Palathingal. 2005. Document Clustering using Particle Swarm Optimization. In *Proceedings 2005 IEEE Swarm Intelligence Symposium*. SIS 2005. 0-7803-8916-6/05.
6. Li, D., Q. Lv, L. Shang, and N. Gu. 2017. Efficient Privacy-Preserving Content Recommendation for Online Social Communities. *Neurocomputing* 219: 440–454.
7. https://github.com/deshanadesai/Symptom-X-/blob/master/dataset_clean1.csv#L2.

Improved Approach to Extract Knowledge from Unstructured Data Using Applied Natural Language Processing Techniques

U. Mahender, M. Kumara Swamy, Hafeezuddin Shaik
and Sheo Kumar

Abstract Extraction of meaningful knowledge from a unstructured data is a complex task. In the literature, efforts have been made using text mining approaches. These approaches employ rich amount of resources in mining the textual datasets. In this paper, we focus on optimization of mining algorithm of text data to generate the automatic text summarization. In this approach, we extract text summaries from the text data carpus using natural language processing techniques. We propose a mining approach in semantic parsing and generate the automatic text summaries. We conduct the experiments on the real-world dataset and show the proposed approach is useful than the existing approaches.

Keywords Data mining · Natural language processing · Text mining · Summarization

1 Introduction

Growth of Internet allows normal people to access the huge amount of online repositories such as Web documents and text documents. On the other hand, people do not have sufficient time to go through the entire text documents for the desired information. To meet this challenge, the area called automatic text summarization

U. Mahender (✉) · M. Kumara Swamy · S. Kumar
Department of Computer Science & Engineering, CMR Engineering College,
Hyderabad, India
e-mail: mahenderudutala@gmail.com

M. Kumara Swamy
e-mail: m.kumaraswamy@cmrec.ac.in

S. Kumar
e-mail: sheo2008@mail.com

H. Shaik
Hyderabad, India
e-mail: hafeezskhbablu@gmail.com

© Springer Nature Singapore Pte Ltd. 2020
K. S. Raju et al. (eds.), *Proceedings of the Third International Conference
on Computational Intelligence and Informatics*, Advances in Intelligent
Systems and Computing 1090, https://doi.org/10.1007/978-981-15-1480-7_12

has been evolved to generate the text summary automatically from the given text document. The summarized text helps the normal people to take a decision whether the document is relevant to their requirement or not. This situation demands the exhaustive research in the area of automatic text summarization. In [1], a summary is defined as a text that is produced from one or more texts, that conveys important information in the original text(s), and that is no longer than half of the original text (s) and usually, significantly less than that.

The process of automatic text summarization produces a concise summary by preserving key information content and overall meaning. In recent years, several approaches have been developed for automatic text summarization and applied widely in various domains. Search engine snippets [2], preview of news articles [3] are some of the examples of automatic text summarization. To meet these requirements, efforts have been made to extract of relevant knowledge from the text dataset using text mining [4] approaches.

In this paper, we propose an approach to extract the text summary from the text data using the semantic-based approach. In the approach, we extract the bag-of-words using the term frequency-inverse document frequency (TF-IDF) [4]. We apply the semantic-based approach to extract the text summaries from the bag-of-words. We conduct the initial experiments using the sample text and the results are encouraging. In this paper, we present the initial experimental results only.

The rest of the paper is organized as follows. In the next section, we present the related work. In Sect. 3, we present about the text mining. In Sect. 4, the proposed approach is explained. We explain the initial experimental results in Sect. 5. In the final section, we present the summary and conclusions.

2 Related Work

In this section, we explain the related work in document summary approaches and also explain how the proposed approach is different from the existing approaches.

A survey has been conducted about the automatic text summarization in [5, 6]. In this survey, various approaches have been summarized.

An important research of these days was [7] for summarizing scientific documents. In [7], introduced a method to extract salient sentences from the text using features such as word and phrase frequency. They proposed to weight the sentences of a document as a function of high-frequency words, ignoring very high-frequency common words.

Edmundson et al. [8] described a paradigm based on key phrases which in addition to standard frequency depending weights, used the following three methods to determine the sentence weight. In [10–15] several other methods have also proposed in this area.

The proposed approach is different from the approaches proposed in [5]. We employ the semantic approach using the Keras library in Python.

3   About Text Mining

In this section, we explain the text mining approach and also explain the stages of knowledge extraction.

Text mining is the procedure of synthesizing information, by analyzing relations, patterns, and rules among textual data semi-structured or unstructured text. Several approaches have been proposed to extract useful knowledge from the text documents. Some of the approaches include clustering, classification, summarization, etc.

Figure 1 illustrates steps to extract the useful knowledge from the text documents. In this figure, the first stage is the problem to be analyzed and goals of mining are proposed. The datasets are gathered from various secures based on the objectives of the mining process. Generally, the data is unstructured format and the data mining approaches cannot be applied directly on these datasets. In the next stage, the data is organized in the required manner using the various approaches such as tf-idf and reprocessing. In the final stage, the knowledge can be extracted based on the objectives and goals defined in the first stage. This knowledge can be used for the different applications within and outside the business.

In this paper, we focus on stage two in Fig. 1. Now, we explain the proposed approach in the following section.

4   Proposed Approach

In this section, we explain the proposed approach.

We propose an optimized approach called exposition approach to extract knowledge from textual data. In this approach, we convert the sentences into a predefined template which gives the sentence structure such as noun, verb, and object. This predefined template structure is given to Keras to find the text summary. Hence, there are two steps in this approach: convert sentences into NLP structure and generate summary. We explain these steps in the following.

Fig. 1 Text mining stages

4.1 NLP Structure

In this step, the text is converted into the NLP structure that is generating the unstructured data into the structured formed as explained in the following. Figure 2 illustrates the segment of feature extraction approach.

Most of the learning models and algorithms are complex, and generally they are based on bag-of-words approach. Some of the approaches exploit the recurrent neural network (RNN) approach to directly translate the sentence into a meaning.

The proposed approach transforms the textual data into a tree-level structure continuously till formal meaning representation is derived. This approach makes the process feature-rich as it breaks the text into a tree-like structure for effective learning and representation. The uniqueness of this methodology is that a single word can be tagged as noun, a named body, and a proper noun, which means that a single word can have multiple features associated with it. This makes the approach a feature reach and lots of scope to study the features further. Here we are converting the textual data into a much simpler logical form understandable by the machine using deep learning techniques (Fig. 3).

In exposition approach of text mining, we focus mainly on word type and order/place of the word in the sentence.

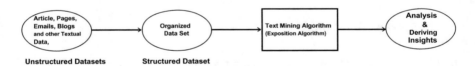

Fig. 2 Optimization levels of text

Fig. 3 Segment of feature extraction

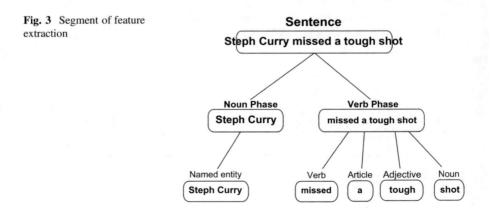

4.2   Generate Summary

Now, explain the process of generating summary using the NLP structure. We shall explore the knowledge extracted to analysis the data. This knowledge extracted is subjected to summarization. It is notable fact that the increase of the knowledge in the data, the importance of semantic destiny is also increase. As the deep learning methodologies in the domain of text mining increased, more sophisticated algorithms such as exposition algorithms are needed.

The automatic text summary is expected to give more accurate and consistent as similar to the human-generated summaries. For example, generating a title for an article based on the content/knowledge. In order to generate the summaries, we exploit the Keras [9] open library. Keras is an open-source neural network library written in Python. Keras contains numerous implementations of commonly used neural network building blocks such as layers, objectives, activation functions, optimizers, and a host of tools to make working with image and text data easier. In addition to standard neural networks, Keras has support for convolutional and recurrent neural networks. It supports other common utility layers like dropout, batch normalization, and pooling. The Keras pretrained model is provided in Fig. 4.

This pretrained model (Keras) takes the input knowledge words from the stream and generates the vector values for the same. These vector values are plotted to derive relevancy.

In Fig. 4, each vector deceits an entity holding a different property which may be anything like gender, age, race, etc. The summarizer generates the vector of words along with the weight of each word using the deep learning methods. Now, the high-weighted words are the candidate words to generate the text summaries. The high-weighted words are taken and generate the senses using the NLP structure as explained in Sect. 4.1. Using this approach, more meaningful automated summaries are generated.

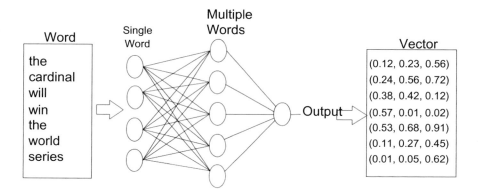

Fig. 4 Pretrained model (Keras)

The following are the applications for the text summaries.

1. Monitoring of social media landscape.
2. Bio-engineering, bio-medical applications, and bio-informatics.
3. Digital advertisements.
4. Building business intelligence.
5. Sentimental analysis.
6. Building of marketing applications.

5   Experimental Results

In this section, we have shown the initial results of the experiments conducted using the football database. We collect the dataset from the Kaggle.

This dataset includes 39,669 results of international football matches starting from the very first official match in 1972 up to 2018. The matches range from FIFA World Cup to FIFI Wild Cup to regular friendly matches. The matches are strictly men's full internationals and the data does not include Olympic Games or matches where at least one of the teams was the nation's B-team, U-23 or a league select team.

The dataset contains the attributes date (date of the match), home team (the name of the home team), away team (the name of the away team), home score (full-time home team score including extra time, not including penalty-shootouts), away score (full-time away team score including extra time, not including penalty-shootouts), tournament (the name of the tournament), city (the name of the city/town/ administrative unit where the match was played), country (the name of the country where the match was played), and neutral (True/False column indicating whether the match was played at a neutral venue).

The preliminary results are shown in Fig. 5.

Here, the magnitude across each x-axis of a particular entity represents relevance property to a word. It means the term 'Woman' is the closest vector to the y-axis which further indicates that the term woman is a suitable word in this context. Sample generated titles based on Keras training model.

Many countries across the world play many sports. But one common thing among all the nations is that almost all nations play football. It is to be noted that football is also the highest watched sport on the air across the world.

6   Conclusion

Extraction of meaningful knowledge from a unstructured data is a complex task. Efforts have been made using text mining approaches. In this paper, we focus on optimization of mining algorithm of text data to generate the automatic text

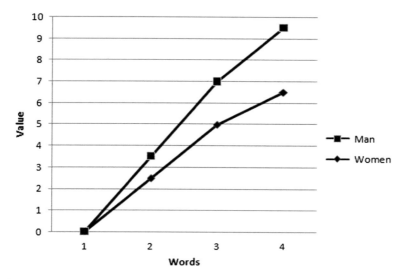

Fig. 5 Text summarizer

summarization. In this approach, we extract text summaries from the text data carpus using natural language processing techniques. We propose a mining approach in semantic parsing and generate the automatic text summaries. We conduct the experiments on the real-world dataset and show the proposed approach is useful than the existing approaches.

As a part of the future work, we are planning to conduct the elaborate experiment in various datasets.

References

1. Murty, M. Ramakrishna, J.V.R. Murthy, P.V.G.D. Prasad Reddy, and Suresh Chandra Satapathy. 2012. A survey of cross-domain text categorization techniques. In *RAIT 2012*, 499–504.
2. Radev, D.R., E. Hovy, and K. McKeown. 2002. Introduction to the Special Issue on Summarization. *Computational Linguistics* 28 (4): 399–408.
3. Turpin, A., Y. Tsegay, D. Hawking, and H. E. Williams. 2007. Fast Generation of Result Snippets in Web Search. In *Proceedings of the 30th Annual International ACM SIGIR Conference on Research and Development in Information Retrieval, ACM*, 127–134.
4. Lahari, K., M. Ramakrishna Murty. 2015. Partition Based Clustering Using Genetic Algorithms and Teaching Learning Based Optimization: Performance Analysis. In *International Conference and Published the Proceedings in AISC*, vol. 2, 191–200. Berlin: Springer. https://doi.org/10.1007/978-3-319-13731-5_22.
5. Ricardo Baeza-Yates, and Ribeiro-Neto Berthier. 2011. *Modern Information Retrieval: The Concepts and Technology Behind Search*, 2nd ed. USA: Addison-Wesley Publishing Company.

6. Edmundson, Harold P. 1969. New Methods in Automatic Extracting. *Journal of the ACM* 16 (2): 264–285.
7. Allahyari Mehdi, Seyedamin Pouriyeh, Mehdi Assefi, Safaei Saeid, Elizabeth D. Trippe, Juan B. Gutierrez, and Krys Kochut. 2017. Text Summarization Techniques: A Brief Survey. *International Journal of Advanced Computer Science and Applications* 8 (10): 397–405.
8. Luhn, Hans Peter. 1958. The Automatic Creation of Literature Abstracts. *IBM Journal of Research and Development* 2 (2): 159–165.
9. Rabinowitz, P. 1980. On Subharmonic Solutions of a Hamiltonian System. *Communications on Pure Applied Mathematics* 33: 609–633.
10. Clarke, F., and I. Ekeland. 1982. Nonlinear Oscillations and Boundary-Value Problems for Hamiltonian Systems. *Archive Rational Mechanics and Analysis* 78: 315–333.
11. Clarke F., I. Ekeland. 1978. *Solutions periodiques, duperiode donnee, des equations hamiltoniennes.* Note CRAS Paris 287, 1013–1015.
12. Michalek, R., and G. Tarantello. 1988. Subharmonic Solutions with Prescribed Minimal Period for Nonautonomous Hamiltonian Systems. *Journal of Differential Equations* 72: 28–55.
13. Tarantello, G. (to appear). Subharmonic solutions for Hamiltonian systems via a ZZ_p pseudoin-dex theory. *Annali di Matematica Pura* .
14. Allahyari, M., S. Pouriyeh, M. Assefi, S. Safaei, E. D. Trippe, J. B. Gutierrez, and K. Kochut. 2017. *A Brief Survey of Text Mining: Classification, Clustering and Extraction Techniques.* ArXiv e-prints.
15. Manjusha, K., M. Anand Kumar, and K.P. Soman. 2017. Reduced Scattering Representation for Malayalam Character Recognition. *Arabian Journal for Science and Engineering.*

Word Sense Disambiguation in Telugu Language Using Knowledge-Based Approach

Neeraja Koppula, B. Padmaja Rani and Koppula Srinivas Rao

Abstract In NLP, many languages will have many ambiguous words, and finding the correct meaning of an ambiguous word is known as word sense disambiguation. This research article is to develop a WSD system for regional Telugu language. Word sense disambiguation system can be developed using three approaches; in this work, we are using knowledge-based approach, where the accuracy is more than unsupervised approaches. In regional Telugu language, research work on word sense disambiguation is not up to the mark. In English and Hindi languages, word sense disambiguation systems are developed using all the three approaches.

Keywords Word sense disambiguation · Telugu language · Knowledge-based approach · NLP

1 Introduction

In natural language processing (NLP), Word Sense Disambiguation (WSD) is an open challenge. Many verbal languages will have many ambiguous words or polysemy words in their language. The process of identifying the appropriate sense of a polysemy word is known as word sense disambiguation. The main application of WSD is that it is used in machine translation. Reasonably less work is reported for regional Indian languages when compared with English language. No work on WSD is reported for Telugu language, which is at infant level.

N. Koppula (✉) · K. S. Rao
Department of CSE, MLR Institute of Technology, Hyderabad, India
e-mail: Kneeraja123@gmail.com

K. S. Rao
e-mail: Ksreenu2k@gmail.com

B. Padmaja Rani
Department of CSE, JNTUCEH, Hyderabad, India
e-mail: Padmaja_jntuh@jntuh.ac.in

© Springer Nature Singapore Pte Ltd. 2020
K. S. Raju et al. (eds.), *Proceedings of the Third International Conference on Computational Intelligence and Informatics*, Advances in Intelligent Systems and Computing 1090, https://doi.org/10.1007/978-981-15-1480-7_13

153

The task is to determine the meaning of an ambiguous word in the given context, using regional Telugu language. For example: బంతి Polysemy word in Telugu language. The word "బంతి", with three distinct senses

sense1 పుష్ప
sense2 గుండ్రని, ఆటవస్తువు
sense3 వరుస

Senses of బంతి in three different contexts

1. పేడతో గొబ్బెమ్మలు చేసి, వాటిపై బంతి పూల రేకులు చల్లి అలంకరిస్తారు. In this context, the appropriate sense is sense1 పుష్ప

2. రాముడికి బంతి తగిలింది. In this sentence, the appropriate sense is sense2 ఆటవస్తువు

3. వివాహ సమయాల్లో ఏర్పాటు చేసే సామూహిక భోజనాలను బంతి భోజనాలు అంటారు. In this context, sense3 వరుస is the appropriate sense.

The process of assigning the correct meaning to the polysemy word (a word having multiple meanings) in the given context is word sense disambiguation. Word sense disambiguation system is developed using three approaches: supervised approach, knowledge-based approach and unsupervised approach. Supervised approach [1] is more accurate when compared with other approaches; this approach uses dataset as corpus. The process of building Telugu corpus is a complex task. In this paper, knowledge-based approach is used to build word sense disambiguation system in regional Telugu language.

In this article, Sect. 2 explains literature survey on the approaches, which is followed by the proposed model, and Sect. 4 explains the proposed algorithm to develop word sense disambiguation system in Telugu Language. Section 5 explains the evaluation process and is followed by future enhancements and conclusion, acknowledgement and references.

2 Literature Survey

Word sense disambiguation [2] is a challenging task in the area of natural language processing. It is the open challenge in the field of artificial intelligence. Many researches are in progress in this field; full-fledged work is not reported in any language. But maximum work is reported in English and Hindi Languages. Less work is reported in Bengali, Nepali and Assamese Languages.

Recently, the graph-based word sense disambiguation system has gained much importance in NLP and is developed for English languages. The drawbacks of the traditional approaches of word sense disambiguation have been overcome in this approach.

In Telugu language, word sense disambiguation is at infant level. The research work in Telugu Language is not up to the mark.

To build word sense disambiguation system [3] Lesk proposed an algorithm in English language which is based on overlapping of content or context words of a target word. WSD system is developed using similarity and semantic metrics between the concepts of a polysemy word.

The traditional approach Walker's [4] algorithm proposed by Walker scientist, he proposed a WSD system based on domain specific or we can say it as categorized methodology. The category for which more no of context words are matched that category sense is assigned to the disambiguated word.

Another traditional approach is conceptual density [5], which is defined as conceptual density between the concepts (senses) of the polysemy word. The WordNet of English language is developed, and this WordNet is represented in the form of a graph; from this graph, we are calculating the conceptual density between the senses of an ambiguous word. The conceptual density is equivalent to the conceptual distance, which is calculated from the formula.

Navigli and Roberto [6] proposed a method for word sense disambiguation system, which is purely based on the graph. WordNet is represented as a graph, and they applied a page rank algorithm on this graph [7], the node for which page rank is more that node is treated as the appropriate sense.

In Regional Telugu language, proposed works are WFS (Word First Sense) finding the correct sense of the polysemy word, by assigning the first sense of the polysemy word from the LKB (Lexical Knowledge Base) as an appropriate sense and WMFS (Word Most Frequent Sense) is the process of identifying the sense, having more no of context words that sense is treated as Most Frequent Sense of the polysemy word. These two methods are context independent, and our proposed method WTSS is context dependent. The WFS and WMFS methods are related to our proposed method Word Total Sense Score (WTSS) in this paper. In the evaluation process, we compared all the three proposed methods WFS, WMFS and WTSS.

3 Proposed Model

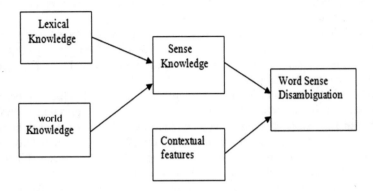

Fig. 1 Word sense disambiguation model [8]

4 Proposed Method

Word Sense disambiguation System in regional Telugu language can be developed using two different categories—one is with context (context-dependent approach) and other is without context (context-independent approach).

In literature survey, the methods WFS and WMFS are discussed which are context-independent approaches, and our present proposed approach is context-dependent Word Total Sense Score (WTSS).

Context-Dependent Approach. This method is known as Word Total Sense Score (**WTSS**) which is our proposed method based on input context words, and surrounding words of a target polysemy word in input sentence are matched with the context words of a particular sense of a target polysemy word in LKB. The context words for each sense of a polysemy word are stored in LKB from training data, with proximity of 2–3 surrounding words of the target polysemy word.

In our proposed method, the input is the sentence with polysemy word. The surrounding words around the target polysemy word are treated as context words, and from the Lexical Knowledge Base (LKB), retrieve the content words of each sense of the polysemy word, and matching is done between the content words and context words. For which sense matching of context words are more, with the input given context words that sense is treated as an appropriate sense. The algorithm is as follows,

Algorithm

Input: Test sentence with target polysemy word.

Output: Appropriate sense of a polysemy word

1. Read input sentence IS

>   *Target word ← null, Context word ← null.*

>   *Score ← 0, Sense ← null*

2. Pre-processing stage   a. Removal of stop words   b. Stemming

3. For each Wj in PSW

>   *If (Wi == Wj)*

>>   *Target word ← Wi*

>   *Else*

>>   *Add Wi to context words*

4. Extract each sense Sk of a target word

>   *Skscore ← 0*

5. Extract context word list PCSk of sense k

6. For each word Wi of context words

>   *If   (Wi in PCSk)*

>>   *Skscore ← Skscore + 1*

7. If (Skscore > score)

>   *score ← Skscore*

>   *Sense ← Sk*

8. Output sense

b. Implementation

Let us consider the following examples

భూమి బంతి వలె ఉంటుంది

నాటకం లో కృష్ణుని పాత్ర మరువలేనిది

రాముని తలకు బంతి తగిలింది

Example: భూమి బంతి వలె ఉంటుంది

Target Word (బంతి)

  Input context words { భూమి }

    Sense ← 0
1. For S1 Sense do
    Sense s1← 0
     For each word Wi in IC
     (భూమి) in PCSI   wrong X

    Sense S1← 0

2. For S2 Sense do
    Sense S2← 0
     For each word Wi in IC
    భూమి in PCS2

    Sense S2← 1

3. For S3 Sense do
    SenseS3← 0
     For each word Wi in IC
     (భూమి) in PCS3 X
    Sense S1← 0
    Max (senseS1, SenseS2, senseS3)   {sense S2   correctly disambiguated}

    = S2   గుండ్రని.

5   Evaluation

The evaluation for word sense disambiguation system is done by F-measure. The precision and recall values are calculated using the formulae,

Precision = No. of correctly disambiguated words/No. of disambiguated words.

It is explained as the number of correctly disambiguated words to the total number of words disambiguated. Some words which are disambiguated may result in incorrect sense.

Recall $=$ No of correctly disambiguated words/No. of tested set of words

It is explained as the number of correctly disambiguated words to the total number of words to be disambiguated.

There is no standard WordNet for Telugu language. Only training data generated from సాహిత్యం, తెలుగు నవలలు is used for evaluation process.

Polysemy words with senses and context words are generated for each polysemy word from training data depending on the senses of polysemy word Telugu online dictionary, and graphs are used for our examples.

Training Datasets: A huge dataset with 1,500 documents, categorized as follows

క్రీడలు , కథలు ,దేవాలయములు ,రాజకీయలు , కవితలు , వేమన పద్యాలు ,తెలుగు సాహిత్యం ,తెలుగు నవలలు.

F1 is the main evaluation in WSD systems. F1 measure is the harmonic mean of precision and recall.

$$F1 = 2 * (\text{precision} * \text{recall}/\text{precision} + \text{recall}).$$

The following Table 1 compares all the three methods' precision and recall values. By using precision and recall [1], we can find accuracy of the WSD system, which can be called as F1 measure (Fig. 2).

Table 1 Comparison of accuracy of three methods

Method	Precision	Recall	Accuracy
WFS	0.56	0.54	0.55
WMFS	0.68	0.65	0.66
WTSS	0.89	0.81	0.85

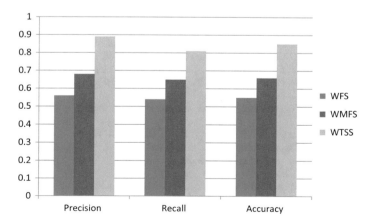

Fig. 2 Accuracy chart for three methods

6   Conclusion and Future Scope

The proposed algorithm word total sense score (WTSS) is using knowledge-based approach for word sense disambiguation in regional Telugu language. Word sense disambiguation is at infant stage; less research work is reported in Telugu language. Word sense disambiguation in Telugu language has more scope than compared to any other language. Future work, word sense disambiguation system for Telugu language is in progress for the refinement of our proposed method.

Acknowledgements I would like to thank Dr. B. Padamaja Rani for her continuous support. I would like to thank Dr. K. Srinivas Rao for his valuable suggestions. Finally, I would like to extend my thanks to the management of MLR Institute of Technology for providing excellent infrastructure to complete this research work.

References

1. Koppula, Neeraja, and B. Padamaja Rani. 2018. Word Sense Disambiguation Using Knowledge Based Approach in Regional Language. *International Journal of Advance Research in Dynamical and Control Systems* 10(05).
2. Neeraja, B. Padmaja Rani, and Koppula Srinivas Rao. 2015. Hybrid Approaches for Word Sense Disambiguation: A Survey. *International Journal of Applied Engineering Research* 10 (23): 43891–43895.
3. Lesk. M. 1986. Automatic sense disambiguation using machine readable dictionaries: How to tell a pine cone from an ice cream cone. In *Proceedings of the 5th annual international conference on Systems documentation, SIGDOC '86*, 24–26. New York, NY, USA: ACM.
4. Walker, D., and R. Amsler. 1986. The use of machine readable dictionaries in sublanguage analysis. In *Analyzing Language in Restricted Domains*, eds. Grishman, and Kittredge, 69–83. LEA Press.
5. Eneko, A., and G. Rigau. 1996. Word sense disambiguation using conceptual density. In *Proceedings of the 16th International Conference on Computational Linguistics (COLING)*. Denmark, Copenhagen.
6. Navigli, Roberto, and Paola Velardi. 2005. Structural Semantic Interconnections: A Knowledge-Based Approach to Word Sense Disambiguation. *IEEE Transactions on Pattern Analysis and Machine Intelligence* 27 (7): 1075–1086.
7. Neeraja, and Dr. B. Padmaja Rani. 2018. Graph based word sense disambiguation for Telugu language. *International Journal of Knowledge Based and intelligent Enginering Systems*. IOS Press
8. Mihalcea, Rada. 2007. Knowledge Based Methods for WSD, Springer. e-ISBN 978-1-4020-4809-2.
9. M. Ramakrishnamurthy, J.V.R. Murthy, P.V.G.D. Prasad Reddy, Suresh C. Satpathy. A Survey of Cross-Domain Text Categorization Techniques. In *International Conference on Recent Advances in Information Technology RAIT-2012*, IEEE Xplorer Proceedings, 978-1-4577-0697-4/12.
10. M. Ramakrishnamurthy, J.V.R. Murthy, and P.V.G.D. Prasad Reddy, et al. 2014. Automatic Clustering Using Teaching Learning Based Optimization. *International Journal of Applied Mathematics* 5(8): 1202–1211, Scientific Research Publishing.

11. M. Ramakrishnamurthy, J.V.R. Murthy, and P.V.G.D. Prasad Reddy. 2011. Text Document Classification Based on a Least Square Support Vector Machines with Singular Value Decomposition. *International Journal of Computer application (IJCA)* 27(7): 21–26.

A Study of Digital Banking: Security Issues and Challenges

B. Vishnuvardhan, B. Manjula and R. Lakshman Naik

Abstract In the extensive sense, mobile banking (M-Banking) is defined as the financial transaction execution through smart devices or electronic devices. The clients are utilizing mobile devices with the combination of mobile communication systems for purpose of electronic transactions (E-transactions). The major reason for the expansion of electronic banking (E-Banking) is due to continuous growth of Internet speed, facilities, personal Digital Assistants (PDAs), Internet-empowered phones, and designing of banking application. This made another new subset of E-Banking and M-Banking. We mainly concentrated on E-Banking and M-Banking aspects of banking area. In this paper, we examined the general structure of Indian banks and identified relevant digital banking types, which are classified into E-Banking and M-Banking. We carried a survey on modern M-Banking and determined five different categories of M-banking services and listed out the functionalities, advantages, disadvantages, and security issues of it. We also examine the security attacks on mobile devices and wireless networks. Finally, a related work is carried on the existing system based on the security concern issues in M-Banking.

Keywords Authentication · Banking · Security · Mobile · Internet · Attacks · SMS · IVRS · GSM · WAP

1 Introduction

A bank is a financial institution licensed to receive deposits from the public and make credits and loans to farmers, traders, and MNCs. In 1694, first bank was established in England which is named "Bank of England." An establishment follows entitle for national or public institutions to mobilize the nation's resources

B. Vishnuvardhan (✉) · B. Manjula
Department of Computer Science, Kakatiya University, Warangal, TS, India
e-mail: vishnuvardhan_phd@kakatiya.ac.in

R. Lakshman Naik
Department of IT, KUCE & T, Kakatiya University, Warangal, TS, India

© Springer Nature Singapore Pte Ltd. 2020
K. S. Raju et al. (eds.), *Proceedings of the Third International Conference on Computational Intelligence and Informatics*, Advances in Intelligent Systems and Computing 1090, https://doi.org/10.1007/978-981-15-1480-7_14

and secure loan for the government. In 1770, Indian government introduced banking sector name as "Hindustan bank" at Calcutta.

Standard banking services for customers include checking accounts, paying checks, collecting check deposits, cash deposits, payments via telephonic transfer, Real Time Gross Settlement (RTGS), wire transfers, automated teller machine (ATM), National Electronic Funds Transfer (NEFT), issuing bonds, securities, money lending, insurance, mortgage, loans, etc.

E-Banking—this service executes the financial administration's task through the Internet; these change the market strategies, which minimize the meantime costs and increase the customer satisfaction. The tremendous expansion of PDAs and Internet-empowered telephones and designing of mobile applications to the mobile devices lead toward E-Banking and M-Banking.

We characterize M-Banking as a sort of implementation of money-related task in electronic strategy. The client utilizes mobile communication procedures in combination with mobile devices. The typical mobile devices such as notepads, sub-notepads, and notebook are usually transferring information through the mobile communication, and these are easily transportable from one location to another location.

We examine the necessities of banking and analyzed wide-range depiction of Indian banks in Sect. 2, and we recognize the importance of digital banking types, which are classified into E-Banking and M-Banking. We carried a study on the futuristic M-Banking models and listed out the functionalities, advantages, disadvantages, and security issues of it in Sect. 3. We examine the security attacks on mobile devices in Sect. 4. In Sect. 5, a related work is carried on the existing system based on the security issues on M-Banking. In the last section, we conclude the significant outcomes and inadequacy of M-Banking applications are determine for further enhancement.

2 Reserve Bank of India (RBI): Types

Before establishment of RBI, there is no central bank in India. It is the highest monitoring and controlling authority of banking system in India. It is called the Reserve Bank of India as it keeps the reserves of all banks. The responsibility of National Bank is to keep up monetary security; generally, a nation's economy will not work legitimately. National Bank goes about as controllers of their nation's interest rates by controlling the measure of cash available for use and purchasing and offering monetary forms.

Some have worked for a long time, and some have gone up against new sorts of business. A few banks are extensive and complete a wide range of capacities, and others are more particular. Based on the several financial institutions, conducting different kinds of business, RBI has been classified into different types of banks (shown in Fig. 1).

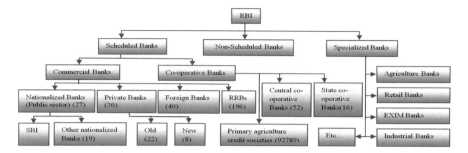

Fig. 1 Structure of Indian banking

According to the Banking Regulation Act of 1965, the bank must fulfill and accompany conditions as per the second schedule.

- The bank ought to paid and hold capital of 5 lakh rupees to the RBI throughout their operational period.
- The bank's affairs are not directly affecting the interest of depositors and must be satisfied by the RBI.

Banks which satisfy the above conditions are called scheduled banks and banks which do not satisfy the above conditions are called non-scheduled banks.

2.1 Scheduled Banks

Under the second schedule of the RBI Act, 1934, scheduled banks have been recorded. This bank has been satisfying certain conditions of Reserve Bank of India. Scheduled banks are additionally grouped into commercial and co-operative banks.

2.1.1 Commercial Banks

Commercial banks (CBs) are one type of scheduled banks. CBs hold major proportion of the business account. CBs afford banking service to small business organization through cooperate organization to coordinate huge partnerships. CBs issue drafts and bank checks and handle cash on term deposits. CBs additionally act as moneylenders, by method for overdrafts and loans installment. These banks makes owned and profit by group of individuals.

In India, CBs are divided into the four groups depending on their ownership and/ or on their identity of operations such as public, private, foreign sector, and regional rural banks which is shown in Table 1.

Table 1 Types of commercial banks

Commercial banks	Key shareholders	Key players
Public sector	Indian Government Government of India holds the 100% ownership.	Canara Bank, PNB, BoB, SBI, BoI, etc.
Private sector	Individuals of private sector Private individuals hold the major share capitals.	HDFC, Axis Bank, Yes Bank, Kotak Mahindra, ICICI, etc.
Foreign sector	Foreign entity Foreign individuals hold the major share.	Deutsche Bank, HSBC, Standard Chartered Bank, Citi Bank, etc.
Regional rural sector	Jointly owned by Indian Government Indian Government, concerned state government and sponsor banks issued capital of an RRB that is shared by the proprietors in the share of 50, 15, and 35%.	Telangana Grameena Bank, APGVB, Prathama Bank, Uttaranchal Gramin Bank, etc.

2.1.2 Co-operative Banks

Co-operative bank is one type of scheduled bank. *Anyonya Co-operative Bank Limited* (ACBL) is first co-operative bank in India, which is located in Vadodara in Gujarat state. Co-operative banks are the key financiers of some small-scale businesses, agricultural farming exercises, and independently employed workers. Co-operative banks work on the premise of "No-Loss No-Benefit." Co-operative banks "for the most part furnish" the individuals with an extensive variety of banking and budgetary administrations (deposits, loans, banking accounts, and so on). Co-operative banks likewise give restricted banking items and experts in farming-related items. In India, the co-operative banking is categorized as three types. They are PACS, State Co-operative Banks, and DCCB.

2.2 Non-scheduled Banks

Under the second schedule of the RBI Act, 1934, a non-scheduled bank has been recorded. All existing Indian banks are scheduled banks and very few numbers of banks are non-scheduled banks. In 2006, only three non-scheduled banks survive in India. In 2011, one more non-scheduled bank is added in India. These banks come under paid-up capital below five lakhs. These four non-scheduled banks in India are Indigenous Banks, Alavi Co-operative bank limited, Amod Nagrik Sahakari bank limited, and Akhand Anand Co-operative bank limited.

2.3 Specialized Banks

These banks are also called development banks. In 1948, Industrial Finance Corporation of India has established specialized banks under State Financial Corporation Act, 1951. More than 60 specialized banks have existed in India at both state and central level. We discuss the major four specialized banks. The specialized banks are those banks that specialize in financing certain economic sectors, and most important types of specialized banks are agricultural banks, industrial banks, retail banks, EXIM banks, etc.

2.3.1 Agriculture Banks

Six decades back, Mahatma Gandhi said that "the agriculture farming is the backbone of the Indian economy." Almost whole economy depends on agriculture. It gives accounts of 52% work for common people and 16% of the GDP of India. This bank runs under the Nationalized Banks, Rural Banks, and Private-Sector Banks. Typical products offered by an agriculture bank include credit loan, subsidy, corp insurance, corp import and export, and extension activities

2.3.2 Retail Banks

All street banks are known as retail banks. Retail banks are also called consumer banks. These banks provide services to individual customers, rather than the companies, corporations, individuals and others. These banks take deposits from people, pay interest on these accounts, and produce saving schemes. Among all the retails banks, ICICI bank is the largest. And other banks which come under retail banks are SBI, HDFC, PNB, etc. The retail bank typical commodities offered saving/current accounts, debit/credit cards, mortgages, and traveler's cheque.

2.3.3 EXIM Banks

The Export and Import Banks of India popularly known as EXIM Banks was started in 1982 under the EXIM Act 1981 mainly for the purpose of enhancing export trade of the country. This bank is started to promote and facilitate the foreign exchange/trade. The major principle of this bank is to engage the financial export and import activities in the country. Typical products offered by EXIM bank include line of credit, import credit, and loans for export units.

2.3.4 Industrial Banks

In 1971, Industrial Reconstruction Corporation of India Ltd. (IRCI) sets up an ailing industrial organization. Later in 1985, this is reconstituted as industrial reconstruction bank under RBI Act, 1984. With a view to changing over the establishment into an undeniable improvement monetary organization, in March 1997, Industrial Reconstruction Bank of India (IRBI) is recorded as Industrial Investment Bank of India Ltd. (IIBI) under the 1956 Companies Act. In 2005, a merger proposal was considered for three banks (Industrial Development Bank of India (IDBI), IIBI, and Industrial Finance Corporation of India (IFCI)) [1].

Industrial bank attempts to make a lively neighborhood economy in different ways including public/private organizations, banking training/financial proficiency workshops, and sponsorships. Industrial bank trusts the educated client is the most valuable client.

Typical products offered by industrial banks include real estate and commercial loans, project finance assistance, short-term loans, or working capital to industries.

3 Digital Banking

The capacity to oversee online bank accounts from PC, tablet, or smartphone is becoming very important to the bank customers. Banks will normally offer digital banking services, for example, online mobile account checking, E-tax pay, text alerts, E-statements, and online bill pays. The digital banking is divided into two parts; the first one is E-Banking and the second one is M-Banking (shown in Fig. 2).

3.1 E-Banking

E-Banking is also called as Internet banking. E-Banking is the electronic banking that gives the money-related support of the individual customer by methods for Internet. E-Banking implies any client with a personal PC, and any browser can get associated with his bank's site to carry out any of the virtual banking functionalities. In E-Banking framework, the bank has an incorporated database that is web-connected. It would be a borderless unit allowing 3A (anyhow, anytime, and anywhere) banking.

Fig. 2 Digital banking types

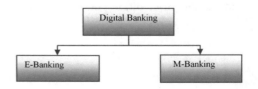

E-Banking is an umbrella term for the method by which a client may perform banking transaction process electronically without going to the bank offices. Online services save the time of customers.

3.1.1 Functions of E-Banking

Account Information: The customer inquires his own account information such as the account's balance and transaction statement on specific period.

Fund Transfer: Customer can fund transfer from one account to other accounts within the same city or out of the city. The client can also transfer fund between his own bank accounts.

Foreign Exchange Transaction: A customer can trade the foreign exchange and inquire about the transaction of foreign exchange based on the bank rates on Internet. It also cancels the orders.

Management of Account: The customer can modify his own account details' limits of his rights, such as changing login password and holding or deleting his card's details.

3.1.2 Advantages and Disadvantages of E-Banking

Advantages

- Opening an account in online is easy, simple to operate, very fast, and efficient.
- Paying bills, transfer funds between accounts, etc.
- Available on all the time, i.e., 24×7. User can process banking operations from anywhere and at any time.
- Managing and monitoring several accounts easily through banking on Internet by keeping an eye on every bank transaction and account details all the time. This helps to know any fraudulent activity details or user account threats.

Disadvantages

- For the first time, beginners might be difficult to understanding the usage of E-Banking.
- Accessing the E-Banking must require Internet connection. Without Internet, it is useless.
- A big issue is security of transactions. Sometimes, user's account details might get hacked or stolen by unauthorized persons on Internet like password, and after getting password, they immediately change and memorize it, and then account may be misused.
- Sometimes, it is difficult to know whether user transaction was successful or unsuccessful. It may be due to net connectivity loss or due to connection slow or due to the bank's server not working properly.
- Sometimes, a face-to-face meeting is required to complex transaction and complicated address problems.

3.1.3   Security Problems in E-Banking

Malware Attacks: Malware is a program that is hurtful to a client's PC. It is consisting of Trojan horses, Worms, Spyware, and Virus. These malevolent programs can perform functions like encrypting, stealing, or deletion of sensitive data, hijacking core computing functions, and monitoring the computer without permission of a user.

Insiders/Disgruntled Employees: An insider threat is cracker or blocker and also called as malicious hacker who is a representative or officer of a business, agency, or institution. The term also can apply to an outdoor one that poses as co-worker or officer by getting false credentials. The cracker gets access to the personal computer or enterprise network to conduct the activities like harm the enterprise.

DoS (Denial of Service) Attack: This attack works on bringing down service, application, or Web site. A DoS attack could be a cyber-attack wherever the offender seeks to form a network resource or machine untouchable to its supposed customers by briefly or indefinitely disrupting a number connected to the web services. A DoS attack floods a site with junk traffic, so it cannot respond to legitimate requests for access—like stuffing a letterbox with junk mail so that genuine letters cannot be received.

Phishing: Phishing is nothing but an attempt to obtain information such as credit card details, passwords, and usernames in an electronic communication by disguising a trustworthy entity.

3.2   M-Banking

M-Banking also called as mobile banking or phone banking. M-Banking is nothing but doing financial transactions through mobile devices like tablets, cell phones, etc. It is a technologically oriented lifestyle in a busy environment. M-Banking allows user to perform many of the same activities as online banking using a smartphone or tablet instead of a desktop computer. M-Banking similar to an ATM on steroids is a very useful and profoundly practical contrasting option to a branch office. Mobile technology facilitates bank to interface with a client in different ways like potentially gainful for both sides without the expenses regularly connected with working a branch office [2].

M-Banking is available in 24/7 basis, and it deals with non-cash deposits and withdrawal transactions. The M-Banking process works based on radio-frequency identification (RFID) and near-field communication (NFC). RFID utilizes electromagnetic field to consequently recognize track labels appended to objects [3]. NFC technique is implemented in Sweden. M-Banking is 5A (anywhere, anytime, any communication device, anyhow, any currency) service [4]. The European banks have started M-Banking in 1999. In the year 1997, banks started SMS alert service to the respective accounts of a customer.

Functions of M-Banking

Mobile Accounting: Mobile accounting is the use of mobile technology—smartphones and tablets—to accomplish business goals in the field of accounting. M-Banking may incorporate account adjusts and arrangements of most recent exchanges, e-charge installments, and exchanges supports between accounts. It is transaction based.

Mobile Brokerage: It means a wireless secure technology for trading. It allows investors to access exchange platforms from their mobile devices rather than traditional trading methods. It is transaction based.

Mobile Financial Information Services: Non-transaction based, for instance, balance inquiries, etc. [5]

Remote Deposits: Remote deposits works take pictures on check front and back to deposits.

M-Banking Models:

  (i)   Short Message Service _(SMS) Based M-Banking
 (ii)   Unstructured Supplementary Service Data (USSD) Based M-Banking
(iii)   Interactive Voice Response System (IVRS/Tele) Based M-Banking
 (iv)   Wireless Application Protocol (WAP) Based M-Banking
  (v)   Application-based M-Banking.

3.2.1 SMS-Based M-Banking

The principal of M-Banking is to offer SMS-based banking service. It deals with plain text message. SMS is a GSM/CDMA-based services and supports up to 140/160 characters to exchange the 7-bit message through Short Message Service Center (SMSC) [6]. SMS-based M-Banking works in two unique modes: pull mode and push mode. Push mode works based on the specific pre-defined code, and it is two-way communications (customer-to-bank and bank-to-customer). Pull mode is a one-way instant message framework where the bank sends an instant message to the clients educating them about certain record circumstance [7]. SMS-based transaction scheme secure messages are used for transfer amount from one mobile user account to another user account through mobile network operator (MNO). It is more useful for small merchants and village consumer's bank transactions [8].

Functions of SMS-Based M-Banking

- Enquiry on account balance, request on mini statement and fund transfers between accounts.
- Electronic bill payment and stopping payment if required.
- Request for a credit/debit cards to be perched.
- Foreign currency exchange rates enquiry and fixed deposit interest rates enquiry.

Advantages and Disadvantages of SMS-Based M-Banking

Advantages

- Anyone can access easily from anywhere, no Internet connection required.
- Familiar technology and cost-effective, virtually available in each and every mobile phone.

Disadvantages

- No security and extra charges.
- Third parties (like MNO) are involved.

Security Problems in SMS-Based M-Banking

Text Messages are not in Encrypted Form: These sorts of exchanges are missing encryption of the data amid the reporting in real-time transmission between MNO and mobile telephone. An encryption of immaculate instant messages is unrealistic [6].

SMS Spoofing Attack: In SMS managing, an account's most hazardous attack is spoofing attack, where attacker can send the message on system framework by controlling sender's numbers. Because of spoofing attack, the vast majority of the associations are not willing to embrace mobile keeping money through SMS [9].

Third-Party Payment Gateway: This process includes third-party agency, or gateway is involved to give service between customer's bank and merchant's bank. So customer trust third party, but it is risk-involved payment process [10].

3.2.2 USSD-Based M-Banking

The service allows the banks and telecom service providers to work together seamlessly. The services of National Unified USSD Platform (NUUP) are based on the USSD method. USSD is a protocol utilized by GSM cell phones to communicate with the service provider's personal computers. A gateway is the combination of hardware and software which is required to interconnect at least two networks, including performing protocol conversion. A USSD is a session-based protocol. USSD messages go over GSM signaling channels and are utilized to trigger services and query data. This protocol establishes a time session among an application (which is handling the service) and mobile handset [7].

The codes which directly communicate with the server of Telecom Company are called as the USSD. User must have noticed that this code starts with "*" (asterisk) and ends with # (hash). As USSD code connects to the telecom operator's server, it also connects to bank's server. Hence, it gives access to user bank account and performs transactions. The entry to user bank account is given on the basis of registered mobile number.

Functions of USSD-Based M-Banking
*99# service can be used by the customers for the following purposes.

(i) Financial

- Send money using Mobile Money Identifier (MMID), using Indian Financial System Code (IFSC) and using Aadhaar number (card).

(ii) Non-Financial

- Know MMID
- Generate and change mobile pin (MPIN) for balance enquiry, mini statement, etc.

(iii) Value-Added Services (VAS)

- Query Service on Aadhaar Mapper (QSAM) or *99*99#

Advantages and Disadvantages of USSD-Based M-Banking

Advantages

- Most of the mobile users are using GSM devices, irrespective of handsets make, operating system working, cost or even the telecom service provider and framework.
- The customers no need to download and install any application on the mobile.
- General Packet Radio Service (GPRS) is not required.
- USSD is application with interactive menu.

Disadvantages

- Not works in CDMA environment
- Third party means mobile network operator involved and no security; encryption is not possible.

Security Problems in USSD-Based M-Banking
Tampering USSD Commands: A malevolent customer can modify USSD commands requests/responses through equipment and programming interceptors provoking false exchanges. Weak scrambled requests/responses messages are prime worries in such threat vectors.

USSD Replay Attacks: When a mobile device is lost, with an installed application of USSD, a rival may perform false transaction on USSD (e.g., IMEI, Unique Message Tracking ID, PIN, and MSISDN).

Prepaid Roaming Access Test on USSD: A rival may bring about loss of direct revenue for specialist organizations by utilizing control of wandering access parameters and getting unapproved access to USSD application prepaid roaming services.

Improper Information Approval (USSD IP Mode Applications): Inappropriate information approval is USSD IP mode application that can prompt cross-site scripting attacks, SQL infusion. A rival may intentionally embed particularly made scripts in client input and may attempt to utilize the same to perform vindictive activities at the database or at another client's dynamic session [11].

3.2.3 IVRS/Tele-Based M-Banking

IVRS takes jump into the banking sector. IVRS is an extreme product of artificial intelligence, which goes about as a communicator and manages to erase the complications involved in the banking process. IVRS is an automatic programmed voice responsive system, which enables a man to extricate, A to Z data of client account by dialing a toll-free number. IVR systems provide world-class services like M-Banking registration, stock quotes, mini-bank statement, bank policy, Sensex trends, investment and loan schemes, credit and debit offering, checkbook request, and check status—stop payment.

Functions of IVRS-Based M-Banking

- Enquiry balance in accounts and bill payments and fund transfer
- Generate and change telepin [12]

Advantages and Disadvantages of IVRS-Based M-Banking

Advantages

- IVRS is a customer-friendly service available 24/7.
- IVR systems are very useful where some of the people served by an organization are illiterate, because they use voice rather than text. It supported all types of mobile phones.

Disadvantages

- Call charges applicable and needs third person's support.
- First-time users may find the system slightly difficult to use.

Security Problems in IVRS-Based M-Banking

- Third party involved like MNO and calls are recorded, so hacking is easy to hacker.
- Difficult for an identity fraudster to defraud user account using M-Banking unless they have a huge volume of information on user including passwords.

3.2.4 WAP-Based M-Banking

Internet is more flexible and easy to access through mobile devices, so in 1999, WAP-based M-Banking has been introduced. WAP is an industry standard wireless application for mobile devices. It is same as the E-Banking and it does not require the installation of a special banking application. Bankers have to create compatible mobile device Web site and can be accessed through the small screen of mobile device [7]. It is based on Internet service and available on all GPRS-enabled GSM/ CDMA mobiles. In transaction/authentication process, banks afford OTP, and this OTP is valid for single transaction process only. So for each transaction, the user receives OTP every time through SMS [13]. The banks are processing sensitive data, so security must be needed. Unauthorized access is not possible in WAP gateway [6].

Functions of WAP-Based M-Banking

- Balance enquiry, mini statement, and fund transfer.
- Check status enquiry and stop payment of checks and service outlet locator.

Advantages and Disadvantages of WAP-Based M-Banking

Advantages

- Real-time send/receive transactions and easy to monitor the accounts.
- Multiplatform functionality (little change is needed to run on any Web site since XMI is used) and mobile-friendly access.

Disadvantages

- WAP-based M-Banking is not having user friendly authentication method.
- Costly to end users as data and roaming charges may apply.
- May require customization to suit different types of mobiles and does not work with all type of mobile devices and require PDA or smart mobile devices.

Security Problems in WAP-Based M-Banking

Mobile Malware: Viruses, Trojans and Rootkits migrating mobile design and malware development is increase in mobile environment with rapid growth of market.

Third Party Application: Users mostly adore their mobile device apps; these apps may not be secure because they are developed by third party. Sometimes, some apps may be developed by fraudsters, and this leads to load malware in the mobile devices.

Wi-Fi Unsecured: The unsecured Wi-Fi network is a toll-free roadway for fraudsters to access cell phones, either to access account data or to grab control.

Client Behavior: Clients are prone to download outsider applications, open, install, and click the links in instant messages or email, and this leads to utilize unsecured network access and data may be lost in the cell phones. This creates a suite of fraudsters and vulnerabilities to exploit the anxious.

Man in Middle Attack: In this attack, the attacker secretly transfers and modifies the communication between two parties who trust they are directly communicating with each other [17].

3.2.5   Application-Based M-Banking

M-Banking is a bank that enables its clients to direct money-related exchanges remotely utilizing cell phone. It uses software called an app. This procedure requires download and establishment of an application into the cell phone. The bank application can give an extensive variety of service to their customers. The app password used in authentication purpose. This password is prescribed to be solid and forestall security attacks [13].

Functions of Application-Based M-Banking

- Balance enquiry, mini statement and fund transfer with mobile to mobile or mobile to account.
- Fund transfer through NEFT or IMPS and Check status request and stop check
- Utility services (like movie ticketing and mobile recharge) and trace ATM/ branch locator (by pin code/location).

Advantages and Disadvantages of Application-Based M-Banking

Advantages

- *Anytime Banking:* M-Banking works anytime and anywhere. One can do most of the banking transaction after banking hours from anywhere, irrespective of whether user is traveling in bus or auto. Easy to use.
- *Free of Access*: The service provided by banks has limited number of transaction to access user account for free [14].

Disadvantages

- It is not available on all mobile phone. It may lead to extra charges in regular use (beyond the limit).
- Mobile device has some limitations in processing the data, speed, battery life, and screen size. This creates the barrier for mobile application.

Security Problems in Application-based M-Banking

DDoS Attacks: The FBI said that "DDoS attack is ranked as third highest treat attack in Application-based M-Banking." It is the most common attack of M-Banking system, and this attacks the target system [15].

Malware Attacks: Malware is the term for malicious crafted software code. It is performed by fake Web site application substitution and account hijacking [14].

TCP/IP Spoofing: IP spoofing is a strategy used to increase unauthorized access to machines and an attacker illegally impersonates another machine by changing IP packets. IP spoofing involves altering the packet header with a spoofed source IP address, the order value, and a checksum [16].

Backdoor Attacks: Backdoor attack is accessed to mobile program that avoided security mechanisms. A programmer may sometimes install some software in the backdoor so that the program can be accessed for troubleshooting or other purposes.

Man in Middle Attack (MIMA): In this process, the attacker secretly relays and possibly manipulates the communication between two users. The client on the customer host wants to make a secure transfer, but it modified by third person in communication is man-in-middle attack [10].

Replay Attacks: The attacker repeats messages sent in past and resend them at a later time to verifier.

Pre-play Attacks: The attacker directs a message from the recorded messages in past communiqué for present communications.

Offline Attacks: The attacks replicate past communication, and after that study, a dictionary reference in pursuit as a secret password is steady with a communication which is recorded. In the event that every secret password is assembled, the attacker brings that this password is appropriate in an attack.

Online Attacks: In these attacks, attackers seek to use password in order to do fraud as the client, which is recorded in dictionary.

Server Compromise Attack: The attacker gets sensitive information saved at the verifier to fraud as a client [18].

4   Attacks on Mobile Devices

Wireless Attacks: These types of attacks target the sensitive data and personal data against mobile phones.

Break-in Attacks: The attacks are enabling the assailant to create programming errors and obtain the power over the targeted device.

Infrastructure-Based Attacks: Infrastructures are the basic essential and provide services by the advanced mobile device utility, such as calls, receiving/placing SMS, and email service. These attacks are GPRS attacks, UMTS attacks, etc. [19].

Worm-Based Attacks: A worm is a collection of instructions (program) that make duplicates of itself, ordinarily starting with one device then onto the next device, utilizing distinctive transport instrument through a current system without client contribution.

Botnets: Group of device is referred to as Botnet; these are polluted by an infection that gives ability to attacker to control remotely [20, 21].

4.1   Wireless Networks Attacks in Mobile Device

Available wireless networks are 2G-5G.

4.1.1   2G Network Attacks

Security Attacks: In the network, no authentication is provided for the end user/client in a given *GSM* system-susceptibility in subscribing the identity secrecy method.

Mimic Attacks: The attacker tends to mimic a legal client to handle an attack.

Gains Anonymity Attacks: The attacker obtains data on the user's behavior, methods of calling, etc., which can be accessed in opposition to the end user.

Secrecy Attacks: The attacker utilizes deficiency in the architecture of GSM design, defects in the GSM network, end user, and the used protocols in between them. These types of well-known attacks are non-cryptanalysis attacks, cryptanalysis-based attacks, and brute force attacks.

Denial of Services Attacks: The attacker surges the network system to do some failures to the end users or disable them from the network by implementing the attack using logical or physical intrusion.

4.1.2   3G Network Attacks

Interception Attacks: This type of attack reads the signaling message and interrupts the information, and it never manipulates the message. Moreover, it infects the privacy of the network subscriber.

Replay Attacks: This type of attacks try to embed fake objects into the system based on physical access type (e.g., fake subscriber data, signaling messages, or fake service logic).

Eavesdropping Attacks: This type of attack intercepts messages without detection.

Resource Modification: In these attacks, resources are harmed and modified by the attackers.

Logical Service Attacks: The attacker roots the critical damages by just assaulting the logical services in the different entities of 3G networks.

4.1.3   4G Network Attacks

These technologies are expected to work totally on the IP architecture and protocol suites; it expands results with respect to the security concern. To achieve the 4G wireless technology, the long-term evolution (LTE) technology is considered.

Interference: The attacker purposely embeds man-designed interference onto a medium which stops functioning communication system due to high signal noise ratio.

Scrambling: This type of interference activates for short-time interims. It focuses on the specific odd parts of frames, which are control information or management to disturb a service. These attacks are extremely hard to launch.

Location Tracking: In this type of tracking, attacker tracks user device in a particular location/cell or across several locations/cells. Even though it is an undirected threat, it creates the security break in the network.

Bandwidth Stealing: By using fraud buffer status reports during the discontinuous reception (DRX) period, the attackers attain the attack by push messages [22].

4.1.4 5G Networks Attacks

In 5G network architecture of mobile as a portion of the key developing advances, those are served to enhancing the design and meeting the requests of customers. Access beam division multiple access (BDMA) and non- and quasi-orthogonal or filter bank multicarrier (FBMC) access [23]. It is important to understand that today in 2016, 5G is not a technical standard. Instead, it is defined by a set of aspirations around desired services intended to be commercially available around 2020. These services are expected to place new requirements on connectivity, flexibility, cost efficiency, and performance. Some companies have already announced that they intend to launch 5G capable networks commercially in 2018 [24].

Femtocell Attacks

- Physical tampering with equipment (interference with other devices)
- Configuration attacks (mis-configuration of analog configuration line (ACL))
- Protocol attacks (MitM during first access)
- From the compromised nodes, the attackers attack on mobile operator's core network
- From open access nodes, the attackers theft the credential, user identification and private information
- Attacker attacks on radio resources [24, 25].

5 Related Work

Digital banking involves several major challenges for the banking transactions and accessing the bank accounts. The challenges and issues in the context of banking security are summarized.

Cyber-security attacks and threats are the common problems of wireless networks. Sabina et al. [22] addressed security issues in wireless mobile networks, such as threats, vulnerabilities, and attacks. These security issues may be covered in cellular networks such as WMANs, WLANs, WMNs, WSNs, Bluetooth, VANETs, and RFID networks and communications. Author did not give solutions to various security issues.

Information security itself is a complicated problem in implementation of wireless communication technology. Xianling and Jin [4] discussed various information security issues like information leakage, loss, virus attacks, and incomplete information and they proposed a framework for WPKI security system. The proposed system performance measures are not practically implemented.

Several entities such as technology platform provider, telecom operator, bank, non-banking financial association, and others are implicating the mobile financial services (MFS) operation. Privacy and information security are a major barrier for adoption of MFS. Effective and secure framework for MFS was proposed [13].

This solution improves the customer satisfaction and confidence. MFS will not provide authentication for more users simultaneously and not suitable to access from anywhere is the major drawbacks. To establish the relation between customers to merchants and merchants to banking sector; lack of infrastructure, high cost and security is the major problem. Soni [8] introduced M-Payment method to solve the secure issues between customers to merchants and merchants to banking sector through the third-party gateway and biometrics authentication was proposed. Third-party gateway is not trustworthy and not secure. In this process, there may be chance of involving third person. This leads to man-in-middle attack, spoofing attack, etc.

In order to create trust to a customer, the E-Banking must provide risk-free services and high-secured services. Hameed [26] identified the attack model and top risk factor faced by E-Banking. For this, author proposed authentication approach to overcome the risk. The proposed system consumes more time, when the number of users is increased, and this leads to extra cost for the banks.

5.1 Mobile Versus Desktops

According to the Graph 1, there is massive change in the utilization of mobile phones. Previously people used to stick to the systems to do any sort of Internet work. As the days are passing by, we can see the usage of mobile phones has tremendous change from year 2009 to 2018 [27]. From 2012, the usage of mobile phones is tremendously increased, whereas desktops are dramatically decreased. Mobile phones are portable and easy to carry. All apps can be downloaded and use randomly. So the people are interested to purchase the mobiles rather than having a desktop at home.

The high-end mobiles are available in the market for very less price, which is a boon to the common man to compete with the world using mobile as a tool. It is very easy to carry. We conclude that the usage of mobiles phones is gradually increasing instead of desktops. So, users are showing interest in making payment through mobile instead of desktop. We can credit the amount and debit the amount through mobile phones. To withdraw/deposit/transfer an amount in the bank,

Graph 1 Year-wise mobile and desktop growth rates

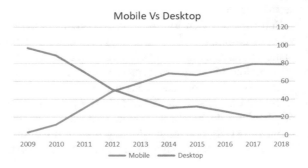

Customers must go to the banks. They must stand in a line and wait for some time to make the transaction. Instead of that, now customer are using the improved technology in such a way that by clicking the button, we can send/receive the money to the assigned person by holding the mobile in any corner of the world.

5.2 Mobile Malware Evolution

A growing number of viruses, worms, and Trojans that targeted the smartphones have been discovered. The evolution of malware on smartphones from 2009 to 2018 is discussed in the below Table 2 [28]. As the technology increases, mobile virus and worms have increased significantly. For every year, a new type of malware is injected in smartphones by third party. Every human has a disease, and you know what even computers and mobile phones also face the same. The table above has given the clear picture of the viruses caused yearly from 2009 to 2018. In 2009, the mobiles were affected with a Spyware software that infesters user computing device, stealing user internet data and sensitive information. Spyware is a malware attack that dramatically branded as Curse of Silence and acts as genuine bug that prevents incoming SMS messages being received. Later, a virus termed worm which is a type of malicious software program whose primary function is to infect other mobile devices while remaining active on infected systems was discovered and termed as Zeus MitMo. As the technology is introducing more and more, mobile users use different software and they got affected with a multifarious malware where they termed it as iSAM. In 2012, they introduced a software named rootkit which is a piece of software that can be installed and hidden in the mobile devices without the user's knowledge and Botnet is a term derived from idea of bot networks. In 2012, the systems were affected with a malware where they termed it as Botnet/rootkit.

Table 2 Year-wise discovery of malware

Year	Type	Name
2009	Spyware	Curse of Silence
2010	Worm	Zeus MitMo
2011	Multifarious malware	iSAM
2012	Malware	Botnet/rootkit
2013	Backdoor attacks	Franco phoned
2014	PoS	Malware
2015	Threats	IoT threat
2016	Malware	CAC (common access card) and Pegasus
2017	Malware	Ransomware
2018	Trojan	Trojan Dropper

A new type of attacks called backdoor attacks are recognized where the attacker's goal is to create a backdoor into a learning-based authentication system, so that the user can easily circumvent the system by leveraging the backdoor. Point of sale software for small and independent retailers with cloud, Pos tracking, management and reporting of the business. Later the Internet of things (IoT)threats are found in the year 2015 which rely on insecure web interface, insufficient authorization etc. Common Access Card is a smart card about a size of a credit card is the standard identification for active duty uniformed service personnel. Pegasus software is based in Kettering, England and develops accounting and financial management and service software applications for small and medium-sized business. Ransomware malware can be spread through email attachments, infected software apps, infected external storage devices and compromised Web sites. A growing number of attacks have used remote mobile protocol and other approaches that don't rely on any form of user interaction. Finally in the year 2018, a software named Trojan Dropper which is a detection name used by Symantec to identify malicious software programs that drop other malware files onto the compromised mobile devices.

5.3   M-Banking Malware Evolution

According to Table 3 [28], Flexi Spy helps to protect children and catch cheating spouses; it requires physical access to a target phone for installation. At times, it may be maliciously exploited by malware where the mobile user unknowingly installs it. The ZeuS Mitmo injects a deceived field into web pages that prompts users for their cell phone number and the type of handset they use this malware automatically sends all mTANs sent to the ZeuS operators. iSAM collects personal info and send malicious SMS. Flamer is capable of infecting myriad computer networks for the purpose of gathering sensitive data. Trojan-spy.html.fraud.gen is a Trojan horse which can easily infect the machine through spam messages. It can be used for collecting sensitive information, performing hazardous actions in the background to install other malware, etc.

Zeus software is a malicious program that is used to steal the user's credentials for accessing various services, such as M-banking. This malware can also allow a cybercriminal to remotely control the infected data servers. Trojan-Spy. Win32. Zbot used the Black Hole exploit kit and Cut wail and Pushto botnets to spread. Gosh push is a malicious DEX file containing compiled android application code, runs and roots the device, causing malicious processes to run upon startup of the app. It can install unwanted apps and programs onto a device. Risk tool malware has number of functions which can be used with malicious intent. It designs to operate on the local devices. A new malicious banking Trojan has been discovered and is targeting over 200 banking apps. for android devices. It is basically designed to steal user's credentials, intercept SMS messages, etc. Due to many cyber-attacks, people are being the victims. Cyber-attacks against banks that exploited the SWIFT

Table 3 Year-wise discovery of M-Banking malware

Year	Type	Mobile banking effect
2009	Flexi spy	Tracking/log of device's usage
2010	ZeuS MitMo	Steal bank account information
2011	iSAM	Collect personal info and send malicious SMS
2012	Flamer	Capable of infecting myriad computer networks for the purpose of gathering sensitive data
2013	Trojan-Spy.html.Fraud.gen/ crypto locker Trojan horse.	Malicious program appears in the form of HTML pages which imitate the registration forms of well-known banks or e-pay systems
2014	Zeus(Trojan-py.Win32.Zbot)	Most widespread banking Trojan although the number of attacks involving this malicious the program, as well as the number of attacked users
2015	Ghostpush/xcodeghost/Gozi/ Swift	Remotely controlled the steal information/ Exploited the SWIFT payment network
2016	Risk tool/Zbot/Swift	Steal banking information by man-in-the-browser keystroke logging and form grabbing
2017	Android.banker.A2f8a and Banker.AndroidOS.Asacub.bj	Steal bank card authentication and money through SMS services, including mobile banking
2018	Zbot Android.banker.A9480	Cryptocurrency apps

payment network, Lazarus Group made headline for hacking an unnamed Philippine bank in October 2015, Vietnam's Tien Phong Bank in late December 2015, and the Bangladesh Central Bank in February 2016.

By consolidating the above-mentioned issues, we examined and highlighted the security attacks on mobile devices, wireless networks, and security issues in E-Banking and M-Banking. We have noticed the change in the functioning of the systems.

6 Conclusions

In the preceding sections, we explain the E-banking and M-Banking. We mainly concentrated on these two aspects of banking system. It allows customers to process many services like enquiry on account, money transfer, get balance sheet information, and many other services. M-Banking can offer 5A services to its customers to enhance the limitation of space and time.

We highlighted the general structure of Indian banks and identify relevant digital banking types, which is classified into E-Banking and M-Banking. Later on, we evaluated five futuristic standard categories of M-Banking services to distinguish their well-known weaknesses, opportunities and talk about the security issues along with their possible impacts. Finally, a related work is carried on the existing system based on the security issues of E-Banking and M-Banking and highlighted malware attacks on mobile phones and banking. In future, we are planning to design an effective and secure framework and authentication system for M-Banking.

References

1. Mega Merger of IDBI, IFCI and IIBI under Consideration. Financial Express. http://www.financialexpress.com/news/71999/1. Retrieved Mar 25, 2012.
2. Krishnan, Sankar. 2014. *The Power and Potential of Mobile Banking*. Wiley.
3. M2 Magazine. 2017. Global RFID Tags Market is Predicted to Grow at Approximately 7% by 2022, M2 Communications. Feb 7, 2017.
4. Hu, Xianling, and Jin Nie. 2008. *Mobile Banking Information Security and Protection Methods*. 03: 587–590.
5. Tiwari, Rajnish, Stephan Buse, and Cornelius Herstatt. 2006. *Customer on the Move: Strategic Implication of Mobile Banking for Banks and Financial Enterprises*. 81–81. https://doi.org/10.1109/cec-eee.2006.30.
6. Key, Pousttchi, and Martin Schurig. 2004. Assessment of today's mobile banking applications from the view of customer requirements. In *2004, Proceedings of the 37th , Annual Hawaii International Conference on System Sciences*, Big Island, HI, 10, HICSS.2004.1265440.
7. SMS, WAP and USSD Information. www.ukessays.com & blog.aujasnetworks.com.
8. Soni, P. 2010. M-Payment Between Banks Using SMS. In *Proceedings of the IEEE*, 98: 903–905.
9. Harb, H., H. Farahat., and M. Ezz. 2008. Secure SMS Pay: Secure SMS Mobile Payment Model. In *2008 Second International Conference on Anti-Counterfeiting, Security and Identification*, ASID, 11–17.
10. Rai, Nitika, and Anurag Ashok et.al. 2012. M-Wallet: An SMS Based Payment System. *International Journal of Engineering Research and Applications (IJERA)* 258–263. ISSN 2248-9622.
11. Desai, Suhas. 2011. Mitigating Security Risks In USSD-Based Mobile Payment Applications. An Aujas White Paper.
12. Sohail M. Sadiq, and Balachandran Shanmugham. 2003. E-Banking and Customer Preferences in Malaysia: An Empirical Investigation. *Information Sciences* 207–217. Elsevier.
13. Dass, Rajanish, and Rajarajan Muttukrishnan. 2011. *Security Framework for Addressing the Issues of Trust on Mobile Financial Services*. IEEE https://doi.org/10.1109/nwesp.2011.6088160.
14. Nosrati, Leili, and Amir massoud Bidgoli. 2015. Security Assessment of Mobile-Banking. In *IEMCON*, IEEE. 7344489.
15. Kumar, Sumeet, Kathleen M. Carley. 2016. DDoS Cyber-Attacks Network: Who's Attacking Whom. IEEE. 7745476.
16. He, Wu. 2012. A review of Social Media Security Risks and Mitigation Techniques. *JSIT* 14 (2): 171–180.
17. Callegati, Franco, Walter Cerroni, and Marco Ramilli. 2009. Man-in-the-Middle Attack to the HTTP Protocol. *IEEE Security & Privacy, IEEE* 7: 78–81. https://doi.org/10.1109/msp.2009.12.
18. Sangjun, lee, SeungBae park. 2005. *Mobile Password System for Enhancing Usability-Guaranteed Security in Mobile Banking* 66–74, Berlin, Heideberg Springer.
19. Neeabh. 2012. Tracking Digital Footprints of Scare ware to Thwart Cyber Hypnotism Through Cyber Vigilantism in Cyberspace. *BIJIT—BVICAM's International Journal of Information Technology, BIJIT* 4 (2).

20. La Polla, Mariantonietta, Fabio Martinelli, and Daniele Sgandurra. 2012. A Survey on Security for Mobile Devices. *IEEE Communications Surveys and Tutorials* 15 (1): 446–471.
21. Zhao, Ting, Gang Zhang, and Lei Zhang. 2014. An Overview of Mobile Devices Security Issues and Countermeasures, ICWCSN. IEEE.
22. Barakovic, Sabina, Ena Kurtovic, Olja Bozanovic, Anes Jasmina, and Barakovic Husic. 2016. Security Issues in Wireless Networks: An Overview. In *2016 XI International Symposium on Telecommunications (BIHTEL)*, 1–6.
23. Akhil, Gupta, Rakesh Kumar Jha. 2015. A Survey of 5G Network: Architecture and Emerging Technologies. *IEEE*, 3: 1206–1232.
24. Bandela, Vishnuvardhan, B. Manjula. 2017. Mobile Communication: Implication Issues. *International Journal of Computer Trends and Technology (IJCTT)* 49(1):9–14. ISSN: 2231-2803. Published by Seventh Sense Research Group.
25. Rodriguez, Jonathan. 2015. Security for 5G Communications. In *Fundamentals of 5G Mobile Networks*, 207–220. London: Wiley. https://doi.org/10.1002/9781118867464.ch9.
26. Khan, Hameed Ullah. 2014. E-Banking: Online Transactions and Security Measures. *RJAS, E&T* 7 (19): 4056–4063.
27. Global Survey. https://www.statista.com/statistics/274774/forecast-of-mobile-phone-users-worldwide.
28. Survey on Mobile & Banking Malware. https://usa.kaspersky.com/resource-center/threats/mobile-malware and https://pwc.com.

Encoding Approach for Intrusion Detection Using PCA and KNN Classifier

Nerella Sameera and M. Shashi

Abstract Intrusion detection is an evolving area of research in the field of cyber-security. Machine learning offers many best methodologies to help intrusion detection systems (IDSs) for accurately identifying intrusions. Such IDSs analyze the features of traffic packets to identify different types of attacks. While most of the features used in IDS are numeric, some of the features like Protocol-type, Flag and Service are categorical and hence calls for an effective encoding scheme for transforming the categorical features into numeric form before applying PCA like techniques for extracting latent features from numeric data. In this paper, the authors investigate the suitability of encoding categorical features based on the posterior probability of an attack conditioned on the feature in the context of IDS. KNN classifier is used for construction of IDS on top of latent features in numeric form. The proposed method is trained and tested on NSL-KDD data set to predict one among the possible 40 distinct class labels for a test instance. Classification accuracy and false positive rate (FPR) are considered as performance metrics. The results have shown that the proposed approach is good at detecting intrusions with an accuracy of 98.05% and a false alarm rate of 0.35%.

Keywords Intrusion · Encoding · Information Gain Ratio · PCA · KNN · CrossTable

1 Introduction

Advancements in the field of computer technology and information age make system security a very challenging task. With the increase of Internet dependency, intrusion rate also increases in both number and complexity. An intrusion can be defined as any unauthorized activity performed by an attacker on computer systems.

N. Sameera (✉) · M. Shashi
Department of CS & SE, Andhra University College of Engineering (A),
Andhra University, Visakhapatnam, India
e-mail: sameerascholar@gmail.com

© Springer Nature Singapore Pte Ltd. 2020
K. S. Raju et al. (eds.), *Proceedings of the Third International Conference on Computational Intelligence and Informatics*, Advances in Intelligent Systems and Computing 1090, https://doi.org/10.1007/978-981-15-1480-7_15

When attacker illegally penetrates through security gates of system infrastructure, it leaves the system in an insecure state. Attackers are becoming smarter, and they even can make an advanced move (attack) on the footsteps of the defenders' action. In order to capture these highly targeted intrusions, information which is tremendously growing in parallel to the technological developments should be monitored carefully. Traditional security mechanisms like firewall are not enough to stand against these consequences. An intelligent system should be needed to secure from these advanced persistent threats. Intrusion detection system (IDS) [1] comes into play in order to handle this situation. IDS is a key to collaborative security. It can be in the form of software or hardware that monitors the activities of a computer or a computer network for suspicious event. Once it finds any such event, it generates an alert. Examples of IDSs include honeypots antivirus, etc.

There exist several data mining methods for implementing IDS like KNN, decision trees, SVM, K-Means, GA, NN, etc. These methods divide data set into training and testing subsets. Training set is used as the basis for learning the model, and testing set is used to evaluate the model to check when it reaches the required level of performance in terms of accuracy and false positive rate.

IDS involves processing the traffic packets represented in terms of numeric and categorical features and relates them to the specific attack detection. K-Nearest Neighbor algorithm has proven its potential for accurate prediction while handling data sets dominated by numeric features and hence is found to be suitable. This paper aims at building an IDS for classifying traffic packets into 40 different attack types by analyzing packets described in terms of 43 features. PCA is applied for extracting latent features while transforming data instances from a high-dimensional space to low-dimensional space in order to eliminate redundancy and noise. Since PCA is directly applicable to numeric data sets, it is required to convert the categorical attributes into numeric form.

In the existing literature, the categorical features were considered as "factors" wherein, each distinct categorical value is mapped to an integer randomly often based on lexicographic order. When such integers are used for data analysis, it results in reduced performance. This paper aims to apply a novel mechanism for converting the categorical features into an equivalent numeric form using encoding approach [2], which considers the posterior probability of an attack conditioned on the feature as equivalent numeric values to replace the corresponding categorical value when occurred in association with a class. The effectiveness of the proposed encoding approach for intrusion detection is investigated using accuracy and FPR metrics.

The remaining part of the paper is structured as follows: Sect. 2 presents related research about handling categorical features and PCA, Sect. 3 presents the algorithm for proposed approach, Sect. 4 gives experimentation of step-by-step procedure for implementing proposed approach and discussion about results and conclusion is given in Sect. 5.

2 Related Work

To handle categorical features, different authors have followed different approaches. Work done by Saxena and Harshit et al. [3] in 2014 uses information gain for selecting the most important features during the process of feature selection. Information gain may not give good results for features with too many labels. To overcome this drawback, Karegowda et al. [4] propose to use IG ratio for feature selection to reduce the bias. Preprocessing in [5] uses direct digitization process for handling categorical features. Here, numbers are directly assigned to the categorical attribute according to the chronological order of attribute values. For example, Tailor, Teacher and Driver are the values of a attribute "Designation"; then this method assigns 1 to the Driver, 2 to the Tailor and 3 to the Teacher. This may not give good results in all situations. Here, PCA is used for dimensionality reduction and clustering is performed by mini-batch K-Means algorithm. In [6], decision tree classification is used which is followed by K-Means clustering. Genetic algorithm is used to optimize the parameters of the clustering algorithm. As a part of prepro-cessing, categorical features are directly converted into numeric values without following any specific approach. Shyu et al. [7] propose two ways for handling categorical features: one is using indicator variable and another way is a scaling method based on MCA. In [8], different techniques are discussed for converting categorical values into numbers. In the first method, numbers are assigned to labels according to chronological order of labels. In the second method, numbers are assigned to numeric bins. Dummy coding is the third method in which presence of a level is indicated with 1 and absence of a level is indicated with 0. Varghese et al. [9] investigated the performance of classification algorithms for IDS with two feature selection methods, PCA and CFS, and finally conclude that dimensionality reduction with PCA will improve detection accuracy. Chabathula et al. in [10] implemented IDS using different machine learning algorithms with and without using PCA and proved that the detection accuracy for all classification algorithms was improved when PCA is applied for dimensionality reduction in preprocessing step. However, work done by Chabathula et al. modeled the IDS problem as a binary classification problem whereas the aim of this work is more finely grained as it classifies a packet into one among 40 possible attack groups rather than declaring it just as an attack and thus requires to extract the inherent information available in the categorical variables also.

All the above said methods use NSL-KDD data set or its variants which include categorical variables also. But no scientific approach for handling categorical features was highlighted. In this paper, the authors propose to use encoding approach of handling categorical features.

3 Proposed Approach

NSL-KDD data set is refined by preprocessing. Preprocessing removes insignificant features and handles all categorical data using encoding approach. Information Gain (IG) ratio is used to decide whether to consider the specific categorical feature or to ignore it. IG ratio of a categorical feature reflects the importance of that feature in finding the class label. Finally, the features with minute IG ratio values can be ignored, and all the remaining categorical features can be sent to encoding approach. Encoding approach then converts these selected categorical features into numeric. With this, NSL-KDD data set becomes numeric but includes features with different ranges. Z-score normalization is applied to neutralize these varied ranges. PCA is applied on the normalized data to transform all the instances into a latent space with features along the principal components identified based on features with high variances in the original data. Classification algorithm KNN is applied on the transformed data to get the required class labels. Block diagram of the proposed approach is given in Fig. 1, and pseudo-code of the proposed approach is shown below.

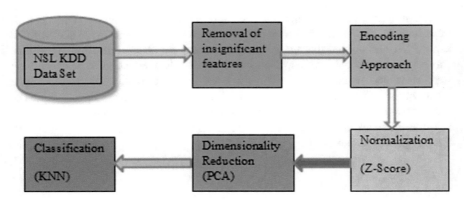

Fig. 1 Block diagram of the proposed approach

Pseudo-code:

Input: The input features F = {F_1, F_2,…, F_n}
       Threshold $\Delta=0.1$
Output: Predicted labels for test data with highest accuracy and minimum FPR
Steps:
1.   for each feature $F_i \in F$ do
2.   if F_i has minimal non zero values
3.   then
4.   F=F-{ F_i }   //Removal of insignificant features
5.   else if F_i is non-numeric
6.   then
7.   estimate X: number of distinct values (F_i)
8.   if X=2   // F_i is a bi-valued categorical feature
9.   then
10. F_i =as.numeric(F_i)   // F_i is converted in to numeric
11.   else   // F_i is a multivalued categorical feature
12.   Calculate IGR_{Fi} =Information_gain_ratio(F_i)
13.   if $IGR_{Fi} < \Delta$
14.   then
15.   F=F-{ F_i }
16.   else
17. F_i =Encoding_approach(F_i)
18.   end if
19.   end if
20.   end if
21. end for
22. F_norm = Z_Score(F)
23. F_transformed = PCA(F_norm)
24. F_train = Random_sample(F_transformed, 60%)
25. F_test = Random_sample(F_transformed, 40%)
26. K=3
27. Repeat for each K until no change in accuracy
28. Prediction = KNN(F_train,F_test,F_train_labels,K)
29. Result = CrossTable(F_test_labels, Prediction)
30. Calculate Accuracy and FPR
31. K=K+2

4 Experimentation and Results

4.1 Data Set

The NSL-KDD data set [11] which is the refined version of KDD-CUP'99 is used to train and test the proposed approach. This data set contains 147,907 instances and 43 features [12]. Each instance belongs to one of the 40 class labels present in the data set. Out of 40 class labels, one is normal activity and the others are attacks. All these attacks in the data set belong to four attack groups, namely (1) DoS,

Table 1 Data distribution in NSL-KDD data set

Label	Count
Normal	76967
Anomaly	70940
Total	147907

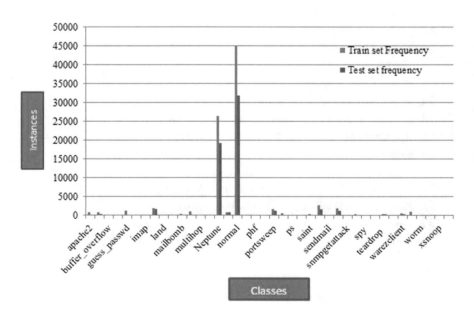

Fig. 2 Feature distribution in training and testing sets

(2) Prob, (3) U2R and (4) R2L. Among the four attack groups present in the data set, majority of the attack instances belongs to DoS category, and a very few instances belong to U2R category. Table 1 gives the data distribution in the data set. From the table, it is observed that half of the data in the data set contain normal activities and remaining part contain attacks.

Data should be preprocessed and prepared before applying classification. Removing insignificant features, encoding and dimensionality reduction by PCA, all these are steps used in this paper to make data ready for classification. Training and testing sets are prepared by considering 60:40 ratio of the preprocessed data set. Attack distribution in the training and testing sets is given in Fig. 2.

4.2 Methodology

Removing insignificant features It was found that certain features are highly sparse and hence do not contribute much to the discriminative information. Hence as a first step of preprocessing, such insignificant features were removed. In the data set, features numbered 7,9,14,15,18,20 and 21 are observed as insignificant as they are highly sparse and contribute no information regarding target class.

Assessing the significance of categorical features If the feature set contains certain bi-values and multi-valued categorical features, they can be handled in such a way that, all bi-valued categorical features were seamlessly converted into binary form while special encoding approach is required for conversion of multivalued categorical features into numeric form. Since the proposed encoding approach requires intensive CPU processing, each of the multivalued categorical attribute is analyzed whether its inclusion is essential or not for predicting the class label. This can be done by calculating IG ratio for each of the categorical attribute and compare its value with a predefined minimum threshold Δ which is set to 0.1. In NSL-KDD data set, IG ratio is calculated for three categorical features, i.e., Protocol-type, Flag and Service, according to the results obtained, the authors have included all the three features. Table 2 gives IG ratio values for the three features.

Table 2 Information gain ratio values for categorical attributes

Feature	Information-Gain Ratio
Protocol-type	0.1291621
Flag	0.4202331
Service	0.218199

Encoding Approach The proposed encoding approach aims to represent the attribute-value pairs related to multivalued categorical features based on their co-occurrence with different class labels. Specifically, the values of the categorical features are converted into the numeric form, reflecting the posterior probability of the corresponding class label given the attribute-value pair. Tables 3a and b illustrate the process of estimating the posterior probabilities, thereby encoding multivalued categorical features. The first two columns of Table 3a depicts the attribute-value pairs of ten instances along with their class labels indicating either normal or attack class. The process of estimating the posterior probabilities for each class label given the attribute-value pairs is shown in Table 3b with three columns. Attribute-value pairs and their co-occurring frequency with different class labels are shown in columns one and two while the posterior probabilities are shown in column three. The values of the categorical features are encoded with the corresponding posterior probabilities as shown in the third column of Table 3a.

Normalization Real-time network traffic data is too complex with different features having different range of values. In order to overcome this, scaling the values of the data set to some specific range is essential, which calls for feature normalization. The proposed approach uses Z-score normalization. Z-score normalization rescales the attribute values based on the mean and standard deviation of the feature, expressing the feature value in terms of its deviation from its mean in the units of its standard deviation and is given by the following formula (1).

Table 3 Process of estimating the posterior probabilities, thereby encoding multivalued categorical features

Table 3a

Prototype	Class	Encoding value for prototype
Tcp	Attack	0.6
Udp	Normal	1.0
Tcp	Attack	0.6
Icmp	Attack	0.33
Tcp	Attack	0.6
Icmp	Normal	0.66
Tcp	Normal	0.4
Udp	Normal	1.0
Icmp	Normal	0.66
Tcp	Normal	0.4

Table 3b

Prototype	Class label		Posterior probability	
	Normal	Attack	Normal	Attack
Tcp	2	3	0.4	0.6
Icmp	2	1	0.66	0.33
Udp	2	0	1.0	0.0

$$z = \frac{(x - \mu)}{\sigma} \tag{1}$$

where x is the attribute to be normalized as z, μ is the attribute mean and σ is the standard.

Dimensionality Reduction In this paper, the authors have used principal component analysis (PCA) for dimensionality reduction [13]. It transforms the original V dimensions to W dimensions in which these V dimensions include the essence of all the W dimensions. These V dimensions are referred with W principal components [14]. PCA concentrates on the variance of the data such that first principal component contains high variance compared to the second principal component and so on. In this paper, $V = 43$ and $W = 10$. Several experiments were conducted to fix the value of W as shown in Fig. 3. Here, accuracy gradually increases up to ten dimensions. Beyond ten dimensions, accuracy seems to be saturated and hence the best number of dimensions (W) considered is ten.

KNN Classifier For each instance in the test data, its K-nearest neighboring instances from training data is considered [15]. Nearness among the instances is calculated based on a similarity metric. The new instance is classified into the class which had got majority votes out of K-nearest instances. Euclidean distance is one of the famous distance metrics used to calculate similarity among instances. The value of K is decided experimentally by supplying different K values to the KNN classifier. The best K value which leads to good classification accuracy is determined by observing the accuracies generated for different K values. The minimum value of K to be considered is 3. The results of several experiments conducted to find the best K value are shown in Fig. 4 and the accuracy is found to be highest at $K = 3$.

Fig. 3 Variation in accuracies for different PCA dimensions

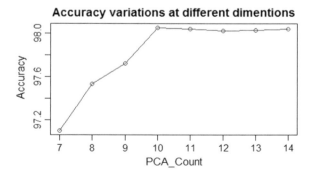

Fig. 4 Accuracy variations at
different *K* values

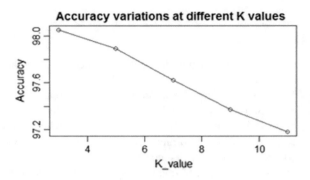

4.3 Results

The proposed approach is implemented using *R*. In this paper, the classification accuracy and FPR metrics are used for performance evaluation of the multi-class KNN classifier. Since the classifier deals with 40 distinct class labels representing different attacks, CrossTable is found to be suitable for performance analysis. Both accuracy and FPR can be calculated from the CrossTable() which is a built-in function offered in R programming that takes actual and predicted class labels as the primary arguments. In this work, the class labels of test set are supplied as first argument to the CrossTable(), and predicted results after applying KNN classification are supplied as second argument. The result of the CrossTable() is a matrix in which actual labels are indicated by rows and predicted labels are indicated by columns. t, prop.col, prop.row, prop.tbl are the arguments of the results generated by CrossTable() in n by m matrix form containing table cell counts, cell column proportions, cell row proportions and cell table proportions, respectively.

Accuracy and FPR are evaluated from the prop.tbl in the following way:

Consider X and Y as the rows and columns of the prop.tbl.

where $X = \{x_1, x_2, x_3, \ldots x_{norm}, \ldots, x_n\}$ and $Y = \{y_1, y_2, y_3, \ldots, y_{norm}, \ldots, y_m\}$

And, x_{norm} is the row containing "normal" label and y_{norm} is the column containing "normal" label

$$\text{Then, Accuracy} = \sum_{i=1}^{n} \text{prop.tbl}[x_i, y_j] \text{ where, label}[x_i] = \text{label}[y_j] \tag{2}$$

$$\text{FPR} = \sum_{j=1}^{m} \text{prop.tbl}[x_{norm}, y_j] \text{ where } y_j \neq y_{norm} \tag{3}$$

Khushboo et al. [16] and Ikram Sumaiya Thaseen et al. [17] in their work for implementation of IDS used the same KDD data set but considered classification into the four attack groups DOS, Prob, R2L, U2R, in addition to normal as class labels instead of considering classification of traffic packets into one among the 40

Table 4 Performance of proposed approach versus other approaches

Method	Accuracy %	FPR %
Proposed method	98.05	0.35
Khushboo et.al [15]	94.17	5.82
Ikram sumaiya et.al [16]	98	0.13

class labels representing each attack as a distinct class. Accordingly, they predict one among the possible 5 distinct class labels for a test instance whereas the proposed approach predicts one among 40 distinct class labels for a test instance. Though the proposed approach considers 40 attack types as individual classes, the results have shown that the proposed approach is much better than [16] approach and competitive with [17] approach. Comparative statement of proposed approach against other approaches is mentioned in Table 4 and the same is presented by graphs shown in Figs. 5 and 6.

Fig. 5 Comparisons of accuracy results

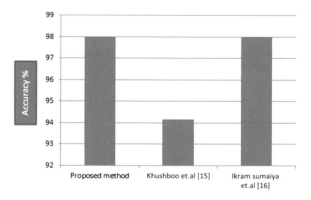

Fig. 6 Comparisons of FPR results

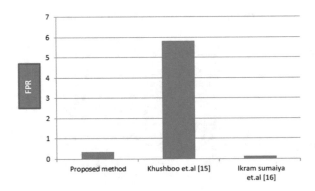

5   Conclusion

While extracting knowledge from data sets dominated by numeric attributes with less categorical features, it becomes essential to transform the categorical features into numeric form without loss of information. In this paper, the authors investigate the suitability of encoding categorical features based on the posterior probability of an attack conditioned on the feature in the context of IDS. Information gain ratio is used to calculate the importance of each categorical feature, and encoding approach was applied only on those features whose gain ratio is above some threshold Δ. Dimensionality reduction is performed using PCA to remove irrelevant and insignificant information and to produce informative and effective high variance transformed data set. KNN classification is applied for identifying attack packets with highest accuracy. Proposed approach is experimented on NSL-KDD data set and compared with existing IDS approaches in terms of classification accuracy and FPR. It is observed that in both [16] and [17], prediction is done against 5 attack groups whereas in proposed approach prediction is done against 40 distinct class labels. So, it is concluded that the proposed approach is clearly better than [16] and competitive to [17] with high accuracy of 98.05% and low FPR of 0.35%.

References

1. Loubna, D., B. Ahmed, E. Hoda, A. Elmoutaoukkil, F. Eladnani, and A. Benihssane. 2015. A Survey of Intrusion Detection System, © IEEE. 978-1-4799-8172-4/15.
2. https.www.youtube.comwatchv=du8YTUgOCJ0. Accessed Sept 2018.
3. Saxena, H., and V. Richariya. 2014. Intrusion Detection in KDD99 Dataset Using SVM-PSO and Feature Reduction with Information Gain 98 (6).
4. Karegowda, A.G., A.S. Manjunath, and M.A. Jayaram. 2010. Comparative Study of Attribute Selection Using Gain Ratio and Correlation Based Feature Selection. *International Journal of Information Technology and Knowledge Management* 2 (2): 271–277.
5. Peng, K., V.C. Leung, and Q. Huang. 2018. Clustering Approach Based on Mini Batch K means for Intrusion Detection System over Big Data. IEEE Access.

6. Elham, A., and K. Rasoul. 2017. IDS Using An Optimized Framework Based on DM Techniques. In *IEEE 4th International Conference on Knowledge-Based Engineering and Innovation (KBEI)*.
7. Shyu, M.L., K. Sarinnapakorn, I. Kuruppu-Appuhamilage, S.C. Chen, L. Chang, and T. Goldring. 2005. Handling Nominal Features in Anomaly Intrusion Detection Problems. In *15th International Workshop on Research Issues in Data Engineering: Stream Data Mining and Applications RIDE-SDMA 2005*, 55–62. IEEE.
8. Simple Methods to deal with Categorical Variables in Predictive Modeling, https://www.analyticsvidhya.com/blog/2015/11/easy-methods-deal-categorical-variables-predictive-modeling/. Accessed Sept 2018.
9. Varghese, J.E., and B. Muniyal. 2017. An Investigation of Classification Algorithms for Intrusion Detection System-A Quantitative Approach. In *International Conference on Advances in Computing, Communications and Informatics (ICACCI), 2017*, 2045–2051. IEEE.
10. Chabathula, K.J., C.D. Jaidhar, and M.A. Ajay Kumara. 2015. Comparative Study of Principal Component Analysis Based Intrusion Detection Approach Using Machine Learning Algorithms. In *3rd International Conference Signal Processing, Communication and Networking (ICSCN)*, 2015, 1–6. IEEE.
11. Dhanabal, L., and S.P. Shantharajah. 2015. A Study on NSL-KDD Dataset for Intrusion Detection System Based on Classification Algorithms. *International Journal of Advanced Research in Computer and Communication Engineering* 4 (6): 446–452.
12. https://www.researchgate.net/post/How_can_I_download_NSL-KDD_Dataset. Accessed Sept 2018.
13. Vasan, K.K., and B. Surendiran. 2016. Dimensionality Reduction Using Principal Component Analysis for network intrusion detection. *Perspectives in Science* 8: 510–512.
14. Chen, Y., Y. Li, X.Q. Cheng, and L. Guo. 2006. Survey and Taxonomy of Feature Selection Algorithms in Intrusion Detection System. In *International Conference on Information Security and Cryptology*, 153–167. Springer, Berlin, Heidelberg.
15. Zhang, M.L., and Z.H. Zhou. 2005. A k-nearest Neighbor Based Algorithm for Multi-label Classification. In *2005 IEEE International Conference Granular Computing*, vol. 2, 718–721. IEEE.
16. Khushboo, S., M. Anil, and K.S. Shiv. 2016. A KNN-ACO Approach for Intrusion Detection Using KDD Cups 99 Dataset. In *International Conference on Computing for Sustainable Global Development (INDIACom)*, 978-9-3805-4421-2/16/$31.00_c IEEE.
17. Ikram Sumaiya, T., and C.A. Kumar. 2017. Intrusion Detection Model Using Fusion of Chi-Square Feature Selection and Multi Class SVM. *Journal of King Saud University-Computer and Information Sciences* 29 (4): 462–472.

Materializing Block Chain Technology to Maintain Digital Ledger of Land Records

B. V. Satish Babu and K. Suresh Babu

Abstract In most of the administrative sections of the government, the enormous amount of digital data is generated. As expected, there will be continuous data insecurity in different forms like tinkering university marks of students, making up of certificates which are bogus, changing or erasing the digital ledger of land records. We need a reliable method that makes this digital data immutable to generate trust between administrative sections of the government and the citizens. To do this, we can materialize the "Block chain, a peer-to-peer software technology" which is an inherent concept from "Bitcoin" crypto currency (Heng in The Application of Block chain Technology in E-government in China, 2017) [1]. To hedge the digital records insecurity, block chain supports a distributed database which clasps encrypted ledger in a decentralized fashion. Upon the occurrence of a transaction, the transaction should be verified and then placed inside distributed encrypted ledger. Afterward, a set of "N" digital transactions are taken into account to calculate hash and kept in a block along with the timestamp. If the hash, timestamp along with transactions placed in a block, it is challenging to change or expunge transactions inside the block, considering that cryptographic block hash is irretrievable. The recently validated block is then connected and appended after the earlier block of block chain using earlier cryptographic block hash value. We have introduced the different schemes for materializing block chain technology in administrative sections of the government, e.g., in the department stamp and registration duties.

Keywords Digital data · Replication · Hash · Cloud · Tamper proof · Distributed ledger · Block chain · Distributed databases · Flask

B. V. Satish Babu (✉)
Research scholar, JNTUH, Hyderabad, India
e-mail: vsatish.825@gmail.com

K. Suresh Babu
Associate Professor of CSE, School of IT, JNUTH, Hyderabad, India
e-mail: kare_suresh@yahoo.co.in

© Springer Nature Singapore Pte Ltd. 2020
K. S. Raju et al. (eds.), *Proceedings of the Third International Conference on Computational Intelligence and Informatics*, Advances in Intelligent Systems and Computing 1090, https://doi.org/10.1007/978-981-15-1480-7_16

201

1 Introduction

Amid of 1990–2010, register books have been utilized to keep handwritten land registration data in pages. These books occupy a pile of physical space and quality is reduced over a lifetime with frequent handling. To deal with a quality reduction of register books, bygone "Stamps and registration department" has begun to use CD's to replica every important document in that book. This precaution is also not feasible to maintain a large number of jpeg images and land document proofs.

In the year 1997, Andhra Pradesh accompanied with other states has begun developing "Computer-Aided Administration of Registration Department (CARD)"—a centralized software system. In the year 2011, the government of Andhra Pradesh came with a resolution of incorporating the CARD online management system in the department of "Stamps and registration duties" by taking assistance from National Informatics Center.

The CARD minimizes many difficulties by changing the regular process of maintaining registry data in books by converting everything into digital. This software system is "Three-layer client–server centralized design [2] (Fig. 1)."

In Layer1, records of entire state will be available and manipulated only by higher officials like commissioner. District-level users are privileged with only read permission.

In Layer2, pertinent master CARD data around the premises of the Sub-Register Officer (SRO) will be available and manipulated only by the registrar at the district level. The Layer2 master CARD data is pertinent to all transactions occurred in the office of SRO. It represents a user data entered and submitted by the corresponding operators. In most of the cases, it is not possible by the operator to modify these transactions.

The next layer, Layer3 of the CARD software system, digital land records are placed in a centralized database that has a variety of privileges. In that environment, we are going to have more chances of tampering [2]. Thus, we need to maintain digital land transactions in an immutable technological environment.

Fig. 1 Components of the card. [2]

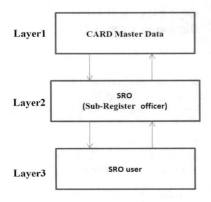

2 Motivation

We have identified some case studies about the manipulating and exploiting of the digital registry using online "WebLand online system."

- "Vizag: In, September,2017, Almost 50 farmers in the tribal area of village Chatterjipuram, in Visakhapatnam district, has made the protest in opposition to change of land details by a tahsildar in the WebLand portal [3]."
- "November 27, 2017 Delhi Officers mutated land registry details using counterfeit court orders to move Rs 600 core money of 30 acres land, to corporate [4]."

Along with the aforementioned incidents, land double registration incidents have already occurred many times. These incidents exploit all possibilities of mutating data in the centralized database domain. We have these situations because the state of our digital data is

- Maintained in a shared database with unified access
- Ability to manipulate database data
- The absence of public transparency.

So, we need to preserve our data under the following attributes

- Database decentralization
- Databases with unchangeable data
- Under the presence of public transparency.

The "block chain technology" is an appropriate technology to provide all these safety attributes.

3 Block Chain Technology

Block chain has a concept of maintaining a public distributed database that contains an encrypted table called **Ledger** [5]. The block chain is managed to keep the consistency of digital ledger among the peers of a distributed network.

Another name of this distributed ledger is "**Block.**" During the creation of a new block, we need to combine a set of transactions in the ledger available at one particular time. This technology is entitled as "Block chain" because a newly generated is always linked to its earlier block by its block hash.

An attempt to change one selected transaction from the ledger; results completely different hash value for the entire block. This factor allows us to identify deception. "Block chain" theoretical concept along with its inner arrangement of the block is represented in Fig. 2.

In Fig. 2, interior of the block is cleaved into two fragments 1) Header 2) Body. The body of the block contains transactions **"T1, T2… T$_N$"** along with their double

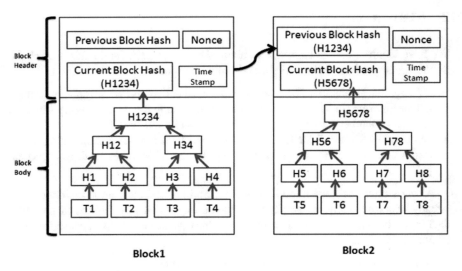

Fig. 2 "Block chain" theoretical concept. [6]

hashes **"H1, H2… H$_{2N-1}$."**The **Nonce** variable sets the challenge to create block [7]. **Timestamp** represents the system time at which block is created.

4 Materialization of Block Chain Technology

Boundaries of this technology are not just limited to crypto currencies. Its characteristics are elongated into different types of applications [8]. Methods of materializing block chain diversify according to the situations. For example, the materialization model of administrative firms is divergent with materialization model of universities. So, irrespective of materialization path, basic semantics of this technology should not be changed.

According to **"Satoshi Nakamoto,"** maintaining every component of block chain is not mandatory, for example, "Miners, Nonce parameter." Therefore, new components are added and needless components are excluded based upon our requirements [7]. This characteristic of block chain made it suitable for a variety of applications. In our proposed method, materialization of block chain technology for land records will be carried out using the following steps

(1) Modeling a prototype to maintain a digital ledger of land records.
(2) Achieving database replication in a distributed fashion.
(3) Creating an immutable ledger by making use of hash algorithms.
(4) All recorded transactions are grouped into a new block. Finally, the new block is linked to the block chain.

(5) The special verification nodes (SV node) are created for verification.

(6) Finally, examine the implementation against real-world cases.

The materialization model of block chain is shown in Fig. 3. In general, block chain does not fit for particular scenarios. So, it is necessary to check the feasibility of applying block chain to those application scenarios.

We have applied the aforementioned materialization model to manage the digital ledger of land records in the block chain. This digital ledger is replicated among peers to their databases in the network. According to Fig. 3, the complete process with the involvement of different actors is sated as follows.

(a) **Administrator:** He is the manager of the digital ledger from the corresponding executive department, for example, Officer of Sub-Registry (SRO). An administrator daily duty consists of adding records to the digital ledger with the help of web application.

(1) The administrator adds submitted details to the digital ledger as one record.

(2) Administrator applies **"SHA-256"** twice to compute the hash of "N" transactions available at one particular time.

(3) Using **"Merkle Tree"** [5], "N" hash values are reduced to a single hash value. This single hash value will be used to compute **"Current block hash."**

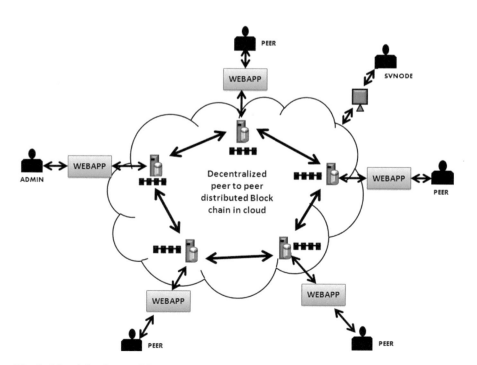

Fig. 3 Materialization model

(4) An administrator creates a new block by comprising **"Block header"** with current block hash, earlier block hash, timestamp. And **"Bock body"** comprises **"N"** number of transactions along with their hash values.

(5) This new block is replicated to every peer in a distributed network (using replication concept of MYSQL).

(6) Now, every peer in the distributed network appends and links new block to their current block chain. In such a way, every peer has an exact copy of block chain.

Any modification to a particular record in the block is simply identified because it invalidates the entire block chain by changing all hash values in the sequence.

(b) **Special verification (SV node)**: SV node downloads only block headers by requesting the administrator. SV node does not preserve the entire block chain due to space constraints [8]. An SV node can be a bank that gives the loan or a court that hears double registration petition, etc.

(1) If it is land registration, special verification node requests the administrator (Sub-Register) to provide particulars about land.

(2) In response, the administrator sends two parameters back to SV node namely **"Merkle root path,"** **"Block Hash value"** of the block that contains requested land records.

(3) Special verification node applies these two parameters to verify land details against block chain. It computes a single hash value using **"Merkle root path,"** **"Block Hash value"** returned by the administrator.

(4) This resultant hash value is compared against block hashes available within all block headers. If SV node finds exact block hash match, then verification is completed successfully. Otherwise, SV node takes it as a fraud.

(c) **Peer nodes:** People who want to purchase and to verify particulars of the land can join as a peer node. These peer nodes simply query block chain to check land records. The administrator is the only person (SRO) privileged with write permissions. Peer nodes and SV nodes are privileged with only read permissions.

5 Configuration for Implementation

Entire block chain is implemented with the assistance of the public cloud **"Amazon web services."** Three **EC2** compute machines have been created with UBUNTU 14.04 Amazon machine image as shown in Fig. 4. These three compute servers are

Master Server1 : Administrator server   ⎫
Slave Server2 : Special verification node ⎬ Three compute machines
Slave Server3 : Peer node server   ⎭

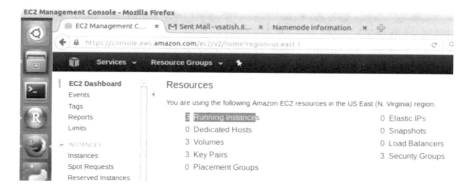

Fig. 4 Three running compute machines in AWS

(1) First log into three running server instances
   "#ssh -i keyfile_path.pem home@Server_PublicIP"
(2) Install **mysql-server** [9] and **mysql-client** [10] in all three machines
(3) Next, install different python libraries required using python package index (PIP).

- Download and install **python-pip** [11] which is used to install other important packages from python libraries.
- Install **python-mysqldb** MYSQL driver for python [12].
- Install package **python-flask** [13] to run web application supported flask server.

(4) Each and every record submitted by the operator should be replicated to the local database of all peers in a distributed network. This replication across master and slave servers is achieved by establishing **"MYSQL replication"** [14]. Stepwise instructions of establishing MYSQL replication are given below

From server of Master

Step 1: From the master server, modify database configuration file "my.cnf." Under **"mysqld"** section of the configuration file set.

- **Binding address to publicIP_address**
- **Server_id to 1**
- **log file to "mysql-bin.log"**
- **Step 2:** Now, at master server, restart and login into MYSQL database server with username "root."
- **Step 3:** Create replication user **"ruser"** and grant privilege of replication to this user. We are going to use this user **"ruser"** from both slave server2, slave server3 from the master.

- **Step 4:** Create the required Tables **T1, T2, T3** under the database called "**blockchaindb**."
- T1 represents the digital ledger of land particulars.
- T2 is the subset of Table T1. It contains all records along with their cryptographic hash values. All these hash values are reduced to Merkle root hash. T2 represents the entire block body of all blocks in the block chain.
- T3 contains block header details of all blocks like timestamp, current block hash, and earlier block hash.
- **Step 5:** Create a dump **blockchaindb.sql** from the database "**blockchaindb**."
- **Step 6:** Master has to Secure copy (scp) that database dump "**blockchaindb. sql**" to all slaves. In our context, they are slave1@publicIP, slave2@publicIP.

From servers of slave: Stepwise instructions in all slaves

Step 1: From both slave servers, modify database configuration file "my.cnf." Under "**mysqld**" section of the configuration file set.

- **Binding address to publicIP_address**
- **Server_id to 1**
- **log file to "mysql-bin.log"**
- **Step 2:** Now, restart both slave servers and then login into MYSQL database server with username "root."
- **Step 3:** From both slaves, set replication master for the user.
- **"ruser"** using master's public IP address and password of MySQL-server.
- **Step 4:** Now, from all slaves, dump **blockchaindb.sql** into their local database and start replication process into all slaves.
- **Step 5:** Finally, test the master–slave replication by adding a record to database **blockchaindb** at the master side. Now, check whether that record is replicated to both slave1 and slave2.
- (5) Create a python script "**blockchain.py**" and include libraries related to flask server and Mysql driver. Design web interfaces using HTML documents and launch them inside the script using flask server. Using the package "**python-mysqldb**" connect to master and slave's database to add records and to perform replication.

6   Web Application Results

We have created web interface with help of flask package. This web application in Fig. 5 will be available at SRO office. The operator at SRO office enters the record into digital ledger database Table "**T1**" using web forms.

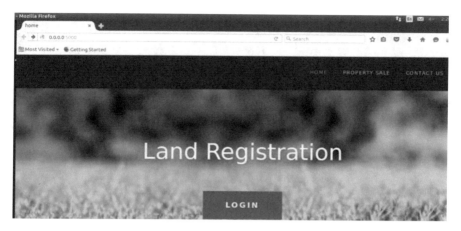

Fig. 5 Web interface for land registration

These records are automatically replicated to peer servers like SV node and peer node. Inside block chain, we have maintained all attributes that are required to register a land like locality of asset, information of seller, SRO office details, etc. Stepwise demonstration of using web application is given as follows

(1) From the master server start flask server.
   home@publicIP: ~ #python3 blockchain.py
   Running on http://0.0.0.0:7000/
(2) Copy and paste the address http://0.0.0.0:7000/ in the address bar of the browser.

At first, the operator logs into the application and then click on "**Registration**" tab to open the registration form. He submits all the details required for land registration provided by the clients (Fig. 6).

As mentioned previously, operator submitted details are stored in three Tables T1, T2, T3 under the database called "**blockchaindb.**" All these three tables together represent the block chain. Entire digital ledger in Table T1 is given below

For every registration, one record is inserted into **T2** along with its hash value. After 10 transactions, all these hash values are reduced to a single "**Merkle root hash**" and placed in Table **T3.**

This arrangement makes block chain data immutable because, if the operator modifies any one record in Table T1 as shown in Fig. 7, it changes the hash value of that particular record in Table T2 as shown in Fig. 8. If the hash value in Table T2 changes, then Merkle root hash stored in Table T3 is affected (Fig. 9).

In general, this Merkle root hash will be used as a previous block hash in the next block. In the next block, current block hash is resultant of **Hash (previous Hash||Time Stamp||Merkle root Hash)**. If the previous block has changed, then current block hash is completely changed. In this way, the signature of all successive blocks linked in block chain will be changed. Therefore, if the operator

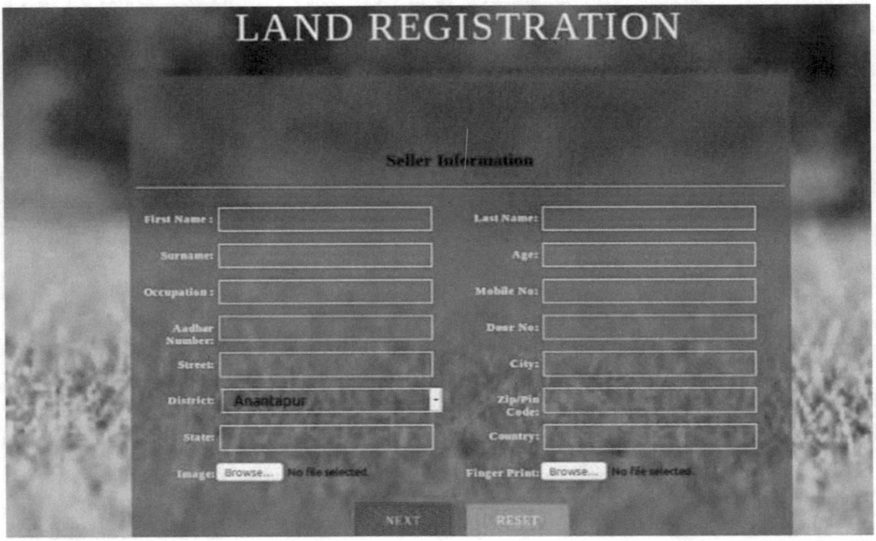

Fig. 6 Registration page

Fig. 7 Entire digital ledger in Table T1

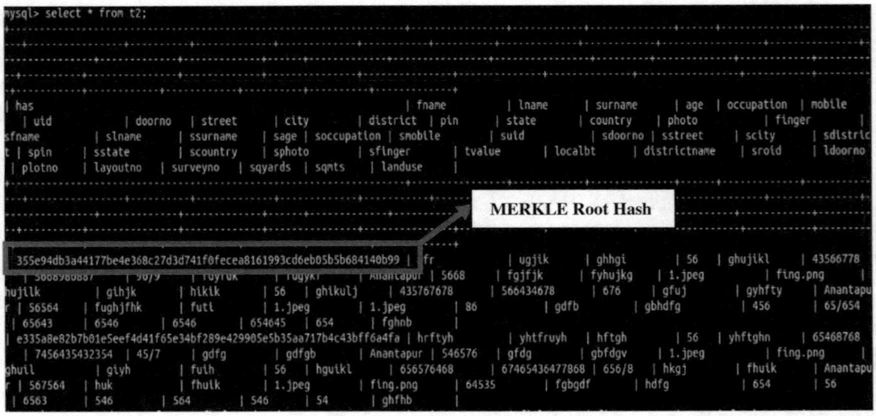

Fig. 8 Block body in Table T2

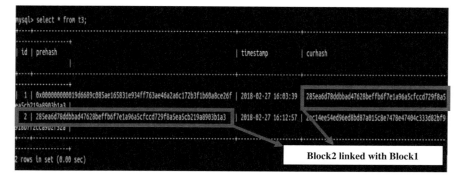

Fig. 9 Block headers in Table T3

wants to tamper one record in block chain, he needs to change signatures of all blocks in block chain which is not practically possible.

7 Conclusion

Block chain technology can be efficiently used to prevent mutation of digital data. In our work, block chain technology is materialized to protect digital ledger of land registry. Further, we need to study the applicability of block chain technology in different scenarios along with storage and time constraints.

Acknowledgements We are really glad to the staff of SRO Office, Patamata, Vijayawada, Andhra Pradesh for giving us information about attributes of land registration.

References

1. Hou, Heng. 2017. *The Application of Block chain Technology in E-Government in China*. Sun Yat-sen University.
2. Card. https://registration.ap.gov.in/CitizenServices/Miscellaneous/NoteOnCARD.pdf.
3. Computer-Aided Registration of Deeds and Stamp. unpan1.un.org/intradoc/groups/public/documents/apcity/unpan019010.pdf.
4. Case Study 1. http://www.thehansindia.com/posts/index/Andhra-Pradesh/2017-09-25/Tribals-DEMAND-probe-into-tampering-of-land-records/328957.
5. Case study 2. https://www.hindustantimes.com/delhi-news/rs-600-crore-property-scam-unearthed-in-delhi-cbi-asked-to-probe/story-FXNlgBOgJwSi328HJCwuiM.html.
6. Zikratov, Igor, Alexander Kuzmin, Vladislav Akimenko, and Viktor Niculichev. 2017. Ensuring Data Integrity Using Block Chain Technology. Lucas Yalansky ITMO University.
7. Nakamoto, Satoshi. 2008. *Bitcoin: A Peer-to-Peer Electronic Cash System*.
8. MIT Block chain. http://blockchain.mit.edu/how-blockchain-works/.
9. Kim, N.H., Kang, S.M. 2017. *Mobile Charger Billing System Using Lightweight Block Chain*.

10. MYSQL Server Installation. https://help.ubuntu.com/lts/serverguide/mysql.html.en.
11. MYSQL Client Installation. https://packages.ubuntu.com/search?keywords=mysql-client.
12. PIP Installation. https://pypi.org/project/pip/.
13. Installing MYSQL Python Drive. https://pypi.org/project/MySQL-python/.
14. Installing Python-Flask. http://flask.pocoo.org/.
15. MYSQL Replication. https://dev.mysql.com/doc/refman/8.0/en/replication.htm.

Review of Optimization Methods of Medical Image Segmentation

Thuzar Khin, K. Srujan Raju, G. R. Sinha, Kyi Kyi Khaing and Tin Mar Kyi

Abstract Medical image segmentation is an important component in medical image analysis and diagnosis which is used as a useful application for medical image processing. Image segmentation of medical images has been implemented and studied by numerous researchers in their various research activities. Robustness of the method is all-time challenge in this type of application of medical image processing. The robustness has been addressed by few researchers but still remains challenging task. The performance of existing research work on medical image segmentation is improved by using optimization techniques. This paper studies and presents a critical review of existing research work that has been used for optimizing the segmentation results. An attempt has also been made to suggest a plan for further formulating a more powerful optimization method to optimize the results that could help in the automated diagnosis of different types of medical images.

Keywords Medical image segmentation · Optimization · Image diagnosis · Robustness

T. Khin (✉) · G. R. Sinha · K. K. Khaing · T. M. Kyi
Myanmar Institute of Information Technology (MIIT), Mandalay, Myanmar
e-mail: thuzar_khin@miit.edu.mm

G. R. Sinha
e-mail: gr_sinha@miit.edu.mm

K. K. Khaing
e-mail: kyi_kyi_khaing@miit.edu.mm

T. M. Kyi
e-mail: tin_mar_kyi@miit.edu.mm

K. Srujan Raju
CMR Technical Campus Hyderabad, Hyderabad, India
e-mail: drksrujanraju@gmail.com

G. R. Sinha
IIIT Bangalore, Bangalore, India

© Springer Nature Singapore Pte Ltd. 2020
K. S. Raju et al. (eds.), *Proceedings of the Third International Conference on Computational Intelligence and Informatics*, Advances in Intelligent Systems and Computing 1090, https://doi.org/10.1007/978-981-15-1480-7_17

1 Introduction

Segmentation of an image is a very important component of automated medical image processing, image analysis, and applications. Actually, in all modern diagnosis centers and hospitals nowadays, modern equipment is used for scanning, analysis, and diagnosis of the medical images such as X-rays, Computed Tomography (CT), mammograms. If a radiologist or physician is interested to extract a certain region, then that portion needs to be brought out; this is possible with the help of segmentation techniques, as highlighted and discussed in several research contributions by Sinha et al. [1–5].

There are various methods in the literature used for this purpose of different categories such as region-based methods, pixel-based methods, and others. The biggest challenge in all these research works is the lack of robustness. There are researches for airborne images, medical images, fingerprints, and their segmentation, which are highly sensitive but again the same problem arises that is robustness and scope of improvement of the performance are always there [6]. Boundaries of anatomical structures, structures of tissues in cancerous images are important things that are required in automatic medical image analysis and diagnosis of images, and in such important and sensitive applications, mistake of few mill inches or micro-inches creates huge difficulties in medical image diagnosis, and therefore, there is need of scope of improvement and robust method so that the method can be applied as versatile method to different types of images for their segmentation and subsequent applications in medical science.

This paper highlights about critical review of what has been done in this field and few major optimization methods that were used by the researchers for optimizing the performance of medical image segmentation.

2 Related Research and Problem Identification

Baatz and Schape [6] suggested about prerequisite method used for object-oriented image processing aiming at the implementation of image segmentation of medical images. Scale parameter and homogeneity definition were involved and controlled in providing and obtaining better results in comparison with current research in the area of medical image segmentation. Chabrier et al. [7] stressed on study of optimization of evaluation parameters and metrics useful for segmentation of medical images. Local ground truth was added with desired level precision along with the contribution of genetic algorithm. Similar to this method could be applied for grayscale images of multicomponent type of medical images successfully.

Ghassabeh et al. [8] attempted in study, analysis, and the implementation of Fuzzy C-Means (FCM) method as clustering method and attempted to minimize the amount of uncertainty level using the concept of fuzzy logic inference. Magnetic Resonance Imaging (MRI) images were subjected to the method, and the method was actually applied in the process of image enhancement that is also very important for obtaining accurate medical image analysis-based results. Computation and accuracy advantages both were achieved slightly better than non-fuzzy methods. Neural network was used in the training method of medical MRI images. Neuro-fuzzy as hybrid method was felt a need for future implementation for improved performance.

Chen et al. [9] discussed a minimization problem concept of optimization for segmentation results of medical images. Shape-related energy and the contour parameters were controlled and optimized. The study of minimization provides a fixed parameter for aligning segmentation and prior shape in medical images. Proper intensity profile of images was created using the method, and it was claimed that the method could be used for all different types of medical images, that is, all modalities of the medical imaging. However, the claim was not substantiated by enough results and proof of optimized performance. Lathen et al. [10] studied about optimization methods that can be used in multiscale filtering and integration-related medical imaging and analysis. The methods were compared for synthetic medical images of blood vessels and structures. No substantial proofs of evaluating robust way of implementation could be established.

Sivaramakrishnan et al. [11] implemented firefly algorithm in tumor detection of brain medical images and combined the approach with enhanced colony optimization method. The detection of breast cancer images found in mammographic images was studied, and few results were obtained. Around 350 different MRI images of a hospital were subjected to the methods, and their results were compared. It was also based on referring few standard databases of images and attempted comparing the results to get some artifacts and tumors present in the images of breast cancer patients. The focus was finding optimum solution for extracting cancerous elements with the help of an important metric called as maximizing a posterior. This metric was used to get the optimization level, and thus, the impact of this was observed in the applications related to segmentation and extraction of tumor in MR images of breast. Then the method was also tested for CT and other medical images, and it was claimed by the researchers that the results obtained are optimal enough that can be relied on the process of diagnosis and prescription.

Valsecchi [12] suggested and rather advocated about hybrid method of region-based segmentation for medical image segmentation. Genetic algorithm using some fine parameters and their tuning was implemented while implementing the process if the segmentation of the medical images. The optimization method was tested for two to three various modalities of medical images. Overall performance was evaluated and fund better than the methods without using such methods related to segmentation tasks. Smistad [13] designed and image-guided surgery method for enhancement as minimal invasive surgery, which is noninvasive but also using computer-aided diagnosis, and with the help of optimization method, the results were improved for various types of diseases. Model-based segmentation with priori knowledge was recommended for such applications. As per the scope to be implemented as future research, GPU computing was also claimed to have been used in ultrasound images where the results were found most satisfactory. Szénási and Vámossy [14] studied the concept of optimization and its impact over the results produced by GPU implemented digital microscopy and diagnosis. Overall, the performance improvement using some optimization methods could be seen but robustness remained a big challenges.

3 Recommendation

Most of the research works that have attempted toward optimization of existing medical image segmentation-based results were applied over-utilize some metrics used in the process of segmentation. This is used as an evaluation criteria with the help of certain metrics. This applies to all medical images irrespective of shapes and structures and especially all the modalities, not even caring about robustness and versatility of the method being implemented or tested. One of the metrics, energy minimization function played important role in obtaining balanced results in image enhancement as well as segmentation related applications. The concept of minimization of energy could be used as versatile method in a number of applications in various stages of implementation to achieve good segmentation results in establishing greater diagnosis findings and outcome.

Major recommendations based on critical review and analysis of the current research on optimization of segmentation for medical images are:

- Different stages of medical segmentation and other components need optimizing their results to get improved and robust findings.
- Robust set of metrics need also to be finalized so that the set of metrics could exactly help in obtaining satisfactory and robust results.
- Focus is required for optimizing color image enhancement method as well as their subsequent stages after enhancement.

For instance, few steps of Sinha et al. [3] are given below to highlight the scope of applying optimization.

- A suitable fitness function is defined.
- Initial population is set or declared.
- A threshold is set.
- The results are compared with respect to threshold using fitness function.
- Rank of the method is evaluated.
- Results are analyzed and performance evaluated.

In the above implementation steps of suitable algorithm, highlighted in Sinha et al. [3], the scope of optimization is very wide that could be applied to all steps so that subsequent stages of implementation provide improved and much satisfying results. A suitable computer-aided diagnosis (CAD) can employ an appropriate optimization using robust set of metrics so as to achieve un-opposed and unique results. This uniqueness could greatly improve the segmentation results in medical images for various applications such as detection of cancers, breast, or any other cancers.

CAD used few important metrics such as true positive, true negative, false positive, and false negative that could also be optimized to achieve robust values of accuracy, sensitivity, and reliability of the CAD system used for medical image segmentation.

4 Conclusions

This paper attempts to bring out major contributions on optimization applied to the segmentation of medical images, and accordingly based on few major problems identified, recommendation has been suggested how one could improve the performance along with addressing the concerns related to robustness of the results so that the methods could be applied as versatile methods in various types of medical image modalities.

References

1. Sinha, G.R., and Bhagwati Charan Patel. 2014. *Medical Image Processing: Concepts and Applications*. Prentice Hall of India.
2. Sinha, G.R. 2017. Study of Assessment of Cognitive Ability of Human Brain using Deep Learning. *International Journal of Information Technology* 9 (3): 1–6.
3. Patel, Bhagwati Charan, and G.R. Sinha. 2014. Abnormality Detection and Classification in CAD for Breast Cancer Images. *Journal of Medical Imaging and Health Informatics* 4 (6): 881–885.
4. Sinha, G.R. 2015. Fuzzy based Medical Image Processing. *Advances in Medical Technologies and Clinical Practice (AMTCP)*, Book Series, 45–61. USA: IGI Global Publishers, IGI Global Copyright 2015.
5. Choubey, Siddhartha, G.R. Sinha, and Abha Choubey. 2011. Bilateral Partitioning based character recognition for Vehicle License plate. In *International Conference on Advances in Information Technology and Mobile Communication—AIM 2011*, eds. V.V. Das, G. Thomas, and F. Lumban Gaol, 422–426, Berlin, Heidelberg: Springer.
6. Baatz, Martin, and Arno Schape. *Multiresolution Segmentation: An Optimization Approach For High Quality Multi-scale Image Segmentation*. http://www.ecognition.com/sites/default/files/405_baatz_fp_12.pdf.
7. Chabrier, S., Christrophe Rosenberger, Bruno Emile, H. Laurent. Optimization Based Image Segmentation by Genetic Algorithms. https://hal.archives-ouvertes.fr/hal-00255987/document.
8. Ghassabeh, Youness Aliyari, Nosratallah Forghani, Mohamad Forouzanfar, and Mohammad Teshnehlab. MRI Fuzzy Segmentation of Brain Tissue Using IFCM Algorithm with Genetic Algorithm Optimization. http://www.cs.toronto.edu/~aliyari/papers/aiccsa.pdf.
9. Chen, Yunmei, Feng Huang, Hemant Tagare, and Murali Rao. 2007. A Coupled Minimization Problem for Medical Image Segmentation with Priors. *International Journal of Computer Vision* 71 (3): 259–272.

10. Lathen, Gunnar. 2010. *Segmentation Methods for Medical Image Analysis Blood vessels, Multi-scale Filtering and Level Set Methods*. PhD thesis, Department of Science and Technology Campus Norrkoping, Linkoping University.
11. Sivaramakrishnan, A., and M. Karnan. Medical Image Segmentation Using Firefly Algorithm and Enhanced Bee Colony Optimization. https://pdfs.semanticscholar.org/3bb5/6bcad9d0d1276761fc1164d7db25383487d8.pdf.
12. Valsecchi, Andrea, Pablo Mesejoyx, Linda Marrakchi-Kacemz, Stefano Cagnoniy, and Sergio Damas. Automatic Evolutionary Medical Image Segmentation Using Deformable Models. https://hal.inria.fr/hal-01221343/file/paperFinal.pdf.
13. Smistad, Erik. *Medical Image Segmentation for Improved Surgical Navigation*. PhD thesis, Norwegian University of Science and Technology (NTNU).
14. Szénási, Sándor, and Zoltán Vámossy. 2013. Evolutionary Algorithm for Optimizing Parameters of GPGPU-based Image Segmentation. *Acta Polytechnica Hungarica*. 10(5).

Automatic Temperature Control System Using Arduino

Kyi Kyi Khaing, K. Srujan Raju, G. R. Sinha and Wit Yee Swe

Abstract Automatic temperature control system is an important application used in almost all modern gadgets and smart homes. The system for controlling temperature automatically is achieved by using Arduino Uno-based microcontroller system. Arduino Uno due to its increased popularity finds its varied range of applications. Temperature sensor LM35 and Arduino Uno are the hardware used interfaced with computer, and the temperature is controlled in the room. Temperature is displayed on LCD display employing A1 pin of hardware with the help of analog pin utilizing pulse width modulation (PWM). We have designed temperature control as an automatic system that has been not attempted before the way it has been implemented.

Keywords Temperature control · Arduino Uno · Temperature sensor · LCD display

1 Introduction

Temperature control becomes an important task in many of automatic operations. There are sensors, right from simple to smart sensors that are used for detecting the temperature. The environmental monitoring application, room temperature control are few of popular examples of temperature control. Now, with the advent of new technologies—hardware and software support—temperature can be controlled, monitored, and recorded more flexibly and with the programmable ways [1–4]. Information and communication technology (ICT) or smart appliances are using

K. K. Khaing (✉) · G. R. Sinha · W. Y. Swe
Myanmar Institute of Information Technology (MIIT), Mandalay, Myanmar
e-mail: kyi_kyi_khaing@miit.edu.mm

K. Srujan Raju
CMR Technical Campus Hyderabad, Hyderabad, India
e-mail: drksrujanraju@gmail.com

© Springer Nature Singapore Pte Ltd. 2020   219
K. S. Raju et al. (eds.), *Proceedings of the Third International Conference on Computational Intelligence and Informatics*, Advances in Intelligent Systems and Computing 1090, https://doi.org/10.1007/978-981-15-1480-7_18

some sort of temperature control; this may be artificial intelligence (AI)-based refrigerator or washing machine.

Microcontroller-based temperature control has become so important that it acts as benchmark for testing and simulation of particular sensors for detection and monitoring of temperature automatically. Various types of projects like minor projects as well as major projects are carried out on suitable hardware and software platform [4–9].

This paper presents an application of control theory using ICT and hardware-based temperature control including design of a circuit (hardware) and implementation and testing on Arduino Uno board. The test results are displayed with the help of LCD display. The program is written in Arduino IDE and facilitates the display of temperature in degree centigrade and also in Fahrenheit. The Arduino Uno board facilitates the temperature measurements input to the fan and cooling system ON/OFF that is automatically done based on varied values of temperature.

2 Related Research and Problem Identification

In the existing literature, there are many research papers that are temperature control but very few of them have used Arduino for automatic control of temperature, especially for monitoring applications. We studied several papers and here few of important contributions are presented.

Atilla et al. presented a case study that the design of heating system controlled by Arduino and has studied the technology, software, hardware used in the heating system, which consists of isolated box, dry resistance, voltage regulator, thermo-couple, air fan, microcontroller, and computer. Proportional–integral–derivative (PID), neural network, fuzzy logic is mainly used for the temperature control of heating systems. The system uses PID controller and exhibits satisfactory value of stability, good reliability, and sensitivity also. Microcontroller-based temperature control was designed with comparing theoretical values of temperature. However, Arduino control and implementation were not done [1].

Abdullah et al. suggested a design of temperature control system and implemented on TudungSaji microcontroller. Hardware implementation, as well as software simulation, was tested and obtained. The purpose of this work includes protection against bacteria after certain value of temperature. The application seems to be very good controlling and rather preventing from bacteria since after certain temperature, the bacteria can be killed. This could be also tested on Arduino IDE system [2]. Wayan et al. proposed the development and design of temperature distribution control for baby incubator application. In this system, it is a very important to maintain a certain temperature inside the room to take care of proper health of a baby. Humidity was also included in the study of experiment using microcontroller-based system for temperature measurement and control. This proved to be very important application for baby care and health [3].

Nagendra et al. presented a design and the implementation of Arduino-based temperature sensor that was also used to measure humidity level [4]. Kanishak et al. brought out a case study on temperature control systems using microcontroller, TRIAC, and bridge rectifiers [5]. Vaibhav et al. implemented speed control system based on change in temperature; for changing temperature measurement, the temperature control system was used. PWM and a simulation software were used to design the hardware and simulate on computer [6]. Theophilus et al. presented a testing of temperature monitoring mechanism with the help of Atmel Atmega 8385 system and LM35 temperature sensor [7].

Kiranmai et al. also proposed a temperature control system, and it was claimed that it is very useful for Internet of things (IoT)-related applications. However, the real-time application for any such application was not tested [8]. Singhala et al. studied a fuzzy-based temperature control system that was completely simulation-based, and no hardware implementation was achieved. The system suggested was very simple and effective but hardware implementation and realization remain as future scope of the work [9].

Muhammad [10] designed a PID controller and implemented on virtual laboratory platform, LabVIEW of National Instruments. It was suggested that Arduino-based hardware realization would produce much relevant and appropriate results for temperature control. Samil et al. employed the concept of PWM and displayed the values of temperature using LCD displays [11]. Masstor et al. suggested a case study for alarm system based on sensing temperature and humidity. Arduino controlled GSM/GPRS module was suggested, but temperature measurement was not accurate in the system [12]. Okpagu et al. suggested the development of temperature control system for egg incubator system utilizing sensors, PID controllers, LCD displays, DC motors, and fan control system. This is a very important type of incubator system because it becomes essential to monitor embryo and its growth, and therefore, temperature control and monitoring played an important role in this system [13].

Christina et al. presented a case study of light and humidity control including temperature control also. Light sensor, temperature sensor, and Arduino hardware interfaced with computer, and the work was implemented. Basically, the work was suggested for environmental monitoring for hospital application [14]. Merean et al. proposed a design and development of remote-controlled temperature monitoring system. Arduino-based work aims at providing a viable solution to the environmental monitoring and care [15].

Summarizing the current literature, temperature control that too automatic way of monitoring has not been attempted by the researchers mainly on temperature in focus of research. We suggest an Arduino control and hardware-based temperature control system, mainly highlighting temperature monitoring and measurement.

3 Hardware Implementation

Hardware implementation was obtained on Arduino IDE interfaced with P-IV computer. Data flow and block diagram of the hardware implementation are shown in Figs. 1 and 2, respectively.

Figures 1 and 2 are simple and self-explanatory where temperature sensors are connected with the help of Arduino and LCD display of 16 × 2 matrix. The fan was additionally connected for cooling mechanism so that automatic control could be achieved which is main objective of the proposed work. The hardware design is very simple without any circuit complexity (Fig. 3).

We used temperature sensor IC LM35 that helps in generating a small voltage for detecting the change in temperature across the temperature sensor.

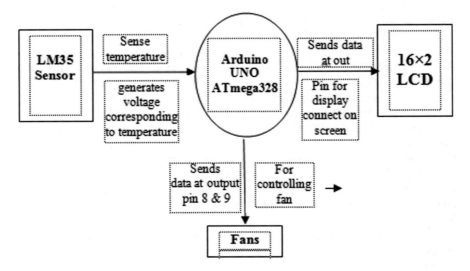

Fig. 1 Data flow in hardware implementation

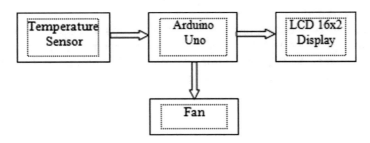

Fig. 2 Block diagram of temperature control hardware

Fig. 3 Flowchart of the
control system

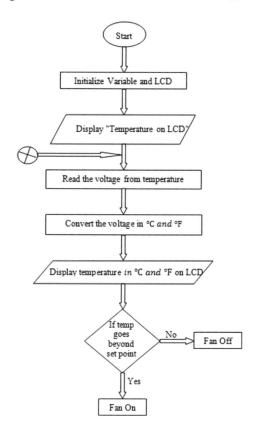

The generated voltage was continuous and analog signal generated through analog
pin of Arduino hardware. Arduino (ATmega328) controller was used, and the
voltage is taken as analog input in port 1 (A1) of Arduino Uno. The hardware reads
the analog signal, and it is then converted into suitable digital output with the help
of appropriate analog-to-digital converter (ADC) circuit that is inbuilt in it. ADC
has maximum capacity of 10-bits output generating from 0 to 1023 (1024
combinations).

Digital data corresponding to analog input received through the port is converted
and multiplied by a coefficient 0.488 just to normalize in centigrade. Similarly,
suitable multipliers such as 1.8 and afterward adding with 32 used to convert the
temperature measurement in Fahrenheit scale. The hardware sends the data to
16 × 2 LCD display which is connected with controller as shown in Fig. 4. Pin 1
and pin 2 are connected to ground and supply VCC, respectively, through Arduino
for activating or switching ON the LCD. Pin 3 enables through 10 K resister as
adjustment of brightness value of LCD display.

So, the display can be made on both the scales of temperature. A control bit,
either 0 or 1, is also sent by the Arduino to port 6 basically for providing control

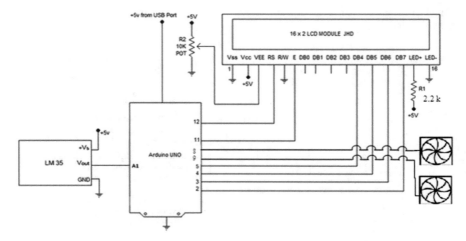

Fig. 4 Temperature control system

application. Pins 8 and 9 can be seen connected with fans just to adjust, normalize, and automatically control the temperate. Whenever temperature goes down after certain value, the fans will be OFF else it will keep running. This is how the control of temperature becomes automatic. Figure 3 shows the flowchart of the system.

4 Results

However, experimental setup was done and lot of temperature measurement was recorded with suitable displays. Few of sample displays and the observations are presented here with brief discussion.

Mainly, we have two outputs in this work: one for displaying the temperature automatically on LCD display and second was even important that is for automatic switching ON/OFF of fans so as to monitor the temperatures on automatic basis.

LCD display produces the output of temperature as well as the status of fans. For example, in Fig. 4, it can be seen that 28 degree centigrade is being displayed by the LCD display along with status of fan as ON. Actually, the running condition of fans depends on the threshold value set. We set 25 degree centigrade as the value, and therefore, it can be seen that for 28 degree centigrade fan is running (i.e., in ON condition). Obviously, for below, for example 23 degree centigrade, the fan will go OFF. The interface of display with bread board and Arduino hardware can be seen in Fig. 5.

Fig. 5 Display of a
temperature and status of fan
in automatic control of
temperature

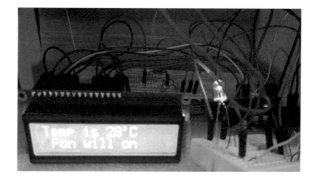

5 Conclusions

We have designed a simple method of temperature control system automatically. Utilizing the concept cooling after certain temperature, Arduino-based hardware along with display was realized in hardware. Few samples are shown in result, but any value of temperature can be generalized in this work. The work is focused mainly on temperature control, and no other parameter is involved. This seems to be robust way of handling only temperature control on automatic basis. This can be extremely useful for persons of physical disability. Soft computing method could be used to make it more robust and fuzzy controlled.

Acknowledgements The successful and final outcome of this paper is required a lot of guidance and assistance from my teacher. I respect and thank Mr. G.R. Sinha, Professor, Myanmar Institute of Information Technology (MIIT), Mandalay (Adjunct Professor, IIIT Bangalore, India).

References

1. Bayram, Atilla, Sulaiman Abdullah Moammed, and Fırat Kara. 2016. Design of Heating System Controlled by Arduino. In *4th International Symposium on Innovative Technologies in Engineering and Science*, 1583–1591.
2. Abdullah, Rina, Zairi Ismael, Rizman, Nik Nur Shaadah Nik, Dzulkefli, SyilaIzawana, Ismail, Rosmawati, Shafie, and Mohamad Huzaimy, Jusoh. 2016. Design an Automatic Temperature Control System for Smart Tudung Saji Using Arduino Microcontroller. *ARPN Journal of Engineering and Applied Sciences* 11(16):9578–9581.
3. Widhiada, W., D.N.K.P. Negara, and P.A. Suryawan. 2017. Temperature Distribution Control for Baby Incubator System Using Arduino ATMega 2560. Bali Indonesia 19 (20) part xv, 1748–1751.
4. Dangi, Nagendra. 2017. Monitoring Environmental Parameters: Humidity and Temperature Using Arduino Based Microcontroller and Sensors.
5. Kesarwani, Kanishak, S.M. Pranav, Tanish Nikhilesh Noah, and K.V.N. Kavitha. 2016. Design of Temperature Based Speed Control System Using Arduino Microcontroller 14 (S3): 753–760. www.sadgurupublications.com.

6. Bhatia, Vaibhav, and Gavish Bhatia. 2013. Room Temperature based Fan Speed Control System using Pulse Width Modulation Technique. *International Journal of Computer Applications (0975–8887)* 81 (5): 35–40.

7. Wellem, Theophilus, and Bhudi Setiawan. 2012. A Microcontroller—Based Room Temperature Monitoring System. *International Journal of Computer Applications* 53 (1): 7–10.

8. Nandagiri, Kiranmai, and Jhansi Rani Mettu. 2018. Implementation of Weather Monitoring System. *International Journal of Pure and Applied Mathematics* 118 (16): 477–494.

9. Singhala, P., D.N. Shah, and B. Patel. 2014. Temperature Control using Fuzzy Logic. *International Journal of Instrumentation and Control Systems (IJICS)*.

10. Muhammad Asraf, H., K.A. NurDalila, A.W. Muhammad Hakim, and R.H. Muhammad Faizzuan Hon. 2017. Development of Experimental Simulator via Arduino-based PID Temperature Control System using LabVIEW. *Journal of Telecommunication, Electronic and Computer Engineering.* 9 (1–5): 53–57.

11. Sami, K., M. Lavanya, M. Arivalagan and Yathrasi Sree Harsha. 2016. Temperature Controlled Based Cooler Pad Using Arduino. *Journal of Chemical and Pharmaceutical Sciences* 216–220.

12. Masstor, Norhidayah Binti. 2015. *Temperature Alert Alarm System (TAAS).*

13. Okpagu, P.E., and A.W. Nwosu. 2016. *European Journal of Engineering and Technology* 2 (7): 13–21.

14. Tan, Christina. 2010. Integrated Temperature, Light and Humidity Monitoring System.

15. Eltrubaly, Meream. 2016. *Remote Temperature Monitoring of Electronic Systems*, 1–29. Sweden: Blekinge Institute of Technology.

Performance Analysis of Fingerprint Recognition Using Machine Learning Algorithms

Akshay Velapure and Rajendra Talware

Abstract This work uses supervised machine learning algorithms to validate the performance of fingerprint recognition. We extracted fingerprint ridge contours, a level 3 feature, from low-resolution fingerprint images that are used for fingerprint recognition. The fingerprint classification is done using support vector machine and logistic regression classifiers. The database of color images taken from the Internet is split into training and test datasets. First, the images are enhanced using Gabor filter and wavelet. The ridge contours are extracted using canny edge detection filter. We create feature vectors from the extracted ridge contours. These features are used for the fingerprint recognition. The performance of the fingerprint recognition system is evaluated and analyzed using different machine learning algorithms. The Python 2.7 programming language using OpenCV, sklearn, Pickle, NumPy packages for image processing is used in this work.

Keywords Fingerprint ridge · Gabor · Wavelet · Ridge contours · Machine learning · Classifier

1 Introduction

Biometric systems using fingerprint recognition gaining popularity because of its uniqueness and permanence. Use of biometric authentication in banking, security systems, personalized computers, etc. is increasing these days. Biometric system can be spoofed easily; we need to provide anti-spoofing measure to prevent from fraudulent attacks. In biometric systems for person recognition using fingerprint, if there are less minutia points present then it is not feasible to use these level 2 features. Instead, we can utilize level 3 features which are dimensional attributes of fingerprint ridge having high discriminating power. Fingerprint level 3 features

A. Velapure (✉) · R. Talware
Department of E&TC Engineering, Vishwakarma Institute
of Information Technology, Pune, India
e-mail: akshay.velapure@gmail.com

© Springer Nature Singapore Pte Ltd. 2020
K. S. Raju et al. (eds.), *Proceedings of the Third International Conference on Computational Intelligence and Informatics*, Advances in Intelligent Systems and Computing 1090, https://doi.org/10.1007/978-981-15-1480-7_19

provide more security as these are finer details of the fingerprint. In this work, we extract fingerprint ridge contours. The ridge contour is outer edge of a ridge that includes width of ridge and edge shape contains valuable information of a finger-print. Ridge contours are more reliable features than pores. For pore detection, a very high-resolution fingerprint images are required typically of 1000 dpi whereas fingerprint ridge contours can be efficiently extracted from low-resolution images. An individual measurable property of our image data is a feature. A set of numerical features can be conveniently described by a feature vector. A machine learning algorithm is applied to generate a model. A model is learned from image data for specific representation. The value to be predicted by our model is called Label (or Target variable). The model is fed feature vectors as input. In training stage, a model is generated from a set of features given and its expected outputs that is Labels. Hence after training, we will have a model that will then map new data to one of the categories trained on. Once our model is trained, a set of new test inputs can be fed to it. This will provide a predicted output (Label). We are going to use supervised machine learning algorithms. In supervised learning process, a function is allotted to some desired category as learnt from supervised training data. The training data consist of a set of training examples where each set consists of a pair consisting of an input object and a desired output value. A supervised learning algorithm learns from this training pair relationship and produces an inferred function. Recognition performance can be analyzed using various machine learning classifiers such as SVM and logistic regression.

2   Previous Work

The authors in [1] have discussed the state of the art in biometrics recognition and have identified key challenges that deserve attention. They summarized the progress in biometric recognition in last 50 years so as to understand how this field emerged, where we are now and where we should go from here. The work in [2] introduces the basic concepts and methodologies for biometric recognition. As discussed in [3], authors introduced a method using wavelet transform and Gabor filters to extract fingerprint ridge contours. Fingerprint level 3 features utility and matching techniques are discussed in [4]. Application of DWT in images processing and wavelet families introduced in [5]. Use of SVM in fingerprint classification and application in pattern recognition have been discussed in [6, 7]. Fusion in bio-metrics using k-NN-based classifiers, decision trees and logistic regression in a multi-modal identity verification application discussed in [8, 9].

3 System Algorithm Flowchart

Figure 1 shows our fingerprint recognition system. We used 'Fingerprint color image database.v1' obtained from 'in.mathworks.com' [10]. We used samples of fingerprint color images of eight different persons. We have taken five samples of fingerprints for training and two samples kept for testing for each person from the fingerprint database. The images are acquired from the database, and further operations are performed.

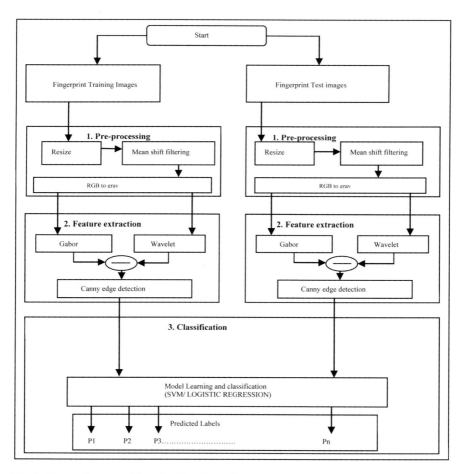

Fig. 1 Fingerprint recognition algorithm flow chart

3.1   Preprocessing

First, the input color image is converted to gray scale by RGB to gray conversion. To make all the images from the database of equal size, we perform image resizing where each image size is changed to [128*128]. Mean shift filtering is used for image segmentation by sharpening the boundaries of the ridges. This process distinguishes the fingerprint ridges from the valleys and ridge edges can be easily detected. Gabor filter is used for ridge enhancement. The equation of the filter is as follows:

$$G(x, y : \theta, f) = \exp\left\{ -\frac{1}{2} \left(\frac{x^2}{\delta_x^2} + \frac{y^2}{\delta_y^2} \right) \cos 2fx_\theta \right\} \qquad (1)$$

where θ is the orientation and f is the frequency of the Gabor filter and δ_x and δ_y are the standard deviations of the Gaussian envelope along the x-axes and y-axes, respectively. The position of a point (e.g., x, y) after it has undergone a clockwise rotation by an angle (e.g., $90° - \theta$) is represented by (x_θ, y_θ). The parameters of the Gabor filter (i.e., $\theta, f, \delta_x, \delta_y$) were determined empirically. The ridges are separated from the valleys after enhancement by removing the noise as shown in Fig. 2b. We have set $\theta = 5.25$.

3.2   Feature Extraction

Wavelet transform is useful in determining at what time interval what frequency bands are present that is they offer a simultaneous localization in time and frequency domain. The main advantage of wavelets is that it can separate the fine details in the image. The basis function of discrete wavelet can be given as:

$$\psi mn(t) = a_0^{-m/2} \psi\left(a_0^m t - nb_0\right) \qquad (2)$$

where 'a' is scale parameter, 'b' is shift parameter and $\Psi(t)$ is the mother wavelet function.

(a) Pre-processed   (b) Gabor Filtering   (c) Wavelet response   (d) (b) – (c)   (e) Ridge contours extraction by Canny
Image   edgedetection filter

Fig. 2   Feature extraction

The discrete wavelet transform for given input signal $x(t)$ is defined as,

$$W_x(m, n;\ \Psi) = a_0^{-m/2} \int x(t)\,\Psi^*(a_0^{-m}t - nb_0)\mathrm{d}t \tag{3}$$

Daubechies (db3) wavelet transform is applied on grayscale image to enhance the ridges by removing the noise presenting in the fingerprint ridges as shown in Fig. 2c. By subtracting wavelet response from the Gabor-enhanced image, fingerprint ridges are further enhanced as shown in Fig. 2d. The canny edge detection filter is applied on the enhanced image obtained in the previous step as shown in Fig. 2e. The algorithm for extraction of the ridge contours is as follows:

a. The image is enhanced using Gabor filter.
b. Apply wavelet transform to the fingerprint image for enhancement of ridge contours. The response of wavelet is subtracted from the Gabor response image hence ridge contours are further enhanced.
c. By applying canny edge detection filter on resulting image, we get ridge contours.

3.3 Machine Learning Model

Once we extract a salient feature from the image, it will be stored in the form of feature vector of 2D matrix wherein each value represents an image pixel at specific location. The training feature vector is given to the classifier and a model is trained. We create a machine learning model using this feature vector in Python programming using Pickle. In pickling process, the Python object is converted into a byte stream known as serializing. For storing a file or transporting data over the network in serialized format, pickling is required. Later, we can load this file to deserialize our model and use it to make new predictions. This process is conversely called unpickling. Any Python object can be converted to string representation and can be reconstructed the object from string representation as Pickle is able to take any object in Python and process it. Several objects can be pickled in same file. For each trained machine learning model, a file is generated in Pickle format.

3.4 Classification

We have used supervised machine learning algorithms for fingerprint classification in this work. From the given dataset, the supervised machine learning algorithm predicts the target variable. To map inputs to the desired outputs, we generate a function from these variables.

3.4.1 SVM

In SVMs, there are two-class problems that are used to determine the separating hyperplane with maximum distance to the closest points of the training set. These points are known as support vectors. If given data in the input space is not linearly separable, then a nonlinear transformation $\Phi(\cdot)$ may be applied. This maps the data points $x \in (R^\wedge n)$ into a high-dimensional space (H) known as feature space. The optimal hyperplane separates the data in the feature space as described above. In the SVM classifier, the mapping $\Phi(\cdot)$ is represented by a kernel function $K(\cdot,\cdot)$ which defines an inner product in H, i.e., $K(x, t) = \Phi(x) \cdot \Phi(t)$. The decision function of the SVM has the form:

$$f(x) = \sum_{i}^{l} a_i y_i K(x_i, x) \tag{4}$$

where $y_i \in \{-1, 1\}$ is the class label of training point x_j and l is the number of data points. The support vectors are the nearest points to the separating boundary.

3.4.2 Logistic Regression

Logistic regression is used for the following:

1. Find the impact of the dependent variables on the response based on the historical data.
2. To predict what can happen in the future using new cases, use generalization in step 1.

Linear regression is used when the response is a continuous variable and logistic regression is used when the response you want to predict/measure is categorical with two or more levels.

4 Results

The output images at each stage are shown in the following figures:

Figure 3 shows all the processes step by step. Figure 3a is input image from the database. Figure 3b is image after resizing. Figure 3c is the image after applying mean shift filter. The color image in Fig. 3c is converted to gray as shown in Fig. 3d. Figure 3e is the Gabor response of image Fig. 3d. The wavelet response of image Fig. 3d is shown in Fig. 3f. Figure 3g shows subtraction of Figs. 3e and 3f. Figure 3h shows extracted ridge contours by applying canny edge detection filter on Fig. 3g.

Fig. 3 Results of all
operations on fingerprint
image

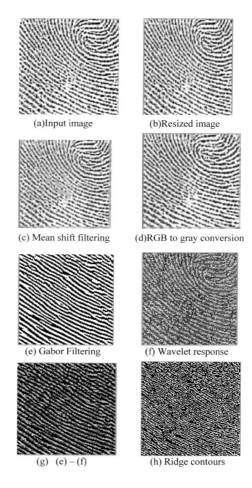

(a)Input image   (b)Resized image

(c) Mean shift filtering   (d)RGB to gray conversion

(e) Gabor Filtering   (f) Wavelet response

(g) (e) – (f)   (h) Ridge contours

Figure 4 shows feature vector of a fingerprint image labeled as Person 1. The values in the matrix indicate the pixel intensities at specific location in the image.

Figure 5 shows the results of the fingerprint recognition with fingerprint image of a person and its predicted label.

Table 1 describes recognition accuracy with utilized fingerprint feature for different classifiers.

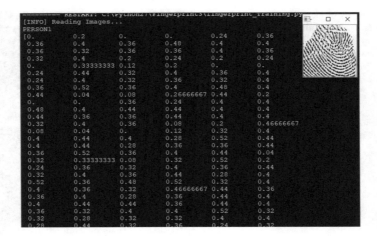

Fig. 4 Fingerprint feature vector

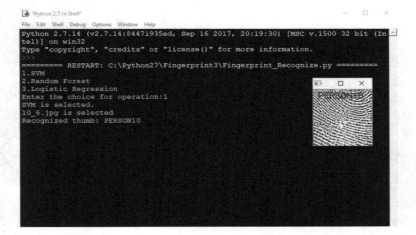

Fig. 5 Fingerprint recognition results

Table 1 Analysis of fingerprint recognition

Classifier	Feature	Recognition accuracy (%)
1. SVM	Ridge contours	87.5
2. Logistic regression	Ridge contours	68.75

5 Conclusion and Future Scope

The main motive of this work is to extract a level 3 feature of fingerprint and verify the performance of the fingerprint recognition system using different classifiers. The proposed algorithm using enhanced ridge contours gives better results. There are two phases in this work. The first phase is preprocessing and feature extraction. The feature extraction involves the Gabor filter and the wavelet. The second phase involves the training of the classifiers and performance analysis. The performance is analyzed using SVM and logistic regression classifiers. Forty fingerprint images were trained, five samples of each person. Out of 16 test images, wrongly predicted labels using SVM are 2 and using logistic regression are 5. Hence, recognition accuracy with SVM and logistic regression is 87.5% and 68.75%, respectively. All image processing operations are carried out in Python 2.7 environment as Python is an open-source software and various Python packages such as OpenCV, sklearn, NumPy, and Pickle are available free of cost. In future work, we may use other level 3 features like pores, incipient ridges, scars and analyze the performance of fingerprint recognition system using unsupervised machine learning and deep learning techniques.

Acknowledgements We would like to show our gratitude to Vishwakarma Institute Of Information Technology, Pune for showing their pearls of wisdom with us during the proposed work. Also thanks to 'Copyright (c) 2015, Ujwalla Gawande, Kamal Hajari and Yogesh Golhar, All rights reserved', for the fingerprint database.

References

1. Jain, Anil K., Nandakumar Karthik, and Ross Arun. 2016. 50 Years of Biometric Research: Accomplishments, Challenges, and Opportunities. *Pattern Recognition Letters* 79: 80–105. https://doi.org/10.1016/j.patrec.2015.12.013.
2. Jain, Anil K., Arun A. Ross, and Karthik, Nandakumar. 2011. Introduction to Biometrics. Berlin: Springer.
3. Jain, Anil K., Yi Chen, and Meltem, Demirkus. 2007. Level 3 Features for Fingerprint Matching. U.S. Patent Application No. 11/692,524.
4. Zhao, Qijun, Jianjiang Feng, and Anil K. Jain. 2010. Latent Fingerprint Matching: Utility of Level 3 Features. MSU Technical Report, vol. 8, 1–30.
5. Gupta, Dipalee, and Siddhartha Choubey. 2015. Discrete Wavelet Transform for Image Processing. *International Journal of Emerging Technology and Advanced Engineering* 4 (3): 598–602.
6. Yao, Yuan, Paolo Frasconi, and Massimiliano, Pontil. 2001. Fingerprint classification with combinations of support vector machines. In *International Conference on Audio-and Video-Based Biometric Person Authentication*. Berlin, Heidelberg: Springer.
7. Byun, Hyeran, and Seong-Whan Lee. 2002. Applications of Support Vector Machines for Pattern Recognition: A Survey. *Pattern Recognition with Support Vector Machines*, 213–236. Berlin, Heidelberg: Springer.

8. Kittler, Josef, et al. 1998. On Combining Classifiers. *IEEE Transactions on Pattern Analysis and Machine Intelligence* 20 (3): 226–239.

9. Verlinde, Patrick, and G. Cholet. 1999. Comparing decision fusion paradigms using k-NN based classifiers, decision trees and logistic regression in a multi-modal identity verification application. In *Proceedings International Conference Audio and Video-Based Biometric Person Authentication (AVBPA)*.

10. https://in.mathworks.com/matlabcentral/fileexchange/52507-fingerprint-color-image-database-v1.

A Survey on Digital Forensics Phases, Tools and Challenges

Sheena Mohammmed and R. Sridevi

Abstract The digital technologies are grown in such way that they are also leading to growth in digital crimes. The aim of digital forensics is to collect, analyze and present evidence related to digital crime and in front of court of law. There are several methods and tools in evidence collection and analysis. This paper gives a survey on digital forensic evidence collection and analysis. Recently, the cloud forensics has become very interesting area of research, as cloud computing is a collection of computer resources and services that can be easily implemented and managed, generally over the Internet. It also discusses about the challenges to be faced in evidence collection and its analysis.

Keywords Digital forensics · Evidence acquisition · Examination · Analysis · Reporting · Cloud forensics

1 Introduction

Digital forensics [1] plays a vital role in detecting, extracting and analyzing the digital evidence collected from a compromised computer and presenting the report in the court [2, 3]. Three important components like hard disk, memory and network in computer forensics are recorded and analyzed to track the behaviors of cyber criminals.

Several tools [4] are available to conduct investigations on computer crimes. These tools are also used to maintain, debug and recovery of computer system. These tools are designed for forensic investigators to help them in investigations. While performing evidence collection and analysis, a huge amount of information

S. Mohammmed (✉)
Vardhaman College of Engineering, Shamshabad, India
e-mail: sheenamd786@gmail.com

R. Sridevi
JNTUH College of Engineering, Hyderabad, India
e-mail: sridevirangu@jntuh.ac.in

© Springer Nature Singapore Pte Ltd. 2020
K. S. Raju et al. (eds.), *Proceedings of the Third International Conference on Computational Intelligence and Informatics*, Advances in Intelligent Systems and Computing 1090, https://doi.org/10.1007/978-981-15-1480-7_20

will be extracted. This big and low-level data is critical to examine and takes more amount of time to analyze. Hence, automated tools need to be developed to handle such type of data. This survey aims to review on various tools and challenges in digital forensic evidence collection and analysis.

2 Digital Forensics Phases

Digital forensics contains mainly four phases [5]: evidence acquisition, examination, analysis and reporting [6]. In addition to these four phases, there are two more phases like identification of crime and preservation of crime and evidence-related information. The digital forensics process can be shown in Fig. 1. The phases are listed as below.

1. **Identification**: In this, the digital crime is identified for which evidences must be collected.
2. **Preservation**: Crime-related information and each evidence once collected are preserved.
3. **Evidence acquisition**: Evidence is collected with proper approval from authority.

Fig. 1 Phases in digital forensics process

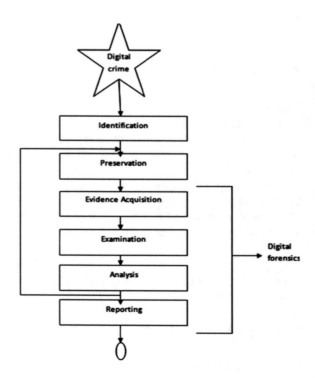

4. **Examination**: The collected evidence is examined to determine whether the identified components are relevant to the case which is being investigated or not.
5. **Analysis**: Determine the type of information which is stored on digital evidence components like hard disk, memory or network and perform a thorough analysis on media.
6. **Reporting**: Once the evidence acquisition, examination and analysis phases are completed for all the evidences, then prepare an official report which can be used to present in front of court of law.

3 Related Study

3.1 Forensic Investigation Models

There are various investigation models proposed such as

1. **Pollitt's Computer Forensic Investigative Process:** It is proposed in 1984, by Pollitt [7, 8]. According to him, there are four phases in investigation process like acquisition, identification, evaluation and admission as shown in Fig. 2,
2. **Digital Forensics Research Workshop (DFRWS) Investigation Model:** This model [9] is proposed in 2001 and consists of six phases like **Identification** of crime, **Preservation** of crime-related information; evidence **Collection**, **Examination** of collected evidence, **Analysis** and final **Presentation** of report (Fig. 3).

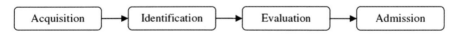

Fig. 2 Pollitt's computer forensic investigative process

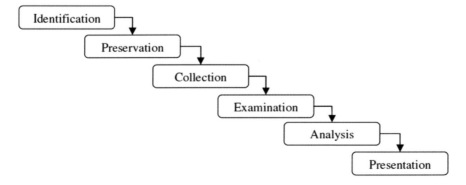

Fig. 3 DFRWS investigation model

3. **Abstract Digital Forensic Model:** This model is extension to DFRWS model which adds three additional phases like preparation, approach strategy and returning evidence to DFRWS model. It is proposed in 2002 by Reith, Carl and Gunsch [10]. The nine phases are shown in Fig. 4.
4. **Integrated Digital Investigation (IDIP) Process:** This model [11] is proposed in 2003 by Carrier and Spafford. It consists of five phases like **Readiness** which indicated the infrastructure must be ready for investigation at any time, **Deployment** which indicated the detection and confirmation of incident happening, **Physical Crime Investigation** which includes investigation of physical evidences, **Digital Crime Investigation** which includes investigation of digital evidence and finally **Review** of evidences investigated and report (Fig. 5).
5. **Enhanced Digital Investigation Process (EIDIP) Model:** This model [12] is proposed in 2004, which is extension to Integrated Digital Investigation Process

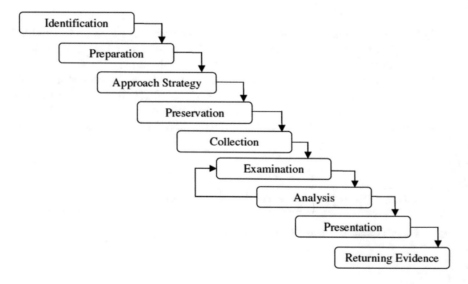

Fig. 4 Abstract digital forensic model

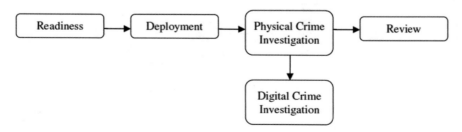

Fig. 5 IDIP model

model and proposed by Carrier and Strafford. It includes **Traceback** to track down the source crime, and Dynamite to conducted investigation at primary crime scene to identify potential culprit (Fig. 6).

6. **Digital Forensic Model based on Malaysian Investigation Process:** This model is proposed in 2009 by Perumal S, [13] which is based on Malaysian investigation process. It is shown in Fig. 7.

4   Tools Used in Digital Forensics

Once the evidence is exacted from data, it must be interpreted to determine the significance of the evidence in the case. There are some techniques of analysis like time frame, information hiding analysis and file and application analysis [14].

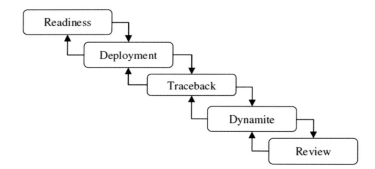

Fig. 6 EIDIP model

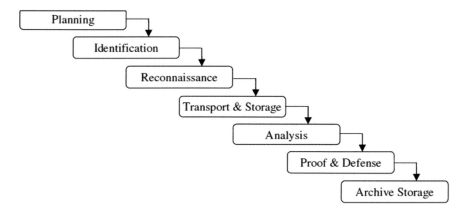

Fig. 7 DFMIP model

Time frame analysis deals with determining time of the event occurrences either by reviewing system and application logs or metadata of file system. Data or information hiding analysis is used to detect and recover data that indicates properties like ownership, purpose, etc., either by examining file headers and extensions. File and application analysis is used to identify what to be done next in analysis processes.

For better investigation, many computer forensic tools are developed [15].

1. **Disk and data capture tools** [16]: Allows the investigator to search the disk and capture data. Some of the disk and data capture tools are developed as given in Table 1.
2. **File viewer:** Allows an investigator to search the computer for different files like images, media, documents, data, etc. Some of the file viewers are listed in Table 2.
3. **File analysis tools:** These tools scan, analyze and report every detail about the file. Some of the tools are listed in Table 3.
4. **Registry analysis tools:** These tools [17] help to extract context and knowledge from a big unused data source and then identifying the context which is modifying contents of the registry. The examples of registry analysis tools are given in Table 4.
5. **Internet analysis tools:** These tools allow us to analyze the incoming and outgoing data to monitor or modify the request–response before they reach the browser. Some of the examples of Internet analysis tools are listed in Table 5.

Table 1 Data and disk capture tools

Name	Description
FAT32	It is used to format the large disks
Magnet RAM capture	Doubtable computer's physical memory can be captured
FTK imager	For mounting the image and viewing the disk
Dumpit	It is used in Windows machines to generate the dump of physical memory
EDD (Encrypted Disk Detector)	This tool is used for checking local drives for encrypted and locked volumes of the system
EWF MetaEditor	This tool is used for removing passwords and modifying metadata
AIM (Arsenal Image Mounter)	For mounting images of disk and to provide access to copies of Volume shadow on windows
Live RAM Capture	It is used for extracting even the protected dump of RAM
OSF Clone	AFF clones or images can be created by using this tool
OSF Mount	Several disk images can be mounted by using this tool
Disk2vhd	This tool is used to created different versions of virtual disks
EnCase forensic imager	It is used to create evidence and logical evidence files

Table 2 File viewer tools

Name	Description
File viewer plus	For opening, editing and converting over 300 file types
File viewer lite	For viewing any file on Windows PC
Free MDB viewer	For opening, viewing and reading MDB/ACCDB files without MS Access
PST file viewer	Outlook PST Viewer for opening PST file without outlook
CDR viewer	For searching, loading, opening and reading CorelDRAW CDR Files
BKF viewer	For viewing contents of BKF (XP backup) files
Microsoft PowerPoint 2007 viewer	For viewing PowerPoint presentations
Microsoft Visio 2010 viewer	For viewing Visio diagrams

Table 3 File analysis tools

Name	Description
eMule reader	For parsing and printing configuration and log files associated with the eMule P2P client
P2P marshal	For discovering and analyzing peer-to-peer files for Windows
Mobius forensic Toolkit	For scanning, retrieving and showing P2P activity information for Ares Galaxy and Shareaza

Table 4 Registry analysis tools

Name	Description
RegEdt	For searching capabilities but not modifying access control lists (ACLs) on registry keys on Windows 2000 and NT
RegsEdt32	Registry editing utility that does not allow you to import or export hive (. reg) files
Prefetch Analyser	This tool can read various prefetch files of different versions of Windows Systems
ForensicUserInfo	User information can be extracted and decrypted from the sequence alignment map files

6. **Email analysis tools:** These tools [18] are used to investigate different e-mail clients and extensions. Some of the examples of Email analysis tools are given in Table 6.
7. **Mobile devices analysis tools:** These tools [19] allow us to recover digital evidence or data from the mobile devices. Some examples are given in Table 7.
8. **Mac OS analysis tools:** These tools allow live data capture, targeted data collection and forensic imaging of Mac OS [20]. Some Mac tools are given in Table 8.

Table 5 Internet analysis tools

Name	Description
Cookie-Cutter	This tool can extract and show the contents of cookies of Google Analytics
Browser history capturer	It can capture history contents of various browsers
Browser history Viewer	It can capture as well as analyze history contents of Internet browsers
Chrome cache viewer	This tool can read and display the contents of the cache of Google Chrome
Chrome session parser	This tools is mainly used to parse the off-line sessions of Chrome such as current and last tabs, sessions etc.

Table 6 Email analysis tools

Name	Description
Mail viewer	This tool is used in email analysis for checking the databases and messages of various mail services
OST viewer	Without involving any exchange server, this tool can capture the OST files of outlook
EDB viewer	Using this tool, we can read EDB files in outlook with no server involvement
MailXaminer	Performs data collaboration and access
MBOX Viewer	This tool is mainly used to check the email contents and attachments of MBOX mails
PST Viewer	PST files can be read without outlook

Table 7 Mobile devices analysis tools

Name	Description
iPhone analyzer	This tool is used for exploring the structure of Internal files in iPhones, iPods, etc.
iPBA2	For Exploring iOS backups
ivMeta	This tool can extract the model of phone, version of software, etc.
SAFT	This tool is mainly used in Android devices to capture the contents line call logs, messages, contacts, etc.

9. **Network forensics tools:** These tools [21] allow us to monitor network to proactive in their security posture. Some network analysis tools are given in Table 9.

10. **Database forensic tools:** These tools [22] allow us to analyze the metadata of the database. Some examples are given in Table 10.

Table 8 MAC tools

Name	Description
Disk Arbitrator	By using this tool, we can block the file system mounting and arbitration of disk
IORegInfo	This tool can identify and locate the devices such as USB drives, RAID devices, etc., connected to the system
Audit	For Auditing Preference Pane and Log Reader for OS X It is used to check the log reader and pane preferences of Mac OS X
PMAP Info	It can be used to display and map the partitions of the drive

Table 9 Network analysis tools

Name	Description
NetworkMiner	It uses packet sniffing or PCAP parsing techniques to detect the ports and host names on the network
Nmap	It is a service provider to discover and audit the network security
Wireshark	Network protocol capturing service and analysis

Table 10 Database forensic tools

Name	Description
Digital forensics framework	This database analysis tool detects the metadata, file systems, hidden data, etc.
SQLite manager	This tools acts as an extension to Firefox browser for checking databases of SQLite

5 Challenges in Evidence Collection and Analysis

Digital forensics is used to identify computer crimes. There are major challenges in these investigations. There are different categories of challenges as listed below [23].

5.1 Technological Challenges

Digital forensic evidence collection faces several technical issues. The major challenge is how to protect the evidence from modifications as it is easy to modify digital evidence than the physical evidence. There are some anti-forensic [24, 25] techniques available which became a big challenge for investigator. The following are some of the anti-forensic tools available.

a. **Encryption:** It makes the evidence unreadable as the evidence is converted to other format. The investigators must decrypt the encrypted evidence to analyze it.
b. **Steganography:** It makes the evidence to be hidden inside other cover file. Investigators have to identify the hidden data to analyze it.
c. **Secret channel:** It allows the evidence to be hidden over network to bypass intrusion detection techniques.
d. **Data hiding in storage space:** It allows the attacker to store the data in the storage space such that it is not visible by regular commands. It makes investigation more complex and more time consuming.
e. **Tail obfuscation:** In this, attacker uses some false information like false file extensions, false headers, etc., to mislead the investigators.
f. **Residual data wiping:** The attackers use the computer for his target task and avoid the risk of hidden processes like temporary files, history of commands by erasing out the tracks made by the process.

5.2 Lawful Challenges

Investigator has to investigate the crime without risking the privacy of organization or victim which is a big challenge for the investigator [26]. The information collected should be stored carefully and legally.

5.3 Resource Challenges

Based on the scenario, a huge amount of data may need to be investigated which may take more time. The investigator can consider only the recent information such as information which is ephemeral, recent user activities which can be available on memory. While collecting the data from the source, the investigator must make sure that the data is not modified or missed.

5.4 Challenges in Cloud Forensics

As the cloud computing is easy to maintain and provides services over Internet, the criminal commit and wiping of evidence are very easy. Hence, the cloud forensics has become a major interest area in research. In some cases, the cloud itself is used as the main tool to perform crime other services hosted by cloud will become the target [27]. In such situation, to identify the crime, the important evidences will be the compromise indicators, abnormal activities on the network.

6 Conclusions

This paper gives a survey on digital forensics. In Sect. 2 and 3, it discussed about various phases involved in digital forensics process. In Sect. 4, it focused on various tools used in digital forensics evidence collection and analysis. In Sect. 5, it gave a brief on challenges to be faced while evidence collection and also it gave some brief review on techniques used in evidence analysis and how cloud is vulnerable to digital crimes.

References

1. Ali, Khidir M. 2012. Digital Forensics Best Practices and Managerial Implications. In *Fourth International Conference on Computational Intelligence, Communication Systems and Networks*, IEEE.
2. Daniel, Larry, Lars Daniel. 2011. *Digital Forensics for Legal Professionals, Understanding Digital Evidence from the Warrant to the Courtroom*. Syngress Publishing.
3. Ieong, Ricci S.C. 2006. FORZA—Digital Forensics Investigation Framework That Incorporate Legal Issues. 3: 29–36.
4. 22 Popular Computer Forensics Tools: https://resources.infosecinstitute.com/computer-forensics-tools/#gref.InfosecResources. 26 Mar 2018.
5. Yusoff, Yunus, Roslan, Ismail, and Zainuddin, Hassan. 2011. Common Phases of Computer Forensics Investigation Models. *International Journal of Computer Science & Information Technology (IJCSIT)* 3 (3).
6. Varol, Asaf, Yeşim Ülgen Sönmez. 2017. Review of Evidence Analysis and Reporting Phases in Digital Forensics Process. In *International Conference on Computer Science and Engineering (UBMK)*, IEEE.
7. Pollitt, M.M. 1995. Computer Forensics: An Approach to Evidence in Cyberspace. In *Proceeding of the National Information Systems Security Conference*. 2: 487–491, Baltimore, MD.
8. Pollitt, M.M. 2007. An Ad Hoc Review of Digital Forensic Models. In *Proceeding of the Second International Workshop on Systematic Approaches to Digital Forensic Engineering (SADFE'07)*. Washington, USA.
9. Palmer, G. 2001. DTR-T001-01 Technical Report. A Road Map for Digital Forensic Research. In *Digital Forensics Workshop (DFRWS)*, Utica, NY.
10. Reith, M., C. Carr, and G. Gunsh. 2002. An Examination of Digital Forensics Models. *International Journal of Digital Evidence* 1 (3): 1–2.
11. Carrier, B., and E.H. Spafford. 2003. Getting Physical with the Digital Investigation Process. *International Journal of Digital Evidence* 2 (2): 1–20.
12. Baryamereeba, V. and F. Tushabe. 2004. The Enhanced Digital Investigation Process Model. In *Proceeding of Digital Forensic Research Workshop*, Baltimore, MD.
13. Rogers, M.K., J. Goldman, R. Mislan, T. Wedge and S. Debrota. 2006. Computer Forensic Field Triage Process Model. *Journal of Digital Forensics, Security and Law* 27–40.
14. Forensic Examination of Digital Evidence: A Guide for Law Enforcement. 2004. Available online at: https://www.ncjrs.gov/pdffiles1/nij/199408.pdf.
15. Patil, Priya S., A. S., Kapse. 2018. Survey on Different Phases of Digital Forensics Investigation Models. *International Journal of Innovative Research in Computer and Communication Engineering*.
16. Cyber Secure India. http://www.cybersecureindia.in/cybergallery/disk-tools-data-capture/.

17. What-When-How, in Depth Tutorials. http://what-when-how.com/windows-forensic-analysis/registry-analysis-windows-forensic-analysis-part-1.
18. Best Forensic Email analysis software. https://www.thetoptens.com/best-forensic-email-analysis-software/.
19. Common Mobile Forensics tools and Techniques. https://resources.infosecinstitute.com/category/computerforensics/introduction/mobile-forensics/common-mobile-forensics-tools-and-techniques/#gref. Infosec Resources. 2018.
20. Hawkings, Peter. 2002. Macintosh Forensic Analysis Using OS X. SANS Institute Reading Room site.
21. Sira, Rommel. 2003. Network Forensics Analysis Tools: An Overview of an Emerging Technology. GSEC. Available online from: https://www.giac.org/paper/gsec/2478/network-forensics-analysis-tools-overview-emerging-technology/104303. SANS Institute.
22. Cankaya, Ebru Celikel, and Brad Kupka. 2016. A Survey of Digital Forensics Tools for Database Extraction, In *2016 Future Technologies Conference (FTC)*, IEEE.
23. Fahdi, M.L., N.L., Clarke, S.M., Furnell. 2013. Challenges to Digital Forensics: A Survey of Researchers & Practitioners Attitudes and Opinions. [Online]. P1. Available from: http://ieeexplore.ieee.org/stamp/stamp.jsp?tp=&arnumber=6641058.
24. Conlan, K., I. Baggili, and F. Breitinger. 2016. Anti-forensics: Furthering Digital Forensic Science Through a New Extended Granular Taxonomy. *Digital Investigation* 18: S66–S75.
25. Rekhis, S., N., Boudriga. 2010. Formal Digital Investigation of Anti-forensic Attacks. [Online]. P34. Available from: http://ieeexplore.ieee.org/stamp/stamp.jsp?tp=&arnumber=5491959.
26. Bui, S., Enyeart, M., and Luong, J. Issues in Computer Forensics. [Online]. P 7. Available from: http://www.cse.scu.edu/~jholliday/COEN150sp03/projects/Forensic%20Investigation.pdf.
27. Pichan, Ameer. 2015. Cloud Forensics: Technical Challenges, Solutions and Comparative Analysis. *Digital Investigation*. 13: 38–57.

A Brief Survey on Blockchain Technology

Vemula Harish and R. Sridevi

Abstract Blockchain technology is based on decentralized architecture. It allows peers to perform transactions without relying on the centralized server or third party but still ensures the privacy and security of users, which is grabbing the attention of many researchers and academicians. It includes cryptographically secured hash functions, public ledgers, consensus algorithms to ensure reliability of transactions and mining algorithms for validating the transactions. This survey is mainly focused on overview of blockchain architecture, basic concepts, applications and Proof of Work, Proof of Stake consensus algorithms.

Keywords Blockchain · Bitcoin · Consensus · Cryptocurrency

1 Introduction

Cryptocurrency is global currency which is grabbing the attention of academicians and researchers. Examples for cryptocurrencies are such as Bitcoin [1], Litecoin, Peercoin, Blackcoin [2], Ripple [3], Namecoin [4], Ethereum [5], Auroracoin, Permacoin, Decred, Dash [6]. Block is an immutable peer-to-peer system. It is based on the decentralized architecture and ensures the privacy and security. The transactions are validated by the peers involved in the network called miners. No centralized authority is responsible for authentication. All transactions in blockchain are tamper proof; transactions are committed using a public ledger, i.e., everyone can view the transactions that are happening over the network. Hence, a peer cannot claim that the transaction is false because public ledger is available to every peer in the network. For example, Ramu wants to send some bitcoins to

V. Harish (✉)
Vardhaman College of Engineering, Shamshabad, India
e-mail: vemula.harish31@gmail.com

R. Sridevi
JNTUH College of Engineering, Hyderabad, Telangana, India
e-mail: sridevirangu@jntuh.ac.in

© Springer Nature Singapore Pte Ltd. 2020
K. S. Raju et al. (eds.), *Proceedings of the Third International Conference on Computational Intelligence and Informatics*, Advances in Intelligent Systems and Computing 1090, https://doi.org/10.1007/978-981-15-1480-7_21

Suresh, available balance in Ramu and Suresh accounts are 100 Bitcoins, 75 Bitcoins, respectively. When Ramu transferred 30 Bitcoins to Suresh, the transaction is validated by miners; transaction is added to block. Now, the available balance in Ramu account is updated to 70, Suresh is 105 Bitcoins.

Miners are the peers who can perform verification and validation of transactions for which they get incentives; few mining algorithms require high computational power and few are needed memory. Every transaction is available to every other peer in the network. But the user addresses are encrypted using cryptographically secured hash functions. No peer in the network has the supreme privileges to control the transactions, i.e., a peer cannot alter or modify the transactions once they are committed. Because every block added to the network is secured by hash keys. Hash keys are generated based on the hash key of its previous blocks in the network. A special data structure is used in blockchain technology known as Merkle tree. Merkle root is constructed based on its two children say left, right. The hash key of left (L_{hk}) and right (R_{hk}) are used to determine the hash key of root (RO_{hk}).

The rest of this paper is arranged as follows: Sect. 2 explains about blockchain architecture, transaction processing in blockchain, In Sect. 3, we discussed types of blockchains with their comparison of attributes, Sect. 4 is about consensus in blockchain includes byzantine node problem, PoW, PoS, Sect. 5 is about most popular cryptocurrencies, and Sect. 6 includes tools and applications in blockchain technology.

2 Blockchain Architecture

Blockchain architecture includes several components such as client, transaction, block, and miner. Client is any node in the network which initiates a transaction, transaction is a process of exchanging currency, assets, etc., using bolckchain technology, and block is a set of transactions represented with a single header. Header includes previous hash, Nonce, timestamp, Merkle root [7]. The block is shown in Fig. 1.

Block can be divided into two half's known as block header and transactions. Header contains hash of its previous block, Nonce which is a random number, Merkle root of a special data structure called Merkle tree used in blockchain, time stamp. Here, hash key is generated by hash function. Hash function takes any sized data as input and can produce a fixed-sized output. It is also known as irreversible encryption technique. Hash function exhibits avalanche effect, i.e., a small change in input makes a significant change in the output.

The transaction processing system of blockchain is shown in Fig. 2.

Fig. 1 A block in blockchain

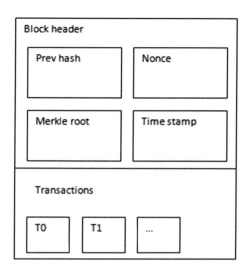

Initially, the client requests for a transaction in blockchain to execute some tasks such as exchanging currency, assets, etc., accordingly a new block is created along with several other transactions happening over the network within certain time period. Then the block is validated by the network participants known as miners. Further, the block is added to existing long chain of the blocks. The added block of transactions is available in blockchain as open public ledger and the transaction is said to be committed. In blockchain, once the transactions are committed, they can never be modified at any cost and no entity has the permission to change the committed transactions. All transactions in blockchain are tampered proof, i.e., it is impractical to alter the transactions.

Public Ledgers: Public ledger in a blockchain is a storage system; it is used to maintain the transactions that are happening over the network. It keeps track of user identities in encrypted format. The transactions that are successful from consensus procedure only included into the public ledger, public ledger is updated at every time a new block is mined and added to the network. As its name indicates, ledger is an account balance of the network participant which is openly available to all other peers in the network.

3 Types of Blockchain

Initially, the bitcoin blockchain is used to perform all kinds of transactions. But later, the private institutions realized and created permission blockchains where miners must be the part of their consortium or legally related to the same organization [8, 9]. Mainly blockchain can be categorized into three; they are (a) Public Blockchain, (b) Consortium Blockchain, and (c) Private Blockchain.

Fig. 2 Transaction
processing

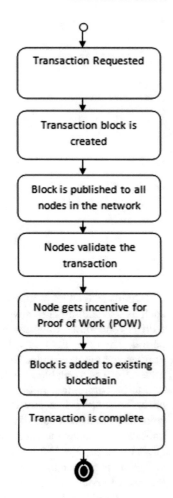

3.1   Public Blockchain

Public blockchains are open source and no one needs permission to participate in it.
The source code is available for free so that anyone can download and use it. Public
blockchain uses Proof of Work (POW)-based algorithms for consensus. Everyone
has access to view the transactions that are happening over the network but the
transactions are anonymous. Examples: Bitcoin, Corda, Ethereum, etc.

3.2 Consortium Blockchain

These are highly scalable and provide high security than public blockchains. Consortium blockchains do not allow everyone to validate the transactions; instead, a group of nodes controls the process of consensus. These kinds of blockchains may restrict the transactions to public participants from viewing. Examples: EWF (Energy), B3i.

3.3 Private Blockchain

In private blockchains, the write permissions are owned by centralized organization; read permissions can be public or may be restricted. Private blockchains exhibit the threat of security breach as similar to the centralized networks. **Examples**: Multichain, MONAX (Table 1).

4 Consensus in Blockchain

Consensus is an agreement to make a decision when required. Consensus algorithms are responsible for authentication, nonrepudiation, integrity, Byzantine fault tolerance, and performance. The most popular consensus algorithms are known as Proof of Work (POW) and Proof of Stake algorithm (PoS) [10]. These algorithms are discussed in the following sessions of the paper.

Table 1 Comparison of various blockchain properties

Property	Public blockchain	Consortium blockchain	Private blockchain
Access	No permission required to read/write	Permission required to read/write	Permission required to read/write
Mining	Every participant can perform mining	Only set of nodes can perform mining	It is specific to one organization
Speed	Performs slower	Performs faster	Performs faster
Immutability	Tamper proof	Not tamper proof	Not tamper proof
Identity	Hidden node addresses	Known node addresses	Known node addresses
Centralized	Decentralized	Partially centralized	Centralized

4.1 Byzantine Node Problem

Byzantine node is a malicious node in distributed network. If a fraudulent individual node is trying to malfunction the network by making false and ambiguous
decisions, then consensus kind of agreement among several nodes is needed to
overcome the problem. According to [11] "The Byzantine Generals Problem" is the
best example for depicting consensus problem, i.e., there are n generals to make
independent decisions to attack a fort. They have to take a decision to either attack
or to retreat; the generals are at far distances, they can only communicate through
messages, they have to take a collective decision, and if one of them makes
ambiguous decision, it leads to a disastrous situation. For example, we say there are
four generals knowing as G1, G2, G3, G4. Assume that G3 is trying to make
ambiguous decision, trying to say "attack contacted to G1"; "retreat contacted to
G4." G2' s decision is to "attack" contacted to G4 and G1. Here, in this situation,
G4 is in confusion because "G3 said to retreat," "G1 said to attack."

4.2 Proof of Work (POW) Consensus

According to Satoshi Nakamoto [1], the Proof of Work system uses SHA-256
algorithm for scanning a hash value begins with number of zero (0) bits; the average
work required can be verified by using a single hash. POW system works based on
the longest chain principle, i.e., in blockchain technology, every block is depended
on its previous blockchain hash values. When new block is to be added, it has to be
added at the longest chain available at that time stamp. PoW solves the consensus
problem by using majority decision, based on one-CPU-one-vote policy. The
attacker who wants to modify the block has to alter all the subsequent blocks, which
is impractical as the complexity grows exponentially.

4.3 Proof of Stake Consensus

Proof of Stake is nothing but proof of ownership; it is similar to Proof of Work
algorithm proposed by Satoshi Nakamoto and is used in minting and security model
of peer-to-peer systems; the concept of coinage is main idea behind the Proof of
Stake (PoS) algorithm design. As similar to POW, the PoS plays a significant role
in minting and the security model of the peer-to-peer system. Coinage is nothing
but currency holding time period; it is used to give priorities for transactions [12].
Example: if Suresh received 20 Bitcoins and held them for 20 days, then it is said to
be Suresh accumulated 400 coin days of coinage. Coinage can be calculated using
block time stamp and transaction time stamp.

5 Cryptocurrencies

It is a virtual currency valid anywhere in the world. There are various cryptocurrencies available as discussed in Sect. 1. The most popular cryptocurrencies are: (1) Bitcoin BTC [1], (2) Ethereum ETH [5], (3) Ripple [3] XRP, (4) Peercoin (PPC) [13].

5.1 Bitcoin (BTC)

Bitcoin [1] is peer-to-peer electronic cash; it is a virtual currency valid over the globe. The current value of the bitcoin is 1 BTC = 4, 83, 06.14 INR. BTC can be represented using eight decimal points. Tax is not applicable to BTC because of its decentralized nature. The Proof of Work (POW) algorithm used for BTC is Hashcash [14].

5.2 Ethereum (ETH)

Ethereum [5] resembles a state machine where a valid state transition comes from a series of transactions. A block is a set of transactions chained using cryptographically secured hashes. Previous block hash is used to construct the current block; network participants called miners validate the transactions, get incentives for validation, and the procedure is called mining. The state transition is represented as the following. The built-in currency for Ethereum is Ether which is also known as ETH. ETH has sub-denominations such as 10^0 is "Wei," 10^{12} is "Szabo," 10^{15} is "Finney," and 10^{18} is "Ether."

5.3 Ripple (XRP)

Ripple [3] is most popular consensus protocol of cryptocurrencies, intended for providing a direct service to exchange goods, currency, etc. Concept of mining is not used for consensus; instead, it uses trust-based system. Consensus involves agreement, correctness, and utility properties in decentralized environment. Ripple protocol contains several components such as Ripple Server, Ledger, Last Closed Ledger, Open Ledger, Unique Node List, and Proposer.

5.4 Peercoin (PPC)

Peercoin [13] is also known as peer-to-peer coin (PPC) which is similar to bitcoin by Satoshi Nakamoto. Proof of Stake (POS) is used for enhancing the security, minting process of the cryptocurrency. It uses coinstake block which is similar to coinbase used for BTC based on Proof of Work (POW) algorithm. POS uses limited space for performing hash operation.

Apart from the above mentioned, there are many other cryptocurrencies available such as Namecoin (NMC) [4], Permacoin [15], Blackcoin (BLK) [16], Auroracoin (AUR) [17], etc.

6   Tools and Applications in Blockchain

6.1   Smart Contracts

Smart contracts are self-executing system executable programs to manage the legal online transactions between network participants. A code is written as an agreement between the buyer and seller. Blockchain technology stores these smart contracts as decentralized distributed ledgers; these contracts are provided by Ethereum using Ether which is a virtual currency. Smart contracts are secure in nature and provide a trusted functionality to users. Currently, they are used in exchanging the assets like bonds, stocks, etc. [18, 19].

Smart contracts are very much useful in voting-based decision making, crowd-funding, and workflow management. These are in essence to provide services to open audience without human intervention in decision making. Blockchain still needed to design and implement these contracts very carefully as they hold million US dollars; miner deviation in automated code execution may lead to a disastrous situation which cannot be tolerated easily. EVM is a stack-based virtual machine runs on bytecode used to execute smart contracts; code is executed in distributed manner to all the peers in the network so that every peer can validate the transaction.

6.2   Blockchain for Digital Identity

In recent times, digital identities of individuals are at risk because of centralized architecture. For example, Unique Identification Authority of India (UIDAI) is maintaining the personal identities of all people over the nation. It is vulnerable to security threats at centralized server; in this case, we can apply the concepts of blockchain for decentralizing the network. So that, no single authority owns the data and no third party can act maliciously. Mainly the identity of network participants has two issues like access control and personal information identity.

Table 2 Tools used in blockchain technology [21]

Name	Description
Blockchain Testnet	It is a system which is alternative for original blockchain for developing decentralized applications (Dapps)
BaaS by Microsoft	Blockchain as a service is created by Microsoft; it is used to develop Dapps supporting multiple chains. Ex: Eris, Storj, and Augur
Mist	Mist is used to develop Ethereum such as to deploy smart contracts, maintain Ether transactions. Ether is considered as a fuel to Ethereum
Coinbase's API	Used to develop bitcoin applications, bitcoin wallets, addresses and it provides SDKs, libraries useful for development.
Tierion	Tierion tool that offers application programming interfaces; tools create database on bitcoin blockchain
Embark	Embark is a framework for Ethereum decentralized applications, to create smart contracts. Embark can modify application based on the modifications in smart contracts
Ether Scripter	Ether scripter tool uses serpent programming language for writing the smart contract code
Solc	Solc stands for Solidity compiler which is used for programming on Ethereum chain; we can use web3.eth.compile.solidity for compiling Solidity files

6.3 Blockchain-Based Voting System

The blockchain technology can add a great transparency in voting system. Using cryptographically secured hash functions, it can enable user privacy. Public ledgers will ensure that the data is available to all network participants transparently. Consensus algorithms can valid the voting transactions and their blocks [20].

6.4 Various Tools Used in Blockchain

See Table 2.

7 Conclusions

The blockchain is a technology in recent years having a significant impact in several industries. Because of its decentralized nature, it is scalable to a large extent and the concepts used in its literature can be applied for many other applications such as to make secure, private, simple, and efficient solutions. In this article, we have given the importance, application areas of blockchain, and several consensus algorithms available for public blockchain.

References

1. S. Nakamoto. 2008. *Bitcoin: A Peer to Peer Electronic Cash System.* [online]. Available https://bitcoin.org/bitcoin.pdf. Accessed Feb 5, 2018.
2. Vasin, Pavel. 2014. Blackcoin's proof-of-stake protocol v2 [online]. Available https://blackcoin.org/blackcoin-pos-protocol-v2-whitepaper.pdf.
3. Schwartz, David, Noah Youngs, and Arthur Britto. 2014. The ripple protocol consensus algorithm, p. 5. Ripple Labs Inc White Paper.
4. Kalodner, Harry, Miles Carlsten, Paul Ellenbogen, Joseph Bonneau, and Aravind Narayanan. 2015. *An Empirical Study of Namecoin and Lessons for Decentralized Namespace Design.* Citeseer: Technical Report.
5. Ethereum, Gavin Wood. 2014. A Secure Decentralised Generalized Transaction Ledger. Ethereum Project Yellow Paper.
6. Mukhopadhyay, Ujan, Anthony Skjellum, Oluwakemi Hambolu, Jon Oakley, and Lu Yu. 2016. Richard Brooks: A Brief Survey of Cryptocurrency Systems. In 14th Annual Conference on Privacy, Security and Trust (PST), Auckland, New Zealand, Dec 12–14, 2016.
7. Blockchain Architecture: [online]. Available https://www.pluralsight.com/guides/blockchain-architecture. October 11, 2017.
8. Voshmgir, Shermin. 2016. *Blockchains & Distributed Ledger Technologies.* [online] Available https://blockchainhub.net/blockchains-and-distributed-ledger-technologies-in-general.
9. Zheng, Zibin, Shaoan Xie, Hong-Ning Dai, Xiangping Chen, and Huaimin Wang. 2017. *Blockchain Challenges and Opportunities: A Survey.* [online] Available https://www.henrylab.net/wpcontent/uploads/2017/10/blockchain.pdf.
10. Seibold, Sigrid. 2016. Samman: Consensus Immutable Agreement for the Internet of Value.
11. Lamport, L., R. Shostak, and M. Pease. 1982. The Byzantine Generals Problem. *ACM Transactions on Programming Languages and Systems* 4 (3): 382–401.
12. S. King, and S. Nadal. 2012. PPCoin: Peer to Peer Crypto-currency with Proof of Stake, 2012. [Online]. Available: https://peercoin.net/assets/paper/peercoin-paper.pdf. Accessed Feb 4, 2018.
13. King, Sunny, and Scott Nadal. Ppcoin: Peer-to-Peer Crypto-Currency with Proof-of-Stake. Self-Published Paper, Aug 19, 2012.
14. Back, Adam. 2003. The Hashcash Proof-of-Work Function. Draft-Hashcash-back-00, Internet-Draft Created, June 2003.
15. Miller, Andrew, Ari Juels, Elaine Shi, Bryan Parno, and Jonathan Katz. 2014. Permacoin: Repurposing Bitcoin Work for Data Preservation. In Security and Privacy (SP).
16. IEEE Symposium on IEEE, 475–490 (2014).
17. Vasin, Pavel. Blackcoin's Proof-of-Stake Protocol v2.
18. Chen, Yi-Hui, Shih-hsin Chen, Iuon-Chang Lin. 2018. Blockchain Based Smart Contract for Bidding system. In IEEE International Conference on Applied System Innovation 2018 IEEE (ICASI), ed. Meen, Prior and Lam.
19. Frantz, Christopher K., Mariusz Nowostawski. 2016. From Institutions to Code: Towards Automated Generation of Smart Contracts. In IEEE International Workshops on Foundations and Applications of Self Systems, 210–215. IEEE.
20. Yavuz, Emre, Ali Kaan Koç, Umut Can Çabuk, and Gökhan Dalkılıç. 2018. Towards Secure e-Voting Using Ethereum Blockchain. In 6th International Symposium on Digital Forensic and Security (ISDFS).
21. Ameliatomasicchio. 2018. *The Best Blockchain Developer Tools.* [Online]. https://blockgeeks.com/-blockchain-developer-tools.

Energy-Efficient based Multi-path Routing with Static Sink using Fuzzy Logic Controller in Wireless Sensor Networks

Subba Reddy Chavva and Ravi Sankar Sangam

Abstract In Wireless Sensor Networks (WSNs), routing protocols manage the communication between wireless sensor nodes. In a multi-path routing protocol (MRP), data can be transferred from source to sink node in a reliable manner based on Received Signal Strength Indication (RSSI) value of a sensor node. This paper improves the life time of WSN by applying Fuzzy Logic Controller (FLC) to construct multi-path routing protocol based on distance, node density and residual energy parameters of wireless sensor nodes. We show the efficacy of our proposed routing protocol by using simulations.

Keywords Energy-Efficient routing · Fuzzy logic controller · Multi-path routing · Sink node · Wireless sensor networks

1 Introduction

Typically, a huge number of sensors are deployed in Wireless Sensor Networks (WSNs). Based on the sensing type of a sensor node, each can sense various actions like vibration, temperature, pressure, etc. These sensor nodes send data directly or indirectly to the base station (sink node). The architecture of a typical WSN is shown in Fig. 1. Due to recent advancements in IoT, WSNs have enormous number of applications like surveillance, agriculture, health care, etc. [1, 2]. The sensor nodes can be deployed in both structural and un-structural networks. The sensor nodes are deployed in a pre-planned manner in structured-based networks. Whereas in unstructured based networks, the sensor nodes are thrown from planes and

S. R. Chavva (✉) · R. S. Sangam
School of Computer Science and Engineering, VIT-AP University,
Amaravati 522237, Andhra Pradesh, India
e-mail: chavvasubbareddy@gmail.com

R. S. Sangam
e-mail: srskar@gmail.com

© Springer Nature Singapore Pte Ltd. 2020

259

K. S. Raju et al. (eds.), *Proceedings of the Third International Conference on Computational Intelligence and Informatics*, Advances in Intelligent Systems and Computing 1090, https://doi.org/10.1007/978-981-15-1480-7_22

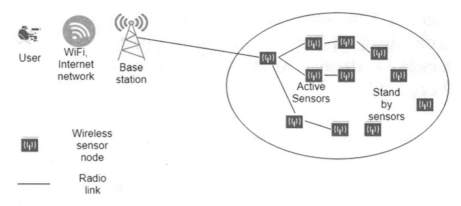

Fig. 1 Architecture of wireless sensor networks

helicopters in a distributed manner [3]. In unstructured networks, replacing batteries to these sensor nodes is very difficult and is of the high cost.

Each sensor has equipped with fixed-limited energy, processing capacity, transmission power and storage capacity. On the other hand, the sink node has high processing speed, storage and energy capacity [4–6]. In unstructured networks, it is very difficult to prolong the life time of network as nodes are physically inaccessible to change their batteries. In this scenario, routing plays a vital role to minimize the energy consumption of sensor nodes by effectively transferring data from sensor nodes to the base station [7]. Typically, in routing, all nodes cooperatively form a route between nodes to the sink node. In this cooperative method, sensor nodes send control packets to entire network by using flooding at a time. However, it is more overhead on the network and consumes more energy of sensor nodes. An alternative routing called, multi-path routing, is used in WSNs to effectively balance the energy consumption. In this routing, sink initiates nearest sensor node to send control packet to form a route to the source node via intermediate nodes. Sink nearest node sends message to neighboring node for constructing route. Selection of nearest sensor node is very important in WSN. There are many methods for selecting node, based on residual energy and distance by using a probabilistic model. It is not an effective model to select node as it takes less number of parameters which often resulting in poor network life time.

It may be noted that routing in WSN is a NP-hard problem. It is a known fact that by using fuzzy logic, we can find effective solutions to NP-hard problems. So, it can be used to effectively select the neighboring sensor node in construction of a path [8]. In this paper, we propose a routing protocol, named it as multi-path routing protocol using fuzzy logic (MRP-FL), to effectively construct a path between source nodes and sink node.

The rest of the paper is organized as follows. In Sect. 2, the related works are reviewed. In Sect. 3, a brief description of System model is provided. Section 4 presents our proposed scheme. Simulations and performance analysis are provided in Sect. 5, and finally, we draw the conclusion in Sect. 6.

2 Previous Work

In this section, we review some important state of art multi-path routing algorithms in WSNs.

Heinzelman et al. [9] proposed Low-Energy Clustering Hierarchy (LEACH) protocol. It is the first hierarchal-based clustering protocol for WSNs. LEACH operates on two stages. In set-up stage, all sensor nodes form clusters, the cluster head in a cluster is selected based on random probabilistic method. In steady-state, data can be transferred from source node to base station. It cannot handle large amount of data.

Lindsey and Raghavendra [10] proposed power-efficient gathering in sensor information systems (PEGASIS) for improvement of LEACH protocol. Similar to LEACH, PEGASIS also operates in two stages, namely chain construction and gathering data. In this protocol, the sensor nodes communicate and transmit data to closest neighbor nodes and from neighbor nodes to base station. It decreases overhead of network but increases the delay of packet delivery in between sensor nodes.

Intanagonwiwat et al. [11] proposed direct-diffusion multi-path routing protocol based on query or on-demand. Sink node or base station initiates flooding message to intermediate nodes in a network. These nodes save the message for later use and calculate gradient in node toward to sink node. In this process, multiple paths are discovered for sending data to sink node. Only selected path source to sink node data can be transferred. If one path fails to transfer data, another path initiated from available alternative paths. This protocol decreases fault tolerance of a routing in a network. However, it affects network life time.

Ganesan et al. [12] proposed highly resilient, energy-efficient multi-path routing. This protocol constructs multiple paths based on partial disjoints paths. It uses two reinforcement messages for path construction. In primary path, sink initiates reinforcement message to best next-hop message for source node. This process repeats till the message reaches source node. The similar path construction approach is followed in alternative path but the only along with primary path next best next-hop toward node origin. The alternative paths preferred only when primary path fails to transfer the data.

Wang et al. [13] proposed energy-efficient collision aware multi-path routing. This routing protocol is reactive or event-driven. The routing protocol creates two collision-free paths between source nodes and sink node using location information of all the nodes. A sensor node finds route message between all other nodes within the communication range based on power and location information. Building of all paths above the communication range decreases the chance of interference. Cost of network increases because of GPS used for location information.

Sharma et al. [14] proposed energy-aware multi-path routing protocol. This protocol avoids flooding. Sink initiates nearest node to send data from source. This message is forwarded to neighbor nodes of initiated node. Each node in the network stores information id, residual energy in neighboring table of its neighboring nodes.

Source node transfers data to nearest node based on RSSI value of node [16]. This protocol sends data to sink node in two paths. In primary path, best next-hop node is selected to construct routing to sink node. If primary path fails, data can transfer via alternative paths. It improves reliable data transfer between sensor nodes. However, this protocol uses less number of parameters which increases the chances of network failure. As a result in this paper, we propose a fuzzy logic-based multi-path routing protocol to increase the life time of WSN.

3 System Model

In this section, for ease of understanding our proposed scheme some assumptions are made to construct an artificial WSN. The constitution network is done using two models viz. network and energy. We emphasize our discussion on network and energy models in Sects. 3.2 and 3.3, respectively.

3.1 Assumptions

The following favorable premises are considered in our proposed protocol.

- All sensor nodes are deployed in a stationary way (including Base Station, i.e., sink node).
- All sensor nodes are uniformly distributed in a network in random way deployment.
- Initially, all sensor nodes have same energy (excluding sink node) and they calculate their residual energy.
- Sink nodes have unlimited energy for computation of receiving data.
- All sensor nodes are homogeneous and have same capabilities.
- Data sent from one node to another in a network is same, i.e., links are symmetric between sensor nodes.

3.2 Network Model

We assume that there are N number of sensor nodes, i.e., $n_1, n_2, n_3, \ldots \ldots n_N$, and one sink node (base station) deployed in a network. The location of each sensor node is n_i $(1 \leq i \leq N)$, is (x_i, y_i). The sensor node communicates using Timeout Medium Access Control (TMAC) method. The sink node has id and location information of all sensor nodes. A route is constructed between sink and source nodes when sink node requires data from source node. Threshold energy is the minimum energy of a sensor node to communicate with other sensor nodes. The nodes with below the specified threshold energy can only perform sensing and relay of data.

3.3 Energy Model

The transmission energy of b bits across c distance is

$$E_{TE}(b, c) = \begin{cases} E_{ele} * b + E_{amp} * b * c^2 \ \text{if}(c \leq c^0) \\ E_{ele} * b + E_{amp} * b * c^4 \ \text{if}(c > c^2) \end{cases} \quad (1)$$

where E_{ele} is energy cost of both transmission and receiving of a single bit.
The receiving energy of b bits wireless sensor node is

$$E_{RE}(b) = E_{ele} * b \quad (2)$$

The sleep time of wireless sensor node is

$$E_{sleep}(t) = E_{low} * t \quad (3)$$

The total energy consumption by wireless sensor node is

$$E_T(t) = E_{TE}(b, c) + E_{RE}(b) + E_{sleep}(t) \quad (4)$$

4 Proposed System

The main goal of multi-path routing protocol with fuzzy logic (MRP-FL) is to select best next-hop node according to decision parameters. It may be noted that fuzzy values are continuous in the range of 0 and 1, and both the values are inclusive. Fuzzy logic uses Fuzzy Inference System (FIS) to make decisions using IF-THEN rules. Besides, it can also append 'OR,' 'AND' connectors as a part of decision-making [15]. As shown in Fig. 2, FIS considers three parameters, viz. residual energy of a node, node density and distance, to choose best next-hop node starting from sink node to the source node. A Fuzzy Logic Controller (FLC) based on Mamdani approach is used to analyze analog values in terms of fuzzy values. Fuzzy controller consists of four modules viz. fuzzification, inference mechanism, rule base and defuzzification.

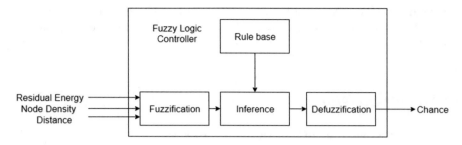

Fig. 2 Architecture of fuzzy inference system

Table 1 Input functions

Input	Membership
Distance	Near, medium, far
Node density	Sparsely, medium, densely
Residual energy	Very low, low, medium, high, very high

Table 2 Output functions

Output	Membership
Chance	Very very weak, very weak, weak, little weak, lower medium, medium, higher Medium, little strong, strong, very strong, very very strong

1. **Fuzzification**: Distance, density of node and residual energy are considered as inputs to fuzzification in the form of crisp values. The inputs are converted in the form of fuzzy sets. Every input has a membership function that gives the degree of membership. The input functions are shown in Table 1. The output chance or probability of selecting node is shown in Table 2.

Remark 1 The process of fuzzification is to change the value from scalar to fuzzy. Different types of membership functions are available to use in fuzzification. These membership functions are representing by triangular and trapezoidal.

Remark 2 The number of neighbor nodes of a node is referred to as node density.

2. **Rule Base**: In rule base, we are going to develop fuzzy rules by using IF-THEN. The fuzzy rules are in the form of if A && B && C then D. Here, A, B and C are the inputs of membership functions, and D is the chance or probability to select a node. Here, we have total 45 rules based on three input membership functions. All these rules are summarized in Table 3.

Table 3 Rules

Residual energy	Node density	Distance	Chance
Very low	Sparsely	Far	Very very weak
Very low	Sparsely	Medium	Very weak
Very low	Sparsely	Near	Weak
Very low	Medium	Far	Very weak
Very low	Medium	Medium	Weak
Very low	Medium	Near	Little weak
Very low	Densely	Far	Very weak
Very low	Densely	Medium	Little weak
Very low	Densely	Near	Lower medium
Low	Sparsely	Far	Very weak
Low	Sparsely	Medium	Weak
Low	Sparsely	Near	Little weak
Low	Medium	Far	Weak
Low	Medium	Medium	Little weak
Low	Medium	Near	Lower medium
Low	Densely	Far	Little weak
Low	Densely	Medium	Lower medium
Low	Densely	Near	Medium
Medium	Sparsely	Far	Little weak
Medium	Sparsely	Medium	Lower medium
Medium	Sparsely	Near	Medium
Medium	Medium	Far	Lower medium
Medium	Medium	Medium	Medium
Medium	Medium	Near	Higher medium
Medium	Densely	Far	Medium
Medium	Densely	Medium	Higher medium
Medium	Densely	Near	Little strong
High	Sparsely	Far	Medium
High	Sparsely	Medium	Higher medium
High	Sparsely	Near	Little strong
High	Medium	Far	Higher medium
High	Medium	Medium	Little strong
High	Medium	Near	Strong
High	Densely	Far	Little strong
High	Densely	Medium	Strong
High	Densely	Near	Very strong
very high	Sparsely	Far	Higher medium
very high	Sparsely	Medium	Little strong
very high	Sparsely	Near	Strong
very high	Medium	Far	Little strong

(continued)

Table 3 (continued)

Residual energy	Node density	Distance	Chance
very high	Medium	Medium	Strong
very high	Medium	Near	Very strong
very high	Densely	Far	Strong
very high	Densely	Medium	Very strong
very high	Densely	Near	Very very strong

3. **Inference Mechanism**: The fuzzification inputs and rules presenting in rule base are aggregated by using Mamdani FLC. The aggregated rules are evaluated using the following equation:

$$\text{chance} = \text{residual energy} \times 2 + \text{node density} + (2 - \text{distance}) \qquad (5)$$

4. **Defuzzification**: In this step, fuzzified values will be converted into crisp values by using center of gravity (CoG) method in centroid methods. The center of gravity is defined as follows [15]:

$$c(i) = \frac{\sum_{k=1}^{n} y_{j*u}(y_j)}{\sum_{k=1}^{n} u(y_j)} \qquad (6)$$

Algorithm 1
ResEng: Residual Energy of sensor node
NeighTable(x): Stores neighbor information (id, distance, ResEng, chance) of node x
Chance (n_i): chance of elect n_i sensor node in the path
Input: N number of sensor node are randomly distributed
Output: one primary and other possible alternative path from source to sink node

1. Neighbor discovery

   (a) Let x is an initiated node, and initially there is no nearest neighbor node.
   (b) x send message to all neighbor nodes and let y is the one of neighbor nodes of x.
   (c) Calculate distance and ResEng of sensor nodes and save in NeighTable(x) in ascending order.
   (d) Repeat above steps (a)-(c) for remaining all neighbor nodes of x.

2. Route discovery

   (a) Sink node initiates route discovery based on location of source node.
   (b) Primary path: Select path from source node to sink node based on chance of a node. Find chance crisp values of all the neighbor nodes of a current node

according to Eq. 6 and store them in neighbor table. The neighbor node with highest crisp value will be considered as best next-hop node. Repeat the same step until reach the sink node

(c) Alternative path: If any node in primary path fails, it goes to the alternative path. The alternative path will be constructed using neighbor table by considering next best next-hop.

3. Sink requests source node for data transmission, and source node sends the data via primary path. If any node fails, it transfers data via alternative path.
4. Sink node initiate reroute if no paths available, it again goes to step 1.

Based on this crisp value, best next-hop node will selected. Based on foregoing discussion, the procedure of MRP-FL is presented in Algorithm 1.

5 Results

Simulations are performed by using FIS in MATLAB to evaluate our methods. All the experiments are executed on Intel processor i7 running on Windows 10 Operating System with 4 GB RAM. Our proposed MRP-FL protocol is compared with MR2 and MRP with the simulation parameters listed in Table 4.

Average Control Packet Overhead: As shown in Fig. 3, average control packet overhead is high in MR2 because of flooding message set to over the network until it reaches the source node. In MRP and MRP-FL, Average Control Overhead is low as compared to MR2. This can be explained by the reason that MR2 is using flooding.

Table 4 Simulation Parameters

Parameter Name	Value
Network area	$500 * 500 \text{ m}^2$
Number of sensor nodes	100
Data packet size	512 bytes
Control packet size	32 bytes
Initial energy	1 J
E_{ele}	50 nJ/bit
$\sum fs$	10 pJ/bit/m^2
$\sum mp$	$0.0013 \text{ pJ/bit/m}^4$
c_o	87 m
E_{low}	0.2 nJ/sec
Simulation time	400 s
MAC protocol	TMAC

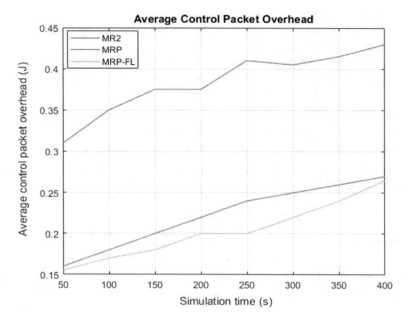

Fig. 3 Average control packet overhead

Average Energy Consumption: It is high in MR2 because of excessive overhead packet transfer over the network. MRP-FL consumes less energy compare to MRP and MR2 (see Fig. 4).

Network life time: As shown in Fig. 5, the network life time of proposed protocol is high as compared to MR2 and MRP due to fewer control packets and load balancing among sensor nodes.

6   Conclusion

It is rattling significant to prolong the network life time in WSNs by using efficient routing protocols as the sensor nodes in WSNs have limited battery power. In MRP, best next-hop node is elected-based RSSI value of sensor node. This approach has considered fewer parameters which result in high energy consumption and further it effects the life time of network. In this paper, we have improved MRP by using fuzzy logic. In our approach, we have considered three parameters, namely residual energy, node density and distance between sensor node to elect best next-hop node in a path. The simulation results have shown that our proposed scheme is outperformed than MRP and MR2 in terms of total network life time.

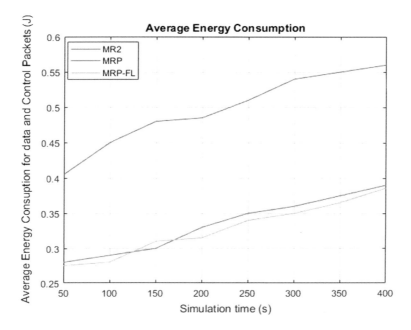

Fig. 4 Average energy consumption

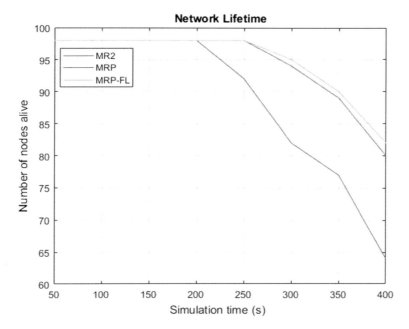

Fig. 5 Network life time

References

1. Shen, C.-C., C. Srisathapornphat, and C. Jaikaeo. 2001. Sensor Information Networking Architecture And Applications. *IEEE Personal Communications* 8 (4): 52–59.
2. Tiab, A., and L. Bouallouche-Medjkoune. 2014. Routing in Industrial Wireless Networks: A Survey. *Chinese Journal of Engineering* 2014.
3. Vu, T.T., V.D. Nguyen, and H.M. Nguyen. 2013. An Energy-Aware Routing Protocol for Wireless Sensor Networks Based on k-Means Clustering. *Recent Advances in Electrical Engineering and Related Sciences* 282: 297.
4. Farjow, W., A. Chehri, H. Mouftah, and X Fernando. An Energy-Efficient Routing Protocol for Wireless Sensor Networks Through Nonlinear Optimization. In *IEEE International Conference on Wireless Communications in Unusual and Confined Areas (ICWCUCA)*, 1–4.
5. Bangash Javed Abdullah Abdul, A.M., and Abdul, K. 2014. A Survey of Routing Protocols in Wireless Body Sensor Networks. *Sensors* 14 (1): 1322–1357.
6. Rault Tifenn, B.A., and Yacine, C. 2014. Energy Efficiency in Wireless Sensor Networks: A Top-Down Survey. *Computer Networks* 67: 104–122.
7. Zhang, D., G. Li, K. Zheng, X. Ming, and Z.-H. Pan. 2014. An Energy-Balanced Routing Method Based on Forward-Aware Factor for Wireless Sensor Networks. *IEEE Transactions on Industrial Informatics* 10 (1): 766–773.
8. Xie, L., Y. Shi, Y.T. Hou, W. Lou, H.D. Sherali, and S.F. Midkiff. 2015. Multi-node Wireless Energy Charging in Sensor Networks. *IEEE/ACM Transactions on Networking* 23 (2): 437–450.
9. Heinzelman, W.R., A. Chandrakasan, and H. Balakrishnan. 2000. Energy-Efficient Communication Protocol for Wireless Microsensor Networks. In *Proceedings of the 33rd Annual Hawaii International Conference on System Sciences*, 10.
10. Lindsey, S., and C.S. Raghavendra. 2002. Pegasis: Power-efficient Gathering in Sensor Information Systems. *IEEE in Aerospace conference proceedings* 3 (2002): 3.
11. Intanagonwiwat, C., R. Govindan, D. Estrin, J. Heidemann, and F. Silva. 2003. Directed Diffusion for Wireless Sensor Networking. *IEEE/ACM Transactions on Networking* (ToN), 11 (1), 2–16.
12. Ganesan, D., R. Govindan, S. Shenker, and D. Estrin. 2001. Highly-Resilient, Energy-Efficient Multipath Routing in Wireless Sensor Networks. *ACM SIGMOBILE Mobile Computing and Communications Review* 5 (4): 11–25.
13. Wang, Z., E. Bulut, and B. K. Szymanski. 2009. Energy Efficient Collision Aware Multipath Routing For Wireless Sensor Networks. In *IEEE International Conference on Communications*, 1–5.
14. Sharma, S., P. Agarwal, and S.K. Jena. 2016. Eamrp: Energy Aware Multipath Routing Protocol for Wireless Sensor Networks. *International Journal of Information and Communication Technology* 8 (2–3): 235–248.
15. Ran, G., H. Zhang, and S. Gong. 2010. Improving on Leach Protocol of Wireless Sensor Networks Using Fuzzy Logic. *Journal of Information & Computational Science* 7 (3): 767–775.
16. Mahapatra, R.K., and N.S.V. Shet. 2017. Localization Based on RSSI Exploiting Gaussian and Averaging Filter in Wireless Sensor Network. *Arabian Journal for Science and Engineering*.

MST Parser for Telugu Language

G. Nagaraju, N. Mangathayaru and B. Padmaja Rani

Abstract Maximum spanning tree (MST) parser is a graph built dependency parser and also finds the k-best spanning trees based on the scores. These spanning trees are constructed based on the global optimization approaches. However, it selects the tree which has the highest score and the method uses Eisner algorithm to find the best spanning tree. Hence, the time complexity has been significantly reduced from $O(n^5)$ to $O(n^3)$. In this paper, we discuss some of the projective and non-projective approaches on how to apply MST parser to Telugu language. In addition to this, we also present a methodology to apply Telugu treebank to MST parsing. Finally, the results are obtained and found to be suggestively improved.

Keywords MST parser · Telugu treebank · Eisner · Chu–Liu Edmonds · Dependency parsing · Graph-based parsing

1 Introduction

Dependency parsing is the process of identifying the relationship between the words of a sentence and it has two types; they are transition-based dependency parser and graph-based dependency parser. Maximum spanning tree (MST) parsing is the graph-based dependency parsing. Dependency parsing is used in many applications like machine translation [1], relation extraction [2], synonym generation [3] and query answering system. The advantage of dependency parsing over phrase structure is that it is efficient and more accurate parsing.

Dependency parsing for the sentence "*Rama is a good boy*" is shown in Fig. 1; in this, each node has a parent node except the root node and a root node is added at the beginning of the sentence. Every sentence is either projective or non-projective.

G. Nagaraju (✉) · N. Mangathayaru
Department of CSE, VNRVJIET, Hyderabad, India
e-mail: nagaraju.gujjeti@gmail.com

B. Padmaja Rani
Department of CSE, JNTUCEH, Hyderabad, India

© Springer Nature Singapore Pte Ltd. 2020
K. S. Raju et al. (eds.), *Proceedings of the Third International Conference on Computational Intelligence and Informatics*, Advances in Intelligent Systems and Computing 1090, https://doi.org/10.1007/978-981-15-1480-7_23

Fig. 1 Projective sentence

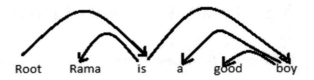

In projective sentence, there is no intersecting edge and in a non-projective sentence, there is an intersecting edge. Most of the parsers will parse projective sentences, but MST parser will parse both the projective sentences and non-projective sentences. The large number of the English sentences is projective and they are automatically generated from Penn Treebank [4].

The majority of the previous methods are focused on projective trees, which are Eisner algorithm, Collins [5], Yamada and Matsumoto [6]. These algorithms are produced more accurate results for projective sentences. Nowadays, several researchers are interested to develop the algorithms for non-projective trees. Non-projective property may frequently occur in free word languages, it means that if we interchange the words position, still it preserves its meaning. An English language is not a free word order language, but some of the English sentences are non-projective like "*Rama saw a girl yesterday she is his schoolmate*" is shown in Fig. 2. Wang and Harper [7] are developed a non-projective parser and Nivre and Nilsson [8] were developed a parsing model for non-projective edges in dependency trees. MST is an alternative method for parsing non-projective sentences and it generates more accurate and efficient dependency trees.

The key concept of MST parsing is that it selects the tree, which has the highest score from a given directed graph. This work is similar to Hirakawa [9]; it reduces the time to find the highest spanning tree from a directed graph, but it uses branch and bound algorithm. So, it has an exponential time complexity in worst case; however, it has better performance in other cases.

In this paper, Sect. 2 explains online large margin learning algorithms, structured classification and margin-infused algorithms. Section 3 explains projective dependency algorithm as Eisner algorithm for parsing projective sentences. Section 4 explains non-projective dependency algorithm as Chu–Liu Edmonds algorithm, Sect. 5 discusses experiments for Telugu language and the last section explains conclusion and future work.

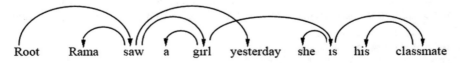

Fig. 2 Non-projective sentence

2 Online Large Margin Learning

In this section, we discuss online large margin learning algorithms. These algorithms are worked on the property of inference. It creates a training model based on the maximum scoring output for a given input.

2.1 Structured Classification

Structured classification algorithms are supervised learning algorithms that map labelled input to its outputs. First, conditional random fields (CRF) [10] were developed and after that, many researchers were developed margin-based algorithms [11]. These algorithms were applied in many real-world applications in natural language processing [12, 13]. These algorithms uses batch learning process, in which all the training instances are optimized simultaneously, whereas in the online learning, algorithms optimize the one input at a time. It seems to be a drawback for online learning as compared to batch learning, but it was overcome with perceptron algorithms [14]; they develop a mathematical model for the given input data. These models handle large amount of data as compared to the batch processing.

Recently, perceptron with online large margin algorithm was popularized because of voted perceptron algorithm [15].Voted perceptron algorithm is the averaged perceptron algorithm. Average perceptron algorithm does not optimize the classification margin; it is required to reduce the generalization error [16]. So that parameter averaging is used. We discuss online large margin algorithms for multiclass classification margin-infused relaxed algorithm (MIRA) for structured outputs. MIRA has many desirable properties to handle accuracy and scalability and it is suitable for handling complex trees.

2.2 Margin-Infused Relaxed Algorithm

Online learning uses the following score function for input and output pairs

$$Score(x, y) = w \cdot f(x, y)$$

Here, score(x, y) is the score function, w is the weight vector and $f(x, y)$ is the high dimensional feature representation and x is the input and y is the output. The main objective of the score function is to increase the score value for correct outputs and decrease the score value for the wrong outputs. Online learning algorithm is described in Fig. 3.

```
Training data  = {(xt , yt)} Here t = 1 to T
W(0)  = 0,  v =0,  i =0
For n = 1 to N
For t=1 to T
W(i+1)  = update w(i)
V = v + W(i+1)
W = V/(N*T)
```

Fig. 3 Online learning algorithm

Minimize w

Such that score(x, y) − score$(x, y') \geq$ Loss(y, y')
For all $(x, y) \in$ training data, $y' \in$ parse(y').

Here, Loss(y, y') is the loss occurred for the parse tree y' as compared to y and that value represents the margin between the correct parse and the incorrect parse. Hence, the weight vector is minimized with respect to the margin. In graph-based parsing, loss is defined as the number of nodes having incorrect incoming edges as it similar to Hamming loss and is used in evaluating the performance.

MRA introduces the weight update equation as the new weight vector is as possible as closer to old weight value. Hence, the following weight update equation is included in the online learning algorithm.

$w^{\text{new}} = \text{argmax}_{w*} \|w* - w^{\text{old}}\|$
such that score(x, y) − score$(x, y') \geq$ Loss(y, y')
for all $y' \in$ parse(x_t).

MIRA is used for multiclass classification; hence, for given random inputs, there are many number of output classes, so the number of output classes is increased exponentially. This is the problem with the sequential classification for dependency parsing.

2.3 K-Best Margin-Infused Relaxed Algorithm

The problem of exponentially increased output classes is reduced by selecting best k highest score trees. Then the weight update equation becomes

$w^{\text{new}} = \text{argmax}_{w*} \|w* - w^{\text{old}}\|$
such that score(x, y) − score$(x, y') \geq$ Loss(y, y')
for all $y' \in$ best$_k$ $(x_t, w^{(i)})$.

3   Projective Dependency Parsing

In projective sentence words, dependent edges cannot cross each other. CKY algorithm [17] generates all possible trees for a given sentence with the time complexity of $O(n^5)$, among all trees, which tree has the highest score is called the maximum spanning tree and it is same as the dependency parser for the given sentence. Eisner [18] was developed a method to parse a projective sentence in $O(n^3)$ time. This method separately parses left dependencies and right dependencies of a word and later it merges. It requires two variables, in which one variable represents the direction to gather its dependents and the other variable represents the item which is complete or not. A complete item and incomplete item were shown in Figs. 4 and 5, respectively.

This algorithm begins with the right-angle triangles, because we assumed that, initially, all the words are complete words. In the next step, the complete items generate incomplete items, which are represented with trapezoidal. This process is repeated so that in the end, an item is completed. This procedure is shown in Fig. 6. Here a, c and d are the start and end indices of the items, and $b1$ and $b2$ are the heads of the items.

Eisner algorithm uses dynamic programming method for parsing. It uses a four dimension array $C[a][d][x][y]$ represents a subtree position from a to d; it uses two binary variables x and y. Here, x is used for the direction of the dependency (Left dependency ($\leftarrow$) or right dependency ($\rightarrow$)) and y represents whether the item is complete or not. If x value is 0, then the item is not complete and 1 for complete. $C[a][d][\rightarrow][1]$ represents the complete item from a to d and the dependency direction is right, shown in below(left figure) and $C[a][b][\leftarrow][0]$ represents incomplete item from a to b and the dependency direction is left, shown in below (right figure).

Eisner cubic parsing algorithm was shown in Fig. 7. The algorithm takes run time complexity of $O(n^3)$. This algorithm always generates maximum spanning tree. Eisner algorithm also generates k-best spanning tree which takes an additional time complexity of O(klogk).

Fig. 4   A complete item

Fig. 5   An incomplete item

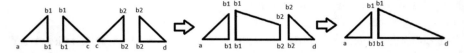

Fig. 6 Eisner cubic parsing procedure [2]

```
Initialization    C[a][a][x][y]=0.0     for all a, x and y
for i: 1 to n
  for j: 1 to n
    K = i + j
    If( k > n) then break
     // create incomplete sub tree
        C[j][k][←][0]  =   max j <= s < k  (C[j][s][→][1]
+    C[s+1][k][←][1]  +  s(k,j))
        C[j][k][→][0]  =  max j <= s < k   (C[j][s][→][1]
+ C[s+1][k][←][1]  + s(j,k))
     // create complete sub tree
        C[j][k][←][1]  =   max j <= s < k   (C[j][s][←][1]
+ C[s][k][←][0])
        C[j][k][→][1]  =  max j < s <= k   (C[j][s][→][0]
+ C[s][k][→][1])
    end for
end for
```

Fig. 7 Pseudo code for Eisner cubic parsing algorithm [18]

4 Non-projective Parsing Algorithm

Many algorithms exist to find a maximum spanning trees for undirected graphs
[19], among that Chu–Liu Edmonds algorithm produces best maximum spanning
tree for directed graphs. [20, 21]. Figure 5 illustrates the Chu–Liu Edmonds
algorithm. First, it greedily selects the maximum incoming edge for each vertex in
the graph and the generated tree is the maximum spanning tree. Suppose it has a
cycle, then it forms a single node with the nodes which were formed a cycle. Then it
calculates the best incoming edge and outgoing edge to the nodes in a cycle with
these equations.

Outgoing edge for a cycle
$S(c, x) = \max_{x' \in c} s(x', x)$
Incoming edges for a cycle
$S(x, c) = \max_{x' \in c} [s(x, x') - s(a(x'), x') + s(c)]$
$S(c) = \sum_{v \in c} s(a(v), v)$ $a(v)$ is the precedence of v in c
Using the above equations, it can remove the cycle from the graph.

Table 1 LA, UAS and LAS results using Telugu treebank

	Results (%)
LA	63.1
UAS	88.5
LAS	62.3

5   Experiments

We applied Telugu treebank to find the performance of the MST parser; it is according to the CONLL format and it was developed by the IIIT Hyderabad. Each word of the sentence has 10 fields like its POS tag for dependency number, root word, universal POS tag, etc., but that the most important fields are POS tag, dependency word position. Telugu is one of the popular Indian languages and it is a free word order language and morphologically rich language [22].

We used different metrics like Labelled Attachment score (LAS), Unlabelled Attachment Score (UAS) and Label Accuracy (LA). LAS is the number of correct labels to its correct heads, UAS is the number of correct heads and LA is the number of correct labels assigned. The following table and graph provide a summary of these accuracies for Telugu treebank (Table 1).

The value of UAS is large as compared with LAS and LA values, as shown in the Fig. 8, because it represents only the correct heads. Labelled Accuracy and labelled complete accuracy for MST parsing are shown in Table 2. Here, x-axis represents the measurements and y-axis represents the accuracy.

Fig. 8 LA, UAS and LAS results for Telugu treebank

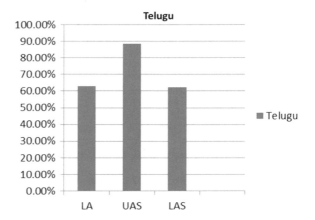

Table 2 Labelled accuracy and labelled complete accuracy for Telugu language

	Results (%)
Labelled accuracy	62.3
Labelled complete accuracy	24.8

6 Conclusion and Future Work

In this work, we discussed a novel methodology for the applications of the projective and non-projective approaches to Telugu language and Telugu treebank fields in MST parser. MST parser can parse both the projective sentences and non-projective sentences. Eisner algorithm is used for parsing projective sentences and Chu–Liu Edmonds algorithm for parsing non-projective sentences. K-best MIRA method is used to find the highest score parse tree and this method reduces time complexity from $O(n^5)$ to $O(n^3)$ [23]. In the end, the results were obtained and found to be positively enhanced on Telugu treebank.

Based on the recorded system results, it has been observed that the labelled accuracy is satisfactory for Telugu treebank. However, this could be improved by incorporating some additional features as part of the future enhancement work. The sentence structure of Indian languages is differing from other language structures in terms of the karakas like karta, karma, kriya, karana, apadaana and sampradana. Hence, these karaka's can be included into the feature set, so as to improve the performance and accuracy of the parser.

References

1. Ding, Y., and M. Palmer. 2005. Machine Translation Using Probabilistic Synchronous Dependency Insertion Grammars. In *Proceedings of ACL*.
2. Culotta, A., and J. Sorensen. 2004. Dependency Tree Kernels for Relation Extraction. In *Proceedings of ACL*.
3. Shinyama, Y., S. Sekine, K. Sudo, and R. Grishman. 2002. Automatic Paraphrase Acquisition from News Articles. In *Proceedings of HLT*.
4. Marcus, M., B. Santorini, and M. Marcinkiewicz. 1993. Building a Large Annotated Corpus of English: The Penn Treebank. *Computational Linguistics* 19 (2): 313–330.
5. Collins, M., J. Hajič, L. Ramshaw, and C. Tillmann. 1999. A Statistical Parser for Czech. In Proceedings of ACL.
6. Yamada, H., and Y. Matsumoto. 2003. Statistical Dependency Analysis with Support Vector Machines. In *Proceedings of IWPT*.
7. Wang, W. and M.P. Harper. 204. A Statistical Constraint Dependency Grammar (CDG) Parser. In Workshop on Incremental Parsing: Bringing Engineering and Cognition Together (ACL).
8. Nivre, J., and J. Nilsson. 2005. Pseudo-Projective Dependency Parsing. In *Proceedings of ACL*.
9. Hirakawa, H. 2019. Semantic Dependency Analysis Method for Japanese Based on Optimum Tree Search Algorithm. In Proceedings of PACLING, 200.
10. Jafferty, L., A. McCallum, and F. Pereira. 2011. Conditional Random Fields: Probabilistic Models for Segmentation and Labelling Sequence Data. In *Proceedings of the International Conference on Machine Learning*.
11. Tasker, B., Guestrin, C., and Koller, D. 2003. Max-Margin Parsing. In Proceedings of Neural Information Processing Systems (NIPS).
12. Tasker, B., D. Klein, M. Collins, D. Koller, and C. Manning. 2004. Max-Margin Parsing. In Proceedings of the Empirical Methods in Natural language Processing (EMNLP).

13. Tsochantaridis, I., T. Hofmann, T. Jaachims, and Y. Altun. 2007. Support Vector Learning for Interdependent and Structured Output Spaces. In *Proceedings of the International Conference on machine Learning*.

14. Rosenblatt, F. 1958. The Perceptron: A Probabilistic Model For Information Storage and Organization in the Brain. *Psychological Review* 68: 386–407.

15. Freund, Y., and R.E. Schapire. 1999. Large Margin Classification Using the Perceptron Algorithm. *Machine Learning* 37 (3): 277–296.

16. Boser, B.E., I. Guyon, and V. Vapnik. 1992. A Training Algorithm for Optimal Margin Classifiers. In *Proceedings of COLT*, 144–152.

17. Younger, D.H. 1967. Recognition and Parsing of Context-Free Languages in Time n^3. *Information and Control* 12 (4): 361–379.

18. Eisner, J. 1996. Three New Probabilistic Models for Dependency Parsing: An Exploration. In *Proceedings of COLING*.

19. Cormen, T.H., C.E. Leiserson, and R.L. Rivest. 1990. Introduction to Algorithms. Cambridge: MIT Press/McGraw-Hill.

20. Chu, Y.J., T.H. Liu. 1965. On the Shortest Arborescence of a Directed Graph. *Science Sinica* 14: 1396–1400.

21. Edmonds, J. 1967. Optimum Branching. *Journal of Research of the National Bureau of Standards* 71B: 233–240.

22. Nagaraju, G., 0N. Mangathayaru, and B. Padmaja Rani. 2016. Dependency Parser for Telugu Language. In Proceedings of the Second International Conference on Information and Communication Technology for Competitive Strategies, Article No. 138.

23. McDonald, Ryan, Koby Crammer, and Fernando Pereira. 2005. Online Large-Margin Training of Dependency Parsers. In *Proceedings of the 43rd Annual Meeting on Association for Computational Linguistics (ACL '05)*. Association for Computational Linguistics, Stroudsburg, PA, USA, 91–98. https://doi.org/10.3115/1219840.1219852.

Design of Low-Power Vedic Multipliers Using Pipelining Technology

Ansiya Eshack and S. Krishnakumar

Abstract This paper proposes a study on how pipelining technology can be used in Vedic multipliers, employing Urdhava Tiryakbhyam sutra, to increase the speed and reduce the power consumption of a system. Pipelining is one of the methods used in the design of low-power systems. Vedic multipliers use an ancient style of multiplying numbers which allows for easier and faster calculations, compared to the regular mathematical method. The concept of pipelining, when used in these multipliers, leads to a system which computes calculations faster using lesser hardware. The study includes the direct use of pipelining in 2×2 bit, 4×4 bit, 8×8 bit and 16×16 bit Vedic multipliers. The pipelining is then further incorporated at different levels to create 8×8 bit and 16×16 bit multipliers. The algorithm of the system is implemented on Spartan 3E field-programmable gate array (FPGA). The designed system uses lesser power and has lower delay compared to the existing systems.

Keywords Pipelining · FPGA · Vedic mathematics · Low-power · Vedic multipliers · Urdhava Tiryakbhyam sutra

1 Introduction

Multiplication is just a collection of additions of partial products, often realized by cycles of shifting and adding. The process becomes complex as the number of partial products increases, due to increase in the number of bits which are being multiplied. This eventually leads to increased use of hardware. As multiplication finds prominence in signal processing and communication applications, it is required that the

A. Eshack (✉) · S. Krishnakumar
School of Technology & Applied Sciences, M. G. University Research Centre,
Edapally, Ernakulam, India
e-mail: ansiya@yahoo.com

S. Krishnakumar
e-mail: drkrishsan@gmail.com

© Springer Nature Singapore Pte Ltd. 2020
K. S. Raju et al. (eds.), *Proceedings of the Third International Conference on Computational Intelligence and Informatics*, Advances in Intelligent Systems and Computing 1090, https://doi.org/10.1007/978-981-15-1480-7_24

process utilizes minimum power and produces maximum throughput. There is always a lookout for systems providing low-power consumption at high speed. Vedic mathematics is a system of mathematics gifted by the ancient sages of India. It contains 16 sutras (algorithms) that show how calculations can be done mentally and in a simple manner. When considering complex problems which involve a large number of mathematical operations to be done, the Vedic mathematical techniques require very little time as compared to the present conventional methods [1]. Vedic multipliers based on the concept of Vedic mathematics prove to be one of the fastest and low-power consuming designs [2–5]. Urdhava Tiryakbhyam (UT) is a simple Vedic Sutra usually employed for multiplication of two numbers. It has been understood that this sutra gives minimum delay during multiplication [6, 7].

Pipelining, on the other hand, is an implementation technique where multiple tasks are divided into subtasks and performed independently, in an overlapped manner. This method of dividing tasks leads to great reduction in the time consumed to obtain the results and also in the amount of energy required by the operation [8]. Pipelining can be used along with the UT sutra, which makes use of the vertical, cross-wise concept, to calculate result of multiplication. It is seen that the end results are achieved at a faster rate when the two techniques are employed together [9]. The present paper proposes the use of the UT sutra and incorporates the pipelining technique in it.

This paper is further organized into five sections; Sect. 2 gives a review of the operating principle of UT sutra. In Sect. 3, a brief description on how pipelining is employed in the multiplier is discussed. Section 4 contains the designed work, and Sect. 5 contains its results and discussions. Conclusion is given in Sect. 6.

2 Vedic Mathematics and the UT Sutra

Vedic mathematics is very efficient in reducing tough and complex calculations to a very simple one. The use of Vedic multiplier using UT sutra leads to lower-power consumption multiplier than the conventional multiplier [10–12].

Consider two numbers $(ax + b)$ and $(cx + d)$. Their multiplication gives '$acx^2 + x(ad + bc) + bd$' as the product. By observation, it can be understood that the first term is got by vertical multiplication of a and c, the middle term is got by addition of the products obtained by cross-wise multiplication of a and d and of b and e, and finally, the independent term is arrived at by vertical multiplication of b and d. This is represented by the line diagram as shown below.

Similarly, for multiplication of any two N-bit numbers, one or more steps of vertical multiplications are required, along with few steps of cross-wise multiplications, and addition of the partial products obtained in each step gives the final result. This forms the basic principle of the UT sutra, and so, this sutra is rightly called vertical and cross-wise multiplication method.

A basic non-pipelined 4×4 bit Vedic multiplier, based on the UT sutra, contains four 2×2 bit multipliers also based on the UT sutra. Similar is the case for an 8×8 bit multiplier which contains four 4×4 bit multipliers based on the same sutra. A 16×16 bit multiplier also contains four 8×8 bit multipliers based on the UT sutra.

3 Pipelining and Partial Product Accumulation in Vedic Multiplier

Consider the implementation of a 4×4 bit multiplier. Let the two numbers be $X = a3a2a1a0$ and $Y = b3b2b1b0$. The partial products are obtained as shown in Fig. 1. The final result of the multiplication is the concatenation of the partial products obtained in the eight steps. It can be seen that Step 2 involves use of a half adder, Step 3 and Step 6 use 4-bit adders, Step 4 and Step 5 use 5-bit adders, and Step 7 uses a 3-bit adder. The partial products are thus obtained in a pipelined fashion with the help of these adders, leading to a reduction in the delay for obtaining the final result [7]. This technique of multiplication, using pipelining, discussed above can be used to obtain results of any two N-bit numbers.

The concept of partial product accumulation is generally used to construct multipliers of wider bit widths using lower width multiplier blocks [13]. Consider two four-bit numbers $A[3:0]$ and $B[3:0]$, their multiplication leads to generation of four partial products PP0, PP1, PP2 and PP3, as shown below.

$$
\begin{array}{rl}
A_{[3:2]}A_{[1:0]} & \\
* \ B_{[3:3]}B_{[1:0]} & \quad \text{Correct to [3:2]} \\
\hline
A_{[1:0]} * B_{[1:0]} & = \text{PP0} \\
A_{[3:2]} * B_{[1:0]} & = \text{PP1} \\
A_{[1:0]} * B_{[3:2]} & = \text{PP2} \\
A_{[3:2]} * B_{[3:2]} & = \text{PP3} \\
\hline
P_{[7:6]} \quad P_{[5:4]} \quad P_{[3:2]} \quad P_{[1:0]} &
\end{array}
$$

The final eight-bit result $P_{[7:0]}$ is the concatenation of $P_{[7:6]}$ & $P_{[5:4]}$ & $P_{[3:2]}$ & $P_{[1:0]}$ where $P_{[7:6]} = \text{PP3}_{[3:2]}$, $P_{[5:4]} = \text{PP3}_{[1:0]} + \text{PP2}_{[3:2]} + \text{PP1}_{[3:2]}$, $P_{[3:2]} = \text{PP2}_{[1:0]} + \text{PP1}_{[1:0]} + \text{PP0}_{[3:2]}$, $P_{[1:0]} = \text{PP0}_{[1:0]}$.

The partial products obtained during each step are:
Step 1: P0 = a0.b0

Step 2: P1= a1.b0 +a0.b1

Step 3: P2 = a2.b0+a1.b1+a0.b2+P1(1)

Step 4: P3 = a3.b0+a2.b1+a1.b2+a0.b3+P2(1)

Step 5: P4 = a3.b1+a2.b2+a1.b3+P2(2)+P3(1)

Step 6: P5 = a3.b2+a2.b3+P3(2)+P4(1)

Step 7: P6 = a3.b3+P4(2)+P5(1)

Step 8: P7 = P6(2:1)

Final result P= P7&P6&P5&P4&P3&P2&P1&P0

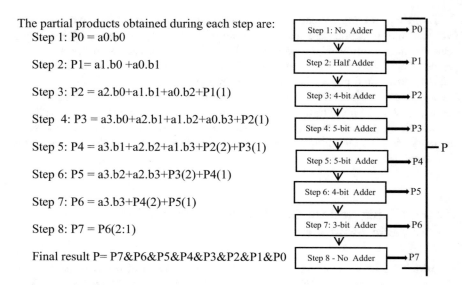

Fig. 1 Pipelining in Vedic multiplier

4 Designed Work

The designed work incorporates the pipelining technology in UT sutra of a Vedic multiplier. Initially pipelining is done in 2×2 bit, 4×4 bit, 8×8 bit and 16×16 bit Vedic multipliers. It is observed that multipliers with pipelining use lesser hardware of the FPGA than those without pipelining. As a further step in understanding the advantage of pipelining, the 4×4 bit multipliers was designed using four 2×2 pipelined multipliers. Also, 8×8 bit multiplier was designed in two ways: The first design contains use of four pipelined 4×4 bit multipliers. The second design uses four 4×4 bit multipliers each of which uses four pipelined 2×2 bit multipliers. Further 16×16 bit multiplier was designed in three different ways: The first design contains use of four pipelined 8×8 bit multipliers. The second design uses four 8×8 bit multipliers each of which uses four pipelined 4×4 bit multipliers. The final design has four 8×8 bit multipliers each of which has four 4×4 bit multipliers employing pipelined 2×2 bit multipliers.

The property of partial product accumulation is also used in the designed work in combination with the pipelining. Consider the example of the 4×4 bit multiplier with the numbers $X = a3a2a1a0$ and $Y = b3b2b1b0$. As shown in Fig. 2, four pipelined 2×2 bit multipliers generate the four partial products PP0, PP1, PP2 and PP3. The values of $P_{[7:6]}$, $P_{[5:4]}$, $P_{[3:2]}$ and $P_{[1:0]}$ are then obtained using the three adders. The same concept is employed in 8×8 bit and 16×16 bit multipliers also.

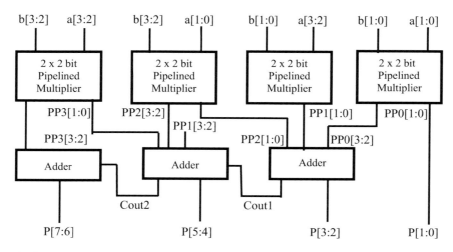

Fig. 2 Modified 4 × 4 bit Vedic multiplier

5 Results and Discussion

Pipelined 2 × 2, 4 × 4, 8 × 8 and 16 × 16 bit multipliers using UT sutra of Vedic mathematics were designed and implemented on the device XC3S500E-5FG320 in Xilinx ISE 8.2i. Table 1 gives the comparison between the hardware used by the pipelined and non-pipelined Vedic multipliers. The 2 × 2 bit multipliers (both pipelined and non-pipelined) show no difference in the values of look-up tables (LUTs) and occupied slices as the mathematical operations performed for both require the same number of adders [13]. But for multipliers with higher bits, there is a difference in the requirements. A reduction in the hardware used is observed when pipelining concept is applied to the higher bit multiplier designs.

Table 2 gives the comparison when pipelining is employed at different levels of the multiplier. It is seen that the level at which the pipelining is done plays an important role in the delay and the power consumption. The values decrease when lower bit pipelined multipliers are used to form higher bit multipliers.

The 4 × 4 bit multiplier, when designed using pipelined 2 × 2 bit multiplier, shows less hardware use than pipelined 4 × 4 bit multiplier. Similar is the case of 8 × 8 bit multipliers. The multiplier uses lower number of LUTs and slices as compared to that designed using pipelined 4 × 4 bit multipliers. Further, these two multipliers occupy less than pipelined 8 × 8 bit multiplier. In the design of 16 × 16 bit multiplier also, the same effect is observed. The number of LUTs and occupied slices decreases as the design employs first pipelined 8 × 8 bit, then pipelined 4 × 4 bit and finally pipelined 2 × 2 bit multipliers.

Table 1 Comparison between non-pipelined and pipelined Vedic multipliers

Type of Vedic Multiplier		2 × 2 bit	4 × 4 bit	8 × 8 bit	16 × 16 bit
No. of input LUTs	Without pipelining	4	42	201	841
	With pipelining	4	35	179	799
No. of occupied slices	Without pipelining	2	22	106	475
	With pipelining	2	21	99	446

Table 2 Comparison between Vedic multipliers employing pipelining at different levels

Type of Vedic Multiplier	Pipelined 4 × 4	4 × 4 with pipelined 2 × 2	Pipelined 8 × 8	8 × 8 with pipelined 4 × 4	8 × 8 with pipelined 2 × 2
No. of input LUTs	35	33	179	174	171
No. of occupied slices	21	17	99	98	93

Type of Vedic Multiplier	Pipelined 16 × 16	16 × 16 with pipelined 8 × 8	16 × 16 with pipelined 4 × 4	16 × 16 with pipelined 2 × 2
No. of input LUTs	799	744	727	705
No. of occupied slices	445	415	411	389

6 Conclusion

Verilog HDL has been used to design the multipliers. It is observed that use of pipelining in UT sutra has led to great reduction in the hardware and time for calculation compared to non-pipelined Vedic multiplier. It is also understood that using lower bit pipelined multiplier, when designing higher bit multiplier, leads to high reduction in the device utilization. This concept can be used in the design of even higher-order bit multipliers like 32 × 32 bit, 64 × 64 bit and so on. Multipliers always form an important part of signal processing and communication applications. The use of designed multipliers leads to design of high speed, low-power multiplier devices.

References

1. Jagadguru, S., and K.T.M. Sri Bharati. 2015. *Vedic Mathematics*, 17th Reprint ed. New Delhi: Motilal Banarsidass Publishers.
2. Amit, K., and P. Hitesh. 2016. Design and Analysis of Faster Multiplier using Vedic Mathematics Technique. In IJCA Proceedings on International Conference on Advancements in Engineering and Technology, 28–31.
3. Mohan, S., and N. Rangaswamy. 2017. Energy and Area Efficient Hierarchy Multiplier Architecture Based on Vedic Mathematics and GDI Logic. *Engineering Science and Technology* 20: 321–331.
4. Saha, P., A. Banerjee, A. Dandapat, and P. Bhattacharyya. 2011. ASIC Design of a High Speed Low Power Circuit for Factorial Calculation Using Ancient Vedic Mathematics. *Microelectronics Journal* 42: 1343–1352.
5. Prabir, S., B. Arindam, B. Partha, and D. Anup. 2011. High-Speed ASIC Design of Complex Multiplier Using Vedic Mathematics. In: Proceedings of 2011 IEEE Students' Technology Symposium (TechSym), 237–241.
6. Challa Ram, G., D. Sudha Rani, Y. Rama Lakshmanna, and K. Bala Sindhuri. 2016. Area Efficient Modified Vedic Multiplier. In Proceedings of 2016 IEEE International Conference on Circuit, Power and Computing Technologies, 1–5.
7. Harish, B., R.N. Satish, D. Bhumarapu, and P. Jayakrishanan. 2014. Pipelined Architecture for Vedic Multiplier. In Proceedings of 2014 IEEE International Conference on Advances in Electrical Engineering, 1–4.
8. Naresh, G., and M.K. Soni. 2012. Reduction of Power Consumption in FPGAs—An Overview. *Information Engineering and Electronic Business* 5: 50–69.
9. Eshack, Ansiya, and S. Krishnakumar. 2018. Implementation of Pipelined Low Power Vedic Multiplier. In: Proceedings of 2018 IEEE International Conference on Trends in Electronics and Informatics, 171–174.
10. Yogita, B., and M. Charu. 2016. A Novel High-Speed Approach for 16×16 Vedic Multiplication with Compressor Adders. *Computers & Electrical Engineering* 49: 39–49.
11. Leonard, G.S., and M. Thilagar. 2010. VLSI Implementation of High Speed DSP Algorithms Using Vedic Mathematics. *Singaporean Journal Scientific Research* 3 (1): 138–140.
12. Bhavani, P.Y., C. Ganesh, R.P. Srikanth, N.R. Samhitha. 2014. Design of low power and high speed modified carry select adder for 16 bit Vedic multiplier. In: Proceedings of 2014 IEEE International Conference on Information Communication and Embedded Systems, 1–6.
13. Information on http://www2.elo.utfsm.cl/~lsb/elo211/aplicaciones/katz/chapter5/chapter05. doc6.html.
14. Poornima, M., et al. 2013. Implementation of Multiplier Using Vedic Algorithm. *Innovative Technology and Exploring Engineering* 2 (6): 219–223.
15. Nizami, I.F., M. Majid, and H. Afzal, et al. 2017. Impact of Feature Selection Algorithms on Blind Image Quality Assessment. *Arabian Journal for Science and Engineering*.

Needleman–Wunsch Algorithm Using Multi-threading Approach

Sai Reetika Perumalla and Hemalatha Eedi

Abstract Needleman–Wunsch algorithm (NWA) is one of the most popular algorithms in the area of bioinformatics to perform pair-wise sequence alignment. It uses dynamic programming to calculate the overall alignment score of two biological sequences. Using multithreading, this algorithm can be parallelized to achieve higher efficiency in terms of speed without compromising on the overall accuracy of the alignment. Multithreading can be implemented in multiple ways on the Needleman–Wunsch algorithm, each of which having different efficiencies in terms of speed and accuracy. The optimal number of threads for varying sequence lengths can also be identified using the results of the experiments performed.

Keywords Sequence alignment · Parallelization · Needleman–Wunsch algorithm · Multi-threading

1 Introduction

The problem of sequence alignment is one of the prominent concerns in the field of bioinformatics [1]. Sequence alignment can provide various insights into how a particular biological component has evolved over the years and how and at what stage the functionality has changed from its predecessors. Sequence alignment can

S. R. Perumalla (✉) · H. Eedi
Department of Computer Science, Jawaharlal Nehru Technological
University, Hyderabad, India
e-mail: saireetika@gmail.com

H. Eedi
e-mail: hemamorarjee@jntuh.ac.in

© Springer Nature Singapore Pte Ltd. 2020
K. S. Raju et al. (eds.), *Proceedings of the Third International Conference on Computational Intelligence and Informatics*, Advances in Intelligent Systems and Computing 1090, https://doi.org/10.1007/978-981-15-1480-7_25

be performed on two biological sequences, termed as pair-wise alignment, and if the number of sequences exceeds three, it is termed as multiple sequence alignment (MSA) [2]. Multiple sequence alignment requires performing pair-wise alignment multiple times before obtaining the final alignment. Hence, optimizing pair-wise sequence alignment has a bigger impact on the overall domain of sequence alignment.

There have been multiple algorithms to perform pair-wise sequence alignment. Some algorithms perform sequence alignment considering the entire sequence at once. These algorithms are termed as global sequence alignment algorithms, and on the other hand, some algorithms perform sequence alignment by considering a part of sequences at once; these algorithms are termed as local sequence alignment algorithms. Needleman–Wunsch algorithm is one of the most popular global sequence alignment algorithms introduced till date. Its popularity is mainly due to its efficiency, simplicity and readability.

1.1 Needleman–Wunsch Algorithm

Needleman–Wunsch algorithm is one of the first applications of dynamic programming in the field of bioinformatics. It was proposed in the year 1970 by Saul B. Needleman and Christian D. Wunsch. It mainly consists of three steps:

- Initialization of scoring matrix
- Filling the scoring matrix based on the given criteria
- Backtracking to obtain the final alignment.

Though the Needleman–Wunsch algorithm is widely popular for its efficiency, its overall time complexity is $O(mn)$ where m and n are the lengths of sequence 1 and 2, respectively. The steps of the Needleman–Wunsch algorithm can be explained via the below algorithm:

Step 1: Initialize the Scoring Matrix, C

Fill the first row and column with the column or row number times the mismatch score and fill all the other cells with zeroes.

Step 2: Calculate the cells of Scoring Matrix

1   **for** $i = 0$ to $len(A) - 1$ do

2   **for** $j = 1$ to $len(B)$ do

3   $q_{diag} = C(i-1, j-1) + S(i, j)$

4   $q_{up} = C(i-1, j) + g$

5   $q_{left} = C(I, j-1) + g$

6   **if** $C[i] = C[j]$ then $S(i, j) =$ m 7

7   **else** $S(i, j) = mm$

8   $C[i][j] = max(q_{diag}, q_{up}, q_{left})$

9   choice $[i][j] = diag$ if $C[i][j] = q_{diag}$

10 choice $[i][j] = up$ if $C[i][j] = q_{up}$

11 choice$[i][j] = left$ if $C[i][j] = q_{left}$

Step 3: Trace Back to Find the Optimal Alignment

1 **for** $i=0$ to $len(A)-1$ **do**

2 **for** $j = 1$ to $len(B)$ **do**

3 **if** $choice[i][j] = diag$ then

4 align the letters of the
sequence 5 C[i] is aligned
with C[j]

5 **else** **if** $choice[i][j] = up$ then

6 hyphen is introduced in the top
sequence (SeqA)

7 **else**
hyphen is introduced in the left of sequence
(SeqB)

1.2 Parallelization

Parallelization or parallel computing is the use of multiple computational resources is a technique to distribute the task at hand to multiple resources and combine their outputs at the end to perform the task quickly and more efficiently. The use of parallel computing can provide scalability to the task. For example, the number of computing resources can be increased or decreased as per need. Parallelization can also be implemented using computing resources that are not physically present at a location, connected through the Internet. This type of parallel computing is called cloud computing or distributed computing based on how the multiple resources are connected. Different computing architectures yield different efficiencies. Thus, the performance of parallel computing is dependent on the architecture used.

1.3 Multithreading

An approach to parallelization is the concept of multithreading, wherein multiple cores of a single computational resource are used instead of multiple computational resources. Multithreading is a way to use the shared-memory architecture, wherein the alignments are stored in a heap memory, and this data can be accessed by multiple threads simultaneously. In a shared-memory architecture, a single copy of data can be accessed by multiple computing resources which is an efficient way to reduce the space requirement. This phenomenon is important as the space occupied by biological data can be huge. Multithreading has in itself a few problems like synchronization of data for multiple read-writes. This has to be taken into consideration in order to preserve the overall accuracy of the algorithm in consideration.

2 Survey

As stated, the Needleman–Wunsch algorithm is a dynamic programming technique used to perform pair-wise alignment of two biological sequences. Various other algorithms for performing pair-wise alignment have been proposed over the decades, but NWA remains unbeaten due to its inherent simplicity and ease of implementation. Needleman–Wunsch algorithm computes the alignment based on the output generated from a scoring matrix. The alignment score can be directly obtained from the scoring matrix. Using the parallelization technique namely multithreading, the areas of parallelization can be identified and implemented to compare the overall efficiency of the algorithm.

2.1 Parallelizing Needleman–Wunsch Algorithm Using Wavefront Framework

In the paper titled "Performance Improvement Of Genetic Algorithm For Multiple Sequence Alignment" by Anderson Rici Amorim et al. [3], the authors proposed that by parallelizing three main phases of MSA, the computational speed of performing MSA can be reduced to a great extent. The three phases include parallelization of Needleman–Wunsch algorithm, parallelization of tournament stage and parallelization of objective function. This paper formed as the basis for the proposed work. Of all the three stages, the parallelization of Needleman–Wunsch algorithm has had a significant impact on the overall computation of MSA.

The above paper uses a third-party framework called wavefront framework in order to achieve a parallel version of the Needleman–Wunsch algorithm. Using a third-party framework is not always reliable as they have a lot of dependencies, instead, the proposed work uses multithreading that is inherent in Java to perform parallelization.

2.2 Multi-threaded Approach of Needleman–Wunsch Algorithm

In the paper titled "A Multithreaded parallel implementation Of A Dynamic Programming Algorithm For Sequence Comparison", by Martins et al. [4], the authors propose various ways of calculating the scoring matrix and propose a new framework called earth framework to parallelize the algorithm in topic.

The proposed work takes their work into consideration and uses multithreading instead of a framework to perform the task at hand.

2.3 Modified Multi-threaded Approach of Needleman–Wunsch Algorithm

In another paper titled A Modified Dynamic Parallel Algorithm For Sequence Alignment In Biosequences by Nirmala Devi and Rajagopalam [5], the authors propose a method to further enhance the efficiency of the multi-threaded Needleman–Wunsch algorithm by calculating the values of three main diagonals namely: the main diagonal, the diagonal above the main diagonal and the diagonal below the main diagonal.

The above method has been taken into consideration in the proposed work but was found to have a toll on the accuracy on the alignment.

The various papers discussed above propose different techniques of parallelizing Needleman–Wunsch algorithm suffer from drawbacks at some stage. On one hand, if the drawback is a strict implementation due to various dependencies, the accuracy is a limitation on the other.

3 Proposed Approach

The Needleman–Wunsch algorithm can be applied in a parallel way by dividing the scoring matrix into parts and by feeding these parts to multiple threads and combining their output alignments at the end which represents the overall alignment of given sequences. Since the initialization and backtracking cannot be parallelized without affecting the accuracy of the algorithm, the second step of the Needleman–Wunsch algorithm, i.e. filling up the scoring matrix is identified to be parallelized.

3.1 Implementation Methodology

The proposed system is implemented in Java using shared-memory architecture to run and execute the multi-threaded versions of NWA. The proposed system runs on Windows Operating system as it facilitates synchronization in managing the resources inherently. The data set is not constant, like regular biological data and changes with time. The data are taken from the NCBI website and are downloaded in FASTA format. Different kinds of data are present in the NCBI website, and the prescribed data format for this experiment is chosen as FASTA as opposed to other data formats as it is easy to interpret and pass on the program as input.

Since the part of the algorithm chosen to be parallelized is Step 2, there are three possible ways of parallelizing the step. The first way is to split the scoring matrix of size $m \times n$ into t rows, where t represents the number of threads. The second way is to split the scoring matrix of size $m \times n$ into t columns, and the third way is to split the scoring matrix t diagonals.

Suppose, there are three threads and the length of sequence 1 and 2 is 6 and 5, respectively, using this method, Thread 1 calculates the rows 0–2, thread 2 calculates 3–5, and thread 3 calculates Row 6.

Figure 1 represents the splitting of the computation of scoring matrix of size $m \times n$ into t threads row-wise.

Suppose, there are three threads and the length of sequence 1 and 2 is 6 and 5, respectively, using the column-division and multi-threaded calculation of NWA, thread 1 calculates the columns 0–3, thread 2 calculates 4–6, and thread 3 calculates column 7.

Figure 2 represent the splitting of the computation scoring matrix of size $m \times n$ into t threads column-wise.

The process of multithreading is chosen as the method of parallelization as multithreading can be used on a single CPU as opposed to using multiple CPUs or computers. Multithreading is a type of computing technique that uses the cores of the CPU in order to perform tasks concurrently. Multithreading is different from multi-processing as in multithreading various computing resources such as memory, I/O devices and translation lookaside buffer (TLB) [6]. Since the memory is shared by multiple threads, this version of NWA can be considered as implementing on shared-memory architecture.

-	-	A	T	C	G	A	C
-	0	-4	-8	-12	-16	-20	-24
C	-4	-3	-7	-3	-7	-11	-15
A	-8	1	-3	-7	-6	-2	-6
T	-12	-3	6	2	-2	-6	-5
A	-16	-7	2	3	-1	3	-1
C	-20	-11	-2	-1	0	-1	8

Fig. 1 Needleman–Wunsch Algorithm implemented row-wise

Fig. 2 Needleman–Wunsch
Algorithm implemented
column-wise

-	-	A	T	C	G	A	C
-	0 → -4 → -8 → -12 → -16 → -20 → -24
C	-4	-3 → -7	-3 → -7 → -11 → -15
A	-8	1 → -3 → -7	-6	-2 → -6
T	-12	-3	6 → 2 → -2 → -6	-5
A	-16	-7	2	3 → -1	3 → -1
C	-20	-11	-2	-1	0	-1	8

4 Experiments

The process of evaluating software during the development process or at the end of
the development process is to determine whether it satisfies the given purpose or
not. The comparisons of the results between the existing system and proposed work
show the improvements and drawbacks if any in the proposed work.

4.1 Results

Experiments have been performed on sequence lengths ranging from 10 to 100,000.
The data have been taken from National Centre for Biotechnology Information
(NCBI)'s website [7]. The data considered for experiments are mostly nucleic acids
[8] of various organisms and sometimes protein sequences [9] of various organisms,
based on the sequence lengths. The number of optimal threads for different
sequence lengths can be calculated by considering the execution time.

Tables 1, 2, 3, 4 and 5 represent the comparison of execution time with a varying
sequence lengths and varying number of threads.

Table 1 shows the execution time of Needleman–Wunsch algorithm applied for
a sequence length of 10. Table 2 represents the execution time for varying number
of threads for the sequence length of 100, Table 3 for sequence length of 1000,
Table 4 for sequence length of 10,000 and Table 5 for the sequence length of
100,000.

Table 1 Execution time
versus the number of threads
for a sequence length of 10

Sequence length	10			
Number of threads	1	2	3	5
Execution time (in s)	0.899	0.657	0.657	0.604

Table 2 Execution time
versus the number of threads
for a sequence length of 100

Sequence length	100			
Number of threads	1	10	20	30
Execution time (in s)	2.120	1.980	1.560	1.960

Table 3 Execution time versus the number of threads for a sequence length of 1000

Sequence length	1000			
Number of threads	1	20	30	40
Execution time (in s)	6.340	5.120	4.320	4.70

Table 4 Execution time versus the number of threads for a sequence length of 10,000

Sequence length	10,000			
Number of threads	1	20	30	40
Execution time (in s)	10.40	8.99	7.20	8.9

Table 5 Execution time versus the number of threads for a sequence length of 100,000

Sequence length	100,000			
Number of threads	1	20	40	60
Execution time (in s)	101.40	98.70	89.9	91.8

The optimal number of threads can be projected taking execution time (time complexity) as the criteria.

The following figure shows the optimal number of threads and the projected equation to find the optimal number of threads. It can be projected by a logarithmic equation given below.

$$y = 21.675 \, In + 3.8464$$

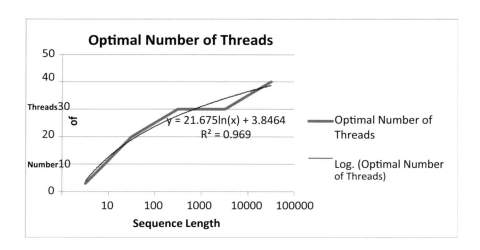

One of the previous research works we took into consideration is titled "Improving the performance of the Needleman–Wunsch algorithm using parallelization and vectorization techniques" by Jararweh, Yaser et al. [3] published in August 2017 in Springer Multimedia Tools and Applications deals with the

Table 6 Comparison between previous work and proposed approach

Results of improving the performance of the Needleman–Wunsch algorithm using parallelization and vectorization techniques					Results of our proposed work				
Sequence length					Sequence length				
Threads	2	4	8	16	Threads	2	4	8	16
10,000	6.414	5.646	5.594	6.027	10,000	6.047	4.898	4.769	5.102
14,992	15.273	13.365	12.193	13.817	14,992	14.978	13.01	11.659	14.023
20,000	36.361	25.148	22.652	30.721	20,000	32.027	24.209	22.578	25.489

diagonal multi-threaded approach and used different architectures to compare the results in terms of execution speed. We tried to use our proposed work with an i7 Processor mentioned in the above paper, and when the results are compared, our method showed an improvement of 7.47% in terms of execution speed. The detailed results are given below (Table 6).

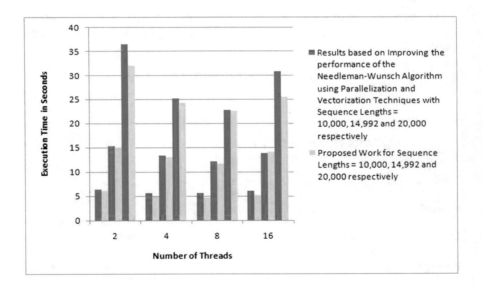

5 Conclusion

The proposed work attempts to enhance the performance of the most-widely used pair-wise alignment technique of bioinformatics called Needleman–Wunsch algorithm. The proposed work uses a single computing resource with multiple cores as opposed to multiple computing resources that are usually used in parallel computing. By using multithreading, the process of calculating the scoring matrix using NWA can be divided into various rows and columns based on the number of threads. Also, the use of multithreading facilitates the use of shared memory, wherein a single copy of data is shared among various threads, reducing the amount of storage space required. The results show the optimal number of threads depends on the sequence length and shows an overall improvement of execution speed by 7.47% compared to the results published in [3].

5.1 Future Work

The experiment has been performed on a PC with a basic configuration of an i5 Processor with a 4 GB RAM and 256 GB memory. This has limited the number of threads and the size of the sequence that can be used to perform the calculations. This version of multi-threaded NWA can be performed on a computer of a higher-end configuration of a network of computers in order to check the performance of the algorithm on them, and the results are to be compared. Also, the multi-threaded NWA can be extended to solve the problem of multiple sequence alignment (MSA), wherein more than two sequences are aligned.

Needleman–Wunsch algorithm can be applied not only to the field of bioinformatics, but it can be applied to other areas of computations as well where there is a need to identify differences and similarity between the text/data [10]. For example, consider the problem of identifying the items that have been bought together the most which is used in the areas of business and marketing, wherein the list of items purchased by each customer is taken into account, and by applying a sequence alignment algorithm like NWA, the sequence of items or the items that have been bought together the most can be identified, and comparisons can be drawn between this approach and the already existing approach of that area.

References

1. Mount, D.M. 2001. Bioinformatics: Sequence and Genome Analysis, 2nd ed. Cold Spring Harbor, NY: Cold Spring Harbor Laboratory Press. ISBN 0-87969-608-7.
2. Pairwise Sequence Alignment. https://gtbinf.wordpress.com/biol-41506150/pairwise-sequence-alignment/.
3. Performance Improvement Of Genetic Algorithm For Multiple Sequence Alignment By Anderson Rici Amorim, JoaMatheusVerdadeiroVisotaky, Allan De GodoiContessoto Leandro Alves Neves, Roge ŕ ia Cristiane Grata De Souza, Carlos Roberto Vale^Ncio, Geraldo Francisco Donega´ Zafalon.
4. Martins, W.S., J.B. Del Cuvillo, F.J. Useche, K.B.Theobald, and G.R. Gao. 2001. A Multithreaded Parallel Implementation of a Dynamic Programming Algorithm For Sequence Comparison.

5. Nirmala Devi, S., and S.P. Rajagopalam. 2012. A Modified Dynamic Parallel Algorithm for Sequence Alignment in Biosequences.
6. Nemirovsky, M., and D.M. Tullsen. 2013. *Multithreading Architecture*. San Rafael: Morgan & Claypool.
7. National Centre for Biotechnology Information. https://www.ncbi.nlm.nih.gov/.
8. Roberts, R. J. 2016. *Nucleic Acid*. Retrieved from https://www.britannica.com/science/nucleic-acid.
9. Biomolecules: Protein Retrieved from https://www.chem.wisc.edu/deptfiles/genchem/netorial/modules/biomolecules/modules/protein1/prot17.htm.
10. Needleman, S.B., C.D. Wunsch. 1970. Needleman-Wunsch Algorithm for Sequence Similarity Searches. Journal of Molecular Biology 48: 443–453.

Emotion-Based Extraction, Classification and Prediction of the Audio Data

Anusha Potluri, Ravi Guguloth and Chaitanya Muppala

Abstract With the rapid penetration of the Internet across the globe and increased bandwidth, the rise of audio data is significant. In one of the recent surveys on the Internet, it is mentioned that video and audio streaming will gobble up more than 50% of the Internet traffic. Moreover, with the recent rise of voice assistants, the significance to the audio data, especially voice data is at the zenith. In this background, there is a need to analyze the audio data for gathering significant insights which will have larger implications in the domains of health, marketing and media. In this project, an open-source approach is proposed to analyze the audio data using the acoustic features like Mel frequency cepstral coefficients (MFCCs), unlike converting the audio to text and performing the analysis on the converted textual data. In this work, a convolutional neural network (CNN) model is developed to predict the emotions on the given audio data.

Keywords Prediction · MFCC · Acoustic features · Audio data · Convolutional neural networks · Emotions

1 Introduction

Rise of audio data in the recent times had caught attention about its implications on many domains like health, media and marketing. The need to analyze and transform this data into meaningful and actionable insights will have significant impact in the way businesses operate or the way customers will be offered services. It will add the

A. Potluri (✉) · R. Guguloth · C. Muppala
Department of Computer Science and Engineering, Malla Reddy College
of Engineering and Technology (JNTUH Hyderabad), Hyderabad, India
e-mail: potluri.anusha1@gmail.com

R. Guguloth
e-mail: g.raviraja@gmail.com

C. Muppala
e-mail: chaitanyamuppala9@gmail.com

© Springer Nature Singapore Pte Ltd. 2020
K. S. Raju et al. (eds.), *Proceedings of the Third International Conference on Computational Intelligence and Informatics*, Advances in Intelligent Systems and Computing 1090, https://doi.org/10.1007/978-981-15-1480-7_26

value to most of the organizations by providing the competitive advantage over others. It can also be extended to the mental health care domain, by offering the capability like early diagnosis of mental conditions and providing better treatments.

One of key analysis on the data that can be done is to extract or elicit the sentiments and more importantly in the form of emotions. This can be achieved with much more sophistication and accuracy by combining with the artificial intelligence and machine learning methods. The application of neural networks and deep learning in the analysis of audio data had significantly increased in the recent times. Convolutional neural networks (CNNs) which have proved its applications in the analysis of image data, for performing the image classification, can be extended to the audio domain [1].

2 Motivation

Detecting emotions from the audio data is going to be marketing strategy in the coming days. With the identification of emotions, we can offer many personalized services to the customers or users. Couple of examples could be to that the call centers can play music when customer on call is angry which can be identified from the tone and voice. The automated self-driving cars can be slowed down when the person in the car is communicating angrily in the car. Businesses can offer the personalized advertisements or information which will be more relevant to the users when using the voice assistants.

However, with the current techniques the key emphasis is to convert the audio to text and then perform the sentiment extraction on the converted text [2]. In this work, the acoustic features of audio are used to predict the sentiments on the audio data. The acoustic features that are used in this work are MFCC [3].

Using MFCC, the speech can be represented in a manner that is elicited from the humans in an accurate manner. The sounds which are generated from human will take a form or shape, which is largely dictated by the shape of vocal tract. This tract includes tongue and teeth etc. The shape will determine what sound will come out of humans. From the proper determination of shape, we can also determine the phoneme that are also produced. This shape represents the short-time power spectrum envelopes and MFCCs will represent these envelopers accurately. Hence, using the acoustic features will offer an advantage in the analysis of the audio data over transforming the data to text.

3 Objective

The objective of the work is to predict the emotions from the audio data. The following are the emotions that are predicted in this work.

1. Angry
2. Calm
3. Fearful
4. Happy and
5. Sad.

Along with the predicting of the emotions, a classification of voices based on genders like male or female is also represented.

The key challenge for this work is to identify the emotional data from different sources. Using this data, an approach and implementation is presented to extract, classify and predict the emotions. In addition to the approach, usage of open-source tools and libraries in implementing the objective is also demonstrated. A clear leverage of machine learning and artificial intelligence methods to build the models is also highlighted.

4 Data

For this work, we have utilized two datasets.

1. RAVDESS Dataset

RAVDESS stands for Ryerson Audio-Visual Database of Emotional Speech and Song [4]. This dataset contains 1500 audio files. 24 different actors contributed voice to the data, out of which 12 are male and 12 are female actors. It contains the short audio recordings in eight different emotions. Each emotion is represented in the form of numbers, viz. 1 = neutral, 2 = calm, 3 = happy, 4 = sad, 5 = angry, 6 = fearful, 7 = disgust, 8 = surprised. For each audio file, the seventh character is consistent with different emotions.

2. SAVEE Dataset

SAVEE stands for Surrey Audio-Visual Expressed Emotion [5]. There are 500 audio files in this dataset. These audios are recorded by 4 different English speaking male speakers. Each speaker is identified by the letters DC, JE, JK, KL as a file name identifier. The speakers constitute of postgraduate students and researchers at the University of Surrey, whose age group falls between 27 and 31 years. The data set comprises of the following emotions—anger, disgust, fear, happiness, sadness and surprise. The emotions are supported by cross-cultural studies performed at Ekman [6] and other studies of automatic emotion recognition which mainly focused on the mentioned emotions. There is also a neural category in the emotions.

5 Approach

One of the key components of the work is to compute the Mel-cepstral coefficients (MFCCs) for the audio files in the data sets. After obtaining the MFCCs, a model is build using the CNNs. Python library, Librosa is used to extract the acoustic features from the audio. Librosa is a popular Python package used for music and audio analysis. This library provides the necessary building blocks that are required to create the audio information retrieval systems. Librosa is used to extract the MFCCs from the given audio data. We also separated out the male and female voices by the file identifiers that are mentioned in the data section.

After the extraction of MFCCs, each audio file contains an array of features with many values. The features are appended with the labels that are extracted in the above step. However, for couple of audio files which are shorter in length there are couple of missing features. For these files, the sampling rate is increased by twice and it resulted in eliciting the unique features for each speech. Any further increase in the sampling rate will attract noise in the collected data and will affect the results. Data is shuffled to introduce the randomness in the collected features and eliminate bias. For cross validation, the data is split as train and test data. A convolutional neural network (CNN) model is built using the train data. The CNN model is built as a sequential model with a linear stack of layers. ReLu, which stands for rectified linear unit is used as activation function. Softmax function is used in the final layer of neural network. The model is finally validated with test data. The approach can be represented on a high level is represented as shown below in the Figs. 1 and 2.

6 Implementation and Results

For implementing the work, Python [7] is used as base programming language. For performing the complex mathematical operations—the popular NumPy [8] package and to build the machine learning models—scikit-learn [9] is used. Tensorflow [10] and Keras [11] are used to build the CNNs. Both Tensorflow and Keras contain many functions out of the box to optimize the model as well. Since we are dealing with the complex conversions and mathematical operations, which are computational heavy, Intel i7 8th generation processors are used along with Nvidia 1080 Ti graphics card and 16 GB RAM. However, the model can also be trained on low-end configuration, but it will take long hours to generate the CNN model.

As discussed in the approach, using the Librosa the audio files are divided into windows and hamming is applied on the small intervals as shown in the Fig. 3.

As represented in the Fig. 4, the MFCCs are calculated using Librosa. The below list contains the sample of features in the form of array.

A convolutional neural network model is generated with multiple layers as represented in the Fig. 5. After successful generation of the model, the test data is used to validate the model. The results are as shown in the following results section.

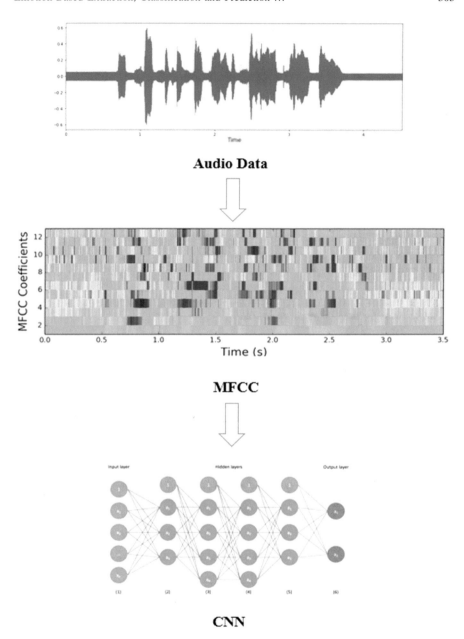

Fig. 1 Statistical approach used in the work, which is to extract the MFCC from the audio data and detect the emotions from the given audio data

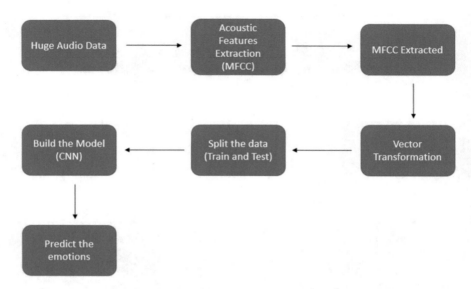

Fig. 2 High-level steps used in the work, which is to extract the MFCC from the audio data and detect the emotions from the given audio data

Fig. 3 Audio divided into shorter windows

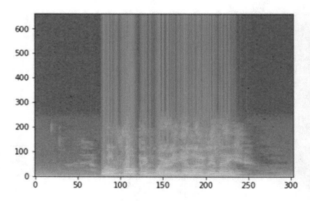

Fig. 4 MFCC features

feature
[-70.2677641611, -70.2677641611, -70.267764161...
[-65.7076524007, -65.7076524007, -63.114722422...
[-65.4824988827, -65.4824988827, -65.482498882...
[-64.5284491035, -64.5284491035, -64.528449103...
[-62.3643105275, -59.9347251381, -61.869599961...

```
:  model.summary()
```

Layer (type)	Output Shape	Param #
conv1d_7 (Conv1D)	(None, 216, 128)	768
activation_8 (Activation)	(None, 216, 128)	0
conv1d_8 (Conv1D)	(None, 216, 128)	82048
activation_9 (Activation)	(None, 216, 128)	0
dropout_3 (Dropout)	(None, 216, 128)	0
max_pooling1d_2 (MaxPooling1	(None, 27, 128)	0
conv1d_9 (Conv1D)	(None, 27, 128)	82048
activation_10 (Activation)	(None, 27, 128)	0
conv1d_10 (Conv1D)	(None, 27, 128)	82048
activation_11 (Activation)	(None, 27, 128)	0
conv1d_11 (Conv1D)	(None, 27, 128)	82048
activation_12 (Activation)	(None, 27, 128)	0
dropout_4 (Dropout)	(None, 27, 128)	0
conv1d_12 (Conv1D)	(None, 27, 128)	82048
activation_13 (Activation)	(None, 27, 128)	0
flatten_2 (Flatten)	(None, 3456)	0
dense_2 (Dense)	(None, 10)	34570
activation_14 (Activation)	(None, 10)	0

```
Total params: 445,578
Trainable params: 445,578
Non-trainable params: 0
```

Fig. 5 Convolutional neural network model summary

6.1 Results

The predictions on the test data have shown more than 70% accuracy in predicting the emotions. The results of actual values versus the predicted values are represented in the Figs. 6 and 7. When predicting the emotions on the recorded audio outside the dataset, we were able to achieve more than sixty percent of accuracy in predicting the emotions.

The classification report of Fig. 7 clearly demonstrates that the accuracy is 60%.

actualvalues	predictedvalues
female_fearful	female_fearful
male_angry	male_angry
male_fearful	male_fearful
male_happy	male_happy
female_happy	female_happy
female_angry	female_angry
female_angry	female_sad
male_sad	male_calm
male_angry	male_calm
male_sad	male_sad

Fig. 6 Results which demonstrates the comparison of actual and predicted values

```
                  precision    recall  f1-score

  female_angry       0.71       0.55      0.61
   female_calm       0.65       0.65      0.65
female_fearful       0.78       0.56      0.65
  female_happy       0.55       0.57      0.55
    female_sad       0.67       0.55      0.60
    male_angry       0.62       0.60      0.61
     male_calm       0.69       0.47      0.56
   male_fearful      0.70       0.65      0.67
    male_happy       0.54       0.57      0.55
      male_sad       0.57       0.59      0.58

   avg / total       0.50       0.47      0.60
```

Fig. 7 Classification report with precision, recall and f1-score values

7   Conclusion

With the current implementation of building the model using the CNN, seventy percent accuracy in predicting the emotions is achieved. The model is also able to clearly classify between male and female voices. Since the model is constructed using the acoustic features, it can easily be extended to other languages apart from English, in predicting the emotions, provided there is a dataset in the intended language. A further extension to the work could be to build a real-time emotion detection system which will have larger implications on many domains.

References

1. Abdel-Hamid, Ossama, Abdel-rahman Mohamed, Hui Jiang, Li Deng, Gerald Penn, and Dong Yu. 2014. Convolutional Neural Networks for Speech Recognition. *IEEE/ACM Transactions on Audio, Speech and Language Processing* 22 (10).
2. Kaushik, Lakshmish, Abhijeet Sangwan, John H. L. Hansen. 2017. Automatic Sentiment Detection in Naturalistic Audio. *IEEE/ACM Transactions on Audio, Speech, and Language Processing* 25 (8).
3. Navya Sri, M., M. Ramakrishna Murty, et al. 2017. Robust Features for Emotion Recognition from Speech by using Gaussian Mixture Model Classification. In International Conference and Published Proceeding in SIST Series, Vol. 2, 437–444. Berlin: Springer.
4. Ezgi Küçükbay, Selver, Mustafa Sert. 2015. Audio-Based Event Detection in Office Live Environments Using Optimized MFCC-SVM Approach. In Proceedings of the 2015 IEEE 9th International Conference on Semantic Computing (IEEE ICSC 2015), 475–480.
5. Livingstone, S.R., and F.A. Russo. 2018. The Ryerson Audio-Visual Database of Emotional Speech and Song (RAVDESS): A dynamic, multimodal set of facial and vocal expressions in North American English. *PLoS ONE* 13 (5): e0196391. https://doi.org/10.1371/journal.pone. 0196391.
6. Haq, S., and P.J.B. Jackson. 2010. Multimodal Emotion Recognition. In *Machine Audition: Principles, Algorithms and Systems*, ed. W. Wang, 398–423. Hershey: IGI Global Press. ISBN 978-1615209194. https://doi.org/10.4018/978-1-61520-919-4.
7. Sauter, Disa A., Frank Eisner, Paul Ekman, and Sophie K. Scott, Cross-Cultural Recognition of Basic Emotions Through Nonverbal Emotional Vocalizations. *Proceedings of the National Academy of Sciences of the USA.*
8. Van Rossum, G. 1995. Python Tutorial, Technical Report CS-R9526, Centrum voor Wiskunde en Informatica (CWI), Amsterdam.
9. van der Walt, Stefan, Stefan van der Walt, and Stefan van der Walt. 2011. The NumPy Array: A Structure for Efficient Numerical Computation. *Computing in Science and Engineering Journal* 13 (2): 22–30.
10. Pedregosa, et al. 2011. Scikit-learn: Machine Learning in Python. *Journal of Machine Learning Research* 12: 2825–2830.
11. Goodfellow, Ian, Derek Murray, and Vincent Vanhoucke. 2015. TensorFlow: Large-Scale Machine Learning on Heterogeneous Distributed Systems. *Google Research*.
12. Keras Documentation—https://keras.io/.
13. Kalyani, G., M.V.P. Chandra Sekhara Rao. 2017. Particle Swarm Intelligence and Impact Factor-Based Privacy Preserving Association Rule Mining for Balancing Data Utility and Knowledge Privacy. *Arabian Journal for Science and Engineering.*

Web Service Classification and Prediction Using Rule-Based Approach with Recommendations for Quality Improvements

M. Swami Das, A. Govardhan and D. Vijaya Lakshmi

Abstract Web-based applications are more popular and are increasing the demand to use of applications. Service quality and user satisfaction are the most significant for the designer. IT provides the best solutions for various applications such as B2B, B2C, e-commerce and other applications. The client demands high-quality service, regarding minimum response time, high availability and more security. The existing approaches and models may not improve the overall performance of web-based application due to not covering all functional, nonfunctional parameters. The proposed model-approach gives the best solution using rule-based classification of web services using QWS dataset, and predictions of quality parameters based on user specifications. The model is implemented in Java, the results of quality parameters will classify and predict the class labels Class-1(high quality), Class-2, Class-3 and Class-4 (low quality), and the system will give recommendations to improve the quality parameters. By using suggested guidelines and instructions to the software developer, that he will meet the client specifications and provides best quality values which will improve the overall performance (in specification parameters including functional and nonfunctional values). The result clearly suggests the improvement of quality parameters by classification and prediction. This paper can be extended to mixed attributes of quality parameters.

Keywords Web service · QoS · Rule-based classification · Prediction · Performance · Knowledge discovery data

M. Swami Das (✉)
Department of CSE, MREC, Hyderabad, India
e-mail: msdas.520@gmail.com

M. Swami Das · A. Govardhan
Jawaharlal Nehru Technological University Hyderabad, Kukatpally,
Hyderabad, Telangana State, India
e-mail: govardhan_cse@yahoo.co.in

D. Vijaya Lakshmi
Department of IT, MGIT, Hyderabad, Telangana State, India
e-mail: vijayadoddapaneni@yahoo.com

© Springer Nature Singapore Pte Ltd. 2020
K. S. Raju et al. (eds.), *Proceedings of the Third International Conference on Computational Intelligence and Informatics*, Advances in Intelligent Systems and Computing 1090, https://doi.org/10.1007/978-981-15-1480-7_27

1   Introduction

Web services are the functional services over networked communication and they are independent for clients to invoke the operations. The development of web services used technologies such as SOAP, UDDI, XML, and WSDL. Clients demand services in high quality. To design a web service is to find the best among the many services, web technology plays an important role, the designer needs to design and develop the web-based application as per client requirements and quality standards. The infrastructural facilities are PCs, communication facilities, programming languages which are used and the management of components by the developer in the organization. Web services can be integrated the components based on usage, for example, EAI, B2B integration, applications, loosely coupled, dynamic interactions, the development of distributed and complex applications meet the industry demand and quality constraint according to user specifications.

The web service quality is two types, namely functional parameters and nonfunctional parameters. Functional parameters include response time, functionalities and nonfunctional parameters such as cost and security. Web-based applications, for example, are B2B, CRM, and reservation etc. Some web services are related to weather, stock, reservation, Google search engine and Amazon web services. The end user seeks web services to provide high availability, minimum response time, high reliability and security. The accessibility of published services are relevant to specific domain(s) and applications. The earlier web pages used HTTP protocols and flow of control operations, where now use SOAP—XML based protocols and flow of transactions by the use of the Internet. The statistics report according to Global Digital Suite on World population in 2018 is 7.59 billion, where the Mobile users 5.13 billion (68%), internet usage people is 4.02 billion (53%), active social is 3.196 (42%) and active mobile social user 2.95 billion (39%) [1]. This shows the application Utilities, Business Services, Internet of Things, apps, and web-based applications, require to meet the requirements and quality parameters such as high availability, minimum response time, and more security. The advantage is to reduce the physical distance between user and web applications, anywhere around the world. The applications with the use of devices like mobile, handheld portable devices use of mobile apps are very user-friendly, portable, and compatible with the users. For example health care, automated applications systems like home appliances, transportation, enterprise, government, and other applications including cloud-based, web applications, and IoT.

The main objective of the paper is to design and develop a web-based application which meets the client specifications and quality parameters with industry standards including functional and nonfunctional parameters. The existing solutions may not resolve the best solution for the development of the application. We propose an enhanced model, the technique is best suitable for designer throughout the software development lifecycle (SDLC) and maintenance with required quality parameters, using machine learning (classification and prediction techniques). These models propose algorithms will improve the performance. For example, the applications which client can use invoke operations with minimum response time, high availability.

However, the new challenge task of quality web service parameter is applying the classification and prediction of web services; these recommendations will help the designer in the initial stages of application and suggested recommendations to be followed in the design and development of web-based applications which will lead to overall improvement. Section 2 deals with related work, Sect. 3 describes the proposed model, Sect. 4 deals with implementation, Sect. 5 results and discussion, and Sect. 6 ends with a conclusion and future scope.

2 Related Work

Weiss et al. [2] suggested that the classification of web services was done by feature modelling approach through requirements, design, and implementation processes. The feature interactions analysis by building the model with features, goals, and analysis of goals, resolving interactions models. The disadvantages are goal conflicts for developing applications due to multiple feature interactions. The goals are ranked satisfied, weakly satisfied, not satisfied, denied, conflict. Zhang et al. [3] applied web service classification with WSDL file using data mining and classification algorithm (i.e., AdaBoost, Decision Trees, Naive Bayes, Random Forest and find the accuracy) by feature attribute using 500 WSDL related dataset and is classified into business, computing, documentation, financial, news, and telephone types values.

Wang et al. [4] used web service classification by SVM and hierarchical classification functional features, UNSPC standards, and taxonomy for finding document type. Use of experiments by OWLS dataset 1007 records into categorical values is as follows: (1) Travel, (2) Education, (3) Weapon, (4) Food, (5) Economy, (6) Communication, (7) Medical. Mohanty et al. [5] applied various machine learning techniques BPNN, PNN, GMDH, J48, CART methods. In these methods, J48 and CART method got more accuracy as compared with other methods. Nonfunctional parameters are not addressed.

Bunkar et al. [6] applied classification rules and techniques (i.e., the techniques used for learning process ID3, CART, and C4.5) to the prediction of performance of graduate students (i.e., final grade) from Vikram University, BA course, 2009, and predictions. Bennaceur et al. [7] used web service classification by machine learning SVM method, text document category into web and news business automatic composition and classifications.

Mohanty et al. [8] applied web service classification using techniques, Markov Blanket: Bayesian network, Tabu search, Markov blanket. Naive Bayes classifier accuracy is 85.6%, markov blanket accuracy is 81.3%, and Tabu Search accuracy is 82.4% which does not improve the overall quality. Syed Mustafa et al. [9] used the QWS dataset with techniques KNN, SVM, FURIA, and RIDOR, and among this, FURIA Fuzzy Unordered Rule Induction Algorithm method got the accuracy of 80.55%. Nonfunctional parameters do not cover and do not improve the overall quality.

Yang et al. [10] applied web service classification using machine learning algorithms C 4.5, SVM, Naive Bayes, Decision Tree, BPNN, and C 4.5 has the highest accuracy. The data source is OWLS-TC4, and the document is classified into nine domains (1) communication (2) economy (3) education (4) food (5) geography (6) medical (7) simulation and (8) weapon, etc. with the small dataset used.

Hamdi et al. [11] classification service behavior, the taxonomy of service, and rules QWS data into platinum(high quality), gold, silver, and bronze(low quality), using web service relevancy function. Behavior of service by {low, medium, high} quality Behavior by formal specifications, maximum 1 and minimum range value 0 (i.e., 0–100%), improvement of characteristics ascending values in between 0 and 1. Kvasnicovaa et al. [12] applied e-service (e-government, e-learning, e-commerce) classification by products and services as high, moderate, and low. For example electronic services, goods, etc.

Reyes-Ortiz et al. [13] used web service classification and ontology with WSDL features text into communication, economy, etc. Here rule-based C 4.5 methods with dataset 899 OLWSTC dataset used. Kumar et al. [14] used web service quality by PCM and SVM techniques, text classification with QoS parameters Availability, response time, latency, compliance, documentation, success ability, throughput, and maintainability ontology words into various types like types like Weapon, communication food, geography, etc.

Xiong et al. [15] goal is web service recommendations, and web application development because of poor services, the techniques used deep learning, hybrid approaches WS functionality and recommendations, Collaborative Filtering techniques. Zhang et al. [16] web service discovery and service goals, into seven types textual descriptions, from discovery SVM, NLP, using 13,520 Web API java services. Hybrid service discovery, keywords, and topic approach. Nonfunctional parameters not addressed. Al-Helal et al. [17] proposed quality of web service selection by re-planning, re-selecting, individual service elements based on response time, cost, availability, reliability, and other quality parameters. Probabilistic distribution, the results are compared the algorithms without replaceability, and with replaceability. Machine learning technique, feature extraction, using a classification technique predicts the label. Input with feature extraction and classification of labels. For example, deep learning: a self-driving car. Lalit et al. [18] proposed a Promethee Plus model for classification of web service nonfunctional parameters, not addressed and not improving the overall quality. The QWS data set [19] was used for experiments. Das et al. [20] proposed a model web service classification using various data mining algorithms and the performance of quality parameters was discussed. The behavior of service is selected, the Hamdi et al. [21] proposed classification service behavior using rough set. Xiao et al. [22] suggested finding service, semantic classification by Naive Bayes method. Kalyani, G et al. [23] used rule mining approach for knowledge discovery. R. Murthy et al. [24] used text document classification, Adeniyi et al. [25] used RSS reader data using KNN method to identify the client stream data. Bakraouy et al. [26] proposed availability of Web services based on classifications.

2.1 Motivation of Work

Service quality issues of web service include, data validations, workflow, service access, availability, easy access, apps used for mobile devices, portable handheld devices for easy access, customers in the organizations functional parameters high response time, low availability, not followed best practices in development of software applications, and maintenance technical (service not available, features, high response time, failed to provider service, non technical issues (failed system, maintenance cost more, error during service invocations). Access financial transactions, security and more cost for development of web-based enterprise and application development integration. Moreover, security issues, Service quality issues, Data validations—workflows, service access—availability, Easy access—apps used in mobiles portability, adaptability, Service providers—consumers—organizations, Design—develop—web service—business systems, automated, real-time applications. An application like e-business, e-government, automatic control systems.

The problem is to classify the web services into Class-1 (High quality), Class-2 (Best quality), Class-3 (Average quality) and Class-4 (Poor quality services). The proposed model applied to this problem is to solve web service classifications using Input dataset, Learning method (Classification Technique) and Training data to classify the data into categorical values.

3 Proposed Model

The proposed Model for web service classification is shown in Fig. 1, which follows KDD steps for classification and predictions, read the data shown in Table 2 (QWS) [24], pre-preprocessing data using Normalization values, and feature extractions. Here we took five parameters response time, availability, best practices, cost and security parameters. Training data(classified labels), testing data to be tested input data set, and learning models (classification methods) are used for classification and predictions using Algorithm 1, Algorithm 2 and Table 1 provides a functional and nonfunctional score and classifications.

Features data, relevant objects in the domain. In this case, QWS data contains various quality parameters, response time, availability, throughput, successability, reliability, compliance, best practices, latency, and documentation. The response time to be minimized is described in Eq. 1.

$$\text{Response time} = \text{Completion time} - \text{Requested time} \tag{1}$$

Completion time depends on CPU scheduling algorithms, load, system, and network bandwidth, number of users, number of applications etc. Availability to be increased as described in the Eq. 2.

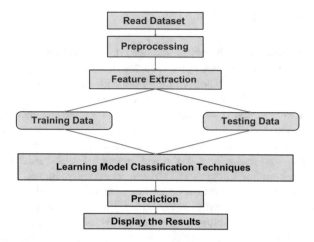

Fig. 1 Web service classification and prediction approach

Table 1 Rules for functional score, nonfunctional score, total score and classification

Functional score	Non functional score	Total score	Classification
if(f_rank=1) then F_score=0.80	If(n_rank=1) then n_score=0.2	if((t_score $\leq$ 1) && (t_score>0.8))	Class-1
If(f-_rank=2) F_score=0.70	if (n_rank=2) then n_score=0.15	If((t_score $\leq$ 0.8) && (t_score>0.7))	Class-2
If(f_rank=3) (F_score=0.60)	if(n_rank=3) then n_score=0.1	if((t_score $\leq$ 0.7) && (t_score>0.6)	Class-3
If(F_score=0.50)	If(N_score=0.05)	If (t_score<0.6)	Class-4

$$\text{Availability} = \frac{\text{uptime}}{(\text{Uptime} + \text{downtime})} \qquad (2)$$

Availability depends on server load, network protocols, services, and FTS systems etc. Throughput to be increased as described in Eq. 3.

$$\text{Throughput} = \frac{\text{Total innovations}}{\text{Period of Time}} \qquad (3)$$

Is the capacity of bits achievable from point to point depends on network bandwidth, server availability, system configurations, and a number of users etc. accessibility to be increased by Successability is described in Eq. 4.

$$\text{Successability} = \frac{\text{Number of responses}}{\text{Number of requests}} \tag{4}$$

Reliability to be increased (i.e., failure is very less) Compliance—WSDL file to locate an information source. Compliance is WSDL specifications. Best practices—web service extensions. Latency -depends on the software component to be minimized is described in Eq. 5.

$$\text{Latency} = \text{Response time} - \text{Requested time} \tag{5}$$

Algorithm 1: Classification of Web Service

```
Input:     Quality of parameters { A₁, A₂,...,Aₙ}
Output:    Classifying the Web service into categorical
           values Cᵢ={C₁, C₂,.., Cₙ} where Cᵢ = 4,
Class-1: High quality, Class-2: medium, Class-3: average
and Class-4: Poor quality.
begin
      Step1. Read the input data
      Step2: Preprocessing the data
      Step3: Apply rule-based classification algorithm
      Step4: Predict the rank and recommendations of the
             designer
      Step5: Based on recommendations proceed for design
      and
             development.
end
```

The most critical parameters for web service design functional parameters are response time, availability and best practices; Nonfunctional parameters are cost and security. The algorithm one which read the required parameters response time, availability, best practices, cost and security parameters for QWS dataset and applying rule-based classification algorithm, rules defined in Algorithm 2, and find the values classification and predictions, recommendations to the user.

Algorithm 2: Classification and Prediction of Web Service—Rule-Based Approach

Input: QWS data parameters related attributes Input data set {Response time (RT), Availability(AV), Best Practices(BP), Cost(CO) and ,Security(SE)} we took normalized values (i.e lying between 0 and 1)

Output: Measure the Classification of Web service and prediction
Categorical values Classification service ={1,2,3,4}, Where class labels indicates, in Class-1 high quality(platinum), Class-2(Gold) Class-3(Silver) and class-4 (Bronze) poor service
Variables:
        Functional_score= 0;
        Non_functional_score =0;
        F_rank // Functional parameter rank
        N_rank // Non Functional Rank
Begin
Step1: Read the Input dataset RT, AV, BP, CO and SE
        // Read normalized values quality parameters lying
            between 0 and 1
        // 0 Means Minimum ,and 1 Means maximum (i.e in
            Percentage 0 % to 100%)
Step2: Apply the Classification Methods to find the
        Classification (Decision Rules)
Step3. Decision Rules for Functional Parameters
        If(RT ≤0.25 && AV ≥0.75)
        then
            F_Rank =1;
        else if(RT ≤0.50 && RT >0.25)&&(AV≤0.75&& AV >0.50)
            F_Rank=2;
        else if((RT≤0.75&&RT >0.5) && (AV≤0.50 && AV >0.25)
            F_Rank =3;
        else F_Rank =4;
        end if Display the Functional Rank(i.e F_Rank)
        // To find the Functional Score
        If (f_rank=1) then
        Functional_score=0.80;
        else if(f-_rank=2)
        {
            Functional_score=0.70; Display the RT < 0.25
            and AV to be improved Above 0.75 to attain
            Functional service Class-1
        }
        else if (f_rank=3)
        {
            Functional_score=0.60; Display RT between 0.25
            to 0.5 and AV to be improved in between 0.5 to
            0.75 to attain Functional service Class-2

```
        }
        else
        {
                Functional_score=0.50; Display the RT beteen
                0.5 to 0.75 and AV to be improved in between
                0.25 to 0.50 to attain Functional service
                Class-3
        }
Step4. Display the Non Functional_score value;
Step5. Decision Rules for Non Functional parameters for
       three variables
// BP - best practices, CO- Cost, and SE- security values
        If (BP≥0.75)&&(CO≥0.75) &&(SE≥0.75)
        Then N_rank=1;
        else If((BP≥0.50)&&(BP<0.75))&&(( CO≥0.50)
             &&(CO<0.75)) &&( SE≥0.50) &&(SE<0.75) Then
        {
                N_rank=2;
                Print the quality value BP, CO and SE to be
                improved more than 0.75 to attain Class-1
        }
        Else if(((BP≥0.25)&&(BP<0.50))&&((CO≥0.25)
        &&(CO<0.50)) &&( SE≥0.25) &&(SE<0.50)then
        {
                N_rank=3;
                Print quality values BP,CO and SE values in
                between 0.5 and 0.75 to attain class-2
        }
        else
        {
                N_rank=4;
                Display the quality values BP, CO and SE
                values in between 0.25 to 0.5 to attain
                class-3
        }
        }
Step.6 Display N_rank of Web service
        // To find the non_functional score
Step.7.Find Non Functional Score for Three variables BP,
       CO, and SE
                If(n_rank=1) then n_score=0.2;
                else if (n_rank=2)then n_score=0.15;
                else if(n_rank=3) then n_score =0.1
                else N_score =0.05;
Step 8. Display the  N_score value
Step 9. Compute the Total score  Functional and
        Non_functional score
                t_score = functional_score+n_score;
```

```
        Dsipaly t_score value
if(t_score≤1 && t_score>0.8) class=1;
else (t_score≤0.8 && t_score>0.7){
        class=2;
        Display in order to attain Class-1
        functional/Non functional RT,AV, BP, CO and
        SE to be improved above 0.75
}
else if((t_score≤0.7 && t_score>0.6){
        class= 3;
        Display in order to attain class-2 functional
        and Nonfunctional RT, AV, BP CO and SE values
        in between 0.5 to 0.75
}
else {
        class=4;
        Display in order to attain class-3 functional
        And Nonfunctional RT, AV, BP CO, and SE
        values in between 0.25 to 0.50
}}
```
Step.10 Display the Web service Classification labels
and predictions.
        The given web service is (Class-1, or 2, or 3 or 4)
end

Algorithm 2 describes classification and prediction read the QWS data response time, availability, best practices, cost and security (including functional and non-functional). Functional score, nonfunctional score used for Functional rank, non-functional rank, make the decision rules based on Eqs. (1)–(5), and display function rank, F-score, find the nonfunctional score in Step 7, compute the total score the sum of the functional and nonfunctional score, based on total score (t_score) classification of web service values class-1, 2, 3 and 4. In step 9 based on classifications and predictions, recommendations to improve the given parameters is described in Algoirthm 2 and Table 1. Finally, display the class labels with predictions.

4 Implementation

The development of web services for business applications with integration on new modules. The applications are easy to change the code and re-use code when it is required to optimize the cost of applications. Growing the need offers the quality web services in terms of standard design, development, and deployment. The quality parameters in the development process are high priority, availability, response time, best practices, security and cost. The implementation of web services

Table 2 QWS dataset parameters

Id	Feature name	Description
A_1	Response time	Time taken to respond and receive a response from web server
A_2	Availability	Number of successful invocations over total invocations
A_3	Throughput	Total number of invocations over a time
A_4	Successability	Number of response over a number of requests
A_5	Reliability	The ratio of number of error messages to total messages
A_6	Compliance	WSDL document specification
A_7	Best practice	Extend Web service WS-I basic profile
A_8	Latency	Time taken to process request at the server
A_9	Documentation	WSDL documentation
A_{10}	WsRF	Web service Relevancy function: Rank of web services
A_{11}	Service name	Web service name
A_{12}	WSDL address	Locate WSDL service

classification and predictions is implemented using Java based on Algorithm 1 by considering these quality parameters QWS data described in Table 2 and, web service classification and Algorithm 2 using rule-based approach. The describes rules in rule-based approach defined in Table 1 for classification and prediction the results shown in Fig. 2a shows the code is implemented in Java which accepts the input parameters, and Fig. 2b displays the results of classifications and predictions to the software developer in development of software which includes analysis, design, implementation, quality analysis, and re-design and implement based on suggestions by the proposed classification model-approach.

(a)   **(b)**

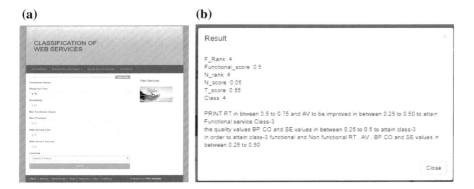

Fig. 2 a Classification of web services, b classification, predictions, and recommendations

5 Results and Discussions

The web service classification labels and predicts the given web service as Class-1 (which is the highest quality), Class-2 (Gold), Class-3 (Silver) and Class-4 (low quality). The Algorithm 1 (Classification) and Algorithm 2 (Classification and Predictions using rule-based approach), the input accepts the text forms (number) values lying between 0 (minimum) and 1(maximum). The classification is based on the Functional and Nonfunctional score of the parameters. Classifying label (Class-2) means the required functional and nonfunctional parameters to be improved/ recommended values will be displayed to attain the highest quality (i.e., Class-1).

6 Conclusion and Future Scope

The growth of web-based application, everyday, the developer follows the best practices and quality parameter recommendations based on analysis and report of classification and prediction of quality parameters, the advancements of security, web, and cloud-based applications which are connected with IoT systems. The existing systems may not improve the overall performance of web service quality applications. The proposed approach (repesented in Fig. 1) does the classification and prediction using Algorithm 1 classification of web services and Algorithm 2. Rule-based approach classification and prediction methods. The application which accepts the functional and nonfunctional parameters based on client specifications and read input data classify and predictions of web services. The model-approach helps the software developer in the prior implementation, that he can choose the best design and implementation of software to improve the overall performance of web applications. In future, this paper can be extended for mixed attributes.

Acknowledgements Thanks to Dr. Eyhab Al-Masri for providing QWS dataset 2507 records. Thanks to Dr. R. K. Mohanty, Professor, KMIT, Hyderabad for his support.

References

1. Global Digital Report. https://wearesocial.com/blog/2018/01/global-digital-report-2018.
2. Weiss, M., B. Esfandiari, Y. Luo. 2007. Towards a Classification of Web Service Feature Interactions. *Computer Networks* 51: 359–381.
3. Zhang, Jing, Dan Pan. 2008. *Web Service Classification*. https://pdfs.semanticscholar.org.
4. Wang, Hongbing, Yanqi Shi, Xuan Zhouy, and Qianzhao Zhou. 2008. Web Service Classification using Support Vector Machine. In International Conference on Tools with Artificial Intelligence, 1–6. IEEE.
5. Mohanty, Ramakanta, V. Ravi, and M. R. Patra. 2010. Web-Services Classification Using Intelligent Techniques. *Expert Systems with Applications* 5484–5490.
6. Bunkar, K., U.K. Singh, B. Pandya, and R. Bunkar. 2012. Data Mining: Prediction for Performance Improvement of Graduate Students Using Classification. In 2012 Ninth

International Conference on Wireless and Optical Communications Networks (WOCN), Indore, 1–5.

7. Bennaceur, Amel, Valerie Issarny, Richard Johansson, Alessandro Moschitti, Romina Spalazzese, and Daniel Sykes. 2012. *Machine Learning for Automatic Classification of Web Service Interface Descriptions*. http://dit.unitn.it/moschitti/articles/2012/ISOLA2011.pdf.

8. Mohanty, Ramakanta, V. Ravi, M.R. Patra. 2012. Classification of Web Services Using Bayesian Network. *Journal of Software Engineering and Applications* 291–296.

9. Syed Mustafa, A., and Y.S. Kumaraswamy. 2014. Performance Evaluation of Web-Services Classification. *Indian Journal of Science and Technology* 7 (10), 1674–1681.

10. Yang, Jie, and Xinzhong Zhou. 2015. Semi-automatic Web Service Classification Using Machine Learning. *International Journal of U and e-service, science and Technology* 8 (4): 339–348.

11. Hamdi, Yahyaoui, Hala Own, and Zaki Malik. 2015. Modeling and Classification of Service Behaviors. *Expert Systems with Applications* 1–24.

12. Kvasnicovaa, Terezia, Iveta Kremenovaa, and Juraj Fabusa. 2016. From an Analysis of e-Services Definitions and Classifications to the Proposal of New e-Service Classification: 3rd Global Conference On Business, Economics, Management and Tourism, 192–196. Amsterdam: Elsevier.

13. Reyes-Ortiz, Jose A., Maricela Bravo, and Hugo Pablo. 2016. Web Services Ontology Population through Text Classification. In Proceedings of the Federated Conference on Computer Science and Information Systems, 491–495.

14. Kumar, Lov, Santanu Rath, and Ashish Sureka. 2017. Estimating Web Service Quality of Service Parameters Using Source Code Metrics and LSSVM. In 5th International Workshop on Quantitative Approaches to Software Quality (QuASoQ 2017), 66–73.

15. Xiong, Ruibin, Jian Wang, Neng Zhang, and Yutao Ma. 2018. Deep Hybrid Collaborative Filtering for Web Service Recommendation. *Expert Systems with Applications* 191–205.

16. Zhang, Neng, Jian Wang, Yutao Ma, Keqing He, Zheng Li, and Xiaoqing Liu. 2018. Web Service Discovery Based on Goal-Oriented Query Expansion. *The Journal of Systems and Software* 73–91.

17. Al-Helal, H., and R. Gamble. 2013. Introducing Replaceability into Web Service Composition. *IEEE Transactions on Services Computing* 1–30.

18. Lalit, Purohit, Sandeep Kumar. 2018. A Classification Based Web Service Selection Approach. *IEEE Transactions on Service Computing* 1–14.

19. QWS Data Set. http://www.uoguelph.ca/~qmahmoud/qws/.

20. Swami Das, M., A. Govardhan, and D. Vijaya Lakshmi. 2018. Web Services Classification Across Cloud-Based Applications. In *Soft Computing: Theories and Applications. Advances in Intelligent Systems and Computing*, vol. 742, ed. K. Ray, T. Sharma, S. Rawat, R. Saini, and A. Bandyopadhyay, 245–260. Singapore: Springer.

21. Hamdi, Yahyaoui, Hala Own, and Zaki Malak. 2014. Modeling and Classification of Service Behavior. *Expert Systems*, 1–27.

22. Xiao, Jian, et al. 2016. An Approach of Sematic Web Service Classification Based on Naive Bays, 1–7. IEEE.

23. Kalyani, G., and M.V.P. Chandra Sekhara Rao. 2017. Particle Swarm Intelligence and Impact Factor-Based Privacy Preserving Association Rule Mining for Balancing Data Utility and Knowledge Privacy. *Arabian Journal for Science and Engineering*.

24. Ramakrishna Murty, M., J.V.R. Murthy, P.V.G.D. Prasad Reddy. 2011. Text Document Classification Based on a Least Square Support Vector Machines with Singular Value Decomposition. *International Journal of Computer Application (IJCA)* 27 (7).

25. Adeniyi, D.A., Z. Wei, and Y. Yongquan. 2016. Automated Web Usage Data Mining and Recommendation System Using K Nearest Neighbor (KNN) Classification Method. *Applied Computing and Informatics* 12: 90–108.

26. Bakraouy, Zineb, Amine Baina, Mostafa Bellafkih. 2018. Availability of Web Services Based on Autonomous Classification and Negotiation of SLAs, 1–6. IEEE.

Location-Based Alert System Using Twitter Analytics

C. S. Lifna and M. Vijayalakshmi

Abstract In today's industry, Enterprises are rigorously blending Social Intelligence with Business Intelligence to achieve Competitive Intelligence. So, the ongoing process of Social Analytics cannot be overlooked. If used judiciously, Social Analytics can even address many sensitive social issues such as violation of Human Rights. The objective of the study was to design a platform for Location-Based Alert System which can aid Government bodies in taking corrective action against violation of Human Rights. The locations extracted from tweets were successfully plotted on to Indian map. This visualization revealed the importance of integrating News Analytics with Social Analytics for deriving precise inferences about event.

Keywords Social Analytics · Twitter · Location · Association Rule Mining · Human Rights

1 Introduction

The digital landscape of India is evolving quickly, with the extensive use of smartphones. The plethora of activities, taking place in the social networking sites [1–6] like Twitter, Facebook, LinkedIn and WhatsApp using these digital devices, generates a humongous amount of data about people preferences, behaviour and sentiments. This scenario has opened up a new arena for the users to express their opinions about a wide range of topics. People today are more interested in social presence. Incidents such as Delhi Zoo incident[1] happened on 23 September 2014

---

[1]Delhi Zoo Incident, http://bit.ly/1B9x2gJ.

C. S. Lifna (✉) · M. Vijayalakshmi
VES Institute of Technology, Mumbai, India
e-mail: lifnajos2006@gmail.com

M. Vijayalakshmi
e-mail: viji.murli@gmail.com

© Springer Nature Singapore Pte Ltd. 2020
K. S. Raju et al. (eds.), *Proceedings of the Third International Conference on Computational Intelligence and Informatics*, Advances in Intelligent Systems and Computing 1090, https://doi.org/10.1007/978-981-15-1480-7_28

were first posted on Twitter prior to any News media catch up with the detailed information. Almost every News channels[2,3,4,5] have Twitter accounts to catch up in the race. Also, social media tools such as GeoTwitter, a distributed sensor system for situational awareness [7–13] have changed society's perspective. This reveals, the relevance of Location-based Alert System using Social Analytics.

Violation against Human Rights has now become a common phenomenon, and the concerned Government Authorities need to be alerted about the upcoming consequences at an earlier stage. But, Government Authorities often fail to extract meaningful information from these social media posts as they are concise. Our main objective is to address the hidden context challenge within a social media post and to utilize it for societal benefit by selecting Human Rights as the application domain. The below-mentioned scenario clearly explains the motto behind selecting the Human Rights domain for the study. In this domain, each tweet represents some real-world events, with some temporal and transitive relationship existing between them. Therefore, the objective was to reveal the implicit transitive property that exists within the identified events. For example, Event A leads to Event B, Event B in turn leads to Event C or D and so on. This implicitly reveals that Event A leads to Event C or D represented as follows:

$$
\left.\begin{array}{c}
\text{Event } A \rightarrow \text{Event } B \\
\text{Event } B \rightarrow \text{Event } C \\
\text{OR} \\
\text{Event } B \rightarrow \text{Event } D
\end{array}\right\}
\quad
\begin{array}{c}
\text{Event } A \rightarrow \text{Event } C \\
\text{OR} \\
\text{Event } A \rightarrow \text{Event } D
\end{array}
$$

This can also be illustrated, with the help of the following example: Let child harassment be a known real-world event, say A and protest is another real-world event say, B. This may lead to another Event C, as criminal arrested, or Event D says child rehabilitated. This clearly explains the transitive property implied from Event A to Event C or Event D. Although there exists transitive property among these events, in some cases B, C or D may be unknown to the social media. In such cases, identification of unknown events turns out to be vital for discovering such temporal rules. Integrating these inferences with location could aid local and central governing bodies towards improving the efficiency of e-Governance. The study conducted over Twitter data streams was able to achieve the following objectives:

1. Extract hidden information from tweets.
2. Establish the presence of temporal association between tweets.
3. Locate events from tweets in Indian context.
4. Visualize the findings using heatmaps on India map.

---

[2]Twitter @BreakingNews, http://bit.ly/2aFxPC1.

[3]CNN BreakingNews @cnnbrk, http://bit.ly/2b6Yubb.

[4]BBC BreakingNews @BBCBreaking, http://bit.ly/2aTLaYi.

[5]Times of India, @timeso_ndia, http://bit.ly/2ayJ913.

2   Literature Review

This section summarizes the related work in the respective domain along with the state-of-art mining techniques which has helped to revamp WUARM into TWUARM. Before diving deep into the current research work, it is imperative to have some knowledge of Twitter.

2.1   Twitter Data Streams

Twitter is undoubtedly one of the popular microblogging sites[6,7,8] (more than 500 million tweets per day) where users search social-temporal information [8, 10, 14, 15] such as breaking news, tweets about celebrities, trending topics, etc. Within the Twitter Developers Community, there exists many real-time Twitter Analytics and visualization tools,[9] which crawl through Twitter streams to deliver some meaningful insights on Twitter activities and trends. Even though these tools exist, they are all proprietary versions.

Papers [14–19] discussed the various supervised learning techniques adopted for classifying and forecasting the tweets in real time. In papers [20–22], authors have attempted to classify tweets into known and unknown events. The attempt of this study was to combine her work with the transitive property that exists between these known and unknown real-world events and come up with a novel initiative in social reform using the active social networking media. In short, there is no existing foolproof state-of-the-art technique which can be applied for mining any Twitter data stream. Since the objective of the study is to discover temporal rules, the next focus of study is upon rule mining techniques.

2.2   Rule Mining Techniques

As mentioned in [3], association rule mining (ARM) is considered as an important technique for the extracting hidden knowledge. In the recent years, many researches have proposed algorithms for mining frequent item-sets as discussed in [23]. In papers [24–27], authors discuss about various temporal association rule mining techniques to efficiently extract global and local frequent patterns without prior domain knowledge. In papers [28–30], authors focus on identifying the utility of the extracted rules and discarding the least significant rules. All the above-mentioned

---

[6]http://www.alexa.com/topsites.

[7]http://www.gurugrounds.com/uncategorized/top-10-microblogging-sites/.

[8]https://about.twitter.com/milestones.

[9]http://twittertoolsbook.com/10-awesome-twitter-analytics-visualization-tools/.

techniques were applied upon an incremental dataset. So, to discover temporal rules, the dataset was treated as temporal, and weighted utility association rule mining (WUARM) was applied upon this dataset. In order to apply temporal WUARM, upon these tweets, there is a need to define domain dictionary. The following subsection discusses the various methods used to create domain dictionaries for specific purposes.

2.3 *Methodologies Used to Create Domain Dictionary*

In paper [31, 32], authors discuss about extracting implicit aspects with the help of NLTK WordNet.[10] But, NLTK WordNet cannot be applied in this context due to the inherent properties of tweets such as:

1. Each tweet is restricted to 140 characters;
2. Tweets are spontaneous;
3. There does not exist a standard lingo for the tweets;
4. Tweets are posted by each community in their agreed upon lingo which is sometimes difficult to decipher.

These properties impose hindrance in designing a fully automated system to create a domain dictionary for a social networking site like Twitter. Thus, the focus of this study is to identify temporal association rules that exist between the tweets in the Human Rights domain. It was inferred that only with the help of human intervention, sub-domains can easily be identified. This justifies how the 13 sub-domains covering all the major concepts of Human Rights were finalized.

2.4 *Natural Language Processing (NLP)*[11]

As tweets were posted by individuals, the location information attached with tweets was of least significance compared to the actual location mentioned in the tweet text. Among the various applications of POS tagging, named entity recognizers (NER) has a major role is extracting named entities from a document. Various educational institutions and organizations are having specialized research centres for developing foolproof NERs and provide APIs for accessing them. Google,[12] IIIT Hyderabad[13] and Stanford University [33] are pioneers in this field.

---

[10]http://www.nltk.org/.

[11]https://en.wikipedia.org/wiki/Naturallanguage_processing/.

[12]https://cloud.google.com/natural-language/.

[13]http://ltrc.iiit.ac.in/.

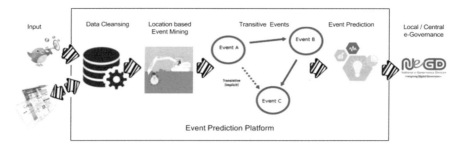

Fig. 1 Proposed architecture for location-based alert system

3 Proposed System Architecture

This section summarizes the procedures, techniques and methodologies developed and adopted during the phases of experimental study. Figure 1 depicts the proposed architecture for location-based alert system. To validate the proposed system, top five Twitter accounts under the Human Rights domain were identified. The proposed system was then experimented upon the user timeline tweets posted on active Twitter accounts of NGOs, Shakti Vahini (India) and Human Rights Watch (Global).

3.1 Data Collection and Preprocessing

For developing any system, deluge of data is required. Due to the inherent properties of tweets as mentioned earlier, data gathering procedure was extensively carried out. Initially, tweets were downloaded, followed by expanding the tiny-URLs in tweets, and finally crawling the web-pages of expanded URLs. Cleansing this raw data obtained after crawling the web pages turned out to be a herculean task. Therefore, the data preprocessing was rigorously performed as follows:

- Removing HTML tags[14]
- Extracting locations [33]
- Extracting meaningful words[15]
- Removing stop words[16]
- Perform stemming[17]

---

[14]http://www.pythonforbeginners.com/python-on-the-web/beautifulsoup-4-python/.

[15]http://pyenchant.readthedocs.org/en/latest/.

[16]See Footnote 10.

[17]http://textminingonline.com/tag/wordnet-lemmatizer.

Table 1 Relevance weightage assigned to 14 sub-domains

No.	Sub-domains	Relevance weightage, R_i
1	Killed	150
2	Harassment	130
3	Kidnapped	130
4	Criminal	110
5	Protestor	110
6	Rescued	100
7	Rehabilitation	100
8	Juvenile Home	100
9	Arrested	80
10	Law	80
11	Women	70
12	Child	70
13	Human	70

- Combine three tweet components into a single Tweet Corpus.
- Group the stem words into 13 identified sub-domains mentioned in Table 1.

3.2 Mining Temporal Association Rules

Upon this refined dataset, experiment was conducted temporal weighted utility association rule mining (TWUARM) using sliding window processing model. The following pseudocode explains the working of TWUARM in brief.

Algorithm: Temporal weighted utility association rule mining (TWUARM)
Input: Refined dataset
Output: Temporal association rules in each window
Parameters: *min_sup = 65%, min_conf = 85%, wt_thres = 20, win_size = 100, win_slide_ratio = win_size/3*;

1. Scan the 100 tuples (*win_size*) from the refined dataset.
2. **Repeat**

   a) Extract frequent item-sets from the window. (based on *min_sup*);
   b) Identify association rules from frequent item-sets. (based on *min_conf*);
   c) Filter out the utility rules from the association rules. (based on *wt_thres*);
   d) Temporal rules were discovered from utility rules;
   e) Slide the window (based on *win_slide_ratio*);

3. **Until** All tuples are scanned at least once;

The algorithm works similar to the exiting Apriori algorithm [23] till step 1(b) except for the processing model. In step 1(c), it extracts utility rules from the derived association rules based on the weight threshold. Then, temporal rules were discovered from these utility rules. Parameters in the algorithm were finalized as follows: (1) Window size, *win_size* set to 100, under the assumption that number of tweets per day shall not exceed 100; (2) Minimum support, *min_sup* and minimum confidence, *min_conf* were set to 65 and 85%, after experimental study and (3) Weight threshold, *wt_thres* was set to 20, to confine the number of rules generated to a specific limit. Similar to the profit-weight assigned to items in the transaction dataset discussed in paper [29], relevance- weight was given to the 13 sub-domains according to their impact upon the topic of discussion in the tweet as shown in Table 1, and the formulas were redefined as follows. Given an item-set $X = \{i_1, i_2, \ldots, i_n\}$. The relevance ratio of item i, Q-factor$_i$ and relevance weight of an item-set X, RW-factor(X) are defined as follows:

$$Q\text{-factor}_i = \frac{R_i}{\sum_{i=1}^{13} R_i}$$

where R_i is the relevance of item i

$$RW\text{-factor} = \text{factor}(X) = \sum \text{freq}(X) * Q\text{-factor}_i$$

where freq(X) is the frequency of item-set, X.

3.3 Experimental Setup

The experiment was performed on Intel® Core™ i3-2350M Processor, 2.30 GHz × 4 with 8 GB main memory running on Ubuntu 16.04 LTS. All programs were coded in Python using Stani's Python Editor (SPE 0.8.4.h). An application named "Mining Concept Drift" was created in Twitter for generating consumer keys and access tokens. These were then used to extract real-time user timeline tweets from Shakti Vahini (@shaktivahini) and Human Rights Watch (@hrw) Twitter accounts.

Tweets were extracted using tweepy Python module downloaded from GitHub. The latest 3200 tweets were downloaded on a day-to-day basis. These tweets were then merged, to extract distinct tweets to construct the final raw dataset. While downloading the tweets, tiny URLs in the tweets were expanded, and their corresponding web pages were crawled simultaneously. For extracting, expanding and crawling the downloaded tweets, Python modules such as urllib, urlclean and Beautiful Soup were used. Upon these three tweet components, cleansing and grouping were performed to derive entries into the merged Tweet Corpus. For cleansing the tweet components, Python modules such as urlclean, Beautiful Soup, urllib, WordNet Lemmatizer, Newspaper, PyEnchant, PyNER, gmaps, pandas, NLTK were used. Finally, temporal association rules were mined from the refined dataset using the sliding window processing model.

4   Results and Discussions

As mentioned in Sect. 3, the entire work was performed in two phases. Even though each one of them could be performed individually, the methodology adopted for this experimental study was sequential. In the first phase, the temporal rules were derived. In the next phase, location information was mapped upon to those temporal rules for extracting inferences. Two distinct datasets were used for the experiments. For the first phase, dataset was created from user timeline tweets downloaded from @shaktivahini to extract location-based temporal rules as shown in Table 2. The dataset for second phase was created from first phase dataset by deriving the frequency of locations to plot them on Google Maps. Table 3 demonstrates the sample dataset used for visualization.

In some extreme cases, only the third tweet component, i.e. the crawled tiny-URLs proved to render some valid stem words relevant to the Human Rights domain. For example, in Tweet #3, there does not exist any stem words associated with Human Rights domain even after expanding the tiny-URL present in it. Table 4 illustrates the data gathering procedure associated with Tweet #3. It was clear that it is difficult to extract meaningful rules from the ocean of stem words relevant to the Human Rights domain. So, related words were identified and grouped under the 13 sub-domains as depicted in Table 1. Upon this resulting refined dataset, temporal weighted utility association rule mining was applied by setting the parameters with the following values: (1) min_sup = 65%, (2) min_conf = 85% and (3) wt_thres = 20. The above values were empirically standardized for the system. It is now time to discover temporal rules. For this, consider the utility rules generated by TWUARM in windows 19, 20 and 21 at min_sup = 65%:

Window 19:   harassment, law $\rightarrow$ juvenile
             killed, juvenile $\rightarrow$ protester
Window 20:   harassment, law $\rightarrow$ juvenile
             killed, juvenile $\rightarrow$ protester
             rehabilitation, juvenile $\rightarrow$ harassment
             harassment, killed $\rightarrow$ protestor
Window 21:   killed, juvenile $\rightarrow$ protester
             rehabilitation, juvenile $\rightarrow$ harassment

There six events have been reported in these windows. They are Event A—harassment; Event B—law; Event C—killed; Event D—juvenile; Event E—protester and Event F—rehabilitation. When the rules are closely examined, the temporal rules in these tweet windows were evident. Also, their transitive property is clearly visible in this illustration.

The location information gathered during the first phase was utilized to plot the following heatmap depicted in Fig. 2. From the heatmap, the following conclusions can be made. Firstly, as per the Tweets from @shaktivahini, New Delhi is the location which is more into discussion during the period as it is more into red

Table 2 Sample dataset with Locations and Domain-specific keywords

Tweet id	...	Location in tweets	Domain-specific keywords
888723904382083073	...	West bengal Kolkata delhi ...	Child law harassment woman...
888721552748855297	...	Sonipat	Woman harassment law...
888711948744785920	...	Mahaguna moderne noida...	Harassment protester woman...
888321262329704448	...	West bengal jharkhand...	Criminal harassment arrested...
...	...	...	...

Table 3 Sample dataset for visualization

Location	Latitude	Longitude	Frequency
Adyar India	13.0011774	80.2564957	10
Agartala India	23.831457	91.2867777	27
Agra India	27.1766701	78.0080745	6
Akkalkuva India	21.5546245	74.0159499	43
...	...	...	...

Table 4 Illustrates the entire data gathering procedure associated with Tweet #3

No.	Methods	Outputs
1	Downloaded tweet	All work, no play http://t.co/vBFJ9lD8TF, http://t.co/xJuyK95kwu
2	Expanded tiny-URL in tweet	All work no play https://bit.ly/2OoEbbU, https://bit.ly/2JZ24Dj
3	Crawled web page of the expanded tiny-URL	https://bit.ly/2DJhOuO

compared to the other locations depicted in yellow (for locations such as Mumbai, Assam and Bengaluru) and in green (for locations in Goa, Kerala, Tamil Nadu, Karnataka). Secondly, to extract precise and all over information, there is need to integrate Social Analytics with News Analytics before integrating the developed system into e-Governance. Thus, the existence of transitive property has a valuable derivative inferences not only to predict events in the Human Rights domain but also to serve as a location-based alert system for concerned Government departments for a smooth and effective e-Governance.

Fig. 2 Location-based heatmap over Indian Terrain for Human Rights violation incidents

5 Conclusion and Future Scope

The experiment started off with downloading the user timeline tweets from @shaktivahini and @hrw in the Human Rights domain. To address the intrinsic properties of tweets, the data gathering, cleansing and grouping steps were performed extensively to derive refined dataset. Thus, at this stage, the first objective and the major contribution mentioned in the paper, i.e. addressing hidden context challenge has been achieved. Since sliding window processing model was efficient as mentioned in paper [34], the refined dataset was processed using the same model. For each window, association rules were mined and refined using TWUARM. The interesting patterns of temporal association between the rules were noticed during the study. Thus, the second major contribution was revitalizing WUARM into TWUARM for such a vibrant data stream from Twitter. Also, the study was able to attain its second objective, i.e. to discover the existence of temporal rules. The third objective was achieved by locating events from tweets in Indian context. And by plotting them, the fourth objective of visualizing the findings using heatmaps was achieved.

As a future scope, we are planning to extend the work towards the goal of integrating the proposed system with e-Governance. Towards the same, the following subgoals have been identified.

1. Creation of Ontology-based domain dictionary for identifying the concepts.
2. Integrate News Analytics with Social Analytics for gathering the exact scenario.
3. Real-time prediction of events.
4. Notify the Government bodies regarding predict events with probability.
5. Improve system performance by incorporating Big Data Tools.

Acknowledgements The authors gratefully acknowledge the support extended by University of Mumbai as Minor Research Project Grant No. 463 (Reference No. APD/237/16/2017 dated 13 January 2017).

References

1. Etlinger, S., and C. Li. 2011. *A Framework for Social Analytics.* San Francisco, USA: Altimeter Group.
2. Lau, R.Y., C. Li, and S.S. Liao. 2014. Social Analytics: Learning Fuzzy Product Ontologies for Aspect-Oriented Sentiment Analysis. *Decision Support Systems* 65: 80–94.
3. Charu, C. 2011. An Introduction to Social Network Data Analytics. In *Social Network Data Analytics*, 1–15. Berlin: Springer.
4. Parsons, T., M. Ratclifie, R. Crumpler, W. Kessler, and K. Freytag. 2014. Social Analytics System and Method for Analyzing Conversations in Social Media. U.S. Patent No. 8,682,723.
5. Ferguson, R., and S.B. Shum. 2012. Social Learning Analytics: Five Approaches. In Proceedings of the 2nd International Conference on Learning Analytics and Knowledge, 23–33. ACM.
6. Nadeem, M.: Social Customer Relationship Management (SCRM): How Connecting Social Analytics to Business Analytics Enhances Customer Care and Loyalty?
7. Amer-Yahia, S., S. Anjum, A. Ghenai, A. Siddique, S. Abbar, S. Madden, A. Marcus, M. El-Haddad. 2012. MAQSA: A System for Social Analytics on News. In Proceedings of the 2012 ACM SIGMOD International Conference on Management of Data, 653–656.
8. Castillo, C., M. El-Haddad, J. Pfeifer, M. Stempeck. 2014. Characterizing the Life Cycle of Online News Stories Using Social Media Reactions. In Proceedings of the 17th ACM Conference on Computer Supported Cooperative Work and Social Computing, 211–223. ACM (2014).
9. Ciulla, F., D. Mocanu, A. Baronchelli, B. Gonficalves, N. Perra, A. Vespignani. 2012. Beating the News Using Social Media: The Case Study of American Idol. *EPJ Data Science* 1 (1).
10. Bandari, R., S. Asur, and B.A. Huberman. 2012. The Pulse of News in Social Media: Forecasting Popularity. *ICWSM* 12: 26–33.
11. Stieglitz, S., and L. Dang-Xuan. 2013. Social Media and Political Communication: A Social Media Analytics Framework. *Social Network Analysis and Mining* 3 (4): 1277–1291.
12. Crooks, A., A. Croitoru, A. Stefanidis, and J. Radzikowski. 2013. # Earthquake: Twitter as a Distributed Sensor System. *Transactions in GIS* 17 (1): 124–147.
13. MacEachren, A.M., A. Jaiswal, A.C. Robinson, S. Pezanowski, A. Savelyev, P. Mitra, X. Zhang, J. Blanford. 2011. Senseplace2: Geotwitter Analytics Support for Situational Awareness. In Visual Analytics Science and Technology (VAST), 181–190. IEEE.
14. Althofi, T., D. Borth, J. Hees, A. Dengel. Analysis and Forecasting of trending topics in online media streams. In Proceedings of the 21st ACM International Conference on Multimedia, 907–916. ACM.
15. Mathioudakis, M., and N. Koudas, Twittermonitor: Trend Detection over the Twitter Stream. In Proceedings of the 2010 ACM SIGMOD International Conference on Management of Data, 1155–1158. ACM.
16. Gaber, M.M., A. Zaslavsky, S. Krishnaswamy. Data Stream Mining. In *Data Mining and Knowledge Discovery Handbook*, 759–787. Berlin: Springer.
17. Lee, K., D. Palsetia, R. Narayanan, M.M.A., Agrawal, A., Choudhary, A. 2011. Twitter Trending Topic Classification. In Data Mining Workshops (ICDMW), 251–258. IEEE.
18. Tare, M., I. Gohokar, J. Sable, D. Paratwar, R. Wajgi. 2014. Multi-class Tweet Categorization Using Map Reduce Paradigm. *International Journal of Computer Trends and Technology (IJCTT)* 9 (2), (2014).

19. Fiaidhi, J., S. Mohammed, A. Islam, S. Fong, and T.H. Kim. 2013. Developing a Hierarchical Multi-label Classifier for Twitter Trending Topics. *International Journal of u-and e-Service, Science and Technology* 6 (3): 1–12.
20. Becker, H., M. Naaman, and L. Gravano. 2011. Beyond Trending Topics: Real-World Event Identification on Twitter. *ISWSM* 11 (2011): 438–441.
21. Psallidas, F., H. Becker, M. Naaman, and L. Gravano. 2013. Effective Event Identification in Social Media. *IEEE Data Engineering* 36 (3): 42–50.
22. Rosa, H., J.P. Carvalho, F. Batista. 2014. Detecting a Tweet's Topic Within a Large Number of Portuguese Twitter trends. In OASIcs-OpenAccess Series in Informatics, Vol. 38.
23. Han, J., J. Pei, M. Kamber. 2009. Data Mining: Concepts and Techniques. Amsterdam: Elsevier.
24. Jain, S., A.P.S. Jain, A. Jain. 2013. An Assessment of fuzzy Temporal Association Rule Mining. IJAIEM.
25. Shirsath, P.A., V.K. Verma. 2013. A Recent Survey on Incremental Temporal Association Rule Mining. *International Journal of Innovative Technology and Exploring Engineering (IJITEE)* 2278–3075.
26. Wang-Wei, Junheng-Huang. 2007. Efficient Algorithm for Mining Temporal Association Rule. *The International Journal of Computer Science and Network Security (IJCSNS)* 7 (4): 268–271, (2007).
27. Chang, C.C., Y.C. Li, J.S. Lee. 2005. An Efficient Algorithm for Incremental Mining of Association Rules. In Proceedings of the 15th International Workshop on Research Issues in Data Engineering: Stream Data Mining and Applications, 3–10. IEEE.
28. Kuthadi, V.M. 2013. A New Data Stream Mining Algorithm for Interestingness-Rich Association Rules. *Journal of Computer Information Systems* 53 (3): 14–27.
29. Dhanda, M. 2011. An Approach to Extract Efficient Frequent Patterns from Transactional Database. *International Journal of Engineering Science and Technology*, 3 (7).
30. Khan, M.S., M. Muyeba, and F. Coenen. 2008. A Weighted Utility Framework for Mining Association Rules. In Second UKSIM European Symposium on Computer Modeling and Simulation, 87–92. IEEE.
31. Lal, M., and K. Asnani. 2014. Aspect extraction and segmentation in opinion mining. *International Journal of Engineering and Computer Science* 3 (05).
32. Falleri, J.R., M. Huchard, M. Lafourcade, C. Nebut, V. Prince, and M. Dao. 2010. Automatic Extraction of a Wordnet-Like Identifier Network from Software. In Program Comprehension (ICPC), 4–13. IEEE.
33. Finkel, J.R., T. Grenager, and C. Manning. 2005. Incorporating Non-local Information into Information Extraction Systems by Gibbs Sampling. In Proceedings of the 43rd Annual Meeting on Association for Computational Linguistics, pp. 363–370. Association for Computational Linguistics.
34. Lifna, C.S., and M. Vijayalakshmi. 2015. Identifying Concept-Drift in Twitter Streams. *Procedia Computer Science* 45: 86–94.

Diagnosis of Diabetes Using Clinical Decision Support System

N. Managathayaru, B. Mathura Bai, G. Sunil, G. Hanisha Durga, C. Anjani Varma, V. Sai Sarath and J. Sai Sandeep

Abstract Medicinal services are one of the prime worries of each individual. This work deals with diabetes, an incessant illness which is exceptionally regular throughout the world. Administration of such complex ailments requires proper diagnosis for which efficient analysis is required. So, extracting the diabetes reports in productive way is an essential concern. The Pima Indian Diabetes Data Set is used for this project, which accumulates the data of individuals who are affected and not affected by diabetes. The work goes for discovering solutions to analyze the illness by looking at patterns found in the information through classification analysis. The altered J48 classifier is applied to enhance the precision rate before which preprocessing and feature selection have been done as this prompts to decisions which are more accurate. The research would like to promote an agile and more proficient method of diagnosing the malady, prompting better treatment of the patients.

Keywords Clinical decision support system · J48 decision tree · Diabetes · Missing values · Normalization · Feature selection

1 Introduction

Repercussions of diabetes are discerned to have a more catastrophic and declining sway for women rather than for men as it is observed that women have lower survival rate and poorer personal satisfaction. World Health Organization reports unveiled the facts that near to around one-third of the women who have experienced the ailing consequences of diabetes have no awareness about it.

N. Managathayaru (✉) · B. Mathura Bai · G. Sunil · G. Hanisha Durga · C. Anjani Varma · V. Sai Sarath · J. Sai Sandeep
Department of Information Technology, VNR Vignana Jyothi Institute of Engineering and Technology, Hyderabad, India
e-mail: Mangathayaru_n@vnrvjiet.in

© Springer Nature Singapore Pte Ltd. 2020
K. S. Raju et al. (eds.), *Proceedings of the Third International Conference on Computational Intelligence and Informatics*, Advances in Intelligent Systems and Computing 1090, https://doi.org/10.1007/978-981-15-1480-7_29

The complication of diabetes is phenomenal in case of women in light of the way that the infirmity is transmitted to their unborn [1]. A diabetic tolerant person fundamentally has very minimal generation of insulin. There are primarily three kinds of diabetes but early stage detection is required to stay away from the complexities related with them. On the off chance that are left untreated, diabetes can cause abundant complexities. Extreme complexities can join diabetic ketoacidosis, non-ketotic hyperosmolar extreme lethargies, or death. Serious whole deal disarrays incorporate coronary illness, stroke, and interminable kidney disappointment and may even be harmful to eyes. When there is an expansion in the sugar level in the blood, it is called pre-diabetes. For such problems, data mining systems such as classifications can be utilized to ponder the well-being states of diabetic patients. These days, humongous amount of data is gathered as patient records by the clinics and this data can be used for knowledge discovery which plays a vital role in data mining. The term knowledge discovery from information (KDD) alludes to the computerized procedure of information disclosure from databases. This research mainly focuses on data transformation, data cleaning, attribute selection, and classification. Data normalization has been a measuring system to scale or a map data or a pre-preparing phase where we get a range of data which is new from the current or existing range of data. It can be supportive for prediction and forecasting.

As we probably are aware that there are a significant number of approaches to anticipate or conjecture in any case, all can fluctuate with each other a ton. So, to keep up the substantial variety of expectation and determine the normalization, procedure is required to make them scale to close values. Our approach leads to Z-score normalization features that will be rescaled with a goal and they will have the properties of a standard normal distribution [2]. This mechanism uses the values of mean and standard deviation to compute the normalized value for each attribute and scale them between the range 0 and 1. In various predictive modeling applications quiet information frequently have missing indicative tests that would be useful for evaluating the probability of analyses or for anticipating treatment adequacy. Instead of expelling factors or perceptions with missing information, another approach is to fill in or "ascribe" missing esteems. An assortment of ascription approaches can be utilized that range from to a great degree easy to rather mind-boggling. These techniques keep the full example measure, which can be worthwhile for inclination and accuracy; notwithstanding, they can yield various types of inclination. Apart from this, numerous insignificant qualities might be available in information to be mined. So, they should be evacuated. Moreover, numerous mining calculations do not perform well with a lot of highlights or characteristics. Accordingly, choice strategies should be connected prior to any sort of mining calculation is connected. The primary targets of highlight choice are to evade overfitting and enhance demonstrate execution and to give quicker and more practical prototypes. The selection of ideal features includes an additional row of entanglement in designing rather than simply finding ideal variables for complete arrangement of highlights, initially ideal element subset is to be found what's more, the prototype variables are to be augmented. Techniques for the choice of variables can be extensively partitioned into channel and wrapper approaches. Here, we use the channel approach, wherein this approach the characteristic choice strategy is autonomous of the data excavation calculation to be connected to the chosen characteristics and survey the importance of highlights by taking a gander at the inborn qualities of the information. As a rule, a highlight pertinence score is computed, and low scoring highlights are expelled. The subset of includes left after element expulsion is exhibited as contribution to the characterization calculation. Then, we have process of classification, a technique used in machine learning based on supervised learning mechanism that allots target classes to various items or groups. It is a bi-step process where the initial process is to construct a classifier model, which is utilized to assess the training set of the database. Second step of this process is model usage, where the developed model or classifier is used for grouping or classifying. As indicated by the level of test dataset that are grouped, the accuracy of the classification is determined. This research mainly focuses on data transformation, data cleaning, attribute selection, and classification which made us to put forward a rapid and proficient method of healing the malady, prompting better treatment of the patients.

2 Literature Survey

Salim Amour Diwanil, Anael Sam used a supervised machine learning algorithm, which is used for predicting the outcome of the dataset from the ML laboratory at University of California Irvine. As part of their work, the affected people's information is trained and later tested using ten cross-validations by using N-Bayes and J48 algorithms [3, 4].

Vijayalakshmi and Thilagavathi used b-coloring approach for clustering analysis. This is a way to predict the diabetic disease for Pima Indian diabetes datasets. The method put forward by them depicts clusters formed by superior objects that ensures the inter cluster disparity in a splitting and used to assess the standard of cluster [5].

Magudeeswaran and Suganyadevimade use of revised radial basis functional neural networks (MRBF), a supervised ML algorithm, which acts as a classifier to estimate whether the patient is diabetic or not. This particular method used the glucose content in blood of diabetic patients for prognosis. The approach that was put forward uses Pima Indian diabetes set of data, and from the results obtained, it is clear that the MRBF obtained greater accuracy than the current radial basis functional (RBF) techniques and other neural networks [6].

3 Methodology

The purpose of this work is to develop a decision support system on the basis of few classification algorithms in the sense to deliver prediction solutions to the existing problem of early detection of diabetes among women. When coming to prediction of diabetes from a given dataset, firstly, we have to identify whether the dataset contains any noisy data, inconsistent data, redundant data, and missing values. So, in our dataset, we have missing values, we must first follow this approach to impute those missing values and predict the solution. This approach mainly uses imputation measures [7], data normalization, feature selection, and classification. Once the above procedure is applied, we get a substituted value for the missing value from the data itself and scaled data. Hence, we get accurate results during the prediction and classification which reduces the risk factor for human lives. In this section, the above approach has been used to build support system for diagnosis of diabetes.

3.1 Data Description

We have used here Pima Indian diabetes data which is taken from Kaggle for experimentation and evaluation. The Pima Indian Diabetes Dataset comprises of data on 768 patients out of which 268 tested_positive occurrences and 500 tested_negative cases. Tested_positive and tested_negative show whether the individual is diabetic or not, respectively. Each row is comprised of 8 attributes, where all numeric and a class are attribute which is factor type. The same has been briefly described in Table 1 [8, 9].

3.2 Data Preprocessing

The data present in the datasets cannot be always used directly for analysis. So, the first and foremost step to be done in any data analysis process is data preprocessing. This dataset contains noise and missing features in most cases and therefore requires significant preprocessing. The essence of the information to a great degree impacts the effect of analysis, which infers data pre-handling takes part an

Table 1 Characteristics of dataset attributes

S. No	Name of the attribute	Value type
I.	No. of times pregnant	N
II.	Plasma glucose conc. at 2 h in an oral glucose tolerance test	N
III.	Diastolic BP (mm Hg)	N
IV.	Triceps skin fold thickness (mm)	N
V.	2-h serum insulin (mm U/ml)	N
VI.	BMI (weight in kg/(height in m)^2)	N
VII.	Diabetes pedigree function	N
VIII.	Age (years)	N

N Numeric type

imperative part in modeling. In this model, we have imputed missing data with median value imputation and multiple imputation measures to advance the dataset. Here, we used median value imputation for less number of missing valued attributes and multiple imputation for large number of missing valued attributes. In the main dataset, the estimations of blood pressure and BMI cannot be zero as it declares that the actual value was absent. To minimize the impact of pointless readings, we replaced 0 s with the value which we got after performing imputation measure. The imputation is done as follows:

Median Imputation:

$$\text{Median} = L + \frac{h}{f}\left(\frac{N}{2} - c\right)\mu.$$ (1)

Multiple Imputation:

$$\bar{Q} = \frac{1}{m}\sum_{i=1}^{m} Q_i.$$ (2)

3.3 Data Standardization

Standardization is a preprocessing procedure used to rescale values of attributes to fit in a particular range. Standardization of the data is imperative when managing characteristics of various units and scales. In our model, we perform Z-score normalization, also referred as Zero-mean normalization. The information here is processed based on the mean and standard deviation. Basically, it converts all values to a common scale with an average of zero and standard deviation of one, i.e., range lies between 0 and 1. Z-score normalization technique is mainly used to compare the difference between normal distributions which lie in a particular range. The normalization is done as follows,

$$z = \frac{x - \mu}{\sigma}.$$ (3)

where μ = mean

$$\sigma = \text{standard deviation}$$

3.4 Dimensionality Reduction

In this work, we performed recursive feature elimination which is a wrapper strategy for feature selection to reduce the dataset dimensionality by analyzing and

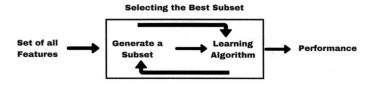

Fig. 1 RFE algorithm

understanding the effect of its features on a model. The primary reason to use feature selection is, it empowers the machine learning algorithm to train quicker and diminishes the complexity of a model and makes it less demanding to interpret, it improves the precision of a model if the correct subset is picked, and reduces over fitting. Figure 1 represents procedure of RFE algorithm.

3.5 Classification

Classification is a supervised learning mechanism which consists of predicting a certain outcome based on a given input. Classification algorithms are broadly utilized in different medicinal applications. Generally, it is a two-stage process in which initial step is the preparation stage which is referred as training phase. Here, the classifier algorithm constructs classifier with the training set of records and the secondary phase is characterization stage also called as classification phase [10]. It is utilized for classifying and its execution is scrutinized with testing set of records. Ten-fold cross-validation is done to prepare training and testing data. The data is split 80% for training and 20% for testing. Then, we train the algorithm on 80% of the dataset keeping the test data aside. This training process will produce the training model based on the logic and the algorithm and the observations of the features in the training set. Then, trial the prototype on the unseen input to assess it. If we trained the model on the whole set of data, at that point it produces a good outcome on the test data as it has seen the pre-dispositions and when we use this model to the real world data, at that point it will perform ineffectively as it is unaware of the biases present in reality. In this manner, we keep the testing data isolated from training data with the goal that it delivers better outcomes.

J48 Decision Tree
In our proposed work, we used J48 decision tree classifier in order to predict the classifier output. J48 is an expansion of ID3. The additional features of J48 are representing values, decision trees pruning, consistent characteristic value ranges, inference of rules, and so on [11].

A decision tree is an insightful machine learning model that picks the objective value (subordinate variable) of another case in perspective of various property estimations of the accessible information. The inner nodes in a decision tree mean the distinctive qualities, the branches linking the nodes discloses the conceivable value that these properties can hold in the recorded instances, where as the nodes at the end reveal us the final characterization of the reliant variable. The attribute that will be anticipated is known as the reliant variable, as the observations are dependent on or decided by, the values of discrete characteristics. The other traits, that support while predicting the records for the reliant variable, are usually referred as the independent factors in a dataset.

The decision tree is generated by calculating entropy value and information gain. The formula is as follows,

$$\text{Entropy}(s) = \sum_{i=1}^{c} -p_i \log_2 p_i. \tag{4}$$

$$\text{Gain}(S, A) = \text{Entropy}(s) - \sum_{v \in \text{values}(A)} \frac{|s_p|}{|s|} \text{Entropy}(s_p). \tag{5}$$

4 Proposed Algorithm

Input: Pima Indian Diabetes dataset which is in .csv format.
Output: It produces J48 decision tree predictive model with leaf node wherein the leaf node that is produced is either tested-positive instances or tested-negative instances and prediction results
Procedure:

Step-1: Load the dataset in a form where records are set as rows and features (attributes) as data columns.
Step-2: Consider the set of data consisting records with 0 s as missing values and assign it as NA except for one attribute.
Step-3: Identify the number of missing values in each attribute and split them into two parts. First part is the attributes that have less number of missing values and second part is the attributes that have more number of missing values.
Step-4: Apply the proposed measure to assign the records missing with the comparing attribute value itself in the dataset. We may likewise think about the recurrence of event if there should be an occurrence of equivocalness.
Step-5: Redo first four steps for each observation where the corresponding values are absent.
Step-6: Standardize the dataset acquired in step-5 reasonably. This includes mapping the column records to fit the requirement for application of the measure which is put forward to identify the absent values.
Step-7: Handled dataset is passed through feature selection wherein set of attributes are removed from the dataset in order to reduce dimensionality reduction of dataset.

Step-8: Now, the preprocessed data is split into two sets where one is used for training the set and the other is used as testing dataset.

Step-9: The J48 decision tree algorithm is employed. For basis of the algorithms, cross-validation and percentage split methods are performed to generate a model.

Step-10: Based on the decision tree generated, we can predict the classifier output.

5   Implementation

The following flowchart (Fig. 2) depicts the implementation of the proposed system.

6   Performance Metrics

We have assessed the performance of the above-mentioned algorithm by applying performance metrics accuracy, precision, root mean square error, and area under curve ROC curve [12]. Accuracy is expressed as the proportion of a number of correct assessments to the total number of assessments. The precision is expressed as the proportion of tp to the sum of tp and fp, where tp is the number of true positives and fp the number of false positives [13]. Mean square error (MSE) indicates the error in prediction. Lesser the MSE value, better the prototype performance. The area covered by the ROC curve is also an efficient performance metric. Higher the AUC value, better the performance.

7   Results and Discussions

The initial data had several missing attribute values. We make use of median and multiple value imputation to handle the missing data. After data imputation, the data is normalized and fed to classification algorithm to check for efficiency. This paper uses J48 to produce performance of model. The efficiency metrics of the above-mentioned algorithm is recorded in Tables 2 and 3. Table 3 records the correctly and incorrectly predicted negative and positive values which give an idea of the accuracy of the model used. As per the decision tree generated glucose is the attribute with highest information gain and is represented using ROC curve w.r.t. class attribute in Fig. 3 (Table 4).

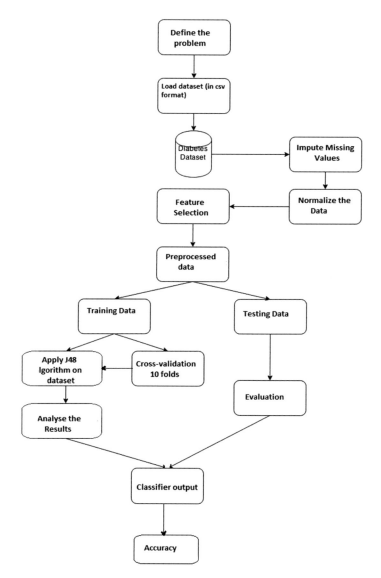

Fig. 2 Flowchart depicting model creation

Table 2 Performance evaluation

Measures	Percentage (%)
Kappa statistic	66.54
Mean absolute error	22.78
Root mean squared error (RMSE)	33.75
Relative absolute error (RAE)	50.4864
Root relative squared error (RRSE)	71.0662

Table 3 Table for performance analysis

	Number of instances	Percentage (%)
Correctly classified instances	521	84.8
Incorrectly classified instances	93	15.14

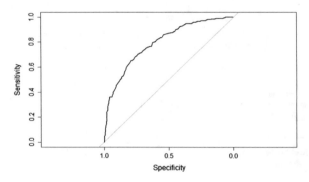

Fig. 3 ROC for class versus glucose

Table 4 Table for comparison of results

	Existing system (I)	Existing system (II)	Existing system (III)	Proposed system
Correctly classified instances (%)	73.8281	76.52	81	84.86
Incorrectly classified instances (%)	26.1719	23.47	19	15.14

I—Improved J48 classification algorithm for the prediction of diabetes
II—Performance analysis of classification algorithms in predicting diabetes
III—A comparative study on the preprocessing and mining of Pima Indian Diabetes Dataset

8 Conclusion and Future Work

From this research, it was discovered that data mining systems can be utilized for anticipating of diabetes. This work proposed deals with accurately predicting diabetes from medical records of patients [14]. The enhanced J48 classifier has been used to improve the rate of accuracy for the data mining procedures [15]. This examination helps the specialists and well-being associations in utilizing the data mining systems in the medicinal field which helps in foreseeing of diabetics and predicting the accuracy of the result using measures which are used to improvise the results. Hence, the proposed show helps in enhancing the determination of the infections which for sure aides in early cure of ailment in the patients.

In future, it is necessary to collect the information from different zones over the world using database frameworks and make a more accurate and judicious prototype for modeling diabetes conclusion [1]. Upcoming analysis will likewise revolve around social affair data from a later day and age. This work can be widened and improved for the mechanization of diabetes examination.

Acknowledgements The proposed research work has been funded under DRDO-LSRB (DRDO-Life Science Research Board)—No. CC R&D (TM)/81/48222/LSRB-284.

References

1. www.ijarcs.info Internet Source.
2. Shirazi, Syed Noorulhassan, Antonios Gouglidis, Kanza Noor Syeda, Steven Simpson et al. Evaluation of Anomaly Detection Techniques for SCADA Communication Resilience. *Resilience Week* (RWS).
3. Kumari, S., and A. Singh. 2013. A data mining approach for the diagnosis of Diabetes Mellitus. In Proceedings of Seventh International Conference on Intelligent Systems and Control, pp. 373–375.
4. Goyal, Anshul, Rajni Mehta. 2012. Performance Comparison of Naïve Bayes and J48 Classification Algorithms. *IJAER* 7 (11).
5. Magudeeswaran, G., D. Suganyadevi. 2013. Forecast of Diabetes using Modified Radial basis Functional Neural Networks. In International Conference on Research Trends in Computer Technologies (ICRTCT). Proceedings Published in International Journal of Computer Applications (IJCA) (0975-8887).
6. Karegowda, A.G., M.A. Jayaram, A.S. Manjunath. 2012. Cascading K-means Clustering and K-Nearest Neighbor Classifier for Categorization of Diabetic Patients. *International Journal of Engineering and Advanced Technology (IJEAT)* 1 (1). ISSN: 2249 – 8958.
7. Mathura Bai, B., N. Mangathayaru, and B. Padmaja Rani. 2015. An Approach to Find Missing Values in Medical Datasets. In Proceedings of the International Conference on Engineering & MIS.
8. Jahangir, Maham , Hammad Afzal, Mehreen Ahmed, Khawar Khurshid, and Raheel Nawaz. 2017. An Expert System for Diabetes Prediction Using Auto Tuned Multi-layer Perceptron. In Intelligent Systems Conference (IntelliSys).
9. Communications in Computer and Information Science, 2016.
10. Vijayarani, S. 2013. Evaluating the Efficiency of Rule Techniques for File Classification. *International Journal of Research in Engineering and Technology*.

11. Submitted to The University of the South Pacific Student Paper.
12. idus.us.es. Internet Source.
13. http://journal.frontiersin.org. Internet Source.
14. Saravanan, N., V. Gayathri. 2017. Classification of Dengue Dataset Using J48 Algorithm and Ant Colony Based AJ48 Algorithm. In International Conference on Inventive Computing and Informatics (ICICI).
15. research.ijcaonline.org Internet Source.

N. Mangathayaru received B.Tech. in Electronics and Communication Engineering from JNTUH, A.P., India in 1997 and received M.Tech. in Computer Science, JNTUH, India in 2003. She is awarded Ph.D. in Computer Science & Systems Engineering from Andhra University, India in 2010. She is working as Professor in the Department of Information Technology at VNR Vignana Jyothi Institute of Engineering and Technology, Hyderabad, India. Her research interests are Data Mining, Network Security, Information Security, and Bio Informatics.

B. Mathura Bai received B.Tech in Computer Science and Information Technology from VNRVJIET, JNTUH, A.P., India in 2003 and received M.Tech in Computer Science, JNTUH, India in 2007. She is working as Associate Professor in the Department of Information Technology at VNR Vignana Jyothi Institute of Engineering and Technology, Hyderabad, India since 2003. Her research interests are Data Mining, Computer Networks, Mobile Computing, and Cloud Computing.

Analysis of Fuel Consumption Characteristics: Insights from the Indian Human Development Survey Using Machine Learning Techniques

K. Shyam Sundar, Sangita Khare, Deepa Gupta and Amalendu Jyotishi

Abstract There are two main factors that need to be considered when using fuel—ecology and economy. Ecologically, the fuels that are clean (fuel that emits less or no CO_2) are more efficient than the ones that are not clean. Economically, such clean fuels are costly compared to their counterparts. The Indian Human Development Survey (IHDS-II) 2011–12 data set provides the usage details on six different types of fuel for over 42000 households in India. This paper shows the details of the requirements and processes taken to classify the data set based on the fuel usage variables. The results are obtained using machine learning techniques on the data set to determine the factors that are responsible for the use of clean fuel over non-clean fuel in households.

Keywords IHDS-II · Fuel switching · Fuel stacking · Clean fuel · Non-clean fuel · ML · Random tree

K. Shyam Sundar (✉) · S. Khare · D. Gupta
Department of Computer Science & Engineering, Amrita School of Engineering,
Bengaluru, Amrita Vishwa Vidyapeetham, India
e-mail: sundarshyam030@gmail.com

S. Khare
e-mail: k_sangita@blr.amrita.edu

D. Gupta
e-mail: g_deepa@blr.amrita.edu

A. Jyotishi
Department of Management, Amrita School of Business, Bengaluru,
Amrita Vishwa Vidyapeetham, India
e-mail: amalendu.jyotishi@gmail.com

© Springer Nature Singapore Pte Ltd. 2020
K. S. Raju et al. (eds.), *Proceedings of the Third International Conference on Computational Intelligence and Informatics*, Advances in Intelligent Systems and Computing 1090, https://doi.org/10.1007/978-981-15-1480-7_30

1 Introduction

Fuel is an important constituent of the livelihood of the people. In the household sector, fuels are consumed for cooking, lighting, heating and other purposes. The major fuels consumed are crop residue, LPG, kerosene, electricity and other non-commercial fuels like firewood and dung cake. Unlike developed countries where electricity is dominating energy source, households in developing countries choose their fuel type based on the availability and socio-economic conditions. Especially, for cooking, it is estimated that around 30% of people rely on biomass [1].

The cost of consuming energy presents itself in different forms, mainly monetary and opportunity costs. The energy capacity of a fuel is defined by its calorific value. But the total potential of the fuel depends on two factors—economy and ecology. Economically, the fuel efficiency is the ratio of the calorific value to the amount spent in acquiring that fuel. Ecologically, the fuel efficiency depends on the amount of direct and progressive effect it has on the natural environment [2]. There are two main classifications of fuel in terms of their effects on the environment—clean and non-clean. This classification is based on the amount of impurities, mainly CO_2 per unit of the fuel consumed. This paper takes six different fuels into consideration— *firewood, dung cake, crop residue, kerosene, LPG* and *coal*. According to sources, LPG [3] and kerosene [4] are classified as clean fuel and the others, coal [5], firewood [6], dung cake [7] and crop residue [8], as non-clean fuel.

This paper is organized as follows—Sect. 2 discusses the literature and theories that have been taken into consideration for the work. Section 3 provides a brief description of the IHDS-II data set. The methodology followed is described in Sect. 4 and the detailed results are presented in Sect. 5. Conclusions derived from the results are described in Sect. 6 along with approaches proposed for future research.

2 Literature Review

There are a handful of theories that depict the nature of household fuel consumption. The most popular among those are the energy ladder model and the energy stacking model. Some studies conclude the effectiveness of the energy ladder [2], while others report the greater efficiency of the energy stacking model over the former [9, 10], especially in the context of developing countries.

This theory puts forth a linear correlation between the household's income and the type of fuel they use. Cleaner fuels, like LPG and kerosene, cost more compared to coal and other non-monetary sources of fuel like firewood and dung cake. This theory explains that households "switch" from using non-clean fuels to clean fuels as their income level increases and vice versa in order to meet their increasing requirements or decreasing savings [11]. This model puts forth a theory contesting the energy ladder that households do not completely switch their fuel preference with increase in income; instead, they use a combination of different fuel types for different purposes.

There are a number of studies that have worked in a similar area. A study in Africa [2] explores the households' fuel choice within the context of the energy ladder hypothesis and tests for the effects of other factors such as asset ownership and the differences in fuel usage between urban and rural areas. A study in Arusha city, Tanzania [12], analyzed the socio-economic factors that are responsible for the urban households' choice for cooking fuel and its toll on the households' expenditure. The study revealed that the choice was influenced mainly by socio-economic and demographic factors like education, marital status, household size and occupation. A study in Pakistan [13] focuses on the effect of poverty on household fuel choice and its effects. The study reveals that the choice is affected not only by poverty but also by other factors like capital, asset ownership, etc. One such study in India [14] focuses on the Uttam Urja initiative that aims to promote cleaner energy options through development of value chains.

The concept of the energy ladder is closely connected with urbanization. The fuel or energy shifts are stimulated by an increase in monetary income mainly and also because of the availability of better and superior fuel sources locally. This shift in many cases becomes a status symbol also in the rural areas. From a purely economic point of view, the shift from a fuel of lower efficiency to more efficient fuel is good, apart from the wood saving which can be achieved if a major percentage of the population can shift away from fuelwood usage. This is based on the assumption that all the households are presented with a variety of choices of fuel types.

For a number of developing countries, this is very important from policy standpoint. Thus, a question arises in this respect, how far are the households free to choose between different energy sources? Are they actually able to exercise the choice between different fuels or is the progress heavily dependent on the economic status? Though studies in India have discussed the methods to promote actual use of cleaner energy in the future [14], no such study has been made that takes into account the actual rate and type of fuel consumption in India, as a whole. Even if a smaller portion of the country was to be taken as a representative, it is not necessary that the results would be befitting the entire nation. India, being a country of diversity not just in culture but also in its resources, will have a diversified usage rate.

As such considering, the country as a whole would be the better option to analyze the consumption pattern among the households. There are many studies that have been conducted, using a survey data as their information source. These studies have taken an unconventional approach of solving the problem [15]. Instead of the study based on the past literature work alone, they have used machine learning approach.

This study takes into account the differing nature of available resources, the nature of the settlements along with the apparent factors contributing to the consumption pattern—income and expenditure. Taking into consideration the similarities and differences between each of the models, a machine learning approach is used to determine the exact factors that contribute to the change in fuel consumption pattern in India, instead of expediting on an assumption that the factors are already known and constant.

3 Data Description

The study is done on the Indian Human Development Survey (IHDS-II) 2011–12 data [16]. It provides different sets of data for different categories like individuals, households, workers, doctors, villages, panchayats, etc. The survey includes 42,152 households in 1503 villages and 971 urban areas. Our focus is directed on the household data set of the survey which has a total of 42,152 records with 758 different variables [17]. From the previous literature on the subject, it is evident that only some of the fields from the data set are relevant to the research. Manual pruning of the fields produced 12 fields relevant to the study. The fields *URBAN2011, INCOMEPC, CO21, POOR* and *FM1* have been selected with reference to previous literature and the other fields have been included in the study to provide a difference in opinion to the existing literature.

4 Methodology

The schematic diagram Fig. 1 represents the overview of the process. The proposed approach is divided into three phases: phase I—data preparation, phase II—training and testing and phase III—result analysis.

4.1 Phase I: Data Preparation

For the data to be used by the machine learning algorithm, it has been free from any discrepancies or other errors present in it. For cleaning, record elimination method has been implemented because records containing missing values are negligble compared to the size of the whole data set ($\sim 0.032\%$). As mentioned earlier, the data consists of six different fields that contain fuel consumption data. However, for a smooth process, it is best if there exists a single distinctive class label. As such, the first step is to derive a single unique class label from the collection of these six fields. Table 1 provides the description of the fields that are taken into consideration for predictors and Table 2 provides the description of the fields considered for the preparation of the class label.

For this process, two temporary fields clean and non-clean are created, which contain the amount of clean and non-clean fuel consumed by the households,

Fig. 1 Overview of the experimental setup

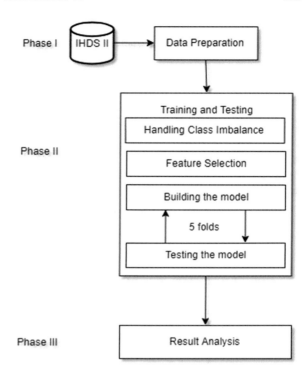

Table 1 Fields selected from the data as predictors

Predictors	Description
STATEID	A unique id for each state
ID14	Main income source
ID18C	Education of the household's head
MG1	Immigrant status
FM1	Land owned or cultivated
POOR	Poverty status
CGVEHIVLE	Vehicle owned
URBAN2011	Settlement region
HHEDUC	Highest education of household
CO21	Total expenditure on fuel
INCOMEPC	Income per capita
NPERSONS	Number of people in the householda

Note The field *CGVEHICLE* is a combination of three fields *CGVEHIVLE, CGMOTORV* and *CG21* present in the original data. The combined field now has four categories (car, other motor vehicles, other vehicles and vehicles not owned)

Table 2 Fields selected from the data for class label

Field	Description
FU7	Consumption of firewood
FU8	Consumption of ding cake
FU9	Consumption of crop residue
FU10	Consumption of kerosene
FU11	Consumption of LPG
FU12	Consumption of coal

Table 3 Class label categories

Class	Condition	Frequency
Clean favoured	Clean > Non-clean	15,462
Non-clean favoured	Non-clean > Clean	5354
Combination	Clean = Non-clean	8087
Only clean	Non-clean = -1	9568
Only non-clean	Clean = -1	2326
Fuel not used	Clean = Non-clean = -1	33

respectively. This is done by changing the value of each of the fields into 1 and -1 each representing whether the household has used that particular fuel or not. Then a sum of each fuel field in the particular category (clean fuel and non-clean fuel) is taken as the values for the temporary fields. Since there are four non-clean fuels and two clean fuels, it is better to first bias the fields by dividing the "non-clean" by four and "clean" by two to get a uniform representation.

Next, we combine these two fields into one single class label. This is done by comparing the numeric values of the fields and the new class label is formed. Table 3 provides a description of the categories in the new class label. For example, a household where the value of the clean variable is greater than that of the non-clean variable is categorized as *Clean Favoured* and so on for each household in the data.

Now, there is a distinct class label which is named "USAGE" with independent categories to work with. This way it is easier to analyze based on the households' inclination to one type of fuel instead of using two different variables and analyzing the data using each of the new fields as a class label. The frequency of each class determines the number of households that follow that particular fuel consumption pattern. A higher value implies a higher number of households following the pattern and vice versa. Since the consideration is given for fuel usage statistics and the fact that the frequency of the class is too low as compared to the others, the class *Fuel not used* has been eliminated from the entire data. As such, a total of 40797 records have been retained as the base for building the model with five different categories present in the class label.

4.2 Phase II: Training and Testing

During this phase, the prepared data is put for testing to determine how much of it is actually contributing to the classification of the data based on the class label. This step is done in three stages, handling the class imbalance, feature selection and building the model.

Handling Class Imbalance

After the errors and inconsistencies are removed from the data, there may be a problem with the distribution of its records over different classes. In such cases, the class with the highest frequency is called the *majority class* and the one with the least frequency, the *minority class*. In these situations, most classifiers are biased towards the majority class. Hence, they produce poorer results than when working on balanced data [18]. One technique for resolving this problem is the *resampling technique* [19]. This method involves two processes—*oversampling* (where samples are taken into account more than once) and *undersampling* (where some samples are ignored). The other technique is the Synthetic Minority Oversampling Technique (*SMOTE*) [20] process where new records are synthesized from existing records instead of creating duplicate copies of the present records. In cases where only one class has a different (high or low) spread compared to others, one of these processes can be used. But, in present scenario, where the difference in spread is not limited to a single class, the combination of the two techniques is used to obtain better results.

Feature Selection

The data contains fields that have been pruned manually based on previous literature. But not all of those fields will be statistically relevant in predicting the class label. This necessitates a selection process to identify which fields matter in predicting the class label.

The CFS [21] evaluation (correlation-based subset evaluation) is one such technique which chooses the fields that have a high correlation value with the class label and have less inter-correlation between each other. That is, it selects the features that provide good enough information about the data to classify it based on the class label. Furthermore, the selected features themselves might still be ranked based on the amount of information they provide to classify the data. This helps to determine which feature of the data is more suitable to determine the value of the class label as compared to others.

The result of the cleaning process produced a total of 40830 records out of 42,152. From Table 2, it is evident that the classes are not evenly spread. After a trial and error process of using these techniques in conjecture, one method proved to be more suitable as compared to the others. Here, the SMOTE and resampling technique has been used in conjecture, by first increasing the minority class to 300% its original size using SMOTE, then applying the resampling technique over the resultant data using the uniform class biasing to even the spread of records through simultaneous oversampling and undersampling. The final distribution shows an equal spread of 9555 records per category. After this process, the data is subjected to the feature selection process using the *cfsSubsetEval* filter provided by Weka.

The *infoGain* or *gainRatio* filter has been used to determine the ranking of the selected features. These fields represent the ones that are statistically relevant in classifying the data based on the class label, i.e. these are the fields that can actually provide information necessary to classify a household based on the factors responsible for following that particular pattern. The features selected by the CFS subset evaluation technique and their corresponding ranks determined by the *infoGain* filter showed, in decreasing order of rank, *INCOMEPC* (1), *STATEID*, *ID14*, *CO21*, *NPERSONS*, *URBAN2011*, *CGVEHICLE*, *ID18C*, *MG1* (9). Higher rank implies higher information contributed to classification by said feature and vice versa, with the highest rank corresponding to *INCOMEPC* and the lowest to *MG1*.

Building the model

The data mostly consists of categorical fields with two or more categories in each field. Only two fields, namely *INCOMEPC* and *CO21,* consist of numeric data. As such, a decision tree model would better suit this classification problem as compared to other techniques, because there is no necessity for any numerical analysis. Hence, the *random tree* [22] classifier is used to build and test the model. Random tree is a straightforward generative classifier. It is not an iterative process. The tree is generated after the first read and hence has a very low building and testing time. While many machine learning algorithms tend to iterate over the initial process to obtain an optimal solution, the random tree classifier works towards generating a tree whose attributes are based solely on the data. The training and testing of the model have been done using a machine learning tool called Weka that contains in-built filters to handle class imbalance, feature selection and model building.

4.3   Phase III: Result Analysis

The overall efficiency of the model is determined based on its accuracy (the fraction of records that have been correctly classified by the model). Higher accuracy implies a higher probability for each of the records to be correctly identified. For example, a model with an accuracy of 1 (100%) will correctly classify each and every record in the data. The individual results of the class label are analyzed based on the *precision* (fraction of selected records that belong to the respective class), *recall* (the fraction of records belonging to the class that has been retrieved) and *f-measure* (the weighted average of precision and recall) [23].

5 Experimental Results

This section provides the results obtained by the implementation of the random tree model over the data set. The data preparation phase has been handled with the help of the R interactive console. The results produced an accuracy value of 82.8864%, i.e. at least 82 records out of a 100 would be correctly classified by the model. Figure 2 shows the distribution of the precision, recall and the *f*-measure for each category in the class label along with their average. As seen in the figure, the recall (ability to identify a record correctly) for each of the categories is varied—some above the mean (only non-clean, 0.906; non-clean favoured, 0.89) and some much below the average (clean favoured, 0.695).

This implies that the records in the categories of higher recall were easily identified correctly by the classifier as compared to those with the lower recall. This is due to the amount of relevancy that each of the fields holds with respect to the class label in that particular category, i.e. the easily classified categories can be explained better in terms of their predictors as compared to the other categories. The differences in the recall measures can be explained by a number of factors—incorrect data, inconsistent data format, insufficient correlation between the fields have with respect to the class label, etc. The variation in the recall measures can be directly related to the varying relevancy of the fields as there are no data discrepancies that can contribute to this variation.

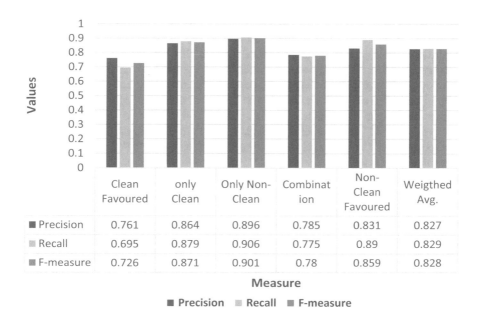

Fig. 2 Variation in the precision, recall and *f*-measure among the categories of the class label

For example, the class with the highest f-measure of 0.901 (only non-clean) is better explained by the information provided by the selected features as compared to a class with a lower f-measure, say clean favoured, which has a value of 0.726. Though this value is not too low to be discredited, it is lower than that of the other classes. Based on the results, it is easier to identify households that use only non-clean fuel as compared to other categories. At the same time, it is a bit tedious to correctly identify if a particular household favours clean fuel over non-clean fuel with higher precision. This implies that the direct relation that exists between the fields and the class label is not constant but actually dependent on the class label itself, i.e. for a category with high recall values, the relationship is stronger between the fields and the class label as compared to the ones with lower recall. The f-measure, being directly dependent on recall and precision, follows a similar rise and fall in correspondence to its independent factors.

6 Conclusion and Future Work

The data that has been used is highly reliable—96.78% records were retained after the cleaning and reduction processes but the data is imbalanced. From the values in Table 2, it can be asserted that the fuel usage in India is not equally distributed among different categories.

The values show that number of people who prefer clean fuel (LPG and kerosene) is greater compared to the number of people who prefer equally distributed usage which is in turn greater than the number of people who prefer non-clean fuel (firewood, ding cake, crop residue and coal). The features selected by the CFS algorithm show a total of eight different fields that are statistically relevant in classifying the data. This shows that, despite the popular belief that the fuel consumption pattern is determined by the economic status of the households, there are other factors that contribute to the determination of the nature of fuel consumption. These factors are described by the respective fields selected during the subset selection process. These include the ones that describe the socio-economic status of the households, like the settlement nature of the household (*URBAN2011*) and the expenditure on fuel consumption (*CO21*), among other factors such as the region of settlement (*STATEID*), number of residents (*NPERSONS*), etc.

This analysis enhances the existing theories that explain the economic factors are responsible for determining the fuel consumption nature of the households. But the findings of this work suggest that it is not the only factor that affects the nature of fuel consumption but it is one among many others. Although this work provides an extended view of the consumption nature of households in developing countries, it is based on the assumptions that the households taken into consideration represent all the households from the same country or state. As such, further works may include not only the data from a single country, but from multiple countries, or across multiple time periods of the same households, to monitor the changes in their fuel consumption pattern over the time period. Further, works could also take into consideration, the individual fuel rates in the particular region, and other factors like availability, taxation, policy laws and other factors that directly or indirectly affect the nature of fuel consumption and observe their effects on the problem.

References

1. International Energy Agency. 2006. *World Energy Outlook*.
2. Toole, R. 2015. The Energy Ladder: A Valid Model for Household Fuel Transitions in Sub-Saharan Africa?
3. A. F. D. Center. Propane Fuel Basics. U.S. Department of Energy.
4. Lam, N.L., K.R. Smith, A. Gauthier, and M.N. Bates, Kerosene: A Review of Household Uses and Their Hazards in Low- and Middle-Income Countries. *Journal of Toxicology and Environmental Health Part B*.
5. Coal, The fuel of the future, unfortunately. *The Economist*, April 16, 2014.
6. Ecofriendly Alternatives to Burning Wood in your Fireplace. *Scientific American*.
7. Pant, K.P. 2010. *Health Costs of Dung-Cake fuel use by the poor in rural Nepal*. Kathmandu: South Asia Network of Economic Research Institues.
8. Crop Residue, Wikipedia. [Online]. Available: https://en.wikipedia.org/wiki/Crop_residue.
9. Mekkonen, A., and G. Kohlin. 2009. Determinants of Household Fuel Choice in Major Cities in Ethiopia. *Working Papers in Economics*.
10. Ado, A., I.R. Darazo. 2016. M. A. 7 (3).
11. Heltberg, R. 2003. *Household Fuel and Energy use in Developing Countrie—A Multicountry Study*. The World Bank: Washington, DC.
12. Thadeo, S.M. 2014. *Economic of Urban households' Cooking Fuel Consumption in Arusha City, Tanzania*. Morogoro, Tanzania: Sokoine University of Agriculture.
13. Nasir, Z.A., F. Murtaza, and I. Colbeck. 2015. Role of Poverty in Fuel Choice and Exposure to Indoor Air Pollution in Pakistan. *Journal of Integrative Environmental Sciences* 107–117.
14. Rehman, I.H., A. Kar, R. Raven, D. Singh, J. Tiwari, R. Jha, P.K. Sinha, and A. Mirza. 2010. Rural Energy Transistion in Developing Countries: A Case of the Uttam Urja Initiative in India. *Environmental Science & Policy* 13 (4): 303–311.
15. Dominic, V., D. Gupta, S. Khare, and A. Agrawal. 2015. Inverstigation of Chronic Disease Correlation Using Data Mining Techniques. In *2nd International Conference on Recent Advances in Engineering and Computational Sciences*, Chandigarh.
16. Narendranath, S., S. Khare, D. Gupta, and A. Jyotishi. 2018. Charateristics of 'Escaping' and 'Falling into' Poverty in India: An Analysis of IHDS Panel Data Using Machine Learning Approach. In 7th International Conference on Advances in Computing, Communications and Informatics (ICACCI), Sept 2018. IEEE.
17. Indian Human Development Survey. [Online]. Available: https://ihds.umd.edu/.
18. Longadge, R., S.S. Dongre, and L. Malik. 2013. Class Imbalance Problem in Data Mining: Review. *Internation Journal of Computer Science and Network*.
19. Brownlee, J. 2018. Need Help with Statistics? Take the FREE Mini-Course. *Machine Learning Mastery*. [Online]. Available: https://machinelearningmastery.com/statistical-sampling-and-resampling/.
20. Chawla, N.V., K.W. Bowyer, L.O. Hall, and P.W. Kegelmeyer. 2002. SMOTE: Synthetic Minority Over-sampling Technique. *Journal of Artificial Intelligence Research*.
21. Hall, M.A. 1999. *Correlation-based Feature Selection for Machine Learning*. New Zealand: University of Wakaito.
22. Random Forest. [Online]. Available: https://en.wikipedia.org/wiki/Random_forest#Algorithm.
23. Joshi, R. 2016. Accuracy, Precision, Recall & F1 Score: Interpretation of Performance Measures. *Exsilio Solutions*.

Efficient Predictions on Asymmetrical Financial Data Using Ensemble Random Forests

Chaitanya Muppala, Sujatha Dandu and Anusha Potluri

Abstract The technological advances in the areas of Big Data and machine learning have led to many useful applications in the financial industry. However, the success of these technologies depends on the analysis of useful information. The financial data is often asymmetrical in nature. It is the nature of information that is crucial in making financial decisions. It is often used to detect the financial frauds, predict the market trends, marketing financial products, and various other use cases. In this work, we are proposing that the ensemble random forests will be able to make better predictions on the asymmetrical financial data. We are taking two cases for making the predictions—one, predicting the customers who will buy the term deposit and two, credit card fraud detection. In both cases, the ensemble random forests were compared with the logistic regression and demonstrated with the results where the random forests performed better than the logistic regression.

Keywords Prediction · Random forest · Asymmetrical financial data · Trends · Logistic regression

1 Introduction

Prediction systems play a prominent role in today's business landscape and are a key proponent in driving business decisions. Machine learning has become a phenomenal technology in providing predictions on the given data. Predictions are elicited from the given data by extracting the patterns and generalizing them to the

C. Muppala (✉) · S. Dandu · A. Potluri
Department of Computer Science and Engineering, Malla Reddy College
of Engineering and Technology, Hyderabad, India
e-mail: chaitanyamuppala9@gmail.com

S. Dandu
e-mail: sujatha.dandu@gmail.com

A. Potluri
e-mail: potluri.anusha1@gmail.com

© Springer Nature Singapore Pte Ltd. 2020
K. S. Raju et al. (eds.), *Proceedings of the Third International Conference on Computational Intelligence and Informatics*, Advances in Intelligent Systems and Computing 1090, https://doi.org/10.1007/978-981-15-1480-7_31

new data. On an abstract level, much of the machine learning prediction can be reduced to build a learning model. The learning model is a function that maps the given inputs to a prediction. Once the model is trained with the given set of training data, it can be used to provide predictions on a set of new and unseen data. However, the major roadblock is to identify the right set of algorithms for a given problem and to use the right set of tools.

2 Motivation

Often, the financial data is asymmetrical in nature [1] and it is important to understand this asymmetry to extract the trends and predictions, which in return will add value to make business decisions. The motivation of this project is to make efficient predictions on asymmetrical financial data. For making efficient predictions, it is very crucial to use the right set of algorithms, tools, and statistical approaches in building the models. It is also important to identify and select the right set of attributes which will contribute to better predictions.

There are many machine learning algorithms and techniques that are proposed and implemented from the past many years to make better predictions. Logistic regression had been a very popular technique that is widely used across. However, in the background of the asymmetrical nature of the data, the ensemble random forests will be the better choice in terms of high precision and accuracy.

3 Objective

The objective of the work is to prove that the random forests are better predictors than the logistic regression algorithms especially in the background of the asymmetrical financial data. To prove the hypothesis, the following two cases are considering.

1. Predicting the term deposit purchase by the existing customers of the bank.
2. Identifying fraudulent credit card transactions to prevent fraud to happen.

One of the useful results of the logistic regression is that it does not provide outright classes as output. It will provide the probabilities that can be associated with each observation. One can apply custom performance metrics on the resulted probabilities and can classify the output for a business problem. Logistic regression is not altered for the multi-collinearity provided, it is implemented with regularization. The problem with logistic regression starts when there are too many features and a good set of missing data. Another problem is that it uses the entire data to come up with its scores [2]. When there are extreme ends in the data, the scores will be highly influenced by these extreme ranges in the data. Moreover, when the selected features are nonlinear, one needs to rely on transformations, which prove to be costly.

Random forest, as the name implies, is used to create forest of decision trees with randomness introduced in it. The more the number of trees, the better will be the result. One significant point to note here is that decision tree generation is quite different from the generation of random forest. Using for both classification and regression is an advantage of random forests. Overfitting can be easily avoided by creating sufficient trees from the given data. Missing values can also be handled, and it can be modeled for categorical values. One of the highlight features is that one can measure the relative importance of a feature, which is responsible for the output or the successful prediction [3, 4].

4   Data

There are two datasets that are used for this project.

1. Bank marketing data from UCI Machine Learning Repository [5].

This data represents the direct marketing campaign of the Portuguese bank. The general marketing campaigns run over telephone calls. The goal of the telemarketing campaign is to make the existing customer purchase the term deposit that is going to be launched by the bank.

There are four datasets under this data:

(a) bank_data_full_addtional.csv contains all examples numbered 41,188 and 20 inputs/attributes. The data in this dataset is sorted by date from May 2008 to November 2010. The data analyzed in Moro et al. [5] is very similar to the selected data.
(b) bank-additional.csv with 10% of the examples numbered 4119, that are randomly selected from (a), and 20 inputs/attributes.
(c) bank-full.csv contains all the examples and 17 inputs, which is ordered by date.
(d) bank.csv contains 10% of the examples and 17 inputs that are randomly selected from (c).

Term deposit purchase or subscription is the final goal of classification from the above data.

2. Credit card transaction data is resulted from the research collaboration of Worldline and the Machine Learning Group of Université Libre de Bruxelles [6].

The dataset holds the transactions that are done using the credit cards during period of September 2013. On a whole, it contains 284,807 transactions, out of which only 492 are the fraud. This clearly depicts that the data is highly imbalanced. The fraudulent transactions, which are positive class in the data, only account for 0.172% of total data.

The data is resultant of PCA transformation; hence there are only numerical input variables. The features $V1$, $V2$, $V3$, ..., $V28$ are resultant of the PCA transformation. The features which are not transformed are "Time" and "Amount." There is also "Class" feature and the fraud transaction is indicated by value 1 and the non-fraudulent transactions are indicated by value 0.

5 Approach

The work begins with building the models for logistic regression and random forests. Figure 1 depicts the statistical method used in the paper. Predictions are made using the models that are constructed. Models are abstractions. It is a mathematical function that generalizes the relationship between predictor variables (X) and a target variable (y).

To address the imbalanced nature of the datasets and to enhance the prediction of the model, a method called SMOTE—Synthetic Minority Over-sampling Technique is used as suggested by Chawla et al. [7]. The data is split as training set and test set and applied with cross-validation, to validate the accuracy of the model.

6 Implementation and Results

Scikit-learn [8, 9] is used as a toolkit to build the models and make predictions. The following are the high-level steps that are used to implement the random forests.

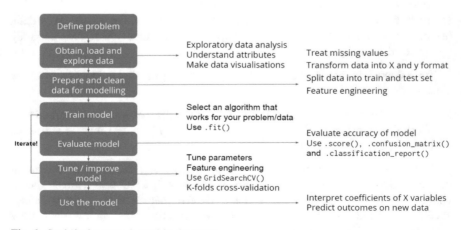

Fig. 1 Statistical approach used in the work

Step (i): From the total of "*f*" features by selecting "*c*" features in such a way that $c \ll f$, which forms a bagging sample.
Step (ii): Using the best split point, calculate node "*d*" among the "*c*" features.
Step (iii): Using the best split, split the node into child nodes.
Step (iv): Repeat the steps (i) to (iii) for predetermined times.
Step (v): Build the random forest by repeating steps (i) to (iv) for "*n*" number times to create "*n*" number of trees and finally average the prediction of each tree as depicted in Fig. 2 [10].

6.1 Predicting the Term Deposit Purchase

Initially, using the logistic regression, the following attributes are identified which have positive effect on the outcome. They are

- consumer price index
- job_retired
- job_student
- default_no
- month_mar
- month_jun
- month_jul

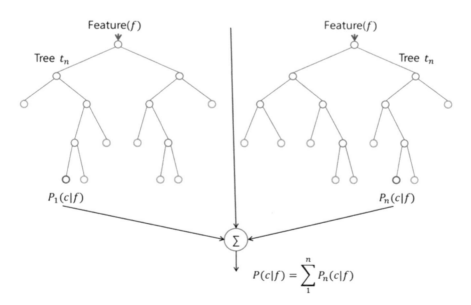

Fig. 2 Sample random forest

- month_aug
- contact_cellular
- day_of_week_tuesday
- poutcome_nonexistent.

The attributes that have negative effect on the outcome are

- emp.var.rate
- euribor3 m
- job_blue_collar
- contact_telephone
- month_may
- month_nov
- poutcome_failure.

The attributes can be plotted as shown in Fig. 3.

The interpretation of the chart can go as such: If there is a unit increase in X, there is a change percentage that the user is more likely to purchase the term deposit from the bank which is $y = 1$. For example, user is predicted to be 16% more likely to purchase a bank term deposit if there is a unit increase in employment variation rate.

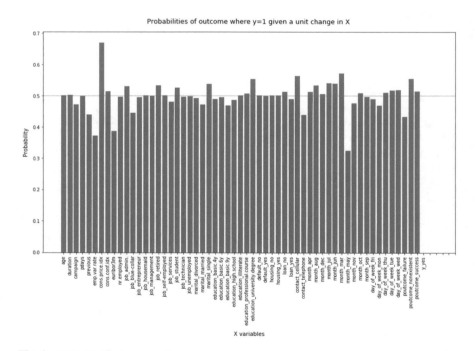

Fig. 3 Attribute effects on results

Results

On the initial run of the model, without any tuning, the logistic regression was able to achieve 90% of the accuracy as shown in Fig. 4.

The confusion matrix represented above denotes the following format.

[[true_negatives, false_negatives]
[false_positives, true_positives]]

In other words, if the business or banks made marketing calls based entirely on this logistic regression model, the following are the implications.

- Marketing people are saved by 35,613 wasted calls
- Accurately, predicted 1855 calls which means 1855 people subscribed to term deposit
- However, the model caused 2785 missed opportunities, which would have said yes, if the marketing team had
- Wasted 935 calls on false negatives who will not subscribe to the term deposit.

Classification report shows four metrics: precision, recall, f1-score, support1. The following are the explanation to the metrics.

1. Precision (also called positive predictive value) is the fraction of relevant instances among the retrieved instances. In other words, precision is the number of correct positive results divided by the number of all positive results. It is calculated as such:
   'true positives/(true positives + false positives)'
2. Recall (also known as sensitivity) is the fraction of relevant instances that have been retrieved over total relevant instances. In other words, recall is the number of correct positive results divided by the number of positive results that should have been returned. It is calculated as such:
   'true positives/(true positives + false negatives)'

Fig. 4 Results of logistic regression

```
COMPARING SCORES OF TRAINING AND TEST SET:

training set score: 0.908906
test set score: 0.912013

CLASSIFICATION REPORT:

                 precision    recall  f1-score   support

            0        0.93      0.97      0.95     36548
            1        0.66      0.40      0.50      4640

avg / total          0.90      0.91      0.90     41188


CONFUSION MATRIX:

[[35613   935]
 [ 2785  1855]]
```

3. F1 score (also F-score or F-measure) is a measure of a test's accuracy in statistical analysis of binary classification. It considers both the precision p and the recall r of the test to compute the score. The F1 score can be interpreted as a weighted average of the precision and recall, where an F1 score reaches its best value at 1 and worst at 0.
4. Support is the number of occurrences of each class in y_true.

With the random forests, 88% of accuracy is achieved as shown in Fig. 5.

On further fine tuning the models, the logistic regression was still able to show only 90% accuracy as shown in Fig. 6.

However, the random forests prediction performance increased to 93% when applied to the tuning parameters as shown in Fig. 7.

From the results, it can be clearly inferred that the random forests performed much better than logistic regression.

In other words, if the business or banks made marketing calls based entirely on this random forest model, the following are the implications.

- Marketing people are saved by 35,883 wasted calls
- Accurately, predicted 2619 calls that converted into successful subscriptions.
- Wasted only 665 calls on false negatives.

Fig. 5 Results of random forests

```
COMPARING SCORES OF TRAINING AND TEST SET:

training set score: 0.907060
test set score: 0.905021

CLASSIFICATION REPORT:

                 precision    recall  f1-score   support

          0         0.91      0.99      0.95     36548
          1         0.78      0.24      0.37      4640

avg / total         0.90      0.91      0.88     41188


CONFUSION MATRIX:

[[36231   317]
 [ 3532  1108]]
```

Fig. 6 Results of logistic regression with model tuning

```
[[35404  1144]
 [ 2506  2134]]
                 precision    recall  f1-score   support

          0         0.93      0.97      0.95     36548
          1         0.65      0.46      0.54      4640

avg / total         0.90      0.91      0.90     41188
```

Fig. 7 Results of random forests with model tuning

```
COMPARING SCORES OF TRAINING AND TEST SET:

training set score: 0.941536
test set score: 0.914538

CLASSIFICATION REPORT:

                 precision   recall  f1-score   support

             0      0.95     0.98      0.96     36548
             1      0.80     0.56      0.66      4640

    avg / total      0.93     0.93      0.93     41188


CONFUSION MATRIX:

[[35883   665]
 [ 2021  2619]]
```

In addition to this, the results on the random forest are improved compared to the work done by Ruangthong and Jaiyen [11], where the accuracy is only 87.54 even after the application of SMOTE [7].

6.2 Identifying Fraudulent Credit Card Transactions

As discussed in the Data Section IV, the data is highly imbalanced, and once again the SMOTE method [7] is used to address the problem of imbalance and to make better predictions. Cross-validation using various splits of the data is performed and the results are as shown below. GridSearchCV and RandomizedSearchCV methods of sklearn are used to identify the optimal hyperparameters in the hyperparameter space.

Results

Logistic Regression

- With undersampled data: 70% accuracy

```
             precision   recall  f1-score   support

         0      1.00     1.00      1.00     284315
         1      0.80     0.63      0.70       492

avg / total      1.00     1.00      1.00     284807
```

- With model tuning: 79% of accuracy

```
              precision    recall  f1-score   support

         0       1.00      1.00      1.00    284315
         1       0.84      0.74      0.79       492

avg / total       1.00      1.00      1.00    284807
```

- With undersampled data and tuning: 95% of accuracy.

```
              precision    recall  f1-score   support

         0       0.93      0.98      0.95       492
         1       0.98      0.92      0.95       492

avg / total       0.95      0.95      0.95       984
```

Random Forests

- With undersampled data: 98% accuracy

```
              precision    recall  f1-score   support

         0       1.00      0.97      0.99    284315
         1       0.05      0.98      0.10       492

avg / total       1.00      0.97      0.98    284807
```

- With undersampled data and 40% test and train data split: 93% of accuracy

```
              precision    recall  f1-score   support

         0       0.91      0.96      0.94       195
         1       0.96      0.91      0.93       199

avg / total       0.94      0.93      0.93       394
```

- With undersampled data and 40% test and train data split and tuning: 98% of accuracy.

```
              precision    recall  f1-score   support

         0       1.00      0.96      0.98    284315
         1       0.04      0.96      0.08       492

avg / total       1.00      0.96      0.98    284807
```

Fig. 8 Comparison of linear regression and random forests with undersampled data and model tuning

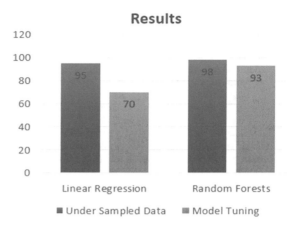

The interpretation of the results can be interpreted in Fig. 8.

One clear inference from the results is that, only with 60 percent as training data, the random forests were able to predict the fraud transactions with an average of 96 percent accuracy.

7 Conclusion

The results clearly demonstrated that the random forests stood out as a better classifier than the logistic regression. When tried out with various validation methods like cross-validation and various splits of the data, the results clearly reiterated that the random forests are the better predictors. The work also demonstrated the methods of addressing the highly imbalanced especially in the credit card fraud detection dataset. However, the application of random forests can be extended to other domains like security, Internet of Things to stand as a testimony of the results that are presented in this work.

References

1. Akerlof, George A. 1970. The Market for "Lemons": Quality Uncertainty and the Market Mechanism. *The Quarterly Journal of Economics* 84 (3): 488–500.
2. Bishop, C., and N. Nasrabadi. 2006. *Pattern Recognition and Machine Learning*, vol. 1. New York: Springer.
3. Biau, G. 2012. Analysis of a Random Forests Model. *The Journal of Machine Learning Research* 98888: 1063–1095.
4. Breiman, L. 2001. Random Forests. *Machine Learning* 45 (1): 5–32.
5. Moro, S., P. Cortez, and P. Rita. 2014. A Data-Driven Approach to Predict the Success of Bank Telemarketing. *Decision Support Systems* 62: 22–31.

6. Dal Pozzolo, Andrea, Olivier Caelen, Reid A. Johnson, and Gianluca Bontempi. 2015. Calibrating Probability with Undersampling for Unbalanced Classification. In Symposium on Computational Intelligence and Data Mining (CIDM), IEEE.
7. Chawla, N.V., K.W. Bowyer, L.O. Hall, and W.P. Kegelmeyer. 2002. SMOTE: Synthetic Minority Over-Sampling Technique. *Journal of Artificial Intelligence Research* 16: 321–357.
8. Pedregosa, et al. 2011. Scikit-learn: Machine Learning in Python. *Journal of Machine Learning Research* 12: 2825–2830.
9. Louppe, Gilles, and Alexandre Gramfort. API Design for Machine Learning Software: Experiences from the Scikit-Learn Project" in European Conference on Machine Learning and Principles and Practices of Knowledge Discovery in Databases.
10. Hastie, Trevor, Robert Tibshirani, Jerome Friedman. 2008. The Elements of Statistical Learning, 2nd ed. Berlin: Springer. ISBN 0-387-95284-5.
11. Ruangthong, Pumitara, and Saichon Jaiyen. 2015. Bank Direct Marketing Analysis of Asymmetric Information Based on Machine Learning. In 12th International Joint Conference on Computer Science and Software Engineering.

Segmentation of Soft Tissues from MRI Brain Images Using Optimized KPCM Clustering Via Level Set Formulation

Kama Ramudu and Tummala Ranga Babu

Abstract An important approach for medical images is image segmentation. Image segmentation is the way toward to eradicate the area of attentiveness by making various segments in an image. This segmentation helps to analyze the representation of image in an easier manner. The segmentation of an image in medical image analysis is considered as one of the challenging tasks in many clinical applications due to noise, poor illumination changes, and also the intensity inhomogeneity. In order to have segmentation of soft tissues from magnetic resonance imaging (MRI) brain images, a new approach had been proposed known as "optimized kernel possibilistic fuzzy c-means (OKPCM)" algorithm using a level set formulation. The proposed algorithm consists of two stages: In order to improve clustering efficiency in the preprocessing, we introduced a hybrid approach which is called particle swarm optimization (PSO) algorithm followed by kernel possibilistic fuzzy C-means (KPFCM) clustering. Firstly, with the help of PSO algorithm, automatically the optimal cluster centers are calculated. Later, these optimum cluster values acted as a cluster centers for KPFCM clustering in order to ameliorate the clustering efficiency. The membership function (MF) of the conventional clustering algorithms, i.e., FCM, PCM along with KFCM, was sensitive to the noise and outliers. The preprocessing segmentation results suffer from boundary leakages and outliers. So, to overcome these drawbacks, it is necessary to use post-processing where we introduce the level set method. The level set method utilizes the efficient curve deformation which is driven by an external and internal force in order to capture the important structures (usually edges) in an image as well as curve with minimal energy function is defined. The combined approach of both preprocessing and post-processing is called as optimized kernel possibility fuzzy c-means (OKPFCM) clustering using level set method. The accuracy and the noise effect in those images can be upgraded by using this method.

K. Ramudu (✉)
Department of ECE, Acharya Nagarjuna University, Guntur, India
e-mail: ramudukama@gmail.com

T. Ranga Babu
RVR&JC College of Engineering, Guntur, India
e-mail: trbaburvr@gmail.com

© Springer Nature Singapore Pte Ltd. 2020
K. S. Raju et al. (eds.), *Proceedings of the Third International Conference on Computational Intelligence and Informatics*, Advances in Intelligent Systems and Computing 1090, https://doi.org/10.1007/978-981-15-1480-7_32

Keywords Image segmentation · Kernel-based possibilistic fuzzy c-mean
algorithm · Particle swarm optimization · Level sets and medical images

1 Introduction

Magnetic resonance imaging or MRI is a radiology system that utilizes radio waves
and magnetism in addition with a computer to deliver the mages of a body structures.
Originally, MRI is also called as "nuclear magnetic resonance imaging (NMRI)" as
well as "magnetic resonance tomography (MRT)" and is the type of a "nuclear
magnetic resonance (NMR)" which is a therapeutic utilization of a nuclear magnetic
resonance (NMR). Nuclear magnetic resonance is a physical marvel in which the
nuclei in a magnetic field assimilate and reproduce electromagnetic radiation which
can likewise be utilized for imaging in other NMR applications, for example, NMR
spectroscopy. NMR is likewise routinely utilized in the advanced medical imagining
systems, for example, in MRI. "Computed tomography (CT)," "magnetic resonance
imaging (MRI)", "digital mammography," as well as other imaging modalities
provide effective means for a noninvasive analysis of the anatomy of a subject. The
brain is the front-most piece of central nervous system. Alongside the spinal cord, it
forms the "central nervous system" (CNS). In a field of medicinal science, an
eccentric cell magnification inside the cerebrum is kenned as a tumor. Brain in the
human body is the most delicate fragment. The most important part in the brain or a
cerebrum MRI can isolate between white matter what's more with the grey
matter-GM and can in like manner be utilized to investigate aneurysms at the same
time tumors also. Since MRI does not use an X-rays or another radiation, it is an
imaging approach of choice where an incessant imaging is required for the assurance
or else treatment, particularly in the cerebrum or in the brain [1–4].

In this paper, we proposed optimized KPCM clustering using level set method,
which is motivated by the KPCM clustering and PSO algorithms to modify and
overcome the drawbacks of the existing techniques such as outlier reduction, spatial
information, noise, and distance metric. In an existing method, the development of
the algorithm "FCM" was pragmatic to the image segmentation concerns particularly simply on the one side, which were often could not be enough to yield the
acceptable outcomes. Therefore, in the proposed process, we do integrate an
algorithm which is single as well as number of improvements that are deemed to be
a relevant one. From the experimental results, we can say that the proposed process
gives an improved segmentation results. For the medical images, cluster boundaries
found through the utilization of optimized KPCM were used as initial contours for a
level set [5–8].

This paper was organized as follows: Sect. 2 is the introduction and how to
calculate the best fitness values in an image by using particle swarm optimization.
Section 3 discusses the algorithm "kernel-based possibilistic fuzzy c-means
(KPFCM)" clustering. An implementation of the proposed optimized KPFCM
clustering using level set formulation is described in Sect. 4. In Sect. 5, we present

experimental results as well as discussion of planned algorithm by utilizing synthetic images, simulated along with the real MR brain images. At last, in Sect. 6 we presented the conclusion and future scope.

2 Particle Swarm Optimization (PSO) Algorithm

2.1 Calculate Optimum Pixel Values by Using PSO Algorithm

Kennedy and in addition Eberhart (1995) [9] stayed enlivened to have an improvement in look system of PSO by the scavenging conduct of groups of fowls along with schools of a fish. Every particle has its own location as well as velocity, at which a quality speaks to the factors of a choice in the present emphasis along with the development vector for the following cycle, separately. A velocity additionally the position of each and every particle as needs be varieties to the information which is partaken in the middle of each particle in the present cycle. The most vital preparing step is that to figure the novel velocity as well as the location of every particle in the following procedure by utilizing the conditions (1) and (2).

$$
\begin{aligned}
v_{id}(t+1) = w * v_{id}(t) + c_1 r_1 (x_{p(id)}(t) - x_{id}(t) \\
+ c_2 r_2 (x_{gd}(t) - x_{id}(t)))
\end{aligned}
\tag{1}
$$

At which, $v_{id}(t)$ indicates the value of velocity of dth.

Dimension' of ith particle in tth-iteration. A variable $x_{id}(t)$ indicates location of dth dimension if ith particle in tth iteration. A variable w is a weight of inertia, c_1 self-cognition acceleration coefficient, and c_2 social cognition acceleration coefficient.

$$
x_{id}(t+1) = x_{id}(t) + v_{id}(t+1)
\tag{2}
$$

Equation (2) represents new location of every particle which is refreshed utilizing a first position along with the new velocity by a condition (1), at which r_1 and additionally r_2 were produced independently. The range of uniform distributed random numbers is (0, 1).The below segment depicts quickly about these two classes of estimations of parameter segment.

The parameters inertial weight (w), acceleration coefficients c_1 and c_2 and r_1 and r_2 are used in Eq. (1). The proper values are choosing for better results. Standard PSO algorithm utilizes the c_1 and c_2 which are equal to 2. For better segmentation results, changes are made in the inertial weight (w) and acceleration coefficients according to the specific application, which is called "adaptive PSO". In this work, we considered the following parameter values for optimum pixel calculation from the input image. The parameter values used in this paper are as shown in Table 1.

Table 1 Parameters used to get optimum pixel values

Parameters	Representation	Values
c_1	Personal particle coefficient	0.8
c_2	Global particle coefficient	0.8
w	Inertial weight	1.2
Particle population	Predefined population	150
Iteration	–	150

3   Kernel Possibilistic FCM (KPCM) Clustering

In our proposed algorithm, changes are made in the preprocessing an image to improve segmentation accuracy. In this paper, some modifications are done in the conventional clustering technique, i.e., possibilistic c-means (PCM) clustering. The improved PCM clustering is called kernel possibilistic c-means (KPCM) clustering by incorporating the Gaussian function as a kernel and modifying the distance metric, i.e., Euclidian distance, which is replaced by Mahalanobis distance to get the smaller distance between pixels and cluster centers. The noise effect in the MRI brain Images is overcome by spatial information of both local and nonlocal, and membership. The objective utility of Kernel-based PFCM is incorporated through this modified distance metric. To overcome all this limitation, the subsequent algorithm is called "kernel-based possibilistic fuzzy c-means (KPFCM)". We utilize a Gaussian function as kernel function to incorporate this function into the PCM clustering, which is called kernel PCM clustering. These results intended for the enhanced segmentation [10, 11].

3.1   Kernel Possibilistic C-Means (KPCM) Algorithm

"Kernel possibilistic clustering algorithm (KPCM)" had taken into consideration to overcome issues of robustness against the noise, outliers as well as the arbitrarily shaped clusters' boundaries. Keller in addition with krishnapuram introduced the new below objective function by utilizing the modified distance metric called mahalanobis distance is as following.

$$J_{\text{PCM}} = \sum_{i=1}^{c} \sum_{k=1}^{n} \mu_{ik}^m d^2(x_k, V_{iopt}) + \sum_{i=1}^{c} \eta_i \sum_{k=1}^{n} (1 - \mu_{ik})^m \tag{3}$$

At which 'd' represents the distance of Mahalanobis:

$$d^2(x_k, V_{iopt}) = (x_k - V_{iopt})^T S_i (x_k - V_{iopt}) \tag{4}$$

$$S_i = \left| \sum_i \right|^{\frac{1}{p}-1} \sum_1$$

(5)

where x_k is pixels of an image, n total number of pixels, 'C' number of clusters, and V_{iopt} optimum cluster center.

$p = 1$, dimension problem, as well as η_i were positive numbers.

An initial term of Eq. (3) campaigns for a distance minimum

In between prototypes in addition with data types, at the same time μ_{ik} should be large as possible as. Here, η_i were preferred as:

$$\eta_i = K_1 \frac{\sum_{k=1}^{n} \mu_{ik}^m \|x_k - V_{iopt}\|^2}{\sum_{k=1}^{n} \mu_{ik}^m}$$

(6)

K_1 was preferred to be a 1, and PCM memberships were modified as below:

$$\mu_{ik} = \frac{1}{1 + \left(\left(\|x_i - V_{iopt}\|^2 \right) / (\eta_i) \right)^{(1)/(m-1)}}$$

(7)

KPCM objective function is represented as:

$$J_{\text{KPCM}}(U, V) = \sum_{i=1}^{c} \sum_{k=1}^{n} \mu_{ik}^m \|\phi(x_k) - \phi(V_{iopt})\|^2 + \sum_{i=1}^{c} \eta_i \sum_{k=1}^{n} (1 - \mu_{ik})^m$$

(8)

By considering KFCM, modified memberships were written as below:

$$\mu_{ik} = \frac{1}{1 + \left[\frac{2(1 - K_1(x_k, V_{iopt}))}{\eta_i} \right]^{(1)/(m-1)}}$$

(9)

$$V_{iopt} = \frac{\sum_{k=1}^{n} \mu_{ik}^m k_1(x_k, v_{iopt}) x_k}{\sum_{k=1}^{n} \mu_{ik}^m k_1(x_k, v_{iopt})}$$

(10)

Here, the Gaussian function was used as kernel function as well as η_i were assessed using:

$$\eta_i = K \frac{\sum_{k=1}^{n} \mu_{ik}^m 2(1 - K_1(x_k, V_{iopt}))^2}{\sum_{k=1}^{n} \mu_{ik}^m}$$

(11)

Usually, K is preferred to be as 1. We here condense a KPCM algorithm as below.

3.2 Algorithm Steps of Optimized KPCM Clustering

1. Initiate (number of a class: c, fuzziness degree: m>1, criterion stopping:ε)
2. Optimum cluster centers are obtained by using Eq. (1) and Eq. (2).
3. Initiate the $U^t = 0$ by utilizing the random values.
4. Assess the η_i, Eq. (11).
5. a. Modify V^t, Eq. (10)

    b. Modify U^t, Eq. (9)

    c. Stop the process if $\left| J_{KPCM}^{t+1} - J_{KPCM}^t \right| < \varepsilon$,

Or else return to stage 3.

The KPCM algorithm undergoes from the issues of an initialization stage, metric distance, spatial data as well as rejection of outlier. So we propose a single algorithm which considers all these aspects which will be taking into account.

4 Proposed Method

Here in this section, we were proposing the implementation of optimized KPCM clustering by using level set method for robustness and reduce outlier rejection. The implementation of post-processing level set segmentation method is described below [12].

4.1 Implementation to Level Set Segmentation

Active contours are a well-defined method by means of image segmenting [13, 14]. Rather than the parametric classification of active contour, the segmentation of level set is embedding them into a period of subordinate "partial differential equation (PDE)" $\psi(a, b, c)$. It is then conceivable toward inexact the evolution of active contour's verifiably through the following level set $\Gamma(a)$. To introduce spontaneously an initial contour of a level set, we do utilize the subsequent clustering accomplished by optimized KPCM clustering.

$$\left\{ \begin{array}{l} \psi(a, b, c) < 0, (b, c) \text{ inside } \Gamma(a) \\ \psi(a, b, c) = 0, (b, c) \text{ at } \Gamma(a) \\ \psi(a, b, c) > 0, (b, c) \text{ inside } \Gamma(a) \end{array} \right\} \qquad (12)$$

Γ might be consist as a contour. It could be effortlessly controlled through checking estimations of a level set function ψ, which adjusts to a topological variations of an implicit interface Γ. An evolution finishing is controlled by:

$$\left\{ \begin{array}{l} \frac{\partial \psi}{\partial t} + F|\nabla \psi| = 0 \\ \psi(0, b, c) = \psi_0(b, c) \end{array} \right\} \tag{13}$$

At which, $|\nabla \psi|$ represents normal direction, $\psi_0(b, c)$ indicates an initial contour, as well as 'F' shows a comprehensive forces. An evolving forces F must be regularized through an edge representation capacity g with a specific end goal to stop a level set evolution near to an optimal solution.

$$g = \frac{1}{1 + |\nabla(G_\sigma * I_{\text{OKPCM}})|^2} \tag{14}$$

where $G_\sigma * I_{\text{OKPCM}}$ is convolution in between the Gaussian Kernel G_σ and the image I_{OKPCM}. ∇ is an gradient operator.

Formulation of the segmentation of level set is treated as follows:

$$\frac{\partial \psi}{\partial t} = g|\nabla \psi|\left(\text{div}\left(\frac{\nabla \psi}{|\nabla \psi|}\right) + v\right) \tag{15}$$

Here, $\left(\text{div}\left(\frac{\nabla \psi}{|\nabla \psi|}\right)\right)$ appropriates a mean curvature k in addition with a customable balloon force.

The algorithm called fast level set was introduced in

$$\frac{\partial \psi}{\partial t} = \mu \zeta(\psi) + \xi(g, \psi) \tag{16}$$

$$\zeta(\phi) = \Delta \phi - \text{div}\left(\frac{\nabla \phi}{|\nabla \phi|}\right) \tag{17}$$

At which, $\zeta(\psi)$ shows a penalty momentum of a ψ as well as $\xi(g, \psi)$ is the image gradient data.

$$\xi(g, \psi) = \lambda \delta(\psi) + \text{div}\left(g \frac{\nabla \psi}{|\nabla \psi|}\right) + vg\delta(\psi) \tag{18}$$

In the above equation, $\delta(\psi)$ represents a function of dirac. v, μ and λ were the constraints to the control of an evolution level set. The parameter ψ in Eq. (16), i.e., $\xi(g, \psi)$ and $\zeta(\psi)$, attracts toward the boundaries of the images and automatically approach nearer to the signed distance function (SDF), espectively.

5   Results and Discussions

Implementation of suggested algorithm is performed on MRI brain T1 image for segmentation of three tissues such as white matter (WM), gray matter (GM) and cerebral spinal fluid (CSF). Dataset is taken from the MRI brain Web science Web site. The proposed results are analyzed and compared with optimized k-means and FCM clustering via level set evolution. The proposed segmentation results are accurate and faster based on the following parameters, i.e., tuning parameter, Dice, Jaccard similarity index, and iterations.

Figures 1 and 2 clearly show that the proposed optimized k-means clustering and optimized FCM clustering and its level set formulation (LSF) which fails to detect and extract the regions of the three tissues compared to the ground truth segmented images So these methods are not superior to an MRI brain T1 as well as coronal brain images compared to the proposed optimized KPCM clustering via LSF. The optimized KPCM clustering via LSF is best suitable for detecting and extracting of three tissues in terms of accurate segmentation of regions by utilizing the particle swarm optimization (PSO) and Mahalanobis distance measured in the preprocessing.

The performance parameters of our proposed model and conventional models are shown in Tables 1 and 2, respectively. So here we can say that optimized k-means clustering via level set formulation (LSF) in terms of Dice similarity (DS) and Jaccard similarity (JS) at which segmentation for the CSF, WM as well as GM tissues could not detected satisfactorily. At the same time, it takes more elapsed time and has less segmentation accuracy. So to improve the drawbacks occurred in Tables 1 and 2, optimized k-means and FCM clustering via level set formulation (LSF) can be done by the proposed method optimized KPCM clustering via level set formulation (LSF). A pixel-based quantitative evaluation approach is used. In this evaluation, approach made a comparison between the final segmented image 'P' and ground truth image 'Q'. The segmentation similarity coefficient (SSC) is measured with the help of Dice and Jaccard coefficients. The proposed final level set segmentation results covered the maximum segmented area compared with conventional models. A pixel-based quantitative evaluation approach is used. In this evaluation, approach made a comparison between the final segmented image 'P' and ground truth image 'Q'. The segmentation similarity coefficient (SSC) is measured with the help of Dice and Jaccard coefficients, true positive fraction (TPF), true negative fraction (TNF), false positive fraction (FPF). For the higher values of the Dice and Jaccard coefficients gives the better performance. The Dice and Jaccard indexes can be defined as

$$\text{Dice} = \frac{2|P \cap Q|}{|P| + |Q|} \quad \text{Jaccard} = \frac{P \cap Q}{P \cup Q} \tag{19}$$

$$\text{TPF} = \frac{P \cap Q}{Q}; \quad \text{TNF} = 1 - \frac{P - Q}{Q} \tag{20}$$

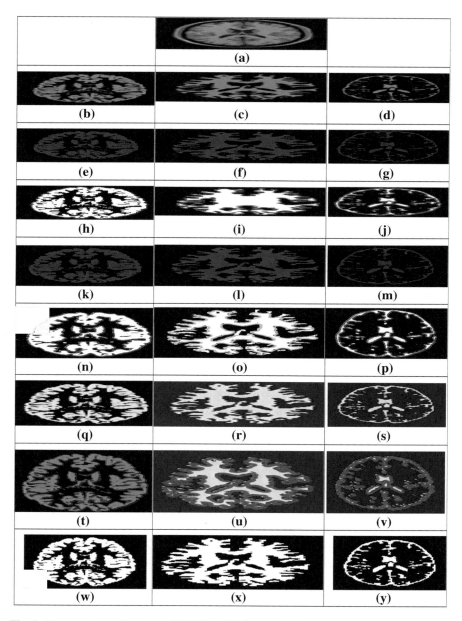

Fig. 1 First row shows that original MRI brain T1 image with 3% noise, slice thickness is 1 mm and non-uniformity of pixels 20%, second row figures (**b–d**) depict the ground truth images, and third row images (**e–g**) and fifth row images (**k–m**) show that optimized k-means and optimized FCM clustered images and its level set evolution in fourth row (**h–j**) and sixth row (**n–p**), respectively; similarly, the seventh and eighth row images such as figures (**q–s**) and (**t–v**) are the proposed optimized KPCM clustered and its level set evolution, respectively. Finally, the last row images (**w–y**) show the extracted segmentation regions from eighth row

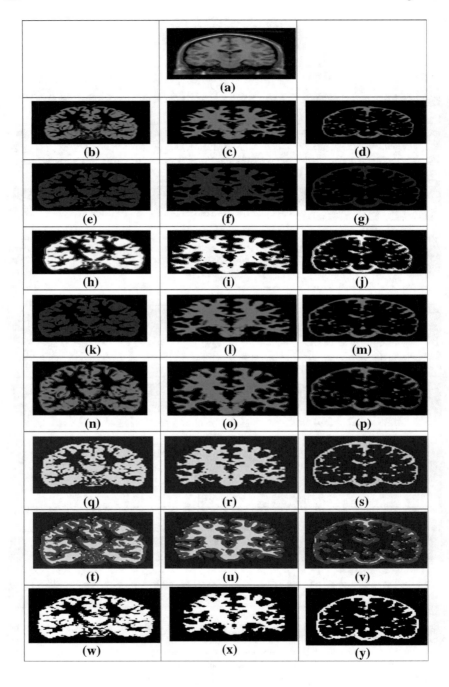

◄**Fig. 2** First row shows that original MRI brain T1 coronal image with 3% noise, slice thickness is 1 mm and non-uniformity of pixels 20%, second row figures (**b–d**) depict the ground truth images, and third row images (**e–g**) and fifth row images (**k–m**) show that optimized k-means and optimized FCM clustered images and its level set evolution in fourth row (**h–j**) and sixth row (**n–p**), respectively; similarly, the seventh and eight row images such as figures (**q–s**) and (**t–v**) are the proposed optimized KPCM clustered and its level set evolution, respectively. Finally, the last row images (**w–y**) show the extracted segmentation regions from eighth row

Table 2 Performance analysis of optimized k-means, optimized FCM and optimized KPCM clustering via level set model in terms of Dice similarity (DS) index, Jaccard similarity (JS) index, true positive fraction (TPF), and false negative fraction (FNF) measures

S. No.	Method	Tissue type	Image 1 (MRI brain T1)				Image 2 (MRI brain T1 coronal scan)			
			DS	JS	TPF	FNF	DS	JS	TPF	FNF
1	Optimized k-means via LSE	WM	0.8380	0.7212	0.721	0.209	0.8715	0.7724	0.7724	0.199
		GM	0.7039	0.3531	0.363	0.196	0.7659	0.3512	0.3512	0.156
		CSF	0.6522	0.4328	0.4328	0.080	0.582	0.3567	0.3567	0.110
2	Optimized FCM via LSE	WM	0.8980	0.4222	0.4222	0.198	0.9225	0.5742	0.5742	0.1053
		GM	0.8869	0.4431	0.441	0.201	0.8959	0.4412	0.4412	0.1966
		CSF	0.7652	0.5398	0.5398	0.345	0.9032	0.4570	0.4570	0.1046
3	Optimized KPCM via LSE	WM	0.9652	0.7328	0.7328	0.001	0.9835	0.7634	0.7634	0.0003
		GM	0.9439	0.5431	0.5431	0.092	0.9695	0.6412	0.6412	0.0187
		CSF	0.9652	0.7328	0.7328	0.015	0.9642	0.8807	0.8807	0.0197

$$\text{FPF} = \frac{P - Q}{Q}; \quad \text{FNF} = \frac{Q - P}{Q} \tag{21}$$

The performance evaluation parameters such as Dice similarity index (DS), Jaccard similarity index (JS), TPF, TNF, FPF, FNF which we considered in our proposed method, the better performance can be achieved by Dice similarity index (DS), Jaccard similarity index, True Positive Fraction (TNF), True Negative Fraction (TNF) for higher values. Moreover, for higher values of false positive fraction (FPF) and false negative fraction (FNF), it leads to the worst performance. Based on the above-mentioned information, Dice similarity index (DS), Jaccard similarity index (JS), true negative fraction (TNF), and false negative fraction (FNF) of the proposed method give improved results when compared with the conventional methods.

From Table 2, the proposed model is superior and segmentation area is nearly to the accurate areas of ground truth images. All the experimentations are done on MATLAB R2017b 64b in Windows 10 OS with Intel(R) dual core(TM) 64-bit processor, CPU @ 1.80 GHz, 2 GB RAM.

6 Conclusion

Optimized kernel possibilistic FCM clustering (KPCM) using level set method is presented in this paper for the efficient segmentation of three tissues' MRI brain images. This is used to suppress the noise effect during the segmentation. In our algorithm, PSO with KPCM is performed in the primary step for improving the clustering efficiency and information of mutually local as well as non-local which are included into the KPCM objective utility which is modified by distance metric. Later, in secondary step, for achieving the robust image segmentation using level set method. The segmentation accuracy is measured based on similarity metrics such as Dice similarity, Jaccard similarity, true positive fraction, and false negative fraction. To demonstrate the superiority of our proposed method, we compared our results with optimized k-means and FCM clustering via level formulation. Thus, our proposed method can show better results than all other two methods.

References

1. Bezdek, J.C., L.O. Hall, and L.P. Clarke. 1993. Review of MR Image Segmentation Using Pattern Recognition. *Medical Physics* 20 (4): 1033–1048.
2. Liu, H.-C., J.-F. Lirng, and P.N. Wang. 2003. Generalized Kohonen's Competitive Learning Algorithm for Ophthalmological MR Image Segmentation. *Magnetic Resonance Imaging* 21: 863–870.
3. Liew, A.W.-C., and H. Yan. 2006. Current Methods in the Automatic Tissue Segmentation of 3D Magnetic Resonance Brain Images. *Current Medical Imaging Reviews* 2 (1): 91–103.
4. Gupta, M.P., and M.M. Shringirishi. 2013. Implementation of Brain Tumor Segmentation in Brain MR Images Using k-Means Clustering and Fuzzy c-Means Algorithm. *International Journal of Computers & Technology* 5 (1): 54–59.
5. Hou, Z., W. Qian, S. Huang, Q. Hu, and W.L. Nowinski. 2007. Regularized Fuzzy c-Means Method for Brain Tissue Clustering. *Pattern Recognition Letters* 28: 1788–1794.
6. Kennedy, J., and R.C. Eberhart. 1995. Particle Swarm Optimization. In *Proceedings of IEEE International Conference on Neural Network* 19-42, Piscataway, NJ.
7. Krishna Puram, R., and J.M. Keller. 1993. A Possibilistic Approach to Clustering. *IEEE Transactions on Fuzzy Systems* 1 (2): 98–110.
8. Pal, N.R., K. Pal, J.M. Keller, and J.C. Bezdek. 2005. A Possibilistic Fuzzy c-Means Clustering Algorithm. *IEEE Transactions on Fuzzy Systems* 13 (4): 517–530.
9. Bezdek, J.C., L.O. Hall, M.C. Clark, D.B. Gold Gof, and L.P. Clarke. 1997. Medical Image Analysis with Fuzzy Models. *Statistical Methods in Medical Research* 6: 191–214.
10. Ramudu Kama, Srinivas A., and Rangababu Tummala. 2016. Segmentation of Satellite and Medical Imagery Using Homomorphic filtering based Level Set Model. *Indian Journal of Science & Technology, 9* (S1): https://doi.org/10.17485/ijst/2016/v9is1/107818, ISSN (Print) 0974-6846 & ISSN (Online) 0974-5645.
11. Ramudu Kama, and Rangababu Tummala. 2017. Segmentation of Tissues from MRI Bio-medical Images Using Kernel Fuzzy PSO Clustering Based Level Set approach. *Current Medical Imaging Reviews: An International Journal, 13* (1), 1–12. ISSN (print):1573-4056 & ISSN (online): 1875-6603 (Indexed by SCI, Impact Factor: 0.933). https://doi.org/10.2174/1573405613666170123124652.

12. Raghotham Reddy, Ganta, Kama Ramudu, A. Srinivas, and Rao Rameshwar. 2016. Segmentation of Medical Imagery Using Wavelet Based Active Contour Model. *International Journal of Applied Engineering Research (IJAER)*, *11* (5), 3675–3680. ISSN 0973-4562.
13. Ramudu Kama, Srinivas, A., and Rangababu Tummala. 2016. Segmentation of Satellite and Medical Imagery Using Homomorphic filtering based Level Set Model. *Indian Journal of Science & Technology 9* (S1). https://doi.org/10.17485/ijst/2016/v9is1/107818, ISSN (Print) 0974-6846 & ISSN (Online):0974-5645.
14. Ramudu, Kama, and Tummala Rangababu. 2018. Level Set Evaluation of Biomedical MRI and CT Scan Images Using Optimized Fuzzy Region Clustering. *Computer Methods in Biomechanics and Biomedical Engineering: Imaging & Visualization*, *7*. https://doi.org/10.1080/21681163.2018.1441074.

Quicksort Algorithm—An Empirical Study

Gampa Rahul, Polamuri Sandeep and Y. L. Malathi Latha

Abstract Quick sort is one of the most sought after algorithms, because if implemented properly quick sort could be much faster than its counterparts, merge sort and heapsort. The main crux of quick sort algorithm is the implementation of the partitioning operation. Nico Lomuto and C. A. R Hoare have put forth partitioning algorithms that have gained prominent significance. Despite this, one can always shed more light on this partially understood operation of partition. Sorting algorithms have been further developed to enhance its performance in terms of computational complexity, memory and other factors. The proposed method is the use of randomized pivot in the implementation of Lomuto partition and Hoare partition algorithms, respectively. It is analysed with and without a randomized pivot element. This paper presents a comparison of the performance of a quick sort algorithm in different cases using contrasting partition approaches with and without randomized pivot. The results provide a theoretical explanation for the observed behaviour and give new insights on behaviour of quick sort algorithm for different cases. The Hoare partition approach with randomized pivot gives best time complexity in comparison to the other cases.

Keywords Quicksort · Analysis of partition algorithms · Randomized pivot · Best-case complexity

G. Rahul · P. Sandeep · Y. L. M. Latha (✉)
CSE Department, Swami Vivekananda Institute of Technology (SVIT),
Hyderabad 500001, India
e-mail: malathilatha99@yahoo.co

G. Rahul
e-mail: gamparahul@gmail.com

P. Sandeep
e-mail: p.sandeep321@yahoo.com

© Springer Nature Singapore Pte Ltd. 2020
K. S. Raju et al. (eds.), *Proceedings of the Third International Conference on Computational Intelligence and Informatics*, Advances in Intelligent Systems and Computing 1090, https://doi.org/10.1007/978-981-15-1480-7_33

1   Introduction

Quicksort, a.k.a partition-exchange sort [1] is essential, as the name suggests used for sorting the elements of an array in a methodical and arranged manner. Developed by C. A. R Hoare in 1959 and published in 1961 [2], it is continually and widely used with variations [3]. The algorithm is simple to execute, works extremely well for various kinds of input data, and is known to utilize fewer resources than any other sorting algorithm [4]. It is based on divide-and-conquer technique. Sedgewick studied quicksort in his Ph.D. thesis [5] and it is broadly described and studied in [6–9]. Its work can be described and implemented by firstly selecting the pivot in the array, then partitioning the array and finally recursively performing the previous steps until the entire array is sorted. The worst-case complexity of quicksort is $O(n^2)$ and the best case/average case is $O(n \log n)$ [3]. Unlike in merge sort where the array of 'n' elements is divided into sub-arrays of $n/2$ elements each, quicksort uses a pivot element to divide the array. The pivot element is to be selected with a motive to cause an equal partition in the given array [10]. Besides the pivot selection, partitioning is the essence and backbone of a good implementation of the quicksort algorithm. By partitioning an array, we are condensing it into two separate sub-arrays and sorting each separately for effective optimization [3].

Several partition algorithms have found their way into the world of computer sciences. But the most commonly used ones are the Lomuto partition and Hoare partition algorithms. The Lomuto partition algorithm's implementation seems simpler and relatively easier [11] but remains less efficient than the Hoare partition algorithm. Hoare partition algorithm primarily seems uncomplicated but the attention to detail adds to its intricacy and makes it a trying task [12]. The fundamental and main objective of the partition algorithm is to necessarily find the middle element [10]. Though these algorithms are accepted and accustomed to, we could benefit from its advancement and refinement in a remarkable manner.

Past survey only studied selected variations of quicksort and used them for sorting small sized arrays. In this paper, we studied and analysed the performance of quicksort using both said portioning methods and proposed randomized pivot implementation of Lomuto partition and Hoare partition, respectively. An empirical study is done on various cases (array size varying from 2000 to 20,000 elements) for a more definitive and optimized result. The rest of the paper is organized as follows: Sect. 2 gives explanation of the partition algorithms and randomized pivot selection. Section 3 describes proposed randomized pivot method. Experimental results are presented in Sect. 4 and finally conclusion is in Sect. 5.

2   Partition Algorithms

2.1   Lomuto Partition Algorithm

This approach was put forth by Nico Lomuto. Its implementation is relatively straightforward and easy to understand, and so, it is rather inefficient when compared to Hoare partition algorithm. In this algorithm, the last element is generally taken as the pivot [1]. It is an in-place partitioning algorithm meaning it does not require auxiliary space or memory for the partitioning process. Lomuto Algorithm picks the last element of the array as its pivot invariably. Variable '*i*' is initialized to low and the array is scanned using another index variable '*j*'. If an element at index '*j*' is less than the pivot, it is swapped with the element at index '*i*' and index '*i*' is incremented. Subsequently, as the entire array is scanned, the element at index 'high' and '*i*' are swapped and '*i*' is returned. Due to its simplicity, it is frequented in the form of study material for beginners [1]. In pseudocode, the algorithm is written as follows:

Algorithm 1
(Lomuto Algorithm)

```
Lomutoquicksort(A, lo, hi)
        {

            if lo < hi then

            p := Lomutopartition(A, lo, hi)

            Lomutoquicksort(A, lo, p - 1 )

            Lomutoquicksort(A, p + 1, hi)
    }

Lomutopartition(A, lo, hi){
            pivot := A[hi]

            i := lo

            for j := lo to hi - 1 do

                if A[j] < pivot then {

                        swap A[i] with A[j]

                        i := i + 1
                }

            swap A[i] with A[hi]
            return i
    }
```

2.2 Hoare Partition Algorithm

This algorithm was put forth by none other than Hoare [2]. It is more efficient than the Lomuto approach as the numbers of swaps are fewer in comparison. Its main aim is to have the pivot in the correct position and sort all the elements lesser than the pivot onto its left side and the elements greater than the pivot onto the right side of the array [12]. It utilizes two index variables with each initialized as low −1 and high +1, respectively. The indexes on either side of the array being partitioned move towards each other. Until they detect an inversion which is a pair of elements one lesser than or equal and the other greater than or equal, to the pivot. The detected elements are swapped to their respective places [1]. This continues as the indices move closer to each other and when they finally close in on each other the algorithm stops and the final index is returned.

The algorithm is written as follows:

Algorithm 2
(Hoare Algorithm)

```
Hoarequicksort(A, lo, hi) {

            if lo < hi then{

                        p := Hoarepartition(A, lo, hi)

                        Hoarequicksort(A, lo, p)

                        Hoarequicksort(A, p + 1, hi)

            }

}

Hoarepartition(A, lo, hi) {

            pivot := A[lo]

            i := lo - 1

            j := hi + 1

            loop forever{
```

```
LquicksortRandomized(A, lo, hi){

        if lo < hi then

                r = Random Number from lo to hi

                swap A[r] and A[hi]

                p :=Lomutopartition(A, lo, hi)

                LquicksortRandomized (A, lo, p-1)

                LquicksortRandomized (A, p + 1, hi)
}
```

3 Proposed Randomized Pivot

Prior to recent developments, the leftmost element in an array was supposedly considered as the pivot element [1] but this led to worst-case time complexity for already sorted arrays which are most frequented in sorting operations. This complication readily resolved itself by taking a random element in the array as the pivot or by considering the middle index element as the pivot [1]. Ideally, the best case for time complexity of quicksort algorithm is when the partitioning is done in the central portion of the array. If partitioning is performed at either end of the array, it results in the worst case for time complexity. So as to avoid the worst case, random selection of a pivot element increases the average time complexity to O (n log n). The worst-case complexity still happens to be O (N^2) [13]. The proposed method chooses the randomized pivot for both partitioning algorithms and their time complexities are analysed.

3.1 *Lomuto Partition Algorithm with Randomized Pivot*

It can be easily identified as Lomuto scheme as A[hi] is set as pivot element.

Algorithm 3
(Lomuto Ranodomized pivot)

```
HquicksortRandomized(A, lo, hi) {

        if lo < hi then

                r = Random Number from lo to hi

                swap A[r] and A[lo]

                p := Hoarepartition(A, lo, hi)


                HquicksortRandomized(A, lo, p)

                HquicksortRandomized (A, p + 1, hi)

}
```

3.2 Hoare Partition Algorithm with Randomized Pivot

It can be identified as Hoare scheme as A[lo] is set as pivot element.

Algorithm 4
(Hoare Randomized Pivot)

```
                do{

                        i := i + 1

                }while A[i] < pivot
                do{

                        j := j - 1

                }while A[j] > pivot
                if i >= j then

                                return j

                swap A[i] with A[j]
                }
}
```

4 Experimental Results

The experiment is carried on Intel(R) Core(TM) i5-8250U CPU @1.60 GHz and 4 GB RAM. Windows 10 and jdk 1.8.0_171 (development kit) Eclipse Java Photon (development kit) are used for testing. The proposed sorting algorithms are implemented in Java programming language and executed on arrays of distinctive sizes. The sorting operations are performed using the partition algorithms and tested in various cases as mentioned below:

Case 1: For array of random elements in range $[1, 10^9-1]$.
Case 2: For array of random elements in range $[0, 9]$ (large no. of repeating elements).
Case 3: For array all elements equal to 0.
Case 4: For array of already sorted elements.
Case 5: For array of reverse sorted elements.

Each case takes into consideration both the partition algorithms and checks for the best case alongside keeping in mind the sub-cases which are, to use or to not use a randomized pivot element. The runtime is estimated by computing the average runtime of a hundred executions for each array given in the data. The array of 'n' elements is generated. Moreover, it is cloned, reason being that all four algorithms are executed on the exact same array.

Let the runtime computed be denoted by $f(n)$, the best-case complexity be indicated by $O(g1(n))$ and the worst-case complexity be symbolized by $O(g2(n))$. Then, the following condition is consistently true.

$$c1 * g1(n) \leq f(n) \leq c1 * g2(n) \tag{1}$$

Constant $c1$ is assumed to be 0.0015

$$gl(n) = n^* \log n.1 \tag{2}$$

$$g2(n) = n^2. \tag{3}$$

substituting (2) and (3) in (1) we get

$$c1 * n * \log n \leq f(x) < = c1 * n^2 \tag{4}$$

Each case, i.e. both the Hoare and Lomuto partition algorithms with and without randomized pivots are analysed and the experimental results are presented. Each case and the values obtained for each specific case is given below in an orderly fashion.

4.1 Results Analysis

Case 1: For array of random elements in range [1, 10^9–1]:

Algorithm 1 and 2 are executed on random elements in range [1, 10^9–1] without randomized pivot and Algorithm 3 and 4 are executed on random elements with randomized pivot. Results are depicted in Table 1 and graph is plotted by taking array size on X-axis and time in millisecond on Y-axis for both partitioning algorithms with and without pivot and shown in Fig. 1. The values seem closer to each other as we go over them. All of the curves in Fig. 1 tend to move towards the average case of time complexity. As the elements are already in random order, selecting a randomized pivot element marginally increases the runtime owing to the extra number of swaps and random selection of a number. Therefore, runtime of all four algorithms is in the order of O (n log n) due to the fact that the elements are randomly placed. Due to this, there is not much change in the performance of each of the algorithms as they all result in a similar form.

Case 2: For array of random elements in range [0, 9] (large no. of repeating elements):

Algorithm 1, 2, 3 and 4 are executed on repeated random elements of size (0–9) with randomized and without randomized pivot and results are presented in Table 2. It is evidently seen in Table 2 that in the case of Hoare partitioning with and without randomized pivot, the result seems as in case 1, whereas in the case Lomuto partitioning in both sub-cases, the values obtained seem very high and may have gone towards the worst case.

Table 1 Hoare and Lomunto partitioning on random elements with randomized and without randomized pivot

Time taken in microseconds				
Size of array (n)	Hoare partitioning		Lomuto partitioning	
	Without randomized pivot	With randomized pivot	Without randomized pivot	With randomized pivot
2000	143	158	135	148
4000	283	297	268	295
6000	437	462	416	464
8000	596	620	559	605
10,000	758	797	697	770
12,000	943	1021	908	978
14,000	1044	1105	993	1079
16,000	1190	1262	1128	1225
18,000	1366	1430	1295	1420
20,000	1565	1627	1442	1571

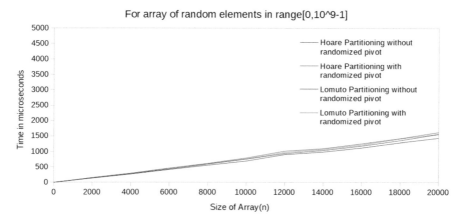

Fig. 1 Hoare and Lomuto partitioning on random elements with and without randomized pivot

Table 2 Hoare and Lomunto partitioning on repeated random elements with randomized and without randomized pivot

Size of array (n)	Hoare partitioning		Lomuto partitioning	
Time taken in microseconds				
	Without randomized pivot	With randomized pivot	Without randomized pivot	With randomized pivot
2000	64	74	340	426
4000	129	150	1242	1575
6000	192	230	2699	3431
8000	262	301	4746	5972
10,000	331	379	7416	9294
12,000	404	461	10,604	13,282
14,000	479	538	14,210	17,983
16,000	540	623	18,542	23,300
18,000	605	691	23,379	29,312
20,000	665	775	28,727	36,226

As the repetition of elements occurs, they are placed in the left side, and sub array can possibly contain elements that are equal to pivot but the right-side sub-array does not contain elements that are equal to the pivot for Lomuto partitioning.

[(**elements less than or equal to pivot**)   **pivot**   (**elements greater than pivot**)]

Whereas in Hoare partitioning, the elements that are equal to the pivot may exist in either of the sub-arrays so there is an opportunity for the occurrence of a more centralized pivot.

[(**elements less than or equal to pivot**)   **pivot**   (**elements greater than or equal to pivot**)]

From Fig. 2, we can infer that the Lomuto partitioning irrespective of using or not using a random pivot time complexity increases exponentially and moves towards the worst case. But in the Hoare partitioning approach with or without a random pivot stays close to the average-case time complexity which is similar to case 1.

Case 3: For array all elements equal to 0:

In this case, for an array with all elements equal to 0, the experimental results are presented in Table 3 and graph is plotted (Fig. 3). From the figure, it is clear that the Hoare partitioning with random pivot and without random pivot stays close to the average case, whereas the Lomuto partitioning algorithm irrespective of the use of a random pivot strays towards the worst case. The Lomuto partitioning invariably chooses the rightmost position or index. Because of this factor, the probability is scarce in acquiring a centralized pivot. Due to all the elements being equal to 0, stack overflow occurs for array of sizes 18,000 and 20,000.

Case 4: For an array of already sorted elements:

In case 4, we have considered array of sorted elements and experimental results of all four algorithms are depicted in Table 4. Both the Hoare and Lomuto algorithms behave similar when using a random pivot element for portioning. This is because the partition almost always occurs at the leftmost position of the array in Hoare

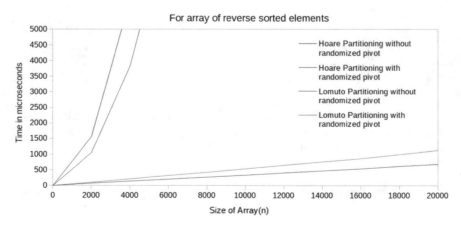

Fig. 2 Hoare and Lomuto partitioning on repeated random elements with and without randomized pivot

Table 3 Hoare and Lomunto partitioning on same random elements with randomized and without randomized pivot

Time taken in microseconds				
Size of array (n)	Hoare Partitioning		Lomuto partitioning	
	Pivot	With randomized pivot	Without randomized pivot	With randomized pivot
2000	29	38	2875	3613
4000	60	81	11391	14,301
6000	103	137	25586	32,157
8000	136	172	45480	56,900
10,000	171	226	71476	89,404
12,000	215	290	102666	128,439
14,000	247	329	140615	175,518
16,000	287	368	183263	228,795
18,000	324	417	−1	−1
20,000	374	480	−1	−1

In the above data, runtime value (−1) indicates that a Stack Overflow has occurred

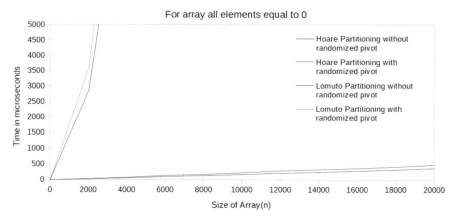

Fig. 3 Hoare and Lomuto partitioning on same random elements with and without randomized pivot

partitioning and at the rightmost position of the array in Lomuto partitioning, the said array of size 'n' is always divided into an empty array and an array of size 'n − 1'. Also, the respective algorithms when implemented by using a randomized pivot remain a bit close to the average case. Stack Overflow occurs for Hoare partitioning without random pivot at 16,000, 18,000 and 20,000. It also occurs for Lomuto partitioning without random pivot at 18,000 and 20,000.

Table 4 Hoare and Lomuto partitioning on sorted array of random elements with and without randomized pivot

Time taken in microseconds				
Size of array (*n*)	Quick sort with Hoare partitioning		Quick sort with Lomuto partitioning	
	Without randomized pivot	With randomized pivot	Without randomized pivot	With randomized pivot
2000	3280	58	2906	74
4000	13,261	120	11,522	150
6000	30,284	179	25,821	227
8000	54,232	239	46,182	310
10,000	90,403	332	78,431	430
12,000	119,620	362	103,600	467
14,000	16,4853	472	144,485	578
16,000	−1	487	184,044	634
18,000	−1	554	−1	719
20,000	−1	609	−1	801

In the above data, the runtime value (−1) indicates that a Stack Overflow has occurred

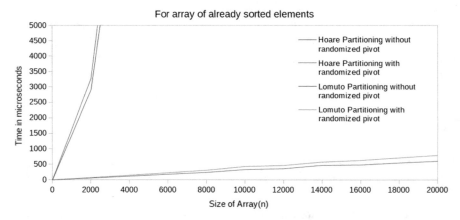

Fig. 4 Hoare and Lomuto partitioning on sorted array of random elements with and without randomized pivot

Figure 4 shows that both the partition algorithms without randomized pivot increase exponentially and move to worst case. This case justifies the use of a random pivot element. Owing to the fact that both the partition algorithms when implemented using a randomized pivot make an opposing change as compared to the other sub-cases by moving towards the average-case time complexity, the runtime is closer to average case and similar to case 1 in its execution.

Case 5: For array of reverse sorted elements:

For an array of reverse sorted elements, we can understand from Table 5 that the results acquired are similar in fashion to that of case 4. Both the partition algorithms when not using a randomized pivot move towards the worst case and also result in Stack Overflows. When a randomized pivot is used, the algorithms are closer to the average case. Stack overflows occur for arrays with size 16,000, 18,000 and 20,000 in Hoare partition and for array with size 18,000 and 20,000 in Lomuto partition algorithm.

Table 5 Hoare and Lomuto partitioning on reverse sorted array of random elements with and without randomized pivot

Time taken in microseconds				
Size of array (n)	Quick sort with Hoare partitioning		Quick sort with Lomuto partitioning	
	Without randomized pivot	With randomized pivot	Without randomized pivot	With randomized pivot
2000	1051	77	1563	104
4000	3826	143	5940	214
6000	8436	204	12,896	323
8000	14,542	272	22,967	424
10,000	22,700	336	35,144	538
12,000	32,753	409	51,107	654
14,000	44,679	476	69,494	764
16,000	−1	543	90,564	870
18,000	−1	624	−1	1002
20,000	−1	691	−1	1136

In the above data, the runtime value (−1) indicates that a Stack Overflow has occurred

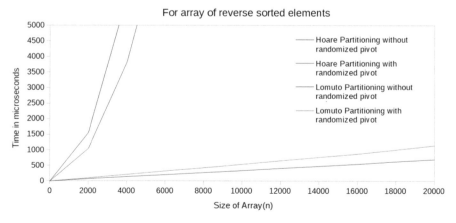

Fig. 5 Hoare and Lomuto partitioning on reverse sorted array of random elements with and without randomized pivot

In the case of using a reverse sorted array of elements, Fig. 5 is similar to that of Fig. 4. That is, the graph lines move exponentially in the cases of both partitioning algorithms when a randomized pivot is not used. And when a random pivot is used, it moves towards the average case as depicted in the graph, Fig. 5. Because it uses the randomized pivot, both the algorithms have the possibility to be in the average-case time complexity.

5 Conclusion

This research adapted a programmatic approach to conduct a comparative study of portioning method of quicksort algorithm for various cases. We have generated random input for our empirical study. Experimental study reveals that: For case 1, the quicksort algorithm using both partitioning methods with and without randomized pivot gives almost similar results, i.e. average-case time complexity. For case 2, 3, 4 and 5, the Lomuto partitioning with and without randomized pivot gives worst-case time complexity, whereas Hoare partitioning with and without randomized pivot gives average-case time complexity. Finally, we can conclude the Hoare partitioning using a randomized pivot is the only algorithm among the four that gave the best complexity in all cases. Both the discussed partition algorithms have a definite capacity for enhancement and advancement as well. The main objective was to bring to light the best case for sorting.

Acknowledgements We thank our mentor and guide Dr. Y. L. Malathi Latha, for her constant support and guidance. Her suggestions have been very helpful in the research and compilation for this paper.

References

1. Quicksort-Wikipedia. https://en.wikipedia.org/wiki/Quicksort.
2. Hoare, C. A. R. 1962. Quicksort. *The Computer Journal* 5 (1).
3. Sedgewick, Robert. 1978. Implementing Quicksort Algorithms. *Communications of the ACM* 21 (10).
4. Sedgewick, R. 1998. *Algorithms in C++*, 3rd ed. New York: Addison Wesley.
5. Sedgewick, R. 1975. Quicksort. Ph.D. dissertation, Stanford University, Stanford, CA, May 1975. Stanford Computer Science Report STAN-CS-75-492.
6. Loeser, R. 1974. Some performance tests of:quicksort: and descendants. *Communications of ACM* 17 (3): 143–152.
7. Bentley, J.L., and R. Sedgewick. 1997. Fast Algorithms for Sorting and Searching Strings. In Proceedings of 8th annual ACM-SIAM symposium on Discrete algorithms, New Orleans, Louisiana, USA, 360–369.
8. Chaudhuri, R., and Dempster, A.C. 1993. A Note on Slowing Quicksort. *SIGCSE* 25 (2).
9. Sedgewick, R. 1977. The Analysis of Quicksort Programs. *Acta Informatica* 7: 327–355.
10. Jaja, Joseph. 2000. A Perspective on Quicksort, Jan/Feb 2000.

11. Knuth, D.E. 1998. The Art of Computer programming, Vol. 3. London: Pearson Education.
12. Abhyankar, D., and M. Ingle. 2011. Engineering of a Quicksort Partitioning Algorithm. *JGRCS* 2 (2).
13. Abdullah, Mirza. 2017. Quick Sort with Optimal Worst Case Running Time. *AJER* 6 (1): 32–36.

Design an Improved Linked Clustering Algorithm for Spatial Data Mining

K. Lakshmaiah, S. Murali Krishna and B. Eswara Reddy

Abstract Recently, various clustering mechanism has been introduced to form data and cluster them into diverse domains. Some of the clustering algorithms clustered the data in proper way for grouping datasets accurately. However, some of the clustering methods roughly merge the categorized and numeric data types. Clustering is a process to identify the patterns distribution and intrinsic correlations in large datasets by separation of data points into similar classes. The proposed system, Improved Linked Clustering (ILC), is introduced to find a number of clusters on mixed datasets to produce results for several datasets. The ILC algorithm helps to prefer which clustering mechanism should be utilized to obtain a coherent and a mixed data significant mechanism for specific character deployment. Moreover, the technique can be used for optimization criteria during cluster formation for assisting clustering process toward better and efficient character interpretable technique. The technique objective is to offer a novel clustering method for data clustering methods evaluation over mixture of datasets including prior spatial information about the relation of elements which represents the clusters. The method provides an idea to estimate the character significance of clustering technique. The method estimates the summarization of the spatial point analysis of clustering technique with respect to coherence and clusters distribution. The proposed method is evaluated in two databases (character, spatial) using three conventional clustering techniques. Based on experimental evaluations, the proposed ILC algorithm improves the 0.04% cluster accuracy and 0.4 s cluster computation time compared to conventional techniques on the spatial dataset.

K. Lakshmaiah (✉)
Jawaharlal Nehru Technological University Hyderabad, Hyderabad, India
e-mail: Klakshmaiah78@gmail.com

S. Murali Krishna
CSE & DEAN-ICT, SV College of Engineering, Karakambadi Road, Tirupati, India

B. Eswara Reddy
Department of Computer Science & Engineering, JNTUA College of Engineering, Ananthapur, Ananthapuram (Dt) 515002, India

© Springer Nature Singapore Pte Ltd. 2020
K. S. Raju et al. (eds.), *Proceedings of the Third International Conference on Computational Intelligence and Informatics*, Advances in Intelligent Systems and Computing 1090, https://doi.org/10.1007/978-981-15-1480-7_34

Keywords Spatial database · Improved linked clustering · Cluster formation · Similarity measure · Character significance

1 Introduction

Spatial dataset clustering is one of the groups which contains the similar spatial object, classes and essential components of spatial data mining. Spatial clustering is used to identify a similar area of spatial information like land usage in an earth observation with merging regions and weather patterns. The spatial cluster is developed for geo-spatial information with respect to corresponding location activities and finding out the appropriate solutions for establishing the relationship between spaces and spatial with observed location attributes of location and events. Still, there is no such method available to solve the distance, neighbors, contiguity, and improper geographic morphology issues. Unfortunately, almost spatial clustering method has some boundary in accuracy, reliability, and the computational time for identifying the clusters in terms of their theoretical statistical finding. It is a very challengeable task to determine which technique will offer the most meaningful solution for specific issues or planning context. The existing spatial clustering methods fully depend on the distance-based similarity estimation and spatial properties. They clustered the objects spatially (globally) close to each other, and they are not suitable for the spatial clustering with constraints. The traditional clustering method is going computationally costly day by day, when the large amount of data should be clustered. There are three ways dataset that can be identified as large: (1) It should contain large amount of elements in the dataset; (2) every element should have various features; and (3) many clusters are required to discover. The existing clustering technique addressed efficiency issues, but did not cover properly. A k-d tree provided an efficient EM-style clustering mechanism to cluster large amount of elements. However, element dimension is too small. The technique performs k-means clustering to find initial clustering spatial points. But it is unable to cluster a large amount of dataset. There are no vast works available which can cluster efficiently. When the dataset becomes extensive in all senses, millions of elements, many features, and many clusters are involved. The existing techniques perform distance measurements only which made among points that occur in a simple canopy. There are several clustering algorithms available to cluster the dataset. However, it cannot be assured the result accuracy and execution time. Most of the clustering algorithms tend to cluster the numeric data and categorical data. But these methods have some complexity and reliability issues. Almost clustering algorithms represent poor performance on grouped and numerical data types. However, k-means algorithm is unable to produce better results on large amount of datasets.

To overcome the above issues, Improved Linked Clustering (ILC) is designed to find a number of clusters on mixed datasets to produce results for several datasets. The ILC helps to prefer which clustering mechanism should be utilized to get a coherent and a mixed data significant technique for specific character deployment. Additionally, the method can be used as optimization criteria during cluster formation to assist the clustering process for better character interpretable approach. The method addresses the spatial data clustering issues. Clustering is a process to presume the spatial data point or characters which are involved in a web process and co-expressed under the control regulatory network. Thus, a detailed evaluation of the categorized patterns to evaluate their memberships to well-known characters pathways. It is used to estimate clusters which are collected from a web perspective. The method addresses the usefulness of the clusters to recognize patterns that change communication and belong to common character pathways regulation. The method performs the objective analysis of clustering technique with respect to coherence and clusters distribution. The paper's contributions are given:

- To design Improved Linked Clustering (ILC) for finding a number of clusters on mixed datasets to produce results for several datasets.
- To optimize criteria for cluster formation to assist the clustering process toward efficient character interpretable mechanisms.
- To estimate the internal character connections of clusters with consideration of common pathways.
- To improve the clustering accuracy and reduce cluster computation time compared to conventional techniques.

The rest of the paper organized as Sect. 2 studies the related work of a proposed system to improve the technique. Section 3 introduces the system methodology, implementation details with workflow. Section 4 discusses the implementation setup, evaluation matrix, and comparative study of proposed technique. Section 5 concludes the overall work with future study.

2   Related Work

In data mining, clustering is an important technique aimed to group similar type of data into several clusters. There exist several clustering algorithms to form clusters. Existing clustering techniques tend to address either numeric data or grouped data. The existing algorithms performed worst on mixture of grouped and numerical data types. Even, k-means algorithm failed to process for large amount of datasets. Asadi et al. [1] developed a clustering technique that mixed the numeric and grouped dataset type that worked based on similarity of weight and filter technique. Shih et al. [2] expressed a two-step clustering method to finding clusters for mixed data types. Categorical attributes are processed to exchange into numerical attributes. The method combined hierarchical and partitioning cluster techniques along with attached attributes for object clustering. Zhang et al. [3] explained a multi-scale semantic zooming method to achieve the scalability and handle the large amount variables. It visualized data relations which spanned by correlated sub-spaces by utilization of variables.

Wiecki et al. [4] identified four levels in computational psychiatry. It indexed various psychological processes by behavioral task process. It identified generative psychological processes by computational models. It defined parameter estimation technique which concerned with quantitatively fitting listed mechanism in subject behavior which concentrates on hierarchical bayesian evaluation. Jacques and Preda [5] illustrated a clustering algorithm for multivariate functional data. It described a parametric mixture model called multivariate functional principal component analysis (MFPCA) which works based on estimation of the principal components normality. It defined and estimated by utilization of an EM-like algorithm. To process large-scale spatial data, GeoSpark, an in-memory clustering technique [6], introduced which contains three layers, namely Apache Spark Layer, Spatial RDD Layer, and Spatial Query Processing Layer for geometrical operations. It helps to create a spatial index for boosting spatial data processing in every spatial RDD partition.

Cao et al. [7] focused on multi-view clustering process by exploration of complementary information among multi-view features. It exposed a diversity-induced multi-view subspace clustering (DiMSC) and multi-view clustering technique. The method used Hibert Schmidt independence criterion (HSIC) to extend the existing subspace clustering into a multi-view domain. Evaluating the mobility data storage is a challengeable task due to the critical spatial situations capability. Von Landesberger et al. [8] described a visual analytics method that addressed the issue by integrated spatial and temporal simplifications. The work designed a graph-based method, called mobility-graphs that revealed movement patterns and occluded in flow maps. The technique enabled the visual presentation of the spatiotemporal variation of moments in series of spatial situations. Nguyen et al. [9] did comprehensive survey on behalf of data stream mining algorithms. The work identified several mining constraints that are available like addressing the spatial-textual objects which are gathered from several areas like location-based services and social networks.

Zhang et al. [10] studied two primary issues in spatial keyword queries called as top k spatial keyword search (TOPK-SK) and batch top k spatial keyword search (BTOPK-SK) to retrieve the closest k objects, where each of the keywords contains all keywords in the query for analyzing and categorizing a large amount of applications. Fahad et al. [11] studied the various existing clustering algorithm with theoretical and empirical perspective difference. The system explained various representative algorithms from each group and given large amount of datasets. Cameron and Miller [12] elaborated a statistical inference on regression during cluster group formation process with errors of regression model independent across clusters which are correlated within clusters. Instead, if several numbers of clusters attained to be large, the statistical inference after such OLS has been based on robust standard errors. It outlined the primary method and several complications which are in practice.

Chen and He [13] introduced a technique for clustering a high-dimensional dataset which is called as Canopies. It is a general approximate distance estimation which is divided the data into overlapping subsets. Canopies method is applied to any domains and used with the variety of clustering technique. Ramakrishna Murty et al. [14] expressed cascaded hidden-space (CHS) feature mapping technique with combination of fuzzy c-regressions (FCR) and classical fuzzy c-means (FCM) techniques. The CHS-FCR method designed with an integration of CHS-FCM with nonlinear switch regressions. Wang et al. [15] developed an end-to-end time series-based YADING clustering algorithm to automatically cluster large scale of time series with faster performance and quality results. YADING included three steps: sampling the input dataset, conducting clustering that turned on the sampled dataset, and assigning the rest of input to clusters that generated on the sampled dataset. The meta-knowledge extracts from every technique learning process to improve the algorithm performance.

Ding et al. [16] expressed a meta-knowledge-based clustering method. It explored characterized clustering issues based on the similarity among objects and combined internal indices for ranking algorithms. Ferrari and De Castro [17] studied data stream model of computation. The data stream mentioned here included multimedia data analysis and web clickstream analysis. The method offered constant-factor approximation technique for k-median issues in data streaming model of computation in a single pass. Guha and Mishra [18] discussed a data clustering method, called as non-dominated sorting genetic algorithm-fuzzy membership chromosome (NSGA-FMC) technique which works based on the L-modes method. It integrated fuzzy genetic algorithm and multi-objective optimization algorithm to solve the clustering formation issues. Yang et al. [19] addressed clusiVAT algorithm designed based on reordered dissimilarity images (RDIs) to cluster the spatial dataset. The technique is used for data clustering and pattern finding. The method estimates the amount of cluster dataset which is too sharply reordered diagonal matrix image for cluster formation of spatial dataset.

3 System Methodology

The section explains Improved Linked Clustering (ILC) algorithm to find the number of clusters from a mixture of the spatial dataset which produces results for several spatial datasets. The ILC decides which clustering technique should be applied to achieve a coherent and a mixed data significant of specific spatial dataset cluster formation. Furthermore, the method is used to optimize the criteria during cluter formation for assisting the clustering process toward efficient character interpretable solutions. Figure 1 expresses the workflow of the proposed ILC algorithm with the implementation procedure and algorithm below in detail.

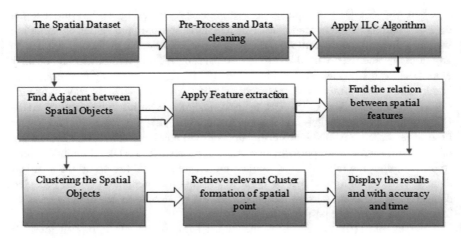

Fig. 1 Workflow of Improved Linked Clustering (ILC) algorithm

3.1   Pre-processing

The module pre-processes the spatial dataset with the spatial object and their reference point. Here, the toy spatial dataset is selected for performance evaluations which permute the dataset, geographic coordinates for the data. The toy spatial dataset contains input details like the neighbor's threshold, distance matrix, and two adjacent spatial relations neighbors to evaluate the connectivity matrix. Then, relevant clustering formation is done to display the clustered result with the best accuracy and minimal computation times. It applies a function that automatically fits multiple theoretical models and ranks the output theoretical models by their sum-of-squared errors selection.

3.2   Feature Extraction

Feature extraction is one of the important stages for spatial data object point identifications. An efficient feature of spatial data investigation is human visual observation and data mining techniques. The module extracts the spatial relationship between objects to find relevant cluster formations. The module used the Jaccard similarity coefficient to find similarity between spatial objects. It is used to improve the clustering accuracy of spatial data objects efficiently by selecting unmistakable constraints. It assures to get the best clustering accuracy with minimal computation times.

3.3   Improved Linked Clustering (ILC) Algorithm

Improved Linked Clustering (ILC) is designed to find a number of clusters on mixed datasets to produce results for several spatial datasets. The proposed ILC algorithm assists to identify which clustering mechanism should be applied to achieve a coherent and mixture data significant method for a specific character deployment. Moreover, the method is used to optimize criteria during cluster designing for helping the clustering process with toward efficient character interpretable approach. The method initiates with an arbitrary starting point that has not been visited. The proposed technique contains three steps. Initially, the spatial adjacent relations among the spatial objects are to be recognized. Hence, the spatial distance is to be estimated according to similarity measurement of the spatial objects, and then fitness matrix will be designed. Next, the constraint-based spatial objects need to be clustered with the different features. The method estimates the spatial adjacent relationship among the objects along with the relevant information of the spatial objects. The technique clusters the objects which are very closest to each other.

The proposed algorithm is used to cluster the spatial information in a specific zone or place. Then the spatial objects are explained by attributes grouping. Categorized data has a higher dimension. Every attribute as a dimension does not perform correctly due to the curve of dimension. Furthermore, the grouping data is not in a metric form. So, it is not necessary to fulfill the triangular inequality. The proposed method works to find the similarity among spatial objects. Jaccard coefficient is used to calculate the similarity among spatial objects.

Let X and Y denote the spatial adjacent relations set which is possessed by two spatial objects, and then the Jaccard similarity coefficient $J(X, Y)$ is represented in Eq. (1):

$$(X, Y) = \frac{|X \cap Y|}{\mathrm{Max}(|X|, |Y|)} \tag{1}$$

where $|X|$ indicates total amount of spatial datasets. The numerator evaluates overlap amount like total number of common adjacent relations among spatial objects, where the denominator standardizes the similarity value between 0 and 1. Thus, the Jaccard coefficient evaluates the fractional overlap between two datasets.

Hence, the **connectivity matrix** is represented on behalf of the link matrix. A connectivity matrix is expressed based on the following information like total volume of connectivity matrix. It contains row number and equivalents cell with the total amount of adjacent relations in network. Connection expresses that each cell has connection between two adjacent spatial objects. **Non-connection** indicates that every cell is not capable to show a direct connection. The connectivity matrix swaps the link matrix, and it displays how the spatial objects are similar to their neighbors. The fitness estimated is utilized to replace the following thing, where n_i denotes the total number of C_i, and two clusters maximum fitness value are integrated into a single group which is expressed in Eq. (2).

$$g\left(C_i, C_j\right) = \frac{\text{Connectivity}[i][j]}{n_i * n_j} \qquad (2)$$

The spatial point neighborhood is obtained, and it has many sufficient points, where cluster is processed. Otherwise, the spatial point is considered a noisy point. If the spatial point is found to consider as a part of a cluster, then all spatial points neighborhood are added. This process continues until all clusters are not found. Then, a new unvisited spatial point is searched and processed to discover the further noisy cluster. The pseudo-code of proposed techniques is explained below in detail:

Input:	Process Toy spatial dataset with spatial object, latitude, and longitude details;
Output:	Display spatial object reference point with the best accuracy and minimal computation time;
Procedure:	
Step 1:	Load the Spatial dataset S;
Step 2:	Extract the spatial dataset with adjacent relations;
Step 3:	Express the spatialAdjacentRelation (doc, minimumAjacentThreshold);
Step 4:	Estimate the spatial distance between objects with fitness matrix using a connectivity matrix;
Step 5:	Compute the connectivity matrix Cm=(S, SmilarityThreshold);
Step 6:	Perform the clustering on the spatial object using similarity easement;
Step 7:	For every spatialAdjacentRelation in S do
Step 8:	q[s]=LocalHeapDesign(Cm, s);
Step 9:	Q=GlobalHeapDesign(S, q);
Step 10:	While |Q|>k do
Step 11:	u=ExtractMax(Q); v=MaxExtract(q[u]);
Step12:	For x in (q[u] or q[v])
Step 13:	M[x,w]=M[x,v]+M[x,u];
Step 14:	Free Memory(q[u]); FreeMemory(q[v]);
Step 15:	Apply clustering spatial object with different features;
Step 16:	Find relevant spatial clustering formation;
Step 17:	Visualizes spatial object reference point with the best accuracy & minimal Computation time;
Step 18:	End;

3.4 Result Visualization

The module evaluates the complete obtained result of spatial object data. The proposed ILC algorithm is used to find a number of clusters on mixed datasets to produce results for several datasets with the best accuracy and minimal computation time. The performance estimation will offer implemented technique performance evaluation metrics and constraints. Finally, the proposed algorithm computes more clear and explorative results which can be easily understood by evaluators.

4 Result and Discussion

4.1 Experimental Setup

The proposed mechanism is deployed with Intel Core i5 7th Generation Processor with 8 GB RAM, 500 GB Memory, and Windows7 professional operating system. The proposed mechanism is implemented in Java programming with Java Development Kit (JDK) 1.8, Netbeans 8.0.2 Integrated Development Environment.

4.1.1 Data

For experimental evaluations, a proposed mechanism used a toy (http://irobotics. aalto.fi/software-and-data/toy-dataset) spatial dataset, permutes the dataset, sets geographic coordinates for the data, and applies a function that automatically fits multiple theoretical models. It ranks the output theoretical models by their sum-of-squared errors select.

4.2 Evaluation Matrix

The section expresses the ILC algorithm evaluation matrix details to estimate efficiency, accuracy reliability, and stability of proposed algorithms in spatial data mining. The proposed technique concentrates to offer spatial object information with best accuracy and minimal computations time. Here, the method illustrates the clustering accuracy and cluster computation time to calculate the performance of proposed ILC algorithms. It expressed that how can best accuracy with minimal computational time be achieved in spatial data clustering.

4.2.1 Cluster Accuracy

In spatial data mining, cluster accuracy is the evaluation criterion of clustering quality. It is the percent of the total amount of spatial objects that were predicted accurately, in the unit range.

The most common evaluation method is classification accuracy which is expressed in Eq. (3).

$$\text{Cluster}_{\text{Accuracy}} = \frac{1}{N} \sum_{i=1}^{k} \max_{j} \left[C_i \cap t_j \right] \times 100 \tag{3}$$

where N = number of spatial objects, k = Total amount of clusters, c_i a group in C, and t_j is the predictions which contain a maximum count for cluster c_i.

4.2.2 Clustering Computation Time (CCT)

The proposed ILC technique illustrates a mathematical model for cluster computation time in Eq. (4). The proposed approach calculated cluster computation time based on time utilization for clustering the spatial objects proposed techniques. Cluster computation time (CCT) is expressed as follows:

$$\text{CCT} = T_{\text{SR}} \times T_{\text{CT}} \tag{4}$$

where T_{SR} is a total number of spatial record and T_{CT} is average computation time for a spatial object clustering in spatial data mining.

Table 1 explains the cluster accuracy (CA), cluster computation time (CCT) for toy spatial dataset. The proposed ILC method displays their average values for respective parameters with the respective spatial dataset. Here, proposed Improved Linked Clustering (ILC) algorithm is evaluated with conventional techniques such as clusiVAT [19], k-means [19], single pass k-means (spkm) [19], and clustering using representatives (CURE) [19] with toy spatial dataset. According to Table 1 observation, it is noticed that the ILC algorithm performs well on respective evaluation matrix and dataset. Hence, it can be said that proposed ILC is the best method compared to conventional techniques on every respective parameter and dataset.

According to Figs. 2 and 3, the proposed ILC algorithm is evaluated with cluster accuracy (CA) and cluster computation time (CCT) for 100,000, 200,000, and 500,000 spatial data points of toy spatial dataset. Here, proposed ILC algorithm is evaluated with clusiVAT [19], k-means [19], single pass k-means (spkm) [19], clustering using representatives (CURE) [19] convention techniques. According to cluster accuracy and clustering computation time, clusiVAT is the closest method of proposed ILC method. ClusiVAT [19] works based on data sampling and recorded image distance to estimate the total amount of clustered data visually.

Table 1 Cluster accuracy and cluster computation time (CCT) for toy spatial dataset

Total no of spatial point	No. of cluster	clusiVAT		k-means		spkm		CURE		ILC	
		CA	CCT	CA	CCT	CA	CCT	CA	CCT	CA	CCT
100,000	3	100	0.05	69.9	0.94	100	0.79	100	4.74	100	0.02
100,000	4	99.8	0.06	85.0	0.96	99.9	0.89	99.9	4.73	99.9	0.04
100,000	5	99.9	0.07	71.8	1.02	94.5	1.12	99.9	4.77	100	0.04
200,000	3	99.9	0.09	92.0	1.74	99.9	4.23	74.2	9.23	100	0.06
200,000	4	99.9	0.12	69.9	2.12	99.9	2.16	98.1	9.35	100	0.09
200,000	5	100	0.14	56.5	2.01	100	1.82	99.2	9.37	100	0.10
500,000	3	100	0.25	78.8	4.66	100	3.93	99.9	23.10	100	0.21
500,000	4	100	0.31	74.5	4.60	100	4.18	96.0	23.27	100	0.28
500,000	5	99.9	0.37	62.2	5.24	96.6	6.61	99.9	23.27	99.9	0.32

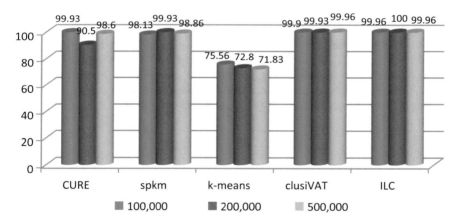

Fig. 2 Cluster accuracy for 100,000, 200,000, 500,000 spatial data points of toy spatial dataset (*X*-axis—Performance and *Y*-axis—Algorithms)

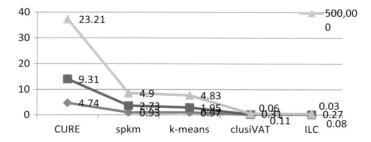

Fig. 3 Cluster computation time (CCT) for 100,000, 200,000, and 500,000 spatial data points of toy spatial dataset (*X*-axis—Computation time and *Y*-axis—Algorithms)

It clusters the sample data by using a relative of single linkage (SL). The method non-iteratively extends the labels of rest dataset using the nearest prototype rule. However, the method consumes more time to execute the clustering process of spatial of the object of a spatial dataset. The spkm [19] algorithm is a crisp adaptation of single pass fuzzy c-means technique. The technique cannot allow for storing the data in main memory, where it represents the number of clustering points that can be stored. It divides the data into multiple points and can be utilized with arbitrarily of large input dataset. However, the method is much slower than other clustering techniques. K-means [19] is one of the most famous clustering techniques due to the simplicity and applicability in various fields. K-means developed independently in different research fields, and it is applied to achieve k clusters. It is generally used for benchmark extension. However, the method is unable to maintain clustering accuracy and computation time efficiently. CURE [19] method is one kind of hierarchical clustering algorithm that randomly samples a fixed amount of points from the hues dataset. So, the representative points retain the geometry of the entire dataset. CURE introduced two kinds of data structures, namely k-d trees and heaps method to precede the cluster formation. However, the method does not assure the accuracy of clustered data and consumes more time to complete the process. Proposed ILC algorithm improves the 0.04% cluster accuracy and 0.4 s cluster computation time compared to conventional techniques on the spatial dataset. Finally, it can be said that the proposed ILC algorithm is the best technique based on the evaluation of respective parameters and dataset.

5   Conclusion

The paper presents Improved Linked Clustering (ILC) algorithm to find a number of clusters on mixed datasets to produce results for several datasets. The ILC algorithm helps to decide which cluster method to be used for achieving a coherent and mixed data significant mechanism for a specific character deployment. The proposed algorithm is used to cluster the spatial information in a specific zone or place. Then attributes grouping explains the spatial objects. Categorized data has a higher dimension. Every attribute as a dimension does not perform correctly due to the curve of dimension. Furthermore, the grouping data is not in a metric form. So, it is not necessary to fulfill the triangular inequality. Jaccard coefficient is used to calculate the similarity among spatial objects. The proposed ILC algorithm is used to find a number of clusters on mixed datasets to produce results for several datasets with the best accuracy and minimal computation time. Proposed ILC algorithm improves the 0.04% cluster accuracy and 0.4 s cluster computation time compared to conventional techniques on the spatial dataset. Finally, it can be said that the proposed ILC algorithm is the best technique based on the evaluation of respective parameters and datasets.

In the future, the paper can be improved in the area of spatial data mining to retrieve and update the spatial data in a cloud environment, where data clustering accuracy and computation time are very challengeable tasks.

References

1. Asadi, S., C.D.S. Rao, C. Kishore, and S. Raju. 2012. Clustering the Mixed Numerical and Categorical Datasets Using Similarity Weight and Filter Method. *International Journal of Computer Science Information Technology and Management* 1 (1–2): 121–134.
2. Shih, M.Y., J.W. Jheng, and L.F. Lai. 2010. A Two-Step Method for Clustering Mixed Categorical and Numeric Data. *Tamkang Journal of Science and Engineering* 13 (1): 1119.
3. Zhang, Z., K.T. McDonnell, E. Zadok, and K. Mueller. 2015. Visual Correlation Analysis of Numerical and Categorical Data on the Correlation Map. *IEEE Transactions on Visualization and Computer Graphics* 21 (2): 289–303.
4. Wiecki, T.V., J. Poland, and M.J. Frank. 2015. Model-Based Cognitive Neuroscience Approaches to Computational Psychiatry: Clustering and Classification. *Clinical Psychological Science* 3 (3): 378–399.
5. Jacques, J., and C. Preda. 2014. Model-Based Clustering for Multivariate Functional Data. *Computational Statistics & Data Analysis* 71: 92–106.
6. Yu, J., J. Wu, and M. Sarwat. 2015. Geospark: A Cluster Computing Framework for Processing Large-Scale Spatial Data. In *Proceedings of the 23rd SIGSPATIAL International Conference on Advances in Geographic Information Systems* 70. ACM.
7. Cao, X., C. Zhang, H. Fu, S. Liu, and H. Zhang. 2015. Diversity-Induced Multi-view Subspace Clustering. In *2015 IEEE Conference on Computer Vision and Pattern Recognition (CVPR)* 586–594. IEEE.
8. Von Landesberger, T., F. Brodkorb, P. Roskosch, N. Andrienko, G. Andrienko, and A. Kerren. 2016. Mobilitygraphs: Visual Analysis of Mass Mobility Dynamics via Spatio-Temporal Graphs and Clustering. *IEEE Transactions on Visualization and Computer Graphics* 22 (1): 11–20.
9. Nguyen, H.L., Y.K. Woon, and W.K. Ng. 2015. A Survey on Data Stream Clustering and Classification. *Knowledge and Information Systems* 45 (3): 535–569.
10. Zhang, C., Y. Zhang, W. Zhang, and X. Lin. 2016. Inverted Linear Quadtree: Efficient top *k* Spatial Keyword Search. *IEEE Transactions on Knowledge and Data Engineering* 28 (7): 1706–1721.
11. Fahad, A., N. Alshatri, Z. Tari, A. Alamri, I. Khalil, A.Y. Zomaya, S. Foufou, and A. Bouras. 2014. A Survey of Clustering Algorithms for Big Data: Taxonomy and Empirical Analysis. *IEEE Transactions on Emerging Topics in Computing* 2 (3): 267–279.
12. Cameron, A.C., and D.L. Miller. 2015. A Practitioner's Guide to Cluster-Robust Inference. *Journal of Human Resources* 50 (2): 317–372.
13. Chen, J.Y., and H.H. He. 2016. A Fast Density-Based Data Stream Clustering Algorithm with Cluster Centers Self-Determined For Mixed Data. *Information Sciences* 345: 271–293.
14. Ramakrishna Murty, M., J.V.R. Murthy, and P.V.G.D. Prasad Reddy et al. 2014. Homogeneity Separateness: A New Validity Measure for Clustering Problems. In *International Conference and Published the Proceedings in AISC*, vol. 248 1–10.
15. Wang, J., H. Liu, X. Qian, Y. Jiang, Z. Deng, and S. Wang. 2017. Cascaded Hidden Space Feature Mapping, Fuzzy Clustering, and Nonlinear Switching Regression on Large Datasets. *IEEE Transactions on Fuzzy Systems* 26: 640–655.
16. Ding, R., Q. Wang, Y. Dang, Q. Fu, H. Zhang, and D. Zhang. 2015. Yading: Fast Clustering of Large-Scale Time Series Data. *Proceedings of the VLDB Endowment* 8 (5): 473–484.

17. Ferrari, D.G., and L.N. De Castro. 2015. Clustering Algorithm Selection by Meta-Learning Systems: A New Distance-Based Problem Characterization and Ranking Combination Methods. *Information Sciences* 301: 181–194.
18. Guha, S., and N. Mishra. 2016. Clustering Data Streams. In *Data Stream Management* 169–187. Berlin, Heidelberg: Springer.
19. Yang, C.L., R.J. Kuo, C.H. Chien, and N.T.P. Quyen. 2015. Non-dominated Sorting Genetic Algorithm Using Fuzzy Membership Chromosome for Categorical Data Clustering. *Applied Soft Computing* 30: 113–122.

An Efficient Symmetric Key-Based Lightweight Fully Homomorphic Encryption Scheme

V. Biksham and D. Vasumathi

Abstract Maintaining the secrecy and privacy of data, generally in cloud scenario, has become an intense challenge for the present day's practical applications. However, transferring private data to any third party consists of large amount of risks of disclosure of private data while computation. This problem can be addressed by performing computations on encrypted data without decrypting it. This utility is evolved as fully homomorphic encryption, which is a great way for securing data which can be manipulated by an untrusted server. In this paper, we have given a fully homomorphic encryption protocol which is lightweight in nature and utilizing symmetric key. Analysis of the scheme confirms that our proposed system is efficient and practical to adopt it in various cloud computation applications.

Keywords Homomorphic encryption · Symmetric FHE · Privacy · Security · Ideal lattices

1 Introduction

In cloud computing scenario, most of the users store and maintain their data in the cloud. Growth in data transmission and storage in the cloud may lead to increase the vulnerability of data [1]. For that, generally, standard encryption mechanisms are utilized for ensuring confidentiality, integrity, and authenticity. Homomorphic encryption is supposed to play a vital role in cloud computing [2]. In homomorphic

V. Biksham (✉)
Department of Computer Science and Engineering, Sreyas Institute of Engineering and Technology, Hyderabad, India
e-mail: vbm2k2@gmail.com

D. Vasumathi
Department of Computer Science and Engineering, Jawaharlal Nehru Technological University, Hyderabad, India
e-mail: rochan44@gmail.com

© Springer Nature Singapore Pte Ltd. 2020   417
K. S. Raju et al. (eds.), *Proceedings of the Third International Conference on Computational Intelligence and Informatics*, Advances in Intelligent Systems and Computing 1090, https://doi.org/10.1007/978-981-15-1480-7_35

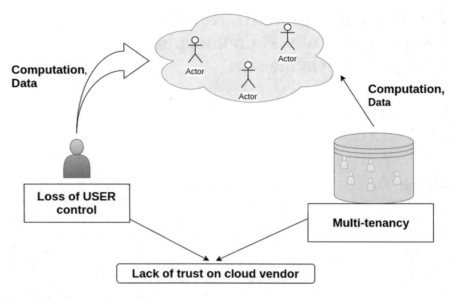

Fig. 1 Specific security concerns

encryption, ciphertext (data in encrypted format) should be sent to the cloud, the computations are made on the ciphertext, and the result of this computation is a ciphertext form itself. If the result of the computation is decrypted, then the correct plaintext result must be obtained. Homomorphic encryption schemes are classified into two types—Partial Homomorphic Encryption (PHE) and Fully Homomorphic Encryption (FHE) schemes. If an HE scheme supports both addition and multiplication, it can also evaluate any arithmetic encrypted data circuit and therefore, we can observe that it is an FHE scheme [3, 4]. Some specific security concerns in cloud computing are—when customers are transferring their private data to any third party, then there is much responsibility for both security and compliance. Therefore, it is necessary that customers should fully faith in their cloud service provider. A small security weakness in any one of the technologies, e.g., databases, network structure, virtualization scenario, resources, transaction management, load balancing factor, memory management, etc. may knock down the complete system. The specific security concerns are presented in Fig. 1.

1.1 Related Work

Cheon [2] presented and examined polynomial analog of the approximate common divisor (poly-ACD) problem. Hemalatha and Manickachezian [5] proposed fully homomorphic encryption (FHE) scheme based on the ring for securing data in cloud computing. Sha and Zhu [6] designed the algorithm which obtains the characteristic of an addition of fully homomorphic encryption. Gavinho Filho et al. [7] proposed a compression method to optimize and reduce the public key sizes by using genetic algorithms (GA). Chen et al. [8] proposed FHE scheme based on the learning with errors (LWE) phenomenon for the optimally efficient key size. Michael and McGoldrick [9] proposed attribute-based FHE scheme. Coron [10] and Howgrave-Graham [11] presented a scheme that remains semantically secure, based on a stronger variant of the approximate-GCD problem. Gentry [3], in 2009, presented three steps procedure. Later, in 2010, Gentry et al. [12, 13] presented fully homomorphic encryption scheme that uses only simple integer arithmetic. Xiao et al. [14] developed a non-circuit-based symmetric key homomorphic encryption scheme. Sharma [15] proposed a "Symmetric FHE" scheme which works with plaintext bits to convert them into integers using linear algebra. Liang [16] developed symmetric and asymmetric quantum fully homomorphic encryption (QHE) and developed four symmetric QHE schemes and distributed key issuing using secret sharing [17–21] that permits all the quantum operators to function. Smart et al. [22] developed FHE scheme which has both relatively small key and ciphertext size. Yagisawa [23] presented new fully homomorphic public key encryption scheme without bootstrapping based on the discrete logarithm and Di e-Hellman assumption on octonion ring. Coron et al. [24] and Cheon et al. [25] presented the improved FHE over integers. Coron et al. [26] given the mechanism for public key compression as well as modulus switching for FHE over integers. In 2016, Dasgupta et al. [27] have given symmetric homomorphic encryption scheme which is based on the polynomial ring.

1.2 Motivation and Contribution

Homomorphism property is a tremendous way in which a problem can be converted from one algebraic system model into another algebraic system. Then the problem is solved in this transformed algebraic system and later the computational result is translated back efficiently. Thus, the cryptosystems having homomorphic properties are in much demand. An application scenario is presented in Fig. 2.

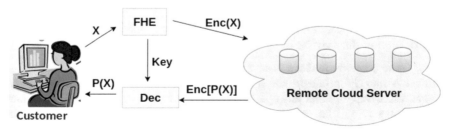

Fig. 2 An application scenario

Main contributions of this paper include,

- to present the existing state-of-the-art methodologies which are explored in the literature.
- in this paper, we have presented our proposed fully homomorphic encryption scheme utilizing symmetric keys.
- we have given the critical security analysis and correctness proof for our proposed methodology.

1.3 Organization of the Paper

In rest of the paper, we begin with the discussion of some required preliminaries in Sect. 2. Our proposed system along with its analysis is given in Sect. 3. Section 4 concludes the paper and presents future research directions.

2 Preliminaries

This section summarizes some preliminaries required. Some desirable and common properties of homomorphic encryption are as below.

2.1 Privacy of Homomorphic Circuit and Computational Variability

This property of HE gives guarantee that the computational function and server's input parameters remain private from client. In other words, the output of func(F, C_1, C_2, C_3, ..., C_t, e_k) is not revealing any information about F to client. Homomorphic encryption phenomenon is significantly considered as a solution to the secure outsourcing of scientific computations, though it is tremendously helpful when computed result can be trusted.

Lemma 1 *It is infeasible to factorizing the N in polynomial time if the integer factorization in the massive scale is termed infeasible.*

Proof Take assumption that x is an adversary who is being able to factorize a large number N into the primes p and q of approximately same bit length in the polynomial time. Suppose that this operation has its probability as p', every factor fact$_i$ of a number N will be possessing at least two number of prime factors. So the probability p_r'' that the attacker could able to factorize it is almost lesser than p'. Thus, the resulting probability that attacker can factorize N is $\prod_{i=1}^{m} p_r'' \leq (p')^m$. Now if p' is negligible, the resultant probability will also be negligible.

Definition 1 A matrix $M \in R^{n,n}$ is known as orthogonal if it is satisfying one of the equivalent conditions—(i) $M.M^T = M^T.M = I_n$ (ii) M is invertible and $M^{-1} = M^T$.

3 Proposed Fully Homomorphic Encryption Scheme

Main Idea: In this section, we have presented our proposed system for fully homomorphic encryption, which is based on symmetric key. Our scheme consists of sub-modules, which are;

1. Key generation step
2. Encryption procedure
3. Decryption procedure
4. Refresh procedure.

3.1 Algorithm

Algorithm 1 KeyGen(l, a)-

1: Randomly choose $2a$ odd number pairs x_i and y_i, $1 \leq i \leq a$, co-prime with each other, consisting the size of l-bits.
2: Compute $S = \prod_{i=1}^{a} n_i$, where, $n_i = x_i \times y_i$
3: Choose an orthogonal matrix K, with it's dimension as 4, in Z_S. Follow approach to choose K - Randomly choose a matrix in the search space of Z_S. Check if it is following the orthogonality property i.e. $K.K^T = K^T.K = I$ or $K^{-1} = K^T$, then search is completed, otherwise repeat until an orthogonal matrix is found.
4: Compute the transpose of matrix K: $K^T \leftarrow tranpose(K)$. mod Z_S
   Note: From matrices orthogonality property, $K^{-1} = K^T$
5: Orthogonal matrix K will act as symmetric key in our cryptosystem.

Algorithm 2 Enc_Procedure(M, n_i, S, K, K^T)-

1: Consider plaintext as M.
2: Choose an integer R randomly, where $R \in Z_S$ and $R \neq M$.
3: Create a matrix Y with it's dimension ($a \times 3$), where each row consists of single occurance of M and rest of two occurances as R.
4: Employ CRT to obtain solution of simultaneous equations - $\alpha_j (1 \leq j \leq 3) \equiv Y_{ij}$. mod n_i; where, $1 \leq i \leq a$.
5: Use Coppersmith-Winograd algorithmic procedure [22] for below step of matrix multiplication computation.
6: Obtained ciphertext $C = K^T \times d(M, \alpha_1, \alpha_2, \alpha_3) \times K$; where, $d(M, \alpha_1, \alpha_2, \alpha_3)$: denotes the diagonal matrix with diagonal elements as parameters.

Algorithm 3 Dec_Procedure(C, S, K, K^T)-

1: Compute plaintext as, $P = K \times C \times K^T$
2: $M \leftarrow [P]_{11}$

Algorithm 4 Refresh_Procedure(S)-

1: Refresh the symmetric key, which is an orthogonal matrix as -
$$K' \leftarrow randortho(t)$$
   Here, randortho() is a randomized function generating new K of dimension t randomly in the space of Z_S.

3.2   Analysis of Proposed Algorithm

Correctness of Decryption algorithm—We can observe that Dec_Procedure(C, S, K, K^T) $\Rightarrow K \times C \times K^T$. Since we know the property of an orthogonal matrix as, $K.K^T = K^T.K = I$ or $K^{-1} = K^T$. So, Dec_Procedure(C, S, K, K^T) $\Rightarrow K \times K^T \times d(M, \alpha_1, \alpha_2, \alpha_3) \times K \times K^T$

$$\Rightarrow I \times d(M, \alpha_1, \alpha_2, \alpha_3) \times I$$

$$\Rightarrow d(M, \alpha_1, \alpha_2, \alpha_3)$$

$$\Rightarrow [P]_{11}$$

$$\Rightarrow M$$

Homomorphic Properties

- *Multiplicative homomorphic property*
  Consider, C_1, C_2 are ciphertexts corresponding to plaintexts M_1, M_2.
  $C_1 = K^T \times d(M_1, \alpha_1, \alpha_2, \alpha_3) \times K$
  $C_2 = K^T \times d(M_2, \alpha'_1, \alpha'_2, \alpha'_3) \times K$
  Now, $C_1 \times C_2 = K^T \times d(M_1, \alpha_1, \alpha_2, \alpha_3) \times K \times K^T \times d(M_2, \alpha'_1, \alpha'_2, \alpha'_3) \times K$
  $= K^T \times d(M_1, \alpha_1, \alpha_2, \alpha_3) \times I \times d(M_2, \alpha'_1, \alpha'_2, \alpha'_3) \times K$
  $= K^T \times d(M_1, \alpha_1, \alpha_2, \alpha_3) \times d(M_2, \alpha'_1, \alpha'_2, \alpha'_3) \times K$
  $= K^T \times d((M_1 \times M_2), \alpha''_1, \alpha''_2, \alpha''_3) \times K$
  Now, Dec_Procedure(C, S, K, K^T) $\Rightarrow K \times K^T \times d((M_1 \times M_2), \alpha''_1, \alpha''_2, \alpha''_3) \times K \times K^T$
  $\Rightarrow I \times d((M_1 \times M_2), \alpha''_1, \alpha''_2, \alpha''_3) \times I$

$$\Rightarrow d((M_1 \times M_2), \alpha_1'', \alpha_2'', \alpha_3'')$$
$$\Rightarrow M_1 \times M_2$$

- *Additive homomorphic property*

Consider, C_1, C_2 are ciphertexts corresponding to plaintexts M_1, M_2.

$C_1 = K^T \times d(M_1, \alpha_1, \alpha_2, \alpha_3) \times K$

$C_2 = K^T \times d(M_2, \alpha_1', \alpha_2', \alpha_3') \times K$

Now, $C_1 + C_2 = (K^T \times d(M_1, \alpha_1, \alpha_2, \alpha_3) \times K) + (K^T \times d(M_2, \alpha_1', \alpha_2', \alpha_3') \times K)$

$= K^T \times d((M_1 + M_2), \alpha_1'', \alpha_2'', \alpha_3'') \times K$

Now, Dec_Procedure(C, S, K, K^T) $\Rightarrow K \times K^T \times d((M_1 + M_2), \alpha_1'', \alpha_2'', \alpha_3'') \times K \times K^T$

$$\Rightarrow I \times d((M_1 + M_2), \alpha_1'', \alpha_2'', \alpha_3'') \times I$$
$$\Rightarrow d((M_1 + M_2), \alpha_1'', \alpha_2'', \alpha_3'')$$
$$\Rightarrow M_1 + M_2$$

Statement 1: *Proposed FHE scheme is lightweight in nature.*

Justification: Our proposed FHE scheme utilizes matrices computational operations, which are lightweight in nature as compared to working with polynomial computations.

Statement 2: *Proposed FHE scheme is parallelizable.*

Justification: As our proposed FHE scheme utilizes matrices, computational operations while encryption as well as decryption, which give the advantage of performing outer product matrix-vector multiplication that is very much parallelizable.

Computational Complexity improvement—in our proposed system, we are utilizing Coppersmith–Winograd algorithmic procedure [28] for matrix multiplication computation in encryption step. This method gives a significantly improved matrix multiplication computational complexity as $O(n^{2.376})$.

4 Conclusion and Future Work

Homomorphic encryption technique enables us to perform the computational functionalities on unoriginal data itself. Still, there exist some reasons for impracticality of the existing schemes. Some of the significant reasons are—(i) inherently slow blueprint (this may be due to ciphertext refreshing algorithm or Bootstrapping procedure) (ii) huge public key (iii) large message expansion (iv) complex circuit evaluation due to bitwise encryption. In this paper, we have presented our proposed fully homomorphic encryption scheme utilizing symmetric keys. We have also given the critical security analysis and correctness proof for our proposed methodology. Our future contribution will perform implementation of proposed symmetric key-based fully homomorphic encryption scheme. Our next consideration could be of designing secure and practical applicable asymmetric FHE scheme.

References

1. Top Threats: Cloud Security Alliance (CSA). https://cloudsecurityalliance.org/group/top-threats/overview.
2. Cheon, Jung Hee. 2016. The Polynomial Approximate Common Divisor Problem and It's Application to the Fully Homomorphic Encryption. *Information Sciences* 326: 41–58.
3. Gentry Craig. 2009. Fully Homomorphic Encryption Using Ideal Lattices. In *STOC*, vol. 9, no. 2009, 169–178.
4. Biksham, V., and D. Vasumathi. 2016. Query Based Computations On Encrypted Data Through Homomorphic Encryption in Cloud Computing Security. In *2016 International Conference on Electrical, Electronics, and Optimization Techniques, Chennai*, 3820–3825. IEEE Digital Library. https://doi.org/10.1109/iceeot.2016.7755429.
5. Hemalatha, S., and Dr. R. Manickachezian. 2014. Performance of Ring Based Fully Homomorphic Encryption for securing data in Cloud Computing. *International Journal of Advanced Research in Computer and Communication Engineering*, 3 (11). ISSN (Print) 2319-5940.
6. Sha, P., and Z. Zhu. 2016. The Modification of RSA Algorithm to Adapt Fully Homomorphic Encryption Algorithm in Cloud Computing. In *4th International Conference on Cloud Computing and Intelligence Systems (CCIS)*, Beijing, 388–392.
7. Gavinho Filho, J., G. P. Silva, and C. Miceli. 2016. A Public Key Compression Method for Fully Homomorphic Encryption using Genetic Algorithms. In *19th International Conference on Information Fusion (FUSION)*, Heidelberg, 1991–1998.
8. Chen, Zhigang, Jian Wang, ZengNian Zhang, and Song Xinxia. 2014. A Fully Homomorphic Encryption Scheme with Better Key Size. *China Communications* 11 (9): 82–92.
9. Clear, Michael, and Ciarn McGoldrick. 2016. Attribute-Based Fully Homomorphic Encryption with a Bounded Number of Inputs. In *International Conference on Cryptology in Africa*. Springer.
10. Coron, Jean-Sbastien. 2011. Fully Homomorphic Encryption Over the Integers With Shorter Public Keys. In *Annual Cryptology Conference*. Berlin, Heidelberg: Springer.
11. Howgrave-Graham, N. 2001. Approximate Integer Common Divisors. In *Proceedings of the Cryptography and Lattices—CaLC* 2001, LNCS, vol. 2146, 51–66. Springer.
12. Van Dijk, Marten, Craig Gentry, Shai Halevi, and Vinod Vaikuntanathan. 2010. Fully Homomorphic Encryption Over the Integers. In *Annual International Conference on the Theory and Applications of Cryptographic Techniques*, 24–43. Berlin, Heidelberg: Springer.
13. van Dijk, M., C. Gentry, S. Halevi, and V. Vaikuntanathan. 2010. Fully Homomorphic Encryption Over the Integers. In *Proceedings of the Advances in Cryptology—EUROCRYPT 2010*, LNCS, vol. 6110, 24–43. Springer.
14. Xiao, Liangliang, Osbert Bastani, and I-Ling Yen. 2012. An Efficient Homomorphic Encryption Protocol for Multi-User Systems. *IACR Cryptology*, 193 (e-Print Archive).
15. Sharma, Iti. 2014. A Symmetric FHE Scheme Based on Linear Algebra. *International Journal of Computer Science and Engineering Technology (IJCSET)* 5: 558–562.
16. Liang, Min. 2013. Symmetric Quantum Fully Homomorphic Encryption with Perfect Security. *Quantum Information Processing* 12 (12): 3675–3687.
17. Kalyani, D., and R. Sridevi, Robust Distributed Key Issuing Protocol for Identity Based Cryptography. *Advances in Computing, Communications and Informatics (ICACCI)*, 16429714, IEEE. https://doi.org/10.1109/icacci.2016.7732147.
18. Tentu, A.N., Banita Mahapatra, V. Ch. Venkaiah, and V. Kamakshi Prasad. 2015. New Secret Sharing Scheme for Multipartite Access Structures with Threshold Changeability. In *International Conference on Advances in Computing, Communications and Informatics, ICACCI 2015*, Kochi, India, 10–13 Aug 2015.
19. Dileep, K.P., A.N. Tentu, VCh. Venkaiah, and Allam Apparao. 2016. Sequential Secret Sharing Scheme Based on Level Ordered Access Structure. *Journal of Network Security* 18 (5): 874–881.

20. Tentu, A.N., Prabal Paul, and V. Ch. Venkaiah. 2014. Computationally Perfect Compartmented Secret Sharing Schemes Based on MDS Codes. *IJTMCC, 2* (4): 353–378.
21. Baby, Vadlana, and N. Subhash Chandra. 2016. Distributed Threshold k-Means Clustering for Privacy Preserving Data Mining. In *2016 International Conference on Advances in Computing, (ICACCI)*, 2286–2289. IEEE, 10.1109.
22. Smart, Nigel P., and Frederik Vercauteren. 2010. Fully Homomorphic Encryption with Relatively Small Key and Ciphertext Sizes. In *International Workshop on Public Key Cryptography*, 420–443. Berlin, Heidelberg: Springer.
23. Yagisawa, Masahiro. 2016. Fully Homomorphic Public-key Encryption Based on Discrete Logarithm Problem. *IACR Cryptology, 54* (e-Print Archive).
24. Coron, J.-S., A. Mandal, D. Naccache, and M. Tibouchi. 2011. Fully Homomorphic Encryption Over the Integers with Shorter Public Keys. In *Proceedings of the Advances in Cryptology—CRYPTO 2011*, LNCS, vol. 6841, 487–504. Springer.
25. Cheon J.H. et al. (2013) Batch Fully Homomorphic Encryption over the Integers. In: Johansson T., Nguyen P.Q. (eds) Advances in Cryptology – EUROCRYPT 2013. Lecture Notes in Computer Science, vol 7881. Springer, Berlin, Heidelberg.
26. Coron, J.-S., D. Naccache, and M. Tibouchi. 2012. Public Key Compression and Modulus Switching for Fully Homomorphic Encryption Over the Integers. In *Proceedings of the Advances in Cryptology—EUROCRYPT 2012*, LNCS, vol. 7237, 446–464. Springer.
27. Dasgupta, Smaranika, and S.K. Pal. 2016. Design of a Polynomial Ring Based Symmetric Homomorphic Encryption Scheme. *Perspectives in Science, 8*: 692–695 (Elsevier).
28. Coppersmith, D., and S. Winograd. 1990. Matrix multiplication via arithmetic progressions. *Journal of Symbolic Computation* 9 (3): 251–280.

MLID: Machine Learning-Based Intrusion Detection from Network Transactions of MEMS Integrated Diversified IoT

Ravinder Korani and P. Chandra Sekhar Reddy

Abstract Human intervene-based automation of the practices related to the appliances involved in human life style is the ecstasy since recent past. In regard to enable such practices by an individual from remote location are realizing because of the concept called "Internet of Things (IoT)". In addition, human intervened automation in industry sector such as manufacturing and health care also equally prioritized. However, the IoT technology advancements, such as MEMS integrated diversified IoT networks, which are giving opportunity to achieve avoidance of human intervention enable fully automated monitoring and controlling practices of devices incorporated in IoT networks. However, the MEMS integrated diversified IoT networks are highly vulnerable to the intrusion scope, which is since the ease of making MEMS such as accelerometers compromised. Hence the increasing dependence on "MEMS integrated Diversified IoT devices and services", the ability to identify malevolent activities and intrusions inside such networks are difficult for flexibility of network infrastructure. Hence, the contribution of this manuscript is a new method for the recognition of intrusion on the basis of classification strategy that trained by n-gram features as optimal features, which are discovered from the training corpus. The classifier that trained by the n-gram optimal features is further used to detect intruded practices like "Remote to Local (R2L)" and "User to Root (U2R) attacks."

Keywords IoT · MEMS · User to root · Remote to local · Machine learning · NIDS · Decision tree · K-means · KNN-algorithm · MLID · TDTC

R. Korani (✉)
ECE Department, Shadan College of Engineering and Technology, Hyderabad, India
e-mail: koraniphd814@gmail.com

P. Chandra Sekhar Reddy
ECE Department, JNTU, Hyderabad, India
e-mail: drpcsreddy@gmail.com

© Springer Nature Singapore Pte Ltd. 2020
K. S. Raju et al. (eds.), *Proceedings of the Third International Conference on Computational Intelligence and Informatics*, Advances in Intelligent Systems and Computing 1090, https://doi.org/10.1007/978-981-15-1480-7_36

1 Introduction

The term which is impressed at every bend of the globe and which became universal by impacting on lives of human in an incredible way is the "Internet." Nevertheless, now, we are inflowing in a period of more extensive connectivity where there is inclusive variety of applications that are associated with Web. And we are inflowing in a period of "Internet of Things (IoT)." Diverse authors defined IoT in diverse ways. Now, the two most prevalent definitions of IoT are discussed. The work [1] presents that an interface between digital and physical worlds is called as IoT. Utilization of plethora of actuators and sensors, the world of digital communicates with physical world.

The work [2] presents that IOT can be defined in another way as model in which networking and computing competences are entrenched in any type of imaginable object. And these competences are utilized to question the object state and if possible, alter the object state. The term IOT refers to novel type of globe where every devices and applications are associated with the network. Utilize them cooperatively to attain intricate tasks which need intelligence in higher degree. To this interconnection and intelligence, the devices of IoT are equipped with the entrenched transceivers, processors, sensors, and actuators. The term IoT is not an individual method; instead, it is collection of several methods which work cooperatively.

The devices actuators and sensors assist in connecting with physical world. The sensors gathered the data and stored and processed brilliantly to define utility implications from them. Sensor is broadly defined as microwave oven or mobile phone can count as sensor when it delivers current state inputs. And the device actuator is utilized to impact the change in environment like air conditioner controlled by a temperature. With the help of remote server, storage and processing of information are done. Data preprocessing is done by either sensor or any other similar device. And the data which are preprocessed are transferred to the server of remote. The capabilities of storing and processing of an IoT are constrained by available possessions that are frequently confined because of the restrictions of energy, computational capability, power, and size.

The work [3] presents that identical to consumer technologies, the technologies of IoT are defenseless intrusion that emerged as important barrier in extensive implementation of "MEMS integrated diversified IoT networks and services". The work [4] presents that the defense mechanisms and intrusion detection are condemnatory in respect to IoT networks.

The different IoT networks are integrated with the MEMS like capacitive accelerometers that can simply be cheated by providing unfair readings to avoid resultant "IoT Network Security." There is a huge growth in the IoT market and wide deployment of the MEMS and sensors which lead to the deficiency of security across corresponding memes and sensors are the defenseless for compromising [5].

2   Related Work

The current intrusion detection methods Bayes theory [6], cluster analysis [7, 8], "Hidden Marko Model (HMM)" [9], and measuring distance [10] have using statistical approaches to identify abnormal activities. The method of anomaly detection can be extensively classified into unsupervised and supervised learning [11]. The work [12] presents that "supervised anomaly detection method," normal conduct of networks is designed utilizing the tagged dataset. The work [13] presents that unsupervised method assumes that ordinary conduct is common, and therefore, the method is constructed on the basis of assumption; therefore, data which are trained are not prerequisite.

The work [14] presents that a suggested unsupervised NIDS on the basis of subspace grouping and identification of outlier and proved that their method functions well in averse to not known attacks. The work [15] presents that "feature section filter module" is suggested that uses "Principal Component Analysis and Fisher Dimension Reduction" to filter the sounds. The "self-organizing maps (SOMs)" a neural method is also utilized to sieve the usual actions. Nevertheless, this method has higher "false positive rate." The work [16] presents that suggested unsupervised framework is on the basis of "K-means clustering method and optimum-path forest algorithm." These framework prototypes are the normal and malevolent conduct of the networks.

The work [17] presents a "supervised anomaly detection" method that influences both measuring of distance and clusters density for the detection of intrusion. The work [18] presents that suggested method is on the basis of "random forest algorithm" to identify anomaly patterns at greater accuracy but with less FNR (false negative rate).

The work [19] presents that suggested "two-level intrusion detection" method at first identifies the misuse and then utilizes the KNN-algorithm to diminish fake alarms. The work [20] suggested "multi-attack classifier" method that implements the combination of fuzzy inference, genetic algorithms, and fuzzy-neural-networks for the detection of intrusion. Regardless of higher rate of accuracy in detecting normal conduct and identifying simpler attacks like probe and DoS attacks, the method functions poorly in identifying less frequency and dissemination attacks like R2L. The work [21] presents that suggested multi-classification method consists of "support vector machine (SVM)" and the "BRICH hierarchical clustering method" to minevital features from dataset of KDD99. Their suggested method has higher rate of detection for probe and DoS attacks yet it is ineffective in averse to R2L and U2R attacks.

The work [22] presents that suggested system for the detection of DoS utilizing "multi-variate correlation analysis (MCA)" to enhance the network traffic characterization accuracy. The work [23] presents that "a two-layer classification module" was utilized to identify the R2L and U2R attacks at less computational intricacy because of the reduction of optimized feature. The work [24] presents that suggested "ensemble based multi-filter feature selection" model to identify disseminated attacks of DoS in the environment of cloud by utilizing four filter models to attain optimum feature selection.

The work [25] presents that suggested "mutual information-based IDS" chooses optimal features for the categorization on the basis of "feature selection algorithm."

The work [26] presents that intrusion detection method have utilized for managing the risks of security in the "industrial control systems." The work [27] presents that suggested automated and systematic method to construct hybrid IDS which learns the specifications based onthe state of temporal for the electric power methods to distinguish accurately among usual control operations, cyber-attacks, and disturbances. The work [28] presents that multi-model and industrial anomaly driven IDS are on the basis of "Hidden Markov" method for filtering the attacks from genuine mistakes.

The work [29] presents that the issues of security will be barrier for the extensive adoption of the IoT devices. The work [30] presents that extensive range of methods might lessen cyber threat aiming at IoT schemes. The work [31] suggested "hierarchical authentication architecture" to offer anonymous information transfer in the networks of IoT. The work [32] presents that emphasized the effect and significance of ghost intrusions on the IoT devices based on ZigBee. The work [33] suggested an "autonomic model-driven cyber security management approach" for the IoT networks that will be utilized to predict, identify, and reply toward the cyber threats with less human interferences. The work [34] suggested a system for the uncomfortable internal intrusions in the networks of IoT by examining the transferring of data of each node of IoT.

Any of these existing models are not well-versed with the intrusion practices observed in MEMS integrated diversified IOT networks, which is since the intrusion practices of these networks are not specific and hard to define the signatures, and majority of the contemporary models rely on the signatures framed from the past intrusion practices noted.

In contrast to the contemporary models explored, a machine learning approach called "two-layer dimension reduction and two-tier classification strategy (TDTC) is devised by Pajouh et al. [35]. However, the TDTC is using the component analysis to reduce the dimensionality of the training data that leads to the minimizing the feature set to train the classifier. The component analysis is often complex and results probabilistic version of the features. Moreover, the said method is intended to identify two categories R2L and U2R of the intrusion practices. The experimental study carried on TDTC is not specific to intrusion practices related to IoT, as it was performed on NSL-KDD dataset. In regard to this argument, the proposal of the manuscript is devising a machine learning approach that selects optimal features using the ANOVA standard called dual tailed t-test, which identifies the features having diversity of the values projected in network transactions labeled as intrude and the network transactions projected as benevolent. In addition, the proposal is using the binary classifier "Decision Trees" that simplifies the complexity of classification strategy.

3 Methods and Materials

3.1 The Data Structure

The dataset DS is of network transactions between MEMS integrated diversified IoT devises and backbone network. Each record of the dataset DS refers a network transaction of an IoT devise with backbone network. Each of these transactions portrayed by sequence of values is observed for set of attributes, and each of these records is labeled as intruded or benign. The said dataset of size |DS| will be considered for training toward defining the fitness function of the classifier. All possible unique subsets of set A of attributes used to frame the records in given corpus are considered as features, which referred further as set F such that each unique subset $\{f \exists f \in F \wedge f \subset A\}$ is an entry in set F. In regard to set theory, the total number of unique subsets (features) is $2^{|A|}$, here $|A|$ is the size of the attribute set A.

Further, for each feature (subset of attributes), let the values portrayed in all of the records from dataset DS as a vector V_f, such that each entry of the vector V_f is the values portrayed for the feature (subset of attributes) f in one or more records and the set V_f is having no duplicate entries. Moreover, the entries in vector V_f are indexed that starts from 1 and increments for each entry of that vector in sequence by 1.

Further, the given training corpus DS is partitioned into two sets, such that the records labeled as intrude are one set I and the records labeled as benign are the other set B.

Later, for each feature $\{f \exists f \in F \wedge f \subset A\}$, the process portrays two vectors v_f^+, v_f^-. Further, for each record $\{r \exists r \in I\}$ from the set I, the vector v_f^+ contains the index (from the vector V_f) of the values portrayed for feature f in the corresponding record r. Similarly, for each record $\{r \exists r \in B\}$ from the set B, the vector v_f^- contains the index (from the vector V_f) of the values portrayed for feature f in the corresponding record r.

3.2 Dual Tailed t-Test

In order to notify the significance of each feature $\{f \exists f \in F \wedge f \subset A\}$, the distribution diversity of the vectors v_f^+, v_f^- corresponding to feature f assesses using the dual tailed t-test that recommended in contemporary literature [36, 37] is as follows.

$$ts = \frac{\left(\langle v_f^+ \rangle - \langle v_f^- \rangle\right)}{\sqrt{stdv(v_f^+) + stdv(v_f^-)}} \tag{1}$$

In (Eq. 1):

- The vector v_f^+ represents the index (from the vector V_f) of the values projected for feature $\{f \exists f \in F \wedge f \subset A\}$ in the records labeled as intrude, and
- The vector v_f^+ represents the index (from the vector V_f) of the values projected for feature $\{f \exists f \in F \wedge f \subset A\}$ in the records labeled as benign
- The notations $\langle v_f^+ \rangle, \langle v_f^- \rangle$ denote the averages of the entries in corresponding vectors v_f^+, v_f^-.
- The notations $stdv(v_f^+), stdv(v_f^-)$ denote the standard deviation of the vectors v_f^+, v_f^- in respective order.

Further, the degree of probability p in regard to the distribution diversity between input vectors v_f^+, v_f^- is assessed by using the t-table [38, 39]. The degree of probability signifies the distribution diversity of the given vectors v_f^+, v_f^-; if the resultant degree of probability is less than the given probability threshold τ, then consider the feature $\{f \exists f \in F \wedge f \subset A\}$ is optimal, otherwise, the distribution of the input vectors said is to be similar and discards the feature $\{f \exists f \in F \wedge f \subset A\}$ from optimal features list.

3.3 The Classifier

In regard to the binary classification (intrude or benign), the classification strategy decision tree is used here in this manuscript. The overall classification process includes branch formation, hierarchical search of the branches to notify the compatible branches in regard to the given records to assign labels, and further assesses the fitness of the given record toward intrusion scope and benign scope, which is based on the compatibility of the branches.

3.3.1 Training Phase

This section explores the training phase of the binary classification process that is carried through decision tree. The training phase builds the decision tree of both labels intrude and benign, such that each decision tree having branches, such that each of these branches referred by an optimal feature (set of attributes as pattern). Further, each of these branches linked to one or more nodes, where node is a unique set of values depicted in the records represented by the corresponding label of the decision tree and representing the attributes of the feature that referring the corresponding branch.

According to the description given about the decision tree formation, the decision trees *MH, BH* are built, which are representing the labels intrude and benign in respective order.

Upon completion of the decision tree formation, the labeling the given records by classification process will be initiated, which is as described in following section.

3.3.2 Classifying

The fitness of the given record estimates based on the number of compatible branches noticed in respective decision trees *MH, BH*. Concerning this, for each branch, any egg of the respective branch is identical to the values observed in given record for the call sequence patterns in corresponding branch representative set, then the fitness of the given record in related to corresponding hierarchy will increment by the 1. This practice delivers the fitness related to malicious and benign state of the given record. Further, the fitness ratio of the given record about both hierarchies will be measured, which is the average of the fitness related to number of branches in corresponding hierarchies. Then, the root mean square distance of the fitness values corresponding to both hierarchies should be measured. Then these fitness ratios and fitness error thresholds corresponding to both hierarchies will use to confirm the state of the given record is malicious call sequence or benign that explored in following section. The mathematical model to assess the fitness follows

Step 1: Let *R* be the given record to be labeled as intrude or benign
Step 2: Let *eR* be the set of value patterns found in given record *R*, such that each pattern refers the values projected for an optimal feature $\{f \ni f \in F \wedge f \subset A\}$.
Step 3:

$$mf = \sum_{j=1}^{|MH_h|} \sum_{i=1}^{|eR|} \left\{ 1 \ni e_i \in nst_j \wedge nst_j \in MH \right\}$$

Add 1 to fitness *mf* of the given record *R* related to malicious scope, if the entry e_i of the vector *eR* is compatible to a node in the branch nst_j of the corresponding tree *MH*.
Step 4:

$$\langle mf \rangle = \frac{mf}{mhnc}$$

Finding the fitness ratio $\langle mf \rangle$ of the given record in related to malicious scope, here, the notation *mhnc* is the number of branches in *MH*.
Step 5:

$$rmsd_{mf} = \frac{\sum_{i=1}^{|mf|} \left\{ \sqrt{(1 - \langle mf \rangle)^2} \right\} + \langle mf \rangle * (mhnc - mf)}{mhnc}$$

The sum of absolute difference of the fitness ratio depicted, and max fitness in regard to each branch, (which is 1) for number of branches having a node that compatible to an entry in set eR and the fitness ratio multiplies by the number of incompatible branches, which is the difference between total number of branches and number of compatible branches that denoted as $mhnc - mf$

Step 6:

$$bf = \sum_{j=1}^{|BH_h|} \sum_{i=1}^{|eR|} \left\{ 1 \exists e_i \in nst_{\{h,j\}} \wedge nst_{\{h,j\}} \in BH \right\}$$

Add 1 to benign fitness bf related to hierarchy BH if the entry e_i of the set eR is compatible to a node of the branch nst_j in decision tree BH.

Step 7

$$\langle bf \rangle = \frac{bf}{bhnc}$$

Finding the fitness ratio $\langle bf \rangle$ of given record R, in related to benign scope, here, the notation $bhnc$ is the number of branches in BH.

Step 8

$$rmsd_{bf} = \frac{\sum_{i=1}^{bf} \left\{ \sqrt{(1 - \langle bf \rangle)^2} \right\} + \langle bf \rangle * (bhnc - bf)}{bhnc}$$

Finding the fitness error threshold of the benign fitness using the similar process defined in step 5.

3.3.3 Discovering the Record State

The fitness ratios $\langle mf \rangle$, $\langle bf \rangle$ and fitness error thresholds $rmsd_{mf}, rmsd_{bf}$ for given input record R, those obtained from decision trees PH, SH built in respective to the labels intrude and benign should use to label the record R. The classification process adapts the following rules to label the given test record:

Rule 1: if the fitness ratio $\langle mf \rangle$ of the given record R that obtained from the decision tree MH is considerably greater than the fitness threshold $\langle bf \rangle$ of the corresponding record that obtained from the decision tree BH, then the record R will be labeled as intrusion.

Rule 2: if the fitness ratio $\langle mf \rangle$ of the given record R that obtained from the decision tree MH is considerably lesser than the fitness threshold $\langle bf \rangle$ of the corresponding record that obtained from the decision tree BH, then the record R will be labeled as benign.

Rule 3: If the fitness ratios $\langle mf \rangle$, $\langle bf \rangle$ obtained for given test record R in regard to the decision trees MH, BH are approximately equal, and the fitness error threshold $rmsd_{mf}$ corresponding to the decision tree MH is considerably less than the fitness error threshold $rmsd_{bf}$ corresponding to the decision tree BH, then the given test record is labeled as intrusion.

Rule 4: If the fitness ratios $\langle mf \rangle$, $\langle bf \rangle$ obtained for given test record R in regard to the decision trees MH, BH are approximately equal, and the fitness error threshold $rmsd_{mf}$ corresponding to the decision tree MH is considerably greater than the fitness error threshold $rmsd_{bf}$ corresponding to the decision tree BH, then the given test record is labeled as benign.

Rule 5: in both cases of fitness ratios $\langle mf \rangle$, and fitness error thresholds $rmsd_{mf}$ found to be approximately equal to their counterpart fitness ratio $\langle bf \rangle$ and fitness error threshold $rmsd_{bf}$ in respective order, then the given record (network transaction) is found to be suspicious and recommends for admin opinion.

4 Empirical Analysis of the Proposed Model

The empirical study is carried out on the Kyoto dataset [40]. The description of the attributes involved in each record of the dataset is explored in [41]. In regard to predict the execution of the suggestion, in the phase of learning, 75% of benevolent and intrude records of the selected dataset were utilized. The remaining 25% of records were untagged and utilized to execute predictive analysis. The outcomes achieved from simulation study were shown in Table 1 that is showing the

Table 1 The metrics and the values obtained from fourfold classification using proposed and contemporary model

		PPV	TPR	TNR	FNR	FPR	ACC
MLID	Fold#1	0.958	0.958	0.947	0.042	0.053	0.953
	Fold#2	0.923	0.962	0.895	0.038	0.105	0.953
	Fold#3	1	0.971	1	0.029	0	0.953
	Fold#4	0.957	0.967	0.947	0.033	0.053	0.93
TDTC	Fold#1	0.88	0.917	0.842	0.083	0.158	0.884
	Fold#2	0.875	0.905	0.842	0.095	0.158	0.86
	Fold#3	0.917	0.917	0.895	0.083	0.105	0.907
	Fold#4	0.84	0.897	0.789	0.103	0.211	0.837

importance of the suggested method toward intrusion detection in MEMS integrated diversified IoT network transactions with backbone networks.

Evaluation results of the model over the test data are illustrated in following description. Effectiveness of the suggested approach involving selecting optimal attribute through dual tailed t-test along with classification using decision tree, the experimental results obtained and compared to the other contemporary model that detects intrusion scope in network transactions of the IoT devises with backbone network, which is done by using "Two-layer Dimension Reduction and Two-tier Classification (TDTC)" [35] method.

The database is subdivided into small groups in fourfold approach. Efficiency and effectiveness of this technique along with comparison to other proposed techniques are determined through different classification criteria [42] which include PPV, TPR, SPC, FNR, FPR, and ACC. Comparative results of this simulation phase are depicted in Table 1.

In the above table, abbreviations represent (in respective order)—PPV, TPR, SPC, FPR, FNR, ACC—precision, recall, specificity, false alarm ratio, miss rate, and accuracy.

The efficiency of the models in correctly estimating the program as either intrusion or normal is assessed in the simulation phase and the corresponding outcomes are provided in Fig. 1. As might be noticed from the figure, MLID model outcome in terms of precisely categorizing the program is highly stable. On the contrary, TDTC model evaluation on similar parameters showed lesser stability. Further, on the precision levels, MLID model posed 93% accuracy, posing superior performance over TDTC which showed precision levels of below 90%.

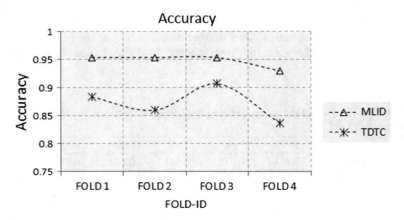

Fig. 1 Classification accuracy ratios of MLID and TDTC observed from fourfold classification

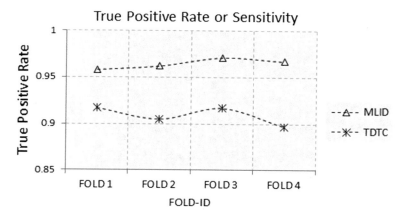

Fig. 2 The sensitivity (malware prediction rate) of MLID and TDTC observed from fourfold classification

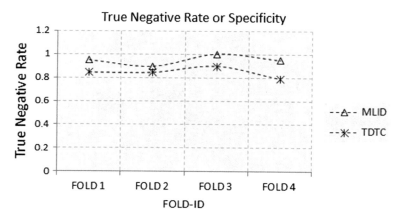

Fig. 3 The specificity (benign state prediction rate) of MLID and TDTC observed from fourfold classification

Figures 2 and 3 present the performance comparison of the MLID with TDTC toward TNR and TPR, which represent prominence of estimating accurately the intrusion program and normal program. The developed technique posed superior performance over TDTC with large margin in this respect.

Figures 4 and 5 depict the FNR and FPR denoting incorrect classification rates of grouping considered dataset into intrusion or normal programs. Comparison of both the approaches on this respect shows that MLID model had much better performance with very low values of FNR and FPR unlike TDTC with relatively higher values.

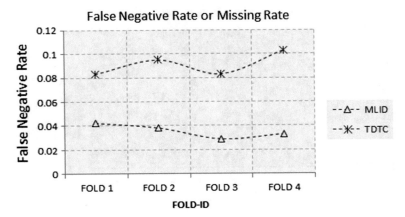

Fig. 4 The malware prediction failure rate (false negative rate) of MLID and TDTC observed from fourfold classification

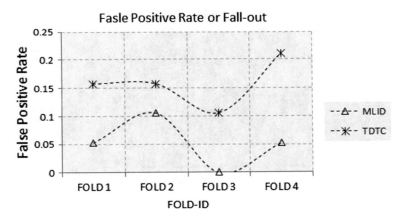

Fig. 5 The benign state prediction failure rate (false positive rate) of MLID and TDTC observed from fourfold classification

The program difficulty assessed during learning stage and experimental stages are presented in Figs. 6 and 7. As can be observed from these illustrations, the duration taken for learning and experimentation in the proposed MLID model is considerably smaller than the duration involved in learning and experimentation in TDTC model.

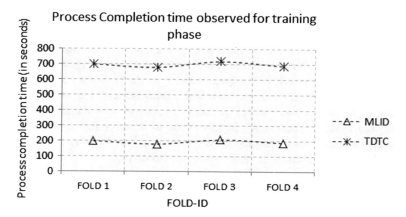

Fig. 6 Process completion time of training phase observed for both MLID and TDTC in fourfold classification

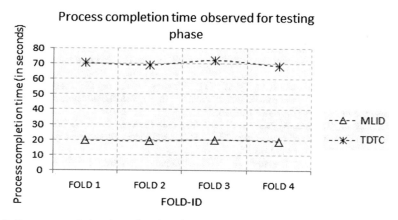

Fig. 7 Process completion time of testing phase observed for both MLID and TDTC in fourfold classification

5 Conclusion

Intrusion detection in MEMS integrated diversified IoT networks using machine learning approach is the contribution of this manuscript. The overall contribution portrayed an optimal technique that built on ANOVA standard called "dual tailed *t*-test" to identify the optimal features, and the values portrayed by each network transaction for the corresponding optimal features are used to train the binary classifier that build by using decision tree. Unlike the contemporary models, the proposal of this manuscript is using the variable length of patterns portrayed as values to represent the optimal features in corresponding network transactions of the given training set. This method of defining variable size patterns of the values related to optimal features escalated the label detection accuracy. The other significance of the contribution is building two decision trees, which are trained by the values projected for optimal features in network transactions labeled as intrude and benevolent in respective order. This practice evinces the performance advantage by considerable reduction in false alarming. The contribution of this manuscript leads to future contribution that enables to define multilevel or ensemble classifier to improve the learning process for high volume of training set with considerable dimensionality in values portraying for the features of the network transaction and to label the network records as intrude and intrusion type.

References

1. Vermesan, Ovidiu, et al. 2011. Internet of Things Strategic Research Roadmap. *Internet of Things-Global Technological and Societal Trends* 1 (2011): 9–52.
2. Peña-López, Ismael. 2005. ITU Internet Report 2005: The Internet of Things.
3. Sicari, S., et al. 2015. Security, Privacy and Trust in Internet of Things: The Road Ahead. *Computer Networks* 76: 146–164.
4. Shakshuki, Elhadi M., Nan Kang, and Tarek R. Sheltami. 2013. EAACK—A Secure Intrusion-Detection System for MANETs. *IEEE Transactions on Industrial Electronics* 3 (60): 1089–1098.
5. Trippel, Timothy, et al. 2017. WALNUT: Waging Doubt on the Integrity of MEMS Accelerometers with Acoustic Injection Attacks. In *2017 IEEE European Symposium on Security and Privacy (EuroS&P)*. IEEE.
6. Koc, Levent, Thomas A. Mazzuchi, and Shahram Sarkani. 2012. A Network Intrusion Detection System Based on a Hidden Naïve Bayes Multiclass Classifier. *Expert Systems with Applications* 18 (39): 13492–13500.
7. Thottan, Marina, and Chuanyi Ji. 2003. Anomaly Detection in IP Networks. *IEEE Transactions on Signal Processing* 51 (8): 2191.
8. Lin, Wei-Chao, Shih-Wen Ke, and Chih-Fong Tsai. 2015. CANN: An Intrusion Detection System Based on Combining Cluster Centers and Nearest Neighbors. *Knowledge-Based Systems* 78: 13–21.
9. Ariu, Davide, Roberto Tronci, and Giorgio Giacinto. 2011. HMMPayl: An Intrusion Detection System Based on Hidden Markov Models. *Computers & Security* 30 (4): 221–241.
10. Weller-Fahy, David J., Brett J. Borghetti, and Angela A. Sodemann. 2015. A Survey of Distance and Similarity Measures Used within Network Intrusion Anomaly Detection. *IEEE Communications Surveys & Tutorials* 17 (1): 70–91.
11. Bhuyan, Monowar H., Dhruba Kumar Bhattacharyya, and Jugal K. Kalita. 2014. Network Anomaly Detection: Methods, Systems and Tools. *IEEE Communications Surveys & Tutorials* 16 (1): 303–336.
12. Theiler, James, and D. Michael Cai. 2003. Resampling Approach for Anomaly Detection in Multispectral Images. In *Proceedings of SPIE*, vol. 5093.
13. Chandola, V., A. Banerjee, and V. Kumar. 2009. Anomaly Detection: A Survey. *ACM Computing Surveys CSUR* 41 (3): 15.
14. Casas, Pedro, Johan Mazel, and Philippe Owezarski. 2012. Unsupervised Network Intrusion Detection Systems: Detecting the Unknown without Knowledge. *Computer Communications* 35 (7): 772–783.
15. De la Hoz, Eduardo, et al. 2015. PCA Filtering and Probabilistic SOM for Network Intrusion Detection. *Neurocomputing* 164: 71–81.

16. Bostani, Hamid, and Mansour Sheikhan. 2017. Modification of Supervised OPF-Based Intrusion Detection Systems Using Unsupervised Learning and Social Network Concept. *Pattern Recognition* 62: 56–72.
17. Pan, Zhi-Song, et al. 2003. Hybrid Neural Network and C4. 5 for Misuse Detection. In *2003 International Conference on Machine Learning and Cybernetics*, vol. 4. IEEE.
18. Zhang, Jiong, Mohammad Zulkernine, and Anwar Haque. 2008. Random-Forests-Based Network Intrusion Detection Systems. *IEEE Transactions on Systems, Man, and Cybernetics —Part C: Applications and Reviews 38* (5): 649.
19. Guo, Chun, et al. 2016. A Two-Level Hybrid Approach for Intrusion Detection. *Neurocomputing* 214C: 391–400.
20. Toosi, Adel Nadjaran, and Mohsen Kahani. 2007. A New Approach to Intrusion Detection Based on an Evolutionary Soft Computing Model Using Neuro-Fuzzy Classifiers. *Computer Communications* 30 (10): 2201–2212.
21. Horng, Shi-Jinn, et al. 2011. A Novel Intrusion Detection System Based on Hierarchical Clustering and Support Vector Machines. *Expert systems with Applications* 38 (1): 306–313.
22. Tan, Zhiyuan, et al. 2014. A System for Denial-of-Service Attack Detection Based On Multivariate Correlation Analysis. *IEEE Transactions on Parallel and Distributed Systems* 25 (2): 447–456.
23. Pajouh, Hamed Haddad, Gholam Hossein Dastghaibyfard, and Sattar Hashemi. 2017. Two-Tier Network Anomaly Detection Model: A Machine Learning Approach. *Journal of Intelligent Information Systems* 48 (1): 61–74.
24. Osanaiye, Opeyemi, Kim-Kwang Raymond Choo, and Mqhele Dlodlo. 2016. Distributed Denial of Service (DDoS) Resilience in Cloud. *Journal of Network and Computer Applications* 67C: 147–165.
25. Ambusaidi, Mohammed A., et al. 2016. Building an Intrusion Detection System Using a Filter-Based Feature Selection Algorithm. *IEEE Transactions on Computers* 65 (10): 2986–2998.
26. Daryabar, Farid, et al. 2012. Towards Secure Model for Scada Systems. In *2012 International Conference on Cyber Security, Cyber Warfare and Digital Forensic (CyberSec)*. IEEE.
27. Pan, Shengyi, Thomas Morris, and Uttam Adhikari. 2015. Developing a Hybrid Intrusion Detection System Using Data Mining for Power Systems. *IEEE Transactions on Smart Grid* 6 (6): 3104–3113.
28. Zhou, Chunjie, et al. 2015. Design and Analysis of Multimodel-Based Anomaly Intrusion Detection Systems in Industrial Process Automation. *IEEE Transactions on Systems, Man, and Cybernetics: Systems* 45 (10): 1345–1360.
29. Whitmore, Andrew, Anurag Agarwal, and Xu Da Li. 2015. The Internet of Things—A Survey of Topics and Trends. *Information Systems Frontiers* 17 (2): 261–274.
30. Ashraf, Qazi Mamoon, and Mohamed HadiHabaebi. 2015. Autonomic Schemes for Threat Mitigation in Internet of Things. *Journal of Network and Computer Applications* 49C: 112–127.
31. Ning, Huansheng, Hong Liu, and Laurence T. Yang. 2015. Aggregated-Proof Based Hierarchical Authentication Scheme for the Internet of Things. *IEEE Transactions on Parallel and Distributed Systems* 3 (26): 657–667.
32. Cao, Xianghui, et al. 2016. Ghost-in-ZigBee: Energy Depletion Attack on ZigBee-Based Wireless Networks. *IEEE Internet of Things Journal* 3 (5): 816–829.
33. Chen, Qian, Sherif Abdelwahed, and Abdelkarim Erradi. 2014. A Model-Based Validated Autonomic Approach to Self-protect Computing Systems.
34. Teixeira, F.Augusto, et al. 2015. Defending Internet of Things Against Exploits. *IEEE Latin America Transactions* 13 (4): 1112–1119.
35. Pajouh, Hamed Haddad, et al. 2016. A Two-Layer Dimension Reduction and Two-Tier Classification Model for Anomaly-Based Intrusion Detection in IoT Backbone Networks. *IEEE Transactions on Emerging Topics in Computing*.

36. Budak, H., and S.E. Taşabat. 2016. A Modified T-Score for Feature Selection. *Anadolu Üniversitesi Bilim Ve Teknoloji Dergisi A-Uygulamalı Bilimlerve Mühendislik* 17 (5): 845–852.
37. Kummer, O., J. Savoy, R.E. Argand. 2012. Feature Selection in Sentiment Analysis.
38. Sahoo, P.R., and T.M. Theorems. 1998. *Functional Equations*. Singapore: World Scientific.
39. http://www.sjsu.edu/faculty/gerstman/StatPrimer/t-table.pdf, 2017.
40. http://www.takakura.com/Kyoto_data/new_data201704/2015/2015.zip.
41. http://www.takakura.com/Kyoto_data/BenchmarkData-Description-v5.pdf.
42. Powers, D.M. 2011. Evaluation: From Precision, Recall and F-Measure to ROC, Informedness, Markedness and Correlation.

A Survey on Deceptive Phishing Attacks in Social Networking Environments

**Mohammed Mahmood Ali, Mohd S. Qaseem
and Md Ateeq Ur Rahman**

Abstract Social phishers continuously adapt the novel trapping techniques of prying usernames and passwords by establishing a friendly relationship through microblog messages using various social networking environments (SNEs) for financial benefits. APWG report of 2018 shows the successive rise in phishing attacks via URLS, fake Web sites, spoofed e-mail links, domain name usage, and social media content (APWG, Anti-Phishing Working Group report [1]). Innumerable defending techniques had been proposed earlier, but are still vulnerable due to exchange of compromised microblogs in SNEs resulting in leaking of confidential information and falling prey to phishers attack. To mitigate the latent fraudulent phishing mechanisms, there is a scope and an immense need to get rid of phishing attacks in SNEs. This paper surveys and analyzes the various social phishing detection and prevention mechanisms that are developed for SNEs.

Keywords Social phishing · Phishing attacks · APWG report · Mitigate · Social network environments (SNEs)

M. M. Ali (✉)
Department of CSE, Muffakham Jah College of Engineering & Technology,
Banjara Hills, Hyderabad, India
e-mail: mahmoodedu@gmail.com

M. S. Qaseem
Department of CSE, Nawab Shah Alam Khan College of Engineering
& Technology, Hyderabad, Telangana, India
e-mail: ms_qaseem@yahoo.com

M. A. U. Rahman
Department of CSE, Shadan College of Engineering & Technology,
Peerancheru, Hyderabad, India
e-mail: mail_to_ateeq@yahoo.com

© Springer Nature Singapore Pte Ltd. 2020
K. S. Raju et al. (eds.), *Proceedings of the Third International Conference
on Computational Intelligence and Informatics*, Advances in Intelligent
Systems and Computing 1090, https://doi.org/10.1007/978-981-15-1480-7_37

1 Introduction

The social phishing is defined as the technique(s) used to trace out the privacy information without users knowledge by sending social messages to gain the victims' trustworthiness and confidentially [2]. Many of research studies were made to detect phishing from e-mails, URLs, Web sites, and domain name usages. But, very less research is paid towards the surveillance of social phishing in social networking environments (SNEs) such as Facebook, Twitter, LinkedIn, Instagram, chatbots, WhatsApp, Teen chat, and many more messaging applications. In SNEs, to open an account, the confidential data are supposed to be uploaded willingly by the users if they want to access and share information. It is well-known that Facebook stores the users' profile data such as account name, place, residential address, contact numbers, current job address, educational details, job description and its experience, personal preferences, hobbies, likes, dislikes, interests, profile photo and videos (optional), and nowadays, to secure the accounts, a two-step-verification strategy of linking mobile numbers to accounts is embedded. Further, e-mail accounts are also merged with Facebook and other social networking accounts opening a hidden channel for social phishing attacks. In mid of 2016, Google thought of linking unique mobile number with the Gmail accounts for restricting the users to limit the opening of multiple accounts up to ten (10), by appending a three-step verification of sending a one-time password (OTP) in the name of high-level secure security. At one end by using OTP, the Google can trace the users' location, latitudes, longitudes through Google maps and can locate the person very easily which is hidden from the users [3].

Ultimately, these social networks trap out the personal and confidential information without the users' knowledge unknowingly. Currently, the accumulated users' information is extensively used by various marketing and other agencies [4]. Simultaneously, the confidential information is known to unauthorized persons such as phishers, and then they make illegal use of data for phishing attacks [5].

This section motivates the social users to carefully make use of personal information while chatting with others on various SNEs for security reasons. Further, alerts the users for the difficulties faced due to negligently sharing of information in SNEs. The remainder of this paper is organized as follows: Sect. 2 highlights the consequences of sharing personal information in SNEs that lead to deceptive phishing attacks and its impact on blocking of online banking transactions. The Sect. 3 elaborates the different phishing strategies developed earlier for overcoming deceptive phishing attacks exclusively in social networking sites (SNS) is discussed, where as Sect. 4 explores the features chosen in generalized architecture of phishing detection system (PDS) to detect deceptive phishing from microblogs which guides in prediction of latent phishing attacks. In Sect. 5, thorough and in-depth comparative analysis among the phishing detection approaches is explained. Section 6 concludes this survey with an outlook to future research challenge for detecting and understanding the psychological intent of phishers in SNSs that needs to be exposed.

2 Problems with Sharing Data in Social Networking Environments

In this section, we explore the problems faced due to carelessly sharing of personal profiles and their data in various SNEs that make them falling prey to phishers. Intelligently, phishers may adopt various SNEs to trap social users for leaking out confidential information. These phishers keep track of users for certain period of time maybe for weeks or months, not only from SNEs but also by making phone calls pretending themselves as official authorized personal requesting to share information such as birth place, date, job, martial status, credit cards, bank accounts, and PAN details through questionnaires pattern.

The sample scenario of the deceptive social message sent by phishers to trace out confidential information is shown in Table 1, that depicts how a phisher plans and forwards spoofed social messages unanimously and then establishes a fraudulent relationship to deceive social users for financial benefits in SNEs.

The bold keywords in Table 1, namely *deposit, amount, account, details, account number, account name, bank, branch IFSC code, mobile number, and DOB*, are the confidential words leaked by IM user to become a social victim due to phishers deceptive microblogs. The above information is enough to block online bank transactions of customers; one of such worse case of misusing of privacy data by phishers is locking the online bank accounts temporarily for a period of 24 h without the bank customers' knowledge. The sample deceptive phishing attack performed by phisher is shown in Fig. 1. The bank customers' accounts are easily blocked by unknown phisher by a simple trick of entering wrong information three (3) times into the online banking Web page. The elliptical marked messages represents that the account is blocked for a day, i.e., 24 h. Further, the leaked private information, such as account number, mobile number, and date of birth of a particular user which is shown in Table 1, is effectively used by a phisher that aids for generation of new password as shown in Fig. 2.

Table 1 Sample microblogs sent by phishers to IM user for trapping via social media

Chatmate-1 (phisher)	Chatmate-2 (IM user)
I wil deposit the amount to ur account, give ur account details	*yes, my **Account number**: 30527854265, **Account name**: Raghu Vamshi, **Bank**: HDFC, **Branch IFSC code**: 74585*
	Received money, Thanxs a lot....
*Pls., **pay back** (50%) **money** after 3 weeks to my **account** as promised by u*	*Ok, sure. Can u pls. send your **account details***
*Ok. My **Account number**: 45565221253, **Account name**: Praveen, **Bank name**: AXIS, Branch **IFSC code**: 58747 ......; whats your phone number and **DOB** please*	*Hi, 9980423222 is **mobile number** 20/2/1987*

Fig. 1 Strategy for locking bank account by entering wrong information three (3) times using the username blocked for 24 h, to unlock see the elliptical marked

Fig. 2 Phisher clicks the "Forgot Login Password" on banks online Web page where account number, account names, country, mobile, and DOB fields are entered

Subsequently, if the intelligent technology that involves deep learning, artificial intelligence, and machine learning techniques are collaboratively used by phishers and hackers on SNEs, then it results in harming and destroying online transactions in different fields of domains which is unimaginable. The third parties are selling the hacked users' profile information obtained through crawling from social networks and making illegal use of private information for their financial benefits to phishers [6].

3 Survey on Deceptive Phishing Detection Approaches in SNEs

Confidential information is both unanimously and sometimes unintentionally exposed through microblogs in social networks environments (SNEs) by various users, because they are interested in exponential exposure and their willingness for establishing good relationships with different social networking groups. Many of these user's of social groups very rarely understand the privacy policies and accepts it blindly by biasing themselves toward leakage of confidential information that traps them into disastrous privacy risks. To detect many strategies and techniques was developed for identifying spim microblogs by using advanced technologies, among them one of machine learning technique is used to detect spam bots in Twitters is explored [7]. Similarly, in another study, machine learning approach is effectively used in detecting and identifying the political goons that misuse the social media [8]. In another approach, analyzing the tweets that constitutes of malicious keywords using Bayesian approach for predicting phishers is explored [9]. Similarly, to detect polluters from SNEs (Facebook and Twitter), a hybrid supervised classifier is proposed that minimizes the burden for service providers [10]. Serious concern of various attacks and its consequences in online social networks (OSN) is briefly discussed, further to reduce OSN attacks the solutions that exists to mitigate is pin-pointed [4, 11].

Many of the machine learning approaches are used to detect phishing only after committing of attacks that comprises of new keywords, new urls, new links, and new domain names which must be frequently updated into the historical phishing database to avoid the future attacks. The drawback in machine learning is that it requires excessive training and mostly works on static OSNs messages. But, most of these are unsuccessful to identify and predict the social phishing attacks in OSNs from online chatting sessions dynamically. Instead of that, a recent study suggested that the use of pre-defined rules can dynamically detect words and is bit faster showing accurate result, further no training is required [12].

One of such systems had been implemented to surveillance the social phishing attack in SNEs named as anti-phishing detection system (APDs) that solely depends on pre-defined words [13]. Similarly, in another approach, an enhanced APD system for detecting the social phishing words at runtime from text and audio is developed [14]. Subsequently, an integrated APD system that uses data mining technique associated with Wordnet Ontology was developed to cross-check and then identify the deceptive spoofed messages from SNEs [15].

4 Generalized Architecture for Phishing Detection System in SNEs

Most of these phishing detection methodologies make use of data mining technique or probabilistic GSHL algorithm and WordNet Ontology for predicting the phishing words in microblogs from social networking sites (SNS) [13–15]. The generalized architecture of phishing detection system (PDS) is depicted in Fig. 3.

The working components of anti-phisihing detection architecture (APD) are illustrated in Fig. 3. This PDS system makes use of databases like TDB (to store user messages), TPDB (stores phishing patterns found), SSPWDB (set of suspicious pre-defined phishing rules), KDB (stores domain with set of phishing words), and ODB (uses OBIE model for predicting phishing words from microblogs using WordNet, which is guided with pre-defined logical rules) [16].

The APD algorithm initiates the steps for capturing the phishing words that are exchanged between the users and then stores them into database for identifying phishing words using pre-defined phishing rules [13]. In PDS, the monitoring system program (APD algorithm) identifies culprit details of phisher and reports to the victim client. The APD algorithmic steps are elaborately discussed in [16].

In this section, we have chosen one of the best architecture out of other approaches for spying of microblogs from SNS that predicts the phishing words dynamically during the chatting session at runtime and then generates an alert pop-up to the users on which this system (APD algorithm) is configured.

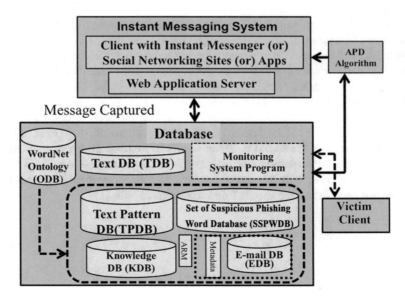

Fig. 3 Architecture of PDS to trap phishing words from social messaging applications

5   Flaws in Earlier Phishing Detection Approaches

In this section, we focus only on the earlier phishing detection approaches that were proposed for identifying the phishing words at runtime when the short messages or posts are shared among the chatmates in SNEs, rather than the static approaches of phishing detection as explored in Sect. 3.

After keenly observing the APWG reports starting from January 22, 2004 till July 31, 2018 [17], to our knowledge APD system was the first attempt made to detect phishing in instant messaging applications from short posts in the year 2011 [18]. In this paper, they have designed an "APD: ARM Deceptive Phishing Detector System," that uses an Apriori algorithm, a data mining technique to detect phishing words from messages but the major drawback of this system is initially it needs an enormous training dataset through phishing messages that constitute of phishing words, which is a time-consuming process. When the test messages are fed to APD for predicting the phishing words, again the Apriori algorithm checks each of the message for phishing words and checks the threshold value which is tuned by the developer, and if it satisfies, then phishing words are detected; otherwise, the messages are ignored.

Similarly, in the year 2012, APD system is improvised by updating the detected phishing words and appending these words to training dataset, which has reduced the processing time of Apriori algorithm by reducing one level of generation of phishing patterns. In 2014, based on the context of words, a phishing detection &and prevention approach for SNS is proposed that analyzes and predicts phishing in instant messages (IM) with relevance to domain ontology (OBIE) and utilizes the classification based on association (CBA) for generating phishing rules and alerting the victims [19].

With the advancement of instant messaging application, the phishers tried to trace out confidential information of the users through voice chatting as well as texting of messages simultaneously. To overcome this phishing detection process, APD system is enhanced to detect phishing words from microblogs that constitutes not only text but also audio (speech) in the year 2012. The speech recognition system is embedded to earlier textual APD system using FFT spectrum analysis and LPC coefficients methodologies [14]. It was time-consuming process as it involves filtering of noise from speech and then converting to voice signals that needs to be checked with wide range of spectrum of voice clips that belongs to phishing words and then subsequently checking the text messages that has phishing words and initiate the integration of phishing words from voice clips and text was bit difficult task.

Later, in the year 2015, the APD system is enhanced and improved a lot with the use of pre-defined logical phishing rules assisted with logical WordNet dictionary using probabilistic GSHL and tree alignment algorithms instead of data mining technique [16]. The results obtained are marvelous giving a good rise in precision rate for the given set of messages. The comparative analysis and the chosen features in each of the strategies are illustrated in Table 2.

Table 2 Illustrates different phishing detection approaches and their deficiencies

Title of paper	Textual detection	Pre-defined rules	WordNet ontology	Text and audio	Approach used	Drawback
APD [18]	Yes	No, but frequent patterns	Not supported	Not supported	Apriori algorithm	Excessive training
APD [13]	Yes	partial, but frequent patterns updated	Not supported	Not supported	Apriori algorithm and updating of frequent patterns	One-level of frequent patterns is reduced
APD [14]	Yes	Initial stage use of pre-defined	Terminological Ontological	Supported	Pre-defined rules mapped and then integrated with speech, using LPC and FFT technique	Processing time is high, accuracy reduced due to inefficiency of pitch signals of voices
APDS [15]	Yes	Yes	Terminological Ontology used	Not supported	OBIE approach used	few phishing words are ignored
CBA [19]	Yes	Context-based rules	Terminological Ontological words are used instead of formal Ontology	Not supported	Classification-based association rules and GSHL algorithm	Classifying and then applying association rules, excepted tree is not formed, time taken is more
APD [16]	Yes	Yes	Supported	Not supported	Probabilistic GSHL and tree alignment algorithms	Short-form words are ignored

6 Conclusion and Future Scope

In this paper, we have surveyed various earlier approaches that were used to detect phishing from microblogs specifically in SNEs. Totally, eight (8) survey papers were thoroughly reviewed that were related to deceptive phishing attacks in SNEs. One of the reasons that can be concluded from APWG reports [1, 17] is that this report has neglected the online social phishing attacks that are yet happening from instant messaging applications [4]. In Sect. 2, we have explored the problems faced with a detailed scenario of tracing out confidential information from SNEs through spoofed social messages by phishers. To overcome, the APD algorithm is discussed in Sect. 4 that can be effectively integrated into SNEs, so that an alarming alert is given to the users of SNEs if it is configured efficiently [16].

Further, we have identified the future scope for enhancing the APD system which is given below as follows:

- Phishing words that are sent using short-form notations are ignored by APD system.
- Multi-lingual phishing words are not identified by APD system.
- Understanding the psychology of phishers is another research area that needs to be explored.
- If deceptive social phishing messages are sent in any other formats such as multimedia (images, audio and video) apart from textual are deferred by APD system.

References

1. APWG. 2018. Anti-Phishing Working Group Report. http://docs.apwg.org/reports/apwg_trends_report_q1_2018.pdf. Accessed 17 Sept 2018.
2. Jagatic, T. 2005. Social Phishing. In *School of Informatics*. Bloomington: Indiana University.
3. Yi-Bing Lin, Min-Zheng Shieh, Yun-Wei Lin, and Hsin-Ya Chen. 2018. MapTalk: Mosaicking Physical Objects into the Cyber World. In *Cyber-Physical Systems*. TandFonline.
4. Kayes, Imrul, and Adriana Iamnitchi. 2017. Privacy and Security in Online Social Networks: A Survey. In *Proceeding of Online Social Networks and Media*, vol. 3–4, 1–21. Elsevier.
5. Shaikh, Anjum N., Antesar M. Shabut, and M.A. Hossain. 2016. A Literature Review on Phishing Crime, Prevention Review and Investigation of Gaps. In: *10th International Conference on Software, Knowledge, Information Management & Applications (SKIMA)*. IEEE.
6. Joseph Bonneau, Jonathan Anderson, and George Danezis. 2009. Prying Data Out of a Social Network. In *Proceedings of Advances in Social Networks Analysis and Mining*.
7. Hai Wang, Alex. 2010. Detecting Spam Bots in Online Social Networking Sites: A Machine Learning Approach. In *Proceedings of ACM*.
8. Ratkiewicz, J., M.D. Conover, M. Meiss, B. Gonçalves, and A. Fla. 2011. Detecting and Tracking Political Abuse in Social Media. In *Proceedings of the Fifth International AAAI Conference on Weblogs and Social Media*. Association for the Advancement of Artificial.

9. Beck, K. 2011. Analyzing Tweets to Identify Malicious Messages. In *IEEE International Conference on Electro/Information Technology (EIT)*, 1–5. IEEE.
10. Park, Byung Joon, and Jin Seop Han. 2016. Efficient Decision Support for Detecting Content Polluters on Social Networks: An Approach Based on Automatic Knowledge Acquisition From Behavioral Patterns. *Information Technology and Management* 17 (1): 95–105.
11. Gupta, Surbhi , Abhishek Singha, and Akanksha Kapoor. 2016. A Literature Survey on Social Engineering Attacks: Phishing Attack. In: *International Conference on Computing, Communication and Automation (ICCCA)*. IEEE.
12. Thelwall, M. 2017. TensiStrength: Stress and Relaxation Magnitude Detection for Social Media Texts. *Journal of Information Processing & Management, 53* (1): 106–121. Elsevier.
13. Mahmood Ali, Mohd, and Lakshmi Rajamani. 2012. APD: ARM Deceptive Phishing Detector System Phishing Detection in Instant Messengers Using Data Mining Approach. In *Global Trends in Computing and Communication Systems, CCIS*, vol. 269, 490–502. Springer.
14. Mahmood Ali, Mohd, and Lakshmi Rajamani. 2012. Deceptive Phishing Detection System: From Audio and Text Messages in Instant Messengers Using Data Mining Approach. In *Proceedings of the IEEE International Conference on Pattern Recognition, Informatics and Medical Engineering*, 458–463. IEEE.
15. Qaseem, Mohd S., Nayeemuddin, M., et. al. 2014. APDs: Framework for Surveillance of Phishing words in Instant Messaging Systems Using Data Mining and Ontology. In: *Symposium of TechSym*, Article No. 486. IEEE.
16. Mahmood Ali, Mohd et al. 2015. An Approach for Deceptive Phishing Detection and Prevention in Social Networking Sites Using Data Mining and Wordnet Ontology. In: *International Conference on Electrical, Electronics, Signals Communication and Optimization (EESCO)*. IEEE.
17. https://www.antiphishing.org/resources/apwg-reports/.
18. Mahmood Ali, Mohd, and Lakshmi Rajamani. 2011. APD: ARM Phishing Detector A System for Detecting Phishing in Instant Messengers. *International Journal of Information Processing*, 5 (2): 22–30. IK Publishers, India.
19. Qaseem, Mohammad S., and A. Govardhan. Phishing Detection in IMs using Domain Ontology and CBA—An innovative Rule Generation Approach. *International Journal of Information Sciences and Techniques (IJIST) 4* (4/5/6). India.

Implementation of Spatial Images Using Rough Set-Based Classification Techniques

D. N. Vasundhara and M. Seetha

Abstract "Artificial neural networks (ANNs)" and support vector networks have been widely used for the purpose of analysis of spatial remote sensory images. Attributes selection plays a major role in reduction of computational time for model processing and improvement in performance accuracy. In this paper, we present rough set-based artificial neural network (RS-ANN) and support vector machine classification (RS-SVM) methods for the classification of hyperspectral images. In both of these techniques, rough set (RS) is used as a feature selection mathematical tool. Further, this diminished dimensionality data is carried to artificial neural network (ANN) and support vector machine (SVM) classifiers correspondingly. The proposed procedure uses spatial information and implements the strategy for efficient processing of the image, and with utilization of the "graphics processing units (GPUs)" successfully implemented with improved efficiency. The experimental analysis achieved efficient outcome that the proposed procedures perform with improved accuracy, improved efficiency with a good amount of reduction in the execution and computing time, and enhancing the accuracy in the categorizing of spatial images in comparison with other conventional strategies.

Keywords Feature extraction · Spatial image data · Artificial neural network · Classification · Rough sets · Support vector machines

D. N. Vasundhara (✉)
Department of Computer Science and Engineering, VNR Vignana Jyothi Institute of Engineering and Technology, Hyderabad, India
e-mail: vasundhara_d@vnrvjiet.in

M. Seetha
Department of Computer Science and Engineering, G. Narayanamma Institute of Technology and Science, Hyderabad, India
e-mail: smaddala2000@yahoo.com

© Springer Nature Singapore Pte Ltd. 2020
K. S. Raju et al. (eds.), *Proceedings of the Third International Conference on Computational Intelligence and Informatics*, Advances in Intelligent Systems and Computing 1090, https://doi.org/10.1007/978-981-15-1480-7_38

1   Introduction

Remote sensing (RS) refers to the passive or active methods used for gathering information of an object without being in physical contact with the object. In relation to Earth observation, RS over the recent decades has proven to be an unprecedented source of data for monitoring the Earth's surface. The spectral, spatial, and temporal characteristics of RS data make it an attractive source of input for image processing. Of particular interest are the images collected by sensors mounted on aircraft or spaceborne platforms as these have the additional advantage of providing with a valuable overview of interrelation between land elements that are not easily obtainable by land surveying.

1.1   Remote Sensing System Overview

The remote sensing system has majorly three phases, viz. image acquisition, preprocessing, and analysis phases as shown in Fig. 1.

The first phase is the acquisition of the image during which the "electromagnetic radiation (EM)" is captured by the sensors. The characteristics of the image are defined by the selection of the sensors that are used for image acquisition. The multispectral images are the images which are acquired at particular frequencies among the EM spectrum. Multispectral radiometers, in contrast to panchromatic sensors that measure the net intensity of radiance, acquire data in broad, wide-spaced spectral ranges commonly referred to as bands. Further, the images are called hyperspectral images when the capturing is done in wider coverage with many number of bands.

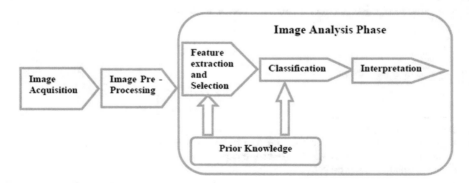

Fig. 1   Block diagram of a remote sensing system

The acquired image data consists of various distortions. Hence, in the second phase i.e., during the preprocessing phase, the distortions of both geometric and radiometric are handled.

The image analysis phase is the final phase and follows the following subphases or steps

i. Identification of the selected features: This step aims at dimensionality reduction of measurement space while retaining optimum salient characteristics. The main objective here is simplifying the amount of resources required to describe the image accurately [1–3].

ii. Classification: During this step, the input space of the features is transformed into feasible classes [4, 5].

iii. Interpretation: In this phase, the required information is finally extracted. The process applied here will vary depending on the final information required [6, 7].

1.2 Hyperspectral Imaging

The latest systems can acquire very high-resolution (VHR) multispectral images characterized by high geometric resolution and superior spectral resolution. In spite of all the advances, multispectral sensors have the inherent limitation of not being able to provide narrow spectral feature discrimination as the data is acquired in a few broad irregularly spaced spectral bands. This limitation makes multispectral sensors less suitable for identifying many surface materials and features [8]. The images from each spectral band when combined form a three-dimensional HS cube that is used for processing and analysis [9–20]. The availability of complete record of spectral responses of materials over the wavelengths considered in HS data sets makes rigorous analysis of surface compositions feasible [21–25].

The roughest theory is used by various scientists applied to ANN classification techniques [26]. The rough set concept is applied and successfully implemented by Vasundhara and Seetha [27] for the classification of spatial image data sets. The same is successfully implemented by various other researchers for hyperspectral images and proposed the decrease in the dimensionality of the feature sets [30–36].

1.3 Motivation and Contribution

Due to the advancement of modern sensory systems such as: AVIRIS, NASA, and DLR sensors, image feature engineering and categorization pertain to the acquisition of some advanced attributes extraction, filtering, attributes selection and classification methods to process those high-resolution images. The key challenges are to deal with data present in high dimension, amount of uncertain pixels presence, and the set of irrelevant attributes present in the imagery data, which may result in high computational time and lesser classification accuracy. Key contributions of this paper are summarized as below:

- Firstly, the ANN and SVM classifiers' overview is presented.
- The overview of proposed RS-ANN classification mechanism along with the experimental results is presented.
- The overview of proposed RS-SVM classification mechanism along with the experimental results is presented.

Experiments are performed in an extensive manner on various standard spatial image data sets: Indian Pines data set and Salinas data set scene. The methods have been efficiently implemented utilizing graphics processing units (GPUs).

2 Existing Techniques

The first is ANN-based classification where the basic backpropagation algorithm is used for the classification of the spatial images. Next, the property of multiplicative weight tuning is introduced to improve the accuracy. The procedure for proposed techniques has been discussed in [27]. The second technique is SVM-based classification for which the procedural steps for spatial image classification are discussed in [38].

3 Proposed Techniques

3.1 RS-ANN Classification [RS-ANN]

In this proposed technique, rough set theory is used for the dimensionality reduction of the features and the "backpropagation" algorithm is used for the image classification (Fig. 2).

3.2 Proposed RS-SVM Classification [RS-SVM]

The proposed technique consists of two phases:

- features of interest identification using rough set theory(Sect. 3.1);
- image classification using support vector machine technique [38].

Image feature selection algorithm using Rough Sets:

Step 1: Since image consists of pixels. In the pre-processing phase, basic image information for each pixel (e.g. RGB values, pixel intensity values, gray code level value etc.) will be obtained. These will be called attribute values for each pixel.

Step 2: Draw the pixel, attribute-value information table. A pair $I = \{P, A\}$, in which $P\{P_1, P_2, \dots, P_n\}$ is meant for finite set of each pixel and non-empty set is called a system of information.

$A\{A_1, A_2, \dots, A_n\}$: meant for attribute set which is finite and non-empty.

The process in which universal set is classified into modules or equivalence classes based on some attribute values, is called Granularization.

Step 3: In this step, the Discernibility matrix for the above pixel, attribute value information table is drawn. Here the explanation of two rough set theory terms is as below:-

Discernibility - When two objects have different attribute values then they can be distinguished based on that attribute value. In RST, this is denoted as term discernibility.

Discernibility matrix - A discernibility matrix of an information table $I = (P, A)$ is a matrix which is symmetric $|P| \times |P|$ in which the entries are given by "$c_{ij} = \{a \in A | a(x_i) \neq a(x_j)\}$" $i, j = 1, \dots, |P|$ Individual c_{ij} consists of those features, who offers the difference between objects i and j

Step 4: For each row of the discernibility matrix, compute discernibility function as $f_1, f_2, \dots, f_{|P|} = \wedge\{\vee c_{ij} | 1 \leq j \leq i \leq |P|, c_{ij} \neq \emptyset\}$

Step 5: Compute resultant discernibility function as $F = \wedge\{f_i, 1 \leq i \leq |P|\}$

Step 6: Attributes shrinking algorithm for computing $(F, \{f_i, 1 \leq i \leq |P|\})$ the algorithm steps are as below:-

– While $(F, \{f_i, 1 \leq i \leq |P|\})$ are not in minimized form

    – remove the super-sets by applying Absorption Rule in $(F, \{f_i, 1 \leq |P|\})iP$)

    – Replace strong shrinkable attributes $(F, \{f_i, 1 \leq i \leq |P|\})$.

Note: strong shrinkability implements where the clause attributes are either together present or missing in all clauses. In that case, those attributes may be substituted by a single temporary attribute.

    – $s \Leftarrow$ most frequent attribute $(F, \{f_i, 1 \leq i \leq |P|\})$

    – apply Expansion Rule $(s, (F, \{f_i, 1 \leq i \leq |P|\}))$

    – substitute the strong shrinkable classes $(F, \{f_i, 1 \leq i \leq |P|\})$

    – $RED \Leftarrow$ calculate the reducts $(F, \{f_i, 1 \leq i \leq |P|\})$

    – return (RED)

Step 7: Set of entire prime implicants in computed discernibility function is obtained. In RST point of view these all are called as Reducts. In general, it is the process of feature selection in rough set theory.

Fig. 2 Rough set algorithm

4 Experimental Results

The existed and proposed algorithms have been implemented on two spatial image data sets, viz. Indian Pines and Salinas data sets.

Setup and Machine Configuration—In our experiments, we used the hardware and software as OS: Windows 7; MATLAB Version: 2014a; RAM size: 16 GB; GPU details: NVIDIA GeForce GTX 1080 GPU; Processor: Octa Core i7 processor @ 3.0 Ghz clock frequency.

Indian Pines data set Scene
Data set description—Indian Pines [28] is a 145 × 145 scene with 224 bands of spectral reflectance, and the wavelength range is 0.4×10^6–2.5×10^6 m. AVIRIS sensor is used to acquire this image set. The Indian Pines scene contains 2/3 agriculture and 1/3 forest or other natural perennial vegetation. The ground truth is designated by 16 classes (*Alfalfa, Cornnotill, Cornmintill, Corn, Grass-pasture, Grass-trees, Grass-pasture-mowed, Hay-windrowed, Oats, Soybeannotill, Soybeanmintill, Soybean-clean, Wheat, Woods, Buildings- Grass-Trees-Drives, and Stone-Steel-Towers*). The original feature bands are 220 in count. Indian Pines data-set is reduced to 200 feature bands after correction and is available for various research purposes.

Results summary—A random input image is shown in Fig. 3. Feature bands obtained in process of RS-based feature selection: total count: 14 (96; 86; 97; 30; 85; 84; 61; 83; 90; 32; 89; 200; 78; 108). We have performed experiments on this data set using eight different methods to achieve multi-class image scene classification. The obtained results are given in Table 1. Final classified images are given as in Fig. 4a–h.

Fig. 3 Input image [28]

Table 1 Methods: I—ANN [without rough sets], II—ANN [with Rough sets], III—ANN with multiplicative weight tuning-based "backpropagation" [without rough sets], IV—ANN with multiplicative weight tuning-based "backpropagation" [with rough sets], V—SVM "RBF kernel" [without rough sets], VI—SVM "RBF kernel" [with rough sets], VII—SVM "POLY kernel" [without rough sets], VIII—SVM "POLY kernel"[with rough sets]

Performance measures	ANN-WOR	ANN-WR	ANNM-WOR	ANNM-WR	SVMR BF WOR	SVMR BF WR	SVM POLY WOR	SVM POLY WR
Overall accuracy (%)	62.60	62.79	63.92	63.90	98.22	98.54	98.66	98.22
Average accuracy (%)	61.02	61.07	62.01	62.04	98.21	98.25	98.18	98.18
Kappa coefficient	0.584	0.583	0.594	0.590	0.978	0.978	0.980	0.980
Exec. time (s)	61.0	29.2	52.0	19.4	401.3	330.2	388.10	319.34

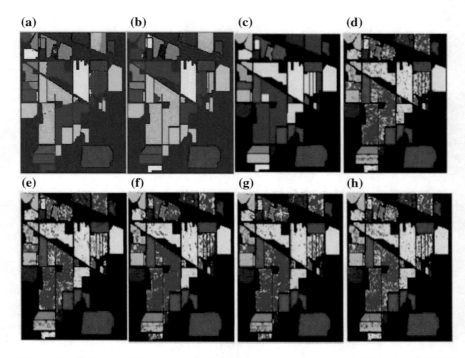

Fig. 4 Classified spatial images of Indian Pines data set. **a** ANN-WOR, **b** ANN-WR, **c** ANNM-WOR, **d** ANNM-WR, **e** SVMRBF-WOR, **f** SVMRBF-WR, **g** SVMPOLY-WOR, **h** "SVMPOLY"-WR

Salinas data set scene

Data set description—This data set scene [29] has number of classes = 16 ("*Brocoli green weeds 1, Brocoli green weeds 2, Fallow, Fallow rough plow, Fallow smooth, Stubble, Celery, Grapes untrained, Soil vinyard develop, Corn senesced green weeds, Lettuce romaine 4wk, Lettuce romaine 5wk, Lettuce romaine 6wk, Lettuc e romaine 7wk, Vinyard untrained, Vinyard vertical trellis*").

Results summary—The input image is shown in Fig. 5, and in the chosen spectral bands, the total number of extracted feature vectors is 63. Each feature vector contains statistical values named: Minima, Maxima, Skewness, Mean, Standard deviation. Gabor filter has been used here for feature extraction. All these feature vectors are combined to form feature space. RS-based feature selection is applied; 17 features are selected. The obtained results are given in Table 2. Final classified images are given as in Fig. 6a–h.

Fig. 5 Input image [29]

5 Comparative Analysis

The experimental results achieved are presented in the above Sect. 3. From these results tabulated in Tables 1 and Table 2, it is clearly observed that the accuracy is improved. Further, this improved accuracy is achieved with very less execution time when the classification utilizes the rough set theory for feature extraction. This is due to the identification of features of interest when the concept of rough set theory is used for feature extraction. As the execution time and accuracy being two major factors of performance for classification techniques, attaining improved accuracy with less execution time implies enhancement in the performance of the classification technique.

The enhancement of the performance for various classification techniques with and without the rough set theory is graphically represented in Fig. 7 for Indian Pines images and in Fig. 8 for Salinas images databases.

From Figs. 7 and 8, we can conclude that the use of rough sets with all the classification techniques is proved to increase the performance of the ANN and SVM. Further, the figures also imply that ANN with multiplicative weight tuning (ANNM) and rough sets gives better performance when compared to ANN, ANNM, and RS-ANN. Similarly, SVM with poly kernel and rough sets outperforms SVMRBF, SVMPOLY, and RS-SVMRBF.

Table 2 Methods: I—ANN [without rough sets], II—ANN [with rough sets], III—ANN with multiplicative weight tuning-based "backpropagation" [without rough sets], IV—ANN with multiplicative weight tuning-based "backpropagation" [with rough sets], V—SVM "RBF kernel" [without rough sets], VI—SVM "RBF kernel" [with rough sets], VII—SVM "POLY kernel" [without rough sets], VIII—SVM "POLY kernel" [with rough sets]

Performance measures	ANN-WOR	ANN-WR	ANNM-WOR	ANNM-WR	SVM RBF WOR	SVMR BF WR	SVM POLY WOR	SVM POLY WR
Overall accuracy (%)	66.56	66.18	67.14	67.16	97.12	97.19	97.64	97.52
Average accuracy (%)	65.14	65.32	67.0	66.02	97.32	97.55	97.33	97.18
Kappa coefficient	0.617	0.621	0.631	0.630	0.962	0.965	0.971	0.971
Exec. time (s)	52.0	18.4	41.0	10.2	380.0	308.04	365.16	296.0

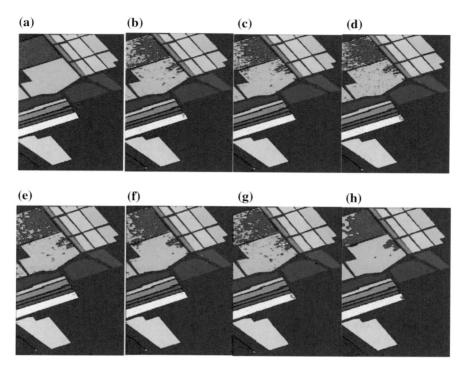

Fig. 6 Classified spatial images of Salinas data set. **a** ANN-WOR, **b** ANN-WR, **c** ANNM-WOR, **d** ANNM-WR, **e** SVMRBF-WOR, **f** SVMRBF-WR, **g** SVMPOLY-WOR, **h** SVMPOLY-WR

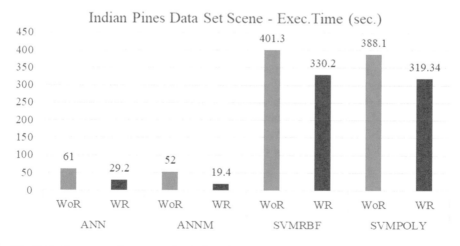

Fig. 7 Performance enhancement for Indian Pines data set scene with improved accuracy and decrease in execution time in sec.

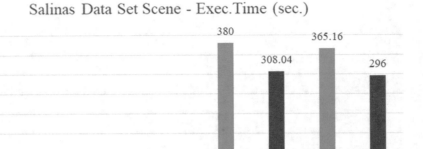

Fig. 8 Performance enhancement for Salinas data set scene with improved accuracy and decrease in execution time in sec.

6 Conclusion

Today, expert systems and machine learning methods are getting much popularity in the classification area because of the effectiveness and accuracy of classification task. This paper presents the rough set-based artificial neural networks and rough set-based support vector machine methods along with an extensive experimental analysis for classification of spatial images. In our proposed mechanism, rough set (RS) is used as a feature selection mathematical tool, which eliminates the redundant features. Further, artificial neural networks (ANN) and support vector machine (SVM) correspondingly are employed for image classification. After successful execution of the corresponding algorithm with rough sets for various combinations of the classification techniques, it can be concluded that the use of rough sets with all the classification techniques is proved to increase the performance of the ANN and SVM. Further, the figures also imply that ANN with multiplicative weight tuning (ANNM) and rough sets gives better performance when compared to ANN, ANNM, and RS-ANN. Similarly, SVM with poly kernel and rough sets outperforms SVMRBF, SVMPOLY, and RS-SVMRBF.

References

1. Chang, C.I. 2007. *Hyperspectral Data Exploitation: Theory and Applications*. New York, NY, USA: Wiley.
2. Li, D., J. Wang, X. Zhao, Y. Liu, and D. Wang. 2014. Multiple Kernel-Based Multi-Instance Learning Algorithm for Image Classification. *Journal of Visual Communication and Image Representation* 25 (5): 1112–1117.
3. Pawlak, Z. 1982. Roughsets. *International Journal of Computer and Information Sciences* 11: 341–356.
4. Petrosino, A., and G. Salvi. 2006. Rough Fuzzy Set Based Scale Space Transforms and Their Use in Image Analysis. *International Journal of Approximate Reasoning* 41: 212–228.
5. Cortes, C., and V. Vapnik. 1995. Support-Vector Networks. *Machine Learning* 20 (3): 273297.
6. Abonyi, J., and F. Szeifert. 2003. Supervised Fuzzy Clustering for the Identification of Fuzzy Classifiers. *Pattern Recognition Letters* 24 (14): 21952207.
7. Boser, B.E., and I.M. Guyon, et al. 1992. A Training Algorithm for Optimal Margin Classifiers. In *Fifth Annual Workshop on Computational Learning Theory*. Pittsburgh: ACM.
8. Chang, C.C., and C.J. Lin. 2001. LIBSVM: A Library for Support Vector Machines.
9. Frohlich, H., and O. Chapelle. 2003. Feature Selection for Support Vector Machines by Means of Genetic Algorithms. In *Proceedings of the 15th IEEE international conference on tools with artificial intelligence* 142–148, Sacramento, CA, USA.
11. Hsu, C.W., C.C. Chang et al. 2003. A Practical Guide to Support Vector Classification. Technical Report. Taipei: Department of Computer Science and Information Engineering, National Taiwan University.
12. Joachims, T., C. Nedellec, et al. 1998. Text Categorization with Support Vector Machines: Learning with Many Relevant. In *Proceedings of the 10th European Conference on Machine Learning* 137–142.
13. John, G.H., R. Kohavi, et al. 1994. Irrelevant Features and the Subset Selection Problem. In *Proceedings of the 11th International Conference on Machine Learning*.
14. Kryszkiewicz, M., H. Rybinski. 1996. Attribute Reduction Versus Property Reduction. In *Proceedings of the Fourth European Congress on Intelligent Techniques and Soft Computing* 204–208.
15. Osuna, E., R. Freund, F. Girosi. 1997. Training Support Vector Machines: Application to Face Detection. In *Proceedings of Computer Vision and Pattern Recognition* 130–136, Puerto Rico.
16. Pawlak, Z. 1982. Rough Sets. *International Journal of Parallel Programming* 11 (5): 341356.
17. Pawlak, Z. 1996. Why Rough Sets. In Proceedings of the IEEE International Conference on Fuzzy System.
18. Pawlak, Z. 1997. Rough Set Approach to Knowledge-Based Decision Support. *European Journal of Operational Research* 99 (1): 4857.
19. Vapnik, V. 1995. *The Nature of Statistical Learning Theory*. New York: Springer.
20. Vapnik, V. 1998. *Statistical Learning Theory*. New York: Wiley.
21. Quinlan, J. R. (1996). Improved use of continuous attributes in C 4.5. Journal of Artificial Intelligence Research, 4(77–90), 325.
22. Scholkopf, B., and A.J. Smola. 2002. *Learning with Kernels: Support Vector Machines, Regularization, Optimization, and Beyond*. Cambridge: The MIT Press.
23. Lin, C.-F., and S.-D. Wang. 2002. Fuzzy Support Vector Machines. *IEEE Transactions on Neural Networks* 13 (2): 415–425.

24. Ahmed, Shohel Ali, SnigdhaDey, and Kandarpa Kumar Sarma. 2011. Image Texture Classification Using Artificial Neural Network (ANN). In *2011 2nd National Conference on Emerging Trends and Applications in Computer Science (NCETACS)*. IEEE.
25. Chen, Degang, Qiang He, and Xizhao Wang. 2010. FRSVMs: Fuzzy Rough Set Based Support Vector Machines. *Fuzzy Sets and Systems* 161 (4): 596–607.
26. Urszula, Stanczyk. 2010. Rough Set-Based Analysis of Characteristic Features for ANN Classifier. In *Hybrid Artificial Intelligence Systems* 565–572. https://doi.org/10.1007/978-3-642-13769-369.
27. Vasundhara, D.N., and M. Seetha. 2016. Rough-Set and Artificial Neural Networks Based Image Classification. In *IC3I*, 14–17 Dec 2016.
28. Remote Sensing Scenes Indian Pines. www.ehu.eus/ccwintco/index.php/Hyperspectral.
29. Remote Sensing Scenes Salinas Scene. www.ehu.eus/ccwintco/index.php/Hyperspectral.
30. Zhang, L., Q. Zhang, L. Zhang, D. Tao, X. Huang, and B. Du. 2015. Ensemble Manifold Regularized Sparse Low-Rank Approximation for Multi View Feature Embedding. *Pattern Recognition* 48: 3102–3112.
31. Zhong, Z., B. Fan, J. Duan, L. Wang, K. Ding, S. Xiang, and C. Pan. 2015. Discriminant Tensor Spectral-Spatial Feature Extraction for Hyperspectral Image Classification. *IEEE Geoscience Remote Sensing Letters* 12: 1028–1032.
32. Zhao, W., and S. Du. 2016. Spectral-Spatial Feature Extraction for Hyperspectral Image Classification: A Dimension Reduction and Deep Learning Approach. *IEEE Transactions on Geoscience and Remote Sensing* 54: 1–11.
33. Chen, Y., H. Jiang, C. Li, X. Jia, and P. Ghamisi. 2016. Deep Feature Extraction and Classification of Hyperspectral Images Based on Convolutional Neural Networks. *IEEE Transactions on Geoscience and Remote Sensing* 54: 6232–6251.
34. Ma, X., H. Wang, and J. Geng. 2016. Spectralspatial Classification of Hyperspectral Image Based on Deep Auto-Encoder. *IEEE Journal of Selected Topics in Applied Earth Observations and Remote Sensing* 9: 4073–4085.
35. Liang, H., and Q. Li. 2016. Hyperspectral Imagery Classification Using Sparse Representations of Convolutional Neural Network Features. *Remote Sens.* 8: 99.
36. Zhang, H., Y. Li, Y. Zhang, and Q. Shen. 2017. Spectral-Spatial Classification of Hyper-Spectral Imagery Using a Dual-Channel Convolutional Neural Network. *Remote Sensing Letters* 8: 438–447.
37. Venkata Sailaja, N., L. Padma Sree, and N. Mangathayaru. New Rough Set-Aided Mechanism for Text Categorisation. *Journal of Information & Knowledge Management*, 17 (2): World Scientific Publishing Co. https://doi.org/10.1142/s0219649218500223.
38. Vasundhara, D.N., and M. Seetha. 2018. Rough Based SVM Technique for Spatial Image Classification. *International Journal of Control Theory and Applications* 9 (44): 365–378.
39. Santhi Sri, T., J. Rajendra Prasad, and R. Kiran Kumar. 2018. SEE: Synchronized Efficient Energy Calculation for Topology Maintenance & Power Saving in Ad Hoc Networks. *Arabian Journal for Science and Engineering (Arab J Sci Eng)*.

Multi-objective Optimization of Composing Tasks from Distributed Workflows in Cloud Computing Networks

V. Murali Mohan and K. V. V. Satyanarayana

Abstract This manuscript proposed and explored a novel strategy for optimizing the composition of tasks from distributed workflows to achieve parallel execution, optimality toward resource utilization. The critical objective of the proposal is to optimize the task sequences from the workflows initiated to execute parallel in distributed RDF environment, which is unique regard to the earlier contributions related to parallel query planning and execution strategies found in contemporary literature. All of these existing models are aimed to notify the tasks from the given workflow, which are less significant to optimize the parallel execution of the multiple tasks located in distributed workflows. In order to this, the multi-objective optimization of composing task (MOCT) sequences from distributed workflows is proposed. The MOCT optimizes the execution of tasks from multiple workflows initiated in parallel. A novel scale called "task sequence consistency score" uses the order of other metric task sequence coverages, dependency scope, and lifespan as input. The experiments are conducted on the proposed model, and other benchmark models are found in contemporary literature. The results are obtained from the experimental study evincing that the MOCT is significant and robust to optimize the task sequences in order to execute distribute workflows in parallel. The comparative analysis of the results is obtained from MOCT, and other contemporary models are performed using ANOVA standards like t-test and Wilcoxon signed-rank test.

Keywords MOCT · ANOVA · Directed acyclic graph · Ant colony optimization · Genetic algorithm · PSO · LMBPSO

V. Murali Mohan (✉) · K. V. V. Satyanarayana
Department of Computer Science and Engineering, Koneru Lakshmaiah Education
Foundation, Guntur, AP 522502, India
e-mail: muralimohan.klu@gmail.com

K. V. V. Satyanarayana
e-mail: kopparti@kluniversity.in

© Springer Nature Singapore Pte Ltd. 2020
K. S. Raju et al. (eds.), *Proceedings of the Third International Conference on Computational Intelligence and Informatics*, Advances in Intelligent Systems and Computing 1090, https://doi.org/10.1007/978-981-15-1480-7_39

1 Introduction

The method cloud computing provides us an effective and simple way to access the resources which are on demand and computable and configurable by an individual which means there will be a less communication of the service provider, user will be accessed and release the services and resources [1]. The term cloud computing has several unique benefits like fault-tolerant [1], high availability, and scalability but it has challenges like variation of performance, scheduling [2], resource provisioning, and security.

Suitable scheduling of resources plays a major role in enhancing the performance and achievement of the cloud computing. Nevertheless, the work [3] presents that the scheduling of workflow is a "NP complete issue," need a maximum consideration, so that there are very few such algorithms in recent literature that optimizing the task sequence composition with linear process complexity, which is having certain limit to optimize the task sequence composition for given workflows. Hence, the optimization of workflow scheduling is considered as research objective. The term workflow is considered as mechanization of the scheme where information or task is the transmission between participants according to the specified guideline which is instructed. By utilizing "direct acyclic graph (DAG)," workflow performs the processes structuring where edge nodes are interdependent and serves to the multiple tasks of divergent processes [4, 5].

Regardless of several researchers are addressing the issue of "scheduling in conventional distributed networks," such as grid, there might be minimum amount of work is executed in the cloud. And it entails a huge determination in scheduling "multi-objective task," specifically when they are reliant.

2 Problem Formulation

Systematic workflows contain variable number of tasks and sizes can also range from the less amount to millions of tasks. If workflow size escalates, then it is useful to disseminate the performance of the tasks among diverse resources of computing in order to attain the sensible time for finishing. The environment of cloud computing provides number of advantages for the scientific performance of application by offering unconfined amount of the resources. And the algorithms of scheduling discover the mapping among the resources and tasks to meet single or multi-objectives of optimization.

The commonly utilized criteria of optimization are minimizing the performance time or cost of application; minimizing the completion time of task which indicates the finishing of final task; maximizing the utilization of resources; and meeting the restrictions of deadline, etc. Besides these, an algorithm for scheduling is essential for deliberating the dependencies of the inter-task in the workflow. The execution of task is done only when the parent tasks have completed its execution in the workflow. Moreover, the environment of cloud computing is advantageous for the workflows as it allows the user to provide the resources directly essential for the implementation and to schedule the mathematical calculation of tasks with the scheduler of user-controlled. This allows the user to assign the resources only if it is essential and once assigned, then the resources will be utilized consequently for the performance of the several tasks.

Therefore, it is obvious that effective task scheduling and provisioning of resources is the most important challenge in attaining maximum performance in the cloud computing [6]. Additionally, the issue of task scheduling is the "NP-hard problem" and will not lend themselves to optimize the solutions in less time, specified a maximum space solution in the environment of cloud computing. Hence, solutions based on the meta-heuristic are utilized to offer the approximate optimum solutions in less time. The methods of "GA-Genetic algorithm," "ACO-Ant colony optimization," and "PSO-particle swarm optimization" are commonly used for improving the algorithms of scheduling for the environment of cloud [7]. And these methods found to execute better when compared with the non-heuristic methods in determining the resource provisioning and effective scheduling algorithms. Nevertheless, there is an opportunity for the future enhancement of solutions.

3 Related Work

The work [8] presents that dynamic model is proposed for the scheduling of workflow on the environment of cloud through an aim to reduce the performance cost on the basis of pricing method of cloud. Several kinds of virtual machines (VMs) will be accessible to lease on the demand of diverse costs. The result lessens the cost; however, is not an approximate optimum solution. The work [9] presents the workflow ensemble improved for the clouds. The range of "static and dynamic algorithms" for scheduling the workflow on the environment of cloud is depicted in [9] through an objective to increase the amount of performed workflows when meeting the budget and deadline confines. And algorithms have deliberated the postponements which are intricate in the leasing of cloud on the basis of VMs, and the results deliberate the changes in the tasks which predicted the time for execution. Nevertheless, algorithms do not deliberate the IaaS clouds heterogeneity as the entire VMs are expected to be of a similar kind.

The work [10] presents IaaS cloud; static algorithm is utilized for scheduling the single workflow. And the algorithm deliberates the difficult route of the workflow and deliberates the VMs heterogeneity and cost methods of the clouds. Objective of the algorithm is to lessen the performance cost deliberating application performance of deadline and the accessibility of resources. Nevertheless, they do not have universal optimization method able to generate the approximate optimum solution. In spite they utilize "task-level-optimization" and therefore unsuccessful to use the entire structure of workflow and the characteristics to produce the best solution.

PSO is used to solve the issue of scheduling the workflow. The work [11] presented a particle swarm optimization based method that aimed the reduction in execution cost and optimal load balancing in work flow processing. However, this algorithm evincing the performance against fixed group of VMs and identical with the method to the algorithms of grid.

The work [12] concentrates on the prerequisite for vigor of the issue of scheduling in the cloud environment. Algorithm schedules workflow tasks on heterogeneous resources of cloud with the aim of lessening the finishing period of the task and the cost. Nevertheless, generally robustness is probable by introducing the idleness in the method, and identically, the depicted algorithm augments the scheduled strength with a resultant augment in the budget.

The work [13] presents that particular cloud computing characteristics have been designed like elasticity and availability of the heterogeneous resources. And a PSO-based method is presented as a method to optimize the resource scheduling for scientific workflows in cloud platforms such as IaaS. Objective of this method is to lessen the entire cost to perform deadline constrained workflows. Nevertheless, the steadiness of optimization is noticed to be probabilistic.

The work [14] suggested the algorithm called S-CLPSO. The work [15] presents a "set-based algorithm" called S-PSO which produces the pair of task resource. The work [16] presents that "CLPSO (comprehensive learning PSO)" is improved on the basis of S-PSO model. The "self-adaptive penalty function" is utilized to develop its execution. The work [17] suggested the "Bi-Criteria Priority Based Particle Swarm Optimization (BPSO)" which allocates the priority to every task to lessen the performance cost and time under specified budget and deadline confines. And the resources are allocated to the works in decreasing sequence of allocated priority. The BPSO does not deliberate the current load.

The work [18] presents that "Load Balancing Mutation a Particle Swarm Optimization (LMBPSO)" is proposed to attain the accessibility and optimization of task finishing time and cost of transmission. This work restructured the PSO method to overwhelm the confines like "failing to map a task with any of the resource available," and "mapping a task with multiple resources" that are frequently noticed in the performance of PSO. The method LMBPSO is bi-phase, where the starting phase maps the resources and tasks by utilizing the PSO, and later stage examines the issues of mapping that tries to solve if it is detected. Nevertheless, the procedure intricacy is not a linear and mapping of resources toward tasks in the second phase is "probabilistic."

The work [19] presents that "a non-dominance sort-based hybrid particle swarm optimization (HPSO)" has suggested bi-phase method. In the starting phase, it schedules workflows toward resources that further utilize PSO in the coming phases which mutate the resulting particles for optimizing the entire cost.

The works [19–24] present that the contributions based on GA targeted to attain optimal finishing time and performance cost of task. Nevertheless, the procedure of mutation is implemented on task ordering and end of resources after the specified threshold evolutions that are not "deterministic." Therefore, deviance in the optimum pairing of the resources and tasks is often showed in concern with these models.

The utilization of ACO in the work scheduling optimization is depicted in [19, 25] that are targeting to optimize performance cost and time. The above-mentioned works are bi-phase models, where primary phase generally composes the resources and workflows, later phase optimizes through restructuring the resource pairing and workflows. Therefore, the procedure intricacy is proportional to the number of resources and workflows. Moreover, the procedure intricacy is not a linear.

The work [26] suggested "Task Scheduling using a multi-objective nested Particle Swarm Optimization (TSPSO)" to simplify the energy and the task end time, which is portrayed on the similar objective of this manuscript contribution. The term DAG is utilized to depict reliance on the task. And the tasks mapped through the processing component from diverse data center and then produce a group of "Pareto optimal solution" for the selection of user. Finishing time of the task is the main objective. The method TSPSO endeavored to lessen the consumption of energy and count of unsuccessful tasks. Nevertheless, this method is showing the procedure intricacy and optimizing the completion of task by a suitable scheduling of resources which is probabilistic. Moreover, the TSPSO is overlooked to optimize the task compositions from the task sequences of the distributed workflows.

In the view of the constraints observed from the contributions of the recent literature, this manuscript aimed to optimize the task composition in regard to distributed workflows that considering the lifespan (deadline), task sequence coverage, and dependency scope as multiple objectives to compose the task sequences from distributed workflows submitted in parallel process. The critical objective of the proposal is to achieve linear complexity in task composition and execution, which is enabled by enabling the independent task sequences to execute in parallel.

4 Multi-objective Optimization of the Task Sequence Composition

This section explores the methods and materials used in multi-objective optimization of the task sequence composition for distributed workflows of the Cloud. The initial step of the MOCT finds all possible task sequences of size 1 to n and lists in descending order of their coverage. Further sort these task sequences in ascending order of the lifespan of the task sequences. On other dimensions, these task sequences enlist in descending order of dependents count, which can often be zero. Further, these task sequences ordered in regard to the coverage of task sequences in respective workflows, lifespan, and the count of dependent tasks are referred further with divergent indices in respect of the ordered list, such that each task sequence will have an index in regard to each of these ordered lists. The index begins at 1 and increased by 1 in the sequence of the task sequence. Further, the process of estimating the consistency of each task sequence in regard to multiple objectives is called task sequence coverage, lifespan, and count of dependent task sequences. The detailed exploration of these phases is depicted in the following sections.

4.1 Notifying the Task Sequence Coverage

This section explores the process of notifying task sequences and their coverage. Initial phase of the proposed model is to identify the task sequences of size 1 and unrelated workflows (no need to search) respective to each discovered task sequence of size 1. Further, the search will be conducted recursively for two-size task sequences to max possible size of task sequences. The detailed process related to identify the task sequences of variable size is explored as follows:

All tasks $E = \{e_1, e_2, \ldots e_m\}$ of size m are a finite set that appears in the distributed workflows given. Consider the distributed workflows $\{q_1, q_2, q_3, \ldots, q_n\}$ as set Q of size n, such that each workflow formed by the subset of tasks in set E, such that each workflow $\{q_r \exists q_r \in Q \wedge 1 \leq r \leq n\}$ is the superset of one or more query tasks QE. Each workflow is notified by a unique workflow id $\{r \exists 1 \leq r \leq n\}$, which referred further as qid. Each task e_j in a given workflow represents task e with id $1 \leq j \leq m$.

Further, the set E that having all unique tasks exists in all of the given distributed workflows is considered as the set of task sequences of size 1. Further, it determines their coverage and list of workflows that are not having the respective task as follows.

$qp_1 \leftarrow E$ //clone the set of all tasks as 1-size task sequences set qp_1
$qps \leftarrow qp_1$ //The set qps retains all possible task sequences discovered that is initialized by moving all task sequences of size 1
step 1.

$$\overset{m}{\underset{i=1}{\forall}} \{p_i \exists p_i \in qp_1\} \qquad \qquad \text{(Begin)}$$

a. $o(p_i) = 0$ //denotes the coverage of task sequence p_i that initialized to 0;

b. $o(p_i) = \sum_{j=1}^{n} \{1 \exists p_i \subseteq q_j \wedge q_j \in Q\}$

c. $tlq(p_i)$ //is an empty list containing the workflows those not having task sequence p_i

d. $\overset{n}{\underset{j=1}{\forall}} \{tlq(p_i) \leftarrow q_j \exists p_i \not\subset q_j \wedge q_j \in Q\}$ //moving all workflows those not having task sequence p_i.

step 2. End //of step 1

The notations used in the algorithm are as follows:

step 1. $s = 2$ //represents the length of task sequences to be discovered in sequence, which is initialized by 2. Since the algorithm uses the task sequences of size 1 as input.

step 2. tlp //Tabu list of task sequences:

step 3. list of unrelated workflows $tlqc$:

step 4. Prepare list of unrelated Workflows tlt: workflows that are having one or zero items from task sequences of length 1

step 5. Main Loop: While (True) Begin

step 6. Pick $s - 1$ size task sequences

step 7. $\overset{|qp_{(s-1)}|}{\underset{j=1}{\forall}} \{p_j \exists p_j \in qp_{(s-1)} \wedge p_j \notin tlp\}$ Begin //for each task sequence p_j in $qp_{(s-1)}$

step 8. $\overset{|qp_{(s-1)}|}{\underset{k=1}{\forall}} \{p_k \exists p_k \in qp_{(s-1)} \wedge j \neq k \wedge p_k \notin tlp\}$ Begin //for each task sequence p_k in $qp_{(s-1)}$, such that $j \neq k$

step 9. $p_{jk} = p_j \cup p_k$ //results new task sequence p_{jk} that contains all unique tasks in p_j and p_k

step 10. $o(p_{jk}) = \sum_{j=1}^{n} \{1 \exists p_{jk} \subseteq q_j \wedge q_j \in Q\}$ //find the occurrence of task sequence p_{jk}

step 11.

$$if(o(p_{jk}) > 0) \hspace{4cm} \text{(Begin)}$$

step 12. $qp_s \leftarrow p_{jk}$ //move task sequence p_{jk} to the task sequence set qp_s

step 13. $\overset{n}{\underset{j=1}{\forall}} \{tlq(p_{jk}) \leftarrow q_j \exists p_{jk} \not\subset q_j \wedge q_j \in Q\}$ find all tabu workflows related to task sequence p_{jk}

step 14. End //of step 11

step 15. Else $tlp \leftarrow p_{jk}$ //move task sequence p_{jk} to unrelated task sequence list tlp, since the task sequence does not exist in workflow

step 16. End //of step 8

step 17. End //of step 7

step 18. $if(|qp_s| > 0)$ begin //If qp_s is not empty

step 19. Update tabu list of Workflows: workflows that are having one or zero task sequences from qp_s list

step 20. $qps \leftarrow qp_s$ //moving all task sequences from qp_s to qps

step 21.

$$s = s + 1$$

step 22. End //of step 19

step 23. Else break the main loop in step 5

step 24. End //of step 5

The process, explored in step 1 to step 24, discovers all possible task sequences as set *qps* and their respective coverage. Further, this set *qps* used as input to allocate discrete ranks to all task sequences about coverage, max search space required, and access cost required respectively. The discrete rank allocation for each task sequence is explored in the following sections.

4.2 Dependent Scope

The count of task sequences that are superset of the given task sequence are referred as dependent scope of the respective task sequence. Estimation of dependent scope is as follows

$$\overset{|qps|}{\underset{i=1}{\forall}} \{qp_i \exists qp_i \in qps\} \ \text{Begin}$$

$\quad\quad \overset{|qp_i|}{\underset{j=1}{\forall}} \{p_j \exists p_j \in qp_i\}$ Begin //for each task sequence of length j

$\quad\quad\quad ds(p_j) = 0$ //dependency scope $ds(p_j)$ of the task sequence p_j that initialized to zero.

$\quad\quad\quad\quad \overset{|qps|}{\underset{k=i+1}{\forall}} \{qp_k \exists qp_k \in qps\}$ Begin /for each set of task sequences of length k

$\quad\quad\quad\quad\quad \overset{|qp_k|}{\underset{l=1}{\forall}} \{p_l \exists p_l \in qp_k\}$ Begin

$\quad\quad\quad\quad\quad\quad if(p_j \subseteq p_l) \ \text{Begin}$

$\quad\quad\quad\quad\quad\quad\quad ds(p_j) + = 1$

$\quad\quad\quad\quad\quad\quad$ End

$\quad\quad\quad\quad\quad$ End

$\quad\quad\quad\quad$ End

$\quad\quad\quad$ End

$\quad$ End

4.3 Task Sequence Lifespan

Another significant factor is the lifespan task sequence, which is denoted by the lifespan of a task that involved in the corresponding task sequence and lesser than the lifespan of the other tasks involved in the corresponding task sequence. The process of evaluating the metric is explored as:

$$\overset{|qps|}{\underset{i=1}{\forall}} \{qp_i \exists qp_i \in qps\} \;\; \text{Begin}$$

$$\overset{|qp_i|}{\underset{j=1}{\forall}} \{p_j \exists p_j \in qp_i\} \;\; \text{Begin //for each task sequence of length } j$$

$ls(p_j) = \infty$ // lifespan of the respective task sequence p_j initialized to the maximum (infinity)

$$\overset{|p_j|}{\underset{k=1}{\forall}} \{t_k \exists t_k \in p_j\} \;\; \text{Begin //for each task sequence of length } j$$

$$if \left(ls(t_k) < ls(p_j) \right) \;\; \text{Begin}$$

$$ls(p_j) = ls(t_k)$$

End

End

End

End

4.4 Task Sequence Consistency Score

This section adopts the deviation between the indices of the task sequences in respective of the multi-objectives of each task sequence to estimate the task sequence consistency score of the respective task sequences. The consistency score for each task sequence can be defined as follows.

The eligible task sequences are being listed as a set *aqps*. Further, sort the list of task sequences *aqps* in ascending order of their task sequence coverage and then register the index of each task sequence in regard to the sorted set *aqps*. Further, sort the task sequences of the set *aqps* in ascending order of the dependency scope and then register the indices of the task sequences in the set *aqps* that sorted in ascending order of the dependency scope. Afterward, sort the task sequences of the set *aqps* in descending order of their lifespan and register the indices of the task sequences in regard to the set *aqps* that sorted in descending order of the lifespan. Further, assess the TSCS of each task sequence from the respective indices assigned to the corresponding task sequences in regard to the coverage, dependency scope, and lifespan.

$$\overset{|aqps|}{\underset{i=1}{\forall}} \left\{ p_i \exists p_i \in aqps \right\} \text{Begin / for each task sequence } p_i \text{ in set } aqps$$

$$I(p_i) = \frac{I_c(p_i) + I_{ds}(p_i) + I_l(p_i)}{3} \text{ // Finding the average of the indices of}$$

the task sequence p_i, which are assigned in regard to coverage, dependency scope, and lifespan

$$d(p_i) = \frac{1}{3} X \left\{ \begin{array}{l} \sqrt{\left(I(p_i) - I_c(p_i)\right)^2} + \\ \sqrt{\left(I(p_i) - I_{ds}(p_i)\right)^2} + \\ \sqrt{\left(I(p_i) - r_i(p_i)\right)^2} \end{array} \right\} \text{ //Finding the deviation of the multi-}$$

objectives assigned to respective task sequence p_i

$$cs(p_i) = \left(\frac{d(p_i)}{I_c(p_i)}\right)^{-1} \text{ //The task sequence consistency score is estimated}$$

as inverse ratio of the mean square distance against the rank assigned for coverage.

End

The task sequence consistency score is estimated as the ratio of deviation against the index assigned to task sequence in regard to coverage. This is due to the equality of the deviation obtained for the distinct lowest indices that is equal to counter measure of deviation from distinct highest indices assigned in regard to multi-objectives.

These task sequences scheduled further in the descending order of their task sequence consistency score. The scheduling of these task sequences is parallel, which is in the context of first-come-first-serve scheduling strategy.

5   Experimental Setup and Empirical Analysis

5.1   Experimental Study

The experimental study is aimed to compare the outcomes of the resource scheduling strategy MOCT that is proposed and other contemporary models of TSPSO (Task Scheduling using a multi-objective-nested Particle Swarm Optimization) [26], which are executing on cloud environment with distributed workflows that simulated with java. The input workflows are synthesized such that no order of priority is applicable to the respective workflows. The experimental study is performed by the proposed and contemporary model of TSPSO.

The proposed model of MOCT and other contemporary models of TSPSO that are considered for performance analysis tend to achieve minimal resource failure ratio and minimal delay in job completion. Hence, in this regard, the fitness of these models is assessed by comparing the results obtained for the metrics: (i) load disparity ratio (LDR), (ii) response failure ratio, and (iii) ratio of workflow completion delay.

5.2 The Performance Analysis

During the simulation process, event log updates carried against each attempt of composing and executing the task sequences from distributed workflows scheduling virtual machines as resources. The values for each of the above-stated metrics were observed at fixed intervals of simulation time, which are assessed from the values listed in event log. The comparative study compares the values obtained for all of these metrics during the composition and execution of the tasks from given distributed workflows by the proposed MOCT and contemporary TSPSO models [26].

The response failure observed for MOCT at fixed intervals is in the range of 1–1.5% that certainly less than the response failure observed for TSPSO, which in range of 1.5–5%. The evinced results indicate that the proposed MOCT is minimal and stable in regard to response failure ratio (see Fig. 1).

The load disparity ratio depicted for MOCT and TSPSO was compared in Fig. 2, which is notified that the MOCT having minimal load disparity ratio than compared to TSPSO. The disparity observed for MOCT is 2.8%, and for TSPSO is 5.4%.

The Fig. 3 portraying the workflow completion delay observed from the proposed model MOCT and the contemporary model TSPSO which is significantly low than compared to the ratio of workflow completion delay observed for the other model TSPSO. The average workflow completion delay observed for MOCT is 7% and that observed for TSPSO is 11%.

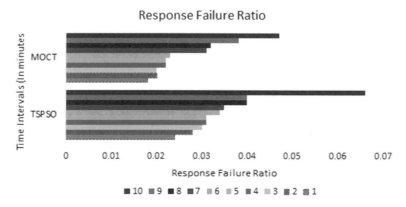

Fig. 1 Response failure ratio observed at different time intervals

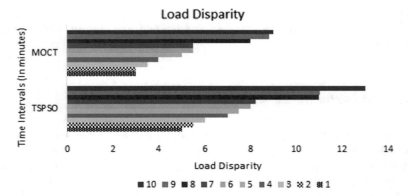

Fig. 2 Load disparity observed at different time intervals

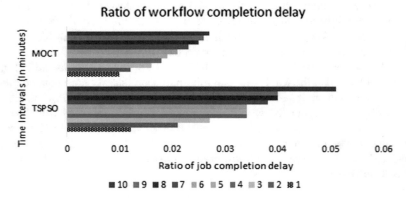

Fig. 3 Ratio of workflow completion delay observed at divergent time intervals

6 Conclusion

The task composition and execution in regard to distributed workflows are the critical research objectives for cloud computing networks. In regard to this, many approaches are depicted in recent literature. However, most of these optimize the task composition in regard to a specific objective such as energy consumption, or deadline based, which is not optimal to reduce the workflow completion failure, and evince the computational complexity as NP-hard. The contribution multi-objective task composition using PSO (TSPSO) [26] is endeavored to achieve the optimality in task composition under multiple objectives such as task priority, delay sensitivity, energy consumption, and deadline constraint. However, the TSPSO is suboptimal to reduce the workflow completion failure and unable to achieve linearity in process complexity. Moreover, the method TSPSO is not intended to compose the tasks from distributed workflows. In contrast to the existing models, the proposed model Multi-Objective Optimization of the Task Composition (MOTC) for distributed workflows has evincing the considerable reduction of the workflow execution failures, which is of by optimizing the task composition under novel objectives, and they are "task sequence coverage, dependency scope, and lifespan" and also evincing the process complexity as linear, which is since the model executes the task compositions in parallel. The future research can eye on evolutionary approaches that use the task sequence consistency score, which is portrayed in this manuscript as fitness function.

References

1. Mell, Peter, and Timothy Grance. 2011. Cloud Computing: Recommendations of the National Institute of Standards and Technology 800–145. NIST, Special Publications.
2. Rodriguez, Maria Alejandra, and Rajkumar Buyya. 2017. A Taxonomy and Survey on Scheduling Algorithms for Scientific Workflows in IaaS Cloud Computing Environments. *Concurrency and Computation: Practice and Experience* 29 (8): e4041.
3. Ullman, Jeffrey D. 1975. NP-complete scheduling problems. *Journal of Computer and System Sciences* 10 (3): 384–393.
4. Yu, Zhifeng, and Weisong Shi. 2008. A Planner-Guided Scheduling Strategy for Multiple Workflow Applications. In *International Conference on Parallel Processing-Workshops, 2008. ICPP-W'08*. IEEE.
5. Yu, Jia, Rajkumar Buyya, and Kotagiri Ramamohanarao. 2008. Workflow Scheduling Algorithms for Grid Computing. In *Metaheuristics for Scheduling in Distributed Computing Environments* 173–214. Berlin, Heidelberg: Springer.
6. Lin, Cui, and Shiyong Lu. 2011. Scheduling Scientific Workflows Elastically for Cloud Computing. In *2011 IEEE International Conference on Cloud Computing (CLOUD)*. IEEE.
7. Zhan, Zhi-Hui, et al. 2015. Cloud Computing Resource Scheduling and a Survey of its Evolutionary Approaches. *ACM Computing Surveys (CSUR)* 47 (4): 63.
8. Mao, Ming, and Marty Humphrey. 2011. Auto-Scaling to Minimize Cost and Meet Application Deadlines in Cloud Workflows. In *2011 International Conference for High Performance Computing, Networking, Storage and Analysis (SC*. IEEE.
9. Malawski, M., et al. 2012. Cost-and Deadline-Constrained Provisioning For Scientific Workflow Ensembles In IaaS Clouds In *Proceedings of the International Conference on High Performance Computing, Networking, Storage and Analysis*, vol. 22.
10. Abrishami, Saeid, Mahmoud Naghibzadeh, and Dick H.J. Epema. 2013. Deadline-constrained workflow scheduling algorithms for infrastructure as a service clouds. *Future Generation Computer Systems* 29 (1): 158–169.
11. Pandey, Suraj, et al. 2010. A Particle Swarm Optimization-Based Heuristic for Scheduling Workflow Applications in Cloud Computing Environments. In *2010 24th IEEE international conference on Advanced information networking and applications (AINA)*. IEEE.
12. Poola, Deepak, et al. 2014. Robust Scheduling of Scientific Workflows with Deadline and Budget Constraints in Clouds. *2014 IEEE 28th International Conference on Advanced Information Networking and Applications (AINA)*. IEEE.
13. Rodriguez, Maria Alejandra, and Rajkumar Buyya. 2014. Deadline Based Resource Provisioning and Scheduling Algorithm for Scientific Workflows on Clouds. *IEEE Transactions on Cloud Computing* 2 (2): 222–235.

14. Chen, Wei-Neng, and Jun Zhang. 2012. A Set-Based Discrete PSO for Cloud Workflow Scheduling with User-Defined QoS Constraints. In *2012 IEEE International Conference on Systems, Man, and Cybernetics (SMC)*. IEEE.
15. Chen, Wei-Neng, et al. 2010. A Novel Set-Based Particle Swarm Optimization Method for Discrete Optimization Problems. *IEEE Transactions on evolutionary computation* 14 (2): 278–300.
16. Liang, Jing J., et al. 2006. Comprehensive Learning Particle Swarm Optimizer for Global Optimization of Multimodal Functions. *IEEE Transactions on Evolutionary Computation* 10 (3): 281–295.
17. Verma, Amandeep, and Sakshi Kaushal. 2014. Bi-criteria Priority Based Particle Swarm Optimization Workflow Scheduling Algorithm for Cloud. In *2014 Recent Advances in Engineering and Computational Sciences (RAECS)*. IEEE.
18. Awad, A. I., N. A. El-Hefnawy, and H. M. Abdel_kader. "Enhanced particle swarm optimization for task scheduling in cloud computing environments." Procedia Computer Science 65 (2015): 920–929.
19. Jain, Richa, Neelam Sharma, and Pankaj Jain. 2017. A Systematic Analysis of Nature Inspired Workflow Scheduling Algorithm in Heterogeneous Cloud Environment. In *2017 International Conference on Intelligent Communication and Computational Techniques (ICCT)*. IEEE.
20. Chandrakala, N., K. Meena, and M. Prasanna Laxmi. 2013. Application of a Novel Time-Slot Utility Mechanism for Pricing Cloud Resource to Optimize Preferences for Different Time Slots. *International Journal of Enhanced Research in Management & Computer Applications*. ISSN 2319-7471.
21. Sindhu, S., and Saswati Mukherjee. 2011. Efficient Task Scheduling Algorithms for Cloud Computing Environment. In *High Performance Architecture and Grid Computing* 79–83. Berlin, Heidelberg: Springer.
22. Butakov, Nikolay, and Denis Nasonov. Co-evolutional Genetic Algorithm for Workflow Scheduling in Heterogeneous Distributed Environment. In *2014 IEEE 8th International Conference on Application of Information and Communication Technologies (AICT)*. IEEE.
23. Visheratin, Alexander, et al. 2015. Hard-Deadline Constrained Workflows Scheduling Using Metaheuristic Algorithms. *Procedia Computer Science* 66: 506–514.
24. Liu, Li, et al. 2017. Deadline-Constrained Coevolutionary Genetic Algorithm for Scientific Workflow Scheduling in Cloud Computing. *Concurrency and Computation: Practice and Experience* 29 (5): e3942.
25. Singh, Lovejit, and Sarbjeet Singh. 2014. Deadline and Cost Based Ant Colony Optimization Algorithm for Scheduling Workflow Applications in Hybrid Cloud. *Journal of Scientific & Engineering Research* 5 (10): 1417–1420.
26. Jena, R.K. 2015. Multi Objective Task Scheduling in Cloud Environment Using Nested PSO Framework. *Procedia Computer Science* 57: 1219–1227.

IP Traceback Through Modified Probabilistic Packet Marking Algorithm Using Record Route

Y. Bhavani, V. Janaki and R. Sridevi

Abstract Denial of Service attack is precarious causing inconvenience to the legitimate user. Distributed Denial of Service (DDoS) attacks can be traced by finding the attack paths from victim to the attacker. We propose a probabilistic packet marking algorithm using record route feature, i.e., options field of a IPv4 packet header is used. In this field, information of the router is marked and the subsequent routers information is XoRed to get the edge information of attack path. Our approach when compared to other methods finds the attackers with less number of packets and without any false positives and false negatives if the network topology is known.

Keywords Denial of service attack · Options field · IP traceback · Packet marking

1 Introduction

In Denial of Service (DoS) attack, the attacker denies the services of a legitimate user by flooding the route with large amount of packets. In case of DoS attack, tracing the attacker is a pressing problem because of IP spoofing. There are many techniques like input debugging [1], controlled flooding [2], ICMP traceback [3], logging [4, 5], and IP traceback to trace the attacker. Among these, IP traceback is the better technique to trace the attacker even in case of IP spoofing.

Y. Bhavani (✉)
Department of Information Technology, Kakatiya Institute of Technology and Science, Warangal, India
e-mail: yerram.bh@gmail.com

V. Janaki
Department of Computer Science, Vaagdevi College of Engineering, Warangal, India
e-mail: janakicse@yahoo.com

R. Sridevi
Department of Computer Science, Jawaharlal Nehru Technological University Hyderabad, Hyderabad, India
e-mail: sridevirangu@yahoo.com

© Springer Nature Singapore Pte Ltd. 2020
K. S. Raju et al. (eds.), *Proceedings of the Third International Conference on Computational Intelligence and Informatics*, Advances in Intelligent Systems and Computing 1090, https://doi.org/10.1007/978-981-15-1480-7_40

481

IP traceback can be performed either probabilistically using probabilistic packet marking algorithm or deterministically using deterministic packet marking algorithm. In our approach, we use the probabilistic packet marking approach because it finds the path from victim to source rather than just the source as in case of deterministic packet marking algorithm. In deterministic packet marking, only ingress router participates in marking the packets and if it gets compromised, then it is difficult to find the attacker, whereas in probabilistic packet marking algorithm, at least a part of the attack path is known.

In this paper, a novel probabilistic packet marking algorithm has been proposed which uses the options field of IPv4 (RFC 791) to store the IP address of the router called as marking information into the packet [6]. This field is chosen compared to Identification field as the former can store more number of bits whereas the latter can store only 16 bits. Our experimental results show that attack path can be constructed with zero false positive and false negative rates and less overhead on the router. The same can also be extended to IPv6 using hop by hop extension header to store the marking information.

The organization of the paper is as follows; Sect. 2 covers the related work. Section 3 elucidates proposed work and Sect. 4 explicates the experimental results. Section 5 concludes our work.

2   Related Work

IP traceback is the process of finding the attack path from victim to source. Many techniques have been proposed with some limitations. Probabilistic packet marking [2, 7–11] is one of the techniques to find the attack path and deterministic packet marking [12–14] is another technique to find the IP address of the ingress router nearer the attacker.

Burch and Cheswick [2] originally proposed probabilistic packet marking algorithm but Savage et al. [7] later modified and called as Fragment Marking Scheme (FMS). In this technique, the IP address of the router and its hash value is fragmented into eight fragments. These fragments are sent through the Identification field probabilistically. The receiver after receiving these fragments generates the full IP address by combining all the partial parts. This procedure results in more false positives due to more combinations. Due to overwriting of marking information by downstream routers the farthest router information may not be received by the victim. This results in false negatives.

Probabilistic packet marking (MPPM) algorithm using Chinese Remainder Theorem [8, 9] is efficient than FMS as it finds a unique value for the IP address of each router using the Chinese Remainder Theorem (CRT). This unique value fragmented is sent as marking information. The receiver after receiving these fragments can combine the corresponding parts with very less false positives and false negatives.

Snoeren et al. [15] in their Source Path Isolation Engine (SPIE) approach finds the attacker with a single packet but requires a large amount of storage space at the router for logging the upstream routers information. This requires change in hardware.

In Fast Internet Traceback (FIT) [16] approach, a hash value of the router is sent instead of the IP address itself. This methodology has less overhead on the router but the disadvantage is the victim needs to find the attack path two times before and after the attack.

Yasin nur et al. [17] in their methodology use the options field of IPv4 to send the marking information to the victim. As this approach appends up to nine IP addresses of the routers in the options field, it may exceed the path MTU limit [RFC 7126] and results in every chance of dropping the packet. In our approach, we use the same field for storing the 32 bit IP address of the router reducing the chance of exceeding the path MTU limit.

Andrey Belenky et al. [12] in their methodology use the deterministic packet marking algorithm. As only ingress router marks the packet, there is more overhead at the ingress router and if it gets compromised, the attacker cannot be identified.

Flexible deterministic packet marking (FDPM) [14] algorithm though have above disadvantages, it is agile as the marking procedure is applied only to the packets when the threshold value crosses the limit value with the packet flow.

3 Proposed Work

In DoS attacks, it is inevitable to find the entire attack path in place of finding only the source of the attack, as the attacker may spoof the IP address. To find the attack path, the victim has to get the information of all the participated routers of the attack path. Hence, the router's identity must be sent through the packets. Savage et al. suggested a methodology to send the router's identity as marking information in the Identification field of the IPv4 packet header. As the Identification field length is 16 bits, 32 bit IP address of the router cannot fit into this field. This results in fragmentation of IP address. At the victim, all these fragments must be properly combined to get the correct IP address of the routers. Failing, it creates false combinations in turn false positives. To overcome the above problem, options field can be used to store the marking information without fragmentation as its length is expandable. The IPv4 packet header is as shown in Fig. 1. According to record route feature (RFC 791), the options field in IPv4 can be used to send the IP address of the router and distance as the marking information to the receiver. The receiver after receiving this information reconstructs the attack path from victim to source.

VER	HLEN	TOS			TOTAL LENGTH	
IDENTIFICATION				FLAGS		FRAGMENT OFFSET
TIME TO LIVE		PROTOCOL		HEADER CHECKSUM		
SOURCE IP ADDRESS						
DESTINATION IP ADDRESS						
OPTIONS(IF ANY) ---						
DATA ----						

Fig. 1 IPv4 Header

3.1 Packet Marking Algorithm

Packet marking procedure is executed at each router in the attack path. Each router when it receives a packet from its upstream router R_j generates a random number

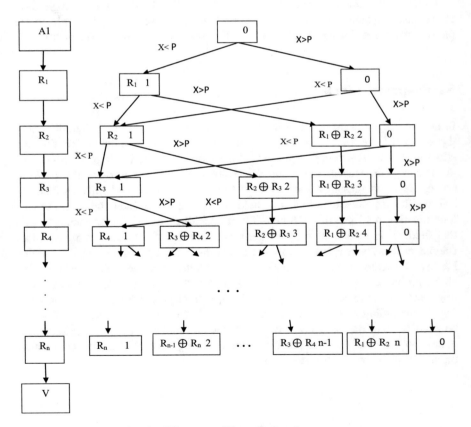

Fig. 2 Procedure showing the different possible marked packets

X between [0..1). If X is less than the marking probability P and flag is equal to 0, then the router address and distance as 1 is stored in the options field of IPv4. If X not less than P and distance d is equal to 0, then R_i is XORed with R_j and flag is set to 1. If flag is set to 1, distance is incremented by 1 which gives the distance of the router R_i to victim V. The procedure is explained in Fig. 2.

Marking procedure at a Router $R_{i:}$

for (each packet p received from its upstream router R_j)

  $X \leftarrow rand()$;

  if $(X < P$ and flag=0) then

           options field $\leftarrow R_i$

           d $\leftarrow 1$

  else

    if (d = 1) then

           $R_i \leftarrow R_i \oplus R_j;$

           set flag to 1;

      endif

    if (flag = 1) then

           increment distance by 1

    endif

  endif

endfor

3.2 Reconstruction Procedure

In this procedure, the victim after receiving the packets separates the marked and unmarked packets. The marked packets are used to construct the path from victim to source and stored in a result table called RT. The packet having distance $d = 1$ contains R_n address. The packet with distance $d = 2$ contains $R_n \oplus R_{n-1}$, so as to obtain R_{n-1}, we perform $R_n \oplus R_n \oplus R_{n-1}$. Similarly, after obtaining R_{n-1} address, this address is XoRed with $R_{n-1} \oplus R_{n-2}$ as $R_{n-1} \oplus R_{n-1} \oplus R_{n-2}$ to obtain R_{n-2}. This process is repeated until the source router R_1 is obtained. In this way, the path from $R_n \leftarrow R_{n-1} \leftarrow R_{n-2} \leftarrow \dots \leftarrow R_2 \leftarrow R_1$ is obtained.

Reconstruction procedure at victim v

for (each packet p received by v)

    store R_i and d in a result table RT

    if (d=1)

   Result ← R_i

    else

    while (there is an entry R_j in RT with distance d-1)

        $R_i = R_i \oplus R_j$

        if (R_i in T)

                append R_i to result

                break

        endif

    endwhile

    endif

endfor

print the result /* Result gives the path from victim to source

4 Experimentation and Results

The following network given in Fig. 3 is considered for experimentation. The router in the attack path marks the packet by placing the IP address of the router and distance as marking information into the options field (as explained in Sect. 3.1). The receiver after receiving this marking information performs the reconstruction procedure and constructs the attack path (as explained in Sect. 3.2).

We compared our results with FMS [7], FIT [16] and Yasin et al. [17] approach. Our approach requires very less number of packets as IP addresses are not fragmented as in FMS and FIT. In FMS, the IP address and the hash value of the IP address are fragmented and sent as eight fragments and FIT requires four fragments, whereas our approach requires only single packet. Though Yasin et al. approach also requires single packet, it stores nine addresses in the options field that may exceed the path MTU limit [18] and has every chance of dropping the packet in the DDoS scenario.

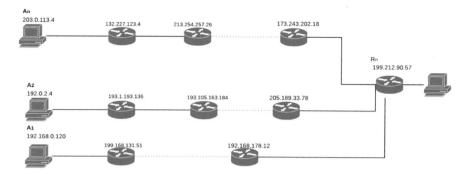

Fig. 3 The network showing the multiple attack paths

In FMS, Savage et al. provided that the victim requires $E(X)$ number of packets to construct an attack path.

$$E(X) < \frac{K * \ln(K * d)}{p(1-p)^{d-1}}$$

where K is the number of fragments, d is the number of hops (number of nodes) from victim to the router, p is the marking probability. In their approach, there is a large deviation in the probability of getting the farthest router's information and nearest router's information. This deviation is given by $\ln(k * d)$[7].

In FIT, the expected number of packets required to reconstruct the attack path is given by

$$E(X) < \frac{K}{p(1-p)^{d-1}}$$

In their approach, the probability of the farthest and nearest routers information is almost the same, only the requirement is the map has to be constructed twice.

In our approach, the excepted number of packets required to reconstruct the attack path is

$$E(X) < \frac{1}{p(1-p)^{d-1}}$$

here, $k = 1$, since we are sending the IP address as a whole in the option field of IPv4 packet without fragmentation. The result is as shown in Fig. 4.

In our approach, the false positive and false negative rates are zero when the network topology is known. The false positives occur due to false combinations but in our approach as the IP address is not fragmented, hence no combination as in FMS [7] and FIT [16].

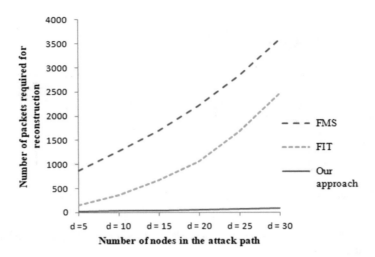

Fig. 4 Comparision of FMS, FIT and our approach in terms of number of packets needed to reconstruct the path

5 Conclusion

This paper presents a novel IP traceback solution to find the attack path from victim to source by storing the IP address of participated routers in its entirety (without fragments) as the marking information in the options field. This facilitated the victim to construct the attack path with a very less number of packets. In our approach, there is less overhead at router as there is no need of fragmentation of IP address and further at victim, no overhead of combining, when compared to FMS and FIT approaches. The false positive and false negative rates are zero if the network topology is known. This can also be extended to IPv6 addressing by storing the marking information in hop by hop extension header without much change done to the packet marking procedure and reconstruction procedure.

References

1. Stone, R. 2000. Center Track: An IP Overlay Network for Tracking DoS Floods. In *9th USENIX Security Symposium*, 199–212.
2. Burch, H., and B. Cheswick, 2000. Tracing Anonymous Packets to Their Approximate Source. In *USENEX LISA. New Orleans*. Lousiana, EUA, 319–327.
3. Bellovin, S., M. Leech, and T. Taylor. 2003. ICMP Traceback Messages. *Internet Draft Draft-Bellovin-Itrace-04.txt*.
4. Wang, Yulong, Sui Tong, and Yi Yang. 2012. A Practical Hybrid IP Traceback Method under IPv6. *Journal of Convergence Information Technology (JCIT)* 7: 173–82.

5. Kamaldeep, Manisha Malik, and Maitreyee Dutta 2017. Implementation of Single-Packet Hybrid IP Traceback for IPv4 and IPv6 Networks. *IET Information Security*, 1–6.
6. RFC 791 Available from: https://tools.ietf.org/html/rfc791.
7. Savage, Stefan, David Wetherall, Anna Karlin, and Tom Anderson. 2001. Network Ssupport for IP Ttraceback. *IEEE/ACM Transactions on Networking*, 9: 226–37.
8. Bhavani, Y., V. Janaki, and R. Sridevi. 2015. IP Traceback Through Modified Probabilistic Packet Marking Algorithm Using Chinese Remainder Theorem. *Ain Shams Engineering Journal* 6: 715–722.
9. Bhavani, Y., V. Janaki, and R. Sridevi. 2017. Modified Probabilistic Packet Marking Algorithm for IPv6 Traceback Using Chinese Remainder Theorem. In *Innovations in Computer Science and Engineering,* ed. H. Saini, R. Sayal, and S. Rawat, 253–263. Singapore: Springer.
10. Vasseur, Marion, Xiuzhen Chen, Rida Khatoun, and Ahmed Serhrouchni. 2015. Survey on Packet Marking Fields and Information for IP Traceback. In *International Conference on Cyber Security of Smart cities, Industrial Control System and Communications (SSIC)*, 1–8.
11. Balyk, Anatolii, Uliana Latsykovska, Mikolaj Karpinski, Yuliia Khokhlachova, Aigul Shaikhanova, and Lesia Korkishko. 2015. A Survey of Modern IP Traceback Methodologies. In: *8th IEEE International Conference on Intelligent Data Acquisition and Advanced Computing Systems*, 484–488.
12. Belenky, Andrey, and Nirwan Ansari 2003. IP Traceback with Deterministic Packet Marking. *IEEE Communications Letters* 7: 162–164.
13. Parashar, Ashwani, and Ramaswami Radhakrishnan. 2013. Improved Deterministic Packet Marking Algorithm. In *15th International Conference on Advanced Computing Technologies*.
14. Yu, S., W. Zhou, S. Guo, and M. Guo. 2016. A feasible IP Traceback Framework Through Dynamic Deterministic Packet Marking. *IEEE Transactions on Computers* 65: 1418–1427.
15. Snoeren, A.C., C. Partridge, L.A. Sanchez, and C.E. Jones, F. Tchakountio, and S.T. Kent. 2001. Hash-based IP Traceback. In ACM SIGCOMM, 3–14.
16. A. Yaar, A. Perrig, and D. Song. 2005. FIT: Fast Internet Traceback. In *Proceeding IEEE INFOCOM*.
17. Yasin Nur, Abdullah, and Mehmet Engin Tozal 2018. Record route IP traceback, Combating DoS attacks and the variants. *Computers & Security*, 72: 13–25.
18. RFC 7126 Available from: https://tools.ietf.org/html/rfc7126#section-4.5.

Necessity of Fourth Factor Authentication with Multiple Variations as Enhanced User Authentication Technique

K. Sharmila and V. Janaki

Abstract Authentication plays a vital role in granting access to a user by any computing system. There are several existing authentication techniques used for different scenarios which include one-factor, two-factor, and three-factor authentication, where the user submits required credentials and gains access to the system. When a legitimate user is unable to provide his credentials at any point of time, the user becomes unauthenticated. At this juncture, we have proposed a fourth factor for authentication where a legitimate user can seek assistance from any of his close associates like a spouse, friend, and colleague and can gain access to the system for a single time. In this paper, we consider multiple variations of fourth-factor authentication and compare them within the framework and explore the mechanism by applying various metrics such as key size, the ease of use, time and space complexity of the algorithms, and security considerations.

Keywords Authentication · Fourth factor · Diffie–Hellman key exchange · AES algorithm · Chinese remainder theorem

1 Introduction

Designing any authentication technique requires the integration of various aspects of computer science and other associated sciences. Depending on the type of technique chosen, it requires design and analysis of protocol, security features, cryptographic calculations, device dependency, etc. The main motivation of any authentication scheme is to ensure that the identification of the user is proved to the computer and access is granted to the user for further transactions. The authentication techniques are mainly available in three categories [1–3].

K. Sharmila (✉) · V. Janaki
Department of CSE, Vaagdevi College of Engineering, Warangal, Telangana, India
e-mail: sharmilakreddy@gmail.com

V. Janaki
e-mail: janakicse@yahoo.com

© Springer Nature Singapore Pte Ltd. 2020
K. S. Raju et al. (eds.), *Proceedings of the Third International Conference on Computational Intelligence and Informatics*, Advances in Intelligent Systems and Computing 1090, https://doi.org/10.1007/978-981-15-1480-7_41

1. Knowledge-based—Something you know: It is also known as one-factor authentication, where the user provides any known information such as PIN or password to the system. If the password given by the user is correct then user is granted access.
2. Possession-based—Something you have: It is also known as two-factor authentication, where the user submits both token and PIN to the system. If both the factors are true, only then the access is granted.
3. Biometric-based—Something you are: It is known as three-factor authentication, where the user has to submit his two factors along with his biometrics like finger print, retina, voice etc.

The efficiency of any authentication mechanism depends on the factors like distribution of keys over the network, space required to store, time required for execution, cost of implementation, and ease of usage by the users.

When the authentication factors are not available with the legitimate user, he becomes unauthenticated and cannot access the system, even though he is legitimate. In such scenario, we propose a novel approach where a trustee helps the unauthenticated legitimate user to enable one emergency transaction using fourth factor for authentication. We have implemented this scenario with multiple variations and algorithms and compared them with some evaluation metrics. The pros and cons of each algorithm are addressed and explained in detail in the next sections.

The paper is organized as follows. Section 2 covers related work, Sect. 3 elucidates our proposed work, Sect. 4 represents evaluation metrics, and Sect. 5 concludes the proposed work.

2   Related Work

Passwords are the most commonly used factors of authentication even today. They are considered easy to use but are easily guessable [3]. These attacks can be overcome by changing the passwords frequently. Moreover, they have to be memorized by the user, which again becomes an overhead. In case, the user is not able to recall the password, backup authentication mechanisms help in regaining them with pre-registered e-mail or phone number [4]. Backup authentication is implemented by posing a unique question known only to the user like his pet name, school name, mother's maiden name, etc. [2]. But since this information is public, it is not considered secure.

To overcome these disadvantages, smart cards were introduced. They can store user's details and when submitted, identifies the user [4]. Smart cards provide authorization, security, and integrity to the user's data. Authentication becomes a problem if the user cannot recollect his PIN or if the card is stolen or lost [2].

The disadvantages of one-factor and two-factor authentication techniques can be defeated by implementing biometrics [4, 5]. Identification of a person is done based on his physiological characteristics and has become an important factor of authentication. Since every person has unique biometric features and do not match with other person's characteristics, it is considered as unique. The system compares these physiological features such as fingerprint, retina, iris, and voice with the database and determines authenticity of the user. The biometric device in few cases may reject a valid user also.

Brainard et al. [2] have proposed and designed fourth-factor authentication mechanism based on social relationships. This protocol provides authentication, based on social relationships. However, the drawback is that the system is prone to social engineering attacks and user impersonation attacks.

Schechter et al. [6] designed a prototype based on trust-based social authentication system [7, 8]. In their proposal, the trustee receives an e-mail from the account recovery system and the trustee visits the respective Web page and vouches for the asker's identity. But accessing an e-mail may not be practically possible by a trustee all the time.

If the authentication factors of the user are unavailable and the biometrics does not match, authentication is not possible. In such situation, we propose a new method for authentication, the fourth factor, somebody the user knows. This known user is termed as helper and will assist the unauthenticated user for making an emergency transaction. We have implemented this mechanism using various algorithms and system requirements. The advantages and disadvantages of each are discussed below.

3 Proposed Work

Human beings continue to maintain social relationships and trust plays a vital role in the society. Authenticating one person by other is possible based on his characteristics, and sometimes with the assistance of a trusted person too. But, when machine fails to recognize an authorized person, the trustee (helper) takes the social responsibility of assisting the user in emergency in case any or all the three existing authentication mechanisms fail. The helper in this regard is termed as fourth factor of authentication. We discuss the necessity of fourth factor in different situations by sharing the secret key using Diffie–Hellman Key Exchange (DHKE), sharing the symmetric key between asker and helper using AES algorithm [9]. The same is also extended to multi-user joint accounts by applying Chinese Remainder Theorem (CRT) with multiple users [10–14]. The user without any authentication factors is referred to as asker and the trustee is known as helper. We compare the requirements, implementation details and the efficiency of the algorithms. The implementation of emergency fourth-factor authentication protocol is described as follows.

A legitimate user performs the following steps to gain access to the system.

1. The user logs into the respective bank Web page or visits any trusted platform.
2. The user tries to make a transaction with his available credentials.
3. If any of them do not match for consecutive three times, the card will be blocked.
4. The asker can select an option to use emergency authentication.
5. The helper's list will be displayed on the trusted platform.
6. When the asker selects one helper, a secret question is sent to both asker (As) and helper (Hp).
7. If the asker and helper submit correct information, the bank understands that asker and helper are genuine and permits one transaction.

Step 1 to step 4 are common to all algorithms but step 5 to step 7 vary from one algorithm to the other (Fig. 1). We present different scenarios, algorithms used, their advantages, and challenges in the coming sections.

3.1 An Alternative Authentication Methodology in Case of Biometric Authentication Failure: AAP Protocol

Diffie–Hellman key exchange is a well-known secret key exchange algorithm. It is mainly used for exchanging authentication signatures. It permits two users to exchange a "key" over an insecure medium such as the Internet [14]. It achieves forward secrecy, as no information about the key is disclosed. Though there is an

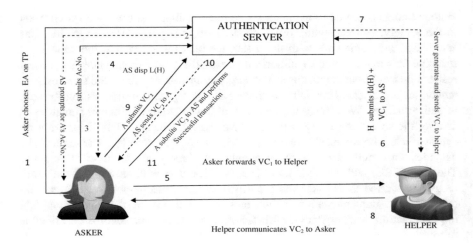

Fig. 1 Model of emergency fourth-factor authentication, "somebody the user knows"

equal chance of interception by the hacker, he will not be able to break the key by applying any mathematical calculations apart from employing the usual brute force method.

We have proposed an alternative solution for authentication by using Diffie–Hellman key exchange algorithm. We make use of a trusted third party who certifies validity of the user, in case of authentication failure. When the valid user fails to access the system due to the failure of third factor authentication with finger injuries, there should be an alternative authentication factor. In this paper, we have proposed a solution where we take help from trusted third party who acts as an interface between the legitimate user and the organization. The third party certifies and authenticates the user in case of failure of biometrics [5, 15].

We have considered a scenario where the user gets injured on his finger or biometrics and does not get authenticated by the system or device. We have proposed an alternative model for authentication with the help of a trusted third party to gain access to the system in emergency for one time.

Implementation of emergency authentication technique using Diffie–Hellman key exchange includes the asker and helper communication with a key exchange. Since exchange of keys is prone to brute force attacks, we have increased the security of key exchange by implementing vouching process as a fourth factor of authentication.

3.2 A Novel Approach for Emergency Authentication Using Fourth Factor

To overcome the disadvantage of key exchange, we have implemented AES algorithm and shared the secret among the users [14]. We have proposed a novel approach for authentication, where a trusted person called helper will assist the legitimate unauthorized user to gain access for one time [16]. In our proposal, all the user's information like names, account numbers, mobile numbers, e-mail address, etc. are stored at the time of creating an account at the organization. The trustees are also informed with the necessary information stating that they are acting as helpers for a specified asker [7]. According to this proposal, the prerequisites are:

- The helper should be a preregistered user of the organization.
- At the time of registration with the organization, it is mandatory that the asker should introduce a helper who is a pre-registered user of the same organization, whom the asker feels trustworthy.
- Both the asker and the helper should share public and private key pairs with the organization.

The process gets initiated only when the asker sends a message to the helper, and in turn the helper acknowledges the asker's request. Then, the helper forwards the request to the organization. Hence, there is no chance for non-repudiation at all the ends since there is a notification to all the parties (asker, helper and organization).

When the asker requests the helper, he on behalf of the asker communicates with the organization by providing his credentials to the bank along with the message sent by the asker. Then, the helper requests the organization to generate a temporary code which can be passed from the helper to the asker to make a transaction. The bank then generates a code which is an encrypted form of combination of private and public keys of the two users (asker and helper) and forwards it to the helper. This code is then forwarded from the helper to the asker to prove his identity. The asker submits this forwarded code to the bank. The bank then compares the code and understands that the asker is indeed a genuine user and allows him to make an emergency transaction by providing an OTP. The asker uses this temporary Vouch Code and continues his transaction. This emergency code given by the organization is an OTP which is valid for a period of 24 h and then expires.

We have used AES algorithm [14] and tested the secure fourth-factor algorithm. The disadvantage here is that the key has to be shared between the two communicating parties, which may be prone to attacks. To overcome these attacks, we have implemented Chinese Remainder Theorem.

3.3 Mutual Consent for Joint Account Transactions Using Chinese Remainder Theorem

In this proposal, we have considered a group of joint account holders. To make a transaction, the power is given to one person of the group only. We are proposing a solution by providing consent by all the users of the group. If all the users accept to the transaction, only then it will be executed. We implement the concept of secret sharing, where the secret is divided into shares and distributed among the number of users. This secret can be constructed only when these shares are combined together and cannot be done for less number [17, 18].

The prerequisites of our proposal are as follows

- A set of relative prime numbers is registered at the key distribution center by all the members of the joint account (with mutual consent) denoted as

$$P = \{m_1, m_2, m_3 \ldots m_N\}$$

- A set of random numbers is registered at the bank denoted as

$$Y = \{Y_1, Y_2, Y_3 \ldots Y_N\} \text{ also called as Pre-OTPs}$$

We have executed our algorithm using Chinese Remainder Theorem in Java environment.

Account holders considered = J_1, J_2, J_3

Account holder 1 logs into his system by providing his credentials

KDC selects 3 prime numbers from $P = \{999, 998, 997\}$

KDC selects 3 random numbers from $Y = \{25, 27, 22\}$
A unique value X is generated by applying CRT
$M = 994010994$
$M_1 = 995006$, $M_2 = 996003$, $M_3 = 997002$
$b_1 = 500$, $b_2 = 997$, $b_3 = 499$
$X = 493518013$
Pre-OTP$_1$ = 25, Pre-OTP$_2$ = 27, Pre-OTP$_3$ = 22 are sent to J_1, J_2, J_3 respectively.

All the members of the joint account must send back the received Pre-OTPs to the KDC. The KDC checks the integrity of the Pre-OTPs and the generated value X. J_1 enters this unique value $X = 493518013$ along with his credentials and the transaction becomes successful. When we have implemented our algorithm by the above inputs, the time taken for the entire transaction was very less.

The main aim of this proposal is to take prior consent of every Joint account holder [10, 19]. Our concept is similar to the existing banking system where OTP is generated randomly. But in our proposal, OTP is generated using Chinese Remainder Theorem. Our scheme provides efficiency and there is no chance for repudiation. One limitation of our proposal is that there is a computational overhead at KDC in generating the final OTP.

3.4 Design of Privileged, Emergency Fourth-Factor Authentication Protocol Using Chinese Remainder Theorem

In this proposal, the emergency fourth-factor authentication protocol is implemented using Chinese Remainder Theorem for only two users, the asker and the helper in a trusted platform like an ATM [11, 14]. The asker chooses the emergency authentication (EA) option from the trusted platform (ATM machine). This alerts the authentication server (AS) of the bank and a new session (s) will be initiated. The AS prompts the asker to submit his account number in the TP. Upon submission, the server validates the account number of asker and generates a temporary vouch code VC$_1$ and transmits it to the registered mobile number of the asker in the form of SMS to assist the asker in authentication process [12].

The asker enters this vouch code VC$_1$ into the TP. The AS generates a list of pre-specified helpers and displays this list to the asker, where he selects one of the helpers from the list specified. The AS alerts the helper with an SMS as notification of the process. The asker requests the helper in this regard over phone or in person and shares his VC$_1$ to strengthen the authentication. The helper receives the vouch code VC$_1$, logs into the webpage by his own credentials user id, password. The AS validates the helper using these submitted credentials, VC$_1$ and then generates another vouch code VC$_2$.

The helper communicates the above generated VC_2 to the asker. The asker upon receiving this vouch code VC_2, from the helper, enters it into the trusted platform. The AS validates both VC_1, VC_2 and generates one time vouch code VC_3. Finally, the asker submits VC_3 to the authentication server and performs one successful emergency transaction. All the above steps create a secure trustworthy platform between authentication server, helper and asker. The final vouch code VC_3 is known only to the asker thus enabling him to continue his transaction (withdrawing money successfully from the ATM machine).

4 Evaluation Metrics

In this paper, we have analyzed several variations of the algorithms on emergency fourth-factor authentication, based on the following metrics. We have spotted some differences based on criteria, tabulated as follows (Table 1).

Table 1 Comparision of multiple variations of emergency fourth-factor authentication

Parameter	Algorithm		
	DHKE	AES	CRT
Scenario of usage at asker's side	In case of biometric mismatch	Any of the three authentication factors are unavailable	Any factor except account number is unavailable
Prerequisites	No prior registration of helper	Helper should be pre-registered	Helper should be pre-registered
Key sharing	Key is publicly exchanged	Symmetric key is shared	Secret key (OTP) is generated
Input	Secret key	Account number of user + secret key	Account number of asker
Helper contact	Helper is contacted by any channel	Helper is contacted over phone	An SMS alert will suffice
Inputs	Asker and helper jointly establish a shared secret key	The input is a combination of three parameters, which cannot be easily hacked	Set of shares (OTPs) are distributed among "n" users
Limitation	Authentication of either party involved is not guaranteed	Difficult to continue the process if secrecy key is lost	If any of the share is lost, secret cannot be recovered
Attacks	Man in the middle attack	Very secure	Very secure and attacks can be traced by log details
Execution time of the algorithm with ready inputs	15 s	12 s	5 s

4.1   Testing and User Studies

We have tested our approach by conducting a user study. As a part of our evaluation strategy, we have simulated our application and installed in a trusted platform. The asker who is an authorized failed user has received the information (VC_3) from their helpers in a stipulated time and was permitted to perform one emergency authentication successfully. The entire procedure is executable in a very less amount of time provided, the asker and helper have a prior communication before the commencement of emergency fourth-factor authentication protocol. The results were encouraging as it is not necessary for the user to remember their helpers since the authentication server displays the list of helpers.

The following table shows the applicability of emergency fourth-factor authentication protocol when the user loses any of the authentication factors (Table 2).

5   Conclusion

Every algorithm is unique and has some advantages and disadvantages. From the experimental results and the comparison, we understand that Diffie–Hellman key exchange is used when a secret key has to be shared between only two users. AES algorithm is selected if the user considers confidentiality and integrity as the important factors. AES algorithm is used to prevent the information from guessing attacks and it can be applied on top of all the Internet protocols that are based on IPv4 and IPv6. Chinese Remainder Theorem is used to generate unique values, every time it is executed and only those set of congruence should be used to regenerate those unique values, which overcomes many attacks. We conclude that a helper can be used as a fourth factor of authentication in case any or all the three existing authentication mechanisms fail. We suggest that when machine fails to recognize an authorized person even though he is a legitimate user, the helper takes the social responsibility of assisting him in crisis.

Table 2 Applicability scenario of emergency fourth-factor authentication protocol

S. No.	Authentication			
	Two factor/three factor			Our algorithm
	Token	PIN	Biometric	Fourth factor
1	✓	✓	✓	Not required
2	✓	✗	✗	Can be applied
3	✓	✓	✗	Can be applied
4	✓	✗	✗	Can be applied
5	✗	✗	✓	Can be applied
6	✗	✗	✗	Cannot be applied

References

1. Guo, Cheng, and Chin-Chen Chang. 2014. An Authenticated Group Key Distribution Protocol Based on the Generalized Chinese Remainder Theorem. *International Journal of Communication System* 27 (1): 126–134.
2. Sobotka, Jiri, and Radek Dolze. 2010. Multifactor Authentication Systems, Elektro Revue, 1–7.
3. Sharmila, K., V. Janaki, and K. Shilpa. 2014. An Alternative Authentication Methodology in Case of BiometricAuthentication Failure: AAP Protocol. *International Journal of Engineering Research & Technology*, 3 (1): 3487–3491.
4. Aloul, Fadi, Syed Zahidi, and Wassim El-Haj. 2009. Two Factor Authentication Using Mobile Phones. In *IEEE/ACS International Conference on Computer Systems and Applications,* 641–644.
5. Cormen, Thomas H., Charles E. Leiserson, Ronald L. Rivest, and Clifford Stein. 2001. *Introduction to Algorithms,* 2nd ed. MIT Press and McGraw-Hill.
6. Schechter, S., S. Egelman, and R.W. Reeder. 2009. It's Not What You Know, But Who You Know: A Social Approach to Last-Resort Authentication. *Human Factors in Computing Systems.*
7. Sharmila, K., V. Janaki, and A. Nagaraju. 2013. Enhanced User Authentication Techniques using the Fourth Factor, Some Body the User Knows. In *Proceeding of International Conference on Advances in Computer Science, AETACS,* 255–262. Elsevier.
8. Ghosh, I., M.K. Sanyal, and R.K. Jana, 2017. Fractal Inspection and Machine Learning-Based Predictive Modelling Framework for Financial Markets. *Arabian Journal for Science and Engineering.*
9. Stallings, William. 2002. *Network Security Essentials.* Prentice Hall, 2nd ed.
10. Navya Sri M,M., Ramakrishna Murty, et al. 2017. Robust Features for Emotion Recognition from Speech by using Gaussian Mixture Model Classification. In *International Conference and Published Proceeding in SIST Series,* vol. 2, 437–444. Springer.
11. Blakley, G.R. 1979. Safeguarding Cryptographic Keys. *National Computer Conference* 4: 313–317.
12. Sharmila, K., V. Janaki, and A. Nagaraju 2017. A Survey on User Authentication Techniques. *Oriental Journal of Computer Science & Technology,* 10 (2): 513–519.
13. Iftene, Sorin. 2001. General Secret Sharing Based on the Chinese Remainder Theorem with Applications in E-Voting. *Electronic Notes in Theoretical Computer Science* 186: 67–84.
14. Huang, Xinyi, Yang Ashley Chonka, Jianying Zhou, and Robert H. Deng. 2011. A Generic Framework for Three-Factor Authentication Preserving Security and Privacy in Distributed Systems. *IEEE Transactions on Parallel and Distributed Systems,* 22: 1390–1397.
15. Sharmila, K., V. Janaki, and A. Nagaraju A. 2017. Novel Approach for Emergency Backup Authentication Using Fourth Factor In: *Innovations in Computer Science and Engineering,* ed. H. Saini, R. Sayal, and S. Rawat, 313–323. Singapore: Springer.
16. Mignotte M. 1982. How to share a secret. In *Beth T, Lecture Notes in Computer Science,* 371–75. Springer-Verlag.
17. Shamir, Adi. 1979. How to Share a Secret. *Communications of the ACM* 22 (11): 612–613.
18. Brainard, John, Ari Juels, Ronald L. Rivest, Michael Szydlo, and Moti Yung. 2006. Fourth Factor Authentication: Somebody You Know. In *ACM Conference on Computer and Communications Security,* 168–178.
19. Asmuth, C.A., and J. Bloom. 1983. A Modular Approach to Key Safeguarding. *IEEE Transactions on Information Theory* 29 (2): 208–210.

Performance Analysis of Feature Extraction Methods Based on Genetic Expression for Clustering Video Dataset

D. Manju, M. Seetha and P. Sammulal

Abstract Clustering of high-dimensional video datasets consists of thousands of features, which encounter problems related to feature selection, computational complexities, and cluster quality. Features are extracted from key frames of videos using feature extraction techniques. Feature selection improves cluster quality by removing irrelevant features from the dataset before applying clustering algorithm. This paper compares the implementation of four feature extraction methods namely texture, speeded up robust feature (SURF), wavelet transform, and scale-invariant feature transform (SIFT). For detecting clusters of different sizes and shapes, density-based clustering algorithm (DBSCAN) is used. It becomes unstable when dealing with high-dimensional data. DBSCAN based on feature selection, new methodology like genetic expression has been adopted. The performance of this approach has been verified on approximately 100 real-world high-dimensional video datasets in terms of accuracy and purity. It has been ascertained that applying genetic expression for feature selection outperforms compared to other feature selection method.

Keywords Clustering · WT · Genetic expression · High-dimensional data · SURF · SIFT · Texture

D. Manju (✉) · P. Sammulal
Department of CSE, JNTUH, Hyderabad, India
e-mail: jadavmanju@yahoo.co.in

P. Sammulal
e-mail: sam@jntuh.ac.in

M. Seetha
Department of CSE, G. Narayanamma Institute of Technology and Science,
Hyderabad, India
e-mail: smaddala2000@yahoo.com

© Springer Nature Singapore Pte Ltd. 2020
K. S. Raju et al. (eds.), *Proceedings of the Third International Conference on Computational Intelligence and Informatics*, Advances in Intelligent Systems and Computing 1090, https://doi.org/10.1007/978-981-15-1480-7_42

1 Introduction

The recent developments in the evolution of digital devices have resulted in an increase of digitally stored images and video datasets, and there is a need for tools which can search and organize these visual data automatically by their content. So, there is a mounting need for a more sophisticated automated system for partitioning the datasets into groups or clusters. In image processing, features tend to be informative and lead to better interpretations. Once a feature extraction has been calculated, it often turns out to be of an extremely high dimensionality. The dimensionality reduction technique applied is FastICA. A FastICA-based unsupervised technique is used to evaluate multispectral images [1]. Selecting the subset features or relevant features and removing the noisy, redundant and irrelevant features is the process of feature selection. The technique adopted for this is called genetic expression. The density-based clustering algorithm (DBSCAN) aims to identify areas of different densities and compares the points by specifying a radius "eps." The performance of a clustering algorithm varies in terms of selection of the relevant features. The results are shown by clustering accuracy and clustering purity that are used as a measure for clustering. Better clustering is assured with high clustering accuracy and purity.

This paper is organized as follows: Sect. 2 overviews existing methods, proposed work is shown in Sect. 3, implementation results are shown in Sect. 4, and conclusion is shown in Sect. 5.

2 Literature Survey

2.1 Existing Methods

Some published literatures well-accomplished different feature extraction techniques where three different image matching techniques like SURF, SIFT, and ORB are compared and observed that performance of ORB is the fastest while SIFT perform is best in most scenarios [2]. GLCM method is used for extracting texture parameters. These features are useful for estimation of videos and in real-time pattern recognition applications [3]. Non-Gaussian feature selection method is compared with varFsGD and MML-based approach [4]. For clustering of videos in heterogeneous feature space, use decoupled semantics clustering tree (DSCT) [5]. Dual-graph sparse non-negative matrix factorization (DSNMF) for discriminative clustering, yields superior performance [6]. On noisy images, the performance of different feature extraction approaches like SIFT, HOG, and SURF were analyzed

[7]. Fuzzy c-means clustering is performed for spatial and spectral features [8]. Compared to other methods, like SVM, CART, and Bayesian classifier, HDMR results are efficient and computational cost is reduced for discriminative features [9]. Performance of adaptive feature selection (AFS) approach is compared with principal component analysis (PCA) method using different number of spectral bands [10]. Fast independent component analysis is used for fast speeding [11]. In case of image classification, it is observed that gene expression algorithms acquire good classification accuracy [12]. Several researchers conducted various studies on high-dimensional datasets, and still, clustering in high dimension is a challenging task. Therefore, a new feature selection method is explored.

2.2 Feature Extraction and Feature Selection

Dimensionality reduction is the process of reducing the numbers of features into few. It has two subcategories called feature extraction and feature selection. A feature extraction is the process of transforming a higher-dimensional space to few dimensions, i.e., attribute that has "n"-dimensional features $(x_1, x_2, \ldots x_n)$ and mapping it to lower-dimensional space $(z_1, z_2, \ldots z_m)$ is "m"-dimensional, where $m < n$ and each of these features is some function, of the original feature set which is $f(x_1, x_2, \ldots x_n)$. Therefore, projection of higher-dimensional feature space to lower-dimensional space is essential for easier processing. There are two main streams to adopt for feature extraction. The first one includes methods like SURF, SIFT, HOG, and another includes PCA, ICA, etc. The feature selection process paves the way toward the accomplishment of selecting only the relevant features leaving out the irrelevant features. When the input data are suspected to be redundant, it' is transformed into a reduced set of features called as feature vector. Feature selection methods are of three types: filter method, wrapper method, and embedded methods.

2.3 Image Matching Methods

SURF Algorithm. One of the key methods to extract points of interest along with the ability to locate, track, and recognize objects is speeded up robust feature (SURF) presented by Herbert Bay at the European Conference on Computer Vision. SURF is a patent local feature detector and descriptor. SURF was developed to increase the speed in every step that is good at handling images with blurring and rotation. The algorithm has three main parts: detection of interest points, feature vector descriptor, and matching.

SIFT Algorithm. David Lowe introduced scale-invariant feature transform (SIFT). SIFT is a successful approach for feature detection, for detecting feature points, SIFT uses cascading filtering approach. The steps followed in this technique are, scale–space peak selection, key point localization, orientation assignment, and key point descriptor. The main advantages of SIFT algorithm is scale invariant, and for small objects, many features can be generated and have good efficiency.

Wavelet Transform. Wavelet is a mathematical function which transforms different frequencies both in frequency and spatial range and consists of infinite set of basis functions. Wavelet transforms produce a few significant coefficients for the signals with discontinuities, and better results are observed for nonlinear approximation. The advantage of wavelet transform is signals that are represented accurately compared to Fourier transform.

Texture Features. Texture is useful for discriminating objects from background. It acts a source for inferring object shapes, orientation, and deformation. Features that are collected through texture analysis process are called as texture feature extraction. First-order statistics, second-order statistics, and higher-order statistics are three different types of statistical-based features. Gray-level co-occurrence matrix (GLCM) extracts second-order texture information from images. In GLCM, the number of distinct gray levels or pixel values in the image is represented by the number of rows and columns. The commonly used features in GLCM are: contrast, correlation, entropy, energy, and homogeneity (Table 1).

3 Proposed Algorithm

To explain the methodology, a sample video is considered. These videos contain rich information, and video clustering techniques are needed that group semantically related videos that are highly scalable and organizes the dataset into structured form. The process converts each video into "n" $M * N$ sized frames; the key frames are then identified and extracted from the "n" frames. Here, each key frame has respective feature vector, once a feature vector has been calculated, which is often of high dimensionality. Efficiency and accuracy of the subsequent representation

Table 1 Texture features extracted using GLCM

S. No.	Contrast	Correlation	Energy	Homogeneity	Entropy
1.	0.248139	0.960507	0.097035	0.892006	7.736384
2.	0.293228	0.951307	0.168323	0.885763	7.247822
3.	0.282543	0.878661	0.208029	0.883159	6.95249
4.	0.25337	0.949314	0.123534	0.902443	7.568491
5.	0.202813	0.981299	0.131387	0.903708	7.389349

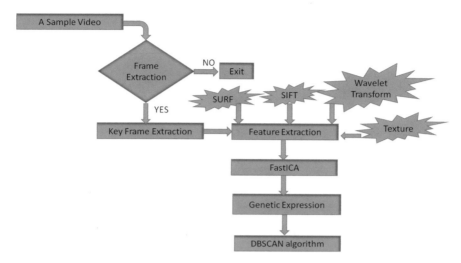

Fig. 1 Flowchart of proposed approach

and clustering can be performed by dimensionality reduction. The dimensionality reduction technique used is FastICA, which is an efficient algorithm consisting of a process called prewhitening the data, single component extraction, and multiple component extraction. Feature selection is done by an algorithm called genetic expression. Here, features are encoded as fixed length genomes or chromosomes. Further, chromosomes are arranged into expression trees which decide the aptness of this solution by adopting fitness function. Finally, for detecting clusters of different sizes and shapes, DBSCAN algorithm is used.

The above procedure is explained in the form of a flowchart as shown below (Fig. 1).

3.1 Genetic Expression

Genetic expression is an evolutionary algorithm that uses population of individuals. Each individual in the population can be represented as a genome or chromosome or expression tree. These programs are expression trees that evaluate the fitness function which determines the best solution for the given problem, by exploring all the paths of the solution space. The higher the fitness value the closer it comes to the best answers [13].

The flowchart of genetic expression is as shown below

The steps used by genetic expression algorithm are:

- For running genetic expression, certain parameters are used, including population size.

$$x = \text{Bird data};$$
$$\text{Population size} = 100;$$
$$\text{No. of selections} = 2;$$

where x is the video dataset loaded, then features are extracted and are later grouped into two groups each.
- Randomly select individuals, where each individual is composed of fitness value or chromosome.
- Fitness function leads to right direction, mark the best and worst individuals.
- Probability is greater for desirable individuals.
- Next, it starts working with two chromosomes chosen randomly. Then, the features are grouped into two groups.
- Evolutionary generation controls the iterations in GE, which is in terms of objective functions.
- When the loop ends, the best individual is the optimum solution that is given as an input to the clustering algorithm.

3.2 DBSCAN Algorithm

DBSCAN is a nonparametric algorithm, which requires two parameters, the "Eps" value and the "Minpts." DBSCAN algorithm cluster regions are based on the concept of reachability. "Eps" is a distance parameter which defines radius for searching nearby neighbors. "Minpts" are minimum number of points in the "eps" neighborhood of a point [6]. The algorithm divides the points into three groups: core point, border point, and noise point.

Core point: is a point that has more points than number of specified points.
Border point: a point with few numbers of "minpts."
Noise point: a point which is neither a core point nor a border point.

4 Implementation and Results

In this paper, experiments were carried out on a sample video dataset VB100. It consists of 1416 video clips of 100 bird species. Each class has on average 14 video clips; the median length of a video is 32 s [14]. The process converted the each video into 1143 frames.

As shown in Fig. 2, the frames were extracted, and key frames were found by using thresholding method. 31 key frames were extracted out of 1143 frames, where a threshold of 900,000 is set and extracted key frames are shown below (Fig. 3).

Clustering algorithm performance varies in terms of selecting the relevant features.

| Frame 0 | Frame 2 | Frame 3 |

| Frame 38 | Frame 500 | Frame 1142 |

Fig. 2 Extracted frames from sample video dataset

| Frame 0 | Frame 500 | Frame 1142 |

Fig. 3 Shows the extracted key frames from sample video dataset

The clustering accuracy r is given as:

$$r = \frac{\sum_{i=1}^{k} a_i}{n} \tag{1}$$

where "k" represents number of clusters, "n" is number of instances in the data set, and a_i denotes the number of instances occurring in both cluster "i" and its corresponding class.

Clustering purity is found using the formula

$$\text{Purity} = \sum_{i=1}^{k} \frac{|C_i^d|}{|C_i|} \Big/ k \tag{2}$$

where c_i is the number of instances in the cluster i, and k is the number of clusters [6].

Table 2 Single-level feature performance evaluation

Feature	Clustering accuracy	Clustering purity
Texture	0.59	0.80
Wavelet	0.63	0.83
SIFT	0.65	0.86
SURF	0.69	0.89

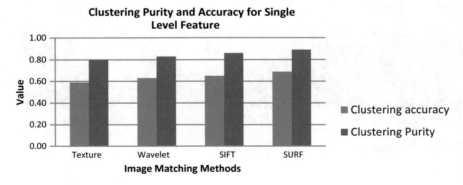

Fig. 4 It is observed that SURF feature extraction method provides better accuracy and purity than SIFT, wavelet, and texture features

Table 3 Multi-level feature evaluation with DBSCAN with and without genetic expression

Method	Clustering accuracy	Clustering purity
DBSCAN (Without genetic expression)	0.63	0.86
DBSCAN (With genetic expression)	0.75	0.95

Fig. 5 Genetic expression based on feature selection process provides better accuracy and purity than other methods. GE feature selection uses search-based optimization principal that improves the performance by using subsets according to fitness function

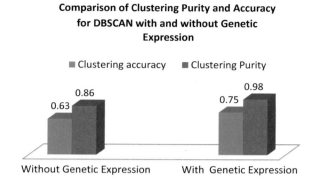

Table 2 shows the clustering accuracy and purity performance of single level and depicts that in the single-level feature performance, SURF performance is high in terms of cluster accuracy and cluster purity (Fig. 4).

Table 3 shows the clustering accuracy and purity performance of DBSCAN genetic expression. On observing the values in the table, it is clear that genetic expression based on optimal feature selection process provides better accuracy and purity than other normal density-based clustering (Fig. 5).

5 Conclusion

The potentials of video clustering related to feature extraction is explored out in this paper. Clustering is a key aspect in the study of video datasets especially for high-dimensional datasets. The four main feature extraction methods, SURF, SIFT, wavelet transform, and texture are implemented and compared. From the implementation results, it is clear that using genetic expression for feature selection, where the process selects a subset of features from a total of original features and improves the performance by using subsets according to fitness function. The DBSCAN based clustering results in highest levels of "Clustering Accuracy" and "Clustering Purity." The effectiveness of this approach becomes further evident from the case study that was carried out with a sample video dataset of 100 data points. The 100 data points were correctly processed to yield 10 different clusters as expected. Finally, it is submitted that the clustering of high-dimensional data is a fast evolving field and is proving to be potential thrust areas and further enhancement of performance is still possible by considering other combinations of techniques in various image and video processing applications.

References

1. Shijie, R., S. Xin, et al. A Wavelet-Tree-Based Watermarking Method Using Fast ICA. In *2009 Second International Symposium on Computational Intelligence and Design*, 162–164. Changsha. ISBN: 978-0-7695-3865-5.
2. Karami, Ebrahim, Siva Prasad, and Mohamed Shehata. 2017. *Image Matching Using SIFT, SURF, BRIEF and ORB: Performance Comparison for Distorted Images*. Canada: Faculty of Engineering and Applied Sciences, Memorial University.
3. Mohanaiah, P., P. Sathyanarayana, and L. GuruKumar. 2013. Image Texture Feature Extraction Using GLCM Approach. *International Journal of Scientific and Research Publications*, 3 (5).
4. Fan, Wentao, Nizar Bouguila, Djemel Ziou, et al. 2013. Unsupervised Hybrid Feature Extraction Selection for High-Dimensional Non-Gaussian Data Clustering with Variational Inference. *IEEE computer Society* 25 (7): 1670–1685. ISSN: 1041-4347.
5. Harit, Gaurav, Santanu Chaudhury, et al. 2003. Clustering in Video Data: Dealing with Heterogeneous Semantics of Features. *Journal Pattern Recognition* 39 (5): 789–811.
6. RamakrishnaMurty, M., J.V.R Murthy, P.V.G.D. Prasad Reddy, et al. 2014. Homogeneity Separateness: A New Validity Measure for Clustering Problems. In *International Conference and Published the Proceedings in AISC and Computing*, vol. 248, 1–10. Springer, (indexed by SCOPUS, ISI proceeding DBLP etc). ISBN: 978–3-319-03106.
7. Meng, Yang, Ronghua Shang, Licheng Jiao, Wenya Zhang, Yijing Yuan, Shuyuan Yang, et al. 2018. Feature Selction Based Dual-Graph Sparse Non-negative Matrix Factorization for Local Discriminative Clustering 87–99. Elsevier.
8. Xianting, Qi, Wang Pan, et al. 2016. A Density-Based Clustering Algorithm for High-Dimensional Data with Feature Selection. In *Conference: 2016 International Conference on Industrial Informatics—Computing Technology, Intelligent Technology, Industrial Information Integration (ICIICII)*, USB. ISBN: 978-1-5090-3574-8.
9. Routray, Sidheswar, Arun Kumar Ray, Chandrabhanu Mishra, et al. 2017. Analysis of Various Image Feature Extraction Methods against Noisy Image: SIFT, SURF and HOG. In *Conference: 2017 Second International Conference on Electrical, Computer and Communication Technologies (ICECCT)*. At Coimbatore, Tamil Nadu, India: IEEE.
10. Ben Salem, M., K.S. Ettabaa, and M.S. Bouhlel. 2016. Hyperspectral Image Feature Selection for the Fuzzy C-means Spatial and Spectral Clustering. In *2016 International Image Processing, Applications and Systems (IPAS)*, 1–5. Hammamet. Electronic ISBN: 978-1-5090-1645-7.
11. Taşkın, G., H. Kaya, L. Bruzzone, et al. 2017. Feature Selection Based on High Dimensional Model Representation for Hyperspectral Images. *IEEE Transactions on Image Processing* 26 (6): 2918–2928. Electronic ISSN: 1941-0042.
12. Rochac, J.F.R., N. Zhang, et al. 2016. Feature extraction in hyperspectral imaging using adaptive feature selection approach. In *2016 Eighth International Conference on Advanced Computational Intelligence (ICACI)*, 36–40. Chiang Mai. ISBN: 978-1-4673-7780-5.

13. Texture feature extraction: shodhganga.inflibnet.ac.in/bitstream/10603/24460/9/09_chapter4. pdf.
14. Mohanaiah, P., P. Sathyanarayana, L. Gurukumar, 2013. Image Texture Feature Extraction Using GLCM Approach 3(5). ISSN: 2250-3153.

Identification of Cryptographic Algorithms Using Clustering Techniques

Vikas Tiwari, K. V. Pradeepthi and Ashutosh Saxena

Abstract Cryptanalysis of cipher text-only attacks is by far the most challenging task in cryptology. This problem has been intimidating researchers as well, and till date no concrete approach has been suggested to address this problem. In this paper, we propose to identify the cryptographic algorithm by analysing the cipher text alone, using the clustering techniques. We have been successful in obtaining results which are able to indicate the cryptographic algorithm used. This methodology of using machine learning algorithms for cryptanalysis is a unique approach, and by looking at our results, it seems to be very promising as well.

Keywords Encryption · Clustering · Cryptanalysis · Machine learning

1 Introduction

Machine learning refers to the ability of machines to perform cognitive tasks like thinking, perceiving, learning, problem solving and decision making. Several methods aim at learning feature hierarchies which are formed by the composition of lower-level features. Automatically, classification features at multiple levels of abstraction allow a system to learn complex functions mapping the feature of input to the output class directly from data, without any dependence on external-crafted features. This is especially important for higher-level abstractions, which humans often do not know how to specify explicitly in terms of raw input making, thus making it increasingly important for a range of applications.

V. Tiwari · K. V. Pradeepthi (✉) · A. Saxena
CR Rao Advanced Institute of Mathematics, Statistics and Computer Science,
Hyderabad, India
e-mail: pradeepthi@crraoaimscs.res.in

V. Tiwari
e-mail: vikas@crraoaimscs.res.in

A. Saxena
e-mail: asaxena@crraoaimscs.res.in

© Springer Nature Singapore Pte Ltd. 2020
K. S. Raju et al. (eds.), *Proceedings of the Third International Conference on Computational Intelligence and Informatics*, Advances in Intelligent Systems and Computing 1090, https://doi.org/10.1007/978-981-15-1480-7_43

Most used types of machine learning algorithm [1] are as follows: (i) classification, (ii) clustering and (iii) reinforcement learning. Of these, classification and clustering are used most frequently to solve many problems. Researchers have found that they are able to solve many tricky problems if there is ample data available to identify patterns.

2 Related Work

Cryptanalysis refers to attacking the encryption system in order to recover key with the knowledge of cipher text [2]. The one who does cryptanalysis is called cryptanalyst. There are various attacks in cryptography, namely

- Known Plain Text Attack: In this attack, the cryptanalyst has the knowledge of cipher text, encryption algorithm and one or more pairs of plain text and the corresponding cipher text. This information is used to reveal the key [3].
- Chosen Plain Text Attack: Here, the cryptanalyst knows cipher text, encryption algorithm and has some chosen plain text and corresponding cipher text pairs.
- Cipher Text-Only Attack: The cryptanalyst has no knowledge of the plain text and must work only using encryption algorithm and the cipher text.

There are a few techniques which proved to be very effective even when targeting modern ciphers and which are based only on the knowledge of the cipher text messages. The most important methods are as follows:

(a) Attack on Two-Time Pad: The fundamental rule of cryptography says the same keystream characters should never be used more than once; otherwise, the cipher would become vulnerable to cipher text-only attack. For example,

$$C1 \ = \ M1 \oplus K1 \, \text{and} \ C2 \ = \ M2 \oplus K1 \tag{1}$$

By taking these cipher texts, the attacker will add these cipher texts and lead to breaking the cipher system.

$$C1 \oplus C2 \ = \ M1 \oplus K1 \oplus M2 \oplus K1 = M1 \oplus M2 \tag{2}$$

The obtained output does not depend on the secret key which we have used. The redundancy in languages and in the ASCII encoding is more; hence, for the attacker, it is possible to extract the original message.

$$M1 \oplus M2 = \ M1, M2 \tag{3}$$

(b) Frequency Analysis: It comes under the category of known cipher text attacks where the frequency of letters or chunk of letters is computed in the cipher text. In any particular language, each letter/alphabet posses its own certain frequency, e.g. in English language, vowels—e, o, a or a consonant—t is frequently used but in another perspective, letters like z and x are used infrequently. These frequencies are intended in an approximate way as different nature of texts (scientific, fiction oriented, etc.) provokes them in a slightly different way. One can differentiate English text from other language's texts by observing any slight patterns, e.g. the presence of some chunks of letters which occur in common pairs like tr, er, on, an, ss, tt and ee. Finding the correct order of letters from mixed words is realizable through this procedure.

Linear cryptanalysis (LC): Linear cryptanalysis is a known plain text attack [4]. "This cryptanalysis is based on finding affine approximations to the action of a cipher, which hold with relatively high probability". The main idea is to find affine relations between bits of the plain text, the cipher text and the key (which hold with relatively high probability) such that we can deduce information about some key bits.

Differential cryptanalysis: Differential cryptanalysis is a chosen plain text attack where the cryptanalyst chooses the plain text pairs which satisfy certain ΔP knowing that it will result in ΔC with high probability [5]. So, the highly likely probable sequence of input differences and output differences is constructed where the output difference of one round is input difference to the next round. This sequence is called differential probability characteristic. Differential cryptanalysis takes advantage of the high probability of certain occurrences of plain text differences and differences into the last round of the cipher text. It is a well-known chosen plain text attack [6]. Two plain texts with a fixed difference are given input to the encryption function, and the difference in cipher text is observed. If the resulting cipher text pair also has a fixed difference, then this can be used to recover the key. Firstly, it determines the partial key bits in the last round. For this, it exploits the plain text pairs with same x-or difference. These pairs are given as input to the target cipher, and the flow of x-or difference through all the rounds is tracked.

These are some works which involve machine learning and cryptography. The primary work in this area is given by Rivest et al. [7]. In this paper authors talk about the machine learning and cryptanalysis are allied fields and can work well together. The possible future collusion between the two has been elaborated in this work. Machine learning can aid in evolution of better compression algorithms, optimization algorithms, etc.

Power analysis has always been a side channel attack that has been exploited by many cryptanalysts. Lerman et al. [8] have captured large amount of power trace data and applied machine learning techniques to the same to identify key information. Barbosa et al. [9] extracted metadata from cipher text performed classification on the metadata to extract patterns. Deep learning has also been used for cryptanalysis of profiling data in side channel attacks.

The analysis of the available literature has shown that there is a strong association between cryptography and machine learning. With this premise, we set out to ascertain that using machine learning, especially clustering algorithms, we can identify the encryption algorithm by using cipher text-based data.

2.1 Ciphers Used for Proposed Methodology

The encryption algorithms that have been used in this paper for analysis are
(i) GOST and (ii) Data Encryption Standard.

GOST block cipher: The GOST block cipher is a standard symmetric key block
cipher of Soviet and Russian government. It has a block size of 64 bits [10, 11].

It is based on Feistel structure with 32 rounds. Round function of gost cipher is
as follows: adding a 32-bit sub-key with modulo 2^{32}. This resultant output traverse
through S-boxes, and then left rotation is performed by 11 bits. Li and Ri represent
32 bits as shown below in Fig. 1. The sub-keys which are 32 bits are taken in a
specified order. Key schedule of GOST cipher is given as follows: 256-bit key is
divided as eight 32-bit sub-keys, in the encryption algorithm, each sub-key which is
derived from 256-bit key is used four times; the initial 24 rounds of the cipher use
the keywords in order and for last eight rounds keys are used in reverse order.

The S-boxes which are used in the cipher are a nonlinear mapping of four-bit
input and four-bit output [12]. Eight 4X4 S-boxes are used in round function.

DES block cipher: Data Encryption Standard (DES) is based on Feistel structure.
DES has 16 rounds [13]. Its structure is shown in Fig. 2. We have 64-bit plain text
block and 64-bit key as input to the encryption algorithm. The 64-bit plain text
block goes through initial permutation (IP). After IP, the plain text is divided into
L and R. This R becomes new L of the next round, and new R is the x-or of the
previous round L and output of the F function. The whole process can be explained
by the following equation:

$$L_i = R_i - 1 \tag{4}$$

$$R_i = L_i - 1 \oplus F(R_i - 1, K_i) \tag{5}$$

The R_i input of 32-bit is expanded to 48-bits using an expansion table. This
48-bit output is xored with key K_i of 48-bits. Now, these resulting 48-bits are
passed through S-box to get 32-bits which are further permuted and xor-ed with L_i
to get $R_i + 1$. The expanded 48-bit output is xor-ed with 48-bit key and given as

Fig. 1 Data encryption
standard round function and
key generation

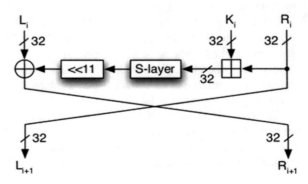

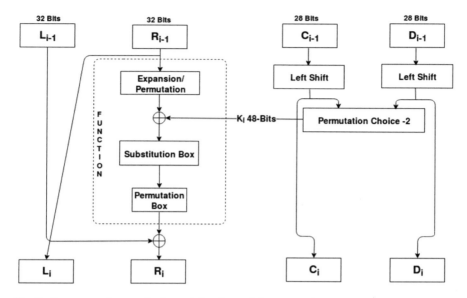

Fig. 2 Data encryption standard round function and key generation

input to the eight S-boxes which takes six-bit input and gives four-bit output. So, this 48-bit xor-ed output is divided into eight blocks of six-bit each, and each of them is given as input to the S-boxes. These 48-bits get transformed into 32-bits as four-bit output is obtained from each S-box. Now, this 32-bit output is permuted and xor-ed with L to get new R, and this permuted output passes through all 16 rounds. The output of the 16th round passes through 32-bit swap and then to the inverse initial permutation to give 64-bit cipher text block. The 64-bit key is mapped into 56-bit using permuted choice 1(PC1). Then in each round, a different 48-bit sub-key Ki is given after passing it through a left circular shift and permuted choice 2 (PC-2).

3 Proposed Methodology

The block diagram of the proposed system is shown in Fig. 3. For the plain text, 20 different text files with sizes varying from 8 kb to 14 kb in English language were considered. These files are encrypted using the encryption algorithms DES and GOST. Then, we extract differences from the cipher text produced. The generation of the difference is explained in algorithm 1.

Fig. 3 Block diagram of the proposed methodology

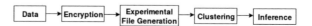

After difference files have been generated, we perform clustering on the difference files by considering each row of the difference file as an instance. Initially, clustering was performed using many algorithms like *K*-Means, Canopy and DBScan. But, we have come to the conclusion that as we do not have any hint about the number of clusters the data will fall into, we cannot use algorithms that need the number of clusters to be specified beforehand. Hence, we consider the results of the expectation–maximization (EM) [14] clustering algorithm because in this algorithm, we need not specify the number of clusters the data is supposed to fall into beforehand.

Algorithm 1 Experimental Setup Algorithm	
1:	i/p: Input data as English text file (say w) is converted into Hexadecimal and partitioned into multiple 64 bits rows in another file (say x)
2:	o/p: Binary Files which is Exor difference of 1 with different levels.
3:	begin
4:	Encryption: X file (n rows) is fed to our encryption algorithm to get the output file Y (n rows). Y file < – Enc Algo[X]
5:	end
6:	begin
7:	Xor Difference: Y file (n rows) is fed to Exor difference algorithm to get the difference files (say Z)
8:	: In file Z, the 1st row & 2nd row is xor-ed, the output of this is stored in file Z1, similarly 2nd row & 3rd row, 3rd row and 4th row,..., (n-1)th and nth row is xored respectively. All xor-ed output is stored in file Z1 row by row.
9:	Now the Z1 file is fed into the Exor difference algorithm and produces an output file as Z2
10:	Similarly generate Z3, Z4 and Z5 files. These files (Z1, Z2, Z3, Z4, Z5) are Difference files.
11:	end

We postulate that if we observe some pattern in the cluster formed by the differences of the subsequent encryption string and then subsequently observing the higher-order differences, we may observe some pattern to deduce nature of the encryption algorithm. Our result validates our hypothesis as we can clearly observe that there exists a pattern in the ciphers considered. The analysis of the result after clustering is done in the next section.

4   Results and Discussion

As mentioned in the previous sections, we perform clustering using the expectation–maximization technique using Weka tool [15, 16]. The purpose for selecting Weka tools are many folds; (i) Open source, (ii) User friendly with interactive GUI,

(iii) Availability of many algorithms to investigate the data and (iv) Import of data sets in varied formats is present.

The results of the first five difference files have been shown for two algorithms DES and GOST. In Fig. 4, we took these two algorithms as both of them are based on the same premise for encryption. We observe that both the graphs work in tandem with negative phase difference in Fig. 4a.

From the subsequent graphs (4b–d), we observe that the negative phase difference is converging, and finally in the difference file 5 (Fig. 4e), we observe the convergence between the two graphs. This validates our hypothesis of using the difference to find the pattern in the cipher text using clustering techniques.

From the different graphs shown in Fig. 4a–e, we can observe that though the initial difference file of the encryption algorithms does not show much of a pattern. However, as we move towards the high-level difference files (especially the

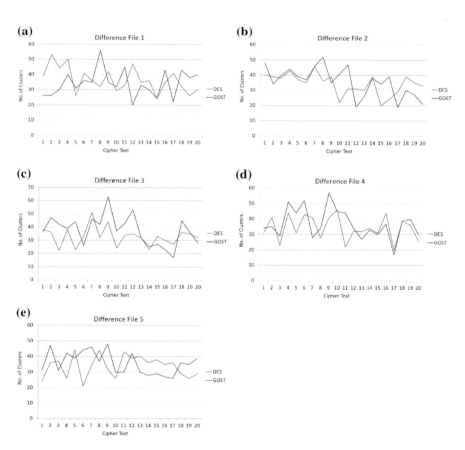

Fig. 4 **a** Clusters formed for difference files 1. **b** Clusters formed for difference files 2. **c** Clusters formed for difference files 3. **d** Clusters formed for difference files 4. **e** Number of clusters formed for difference files 5

difference file 5-Fig. 4e), both the encryption algorithms have the same pattern in the number of clusters that are being generated. This is allowing us to identify the type of algorithm that is being used for encryption as Feistel-based block cipher.

5 Conclusion and Future Works

From the results, it can be observed that though the initial results of difference file 1 do not show any clear patterns. From the subsequent graphs (Fig. 4b–d), we observe that the negative phase difference is converging, and finally in the difference file 5 (Fig. 4e), we observe the convergence between the two graphs. This validates our hypothesis of using the difference to find the pattern in the cipher text using clustering techniques.

With these results as a base, we will be further investigating the various metadata that be extracted from the cipher text and applying the different machine learning algorithms of classification and clustering on them.

References

1. Witten, Ian H., Eibe Frank, Mark A. Hall, and Christopher J. Pal. 2016. Data Mining: Practical machine learning tools and techniques, Morgan Kaufmann.
2. Stallings, William. 2006. *Cryptography Theory and Network Security.* Pearson Education.
3. Stinson, Douglas R. 2006. *Cryptography Theory and Practice.* Chapman Hall/CRC.
4. Heys, H.M., and S.E. Tavares. 1996. Substitution-Permutation Networks Resistant to Differential and Linear Cryptanalysis. *Journal of Cryptology* 9 (1): 1–19.
5. Biham, Eli, Adi Shamir. 1990. Differential Cryptanalysis of DES-like Cryptosystems, The Weizmann Institute of Science Department of Applied Mathematics.
6. Heys, Howard M. 2002. A tutorial on Linear and Differential Cryptanalysis. *Journal Cryptologia* 26 (3).
7. Rivest, Ronald L. 1991. Cryptography and Machine Learning. In *International Conference on the Theory and Application of Cryptology*, 427–439. Springer.
8. Lerman, Liran, Gianluca Bontempi, and Olivier Markowitch. 2011. Side Channel Attack: An Approach Based on Machine Learning. *Center for Advanced Security Research Darmstadt*, 29–41.
9. Barbosa, Flávio, Arthur Vidal, and Flávio Mello. 2016. Machine Learning for Cryptographic Algorithm Identification. *Brazilian Journal of Information Security and Cryptography* 3 (1): 3–8.
10. Madhuri, R., M. RamakrishnaMurty, J.V.R. Murthy, P.V.G.D. Prasad Reddy, et al. 2014. ClusterAnalysis on Different Data sets using K-modes and K-prototype algorithms. In *International Conference and Published the Proceeding in*, vol 249, 137–144. ISBN 978-3-319-03094-4.
11. Zabotin, I.A., G.P. Glazkov, and V.B. Isaeva. 1989. Cryptographic Protection for Information Processing Systems, Government Standard of the USSR, GOST 28147-89, Government Committee of the USSR for Standards.
12. Dolmatov, Vasily (ed.). 2010. RFC 5830: GOST 28147-89 encryption, decryption and MAC algorithms, IETF. ISSN: 2070-1721.

13. Leander, G., and A. Poschmann. 2007. On the Classification of 4-bit S-boxes, WAIFI, 159–176. SpringerVerlag.
14. National Bureau of Standards. 1977. Data Encryption Standard, G.S. Department of Commerce, FIPS pub. 46.
15. Jin, Xin, and Jiawei Han. 2016. Expectation Maximization Clustering, Encyclopedia of Machine Learning and Data Mining, 1–2.
16. Weka. 1999–2008. *Waikato Environment for Knowledge Analysis Version 3.6.0.* The University of Waikato, Hamilton New Zealand.

A Framework for Evaluating the Quality of Academic Websites

Sairam Vakkalanka, Reddi Prasadu, V. V. S. Sasank and A. Surekha

Abstract The main goal of this paper is to design a tool for the evaluation of academic website, taking into account perspectives of different user groups. A literature review was conducted on the existing models, and a list of the factors affecting the quality of academic websites was identified. A framework was developed based on the identified quality factors, to evaluate the new framework, a questionnaire was devised, and a survey was conducted on the reliability of this questionnaire. To assess the effectiveness of the framework, an experiment was conducted, considering six academic websites and 6300 people from different user groups. The threats encountered during the study were also discussed with recommendations for future work.

Keywords Website quality · Academic website · Framework · Quality evaluation

1 Introduction

The growth of Internet rose to great heights in this era; it has changed into one of the most powerful information media of this decade. Everyday, users search websites in order to find the most convenient, relevant, and up-to-date information they need in domains such as business, health, education, and governance [1–4]. Academic websites serve as an effective tool for communication between the users and the management in the domain of education. Only a few academic websites satisfy their intended use. There are several factors contributing to this problem; one such problem is the limited knowledge of the developer or lesser resources such as time and work force [3]. If the academic website does not meet the expectations of users or of the academic website does not provide the users with quality of information and quality in look and feel, the effort and cost put into the maintenance and hosting of that website become useless and waste. Academic websites which

S. Vakkalanka (✉) · R. Prasadu · V. V. S. Sasank · A. Surekha
ANITS, Visakhapatnam, India
e-mail: sairam.vakkalanka@gmail.com

© Springer Nature Singapore Pte Ltd. 2020
K. S. Raju et al. (eds.), *Proceedings of the Third International Conference on Computational Intelligence and Informatics*, Advances in Intelligent Systems and Computing 1090, https://doi.org/10.1007/978-981-15-1480-7_44

provide the user with most intuitive and quality experience are likely to be visited by the user number of times [5]. Evaluation of academic websites helps to know whether the website meets the requirements of the intended users. Though there are a few models to evaluate the quality of academic websites, they do not take into regard the perspectives of different users who visit academic websites.

There are different users who visit an academic website such as the students, parents, and faculty. Students use the academic website frequently to view their academic information. Parents use the website to know the progress of their ward (This differs from country to country. In a country like India, where parents are worried about their ward's progress, they cannot make a visit to the college every time instead they can make use of this academic website), and faculty also use the academic website frequently to post the information regarding their course updates, etc. Similarly, there is no quality assessment model which focuses on quality factors such as the tastes and liking of a gender (as men may like bold colors or women may like pale colors, etc.), symbols and images (A swastika symbol which might be the brand image of an educational society, when published in a website in India represents holiness and may not be treated illegal but it may be not the same everywhere in the world), multiple language support, last update of content of the website, etc. Also, the prioritization of quality factors and opinions based on their priorities is not dwelled much upon. Hence, there is a great need for a comprehensive framework which evaluates the quality of academic websites, taking into consideration the above stated.

1.1 Background and Related Work

McCall's model was developed by Jim Mc Call in 1977 for the US Air Force. This model is divided into three parts, namely product revision, product transition, and product operation. Product revision deals with quality factors which change a software product. Product transition deals with the quality factors which adapt when there is a change in the environment. Product operation deals with those quality factors which fulfill the user specification.

ISO-9126-1 is designed and documented by International Standards Organization in 1991 and is continuously being updated [6]. This standard is divided into three parts consisting of internal quality, external quality, and quality in use. Internal quality deals with internal attributes such as specification, architecture, and design. External quality deals with the dynamic properties which can be viewed while the execution of the software is made. Quality in use deals with the effectiveness and satisfaction of the users, etc. Though there are several other models, they are mostly based on the above-stated quality evaluation models. We have chosen the following research questions to guide our research.

1.2 Research Questions

- What are the factors of quality present in a website?
- Which quality factors of websites act as characteristics to academic websites too?
- How can we derive the level of user experience based on these quality factors?
- How effectively does our evaluation method provide suggestions for improving the website?

2 Research Methodology

Research is nothing but a systematic way of collecting, analyzing, and finding a solution to a problem [7–9]. To attack our problem and to answer our research questions, we have used the mixed method approach [10], which includes both qualitative and quantitative methods. *To answer research questions* (1) *and* (2), we chose to conduct a literature review.

2.1 Literature Review

The main aim of this literature review is to know about the factors which influence the quality of academic websites. We have chosen different databases such as ACM, IEEE XPLORE, Inspec, and Compendex to conduct our search. We have also searched on search engines such as Google Scholar, citation databases such as the ISI and Scopus for the sorting and analysis. To define our search strategy, we have formulated a search string with all possible combinations of different keywords.

(((((website OR web site) WN All fields) AND((academic) WN All fields))AND ((evaluation) WN All fields))

We have included a study if it on evaluation of website, evaluation of website quality and assessing the quality of academic websites. We have also included those papers which published only in the last decade. Excluded those articles which are based on e-commerce websites evaluations and if the article is a letter or a review. We have also excluded those articles which are under learning management systems (LMS) though these falls under educational websites. Before using the inclusion and exclusion criteria, we had 539 papers after step by step exclusion and inclusion, and we have cut the number to 232 and then to 64 and then to 34, finally we are left with eight articles.

The existing quality evaluation models such as Web-QEM [11], Web-Qual [12], and 2QCV3Q [13] are brought into light. Web-QEM [11] is based on ISO-9126-1 software model, and it uses logic score preference (LSP) [14] for the evaluation of the academic websites. Here, based on the opinions of expert groups, a logic score is calculated. Web-Qual [12], four universities in Europe developed this model which is an Internet-based questionnaire. Feedback of the questionnaire is collected; means and variances are calculated. Based on the results obtained from the questionnaire, the evaluation is made. 2QCV3Q [13], this model discusses about Cicirones seven points and gives a description of quality factors such as the management, usability, feasibility, identity, services, location, and maintenance.

2.2   Data Analysis

We have a list of the quality factors obtained from three different models, namely MINERVA [15], Web-QEM [11], and 2QCV3Q [13]. The quality factors are mainly divided into two types such as the high-level quality factors and the sub quality factors. Table 1 shows a list of high-level quality factors obtained from literature review.

Apart from these, we also chose to consider quality factors from the ISO 9126 and also Mc Calls quality model. High-level quality factors such as testability, integrity, flexibility, correctness, and reusability are found in Mc calls quality model.

Table 1 List of quality factors from literature review

S. No.	Quality factor	MINERVA	2QCV3Q	WEB-QEM
1	Interoperability	X		
2	Feasibility		X	X
3	Content	X	X	X
4	Usability	X	X	
5	Portability			X
6	Functionality	X	X	X
7	Efficiency	X		X
8	Maintainability	X	X	X
9	Understandability	X	X	X
10	Navigation	X	X	X
11	Presentation	X		
12	Transparency	X		
13	Reliability			X

2.3 Listing the Quality Factors

Here, in the quality evaluation models, a high-level quality factor in one model acts as a sub quality factor in another model. Such as *understandability* which is a high-level quality factor in all three models in MINERVA, Web-QEM and 2QCV3Q is a sub quality factor in ISO-9126-1 model. We considered quality factors as high-level quality factors if their occurrence as a high-level quality factors is more in number. A cut-off value of 4 is considered. Figure 1 shows the frequency of quality factors in the evaluation models identified from the literature review. Here, we formulated a new list of quality factors with their sub quality factors based on their occurrences. The quality factors include

- **Content**: Content is considered as one of the most important part of a website [16]. It contains information which needs to be accurate, attractive, and updated. The content present in a website conveys the identity of the organization.
- **Usability**: It is defined as the ease of use of the user interface to a user [17]. The website should always be accessible, interactive, easily understandable and also must be easily operable. It should also have the feature of multilingual support.
- **Reliability**: It is about the availability of the website to the users and also its ability to recover quickly from problems [18].
- **Portability**: The ability of a website to work in different environments.
- **Maintenance**: It is to the extent to which the website can accept changes made to it.
- **Functionality**: It is the capability a product can perform based on the user needs [19]. The functionality of a website includes the security, navigation, help, search, site maps, etc.

Based on the above-mentioned observations, a list of high-level quality factors and their corresponding low-level quality factors are identified. Table 2 shows the list of high-level factors and their sub quality factors.

Fig. 1 Occurrence of quality factor in existing evaluation models

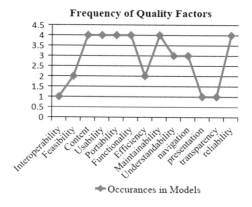

Table 2 Categorization of high-level quality factors and low-level quality factors

#	High-level factors	Low-level factors
1	Content	Accuracy
		Attractiveness
		Identity
		Updates
		Relevance
2	Functionality	Navigation
		Load time
		Security
		Search
		suitability
		Help
		Maps
3	Usability	Accessibility
		Interactivity
		Operability
		Understandability
		Global language
		Aid to the challenged
4	Reliability	Recoverability
		Availability
5	Portability	Adaptability
		Coexistence
6	Maintenance	Testability
		Changeability

To answer *research question* (3), we chose to conduct a survey on the user experience of a visitor using the academic website, and a questionnaire was devised for this purpose..

2.4 Formulating the Questionnaire

We have formulated a questionnaire in such a way that different user groups who use a website can respond and give the feedback. To achieve this, we tried to at least form three to four questions on each high-level quality factor.

2.5 Reliability of the Questionnaire

To know how reliable the questionnaire is, the questionnaire is sent to an expert group. The expert group consists of 511 individuals from different places such as

Fig. 2 Reliability score of questionnaire

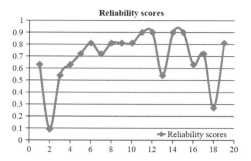

Germany, India, the UK, the Netherlands, Australia, Sweden, and California. They are well experienced in the areas of development and maintenance of academic websites and quality assurance for websites. We have sent the questionnaire to the expert's groups asked them to poll their response to the deletion of a certain question on a scale of Yes/No. Figure 2 shows the reliability scores of the questionnaire.

If the value is Yes, a value of 1 is assigned, and if the answer is no, a value of zero is assigned. The mean of their response is calculated, and changes to the questionnaire are made. The items from the questionnaire 2 and 18 are deleted, and the questionnaire is reframed again as shown in Table 3.

2.6 Analyzing the Feedback

The opinions of the user groups are polled on a scale consisting of values from 0 to 1, respectively. To analyze the feedback of the users, we chose to devise the following method. The means opinions of the entire high-level factor are calculated.

$$\text{Quality of each high level Factor(Fq)} = \sum_{1}^{n} \left(\frac{\text{opinions of low level factors}}{\text{No. of Questions attributed to a quality factor}} \right)$$

The total quality of the website can be calculated using

$$\text{Total Quality(Tq)} = \frac{\sum (Fq * W)}{Nq}$$

- Weight (W) is the number of questions per a high-level quality factor
- Nq is the total number of questions in the questionnaire = 17

If the factor quality or total quality (Tq) lie between the following ranges, the following quality outcomes can be attributed to the website shown in Table 4.

Table 3 Improved questionnaire based on reliability scores

No	Questions	High-level factor
1	Is the information provided in the academic website accurate?	Content
2	Is the text and content in the academic website attractive?	
3	Does the academic website portray the identity of the organization?	
4	Are the updates timely?	
5	Does the academic website provide you with your relevant information?	
6	Is the website navigation ok?	Functionality
7	Is the load the time of web pages quick enough?	
8	How secure is the content in this academic website?	
9	Is the search mechanism sufficient?	
10	Is the help functionally in this academic website handled well?	
11	Are there site maps provided to aid the users?	
12	Is the feedback and support mechanism functioning well?	usability
13	Does the website provide any features to aid the challenged?	
14	Does the academic website provide multi language support?	
15	Is the website available all the time?	Reliability
16	Does the website work in all browsers?	Portability
17	Does the website respond after maintenance well?	Maintainability

Table 4 Mapping of user opinions and quality outcomes

Range	Quality outcome
$0.00 <= Fq$ or $Tq < 0.20$	Needs major improvement
$0.20 <= Fq$ or $Tq < 0.40$	Needs minor improvement
$0.40 <= Fq$ or $Tq < 0.60$	Ok
$0.60 <= Fq$ or $Tq < 0.80$	Sufficient
$0.80 <= Fq$ or $Tq <= 1.00$	Good

To answer the research question (4), we chose to conduct a survey-based experiment.

2.7 Experiment Planning

We chose six academic websites of some reputed institutions from different places around the world. Namely Indian Institute of Technology IIT-D [20], Missouri University of Science and Technology MST [21], Aachen University of Technology AUT [22], University of New South Wales UNSW [23], University of Pretoria UVP [24], and Royal Institute of Technology KTH [13]. The questionnaire is placed to them, and a time of one hour is given to answer the questionnaire.

We have collected the feedback given to website hosts on the user experience of visitors before using this tool and the feedback after using this tool. The feedback comparison is made to know how well this tool helps website administrators in making changes to the website to improve its quality.

2.7.1 Population

To conduct this experiment, we chose population of 6300 people which included 2100 students, 2100 parents, 2100 faculty, and staff members. We have chosen equal population because it would give equal prominence to all the user groups. To select the sample students, parent and faculty, around 5000+ invitations via email were sent to students and faculty from top 1000 academic institutes in the globe listed by Times Higher Education Rankings [25]. To find the parent sample was an arduous one, and parents email ids were hard to obtain. Hence, we have considered the parents/guardians of the 2100 interested students and also sent mails. Feedback evaluation was conducted thrice. To achieve this, the 2100 population sample from each group was equally divided into three subgroups, with each group consisting of 700 people. Questionnaire was sent to each subgroup once, and the average of responses of all subgroups was collected and analyzed.

3 Results and Analysis

The following are the results obtained from the experiment conducted. The table below lists the means of higher-level quality factors for the following university websites. S. No 1–6 represents content, usability, functionality, reliability, portability, and maintenance. The user opinions on the total quality factors of these websites are shown in Table 5.

Table 5 User opinions on selected academic websites

Quality factors	IIT-D	MST	AUT	UNSW	UVP	KTH
Content	0.96	0.95	0.98	0.93	0.89	0.97
Usability	0.78	0.43	0.71	0.79	0.43	0.83
Functionality	0.61	0.57	0.57	0.62	0.39	0.54
Reliability	0.67	0.68	0.66	0.58	0.67	0.62
Portability	0.62	0.67	0.54	0.45	0.65	0.63
Maintenance	0.75	0.71	0.73	0.74	0.76	0.77
Overall Quality	0.731	0.668	0.698	0.685	0.631	0.726

Table 6 Quality outcomes based on user opinions

Quality factor	IIT-D	MST	AUT	UNSW	UVP	KTH
Content	Good	Good	Good	Good	Good	Good
Usability	Sufficient	Ok	Sufficient	Sufficient	Ok	Good
Functionality	Sufficient	Ok	Ok	Sufficient	Needs minor improvement	Ok
Reliability	Sufficient	Ok	Sufficient	Ok	Sufficient	Sufficient
Portability	Sufficient	Sufficient	Ok	Ok	Sufficient	Sufficient
Maintenance	Sufficient	Sufficient	sufficient	sufficient	sufficient	Sufficient
Overall Quality	Sufficient	Sufficient	Sufficient	Sufficient	Sufficient	Sufficient

Based on the results obtained, the following analysis was made. The results of high-level quality factors analysis are obtained, the opinions are generalized, and the quality outcomes are shown in Table 6.

The overall quality of the selected websites is sufficient as all of them fall under the range of $0.60 <= TQ < 0.80$. Though these are of sufficient quality, there are places of improvement which can be seen in the high-level quality factors. These results show that formulated tool can accurately locate the areas of improvement, easier to use and help in improving the quality of academic websites.

4   Threats to Validity

- Lack of interest in participants can prove fatal in a questionnaire-based survey such as this. The results of the survey and the experiment may go wrong due to the lack of interest. To avoid this, we have selected those participants who have shown enthusiasm to participate.
- Population validity may cause a threat when we generalize an issue to larger set of population while only considering a smaller size. To avoid this, we chose to take a considerable sample size to make our generalizations.
- Bias is a very dangerous threat; if the response is biased, the results of the evaluation may not be of any use, and to avoid this threat, we chose to take the sample from the population in such a way that the bias is minimized.

5   Conclusion and Future Work

We have designed a tool for the evaluation of academic websites considering the different user groups who visit the academic website. A literature review was conducted, listed the different quality factors and has thrown light on the existing models for evaluation. A questionnaire was devised based on these quality factors and have conducted a survey to know the user experience of a visitor of academic website. To establish the effectiveness of this, we chose to conduct and experiment and deduced results from the experiment. We have also made a discussion on the threats encountered during the study.

Acknowledgements We thank all the numerous participants who participated in evaluating the different websites, and thank those reviewers for suggesting changes to the questionnaires. No data was collected regarding the details of participants for website evaluations and all of the participants were anonymous.

References

1. Mendes, E. 2006. *Web Engineering*. Berlin, Heidelberg: Springer-Verlag.
2. Alexander, J., and M. Tale. 1999. *Web Wisdom: How to Evaluate and Create Information Quality in the Web*. Lawrence Erlbaum Associate Inc.
3. Dragulenscu, Nicolae-George. 2002. Website Quality Evaluations: Criteria and Tools. *The International Information & Library Review* 34 (3): 247–254. ISSN 1057-2317. https://doi.org/10.1006/iilr.2002.0205.
4. Wu, Y., and J. Offutt. 2002. *Modeling and Testing Web-based Applications*. George Mason University.
5. Krug, S. 2006. *Don't Make Me Think: A Common Sense Approach to Web Usability*, 2nd ed. Berkeley, CA: New Riders.
6. Abran, A., A. Khelifi, A. Seffah, and W. Suryn. 2003. Usability Meanings and Interpretations in ISO Standards. *Software Quality* 11: 325–338.
7. Dyba, Tore, Erik Arisholm, Dag I. K. Sjoberg, and Jo E. Hannay. *Are Two Heads Better than One? On the Effectiveness of Pair Programming*. IEEE Computer Society.
8. Kitchen ham, B.A., and S. Charters. 2007. Procedures for Performing Systematic Literature Reviews in Software Engineering. In *EBSE Technical Report*, Software Engineering Group, School of Computer Science and Mathematics, Keele University, UK and Department of Computer Science, University of Durham, UK.
9. Web link to Social Research. www.socialresearchmethods.net/kb/concthre.php.
10. Creswell, John W. *Research Design: Qualitative, Quantitative and Mixed Methods Approaches*. Sage Publications, Second.
11. Olsina, L., and G. Rossi. 2002 *Measuring Web application quality with WebQEM*, vol. 9, no. 4, 20–29. Multimedia, IEEE. https://doi.org/10.1109/mmul.2002.1041945.
12. Longstreet, P. 2010. Evaluating Website Quality: Applying Cue Utilization Theory to WebQual. In: *43rd Hawaii International Conference on System Sciences (HICSS)*. vol. no., 1–7, 5–8 Jan. 2010. https://doi.org/10.1109/hicss.2010.191.
13. Olsina, L., D. Godoy, G. Lafuente, and G. Rossi. 1999. Source: Assessing the Quality of Academic Websites: A Case Study. *New Review of Hypermedia and Multimedia* 5: 81–103.
14. Yip, C.L., and E. Mendes. 2005. Web Usability Measurement: Comparing Logic Scoring Preference to Subjective Assessment. In *ICWE: International Conference on Web Engineering*, vol. 3579, 53–62. Sydney, Australia: Springer.
15. Web link to Minerva http://www.minervaeurope.org/publications/qualitycommentary/qualitycommentary050314final.pdf.
16. Nielsen, J. 2000. Is Navigation useful?. In *Jakob* Nielsen's Alert Box.
17. Nielsen, J. 2002. *Introduction to Usability*.

18. Micali, F., and S. Cimino. 2008. Web Q-Model: A New Approach to the Quality. In *The 26th Annual CHI Conference on Human Factors in Computing Systems Florence*, Italy.
19. Burris, E. 2007. *Software Quality Management.*
20. Web link to IIT- D http://www.iitd.ac.in/.
21. Web link to MST http://www.mst.edu/.
22. Web link to AUT http://www.rwth-aachen.de/go/id/bdz/.
23. Web link to UVP http://www.cs.up.ac.za/.
24. Web link to UNSW http://www.unsw.edu.au/.
25. Web link to Times Higher Education Rankings. https://www.timeshighereducation.com/.
26. Web link to KTH. http://www.kth.se/en.
27. Web link to Mc Calls. http://www.sqa.net/softwarequalityattributes.html.
28. Trochim, W. 2000. *The Research Methods Knowledge Base*, 2nd ed. Cincinnati OH: Atomic Dog Publishing.

A Survey on Analysis of User Behavior on Digital Market by Mining Clickstream Data

Praveen Kumar Padigela and R. Suguna

Abstract Data stream mining has emerged as one of the most prominent areas with its applications in various areas like network sensors, stock exchange, meteorological research and e-commerce. Stream mining is potentially an active area in which the data is continuously generated in large amounts which are dynamic, non-stationary, unstoppable, and infinite in nature. One of such streaming data generated with the user browsing tendency is Clickstream data. Analyzing the user online behavior on e-commerce Web sites is helpful in drawing certain conclusions and making specific recommendations for both the users and the electronic commerce companies to improve their marking strategies and increase the transaction rates effectively leading to enhance the revenue. This paper aims at presenting a survey of different methodologies and parameters used in analyzing the behavior of a user through Clickstream data. Little deeper, this article also outlines the methods used so far for clustering the users based on mining their interests.

Keywords Clickstream · Behavior · Digital market · Collaborative filtering

1 Introduction

It has become a common tendency for every user to make an online purchase where this is being put up as a trend in making digital purchases. The usage of e-commerce market has grown exponentially for the last few years. As per the statistics stated by a statistics analyzer company, it is to be known that estimated people of about 1.66 billion around the world have made their purchases online in 2017. It is also estimated that there could be a rise of 4.48% in e-commerce purchases by 2021.

P. K. Padigela · R. Suguna (✉)
Department of Computer Science, Vel Tech Rangarajan Dr. Sagunthala R&D Institute
of Science and Technology, Chennai, Tamilnadu, India
e-mail: drsuguna@veltechuniv.edu.in

P. K. Padigela
e-mail: praveen.padigela@gmail.com

© Springer Nature Singapore Pte Ltd. 2020
K. S. Raju et al. (eds.), *Proceedings of the Third International Conference
on Computational Intelligence and Informatics*, Advances in Intelligent
Systems and Computing 1090, https://doi.org/10.1007/978-981-15-1480-7_45

This indicates that there is lot more knowledge to be extracted from the access behavior of a user online and predict the interest of the user so as to recommend the type of product that he or she may show interest toward.

Streaming data is a type of data which is generated in large amounts from different sources at high continuous rates. Data streams may have their applications in a wide variety of fields and data being produced from resources such as sensors, ATM machines, and e-commerce Web sites. This data needs incremental processing by using the stream processing techniques of big data analytics.

As the tendency of buying smart has been on the rise, it is very important for any e-commerce company to analyze the behavior of a user. To make this part of analysis, web data mining is a prominent approach through which track of Clickstreams can be recorded.

Clickstream data is a data which is generated in large amounts. This data acts as the 'digital footprints' of any customer browsing a Web site or an e-commerce application in particular to this survey. Clickstream, also known as Clickpath, defines the information about the path chosen by an e-customer while navigating through the Web site. Whenever a user takes a chance to access a site to look into some information or to know the details or purchase the product, this data of records of the clicks performed by the user is the base of analyzing the behavior of a customer [1].

Data stream clustering determines clusters in the continuous amount of data streamed. Clustering this relative high speed and continuous data is different from traditional clustering. As for the part of traditional clustering, data sets are static in nature, but the data stream is highly dynamic in nature. Storing data streams in memory and scanning it multiple times are a very challenging task because of their massive size. But, for traditional clustering, data sets are stored in memory and can be scanned multiple times [2]. The clustering results of data streams will often change over time, but not so for traditional clustering.

This paper aims at carrying out a survey of the known approaches used for clustering Clickstream data. Rest of this paper is organized as follows: Sect. 2 consists of discussion about web mining, Sect. 3 describes Clickstream data, Sect. 4 focuses on data stream clustering techniques, Sect. 5 includes the characteristics of user behavior, Sect. 6 presents literature survey on different applications and developed tools based on data stream clustering, and Sect. 7 draws certain conclusion and discusses the future scope observed from the literature available.

2 Web Mining

People find a few things on the web as most significant and influential when browsing a Web site or making a purchase. Details such as the advertisements displayed, offers highlighted, rating and comments of the users on the Web site are the areas where a user is most likely to have a click in the given site.

Fig. 1 Relation between web mining categories and DM techniques

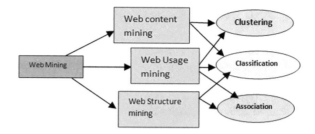

It is not only important for the e-commerce sites to pay attention in understanding or collecting the customer data; rather understating the ways of extracting knowledge and paving a way to attract customer behavior seem to be a point of strategic business improvisation.

Web mining is one aspect which technically supports in searching the needful information from the web. It is one such techniques of data mining which is used in discovering patterns from the WWW [3]. It is an automated process of extracting structured and unstructured information of user indirectly from the sources such as Web site, access page information, and activity web and server logs.

Web data mining is of three types categorized into: web usage mining, web content mining and web structure mining, and this area of research to understand the user behavior focuses on web usage mining (Fig. 1).

3 Clickstream Data

Data at the weblog is continuously streamed which is produced by the activities of the customer on a Web site. This is a process aimed at giving practical solution for data collection and aggregated reporting of pages in which a user visits the Web site. The flick through path of the Web site navigation by the user is called as Clickstream data. Clickstream data is the aggregation of data about what part of a Web site a user visits and in what order. This data is further useful for analysis and reporting of the user behavior on a Web site.

Clickstream analytics falls into two levels where traffic analytics is one kind and the other is e-commerce analytics. Traffic analytics is a server level operation and keeps track of number of pages produced to the user, time taken for each page to be loaded, frequency of crawling back to the previous web page or hit on a stop button, and amount of data transmitted prior to the user movement on to the next Web site.

On the other side, e-commerce analytics of Clickstream data is about the web page on which a user lingers on, add-ons of the user to the cart, details of the purchases made, and data about the mode of payment made detailing about promos, if any, applied.

4 Data Stream Clustering

Data mining focuses on the usage of machine learning and pattern recognition. The computing devices used in current generation generates large amount of data and for effective discovery of patterns or finding the correlation with the hidden data, and analytics of big data or machine learning is useful [4].

In the opinion of a senior industrial data analyst Doug Laney, big data is a study related to the data sets that are very large in volume, velocity, veracity, variety, and variability [5]. All data which is characterized by these Vs is inadequate to be dealt with the software of traditional data processing applications, and there is a requirement of machines to be trained for fulfilling the processing needs of these data.

Machine learning algorithms are categorized as either supervised or unsupervised learning mechanisms [5].

Supervised machine learning mechanism is a kind of learning in which some part of the data is already labeled or a prior knowledge about the output for sample data is already predicted. Regression and classification are the types of algorithms which fall under supervised learning.

On the other hand, the area of pattern recognition used in predicting the behavior is clustering. Clustering is a form of an unsupervised learning mechanism in which the outputs are unlabeled/unstructured or there is no output variable corresponded. For any e-commerce organization, the tedious task of analyzing a user behavior is unknown as the preferences of an individual user differ from one another. It is tedious to classify each user as there is no predictable nature of a person toward a product or in browsing a Web site.

Hence, grouping of unknown behavior of users in which users do not have anything in common can be easily achieved by using clustering mechanism. This paper outlines the survey of research carried out by different scholars aimed at grouping the online purchase users as per different characteristics.

5 Characteristics of User Behavior

Every user who wishes to access any e-commerce commodity Web site owns a set of characteristic features. Few of these parameters can be a part of the user social profile, and the other can be extracted or retrieved by using various methodologies available in log files.

The different parameters that can be used in analyzing an online user behavior are as follows:

- Demographic content
- Contents of a weblog file
- User behavior variables.

Demographic Content:

Demographic contents are the features owned by an e-commerce user. These characteristics typically include information regarding various parameters. Age, race, gender, ethnicity, education, profession, marital status, and income level are the examples of demographic attributes that are used in surveying the behavior of a user [6–8].

While demographic information is a mere concern for every individual, this information is more crucial to the e-commerce commodity.

A recent survey suggests that accumulation of information related to a personal web user would be useful in drawing a conclusion of making a decision of user preferences.

The information given above in Table 1 is a part of a survey conducted by civicscience.com. This sample information suggests that consumers who prefer to shop online are most likely to be the youngsters ranging from age 18 to 34. The statistics of this survey can be a point for making the analysis on kind of users using e-commerce Web sites.

Contents of Log File:

Besides the demographic information of a user, another way of computing which records the events or communication between the user, software, and the operating system is web log file [9]. Web log file is a file generated on the server side for recording the following information:

- **User ID**: Any user who has visited the Web site is identified by an id. Every user's identification depends on the unique address provided by ISP. This unique temporary id is called as IP address. As the id would be temporary, this may not be optimal for analysis.
- **User Profile**: Most Web sites prefer to identify the user based on the information of their profile. This is the information such as username and privacy

Table 1 Demographics of in-store versus online shoppers

	In stores	Online through store specific web sites	Online through web only retailers
Gender	55% women, 45% men	61% women, 39% men	No gender difference
Age: 18–34	30	26	34
Age: 35–54	37	42	39
Age: 55+	33	32	26
Income: $75k and under	68	63	60
Not a present	37	33	46
Education: degree or higher	50	49	62

credentials required to access the Web site would render a unique feature in identifying the user.

- **Access Path**: Also called as the visiting path. This is the path chosen by the user in visiting and accessing the Web site. A user may access the Web site by directly using the URL of the Web site, or he may also prefer to browse through a search engine or click the link on any other Web sites.
- **Traversal Path**: This is useful for fetching the information related to identify the path chosen by the user while accessing the Web site by using various access links.
- **Request Type**: The type of request made to the server which is used for retrieving the web page. Methods such as GET and POST can be used.
- **Page Last Visited**: This details the information of the page which was visited for the last time before the user switches to another web page.
- **Time Stamp**: The amount of time spent by the user on each page at the time of accessing the Web site. Timestamp can be indirectly treated as a session.
- **User Agent**: It is a type of software which acts on the user side and tells the information about the type of operating system and browser being used. This information is basically sent to the Web site.
- **URL**: A uniform resource locator or a web address that acts as a reference to the web resource which specifies the mechanism in locating and retrieving a web resource on computer network
- **Success Rate**: The number of clicks against the downloads, things, or products purchased and number of users visited the Web site determine the success rate of a Web site

User Behavior Variables:

A user behavior is always affected by different factors which are difficult in some cases to be measurable.

Kimar and Dange propose their identification of customer behavior motivated into two different categorical factors: (1) External and (2) Internal factors.

All the factors that are beyond the control of the user and which could stand as user behavior influential factors are treated as external factors. These are broadly divided into different sectors such as socio economics factors, demography factors, public, private, and reference cultural groups, while the factors which are solely dependent on the personal traits and user behavior are treated as internal factors. These factors include user's attitude toward a product, learning and usage experience, motivation, etc.

While internal and external factors are part of human behavior, there are motives which may push the user behavior leading to a transaction or which may deny the user from fulfilling his intentions. The functional motives include time, ease of online shopping, product price, and selection. The non-functional motives include the warehouse or the brand of the product or the e-commerce site where the product is aimed at purchase. These motives are solely responsible for the customers who add the product to the cart with the intention to purchase and may not buy it as the customer may look for benefits in the price reduction or seasonal or other exiting hourly or daily deals.

Final defining feature of any e-commerce Web site is 'Add to Cart' feature—the penultimate step before making a purchase. The motives of user behavior may sometimes lead the product to be left in the cart. Though the user behavior on different aspects has been well researched for clustering, one typical scenario of user behavior where the product is left in the cart without purchase has a larger scope for carrying out the research.

With the evolution of technology, browsing e-commerce Web sites has become a prioritized activity, and for better experience e-commerce applications should be made capable of interacting as per the customer preferences and choices [10]. There is a need for identification of a group of customers whose choices or product preferences are similar so that recommending the product in that group can be made easy.

6   Survey

Many efficient methods were proposed recommending e-commerce products to the user as per their relevance and likeliness. Different techniques have used different parameters based on the type of analysis used for clustering. Most of these techniques in existence have employed the usage of customer ratings to measure the interest of a user.

Ratings from a user have been the important measurement property over a longer period of time to analyze whether the user is interested in it or not. The analysis obtained from ratings indicates how much the user likes the item [10].

Collaborative Filtering

Collaborative filtering (CF) is an important method to be used in recommended systems by making automatic predictions about the user interests. CF methods are classified into two types user-based CF and item-based CF [3].

User-based CF approach is to find out a set of users who have similar patterns in favor to a user given (i.e., user neighbors) and recommend to the user with those items that are liked by other users in the same group. On the other hand, in item-based CF approach, recommendation of a product is aimed at an item for which the items with high correlations are considered. Currently, most CF methods measure the user similarity or item similarity based on common users of items [11].

For example, Sarwar et al. have discussed various techniques used for measuring the item similarity and then obtained recommendations for item-based collaborative filtering; Deshpande and Karypis have presented and evaluated a class of model-based top-N recommendation algorithms that use item-to-item or set-to-item similarities for recommendation.

In either of the collaborative filtering methods, similarity measurement between the customers or the product is a prominent step, and vector space similarity, Pearson correlation coefficient, and cosine-based similarity are some of the common similarity measurements.

To calculate the correlation as per the Pearson method, let us consider the ratings of person X and Y of the item k are written as X_k and Y_k, while $\overline{X}$ and $\overline{Y}$ are the mean values of the ratings, respectively. The correlation between X and Y is then given by

$$r(X, Y) = \frac{\sum_k (X_k - \overline{X})(Y_k - \overline{Y})}{\sqrt{\sum_k (X_k - \overline{X})^2 \sum_k (Y_k - \overline{Y})^2}} \tag{1}$$

A prediction computed for showing the rating of person X of the item i based on the ratings of people who have rated item i is as follows: $\overline{X}$

$$p(X_i) = \frac{\sum_k Y_i - r(X, Y)}{n} \tag{2}$$

However, the information which can be obtained from ratings is too limited as it cannot describe the Web site navigation process of a user. Besides, the ratings given by new users are insufficient for analysis, while ratings from experienced customers though maybe useful. The user may not be ready to provide ratings every time they visit a Web site [1].

Although the use of recommendation methods is widespread in e-commerce, a number of inadequacies have been identified which include data sparsity and inaccuracy of the recommendation [3].

To overcome this problem of this sparsity, King [12] and Huang et al. have used a few hybrid methods such as associative retrieval techniques. Hu et al. [13] have explored the algorithms which are suitable in implicit feedback processing.

Recent popular model called latent factor associates each user u with a user-factor vector p_u, and each item i with an item-factor vector q_i. The prediction is done by taking an inner product and involves parameter estimation.

$$\hat{r}_{ui} = \bar{b}_{ui} + \bar{p}_u^T q_i \tag{3}$$

Some other research works in the recent developments suggest other methods. Leung et al. [14] have proposed clustering-based collaborative location recommendation framework.

The widespread availability of Internet and its role in the daily life of most people creates huge amounts of data in parallel to their web usage. Such data is captured using Clickstream data which contains a wide variety of complete information. This data adds a little more information to the research extension in user behavior analysis. It records the user activities such as browsing paths, purchased products, and clicked banner ads.

Till now, research has been explored using Clickstream data from Web sites that sell a single product to the users who have browsed multiple Web sites for a single product. Substantial increase in online purchases has demanded the study of analyzing user behavior and thus the intention of clustering user behavior has increased, and various researchers have proposed various methods and tools for recommendation [14].

In this connection, Lu Chen et al. have proposed a leader clustering algorithm with cosine similarity to discover the interest of a user at digital business sites. The user behavior analysis was based on three main categories: Visiting path, browsing frequency, and relative length of access time were key considerations refined from Clickstream data.

These indicators were part of analysis in proposing an improved clustering algorithm with rough set theory. This was used to make the clusters of users with similar interest.

While identification of common user behavior was a challenging task, Qiang, proposal on a Chinese e-commerce Web site with 3 million Clickstream data which includes crucial parameters such as browsing path, frequency of page visits, time spent on each category, visiting sequence proposed novel rough leader cluster algorithm is expected to outperform K. Mediods algorithm in terms of efficiency [13].

Leung [15] has proposed clustering using K-means and hierarchical clustering (HC) method in development of recommender tool. The proposed technique was useful in partitioning customers using a similarity graph of higher-level (general behavior) and lower-level clusters (key behavior). Other techniques were also proposed by Cai [12], Hu et al. [14], Chen [16] to make a group of people with similar categorical interest.

S. No.	Author name	Data set	Parameters	Algorithm	Outcomes
1	Lu Chen IEEE 2013	E-commerce Web site, 20,000 sessions + Large data set	Visiting path, browsing freq, access time, sequence of pages	Leader clustering algorithm with cosine similarity	1. Can be applied to support decision making for E-commerce sites. 2. Similarity between two users is analyzed
2	Qiang. SU ELSEVIER 2014	Chinese E-commerce Web site 3million clickstream data	Browsing path, freq of page visits, time spent on each category, visiting sequence	Novel rough leader cluster algorithm	1. Improved efficiency over K. Mediods algorithm. 2. Provides assistance in personalization of Web site. Layout & navigation structure

(continued)

(continued)

S. No.	Author name	Data set	Parameters	Algorithm	Outcomes
3	Hernandez Sergio, IEEE 2016	UP&SCRAP Spanish E-commerce Web site Large data	Users session trace Product category Visiting product, Product WishList, page visits, time spent	Temporal Logic. Model Checking. Text mining	1. Navigational behavior and user interests are detected 2. Recommender for e-commerce and web structuring 3. Buying process is focused on
4	Sahana Raj IEEE 2016	Large Date set of web clicks	Session level characteristics. 1. Visit duration/ amount of time spent, user add, Time stamp. 2. Session details-links clicked, navigated, items viewed, product details, sequence of products viewed	Markov chain model. Threshold nearest neighbor clustering algo	Predicting the prospect of customer to buyers
5	Gang Wang ACM 2016	Large Data set (Renren) 142 million clicks	Demographic parameters related to social networking site	Hierarchical clustering approach applying iterative feature pruning tech	Identifies customers of similar interests. Partitions customers using a similarity graph of higher-level (general behavior) and lower-level clusters (key behavior)

7 Conclusion and Future Scope

From the survey carried out, it is to draw conclusions that many researchers have recommended different parameters in analysis of user behavior preferences. This factor of analysis when turned into a feature of e-commerce application results in increasing business levels as the preference of user is very much in track.

Besides this, a user always navigates with a dual mindset; often cross browsing option will always divert the preference of a user in which there is a chance that user may look at offers, coupons, payback.

There is a chance that the user may add some items into cart which may not be purchased because of change in his preferences. So, there is a bigger scope in this regard for an e-commerce company to analyze the behavior of a user who tends to change the preference having the product un-transacted once added to cart may be turned into a business.

References

1. Su, Qiang, and Lu Chen. 2014. A Method For Discovering Clusters Of E-Commerce Interest Patterns Using Click-Stream Data, 1–13. Elsevier.
2. Constantine, J. Aivalis. 2011. Log File Analysis Of E-Commerce Systems. In *Rich Internet Web 2.0 Applications, Panhellenic Conference on Informatics*. IEEE.
3. Zhao. 2013. Interest Before Liking: Two-Step Recommendation Approaches, 46–56. Elsevier.
4. Anto Praveena, M.D. 2017. A Survey Paper on Big Data Analytics. In *International Conference On Information, Communication & Embedded Systems (ICICES)*. IEEE.
5. Venkatkumar, Iyer Aurobind. 2016. Comparative Study Of Data Mining Clustering Algorithms. In *International Conference On Data Science And Engineering ICDE*. IEEE.
6. RamakrishnaMurty, M., J.V.R. Murthy, P.V.G.D. Prasad Reddy, Suresh. C. Sapathy. 2012. A survey of Cross-Domain Text Categorization Techniques. In *International conference on Recent Advances in Information Technology RAIT-2012*. ISM-Dhanabad, IEEE Xplorer Proceedings. 978-1-4577-0697-4/12.
7. Wang, Gang. 2016. Unsupervised Clickstream Clustering for User Behavior Analysis. In *Proceedings of the 2016 CHI Conference on Human Factors in Computing Systems*. ACM.
8. Abhaysingh. 2017. Predicting Demographic Attributes from Web Usage: Purpose and Methodologies. In *International conference on I-SMAC*. IEEE.
9. Joshila Grace, L.K. 2011. Analysis of Web Logs And Web User In Web Mining. *International Journal Of Network Security & Its Applications* (IJNSA), 3 (1), January.
10. Sergio, Herna´ndez. 2016. *Analysis of users' behaviour in structured e-commerce websites*. IEEE.
11. Ben Schafer, J. 2007. *Collaborative Filtering Recommender Systems*, 291–324. ACM DL.
12. Cai, Yi. 2013. *Typicality-based Collaborative Filtering Recommendation*. IEEE.
13. Ma, H., I. King, and M.R. Lyu. 2007. Effective Missing Data Prediction for Collaborative Filtering. In *SIGIR '07: Proceedings of the 30th annual international ACM SIGIR conference on Research and development in information retrieval*. New York, USA: ACM.
14. Hu, Y., Y. Koren, and C. Volinsky. 2008. Collaborative Filtering for Implicit Feedback Datasets. In *Proceedings of ICDM '08*. Washington, DC, USA: IEEE Computer Society.
15. Leung, K.W.-T., D.L. Lee, and W.-C. Lee. 2011. Clr: a collaborative location recommendation framework based on co-clustering. In *Proceedings of SIGIR '11*. ACM.
16. Chen, Lu, and Qiang Su. 2013. *Discovering User's Interest At E-Commerce Site Using Clickstream Data*. Hong Kong: IEEE.

Optimal Resource Allocation in OFDMA-LTE System to Mitigate Interference Using GA rule-based Mostly HBCCS Technique

Kethavath Narender and C. Puttamadappa

Abstract LTE has turned into true innovation for the 4G systems. It offers phenomenal information transmission and low dormancy for a few kinds of uses and administrations. In remote, expansive band gets to the systems, and a greater part of the indoor situations experiences genuine inclusion issue. The aim is to enhance the OFDMA-LTE execution, Genetic Algorithm (GA) and Hybrid Bee Colony and Cuckoo Search (HBCCS) strategies based Optimal Resource Allocation (RA) in OFDMA-LTE System presented. The ideal power esteems are refreshed to assign every one of the clients in the femto-cell and large-scale cell. Likewise, this enhanced streamlining procedure guarantees to uncover the relieved system framework as far as to flag the obstruction clamor proportion (SINR), ghastly proficiency, throughput and blackout likelihood (OP) over the ordinary strategies.

Keywords OFDMA-LTE · Signal-to-interference noise ratio · Bee colony algorithm · Genetic algorithm optimization · Femto-cell · Macro-cell · Cross-tier interference

1 Introduction

As of late, the world saw quickly developing remote innovation and expanding request of the remote correspondence administrations [1]. The remote transmission strategies that advances range utilization proficiency and empower high information rate correspondence over the multipath radio, for example, symmetrical recurrence division numerous entrance (OFDMA) have discovered across the board sending in

K. Narender
ECE Department, Visvesvaraya Technological University, Belagavi, Karnataka, India

C. Puttamadappa (✉)
ECE Department, Dayananda Sagar University, Karnataka, India
e-mail: puttamadappa@gmail.com

© Springer Nature Singapore Pte Ltd. 2020
K. S. Raju et al. (eds.), *Proceedings of the Third International Conference
on Computational Intelligence and Informatics*, Advances in Intelligent
Systems and Computing 1090, https://doi.org/10.1007/978-981-15-1480-7_46

current remote transmission advances [2]. Despite the fact that the OFDMA idea is basic in its essential guideline, fabricating a viable OFDMA framework is a long way from being an unimportant undertaking without a very much conceived RA calculation [3]. An effective calculation for subcarrier determination can fundamentally build the SINR that is important to improve the throughput in a dynamic situation [4]. The water-filling (WF) rule is inferred for Digital Multitude (DMT) weak frameworks, which assigns the data bits to the clients with the most astounding sign to-clamor proportion (SNR) transporters [5]. In an OFDMA arrange, the BS should ideally allot power and bits over various sub-transporters in view of prompt channel states of various dynamic remote terminals [6]. By applying a GA-HBCCS strategy, the ideal power esteems are refreshed to assign every one of the clients in the femto-cell and large-scale cell. The GA-HBCCS strategy enhances alleviated system framework execution as far as SINR, otherworldly productivity, throughput, and OP.

2 Literature Review

Xiao et al. [7] augmented the throughput of the femto-cell while staying away from extreme between level impedance with the full-scale cell by means of joint sub-channel task and power potion (PA). The sunken arched methodology changes the non-raised primal issue into a tractable frame through consecutive curved approximations and after that utilized the sub-slope strategy to tackle the double issues.

Tehrani et al. [8] considered a general RA issue in a heterogeneous OFDMA-based system comprising of defective FD full-scale BS and femto-BSs and both HD and blemished FD clients. The creator augmented the down-interface and up-connect weighted entirety rate of femto-clients while securing the large-scale client's rates. The weights are considered to use separate classes of administration, oblige both recurrence and time division duplex for HD clients, and organize up-interface or down-connect transmissions. The real impediment of this strategy is high computational complexity.

Zhao et al. [9] exhibited a joint confirmation control and a RA methodology for a symmetrical recurrence division with different access-based femto-cell arrangements. In this framework, clients were ordered into two kinds for an OFDMA-based femto-cell arranges: high-need (HP) and low-need (LP), of which HP clients qualified to appreciate a more recognized QoE; furthermore, a HP client has higher need to get to the system over all the LP clients in a similar dispute district. The significant confinement of this method is computational many-sided quality.

3 GA-HBCCS-ORA Method

The goal of the GA-HBCCS-ORA framework is to diminish the impedance happened between the femto-cell and large-scale cell. In correspondent frameworks, it is the normal rate of fruitful message conveyance over a channel. In view of the framework, the throughput and the obstruction are associated and consider the system of multi-client OFDMA with mitigate interference, which is appeared in Fig. 1.

The resource of the framework has been designated by GA-based HBCCS technique with the assistance of existing allotted power, distance and BER. Both the algorithms separately optimize the power value. At the time of PA, this system compares the GA output and HBCCS algorithm output, and the lesser value will get allocated to the transmission. This method reduces the BER as well as improves the spectral efficiency and throughput.

3.1 Modulation

In MIMO-OFDMA, the QPSK modulation is employed and it is a digital modulation technique. Quadrature means the signal shifts between the states of phases that are divided into 90°. QPSK increases the signal to 90° from 45° to 135°, −45° (315°), or −135°(225°). The constellation diagram for the QPSK modulation is given in Fig. 2.

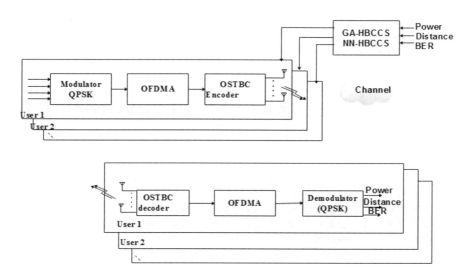

Fig. 1 Block diagram of GA-HBCCS-ORA methodology

Fig. 2 Constellation diagram
for QPSK modulation

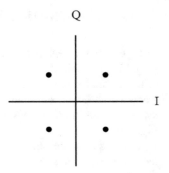

3.2 Orthogonal Space–Time Block Code Scheme

After performing the modulation process, the data packets are sent to the OSTBC encoder and OSTBC decoder for encoding and decoding the data packets, respectively. Assume that OSTBC has N_T of transmit antennas and N_R of receive antennas. The following matrix rows represent different time instant, and the following matrix columns denote the transmitted symbol along with each different antenna.

$$\begin{bmatrix} s_{11} & \cdots & s_{1nT} \\ \vdots & \ddots & \vdots \\ s_{T1} & \cdots & s_{TnT} \end{bmatrix} \tag{1}$$

OSTBC encoder is used for mapping the modulated symbols of transmission matrix. Consider the OFDM system has N amount of sub-carriers and within that sub-carriers only N_A amount of sub-carriers are active. The remaining sub-carriers are called as virtual sub-carriers derived from Eq. (2).

$$N_V = N - N_A \tag{2}$$

The OFDM symbols transmitted through the OSTBC are expressed in Eq. (3).

$$S_j = \left[O_{N_V/2 \times 1} S_j^{-T} O_{N_V/2 \times 1} \right] \tag{3}$$

where the guard band is represented as O's and the data vector of length is j, and it is calculated from Eq. (2). In first OFDM symbol period, s_1 and s_2 are transmitted from the transmit antenna, respectively. Time-domain signal is obtained by using the IFFT operation in frequency domain signal. The encoding matrix of OSTBC encoder is given in Eq. (4).

$$S = \begin{pmatrix} s_1 & s_2 \\ -s_2 & s_1 \end{pmatrix} \tag{4}$$

The obtained OFDM symbols are given in Eqs. (5) and (6).

$$y_1^* = h_{1,1}s_1 + h_{2,1}s_2 + n_1 \tag{5}$$

$$y_2^* = -h_{1,1}s_2 + h_{1,2}s_1 + n_2 \tag{6}$$

where, y_1^* and y_2^* are the received 1 symbol period from antennas 1 and 2, $h_{j,k}$ describes as the NN transmission matrix from the kth transmit antenna to the jth receive antenna, n_1 and n_2 are $N1$ complex Gaussian random noise.

$$y_1 = Fy_1^* + Fn_1^* \tag{7}$$

$$y_2 = Fy_2^* + Fn_2^* \tag{8}$$

Equation (9) shows that the symbols from the OFDM.

$$\begin{bmatrix} y_1 \\ y_2 \end{bmatrix} = \begin{bmatrix} h_{1,1} & h_{2,1} \\ h_{2,2} & -h_{1,2} \end{bmatrix} * \begin{bmatrix} s_1 \\ s_2 \end{bmatrix} + \begin{bmatrix} n_1 \\ n_2 \end{bmatrix} \tag{9}$$

The received signal of Eq. (10) is achieved by rewriting Eq. (9).

$$y = Hs + n \tag{10}$$

4 Genetic Algorithm

GA was introduced by John Holland in the 1970s. Throughout the years, GA ended up a standout among the most well-known metaheuristics in critical thinking. GA is a populace-based metaheuristic enlivened by organic procedures and regular advancement. In GA, we make a populace of people. Every one of them speaks to an answer for the specific issue. This is one of the primary qualities of GA. At each progression, GA can manage an arrangement instead of a solitary one. This methodology permits a superior investigation of the hunt space (Fig. 3).

4.1 Hybrid Bee Colony and Cuckoo Search (HBCCS)

HBCCS is thought to be a standout among the latest metaheuristic as well as swarm intelligent calculations (SI) like GA, particle swarm enhancement, ant colony optimization (ACO), and differential evaluation (DE). It depends on rearing and

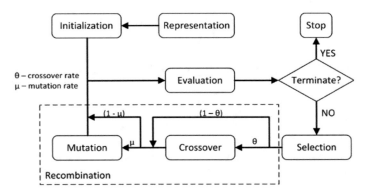

Fig. 3 Genetic algorithm working

exact flight searching conduct of the cuckoo flying creatures. The working of the HBCCS is appeared in Fig. 4.

5   Results and Discussion

GA-HBCCS-ORA framework was executed by utilizing MATLAB 2018a programming device for the reproduction reason in the Intel i5 work area registering condition with 8 GB RAM memory limit. In this GA-HBCCS-ORA framework has

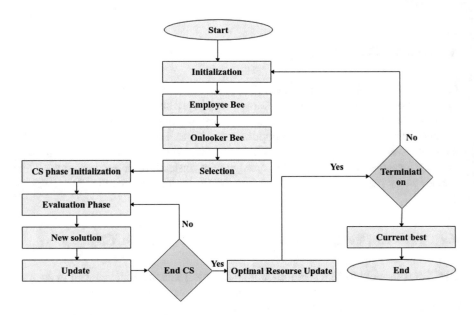

Fig. 4 HBCCS algorithm working

significantly comprised of resource assignment with GA and HBCCS plots in femto-cell arrange-based OFDMA-LTE framework (Table 1).

In the recreations, the full-scale cell has an inclusion sweep of 500 m. Each femto-cell has an inclusion range of 10 m. K FBSs and 50 large-scale clients are arbitrarily circulated in the full-scale cell inclusion region. The base separation between the MBS and a full-scale client (or a FBS) is 50 m. The base separation between FBSs is 40 m. Femto-clients are consistently dispersed in the inclusion region of their serving femto-cell. Both full-scale and femto-cells utilize a transporter recurrence of 2 GHz, $B = 10$ MHz, and $N = 50$. The AWGN fluctuation is given by $\sigma_2 = BNN0$, where $N_0 = -174$ dBm/Hz.

OFDMA-LTE environment with the femto-cell interfering users defines the user in the femto-cell base station is conflict with macro-cell base station. Figure 5 shows the convergence rate of the HBCCS-ORA and GA-HBCCS-ORA methods. Figure 6 shows the capacity of the macro-cell when the number of femto-users per femto-cell increases from 1 to 6, for $K = 20$, 30, and 50.

It can be observed that the GA-HBCCS-ORA algorithm outperforms and HBCCS-ORA algorithm by up to a 23‰ increases in macro-cell capacity so the number of femto-cell users also increase in Fig. 7. Therefore, the GA-HBCCS-ORA algorithm is more and more superior compared to the UHBCCS-ORA algorithm. Figure 8 shows the SINR of the connection between the base station k and user i (SINR$_i$) and OP (P_{out}). It describes when the SINR$_i$ rapidly increased; the outage probability was reduced rapidly in our work compared with HBCCS methods.

Table 1 Simulation parameters

GA-HBCCS-ORA system testing	
Data bits	5000 bits data's (with pocket size of 25)
Sampling rate	1e6
Path delays	0 to 2e−6
Path gain	0 to −10
Modulation and demodulation	QPSK
Channel encoding and decoding	OSTBC
Data encoding	Turbo coding
SNR value for analysis	−35:10:45
Total available bandwidth	50 MHz
Maximum transmitted power	23 dBm
Macro-cell coverage radius	500 m
Number of macro-cell base station	3
Maximum number of users in FC	5
Sub-carrier bandwidth	15 kHz
Maximum number of mobile users	50

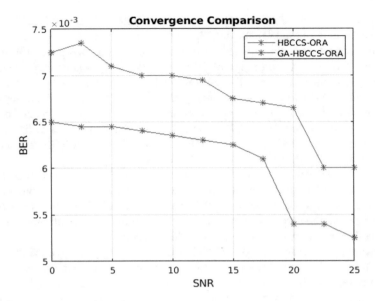

Fig. 5 Convergence rate comparison

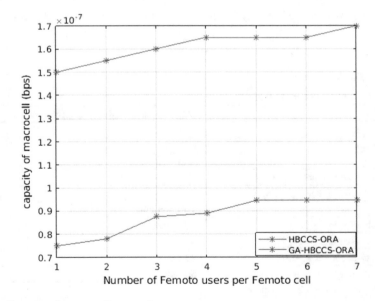

Fig. 6 Capacity of macro-cell versus femto-users

Fig. 7 Outage probability comparison

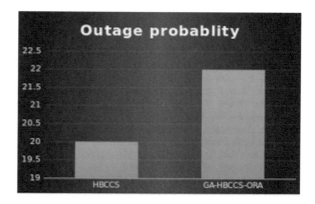

Fig. 8 Signal-to-interference and noise ratio comparison

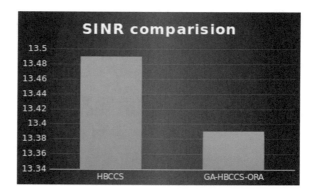

6 Conclusion

In this paper, we examined the RA in OFDMA-LTE condition. In this sense, we proposed another system in light of GA and HBCCS. We thought about a situation with information transmission and assessed the GA-HBCCS-ORA calculation utilizing MATLAB and looked at its execution against the HBCCS booking strategy. From the results, we trusted that GA-HBCCS-ORA can be an imperative apparatus in LTE up-interface RA. In our assessment, the GA-HBCCS-ORA calculation displayed an unrivaled execution in terms of SINR, throughput, spectral effectiveness, and OP.

References

1. Govil, Jivesh, and Jivika Govil. 2008. An Empirical Feasibility Study of 4G's Key Technologies. In *IEEE International Conference on Electro/Information Technology, EIT 2008*, 267–270. IEEE.
2. Snow, Chris, Lutz Lampe, and Robert Schober. 2007. Performance Analysis and Enhancement of Multiband OFDM for UWB Communications. *IEEE Transactions on Wireless Communications* 6(6).
3. Choi, Young-June, Kwang Bok Lee, and Saewoong Bahk. 2007. All-IP 4G Network Architecture for Efficient Mobility and Resource Management. *IEEE Wireless Communications* 14 (2).
4. Martinek, Radek, and Jan Zidek. 2014. The Real Implementation of ANFIS Channel Equalizer on the System of Software-Defined Radio. *IETE Journal of Research* 60 (2): 183–193.
5. Shams, Farshad, Giacomo Bacci, and Marco Luise. 2014. A Survey on Resource Allocation Techniques in OFDM (A) Networks. *Computer Networks* 65: 129–150.
6. Cover, Thomas M., and Joy A. Thomas. 2006. Introduction and Preview. In *Elements of Information Theory*, 2nd edn., 1–12.
7. Xiao, S., X. Zhou, Y. Yuan-Wu, G.Y. Li, and W. Guo. 2017. Robust Resource Allocation in Full-Duplex-Enabled OFDMA Femtocell Networks. *IEEE Transactions on Wireless Communications* 16 (10): 6382–6394.
8. Tehrani, Peyman, Farshad Lahouti, and Michele Zorzi. 2018. Resource Allocation in Heterogenous Full-duplex OFDMA Networks: Design and Analysis. arXiv preprint arXiv:1802.03012.
9. Zhao, Feifei, Wenping Ma, Momiao Zhou, and Chengli Zhang. 2018. A Graph-Based QoS-Aware Resource Management Scheme for OFDMA Femtocell Networks. *IEEE Access* 6: 1870–1881.

Review of Techniques for Automatic Text Summarization

B. Shiva Prakash, K. V. Sanjeev, Ramesh Prakash, K. Chandrasekaran, M. V. Rathnamma and V. Venkata Ramana

Abstract Summarization refers to the process of reducing the textual components such as words and sentences but conveying most of the information in the input text. Research in summarization is very prominent in the current scenario where the textual data available is enormous and contains valuable information. People have been interested in summarization since time immemorial. The methods adopted in the past relied on manually reading the text and based on one's understanding of the text, manually generating the summary. In the current world, due to the explosion of data from Internet and social media, the manual process is very tedious and time-consuming. As a result, there is a great need to automate the process of summarization. In this paper, we summarize most of the researches in the field of summarization which is unique and path-breaking.

Keywords Automatic text summarization · Extraction-based summarization · Abstraction-based summarization

B. Shiva Prakash · K. V. Sanjeev · R. Prakash · K. Chandrasekaran (✉)
Department of Computer Science and Engineering, National Institute of Technology
Karnataka Surathkal, Mangalore, India
e-mail: kch@nitk.ac.in; kchnitk@ieee.org

B. Shiva Prakash
e-mail: shiva96b@gmail.com

K. V. Sanjeev
e-mail: sanjeev.vadiraj@gmail.com

R. Prakash
e-mail: rameshprakash6196@gmail.com

M. V. Rathnamma
Kandula Srinivasa Reddy Memorial College of Engineering, Kadapa,
Andhra Pradesh, India
e-mail: rathnamma@ksrmce.ac.in

V. Venkata Ramana
Chaitanya Bharathi Institute of Technology, Proddatur, Andhra Pradesh, India
e-mail: ramanajntusvu@gmail.com

© Springer Nature Singapore Pte Ltd. 2020
K. S. Raju et al. (eds.), *Proceedings of the Third International Conference on Computational Intelligence and Informatics*, Advances in Intelligent Systems and Computing 1090, https://doi.org/10.1007/978-981-15-1480-7_47

1 Introduction

Any process which reduces human effort by removing his/her involvement in the task is called an automatic process. In any language, in general, a text is a sequence of symbols from its character set, a subsequence of which form constructs like words and sentences which have individual existence and meaning. Summarizing is the process of generating a text whose length is lesser than the original text compromising as little as possible on the information conveyed in the original text. Automatic text summarization refers to summarizing without human involvement.

Baby steps in the field of automatic text summarization started with the research paper in 1952 by H P Luhn of IBM [1, 2]. Many approaches and models have been proposed since then in this field, but most of them were very basic with stress on bag-of-words approach, word ranking, and sentence ranking among many other similar approaches till the year 1990. From the 1990s, the research was based on many statistical and machine learning algorithms.

All the algorithms in automatic text summarization (ATS) can be classified broadly under two main approaches, namely extractive and abstractive summarization. Text compression is another domain, which is generally included under extractive summarization. Choosing a subset of the sentences from the input text based on some algorithm is called extraction, whereas generating a summary that does not have the same structure of the input sentences but conveys the overall information present in the input is called abstraction. A human-generated summary would be an abstractive summary. Most of the research is on extractive summarization due to the complexities involved in generating coherent and grammatically correct sentences in abstractive summarization.

Automatic text summarization will play a huge role in improving analytics. Conveying all the opinions of expert business analysts in a small piece of text would help the investors in making better and profitable decisions. In many contexts where the text contains valuable information hidden in huge volume, ATS is helpful in finding those gold bits from the haystack. It would increase the efficiency of processing and assimilating any text, as a result, being of great help in many activities.

2 Key Concepts in ATS

Automatic text summarization working can be broken down in terms of many modules with the different set of tasks. One of the most important modules in the text preprocessing module includes lexicon reduction and word normalization [3, 4]. Preprocessing results in a text which is significantly easy to manage and occupies lesser space. This stage includes the routine operations applied on a textual data such as removing stop words, lemmatization, and tokenization. Once preprocessing is done, the task is to apply the algorithm on the processed text and then evaluate the performance of the suggested approach (Fig. 1).

Fig. 1 Preprocessing text [1, 6]

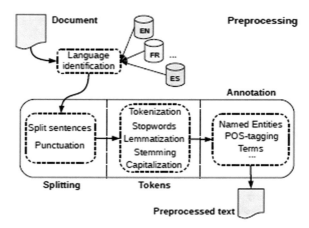

2.1 Extraction

Picking the important sentences from the given input text without any changes to the structure or composition of that sentence is called extraction. The only task of extractive algorithms is to rank the sentences appropriately according to certain heuristics. Once the ranking is done, the top K sentences are picked from the list as a summary to the input text. The value of K can either be user-defined or a default value. Extractive summarization algorithms are the easiest to develop and implement. However, the sentences in the output summary have many shortcomings. Sentences may not be coherent and might be confusing for the reader. Cohesion and coherence being one of the frequently used metrics to evaluate a summary and extraction suffer in this context [5].

2.2 Abstraction

Identifying the important information present in the input text and representing the same in a different and concise manner are called abstractive summarization. The output of this approach would be very similar to the ones generated by a human and work really well for multi-document summarization.

2.3 Aided Summarization

Machines have helped man in many tasks, and aided summarization is one such example. This method is an ensemble of both extractive and abstractive summarization. Given a huge input text, the machines perform a very efficient extractive summarization.

3　Literature Survey

H P Luhn (1958) [2] is the main paper that started the examination in automatic text summarization. Despite the fact that the algorithm proposed in the paper is exceptionally fundamental, it is a noteworthy gradual step. The creator positions the words in light of their frequencies. As the most successive words are the stop words, the upper bound (C) for the consideration of words is chosen. In the same fashion, a lower bound (D) is chosen to disregard certain words in view of their low recurrence in the record. Sentence positioning is done utilizing the words in the sentence and their relative position. The paper suggests that noteworthy words which are near each other in a sentence frame a sentence which is of high significance. As the separation between the words builds, the significance of the sentence diminishes. The paper considers just the physical structure of the record and not the scholarly limit of the creator of the archive. The summary produced is subsequently, exceedingly reliant on the recurrence and no other parameters [2].

Table 1 shows a brief description of different ATS approaches, describing their uniqueness and advantages over the previous algorithms.

Table 1　Automatic text summarization algorithms

Algorithm or author	Year	Description	Advantage
Hans Peter Luhn [2]	1958	Frequency of important words as a metric for determining the significant factor of a sentence. No given to the meaning of words	First published work; low production cost; reduction of problems relating to subjectivity and variability; stemming (normalization)
Baxendale [1, 3]	1958	Position of the sentence	Naive but effective, as many texts have the first and last sentence important
Edmundson's linear combination [1]	1959–68	Scoring sentences in a corpus in order to rank linear combination	First approach to removal of stop words
Extracts by elimination (rush) [1]	1971	Keyword-based search	Sentence rejection rather than selection (change of approach); ADAM: First commercialized summarizer
Vasiliev [1, 3]	1963	Statistical approach, descriptor approach and a semantic-logical approach to ATS; fixed phrase feature	First report on the state of abstraction research
Bayes classifier [7, 8]	Late 1960s	Disadvantage: learning set was very small, domain-specific	First machine learning approach

(continued)

Table 1 (continued)

Algorithm or author	Year	Description	Advantage
Rush, Pollock, and Zamora [1]	1971	Specialized corpus to generate the summary	ATS is more successful when the model is trained on a specific genre
Yale Artificial Intelligence Project [1, 3]	1978	Linguistic knowledge along with a pragmatic and semantic knowledge frame; used a structure called sketchy script	High-level reasoning system with low-level text analyzer
Sparck-Jones [1, 3]	1990	Uses rhetorical structure theory, linguistic sources, domain sources, and communicative sources	Based on human summarizing
Multi-document summarization [1]	1998	Single summary from many documents	All the important information in a concise, single document
Graph-based approach [1]	1991	Parts-of-speech (POS) tagging, information extraction, bag-of-words	–
Marcu [9]	Mid 1990s	Rhetorical parsing algorithm, rhetorical relations, discourse trees	–
Pagerank [10]	1999	User search space and directional hyperlink	Google's webpage ranking
Lexrank and Textrank [7, 11]	2004	Constructing the graph; Implementing link analysis algorithms; concept of eigenvector centrality in a graph representation of sentences	–
Barzilay and Elhadad [12]	1998	Topic progression, WordNet, POS tagger, shallow parser, and segmentation algorithm	Summarize texts without requiring full semantic interpretation
Carbonell and Goldstein [13]	2000	Maximum marginal relevance (MMR), multi-document summarization	Reducing redundancy while maintaining relevance
Witbrock and Mittal [14]	2000	Statistical model. Term selection, term ordering, and style learned from the training corpus	–
Hovy and Lin [15]	1998	Three-phase summarization system	–
Knight and Marcu [16]	the Early 2000s	Decision-based model to reduce text and compress sentences	Coherent output
Radev, Jing, and Budzikowska [17]	2000	Multi-document summarizer (MEAD). Cluster centroids, topic detection, and topic tracking	–

(continued)

Table 1 (continued)

Algorithm or author	Year	Description	Advantage
Saggion and Lapalme [18]	2002	Technical text to indicative, informative summary	Elaborates topics based on reader's interests
Erkan and Radev [11]	2004	LexRank for relative importance of textual units. Graph-based centrality score and similarity graph of sentences	Better view of important sentences compared to centroid approach
Barzilay and McKeown [19]	2005	Bottom-up local multi-sequence alignment to identify phrases conveying similar information and statistical generation to combine common phrases into a sentence	Combine the information in all the documents with sentences, not in any of the original documents
Fernandez, SanJuan, and Torres-Moreno [20]	2007	Neural network based on textual energy	Good results in ATS and topic segmentation
Svore, Vanderwende, and Burges [21]	2007	Net sum for single doc summarization. Both neural network and third-party datasets for features	–
Saggion [22]	2008	Toolkit called SUMMA, which presents a set of adaptable summarization components together with well-established evaluation tools	Resources for computation of summarization features
Filippova [23]	2010	Multi-sentence compression	Compressed and grammatical sentences without any parser or handcrafted linguistic rules
Torres-Moreno [24]	2012	ARTEX—inner product, document vector and lexical vector between sentences	Not require any linguistic knowledge
Litvak and Vanetik [9]	2014	Tensor-based representation	–

4 Analysis

4.1 Classification

There are a plethora of automatic text summarizers available. A complete understanding of each summarizer along with its variations from its adversaries is the elementary step in deciding the appropriate summarizer for the text. Therefore, the art of summarizing is still a field that is appealing. Some of the parameters that can be used to distinguish and classify summarizers are listed below.

- Approach based.
- Based on detail.
- Based on information retrieval.

4.2 Evaluation Techniques

Evaluation is not a general (fully automated) task in the present research community but a task-based, goal-based, semiautomatic approach. Mainly there are two types of evaluation, viz. intrinsic and extrinsic. Intrinsic evaluation deals with the metrics which are already present in the text, whereas extrinsic evaluation deals with satisfying a particular task or a goal. An example of extrinsic evaluation would be a requirement—'The summary should be as positively worded as possible.' The summarizer which comes up with a summary which conveys the same meaning in a positive way will be evaluated better in this extrinsic evaluation.

- Intrinsic.
- Extrinsic.

5 Open Research Problems

Until very recently most of the research in the field of summarization was extraction based as it is easier to conceptualize and also implement. Another main reason for not favoring abstractive summarization is that natural language processing was just in its infancy. Automatic text summarization has a great potential to impact the entire world. Research in this field should now be more focused on generating coherent abstractive summaries. In the generated summary, the text quality is the most important factor. While generating the summary using abstractive methods, there are many problems in understanding the text to generate the intermediate representation, like the word sense disambiguation. Domain-independent abstractive summarization is thus wide open and a very interesting topic to research upon.

6 Conclusion

The huge growth of data has created an immense need for automating the process of text summarization in today's world. There are still so many issues which should be addressed in this field through the research started about six decades back. Unlike those days where summarization was of interest only for scientific journals, today summarization is important in almost all the domains of knowledge and record. The research in this field has been appreciable all these years and has picked up pace in the past decade or two because of well-established statistical and machine learning approaches. Contribution to this field of research will improve the efficiency of information assimilation and aid many generations to come.

References

1. Torres-Moreno, Juan-Manuel. 2014. *Automatic Text Summarization*. Wiley.
2. Fattah, Mohamed Abdel, and Fuji Ren. 2008 Automatic Text Summarization. *World Academy of Science, Engineering and Technology* 37 (2008).
3. Das, Dipanjan, and André F.T. Martins. 2007. A Survey on Automatic Text Summarization. *Literature Survey for the Language and Statistics II course at CMU* 4: 192–195.
4. Church, Kenneth, and William Gale. 1999. *Inverse Document Frequency (idf): A Measure of Deviations from Poisson. Natural Language Processing Using Very Large Corpora*, 283–295. Netherlands: Springer.
5. RamakrishnaMurty, M., J.V.R Murthy, P.V.G.D. Prasad Reddy, and Suresh. C. Sapathy. 2012. A Survey of Cross-Domain Text Categorization Techniques. In *International conference on Recent Advances in Information Technology RAIT-2012*, ISM-Dhanabad, IEEE Xplorer Proceedings. 978–1-4577-0697-4/12.
6. Mani, Inderjeet. 2001. *Automatic Summarization*, vol. 3. John Benjamins Publishing.
7. Erkan, Günes, and Dragomir R. Radev. 2004. LexRank: Graph-based Lexical Centrality as Salience in Text Summarization. *Journal of Artificial Intelligence Research* 22: 457–479.
8. Mihalcea, Rada. 2004. Graph-Based Ranking Algorithms for Sentence Extraction, Applied to Text Summarization. In *ACL 2004 on Interactive Poster and Demonstration Sessions*. Association for Computational Linguistics.
9. Carbonell, Jaime, and Jade Goldstein. 1998. The Use of MMR, Diversity-Based Reranking for Reordering Documents and Producing Summaries. In *21st Annual International ACM SIGIR Conference on Research and Development in Information Retrieval*. ACM.
10. Radev, Dragomir R., Hongyan Jing, and Malgorzata Budzikowska. 2000. Centroid-Based Summarization of Multiple Documents: Sentence Extraction, Utility-Based Evaluation, and User Studies. In *NAACL-ANLP Workshop on Automatic summarization. Association for Computational Linguistics*.
11. Page, Lawrence, et al. 1990. *The PageRank Citation Ranking: Bringing Order to the Web*.
12. Lin, Chin-Yew, and Eduard Hovy. 2002. Manual and Automatic Evaluation of Summaries. In *ACL-02 Workshop on Automatic Summarization-Volume 4*. Association for Computational Linguistics.
13. Litvak, Marina, and Natalia Vanetik. 2014. Multi-document Summarization Using Tensor Decomposition. *Computación y Sistemas* 18 (3): 581–589.
14. Marcu, Daniel. 1998. Improving Summarization Through Rhetorical Parsing Tuning. In *The 6th Workshop on Very Large Corpora*.

15. Robertson, Stephen. 2004. Understanding Inverse Document Frequency: On Theoretical Arguments for IDF. *Journal of documentation* 60 (5): 503–520.
16. Banko, Michele, Vibhu O. Mittal, and Michael J. Witbrock. 2000. Headline Generation Based on Statistical Translation. In *Proceedings of the 38th Annual Meeting on Association for Computational Linguistics*. Association for Computational Linguistics.
17. Hovy, Eduard, and Chin-Yew Lin. 1998. Automated Text Summarization and the SUMMARIST System. In *Proceedings of a workshop on held at Baltimore*, Maryland: October 13–15, 1998. Association for Computational Linguistics.
18. Knight, Kevin, and Daniel Marcu. 2000. Statistics-Based Summarization-step One: Sentence Compression. *AAAI/IAAI* 2000: 703–710.
19. Torres-Moreno, Juan-Manuel. 2012. *Artex is Another Text Summarizer*. arXiv preprint arXiv:1210.3312.
20. Barzilay, Regina, and Michael Elhadad. 1999. Using Lexical Chains for text Summarization. *Advances in Automatic Text Summarization* 111–121.
21. Fernandez, Silvia, Eric SanJuan, and Juan Manuel Torres-Moreno. 2007. Textual Energy of Associative Memories: Performant Applications of Enertex Algorithm in Text Summarization and Topic Segmentation. In *Mexican International Conference on Artificial Intelligence*. Springer, Berlin Heidelberg.
22. Svore, Krysta Marie, Lucy Vanderwende, and Christopher JC Burges. 2007. *Enhancing Single-Document Summarization by Combining RankNet and Third-Party Sources*." EMNLP-CoNLL.
23. Saggion, Horacio. 2008. A Robust and Adaptable Summarization Tool. *Traitement Automatique des Langues* 49 (2).
24. Filippova, Katja. 2010. Multi-Sentence Compression: Finding Shortest Paths in Word Graphs. In *Proceedings of the 23rd International Conference on Computational Linguistics*. Association for Computational Linguistics.

Data Mining Task Optimization with Soft Computing Approach

Lokesh Gagnani and Kalpesh Wandra

Abstract The data mining task optimization (DMTO) is one of the emerging research areas in the branch of data mining. We discuss the data mining tasks as classification and clustering. The optimization of these tasks is done using soft computing methods. The soft computing is also a new term coined for optimization problems. The classification task is achieved using the support vector machine and results optimized with particle swarm optimization and simplified swarm optimization with exchange local strategy (SSO with ELS). The clustering task is achieved using the kernel fuzzy c-means (KFCM) and results optimized using particle swarm optimization (PSO) and intelligent firefly algorithm (IFA). The results of both the tasks are worked out in the same paper. The results obtained outperform the existing approach SVM with PSO with cuckoo search (CS) and PSO with social spider optimization (SSO) for classification, KFCM with PSO with Bacteria Foraging Optimization (BFO) for clustering in data mining task optimization framework.

Keywords Data mining task optimization · Soft computing · Classification · Clustering · Support vector machine · Kernel fuzzy c-means

---

Please note that the LNCS Editorial assumes that all authors have used the western naming convention, with given names preceding surnames. This determines the structure of the names in the running heads and the author index.

L. Gagnani (✉)
Department of Computer Engineering, C U Shah University,
Wadhwan, Gujarat 363030, India
e-mail: gagnani.lokesh@gmail.com

K. Wandra
Department of Computer Engineering, GMB Polytechnic, Rajula, Gujarat, India

© Springer Nature Singapore Pte Ltd. 2020
K. S. Raju et al. (eds.), *Proceedings of the Third International Conference on Computational Intelligence and Informatics*, Advances in Intelligent Systems and Computing 1090, https://doi.org/10.1007/978-981-15-1480-7_48

1 Introduction

Data mining task optimization has recently emerged as a hot topic in data mining. Data mining is the extraction of useful patterns from data repositories. In this, the extraction process may not give interesting patterns and so various optimization methods are employed [1].

Soft computing is also one of the topics in optimization that has gained importance nowadays. It is a multi-disciplinary term that combines or hybridizes the various technologies like artificial intelligence, fuzzy logic, evolutionary computing, machine learning for optimization [2].

In this paper, the DMTO framework takes the data mining tasks of classification and clustering. Classification and clustering of data in DMTO framework with soft computing approach give better results compared to existing techniques in terms of optimized values. In context of machine learning, classification is supervised learning and clustering is unsupervised learning.

The paper is organized, as Sect. 1 gives brief overview of data mining tasks in DMTO. Section 2 gives the architecture of DMTO framework. Sections 3 and 4 summarize the classification and clustering techniques as data mining tasks, respectively. Section 5 shows the experiments conducted on UCI datasets. Section 6 concludes the paper.

2 DMTO Framework

The data mining task optimization framework (DMTO) optimized with soft computing (SC) approach is shown in Fig. 1. It consists of four steps and then the results are compared using benchmark functions, evaluation parameters, or non-parametric statistical tests. Here classification and clustering are considered as data mining tasks, and their results are optimized using machine learning and fuzzy logic, respectively.

The four steps can be summarized as:

1. Input the dataset.
2. Preprocess/filter data (normalization to numerical data). This step is optional depending on the dataset considered.
3. Consider any task—classification/clustering.
4. For clustering optimization is done with fuzzy system and soft computing techniques.
   For classification optimization is done with machine learning and soft computing techniques.
5. Evaluation of results. evaluation is done based on parameters of task being considered.

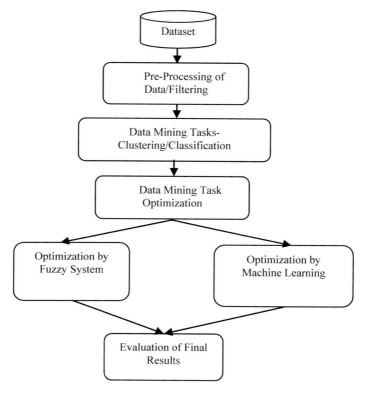

Fig. 1 DMTO framework with soft computing approach

The SC approaches employed in clustering comprise of kernel-based fuzzy c-means (KFCM) [3], in the classification is support vector machine (SVM). The results are further optimized by hybridizing the particle swarm optimization–intelligent firefly algorithm (PSO-IFA) method in clustering and by simplified swarm optimization with exchange local search (SSO-ELS) [4] and particle swarm optimization (PSO) in classification. The evaluation is done on the basis of objective function (OF) in clustering task and four parameters—accuracy, recall, F1 measure, and precision in the classification task. It is done by comparing with existing methods such as bacteria foraging optimization–particle swarm optimization (BFO-PSO) for clustering task and cuckoo search–particle swarm optimization (CS-PSO), social spider algorithm–particle swarm optimization (SSA-PSO) for the classification task.

The DMTO framework produces optimized results when compared with existing methods for classification as well as clustering tasks.

3   Classification Optimization

Classification is prediction of output based on some trained input. In classification, there are training and testing data. If the data is linearly separable classification becomes easy but if the data is not linearly separable, then classification becomes a bit typical task. For this, optimization of support vector machine (SVM) can be done to convert this nonlinearly separable data into linear in some high-dimensional space. The nonlinearly separable data for this is shown in Fig. 2.

In SVM, main criteria are proper selection of hyper-parameters such C and Epsilon. They are referred to as penalty factor and insensitive parameter, respectively. The empirical risk becomes more on effect of larger penalty and precision decrease with larger value of sensitive parameter. Hence, hyper-parameters are very sensitive to initial values.

3.1   Existing Method

In classification using DMTO framework, two existing techniques are employed.
Approach1 [5]:
This method of classification employs three stages:

1. Finding the initial values of SVM hyper-parameters by cuckoo search (CS) method.
2. Finding the best value of optimization through the particle swarm optimization (PSO).
3. The best values are fed to SVM for getting accuracy.

Cuckoo search is a novel approach developed by Yang [6] which has behavior similar to obligate blood parasitism. It lays the eggs in the nests of other species—host birds.

Fig. 2 Non-linearly separable data for classification

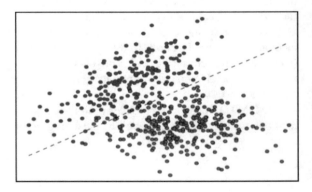

The Levy flight is essentially a random walk with random step using a Levy distribution:

$$\text{Levy} \sim u = t^{-\lambda}(1 < \lambda \leq 3). \tag{1}$$

PSO is a swarm intelligence technique mainly for optimization purposes. A particle is represented as a single solution in PSO search space. Fitness function is indicated by fitness values of all particles and velocities by the flying of particles. Velocity and position of particles are updated by following equations, respectively, till the optimal solution is reached.

$$V_i(i+1) = \omega \cdot V_i(i) + C1 \cdot \emptyset_1 \cdot (P_{\text{best}} - X_i(i)) + C2 \cdot \emptyset_2 \cdot (G_{\text{best}} - X_i(i)) \tag{2}$$

$$X_i(i+1) = X_i(i) + V_i(i+1) \tag{3}$$

Approach 2:
This method also employs three steps as follows:

1. Finding the initial values of SVM hyper-parameters by social spider optimization algorithm (SSA).
2. Finding the best value of optimization through the particle swarm optimization (PSO).
3. The best values are fed to SVM for getting accuracy.

In SSA [7], we formulate the search space of the optimization problem as a hyper-dimensional spider web. Hence, a spider moves to a new position, and it generates a vibration which is propagated over the web. Each vibration holds the information of one spider, and other spiders can get the information upon receiving the vibration. In SSA, all spiders follow positions constructed by other's current positions and their own historical positions.

3.2 Proposed Method

The three stages are:

1. Finding the initial values of SVM hyper-parameters by simplified swarm optimization (SSO) with exchange local strategy (ELS).
2. Finding the best value of optimization through the particle swarm optimization (PSO).
3. The best values are fed to SVM for getting accuracy.

The optimal parameters are found from simplified swarm optimization with exchange local search strategy (SSO-ELS) [4, 8] in which range of parameters is taken as lower bound and upper bound. These optimal values taken as input in

Attribute 1	LB	UB	..	Attribute N	LB	UB	Class X

Fig. 3 Particle position encoding

particle swarm optimization (PSO) for finding best parameters and that are entered into support vector machines (SVM) for classification accuracy. The encoding of particle's position in SSO-ELS is illustrated in Fig. 3.

4 Clustering Optimization

Clustering is an unsupervised form of technique in data mining. Its aim is to divide the data into various clusters depending on the property of high similarity in the same cluster data and low similarity in different cluster data. This encounters the problem that some data cannot be grouped into cluster due to overlapping of clusters [9, 10]. Hence, kernel-based fuzzy c-means (KFCM) [3] solve this problem as it is insensitive to noise and shape of clusters. It achieves this by converting the space into some high-dimensional space with the use of kernel functions such as Gaussian radial basis function (RBF). It is given by the following equation.

$$K(x_j, c_i) = e^{\frac{-\left\|x_j - c_i\right\|^2}{2\sigma^2}} \tag{4}$$

The kernel functions map the input space into some high dimensionality so that data is linearly separable and that overcomes the data in overlapping clusters. The main aim is to minimize the objective function which indicates increasing similarity of data in the same cluster and decreasing that in data of different clusters. It is given by:

$$J = 2\sum_{i=1}^{c} * \sum_{j=1}^{n} u_{ij}^m \left(1 - K(x_j, c_i)\right) \tag{5}$$

The membership u_{ij}^m is calculated as:

$$u_{ij} = \frac{\left(1 - K(x_j, c_i)\right)^{\frac{-1}{(m-1)}}}{\sum_{i=1}^{c} \left(1 - K(x_j - c_i)\right)^{\frac{-1}{(m-1)}}} \tag{6}$$

The center c_i is calculated as:

$$c_i = \frac{\sum_{j=1}^{n} K(x_j, c_i) x_j}{\sum_{j=1}^{n} u_{ij}^m K(x_j, c_i)} \tag{7}$$

The centers are optimized by the hybridization of PSO-IFA method and compared with the existing BFO-PSO method [11, 12]. Accuracy is calculated based on the membership values and the confusion matrix. Accuracy is the ratio of summation of true positive (TP) and true negative (TN) to total rows. Accuracy is to be higher, and objective function is to be lower.

4.1 Existing Approach (Hybrid BFO-PSO)

BFO is an inspired by the bacteria's searching food behavior present in human body. The solution towards optimality is achieved by processes such as Chemotaxis (cost is calculated), and other three.

The tumble direction is given by:

$$\emptyset(j+1) = V \cdot \emptyset(j) + C_1 \cdot R_1 \cdot (P_{\text{Lbest}} - P_{\text{current}})$$
$$+ C_2 \cdot R_2 \cdot (P_{\text{Gbest}} - P_{\text{current}}) \tag{8}$$

4.2 Proposed Approach (Hybrid IFA-PSO)

The main logic of our intelligent firefly algorithm (IFA) [13] is to make use of the ranking information such that every firefly is moved by the attractiveness of a fraction of fireflies only and not by all of them. Ø is a highest portion of the fireflies based on their rank. It is a modification of firefly algorithm [14].

The tumble direction is given by:

$$x_i(t+1) = w \cdot x_i(t) + c_1 e^{-r_{px}^2}(pbest_i - x_i(t))$$
$$+ c_2 e^{-r_{gx}^2}(gbest_i - x_i(t)) + \alpha(\gamma - 0.5) \tag{9}$$

5 Experimental Results

Experiments of DMTO tasks were programmed in Java in Netbeans 8.0 platform and executed on Intel i3 processor, 2.20 Ghz, 6 GB RAM.

5.1 Dataset Description

Two datasets, namely cancer and heart disease, from UCI [15] are used for classification experiment. The details of it are shown in Table 1. Tr are number of training rows and test is number of testing rows.

Table 1 Dataset description for classification task in DMTO framework

Name	Rows	Attributes	Class attribute	#Tr Rows	#Test rows
Cancer	699	10	{2,4}	550	149
Heart	270	13	{1,2}	190	80

Table 2 Dataset description for clustering task in DMTO framework

Dataset	# of Tuples	# of attributes	Class attribute
Heart	270	13	{1,2}
Liver	345	6	{1,2}
Diabetes	1151	19	{1,2}
Bank Loan	45,211	16	{1,2}
Iris	150	4	{1,2,3}
Car	1728	6	{1,2,3,4}

Clustering experiments were conducted on six datasets (4 binary and 2 multiclass) from UCI. The details of it are shown in Table 2.

5.2 Parameter Settings

Various parameters that give optimum results for DMTO with classification and clustering tasks are given as follows:

Parameters for Classification Model:

Gamma = 0.5, Nu = 0.5, Swarm Size = 20, $C1 = 1.5$, $C2 = 1.7$, Nests = 25, Pa = 0.25, Cg = 0.55, Cp = 0.75, Cw = 0.95, Nsol = 150, Ngen = 100, Nrun = 30.

Parameters for Clustering Model:

Inertia (w) = 0.9, $c1 = c2 = 0.49$, $r1 = 0.5$, $r2 = 0.7$, limit = 0.1, Iterations = 5, $w = 0.9$, $c1 = c2 = 0.49$, limit = 0.1, $\alpha = 0.7$, $\varnothing = 0.49$.

Table 3 depicts the values of parameters of clustering approach in DMTO framework which gave good results compared to other values. The values are not restricted to this, but optimal results were obtained when tried for variable number of iteration, fuzziness coefficient (m), and kernel parameter (σ).

5.3 Results

Experiments conducted with DMTO framework in classification and clustering model (existing and proposed) are shown in Tables 4 and 5, respectively. The code

Table 3 Value of m and RHO

Class	Dataset	m	σ
Binary	Heart	2.0	600
	Liver	1.8	600
	Diabetes	1.8	600
	Bank loan	1.2	0.1
Multi class	Iris	2.0	600
	Car	2.0	600

Table 4 Results for DMTO in classification approach

	Approach 1 (CS-PSO-SVM)				Approach 2 (SSA-PSO-SVM)[*]				Proposed approach (SSO-PSO-SVM)			
	A	R	P	F1	A	R	P	F1	A	R	P	F1
D1	73	100	73	84.3	97.1	97.9	97.9	97.9	**100**	100	100	**100**
D2	75	100	75	86	**94**	96	94	**95**	**93**	96	94	**95**

The best results are indicated in bold
[*]Additional method compared with existing and proposed method

Table 5 Results for DMTO in clustering approach

Dataset	Existing approach (KFCM with Hybrid BFO-PSO)		Proposed approach (KFCM with Hybrid IFA-PSO)		Change (%)
	OF	Accuracy (%)	OF	Accuracy (%)	OF (%)
Heart	521	66	402	69	23
Liver	610	49	427	50	30
Diabetes	5573	56	3700	57	34
Bank Loan	878,118	79	103,359	80	88
Iris	0.40	89	0.214	90	46
Car	19.5	35	9.54	35	51

was run for several iterations ranging from 100 to 500, but the best results here are obtained in 100 iterations in classification case.

In classification, the evaluation parameters are precision, recall, accuracy, and F1 score. Accuracy is to be high, and F1 score is always a value that lies between the recall and precision.

Precision is well known as the positive predicted value and recall as sensitivity.

In Table 4, the notations D1 and D2 in the first column are the datasets considered for classification tasks, where D1 is cancer dataset and D2 is heart dataset. Both are easily available from UCI repository. Here 2 approaches are compared in which the hyper-parameters of SVM are optimized by (1) CS-PSO (Approach 1) (2) SSA-PSO-SVM (Approach 2) with the proposed approach (SSO-PSO) which is the SSO-ELS approach.

The approaches are compared with four parameters. The parameters are denoted as: A = Accuracy, R = Recall, P = Precision and F1 = F1 measure/F1 score

6 Conclusion

The paper concludes that DMTO framework obtains better results in both classification and clustering tasks when associated with soft computing approach. For classification, the proposed approach of SSO- and PSO-based SVM obtained better results in terms of accuracy and F1 measure than the existing approaches as inferred from Table 4. Further, the proposed approach of KFCM based on IFA and PSO obtained minimized OF and better accuracy than the existing approaches as inferred from Table 5.

References

1. Fayyad, U., G. Piatesky-Shapiro, and P. Smyth. 1996. From Data Mining to Knowledge Discovery in Databases. *AI Magzine* 17 (3): 37–54.
2. Beni, G., and J. Wang. 1989. *Cellular Robotic Systems*. http://en.wikipedia.org/wiki/Swarm_intelligence.
3. Graves, D., and W. Pedrycz. 2010. Kernel-Based Fuzzy Clustering and Fuzzy Clustering: A Comparative Experimental Study. In *Fuzzy Sets and Systems*, 522–543.
4. Bae, Changseok, Wei-Chang Yeh, Noorhaniza Wahid, Yuk Ying Chung, and Yao Liu. 2012. A New Simplified Swarm Optimization (SSO) Using Exchange Local Search Scheme. *International Journal of Innovative computing, Information and Control (IJICIC)* 8 (6): 4391–4406.
5. Liu, Xiaoyong, and Hui Fu. 2014. *PSO-Based Support Vector Machine with Cuckoo Search Technique for Clinical Disease Diagnoses*, 1–7. Hindawi Publishing Corporation.
6. Yang, Xin-She, and Suash Deb. 2009. *Cuckoo Search via Levy Flights*. IEEE, 210–214.
7. James, J.Q., and O. Victor. 2015. A Social Spider Algorithm for Global Optimization. *Applied Soft Computing* 30: 614–627.
8. Liu, Yao, Yuk Ying Chung, and Wei Chang Yeh. 2012. Simplified Swarm Optimization with Sorted Local Search for Golf Data Classification. In *IEEE World Congress on Computational Intelligence*.
9. Marr, J. 2003. Comparison of Several Clustering Algorithms for Data Rate Compression of LPC Parameters. In *IEEE International Conference on Acoustics Speech, and Signal Processing*, vol. 6, 964–966.
10. Wong, K.C., and G.C.L. Li. 2008. Simultaneous Pattern and Data Clustering for Pattern Cluster Analysis. *IEEE Transaction on Knowledge and Data Engineering* 20: 911–923 (Los Angeles, USA).
11. Kora, P., and S.R. Kalva. 2015. *Hybrid Bacterial Foraging and Particle Swarm Optimization for detecting Bundle Branch Block*, SpringerPlus, vol. 4, no. 1.

12. Korani, W.M. 2008. Bacterial Foraging Oriented by Particle Swarm Optimization Strategy for PID Tuning. In *Proceedings of the 2008 GECCO Conference Companion on Genetic and Evolutionary Computation—GECCO '08.*
13. Fateen, S.E.K., and A. Bonilla-Petriciolet. 2013. Intelligent Firefly Algorithm for Global Optimization. In *Studies in Computational Intelligence*, 315–330.
14. Yang, X.-S. 2014. *Firefly Algorithms, Nature-Inspired Optimization Algorithms*, 111–127.
15. UCI Machine Learning Repository. http://archive.ics.uci.edu/ml.

Dog Breed Classification Using Transfer Learning

Rishabh Jain, Arjeeta Singh, Rishabh Jain and Praveen Kumar

Abstract This research focuses on the use of machine learning and computer vision techniques to predict the dog breed using a set of images. In this, the convolutional neural network is used as a base which is responsible for the prediction of dog breed. For making this model effective, there was further inclusion of two cases. The first case was that to give the human image as an input, and it will provide the resemblance breed of the dog as output, and the second was to give an image of things other than a dog or a human, and it will provide "something else" as an output. The test accuracy of the model is 84.578%.

Keywords Convolutional neural network · Transfer learning · OpenCV · ResNet-50

1 Introduction

According to the reports of [1] Fédération Cynologique Internationale, one of the renowned organizations that studies about the dog breeds recognizes 344 breeds officially. But, during the research, it was found that there were multiple reports such as [2] American Kennel Club which identified 202 breeds of the dogs, and the reason behind it being that the recognition process varied from country to country.

Thus, it is a complicated process to recognize the right breed of a dog picked at random which motivated us to do something regarding this issue. There are few

R. Jain · A. Singh · R. Jain · P. Kumar (✉)
Amity University Uttar Pradesh, Noida, Uttar Pradesh, India
e-mail: pkumar3@amity.edu

R. Jain
e-mail: rishabhjain6416@gmail.com

A. Singh
e-mail: arjeetasingh21@gmail.com

R. Jain
e-mail: jain.rishabh@gmail.com

© Springer Nature Singapore Pte Ltd. 2020
K. S. Raju et al. (eds.), *Proceedings of the Third International Conference on Computational Intelligence and Informatics*, Advances in Intelligent Systems and Computing 1090, https://doi.org/10.1007/978-981-15-1480-7_49

breeds of dogs which are the result of crossbreeding, and hence due to this, some have similar looks, behavior, and also, they have similarity in other aspects. So, it becomes challenging for humans to recognize them by naked eyes. For example:- Alaskan Malamute (Fig. 1a) looks similar to Siberian Husky (Fig. 1b), Belgian Malinois and German Shepherd Dog, Whippet and Italian Greyhound, etc.

This research will help the veterinary doctors and other researchers to identify the special breeds among all the present breeds which may need some different medical assistance. It also helps us to know the diversity present in the ecological system. So this model will act as a base and can be further used to identify breeds of not only dogs but cats, horses or birds, etc. can also be a part of the same model.

Rest of the research paper is ordered as follows: In Sect. 2, Literature Review is mentioned. In Sect. 3, a brief description of Convolutional Networks and Transfer Learning is given. In Sect. 4, the implementation of Transfer Learning in dog breed classification is given. Results and Concluding Remarks are addressed in next section that is Sect. 5. Future Prospects is mentioned in Sect. 6 followed by the References in Sect. 7.

2 Literature Review

As it can be shown there are plenty of research papers available in the field of fine-grained classification which is a key concept used in the model, but the model mainly revolved on the concept of dog breed classification for which the references were taken of the following research papers: The Stanford University researchers, namely [3] Whitney, Brian, and Vijay, made a model using deep learning named "Dog Breed Identification" to classify the breed of the dogs. At the end, they were able to achieve 50% of the accuracy working on 133 different breeds of dogs. Also, in 2016 [4] Pratik Devikar researched on the model that uses Transfer Learning as base for the classification of images of various dog breeds. In 2012 [5] Jiongxin, Angjoo, David, and Peter proposed an approach for the dog breed classification using Part Localization. They achieved 67% recognition rate in their research. Dąbrowski and Michalik [6] made a research on "How effective is Transfer

Fig. 1 a Alaskan Malamute. **(a)**   **(b)**
b Siberian Huskey

Learning method for image classification." In this, they showed that how much Transfer Learning is effective for increasing the accuracy of the model by retraining neural network-based image classification using the approach known as the Transfer Learning.

3 Theory

For making a classifier like what is made, there have to be different techniques implemented together. It was decided to start with the conventional convolutional neural network [7–9] for building a machine which can act as the classifier. But at first, when the process of testing was started with the trained model, the accuracy was pretty low what was not acceptable in the perspective. At last, to have the accuracy of the model up to the mark of the expectations, the implementation of the Transfer Learning [10, 11] concept was used, and it gave 81.578% accuracy. Later, it checks whether if it can predict a dog breed that resembles a human by detecting the human face [12] from the collection of photographs that were provided in the training set. The brief explanation of each of the basic terms used in this model is as follows.

3.1 *Convolutional Neural Networks*

These neural networks are composed of different kinds of layers. It works on a basic algorithm which takes the small windowed like structure (square) and starts applying it on the image. The function of the filter is that it helps the CNN to identify the different patterns present in the picture. The neural network will look for the parts of the image where a filter meets the content of the image. The starting layers of the networks detect the simple features or basic geometrical shapes like circles, lines, edges, etc. In all the layers, Convolutional Neural Network associates these findings and continuously learn more complicated concepts as it goes deeper and deeper into the segments of the network. It mainly constitutes of the following layers:

(i) Input layer: In this layer, the network took all the data into the system for sending it for further processing by successive layers of the net.
(ii) Hidden layer: This layer present in between the input layer and the output layer. This layer takes the weighted sets as an input and produces an output with the help of the activation function.
(iii) Output layer: It is the last layer of the network and generates the given outputs after processing.

Also, there are several other layers present, i.e., the activation layer, convolutional, pooling, dense, dropout and SoftMax layer. Conv layer is at the core of the network. Pooling layer helps to avoid overfitting. It is used to apply nonlinear downsampling on the activation maps (Figs. 2 and 3).

Max pooling helps to reduce the amount of memory required, and the output of the max pooling is calculated as:

Let d_{out} be the dimensions of output, din be the dimensions of the input, w be the window size, and s is the stride.

Now $d_{out} = \text{floor } ((d_{in} - w)/s) + 1$

Dropout layers are also used to avoid overfitting, but it randomly ignores the certain activation functions. Dense layers are the fully connected layers in which the input portrayal is flattened into a feature vector, and then, it gets passed through a network to predict the output probabilities.

SoftMax function is used in the output layer, and it helps in normalizing; it generates a discrete probability vector. For checking the performance and

Fig. 2 Nonlinear downsampling on activation map using pooling layers

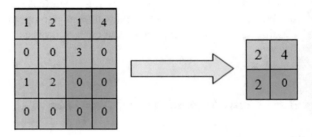

Fig. 3 Flattening operation

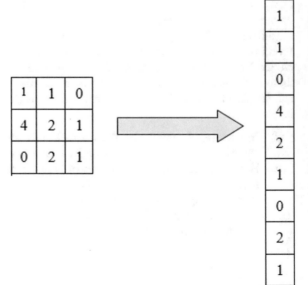

evaluating the model, the loss function is used at an end. A common loss function is used while predicting the multiple output classes, and it is termed as the categorical cross-entropy loss function.

3.2 Transfer Learning

It is the technique in which the pre-trained model is used on a new classification problem. It helps the user to train the deep neural network with relatively low data, and due to this, the accuracy and efficiency of the model increase. It is beneficial because the real-world problems commonly do not have labeled data points to train such complex models. In this, it tried to transfer as much knowledge as achievable from the previous model. It helps to reduce the training time. Here is the simple flowchart given to summarize the whole process in the Transfer Learning (Fig. 4).

In the next section, the implementation of these terms in the model can be seen, and also it shows how to make the efficient use of Transfer Learning in increasing the accuracy of the model.

4 Implementation of the Transfer Learning in Dog Breed Classification

In order to get started with the implementation, it was important to have the environment set as per the requirements of work that was about to begin. To have it done, it was important to have installation and setup of various environments and libraries such as TensorFlow [13], OpenCV [14], and Keras [15]. After that,

Fig. 4 New classifier using transfer learning

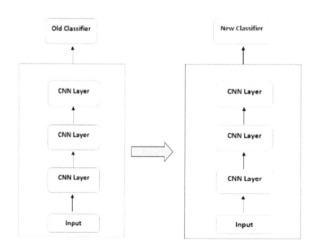

preparation of the dataset was to be done and used. Now it was possible that the dataset could contain pictures of humans also, so it was important for the model to detect these photographs with human faces. To get this done, the face detection algorithm [16, 17] which works using OpenCV and has several inbuilt functions was used. At first, it takes an image and converts into grayscale because for any face, detection algorithm is essential for an image to be in grayscale, after which the detectMultiScale function initializes the classifier in the face cascade which accepts only grayscale images (Fig. 5).

After it was essential to have a check whether an image had a dog face or not, a pre-trained model was used on the dataset which can classify from around 1000 categories. It uses TensorFlow in the backend, and being a Keras CNN model, it needed a 4D array as its input after which it returns us the prediction of the type of object contained in the image. An inbuilt function known as "paths_to_tensor" is responsible for taking the path of collected images and returning them in required size (usually what is 224×224 pixels) in the form of the 4D array (also called as "Tensor") which is what a Keras CNN requires. As it is a pre-trained model, it does have an additional step for the normalization that means that it subtracts pixel from each pixel of every image provided to it. This process is done with the help of an imported function known as "preprocess input." This returns an integer value as its final prediction which corresponds to one of the object classes of which the "argmax" was taken, and then, mapping up of the objects can be done by looking into the dataset dictionary. Now after getting ready with functions that will help us detect both the humans and with such a big set of images, it was essential to reduce the time of training of the CNN, and this should not be done by losing accuracy, so here the concept of Transfer Learning helps us to do so with the model. Transfer Learning allows us to work with some pre-trained learning models that are provided by Keras. These models are pre-trained on a huge dataset, and they retain their knowledge of what they learnt which can be used alongside the dataset layers which get added with the existing ones. This kind of prediction is more accurate and can give fine-tuned results. As it has been discussed that the layers could be added

Fig. 5 Basic work flow

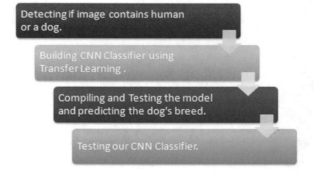

alongside the pre-existing layers of the model that is being used for the Transfer Learning, that means the layers that are going to get added will get treated as model's output also known as "bottleneck features." In this way, the extraction of all the things was achieved, and the model was ready to get trained. Now it was set that the model will receive the last output as the input for the model, and then, it will add several layers such as the max pool layer, the dense layer, and activation function layer of type SoftMax. At last, the compilation and checking of the CNN were done up to what extent it can predict the kind of results that were expected, but in early stages to get better accuracy, it had to do many iterations for obtaining higher efficiency and for optimizing the loss function.

Now once the things were in order, it was important to have an automated function which can do the tasks for us whenever it is called and can return us with the prediction of the dog breed that was needed, so in order to get this, a function was made which took input of images and returns us with the required result. Finally, it was achieved what was needed with an accuracy high as 84.578% which gave near-perfect results, and this accuracy what it achieved was satisfying; as an accuracy above 80% is always considered to predict accurate results. This model was not only able to predict dogs breed when it has been given an image of dog but was also able to return the closest resemblance of a breed if a picture had a human in it. It was designed to return back an error if an image had neither of them.

Working environment for this model was implemented on Jupyter Notebook, and while the TensorFlow was running in the background, it takes an image using the custom-made function named as "predicts_dog_breed_by_image" which needs the path of the image as an argument. Then, once this argument is passed to classifier, it gets initiated at that moment. At last, the classifier generates the required results using CNN (Fig. 6).

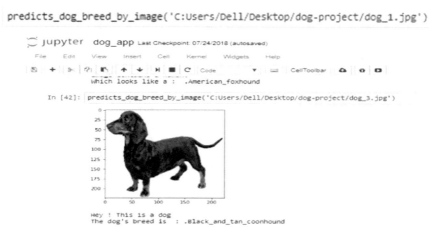

Fig. 6 Implementation on Jupyter Notebook

5   Result

The model was able to predict accurately from the data that was given to it as an input. During the testing, the model was tested with all kinds of test cases, and it was able to predict precisely. The following results of test cases can be shown in the images given down below:

5.1   Case-I

In this, the bulldog image and the German shepherd image are provided as an input, and the classifier model gives the breed of the particular dog as the output (Figs. 7, 8 and 9).

5.2   Case-II

In this, the cycle and the waterfall images are provided as an input, and the classifier model gives "*This is something else*" as the output (Figs. 10 and 11).

5.3   Case-III

In this, the human images are provided as an input; the classifier model gives "*Hey!*" "*This is a human,*" and it gives the corresponding dog breed as the output, i.e., beagle and bull terrier (Figs. 12 and 13).

Fig. 7 Output of German Shepherd_1. The image of a dog was collected from the dataset present on the web for testing of the output of the model

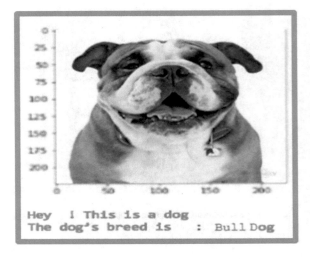

Fig. 8 Output of German Shepherd_2. The image of a dog was collected from the dataset present on the web for testing of the output of the model

Fig. 9 Output German Shepherd_3. The image of a dog was collected from the dataset present on the web for testing of the output of the model

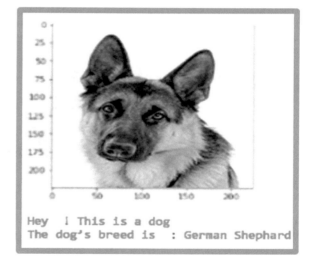

6 Conclusion

The research concludes that the Transfer Learning boosted the performance of the classifier model. It gave a significant increase in accuracy from 45.34% to a remarkable 84.578%. This research proves that Transfer Learning can be significant in the field of computer vision and machine learning.

Fig. 10 Output of cycle image. The image of a cycle was collected from the dataset present on the web for testing of the output of the model

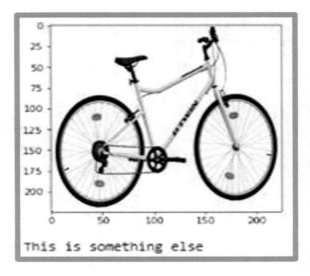

Fig. 11 Output of the Waterfall image. The image of a waterfall was collected from the dataset present on the web for testing of the output of the model

7 Future Prospect

For further improving the results, the images present in the dataset should be treated so that the subject can be clear and the noise level of an image can be reduced. But, as all know that today, the size of the datasets is huge so it is not possible for someone to do it manually and thus need to develop algorithms to do it

Fig. 12 Output of human images_1 .The image of a human was collected from the dataset present on the web for testing of the output of the model (Taken from Google)

Fig. 13 Output of human images_2. The image of a human was collected from the dataset present on the web for testing of the output of the model (Taken from Google)

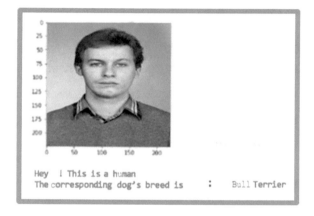

automatically is important. So, the dataset of the images should get converted by these algorithms to "treated images." The dataset of these images will now be used by the classification model.

References

1. FCI Annual Dog Breed Report. 2018. *FédérationCynologiqueInternationale.*
2. AKC Annual Dog Breed Report. 2018. *American Kennel Club.*
3. Dog Breed Identification. *Stanford University.*
4. Devikar, Pratik. 2016. Transfer Learning for Image Classification of Various Dog Breeds. *International Journal of Advanced Research in Computer Engineering & Technology (IJARCET)* 5 (12). ISSN: 2278-1323.
5. Liu, Jiongxin, Angjoo Kanazawa, David Jacobs, and Peter Belhumeur. *Dog Breed Classification Using Part Localization.*
6. Dąbrowski, Marek, Tomasz Michalik. How Effective is Transfer Learning Method for Image Classification. Position Papers of the Federated Conference on Computer Science and

Information Systems, vol. 12, 3–9 https://doi.org/10.15439/2017f526 ISSN 2300-5963 ACSIS.

7. Hu, Baotian, Zhengdong Lu, Hang Li, and Qingcai Chen. 2014. Convolutional Neural Network Architectures for Matching Natural Language Sentences. *Part of: Advances in Neural Information Processing Systems* 27.

8. Krizhevsky, Alex, Ilya Sutskever, Geoffrey E. Hinton. 2012. ImageNet Classification with Deep Convolutional Neural Networks. *Advances in Neural Information Processing Systems* 25.

9. Ciresan, Dan C., Ueli Meier, Jonathan Masci, Luca M. Gambardella, and Jurgen Schmidhuber. Flexible, High Performance Convolutional Neural Networks for Image Classification. In *Proceedings of the Twenty-Second International Joint Conference on Artificial Intelligence.*

10. Yosinski, Jason, Jeff Clune, YoshuaBengio, and Hod Lipson. 2014. How Transferable are Features in Deep Neural Networks? *Part of: Advances in Neural Information Processing Systems* 27.

11. Hoo-Chang, Shin, Holger R. Roth, Mingchen Gao, Lu Le, Xu Ziyue, Isabella Nogues, Jianhua Yao, Daniel Mollura, and Ronald M. Summers. 2016. Deep Convolutional Neural Networks for Computer-Aided Detection: CNN Architectures. *Dataset Characteristics and Transfer Learning, IEEE Trans Med Imaging.* 35 (5): 1285–1298.

12. Lawrence, S., C.L. Giles, Ah Chung Tsoi, and A.D. Back. 1997. Face Recognition: A Convolutional Neural-Network Approach. *IEEE Transactions on Neural Networks* 8 (1).

13. Goldsborough, Peter. 2016. A Tour of TensorFlow. *Proseminar Data Mining* (Submitted on 1 Oct 2016).

14. Čuljak, Ivan, David Abram, Tomislav Pribanic, HrvojeDžapo, and Mario Cifrek. *A Brief Introduction to OpenCV.*

15. Chollet, François. 2018. Keras: The Python Deep Learning library, Astrophysics Source Code Library, record ascl:1806.022, 06/2018.

16. Wang, Yi-Qing. An Analysis of the Viola-Jones Face Detection Algorithm. Image Processing on Line on 2014–06–26. Submitted on 2013–08–31, accepted on 2014–05–09. ISSN 2105–1232 © 2014 IPOL & the authors CC–BY–NC–SA.

17. Hsu, Rein-Lien, M. Abdel-Mottaleb, and A.K. Jain. 2002. Face Detection in Color Images. *IEEE Transactions on Pattern Analysis and Machine Intelligence* 24 (5).

A Framework for Dynamic Access Control System for Cloud Federations Using Blockchain

Shaik Raza Sikander and R. Sridevi

Abstract A cloud federation is a network of cloud computing environment consisting of two or more providers. In this framework, the user classification is done dynamically depending on their user's blockchain transactions. The data classification is done in the cloud using the suitable methods. The data access from the cloud will be given to the users who have the access or permission to use the data depending on their access/privilege rights. For the new user, the access will be through voting based on the new users from different organizations. The votes are given by the trusted parties, and their votes are valid only after their blockchain transaction verification. For a new user to access the data, he requires minimum of votes from the trusted parties.

Keywords Blockchain · Data classification · User classification · Cloud computing · Access control · Cloud federation

1 Introduction

There has been an exponential growth in the cloud computing in recent years. Cloud computing has created a platform for cross-organization collaborations called cloud federations. In the cloud federations [1], there are different organizations who host their data on the cloud, a private shared cloud. So, the data is available to not only the user of the same organization but also the users from the different organizations that are part of the cloud federation. However, the realization of federations of private clouds is hindered by partner organizations' privacy concerns sharing its data across private cloud infrastructures [2]. So, there came the requirement for a framework for the identity and the access management to allow

S. R. Sikander · R. Sridevi (✉)
CSE Department, JNTUHCEH, Hyderabad, Telangana, India
e-mail: sridevirangu@jntuh.ac.in

S. R. Sikander
e-mail: raza.sikander.s@gmail.com

© Springer Nature Singapore Pte Ltd. 2020
K. S. Raju et al. (eds.), *Proceedings of the Third International Conference on Computational Intelligence and Informatics*, Advances in Intelligent Systems and Computing 1090, https://doi.org/10.1007/978-981-15-1480-7_50

users from other organizations in the cloud to access the data which is stored in the private cloud server.

The framework proposed in this paper is for the user classification, data classification, data authentication to the new users from other organizations and trusted party verification.

The user classification in this framework is done dynamically with the help of the blockchain [3] technology. Here, the transaction stored in the blockchains is being used to validate the user. Depending upon the classification, they will be given rights/priorities in the cloud federation. The data classification in the cloud can be done through using the machine learning algorithms like support vector machines [4], random forest [5] or logistic regression [6].

The data authentication to the users will be done with the help of a proposed protocol which is a type of voting protocol which uses blockchain technology to check the authenticity of the users who are giving votes for the user of one federated organization to get him access to the data of other federated organizations.

2 Background

In this section the basic background to the main technologies which are being used in this framework are provided.

2.1 Blockchain

A blockchain is a growing list of records called blocks which are linked cryptographically. Each block contains a cryptographic hash from previous block, a time stamp and transaction data.

Blockchain was invented by Satoshi Nakamoto in 2008 to serve as the public transaction ledger of the cryptocurrency bitcoin [7].

The invention of the blockchain for bitcoin made it the first digital currency to solve the double-spending problem without the need of a trusted authority or central server. The bitcoin design has inspired other applications [7, 8], and blockchains which are readable by the public are widely used by cryptocurrencies. Private blockchains have been proposed for business use.

2.2   Data Classification Algorithms

The machine learning has basically four types of learning algorithms

(1) Supervised learning algorithms.
(2) Unsupervised learning algorithms.
(3) Semi-supervised learning algorithms.
(4) Reinforcement algorithms.

In this framework, the supervised learning algorithms like support vector machines [4], random forest [5] or logistic regression [6] are used for the data classification.

3   The Proposed Protocol

The protocols proposed in the framework are for the user classification, data access and voting-based access.

The main intention of this protocol is to make classification of the users simple and granting them data access and keep a check on their transactions in the cloud federation and depending on their transaction classifying the users.

3.1   Procedure

3.1.1   User Classification

The cloud federation has a blockchain for recording the transactions of the user in the cloud federation. The success and failure transactions are all recorded in the blockchain. There are two flags/counters in the federation for each individual user which keeps track of the user transactions. The positive flag/counter is incremented by one after one successful transaction, and the negative flag/counter incremented by one after one failure transaction. Depending on the flags/counters, the users are classified automatically.

Example of transaction:

1. If User A has access to the data in the cloud, he can directly access it without any issue. This is recorded as a successful transaction in the blockchain, and the positive flag/cloud is incremented.

2. If User A does not have the access for the data but still tries to access the data without requesting the access, it results in a failure, this is a failure transaction and recorded in the blockchain and the negative flag/counter is incremented.

There are three types of users in the cloud federation. They are:

1. Type-1 (new users)
2. Type-2 (moderate users)
3. Type-3 (high-level users).

The users having no successful transaction or less than 500 successful transactions are present in the Type-1.

The users having above 500 transactions and less than 1500 successful transactions are present in the Type-2.

The users having above 1500 transactions are present in the Type-3.

The Type-1 users should ask access for any data as they are new and do not have any access rights in the cloud federation. The access is provided to them only after the voting protocol.

The Type-2 users have access to some data and want access of same organization they can get it by requesting their organization authorities. If they want data from other organizations, then he also needs to go through voting phase as the data is of other organizations.

The Type-3 users have all the access to the data present in the same organization. But if they need data of other organizations, they can get it by requesting authorities of other organizations.

3.1.2   Voting Protocol

1. User 'A' from organization 'X' requests the data which belongs to other organization 'Y' which is present in the cloud federation.
2. Then, the data which is requested by the user 'A' type is checked. If it is a confidential data, then the request is not proceeded and a notification is given to the user.
3. If the data requested is sensitive or unrestricted, the protocol goes to next step.
4. Now, the voting phase began where the users of the same organization will cast the vote for the data access.
5. Not all the users of the organization 'X' can cast the votes. Only the users who are trusted only can cast the vote.
6. The user is trusted or not is checked using the blockchain transactions of the user. After the verification of the details from the blockchain, then only the user can cast the vote.
7. If the votes are in favor of the user 'A', then the protocol goes into further stage or else it stops and intimates the user and stores this as a block in the blockchain.

8. Then, the request and votes from the organization 'X' are forwarded to the organization 'Y'.
9. In the organization 'Y', again the voting is taken by the higher officials (users) whether to provide access to the user from organization 'X'.
10. If votes are in favor, then the access to the user is provided or else the user 'A' of organization 'X' is notified and this is stored as a transaction in the blockchain.

In this method, the user 'A' details are known only to the organization 'X' and unknown to the organization 'Y' as the user details will be kept confidential; only their id and their type are known to other organizations.

3.2 Algorithms

Algorithm: 1 User Classification

```
Initialize
        Int Positive_flag
        Int Negative_flag
        Blockchain B
Processing

        Type-1 user
                If 0<Positive_flag<500
                        User=Type-1
        Type-2 user
                If 500<Positive_flag<1500
                        User=Type-2
        Type-3 user
                If 1500<Positive_flag
                        User=Type-3
End
        If Positive_flag<Negative_flag
                User is moved to lower level
Note: if user is at the lower level and if the negative
flag keeps on increasing then his access is revoked and
he is classified as malicious user
```

Algorithm: 2.1 Checking the type of data requested by user(Data_req())

```
Initialize
     Receive Req1 from User-A
     Check the data type
     If(Data type!=Confidential)
          Req1= 1
     Else
          Req1= 0
End
     If(Req 1=1)
          Go to Votephase
     Else
          Go to Rejectphase with Rej1=0
```

Algorithm: 2.2 Notification_phase

```
     Initialize
          Receive Rej1 from Data_req()
     Processing
          If(Rej1==0)
               Reject  the  access  to  the  User-A  as  data
          confidential.
               Create_trans  (Create  a  transaction  in  the
          Blockchain for future use)
          If(Rej1==1)
               Reject access to  User-A as not enough votes
          from organization-X.
          Create_trans  (Create  a  transaction  in  the
     Blockchain for future use)
     If(Rej2==2)
          Reject  access  to  User-A  as  not  enough  votes
     from organization-Y.
               Create_trans  (Create  a  transaction  in  the
          Blockchain for future use)
End
     Notify the user with the reason.
```

Algorithm: 2.3 Vote phase for processing

```
Initialize
      Request received after the data type verification
Processing
      Vote_phase-1 Initialized
            If vote from a trusted user to grant access
      then grant=+1
            If vote from a trusted user not to grant then
      No_grant=+1
            If(grant>No_grant)
                  Go to Vote_phase-2
            Else
                  Return Rej1=1 to Notification_phase
      Vote_phase-2 Initialized
            Votes will be from higher authorities of
      organization-Y
            If vote  to grant access then grant=+1
            If vote  not to grant then No_grant=+1
            If(grant>No_grant)
                  Go to Grant_phase
            Else
                  Return Rej1=2 to Notification_phase
End
      Returns   value   to   either   Grant_phase   or
      Notification_phase
```

Algorithm: 2.4 Grant_phase

```
Initialize
      Receive the grant request from the Vote_phase
      Processing
      Now  the  data  access  to  the  User-A  from
      organization-X is provided by the organization-Y
End
      Create_trans  (Create  a  transaction  in  the
      Blockchain for future use)
```

3.3 Flowchart

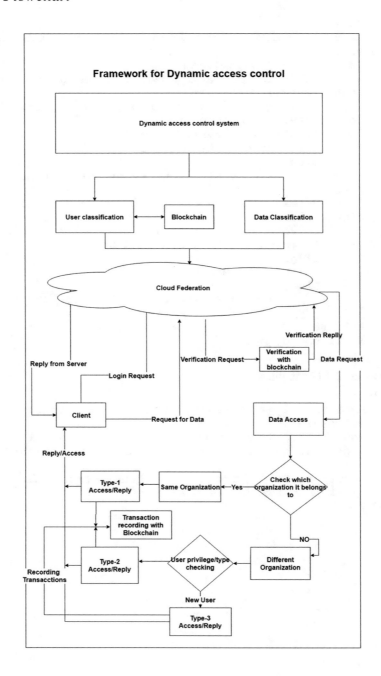

4 Results

The stimulation of the framework has been done with 200 users of different user types and organizations. Depending upon the transactions in the cloud federation, the users were dynamically classified always and the user type was automatically changed depending on the transactions. Different college data sets have been used in this for the implementation. The data set used for this framework consisted of the students' data, faculty data, higher authorities' data of different organizations. The data classification was done accordingly, and the type of data for each user access was successfully classified. The Type-1 users were able to access the data like name of faculty, students and higher authorities. The Type-2 users were able to access

Table 1 Sample user data

User name	User type	Positive Flag	Nagative Flag	Flag = Positive-Nagative	Result
John	Type-3	1700	200	1500	Stay in Type-3
Smith	Type-2	1100	300	800	Stay in Type-2
Sara	Type-3	1570	200	1370	Demoted to Type-2
Steve	Type-1	400	100	300	Stay in Type-1
Arnold	Type-1	550	20	530	Promoted to Type-2

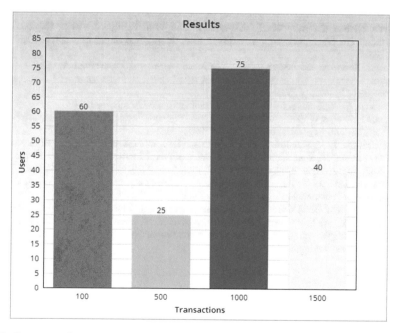

Fig. 1 Summary of total user classification

some more details like photograph of the faculty and his designation. The Type-3 users could access the more personal details like contact number of the student, faculty, higher authorities and their biodata stored in portable document format (PDF) (Table 1; Fig. 1).

5 Conclusion

In the framework for the cloud federations, user classification and voting-based protocol for the data access are implemented successfully. The blockchain technology is used for tamper proof access control and dynamic user privileges which are used to do user classification dynamically in the cloud federation.

Acknowledgements Data was collected from JNTUH College of Engineering Hyderabad's CSE Department and MVSR Engineering College's CSE-3 (2013-17) batch students. The data was willing given by the HOD of CSE JNTUH CEH and students of MVSR college BE CSE-3 2013-17 batch. None of ethical committee was involved in it. No persons were included in the framework testing.

References

1. Popper, Nathan (21 May 2016). A Venture Fund With Plenty of Virtual Capital, but No Capitalist. The New York Times. Archived from the original on 22 May 2016. Retrieved 23 May 2016.
2. Alansari, Shorouq, and Federica Paci. Vladimiro Sassone: A Distributed Access.
3. Control System for Cloud Federations. In: *2017 IEEE 37th International Conference on Distributed Computing Systems*.
4. Nakamoto, S. 2008. Bitcoin: A Peer-To-Peer Electronic Cash System.
5. Cortes, Corinna, and Vladimir N. Vapnik. 1995. Support-Vector Networks. *Machine Learning* 20 (3): 273–297. https://doi.org/10.1007/BF00994018.
6. Ho, Tin Kam. 1995. Random Decision Forests (PDF). In *Proceedings of the 3rd International Conference on Document Analysis and Recognition*, Montreal, QC, 14–16 August 1995. 278–282. Archived from the original (PDF) on 17 April 2016. Retrieved 5 June 2016.
7. Kleinbaum, D.G. 1994. Introduction to Logistic Regression. In: *Logistic Regression. Statistics in the Health sciences*. Springer, New York, NY.
8. Blockchains. The Great Chain of Being Sure About things. The Economist. 31 October 2015. Archived from the original on 3 July 2016. Retrieved 18 June 2016. The technology behind bitcoin lets people who do not know or trust each other build a dependable ledger. This has implications far beyond the cryptocurrency.

ESADSA: Enhanced Self-adaptive Dynamic Software Architecture

Sridhar Gummalla, G. Venkateswara Rao and G. V. Swamy

Abstract With the advent of new technologies and the trend in integration of related business, the software development has become very complex. However, complex systems are realized due to distributed computing technologies like Web services. With machine-to-machine (M2 M) interaction, human intervention is greatly reduced in distributed applications. Nevertheless, there is need for continuous changes in complex software systems. Manual incorporation of changes is both time consuming and tedious task. The self-adaptive features of software can cater to the needs of ad hoc demands pertaining to changes. Therefore, it is desirable to have a self-adaptive software architecture for distributed systems to adapt to changes automatically without traditional reengineering process involved in software update. The existing solutions do have limitations in self-adaptation and need human intervention. Rainbow is one of the examples for self-adaptive dynamic software architecture. However, it does not have knowledge mining and quality of software analysis for further improvements. It is essential to have such enhancements in the wake of self-adaptive systems of enterprises producing huge amount of data related to operations, service quality and other information required for analysing the architecture. We proposed a self-adaptive dynamic software architecture named enhanced self-adaptive dynamic software architecture (ESADSA) which is influenced by Rainbow. It incorporates modules such as QoS analyser and knowledge miner with two data mining algorithms for enhancing capabilities of the architecture. ESADSA decouples self-adaptation from target

S. Gummalla (✉)
Department of Computer Science & Engineering, Shadan College
of Engineering & Technology, Hyderabad, Telengana, India
e-mail: sridhargummalla1975@gmail.com

G. Venkateswara Rao
Gitam Institute of Technology, GITAM (Deemed to be University),
Visakhapatnam, A.P, India
e-mail: vrgurrala@yahoo.com

G. V. Swamy
Gitam Institute of Technology, GITAM (Deemed to be University),
Visakhapatnam, A.P, India
e-mail: drgv_swamy@yahoo.co.in

© Springer Nature Singapore Pte Ltd. 2020
K. S. Raju et al. (eds.), *Proceedings of the Third International Conference
on Computational Intelligence and Informatics*, Advances in Intelligent
Systems and Computing 1090, https://doi.org/10.1007/978-981-15-1480-7_51

system by preserving cohesion of target system with loosely coupled interaction. A real-time case study is considered for proof of the concept. The experimental results revealed significant improvements in dynamic self-adaptation of the proposed architecture.

Keywords Self-adaptation · Dynamic software architecture · Reusability · Maintainability

1   Introduction

Nature has inspired many engineering solutions. Adaptation to situations is one of the inspirations of nature. Living organisms also evolved over time with adaptations. Self-adaptation can occur over time or instantly. Adaptations to software are also required as the software needs to handle runtime situations instead of making explicit changes to software to meet such requirements. Evolution is therefore important to software systems also. It is more so in the wake of distributed computing technologies and the availability of reusable components. The component architecture made it possible to make decisions on the adaptation to components based on the runtime requirements of software in question.

As software ages, it needs to adapt to new environments. Since software maintenance is costly, it is important to have self-adaption capabilities to reduce maintenance problems. Managing software changes, making the changes easier and reducing downtime of software are important aspects. Self-adaptive software (SAS) is the promising solution to have software that can adapt to runtime environments without the need for developers to make them explicit changes. High availability needs and changing requirements argue for the need of appropriate adaptive behaviour. It is good to manage certain changes without explicitly making changes to software through self-adaptation.

It is very challenging to have self-adaptation possible. In the literature, some important solutions are found that can be used as reference implementations of self-adaptive software. Software architecture-based self-adaptation is explored in [1, 6, 14] and [20]. The concept of self-healing with respect to software architectures is studied in [3, 5] and [19]. The need for different architectural styles and the role of them in self-adaptive software architecture are found in [4] and [5]. Most of the research was carried out in distributed environment where software complexity is more due to the components involved and the location of different software components across the globe. Since a distributed application can pave way for inter-operability due to innovative technologies like Web service and service-oriented architecture (SOA), it is found that such application needs to cope with different runtime situations.

Many solutions came into existence for self-adaptation. However, it is understood from the literature that there is much score for further exploration of the area towards optimizing self-adaptive software architectures. IBM's MAPE is a reference architecture that is used in one of the frameworks known as Rainbow [1]. Our work in this paper is inspired by this architecture which has all ingredients to cope with runtime changes. Though it is one of the good architectures for dynamic self-adaption, it lacks two important features. They are known as quality of service analysis and knowledge management. These two modules are incorporated in our self-adaptive dynamic software architecture as shown in Fig. 1. Two algorithms are proposed to realize the new modules. The remainder of the paper is structured as follows. Section 2 provides review of the literature on self-adaptive software architectures. Section 3 presents the proposed dynamic self-adaptive software architecture and algorithms. Section 4 presents a case study considered for dynamic self-adaptation. Section 5 presents experimental results, while Sect. 6 concludes the paper besides giving directions for future enhancements.

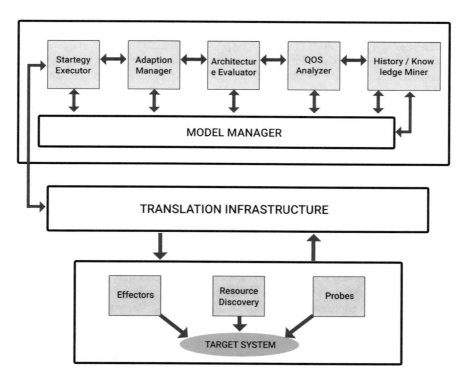

Fig. 1 Proposed enhanced self-adaptive dynamic software architecture (ESADSA)

2   Related Work

Many researchers contributed towards dynamic self-adaptive software architectures. For instance, architecture-based solutions [2–4] and self-healing systems [5] were explored by characterizing the style requirements of systems. The following sub-sections provide more details of the review.

2.1   Adaptation in Distributed Environment

Self-organizing systems have provision for self-managing units that work with a common objective based on the software architecture [6]. They are based on the concept known as architectural formalism of Darwin [7]. There are many compo-nents in the self-organizing system. Each one takes care of its own adaptation which is part of the whole system. To achieve this, each component remembers the architecture model. It can get rid of single point of failure besides having distributed control, consistent model. However, it causes significant overhead with respect to performance and needs an algorithm for global configuration. In our work, we could overcome this problem by supporting global reorganizing and trade-off between local and distributed controls while making adaption decisions.

2.2   Dynamic Architectures

Formal dynamic architectures can achieve self-adaptation. Towards this end, many approaches came into existence. [8] proposed a high-level language to describe an architecture formally in such a way that it can reconfigure when changes are made to the components in the system and interconnections. The K-component model proposed in [9] addresses two issues of the dynamic architectures. They are safety of evolution and integrity. They also took care of meta-models required for graph transformations. Many researchers focused on the programs called adaptation contracts to build applications that are self-adaptive [10]. Darwin is an example for ADL and used for describing distributed systems with self-adaptive architecture including operational semantics that take care of runtime dynamics of the system and reconfigurations [7]. Since the components and their organizations can change at runtime, there is mechanism to analyse changes. Thus, Darwin can help to be used as general-purpose configuration language in distributed environment. Other known examples for ADL are Pilar [11] and ArchWare [12]. These are used to model architectural layers that can separate concerns and improve performance. These approaches are based on the reflective technologies for dynamic evolution.

However, they assume that the implementations of systems are based on the formal architectural descriptions. Our approach in this paper decouples target system and external mechanisms for improving the adaption features.

2.3 Self-adaption with Quality Attributes

Many architecture-based solutions came into existence for self-adaptation. They are based on the quality attributes identified. For instance, they focused on performance as explored in [13, 14] and [15]. Survivability is another quality attribute used [16]. There are some researchers who focused on the architectural styles while building self-adaptive architectures [17, 18]. The style-based architectures had formal specifications that reflect different styles in self-adaptation. Many works in the literature are close to our work in this paper. They include the work done by UCI research group [19] and the work of Sztajnberg [20]. An architecture-based solution for dynamic adaption at runtime was explored as an extension to [18]. This was achieved using planning loop, execution loop and monitoring. Its focus was on the self-adaption of C2-style systems. Merging and architectural differencing techniques were used in the implementation. Style-neutral ADL was explored by Sztajnberg and Loques besides many other aspects such as architectural contracts and architectural reconfigurations. The model had support for formal verification but lacked in automated adaption of multiple objectives. In this paper, our approach addresses this problem. Rainbow framework [1] tried to provide far better solution for dynamic self-adaptive software architecture. However, it can be further improved with two additional modules as done by us in this paper. The modules are known as QoS analysers and knowledge mining. These modules can improve the performance of the architecture.

The approaches found in the literature have certain limitations and issues that need to be resolved. Many were addressed by Rainbow framework [1] such as exception handling and balance between local and global perspectives. It also focuses on quantity of adaption with many building blocks to achieve self-adaption. Customizable elements and reusable infrastructures are the two good features of Rainbow. However, the Rainbow framework has certain limitations as described here. Utility-based theory was used for best adaptation path under uncertainty. The utility-based frameworks are not fully quality-aware. Quality of service (QoS) analysers can be built to continuously monitor for improvement opportunities. The historical information usage is not sophisticated in the existing frameworks. It can be improved with the state-of-the-art data mining techniques for improving decision-making.

2.4 Dynamic Software Architectures Versus Systems of Systems

Silva et al. [21] studied systems of systems (SoS) that are made of many complex and independent systems that are integrated to form a more complex system with its mission to be accomplished. Especially, they focused on characterization of missions of SoS. They made systematic mapping of missions of SoS to discover elements associated with SoS. Cavalcante et al. [22] studied dynamic software architectures. They introduced dynamic reconfiguration support provided by π-ADL which is a language used to formally describe dynamic software architectures. The π-ADL language has main architectural concepts such as components, connectors, component behaviour and connector behaviour. Cavalcante [23] studied on software-intensive SoS architectures. Formal specification of SoS and its dynamic reconfiguration are explored. In their proposed approach, both architectural-level and execution-level descriptions are given. At architectural level, formal ADL is used. At execution level, there are runtime aspects of different components provided. There is integration between architecture-level and execution-level components. Guessil et al. [24] characterized ADLs for software-intensive SoS. The main purpose of their study was to identify the features of ADLs in effectively describing SoS architectures.

Quilbeuf et al. [25] proposed logic for verification of reconfiguration and other system functionalities of dynamic software architectures through statistical model checking. They built new logic named DynBLTL for expressing behavioural and structural properties of the architectures. They also proposed a statistical model checking approach for analysing those properties. Silva et al. [26] studied missions of software-intensive SoS and introduced mobile-based refinement process and support for model-based transformations. Cavalcante et al. [27] introduced statistical model checking (SMC) as the runtime behaviour of dynamic software architectures is unpredictable. SMC helps in analysing software architectures with reduced time, computational resources and effort needed. Silva et al. [28] studied local and global missions and introduced M2Arch which is a model-based process for analysing mission models and refining them.

3 Proposed Dynamic Self-adaptive Software Architecture

In this section, we describe the proposed self-adaptive dynamic software architecture. A self-adaptive software architecture can cater to the dynamic needs of the software at runtime. It can adapt to runtime situations. Select adaptation needs to work for different kinds of systems and quality requirements. The adaptation is to be made with explicit operations that are chosen at runtime. Such architecture should provide an integrated solution that saves time and effort of engineers as it can adapt to situations without the need for writing code and update the software explicitly. Our framework satisfies these requirements by supporting many mechanisms that lead to dynamic self-adaptation. Our architecture is known as enhanced self-adaptive dynamic software architecture (ESADSA). This work has been influenced by Rainbow framework [1]. Our architecture shown in Fig. 1 extends Rainbow framework with two additional modules. They are known as quality of service (QoS) analysers and history/knowledge miner.

The framework has two layers known as architecture layer and system layer. Broadly, architecture layer represents self-adaptive mechanism which is made up of many components that work in tandem with each other. The system layer represents the target system and the plumbing components that are used to realize self-adaptation. There are two mechanisms for monitoring the target system. They are known as probes and gauges. The observations are reported to model manager. The architecture evaluator component is responsible to evaluate the model when model gets updates. It checks architectural constraints within acceptable range. When evaluation concludes that there is a problem in the system, the evaluation management invokes adaptation manager. The adaptation manager is responsible to initiate adaptation process and select a suitable adaptation strategy. Then, the strategy executor comes into picture in order to execute the strategy on the runtime system. This is achieved through system-level effectors that ensure realization of changes to the target system through self-adaptation.

There are three important aspects that are used to realize the software architecture. They are known as software architecture, control theory and utility theory. Self-adaptation is possible with proposed software architecture that makes it cost-effective. Control theory and mechanisms ensure smooth adaptation. Utility theory helps in finding best strategy in self-adaptation. Moreover, the proposed architecture has two components for optimization. These components are known as QoS analysers and knowledge miner.

3.1 Translation and Monitoring

The translation infrastructure provided in architecture takes care of monitoring and action. This will help in bridging the gap between the target system and architecture layer. It has monitoring mechanisms like probes and gauges. These mechanisms get system states and update the model from time to time. The probe is responsible to measure target system, while the gauge is responsible to interpret the measure identified. The probing is done on the attributes such as process runtime and CPU load.

The solution given in [1] has certain limitations. They are described here. Utility-based theory was used for best adaptation path under uncertainty. The utility-based frameworks are not fully quality-aware. Quality of service (QoS) analysers can be built to continuously monitor for improvement opportunities. The historical information usage is not sophisticated in the existing frameworks. It can be improved with the state-of-the-art data mining techniques for improving decision-making. The modules incorporated in the framework for overcoming the drawbacks of [1] are as follows. These modules improve the performance of the architecture in making it robust and dynamic and self-adaptive software architecture.

3.2   QoS Analysers

- These are the components in the architectural layer of the proposed framework.
- They take care of quality-aware analysis to exploit opportunities at runtime based on the QoS parameters like resource utilization, response time, etc.

Algorithm 1. QoS-Aware analysis algorithm

```
Algorithm: QoS-Aware Analysis (QAA)
Input: Set of QoS parameters P, set of resources R
Output: Self adaptation decisions D, knowledge update K
Step 1: Initialize decisions vector D
Step 2: Initialize knowledge base vector K
Step 3:  For each p in P
                 For each r in R
                     IF r or R does not satisfy p THEN
                         IF resource available THEN
                 Add decision to D
                         END IF
                 END IF
Step 4:     Updated K
   End
End
Step 5: Return D
```

The algorithm takes the QoS parameters required by the software under test and takes the stock of resource available. After analysing the resources and parameters, it makes set of decisions that help in further improvement in quality adaptations.

3.3   Building and Mining Knowledge Base

A holistic historical information maintenance and analysis are made using the state-of-the-art data mining techniques for making well-informed decisions towards best adaptation path.

Algorithm 2. Knowledge mining algorithm

Algorithm: Knowledge Mining (KM)

Input: Set of QoS parameters P, knowledge base K

Output: Self adaptation decisions D

Step 1: Initialize decisions vector D

Step 2:   Initialize self adaptation decisions D

Step 3: For each p in P

  For each k in K

    IF k or K does satisfies p THEN

      Add decision to D

    END IF

   END IF

 End

    End

Step 4:   Return D

The algorithm takes the QoS parameters required by the software under test and knowledge base and then mines the database for discovering optimal decisions.

4   Self-adaptation Case Study

The case study is a Web-based application with distributed architecture. The application is built using multi-tier architecture. It can be accessed by thousands of clients across the globe. A set of clients can simultaneously make calls to Web server. Then, the servlets running in Web server invoke business components running in application servers. RMI technology is used to have remote method invocations and also integrate with effectors to ensure self-adaptation. The application is made up of thin clients or browsers, Web servers, application servers and database servers. The business components deployed in application server can access database servers in order to provide requested data to clients. The application scenario is shown in Fig. 2.

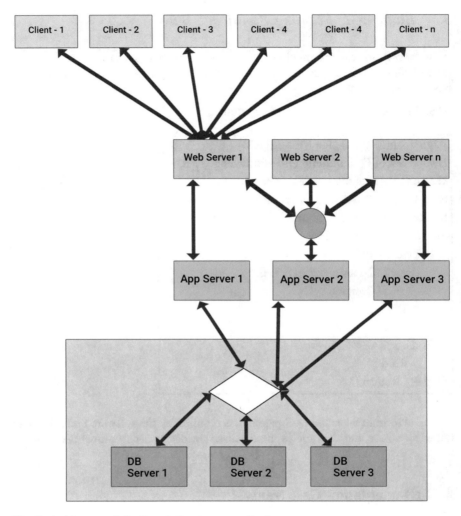

Fig. 2 Architecture of distributed client/server application

This application is in the system layer of the proposed adaptation framework. It is the system which is monitored for dynamic adaptation based on runtime experiences. This application is deployed in Amazon Web Services (AWS) which is one of the widely used public cloud platforms. The rationale behind this is that AWS can provide the state-of-the-art infrastructure in rendering hosting and execution services in pay per use fashion. As it is cloud-based application, it can be accessed by millions of users across the globe. Here comes the need for performance. Before discussing about performance issues and the dynamic adaption, Fig. 3 shows the reference architecture provided by Amazon for deploying Web-based and distributed applications.

The reference architecture takes concurrent requests, and there is load balancer to determine which Web server needs to be used for processing request. There are plenty of Web servers that contain replicated Web application deployed. The Web servers invoke business components running in application servers. The application servers in turn invoke database servers. Replication is the feature in servers so as to serve clients efficiently. As the cloud environment is accessed in pay as you use fashion, it is important to understand the need for using resources effectively. The proposed self-adaptation framework is deployed along with the application. The self-adaptation framework needs to monitor the state of the distributed application and take care of self-adaptation. As mentioned earlier, performance is important from user point of view. Therefore, response time is considered to be the important adaptation factor. The response time is to be maintained as per the service level agreements (SLAs) provided to end users. This is the key factor that needs self-adaptation.

Before deploying the distributed application, AWS account is taken, and cluster is created with commodity servers that act as Web servers, application servers and database servers. Amazon Relational Database Service (RDS)-based MySQL database is preferred. Initially, ten Web servers, ten application servers and five database servers are used by the distributed application. Based on the response time, adaptation framework determines the usage of additional servers. A threshold for response time is taken. This maintained consistency. When response time is dete- riorated, the adaptation framework takes necessary steps to utilize more servers in order to ensure that the response time is complying with SLAs. The response time of a request depends up on 1) heavy load on servers, 2) internet speed i.e avaliable bandwidth of internet connection. Based on the analysis of these parameters, decisions are made dynamically for self-adaptation.

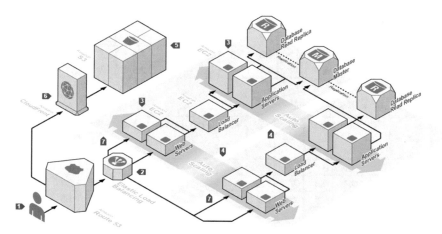

Fig. 3 Reference architecture of AWS Web hosting used for deploying distributed C/S application

5 Experimental Results

This section provides experimental results with the case study application deployed in AWS cloud computing environment. The performance is evaluated in terms of latency without self-adaptation, with self-adaptation and self-adaptation along with BI.

5.1 Without Self-adaptation

See Tables 1, 2 and 3.

As shown in Fig. 4, it is evident that the elapsed time is taken in horizontal axis and latency in vertical axis. The latency or response time should be 1 or below 1 s. This is the threshold considered for self-adaptation. As there is no self-adaptation involved, the results revealed that the SLA with respect to response time is not honoured as elapsed time is increased gradually.

Table 1 Latency without self-adaptation

Elapsed time in seconds	150	300	450	600	750	900	1050	1200	1350	1500	1650	1800
Latency in seconds for randomly chosen requests	0	0.3	0.1	0.1	0.1	1.5	1.6	2	2.5	2.9	3.2	4.6
	0.3	0.7	0.3	0.3	0.3	1.5	1.3	2	2.3	3	3.2	4.6
	0.5	0.6	0.5	0.5	0.5	1.3	1.6	2	2.4	2.9	3.1	4.6
	0.7	0.5	0.7	0.1	0.1	1.5	1.4	2	2.5	2.8	3.2	4.6
	0.8	0.9	0.8	0.6	0.6	1.5	1.6	2	2.5	2.9	3.2	4.6
	0.3	0.7	0.3	0.3	0.3	1.8	1.7	2	2.1	2.7	3.3	4.6
	0.5	0.6	0.5	0.5	0.5	1.5	1.6	2	2.5	2.9	3.2	4.6
	0.7	0.5	0.7	0.7	0.7	1.5	1.6	2	2.5	2.9	3.2	4.6

Table 2 Latency with self-adaptation

Elapsed time in Sec's	150	300	450	600	750	900	1050	1200	1350	1500	1650	1800
Latency in sec's for Randomly chosen Requests	0	0.9	0	0.9	3	3	0	0.9	0.8	0	0.9	0.9
	0.3	0.7	0.3	0.7	3.5	2.4	0.3	0.7	0.8	0.3	0.7	0.9
	0.5	0.6	0.5	0.6	3.5	3	0.5	0.6	0.8	0.5	0.6	0.9
	0.7	0.5	0.7	0.5	2.9	2.3	0.7	0.5	0.8	0.7	0.5	0.9
	0.8	0.9	0.8	0.9	3	3	0.8	0.9	0.8	0.8	0.9	0.9
	0.3	0.7	0.3	0.7	2	2	0.3	0.7	0.8	0.3	0.7	0.9
	0.5	0.6	0.5	0.6	3.2	3.1	0.5	0.6	0.8	0.5	0.6	0.9
	0.7	0.5	0.7	0.5	3	3	0.7	0.5	0.8	0.7	0.5	0.8

Table 3 Latency self-adaptation with BI

Elapsed time in sec's	150	300	450	600	750	900	1050	1200	1350	1500	1650	1800
Latency in sec's for randomly chosen requests	0.6	0.9	0.3	0.8	0.4	0.4	0.6	0.75	0.8	0.3	0.9	0.9
	0.35	0.7	0.3	0.7	0.6	0.6	0.3	0.7	0.8	0.3	0.7	0.7
	0.5	0.6	0.5	0.6	0.3	0.3	0.5	0.6	0.8	0.5	0.6	0.8
	0.7	0.5	0.7	0.5	0.3	0.3	0.7	0.5	0.8	0.7	0.5	0.5
	0.65	0.9	0.8	0.9	0.7	0.7	0.8	0.9	0.8	0.8	0.9	0.3
	0.3	0.7	0.3	0.7	0.4	0.4	0.3	0.7	0.8	0.3	0.7	0.9
	0.5	0.6	0.5	0.6	0.3	0.3	0.5	0.6	0.8	0.5	0.6	0.9
	0.4	0.5	0.7	0.5	0.6	0.6	0.7	0.5	0.8	0.7	0.5	0.8

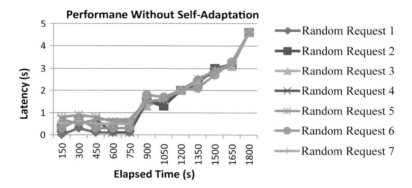

Fig. 4 Performance of the system without self-adaptation

5.2 With Self-adaptation

As shown in Fig. 5, it is evident that the performance of the system started with good response time that is in the limits of SLA and then after reaching 600 s elapsed time, it violated SLA. This has triggered self-adaptation immediately, and the rest of the period, the response time is within acceptable threshold.

5.3 Self-adaptation with Business Intelligence

As shown in Fig. 6, it is evident that the performance of the system started with good response time that is in the limits of SLA continued to do so as the self-adaptation is employed immediately with the BI algorithms in place. Thus, it is almost following real-time self-adaptation strategy.

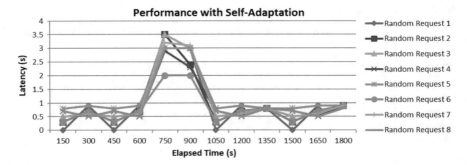

Fig. 5 Performance of the system with self-adaptation

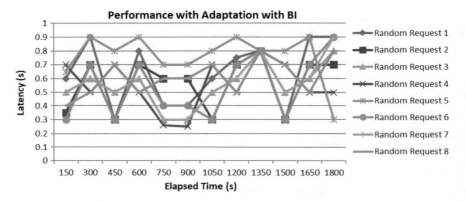

Fig. 6 Performance of the system with self-adaptation and business intelligence

5.4 Performance Comparison

As shown in Table 4, the latency is recorded at different elapsed times for some requests processed by both (E) existing and (P) proposed approaches.

When compared with self-adaptation, details of existing and proposed as given in Fig. 7, the results revealed that there is certain issue with service level agreement that is latency should be less than one second for both approaches. When there is the issue encountered, both performed adaptation. However, the proposed adaptation strategy appeared better than the existing one. The results of existing system are taken from [1]. The existing system does not have business intelligence approach, but self-adaptation strategy is there. The results show the performance improvement over the existing system.

Table 4 Latency comparison with self-adaptation and self-adaptation with business intelligence approaches

Elapsed time in seconds	150 E	150 P	300 E	300 P	450 E	450 P	600 E	600 P	750 E	750 P	900 E	900 P
Latency in seconds for randomly chosen requests	0	0.6	0.9	0.9	0	0.3	0.9	0.8	3	0.4	3	0.4
	0.3	0.4	0.7	0.7	0.3	0.3	0.7	0.7	3.5	0.6	2.4	0.6
	0.5	0.5	0.6	0.6	0.5	0.5	0.6	0.6	3.5	0.3	3	0.3
	0.7	0.7	0.5	0.5	0.7	0.7	0.5	0.5	2.9	0.3	2.3	0.25
	0.8	0.7	0.9	0.9	0.8	0.8	0.9	0.9	3	0.7	3	0.7
	0.3	0.3	0.7	0.7	0.3	0.3	0.7	0.7	2	0.4	2	0.4
	0.5	0.5	0.6	0.6	0.5	0.5	0.6	0.6	3.2	0.3	3.1	0.3
	0.7	0.4	0.5	0.5	0.7	0.7	0.5	0.5	3	0.6	3	0.6

Elapsed time in seconds	1050 E	1050 P	1200 E	1200 P	1350 E	1350 P	1500 E	1500 P	1650 E	1650 P	1800 E	1800 P
Latency in seconds for randomly chosen requests	0	0.6	0.9	0.75	0.8	0.8	0	0.3	0.9	0.9	0.9	0.9
	0.3	0.3	0.7	0.7	0.8	0.8	0.3	0.3	0.7	0.7	0.9	0.7
	0.5	0.5	0.6	0.6	0.8	0.8	0.5	0.5	0.6	0.6	0.9	0.8
	0.7	0.7	0.5	0.5	0.8	0.8	0.7	0.7	0.5	0.5	0.9	0.5
	0.8	0.8	0.9	0.9	0.8	0.8	0.8	0.8	0.9	0.9	0.9	0.3
	0.3	0.3	0.7	0.7	0.8	0.8	0.3	0.3	0.7	0.7	0.9	0.9
	0.5	0.5	0.6	0.6	0.8	0.8	0.5	0.5	0.6	0.6	0.9	0.9
	0.7	0.7	0.5	0.5	0.8	0.8	0.7	0.7	0.5	0.5	0.8	0.8

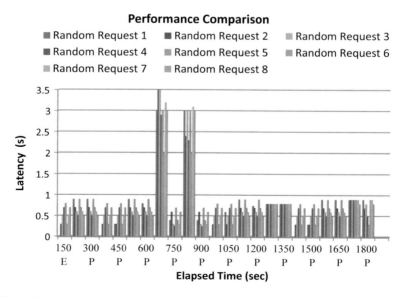

Fig. 7 Performance of the system with self-adaptation and self-adaptation with business intelligence

6 Conclusions and Future Work

Software systems have been evolving. The contemporary applications are able to drive businesses of multiple organizations that form as an integrated set of businesses. Applications in such environment need to adapt to changes dynamically. The changes are to be incorporated in traditional approach by software engineers. However, this approach is costly and time consuming. To overcome this problem, many self-adaptive frameworks came into existence. Rainbow is one such architecture which enables dynamic self-adaptive software. However, the Rainbow architecture has specific limitations in terms of QoS analysis and knowledge mining in making decisions for adaptation. In this paper, we proposed and implemented a self-adaptive dynamic software architecture named ESADSA which is based on a holistic approach with focus on QoS and knowledge mining for making expert decisions as part of self-adaptation. We built a prototype application to demonstrate proof of the concept. Our results revealed that the proposed architecture significantly improves the self-adaptation performance. In future, we focus on defining self-adaptive dynamic software architecture for cloud-based systems.

Acknowledgements We thank Dr. P.V. Nageswara Rao, Dr. G. Appa Rao, Dr. Thammy Reddy and Dr. K Srinivasa Rao for their comments which helped to improve this paper.

References

1. Garlan, D., B. Schmerl, and S. Cheng. 2009. *Software Architecture-Based Self-Adaptation. Automatic Computing and Networking*, 31–55. Berlin: Springer.
2. Bradbury, Jeremy S., James R. Cordy, Juergen Dingel, and Michel Wermelinger. 2004. A Survey of Self-management in Dynamic Software Architecture Specifications. In *WOSS '04: Proceedings of the 1st ACM SIGSOFT Workshop on Self-managed Systems*, ACM, New York, pp. 28–33.
3. Ghosh, Debanjan, Raj Sharman, H. Raghav Rao, and Shambhu Upadhyaya. 2007. Self-healing Systems—Survey and Synthesis. *Decision Support Systems* 42 (4): 2164–2185.
4. Kim, Jung Soo, and David Garlan. 2006. Analyzing Architectural Styles with Alloy. In *Workshop on the Role of Software Architecture for Testing and Analysis 2006 (ROSATEA 2006)*, Portland, ME, July 17, 2006.
5. Mikik-Rakic M., Mehta, and Medvidovic N. Architectural style requirements for self-healing systems. In Garlan et al, Apr 30, 2009, pp. 49–54.
6. Georgiadis I., J Magee, and J Kramer. Self-organizing software architectures for distributed systems. November 01, 2004, Newport Beach, California pp. 33–38.
7. Magee, Jeff, and Jeff Kramer. 1996. Dynamic structure in software architectures. In *SIGSOFT '96: Proceedings of the 4th ACM SIGSOFT Symposium on Foundations of Software Engineering*, ACM, New York, pp. 3–14.
8. Wermelinger, Michel, Antonia Lopes, and Jose Luiz Fiadeiro. 2001. A Graph Based Architectural Reconfiguration Language. *SIGSOFT Software Engineering Notes* 26 (5): 21–32.
9. Dowling, Jim, and Vinny Cahill. 2001. The k-component Architecture Meta-Model for Self-adaptive Software. In *REFLECTION '01: Proceedings of the 3rd International*

Conference on Metalevel Architec-tures and Separation of Crosscutting Concerns, Springer, London, UK, pp. 81–88.

10. Le Metayer, Daniel. 1998. Describing Software Architecture Styles Using Graph Grammars. *IEEE Transactions on Software Engineering* 24 (7): 521–533.

11. Cuesta, Carlos E., Pablo de la Fuente, and Manuel Barrio-Solarzano. 2001. Dynamic Coordination Architecture Through the Use of Reflection. In *SAC'01: Proceedings of the 2001 ACM Symposium on Applied Computing*, ACM, New York, pp. 134–140.

12. Morrison, Ronald, Dharini Balasubramaniam, Flavio Oquendo, Brian Warboys, and R. Mark Greenwood. 2007. An Active Architecture Approach to Dynamic Systems Co-evolution. In *ECSA*, vol. 4758 of LNCS, Springer, New York, pp. 2–10. 24–26 Sept 2007.

13. Batista, T., A Joolia, G Coulson. 2005. Managing Dynamic Reconfiguration in Component-Based Systems. In *EWSA*, vol. 3527 of LNCS, Springer, pp. 1–17, 13–14 June 2005.

14. Hinz, Michael, Stefan Pietschmann, Matthias Umbach, and Klaus Meissner. 2007. Adaptation and Distribution of Pipeline-Based Context-Aware Web Architectures. In *WICSA'07: Proceedings of the 6th Working IEEE/IFIP Conference on Software Architecture*, IEEE Computer Society, Washington, DC, pp. 15.

15. Liu, Yan, and Ian Gorton. 2007. Implementing Adaptive Performance Management in Server Applications. In *Proceedings of the 2007 International Workshop on Software Engineering for Adaptive and Self-Managing Systems (SEAMS'07)*, IEEE Computer Society, Washington, DC, pp. 12.

16. Wolf, A. L., D Heimbigner, A Carzaniga, K. M. Anderson, and N Ryan. Achieving Survivability of Complex and Dynamic Systems with the Willow Framework. In 8. Proc. of the Working Conf. on Complex and Dynamic Systems, Apr 30, 2009.

17. Gorlick, Michael M., and Rami R. Razouk. 1991. Using Weaves for Software Construction and Analysis. In *Proceedings of the 13th International Conference of Software Engineering*, IEEE Computer Society Press, Los Alamitos, CA, USA, pp. 23–34, May 1991.

18. Oreizy, Peyman. 2000. Open Architecture Software: A Flexible Approach to Decentralized Software Evolution. Ph.D. Thesis, University of California, Irvine.

19. Dashofy, Eric M., Andre van der Hoek , Richard N. Taylor. Towards Architecture-Based Self-healing Systems, 26th International Symposium on ... Autonomic Computing: Applications of Self-Healing Systems. DeSE, pp. 21–26, 2002.

20. Sztajnberg, Alexandre, and Orlando Loques. Describing and Deploying Self-adaptive Applications. In *Proceedings of 1st Latin American Autonomic Computing Symposium*, July 14–20, 2006.

21. Silva, Eduardo, Everton Cavalcante, Thais Batista, Flavio Oquendo, Flavia C. Delicato, and Paulo F. Pires. (2041). On the Characterization of Missions of Systems-of-Systems. ACM, pp. 1–8.

22. Cavalcante, Everton, Thais Batista, and Flavio Oquendo. (2015). Supporting Dynamic Software Architectures: From Architectural Description to Implementation. *WICSA '15 Proceedings of the 2015*. IEEE, pp. 1–10.

23. Cavalcante, Everton. (2015). On the Architecture-Driven Development of Software-Intensive Systems-of-Systems. IEEE, pp. 1–4.

24. Milena, Guessi, Everton Cavalcante and Lucas B. R. Oliveira. (2015). Characterizing Architecture Description Languages for Software-Intensive Systems-of-Systems. ACM, pp. 1–7.

25. Quilbeuf, Jean, Everton Cavalcante, Louis-Marie Traonouez, Flavio Oquendo, Thais Batista, and Axel Legay. (2016). A Logic for the Statistical Model Checking of Dynamic Software Architectures. ACM, pp. 1–17.

26. Silva, Eduardo, Everton Cavalcante, Thais Batista, and Flavio Oquendo. (2016). Bridging Missions and Architecture in Software-intensive Systems-of-Systems. IEEE, pp. 1–6.

27. Cavalcante, Everton, Jean Quilbeuf, Louis-Marie Traonouez, Flavio Oquendo, Thais Batista, and Axel Legay. (2016). Statistical Model Checking of Dynamic Software Architectures. IEEE, pp. 1–18.

28. Silva, Eduardo, Everton Cavalcante, and Thais Batista. (2017). Refining Missions to Architectures in Software-Intensive Systems-of-Systems. ACM, pp. 1–7.

Detection of Parkinson's Disease Through Speech Signatures

Jinu James, Shrinidhi Kulkarni, Neenu George, Sneha Parsewar, Revati Shriram and Mrugali Bhat

Abstract Parkinson's disease is a very common neurodegenerative disorder and movement disorder. Two types of symptoms are observed in Parkinson's disease which are motor and non-motor symptoms. Out of these, the non-motor or dopamine non-responsive symptoms have a major impact on the patients. Some of the non-motor symptoms are cognitive impairment, depression, REM sleep disorder, speech and swallowing difficulties, loss of smell and change in the body odor. It becomes difficult to perform basic tasks in daily routine as the symptoms aggravate. The symptoms and the rate at which the disease worsens vary from individual to individual. Patients suffering from this disease also have soft speech, impaired voice or voice box spasms. The objective of our project work is to explore this symptom and its detection. The voice signals will be captured using MATLAB. Comparison of the signals obtained with the corresponding signals of a healthy person will determine whether the individual is affected by the disease.

Keywords Parkinson's disease · Soft speech · MATLAB

1 Introduction

Parkinson's is a disorder which affects the nervous system slowly with time which leads to gradual destruction in the movement. According to the studies, over 10 million people globally are detected with this disorder [1].

Dopamine plays a very important role inside the brain as it acts as a chemical messenger for communication, which is produced in the substantia nigra. Parkinson's leads to breaking down of nerve cells which leads to reduction in the amount of dopamine, thus causing abnormal brain activity [2].

Parkinson's disease has two types of symptoms, which are motor and non-motor. Motor symptoms consist of postural instability, rigidity, bradykinesia (slowness of

J. James (✉) · S. Kulkarni · N. George · S. Parsewar · R. Shriram · M. Bhat
MKSSS's Cummins College of Engineering, Pune, India
e-mail: jinu.james@cumminscollege.in

© Springer Nature Singapore Pte Ltd. 2020
K. S. Raju et al. (eds.), *Proceedings of the Third International Conference on Computational Intelligence and Informatics*, Advances in Intelligent Systems and Computing 1090, https://doi.org/10.1007/978-981-15-1480-7_52

movement), gait analysis, and non-motor symptoms consist of mood disorders, cognitive changes, slowing of thoughts, etc. In the proposed work, speech signal is used for analysis of Parkinson's disease. A person diagnosed with Parkinson's disease has variation in voice as it becomes softer, slurred and breathy which leads to difficulty in understanding [3].

The time-frequency-based extracted features for detection of Parkinson's disease are pitch, jitter, shimmer, SNR and formant frequency [4].

Pitch:

Pitch is a non-cognitive property that allows the ordering of sounds on a frequency-related scale. It is known as the rate at which vibrations are produced. Pitch period is the time duration of one glottal cycle. They are usually expressed in Hertz. One cycle represents the complete vibration of speech signal back and forth. The pitch is high when frequency of the tone is high [5].

Formant Frequency:

The formant is known as concentration of acoustic energy around a particular frequency or resonant frequencies of the vocal tract. The shape and dimension of the vocal tract lead to changes in formant frequencies. The major resonances of the vocal tract can be approximately characterized by the first four resonant frequencies. These resonant frequencies are denoted by F_1, F_2, F_3, F_4 formants. The fundamental frequency F_0 and the formant frequency are correlated. The relation between the nth formant frequency F_n and the fundamental frequency F_0 can be approximated as

$$F = a_n(F_0 + b_n) \tag{1}$$

where a_n and b_n are vowel dependent constants [6].

Jitter:

Jitter is the alteration of a periodic signal from its true periodicity. It attributes to the variations of fundamental frequency between cycles of vibration. According to researchers, from 0.5 to 1% is the normal range of jitter. Due to the loss of control of vibration of the vocal cords, the jitter is affected adversely. High percentage of jitter is seen amongst patients affected by Parkinson's because this disorder interferes with the normal vocal fold [7].

Shimmer:

Shimmer acts toward amplitude change in the voice. Due to the reduction of glottal resistance on the vocal cord, shimmer differs. Shimmer deals with the percentage distortion in the amplitude of the vocal cord. The fluctuation in the intensity of the neighboring vibratory cycles of the vocal folds is calculated by shimmer. Therefore, the noises inside the human body can be calculated by jitter and shimmer. Higher jitter and shimmer levels reflect neuromuscular problems [7].

2 Methodology

The objective of this system is to anticipate Parkinson's disease, based on speech signal analysis.

Main components of this system are as follows:

  (i) MATLAB toolbox
 (ii) MATLAB programming.

Figure 1 displays the block diagram of the proposed system used for speech signal.

The proposed system aims to record speech signal using MATLAB toolbox and thereafter calculating parameters.

  (i) MATLAB toolbox (Recorder):
      In the experiment that has been conducted, MATLAB toolbox has been used as an equipment to record the speech signal. To record data, audio input device (microphone) is connected to the system.
 (ii) MATLAB Programming:

    (A) Windowing:
        After reading the recorded signal, it has been seen that there are a lot of spurious high frequencies. Windowing helps us taper the signal smoothly at the start and the end and also suppresses the discontinuity in the signal. Change in properties occurs in a speech signal over time. Some properties of the speech are of short or of long period of time. With the help of DFT, hamming or autocorrelation signal processing methods can be carried out for short period of time. Speech processing is done by considering short windows called frame and then processing them. Long signal of speech is given a finite length of the original signal by multiplying it with a finite length of window function [8].

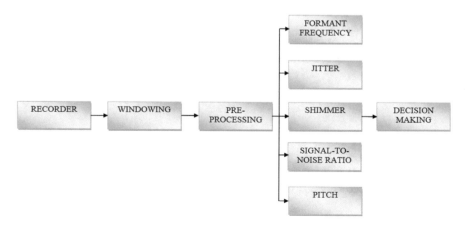

Fig. 1 Block diagram

(B) Pitch Determination:
    To determine the pitch or fundamental frequency of a speech signal, a
    pitch detection algorithm (PDA) is used. Different techniques used for
    pitch period estimation are autocorrelation, cepstrum and SIFT method
    [9]. Due to the convolution of the vocal tract and vocal source, there is an
    existence of larger peaks comparing to the peak of the pitch period giving
    us the wrong estimation, and this is the main disadvantage of estimating
    pitch with the help of autocorrelation method. The cepstrum of speech is
    stated as inverse Fourier transform of the log-magnitude spectrum.
    Slowly varying and fast-varying components are given out to the low
    frequency and high-frequency region, respectively, by the cepstrum in
    log-magnitude spectrum. The envelope corresponding to the vocal tract is
    indicated by slowly varying component, and the excitation source is by
    fast-varying components. Hence, in the spectrum of speech, the vocal
    tract and excitation source components get indicated naturally [10].
(C) Signal-to-Noise Ratio (SNR):
    It is defined as the ratio of signal intensity to noise intensity, expressed in
    decibels.

$$\text{SNR} = 20 \log_{10}\left(\frac{S_{\text{rms}}}{N_{\text{rms}}}\right) \tag{2}$$

    To determine the quality of audio data, signal-to-noise ratio is an
    important feature. Recognition performance is strongly influenced by the
    SNR which is why this parameter is important [11].
(D) Formant Estimation
    Linear prediction method predicts the output of the linear system based
    on its input and previous output of a linear system. Linear predictive
    coding is used for determining formants, and the intensity and the fre-
    quency are estimated by extracting their effects from the speech signal.
    This technique of formants extraction is known as inverse filtering. The
    formant frequencies rely on the size and shape of the vocal tract [12].
(E) Jitter Determination
    Jitter (absolute) is the cycle-to-cycle variation of fundamental frequency,
    i.e., the average absolute difference between consecutive periods [13].

$$\text{Jitter (absolute)} = \frac{1}{N-1}\sum_{i=1}^{N-1}|T_i - T_{i+1}| \tag{3}$$

F. Shimmer Determination
    Shimmer (dB) is expressed as the variability of the peak-to-peak
    amplitude in decibels, i.e., the average absolute base-10 logarithm of the
    difference between the amplitudes of consecutive periods, multiplied by
    20 [13].

$$\text{Shimmer (absolute)} = \frac{1}{N-1} \sum_{i=1}^{N-1} \left| 20 \log \frac{A_{i+1}}{A_i} \right| \qquad (4)$$

3   Result and Analysis

Table 1 lists the parameters calculated for 20 samples that are not affected by Parkinson's disease. The speech signal was recorded with the help of MATLAB toolbox. These readings are taken during the period when these subjects were not sick. The plots obtained for a normal subject for the parameters of the speech signal are as shown in Figs. 2 and 3.

Figure 2 shows the fundamental frequency or pitch estimated by cepstrum method. The figure consists of three graphs of which the first graph is plot of the speech signal recorded. Second graph is the plot of magnitude spectrum of speech signal. Third graph is plot of cepstrum, i.e., DFT of log spectrum.

Table 1 Parameters calculated using MATLAB

Name	Pitch	SNR	Formant frequency	Jitter	Shimmer
Abraham	905.66	−19.46	382.3, 1581.2, 2405.2	7.67×10^{-10}	2.72×10^{-6}
Aishwarya	827.58	−25.56	470.9, 1320.1, 2005.1	8.25×10^{-10}	1.69×10^{-7}
Ajoe	923.07	−18.42	441.2, 1303.3, 2557.9	7.52×10^{-10}	1.24×10^{-6}
Ameya	923.07	−19.15	495.8, 1266.3, 1826.0	7.21×10^{-10}	-9.93×10^{-7}
Neethu	905.66	−23.62	341.0, 717.5, 2147.6	7.20×10^{-10}	-2.58×10^{-7}
Elsa	1000	−22.61	519.1, 1131.6, 1721.8	6.73×10^{-10}	-3.48×10^{-6}
Neenu	786.88	−21.92	444.1, 783.9, 1882	8.61×10^{-10}	-2.44×10^{-6}
Shrinidhi	923.07	−23.05	420.2, 500.5, 1844.2	7.32×10^{-10}	5.18×10^{-6}
Jinu	738.46	−22.73	415.7, 776.3, 1947.1	9.40×10^{-10}	3.25×10^{-7}
Jease	923.07	−17.09	463.3, 713.1, 1632.3	7.60×10^{-10}	-2.65×10^{-6}
Likhita	750	−22.48	371.7, 781.8, 2102.7	9.10×10^{-10}	4.21×10^{-6}
Niyanta	685.71	−19.7	426.4, 918.6, 1918.9	9.95×10^{-10}	-1.16×10^{-6}
Pallavi	827.58	−19.59	398.5, 964.0, 1976.8	8.29×10^{-10}	-3.46×10^{-6}
Roshan	1000	−19.89	532.1, 1726.8, 2575.9	7.56×10^{-10}	8.83×10^{-7}
Rucha	923.07	−17.93	488.8, 1911.6, 1928.9	7.53×10^{-10}	1.90×10^{-6}
Shreya	1000	−21.19	570.7, 709.1, 2000.5	6.94×10^{-10}	-1.29×10^{-6}
Simi	872.72	−24.1	384.8, 796.7, 1656.4	7.88×10^{-10}	-2.16×10^{-7}
Sneha	786.88	−18.94	465.1, 1733.2, 2790.6	8.78×10^{-10}	-3.59×10^{-6}
Vaidehi D	738.46	−19.43	363.9, 660.3, 2016.3	9.41×10^{-10}	4.30×10^{-6}
Vincy	872.72	−19.72	390.5, 961.9, 1945.8	7.44×10^{-10}	2.15×10^{-6}

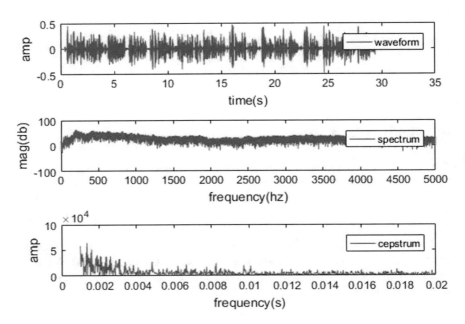

Fig. 2 Fundamental frequency estimation

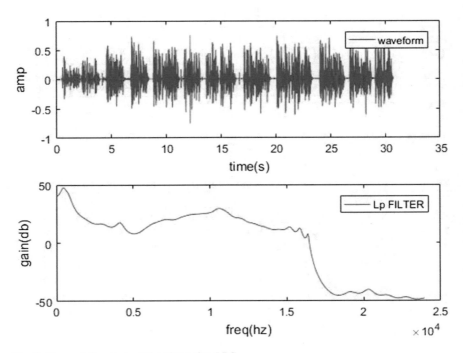

Fig. 3 Formant frequency estimation using LPC

Figure 3 shows the formant frequency estimation of the recorded signal. It consists of two graphs, first graph is the plot of the recorded signal, and the second graph is the plot of the filter's frequency response using linear predictive coding.

4   Discussion

The main reason behind such a study is early diagnosis of Parkinson. It was observed that for early detection of Parkinson's disease, more concentration must be given on non-motor symptoms only rather than motor symptoms as they start occurring after significant years. In the system, proposed speech signal analysis has been considered for Parkinson's disease detection. There are many parameters that show variation when the subject is affected with Parkinson's disease. Some of these are jitter, shimmer, signal-to-noise ratio, formant frequency, pitch, energy, intensity, phase differences, correlation dimension, harmonicity and many more. This is an ongoing work. Amongst these, work was done on five parameters. After analyzing the parameters of the healthy patients, the amount of deviation can be measured for a person affected with Parkinson's disease. Many other works have been carried out based on non-motor symptoms of Parkinson's disease which are tremor analysis, skin impedance, change in facial expression and multiple characteristics of finger movement while typing [14, 15].

5   Conclusion

The detection and diagnosis of Parkinson's disease at an early stage are extremely helpful. The current treatments available for Parkinson's disease are more potent if the disease is detected at the starting phase. Along with this, the physiotherapies and exercises provided to delay the disease advancement and improve the symptoms are much easier to perform at the onset of the disease. This method of detection of Parkinson's disease is reliable and non-invasive. Various parameters have been discussed in the proposed work. According to the research carried out, the values for jitter and shimmer are higher for Parkinson's disease patients than normal subjects, while the values for pitch and formants are lower. This is basically due to changes in the muscle contraction as a result of which the vocal cords do not function properly leading to voice disturbance. Researches in the areas of biomarkers are currently being pursued. These markers in the blood, urine or CSF which can reliably detect Parkinson's disease will be the biggest assets in early detection of the disease. Brain imaging techniques which are highly sensitive in detecting the disease at an early stage are also under study.

Acknowlegements Data was collected from Cummins college of engineering for women and the Parkinson's Mitra Mandal Association, Pune. All the subjects were willing volunteers who participated in the data collection and have given verbal consent for using the data for research purpose and for further publication. None of the ethical committee was involved in it.

References

1. Parkinson's disease Foundation. www.Parkinson's disease f.org/en/parkinson_statistics.
2. WebMD. Leading source for trustworthy and timely health and medical news https://www. webmd.com/parkinsons-disease/default.htm.
3. ParkinsonsDisease.net. Parkinson's disease Health Information & Community. https:// parkinsonsdisease.net/symptoms/speech-difficulties-changes/.
4. Dixit, Vikas Mittal, and Yuvraj Sharma. Voice Parameter Analysis for the Disease Detection. *IOSR Journal of Electronics and Communication Engineering (IOSR-JECE)*.
5. Rabiner, L.R., and R. W. Schafer. *Digital Processing of Speech Signals*. Pearson Education India.
6. Jayan, A.R. *Speech and Audio Signal Processing*. PHI Learning.
7. Teixeira, Joao Paulo, Carla Oliveira, Carla Lopes. Vocal Acoustic Analysis—Jitter, Shimmer and HNR Parameters. CENTERIS 2013—Conference on ENTERprise Information Systems/ HCIST 2013—International, Conference on Health and Social Care Information Systems and Technologies.
8. Dhanalakshmi, P., S. Palanivel, and V. Ramalingam. 2009. Classification of Audio Signals Using SVM and RBFNN. *Expert Systems with Applications* 36 (3): 6069–6075.
9. Rabiner, A.R. 1997. On the Use of Autocorrelation Analysis for Pitch Detection. *IEEE Transactions on Acoustics, Speech, and Signal Processing* 25 (1): 24–33.
10. Amrita Vishwa Vidyapeetham. Estimation of Pitch from Speech Signals. http://vlab.amrita. edu/?sub=3&brch=164&sim=1012&cnt=6.
11. ICSI Speech. Chapter 4.1 Signal processing and audio. http://www1.icsi.berkeley.edu/Speech/ faq/speechSNR.html.
12. Snell, Roy C. 1993. Fausto millinaso, Formant Location from LPC Analysis of Data. *IEEE Transactions on Speech, and Audio Processing* 1 (2).
13. Farrús, Mireia, Javier Hernando, Pascual Ejarque. *Jitter and Shimmer Measurements for Speaker Recognition*. TALP Research Center, Department of Signal Theory and Communications Universitat Politècnica de Catalunya, Barcelona, Spain.
14. Mrugali, Bhat, Sharvari Inamdar, Devyani Kulkarni, Gauri Kulkarni, and Revati Shriram. 2017. Parkinson's Disease Prediction Based on Hand Tremor Analysis. 2017 International Conference on Communication and Signal Processing (ICCSP).
15. Akshada Shinde, Rashmi Atre, Anchal Singh Guleria, Radhika Nibandhe, and Revati Shriram. Facial Features Based Prediction of Parkinson's Disease. 2018 3rd International Conference for Convergence in Technology (I2CT).

Rough Set-Based Classification of Audio Data

T. Prathima, A. Govardhan and Y. Ramadevi

Abstract For effective multimedia content, retrieval audio data plays an important role. Recognising classes of audio data which is neither music nor speech is a challenging task; in this aspect, the authors proposed to work on environment sounds. To represent the audio data, low-level features are extracted. These low-level descriptors are computed from both time domain and frequency domain representation of audio data. From the extracted descriptors, midterm statistics are computed and an information system (IS) is built with class labels. From this IS using the concept of rough set theory, reducts are computed, and from the reducts, rules are generated. The rules obtained are tested against the test set sampled from ESC-10 dataset.

Keywords ESC-10 · Information system · Discretisation · Discernibility relation · Reduct · Rough set classifier

1 Introduction

With humungous amount of multimedia archives available at the click of a button, high-speed and high-accurate algorithms are required for retrieving the user interested data. Audio data plays a vital role in retrieving the data of interest. Digitised CD-quality audio recording requires 44,100 samples of real data to store one second

T. Prathima (✉)
Department of IT, CBIT, Hyderabad, India
e-mail: tprathima_it@cbit.ac.in

A. Govardhan
Department of CSE, Jawaharlal Nehru Technological University Hyderabad, Kukatpally, Hyderabad, Telangana State, India
e-mail: govardhan_cse@jntuh.ac.in

Y. Ramadevi
Departemnt of CSE, CBIT, Hyderabad, India
e-mail: yrdcse.cbit@gmail.com

© Springer Nature Singapore Pte Ltd. 2020
K. S. Raju et al. (eds.), *Proceedings of the Third International Conference on Computational Intelligence and Informatics*, Advances in Intelligent Systems and Computing 1090, https://doi.org/10.1007/978-981-15-1480-7_53

of audio data [1]. Using the raw audio data in its direct form by an algorithm for processing will not yield commendable results. Audio data must be represented in a more succinct form for further processing. Low-level audio features are extracted from both frequency and temporal domains. From these low-level features, midterm statistics are computed and an information system is created. Each audio snippet is represented as a row in the information system along with its class label. Using this information system based on rough set theory, a reduct is computed; reduct is nothing but subset of attributes or features that are sufficient to represent a class of audio samples. Reduct helps in reducing the information system, which is used for training the rough set-based classifier to label the audio snippets. The dataset chosen is ESC: Dataset for Environmental Sound Classification [2].

This paper is organised as detailed: Sect. 2 discusses about the audio media and its representation, Sect. 3 throws light on the existing work, Sect. 4 discusses about ESC dataset, Sect. 5 presents the rough sets, Sect. 6 details the algorithm, Sect. 7 elaborates the results and Sect. 8 concludes the paper.

2 Audio and Its Representation

The tasks that can be performed on multimedia data range from retrieving user interested content, summarisation, classification, grouping based on semantic similarity, etc. Audio media plays an important role in the analysis of multimedia data for any of the listed tasks.

To analyse the audio data for its content, low-level features such as zero-crossing rate, mean crossing rate, maximum signal amplitude, minimum signal amplitude, energy, root mean square energy, phase and magnitude from FFT spectrum, bands 0-N_{Mel} from Mel/Bark spectrum, FFT based and filter based from semitone spectrum, MFCC and PLP-CC from cepstral feature group, autocorrelation function, formants, pitch, F0 harmonics, HNR, jitter, shimmer from voice quality, chroma, and CENS from Tonal group can be extracted [3].

Audio signal is quasi-stationary in nature; to exploit this quasi-stationary nature of audio data, a hamming window function of size 0.05 s is used; this window length is the frame size. Consecutive windows are overlapped by 50%, and each window will be of length 2205 samples. From every window, 35 low-level features are extracted. The number of frames for each audio sample is calculated as shown below [4]:

$$\text{Number of Frames} = \text{floor}((L - \text{window length})/\text{step}) + 1 \qquad (1)$$

where L is length of the entire signal, window length is the product of window size and sampling frequency of the audio signal, and step size is 0.025 s.

For example, if the audio signal size is of five seconds length, sampling frequency is 44,100, window size is 0.05 s, step size is 0.025 s, and it results in 199 short-term frames.

3 Related Work

Different approaches based on machine learning were evaluated on both ESC-10 and ESC-50 [5] by various authors who reported accuracy ranging from 32.2 to 94%. The different methodologies reported in the literature are convolutional neural networks [6–9], deep learning [10], mixture models [11], convolutional recurrent neural networks [12], random forest, support vector machines, k-nearest neighbour [2], etc.

Hardik et al. [6] trained convolutional restricted Boltzmann machine in an unsupervised method to model arbitrary length audio snippets with a learning rate of 0.001. They have trained the model using 60 sub-band filters with convolutional window lengths varying from 132 to 220 samples. The authors reported that Adam optimization gave better accuracy than stochastic gradient descent. They compared the performance of convolutional RBM's with filterbanks and convolutional RBMs with Gamma tone spectral coefficients and reported that convRBM's with filterbanks outperformed GTSC convRBM's. They have achieved an overall accuracy of 86.5%.

Rishabh et al. [7] used phase encoded filterbank energies with Mel filterbank and convolutional neural network. They achieved better performance by combining filterbank energies and phase encoded FBE's features sets. They have used a convolutional neural network proposed in [9]. The authors reported an accuracy of 84.15% on ESC-50 dataset.

Yuji et al. [10] proposed a deep learning method for sound recognition. They have used the concept of between-class learning by giving as input signals of two audio signals belonging to different classes with random ratio to train the model to identify the mixing ratio. They claimed that between-class (BC) learning shuns the possibility of decision boundary of a class appearing between any other classes. They constructed a deep net EnvNet-v2 based on EnvNet [8] and reported error rates of 18.2% on ESC-50 dataset and 8.6% on ESC-10 dataset.

Baelde et al. [11] built a dictionary from mixture models of sound spectrum. Classification is carried out by computing the likelihood of the test sample to the models in dictionary, and the closest sample's class label is assigned to the test sample. The authors carried out a five-fold test and reported an accuracy of 94% for ESC-50 dataset and 96% for ESC-10 dataset. The authors claimed that their proposal outperformed parametric and non-parametric methods as well as human scores.

Table 1 ESC-50 dataset categorisation

Human, non-speech sounds	Animals	Natural soundscapes and water sound	Exterior/ urban noises	Interior/ domestic sounds
Crying baby	Dog	Rain	Helicopter	Door knock
Sneezing	Rooster	Sea waves	Chainsaw	Mouse click
Clapping	Pig	Crackling fire	Siren	Keyboard typing
Breathing	Cow	Crickets	Car horn	Door, wood creaks
Coughing	Frog	Chirping birds	Engine	Can opening
Footsteps	Cat	Water drops	Train	Washing machine
Laughing	Hen	Wind	Church bells	Vacuum cleaner
Brushing teeth	Insects	Pouring water	Air plane	Clock alarm
Snoring	Sheep	Toilet flush	Fireworks	Clock tick
Drinking	Crow	Thunderstorm	Hand saw	Glass breaking

4 Datasets

For experimentation, ESC-10 which is a subset of ESC-50 dataset was considered [2]. Environmental sound classification dataset is extracted from Freesound project [12]. This dataset comprises of 50 different categories of environmental sounds, and each category has about 40 samples. A total 2000 audio files of each five-second length are available. All the audio samples in the dataset are neither speech nor music, which are everyday sounds that are heard. Labelling these sound snippets is important in tasks such as identifying the presence of vehicles, and animals. Table 1 shows the categorisation of 50 different classes of audio samples divided into five major groups with ten classes in each category.

ESC-10 dataset is a subset of ESC-50, 400 audio files belonging to classes person sneeze, baby cry, dog bark, rooster, rain, sea waves, fire crackling, clock tick, helicopter and chainsaw are included.

5 Rough Sets

The concept of set theory says whether an object belongs or does not belong to a set, based on boundaries and regions. Rough set concept can be used to find hidden interesting patterns, identify reducts, evaluate importance of data, generate decision rules from the input data, in parallel processing, etc. Defining the information system, the indiscernibility relation, lower approximation, upper approximation, boundary region, positive region, negative region, reduct and core are the fundamentals of rough set theory [13].

6 Algorithm

Algorithm: *Rough set-based classifier for the classification of environmental sounds.*
Input: *Audio samples from ESC-10 dataset.*
Output: *Classification accuracy.*
Step 1: Read the frames from the audio sample.
Step 2: Extract 35 low-level descriptors for every short-term frame.
Step 3: Compute midterm statistics for all the short-term frames of the sample.
Step 4: Build an information system.
Step 5: Discretise the data in the information system.
Step 6: Vertically split the information table into three sub-tables.
Step 7: Partition the dataset into train and test sets and compute the reduct on training sets.
Step 8: Generate rules based on the computed reduct.
Step 9: Classify test set using the rules generated in Step 8.
Step 10: Repeat Steps 7–9 for all the sub-tables.

Step 1: In this step, the input audio samples are read which are of length five seconds.

Step 2: Extract 35 low-level descriptors for every short-term frame [4].

In this step, short-term frames of length 0.05 s are considered with 50% overlap for feature extraction considering the quasi-stationary characteristic of audio signal. So the actual frame size with respect to number of samples will be 2205 if the sampling frequency is 44,100. For every frame, the following low-level descriptors are extracted: ZCR, energy, energy entropy, spectral centroid, spectral entropy, spectral flux, spectral roll-off, harmonic features hr, F0, Mel frequency cepstral coefficients and chroma features. The output of this step results in a 198×35 size table for a five-second audio snippet as shown in Table 2.

Step 3: Compute midterm statistics for all the short-term frames of the sample.

In this step, statistics are computed for all the 35 features extracted for 198 short-term frames. The six statistics that are computed are mean, median, standard deviation, standard deviation/mean, minimum and maximum. These six statistics are computed over 35 low-level descriptors for all the frames of a single audio file which results in a 210×1 row. For 400 audio files of the dataset, the table size will be 210×400.

Table 2 35 low-level descriptors extracted for 198 frames of five-second audio sample

S. No.	ZCR	Energy	Energy entropy	Spectral centroid	Spectral entropy	Spectral flux	Spectral roll-off	.....	Chroma12
F# 1	0.042857	0.000156	2.87805	0.210898	0.225283	0.28543	2.30E−25	...	1.059107
.	...	...	...	...	...	...	...	...	...
F#198	0.017007	0.001241	1.627155	0.104499	0.185777	0.029434	0.034111	...	1.220032

Table 3 Information system—audio data [14]

File No.	Mean of *fe#1	...	Median of fe#1	...	.....	Minimum of fe#35	Class label
^Fi# 1	0.007472	.	0.007472	.	.	3.625207	Dog
Fi#2	0.158657	.	0.158657	.	.	7.933208	Sneezing
.	.	.	.	.	.	.	.
Fi#400	0.096439	.	0.096439	.	.	5.246916	clock_tick

*—fe is feature; ^—Fi is file number

Step 4: Build an information system.

In this step, the class labels are appended for every row of the midterm statistics that are obtained from Step 3. This step will result in a 211×400 size data table, where rows correspond to the objects of the universe (U); each object represents a single audio file. Hence, the information system I = (U, F), where F is a set of conditional features F_c fe#1 to fe#210 and decision attribute F_d class label (Table 3).

Step 5: Discretise the data in the information system.

This is a preprocessing step where all conditional attributes (F_c) which are numeric need to be discretised for further analysis, as rough sets can handle discretised data. In this step, all 210 conditional attributes are discretised, where binning is done and intervals will replace the continuous values.

Step 6: Vertically split the information table into three sub-tables

Once done with preprocessing, classifier can be trained. The entire table is now vertically split into three sub-tables. The first table consists of the first nine features, i.e. ZCR, energy, energy entropy, spectral centroid, spectral entropy, spectral flux, spectral roll-off, harmonic features hr, F0 along with their statistics, the second table consists of 13 MFCC features and their respective statistics, and the third table consists of 12 chroma features along with their statistics.

Step 7: Partition the dataset into train and test sets and compute the reduct [15] on training sets

The dataset was split into training and testing, where 75% of the objects were considered for training and 25% for testing. From the training set, discernibility table was built and reduct was computed from the discernibility table. For all the sub-tables from Step 6, the lengths of computed reducts varied from 8 to 12. In one of the sub-table, the reduct was {"std of fe#2", "std of fe#3", "std of fe#4", "std of fe#5", "std of fe#6", "std of fe#7", "std of fe#8", "std of fe#9"} of length 8.

Step 8: Generate rules based on the computed reduct.

From the reducts, the rules are generated and the rule set size of a sub-table varied from 232 to 242. One of the sample rules for audio class rooster is of the form ("std of fe#2" = \'(-inf-0.1\'')&("std of fe#3" = \'(0.9-inf)\'')&("std of

Results of experiments by train&test method: stdFe1#9_0.25

		Predicted												
		sneezing	sea_waves	rooster	clock_tick	chainsaw	fireworks	helicopter	dog	rain	crying_baby	No. of obj.	Accuracy	Coverage
Actual	sneezing	2	0	0	1	0	0	0	0	0	0	14	0.667	0.214
	sea_waves	0	6	0	0	2	0	0	0	0	0	10	0.75	0.8
	rooster	1	0	1	0	0	0	0	0	0	0	5	0.5	0.4
	clock_tick	0	0	0	0	0	0	0	0	0	0	7	0	0
	chainsaw	0	3	0	0	0	0	1	0	0	0	7	0	0.571
	fireworks	0	0	0	1	0	3	0	1	0	0	16	0.6	0.312
	helicopter	0	0	0	0	0	0	3	1	1	0	10	0.6	0.5
	dog	1	0	0	0	0	1	0	3	0	1	10	0.5	0.6
	rain	0	2	0	0	2	0	1	0	1	0	11	0.167	0.545
	crying_baby	0	0	0	0	0	0	0	0	0	4	10	1	0.4
	True positive rate	0.5	0.55	1	0	0	0.75	0.6	0.6	0.5	0.8			

```
Total number of tested objects: 100
Total accuracy: 0.535
Total coverage: 0.43
```

Fig. 1 Confusion matrix from standard deviation statistic obtained from the first sub-table

fe#4"=\'(0.1-0.2]\")&("std of fe#5"=\'(0.3-0.4]\")&("std of fe#6"=\'(-inf-0.1] \")&("std of fe#7"=\'(0.2-0.3]\")&("std of fe#8"=\'(-inf-0.1]\")&("std of fe#9"=\'(0.2-0.3]\")=>("class Label" = {rooster[1]}).

Step 9: Classify test set using the rules generated in Step 8.

In this step, rules that are generated from Step 8 are used for classifying the test set.

Step 10: Repeat Steps 7–9 for all sub-tables.

Steps 7–9 are repeated for all folds and accuracy obtained varied from 41.9 to 76.5% in the total covered objects (Fig. 1).

7 Results and Discussions

In the existing work, deep networks either outperform or match the accuracy of convolution neural network-based classification methods. The performance of CNN-based classifiers [6, 7, 9, 10, 12] varied from 64.5 to 94%. The traditional

Table 4 Summary of the results obtained on the test sets of the sub-tables

Statistic	Results obtained on a test set size of 100 objects					
MFCC features			Features 1–9		Chroma	
	Accuracy (%)	Coverage (%)	Accuracy (%)	Coverage (%)	Accuracy (%)	Coverage (%)
Mean	61.9	21	45.7	35	68.4	19
Median	75	12	41.9	31	63.2	19
Std	60	10	53.5	43	55.6	18
Stdbymean	58.8	17	48.8	41	62.5	8
Maximum	50	8	63.9	36	76.5	17
Minimum	57.1	7	55.6	36	60	20

Table 5 Comparison of the parameters for all sub-tables

Sub-table	Size of the fold		Size of reduct	No. of rules generated from the reduct	Minimum and maximum accuracy observed across folds (%)
	Training	Testing			
MFCC	300	100	10–11	291–2568	57.1–75
First 9 features	300	100	8–9	232–242	41.9–63.9
Chroma	300	100	11–12	259–540	55.6–76.5

classifiers k-NN, SVM and random forest [2] reported accuracies of 66.7%, 67.5% and 72.7%, respectively. The reported accuracy when classification was done manually was 95.7% [2]. The authors carried out the experiments on ESC-10 dataset. The observed accuracies are presented in Table 4.

In Table 5, the lengths of the reducts, number of rules generated, etc., can be observed for all the sub-tables.

It is noticed whenever the coverage is less the accuracy is more and when more than one reduct is computed the number of rules generated is substantially more. Best performance in terms of both coverage and accuracy was obtained for the second sub-table with nine features and standard deviation statistic. The most influencing statistics are found to be standard deviation, and the important attributes are energy, spectral centroid, spectral spread, spectral entropy, spectral flux, spectral roll-off, eight MFCC features and harmonic feature.

8 Conclusion

The aim of this work was to train a rough set-based Classifier for environmental sound classification using the benchmark dataset ESC-10. In the process, an information system was built from the midterm statistics computed against the low-level descriptors extracted from the 400 audio snippets of ESC-10. The continuous data was discretised, and reduct was computed using rough sets. The length of the computed reduct varied from 8 to 12 for ESC-10 dataset. Using these reducts, rules were generated. These generated rules are used for testing and the rough set-based classifier accuracies for ESC-10 dataset varied from 41.9 and 76.5%. Rough set-based classifier reported average accuracies when compared to the existing work.

References

1. Schuller, Bjorn W. 2013. *Intelligent Audio Analysis*. Berlin Heidelberg: Springer.
2. Piczak, K.J. 2015. ESC: Dataset for Environmental Sound Classification. In: *23rd Annual ACM Conference on Multimedia*, Brisbane, Australia. pp. 1015–1018.
3. Florian, Eyben. (2016). *Real-time Speech and Music Classification by Large Audio Feature Space Extraction*. Springer Theses, Springer International Publishing.
4. Giannakopoulos, Theodoros, and Aggelos Pikrakis. 2014. *Introduction to Audio Analysis: A MATLAB® Approach*. Academic Press.
5. Piczak, K.J. https://github.com/karoldvl/ESC-50.
6. Sailor, Hardik B., Dharmesh M. Agrawal, and Hemant A. Patil. 2017. Unsupervised Filterbank Learning Using Convolutional Restricted Boltzmann Machine for Environmental Sound Classification. In: *INTERSPEECH*, 2017 Aug 2017, Stockholm, Sweden, pp. 3107–3111.
7. Tak, Rishabh N., Dharmesh M. Agrawal, and Hemant A. Patil. 2017. Novel Phase Encoded Mel Filterbank Energies for Environmental Sound Classification. In: *Pattern Recognition and Machine Intelligence: 7th International Conference, PReMI 2017*, Kolkata, India, 5–8 Dec 2017, pp. 317-325.
8. Tokozume, Yuji, and Tatsuya Harada. 2017. Learning Environmental Sounds with End-To-End Convolutional Neural Network. In *2017 IEEE International Conference on Acoustics, Speech and Signal Processing (ICASSP)*, New Orleans, LA, pp. 2721–2725.
9. Piczak, K.J. 2015. Environmental Sound Classification with Convolutional Neural Networks. In 25th International Workshop on Machine Learning for Signal Processing (MLSP), pp. 1–6. Boston, MA, USA.
10. Tokozume, Yuji, Yoshitaka Ushiku, and Tatsuya Harada. 2018. Learning from Between-Class Examples for Deep Sound Recognition. In *Sixth International Conference on Learning Representations*, ICLR, Vancouver.
11. Baelde, Maxime, Christophe Biernacki, and Raphael Greff. 2017. A Mixture Model-Based Real-Time Audio Sources Classification Method. In *IEEE International Conference on Acoustics, Speech and Signal Processing (ICASSP)*, New Orleans, LA, 2017, pp. 2427–2431. (2017).
12. Freitag, Michael, et al. 2017. auDeep: Unsupervised Learning of Representations from Audio with Deep Recurrent Neural Networks. *The Journal of Machine Learning Research* 18 (1): 6340–6344.
13. Audio Files. Retrieved from https://freesound.org/.

14. Zdzisław, Pawlak. (1982). Rough Sets. *International Journal of Computer and Information Sciences* 11 (5).
15. Li, Xiao-Li, Zhen-Long Du, Tong Wang, and Dong-Mei Yu. Audio Feature Selection Based on Rough Set. *International Journal of Information Technology* 11 (6).

Workload Assessment Based on Physiological Parameters

Tejaswini Dendage, Vaidehi Deoskar, Pooja Kulkarni,
Revati Shriram and Mrugali Bhat

Abstract This paper deals with the various physiological parameters like ECG, EEG, PP and PPG that show deviation from normal values when a person is under stress. Electrocardiography (ECG) is the process of capturing the electrical activity of the heart for a period of time using electrical conductors placed over the skin. ECG waveform tells us about the electrical activity of the heart. Electroencephalography (EEG) is the process of capturing electrical activity of the brain. EEG measures changes in voltage fluctuations resulting from ionic current present inside the neurons of the brain from the scalp and the difference between systolic blood pressures also known as pulse pressure (PP). The stress can be of many types. These signals are monitored in order to comment on the overall stress life of humans. Real-time biofeedback may help us to understand an individual's progression towards acute stress-induced performance decrement. We have recorded and analysed data of four prominent signals from the frontal region of brain to understand the brain activity, and we have also learnt its presentation on various time schedules. Participants are connected to various electrodes. The signals will be taken at various times. For example, the first set of signals will be captured in the morning when the person is fresh and the next set of readings will be taken in the evening after the completion of the day. Towards the end of the day, the person is mentally tired and hence the two signal sets will show the required deviations. In this project, we intend to capture biomedical signal of human and use two techniques for signal conditioning. After data acquisition, MATLAB and Arduino programming will help us in proper analysis of workload.

Keywords ECG · EEG · PP · PPG · Signal analysis · Stress

T. Dendage (✉) · V. Deoskar · P. Kulkarni · R. Shriram · M. Bhat
MKSSS's Cummins College of Engineering, Pune, India
e-mail: tejaswini.dendage96@gmail.com

© Springer Nature Singapore Pte Ltd. 2020
K. S. Raju et al. (eds.), *Proceedings of the Third International Conference on Computational Intelligence and Informatics*, Advances in Intelligent Systems and Computing 1090, https://doi.org/10.1007/978-981-15-1480-7_54

1 Background and Need for This Topic

Workload is an important parameter for a healthy life. Workload causes stress which leads to other health problems such as heart problems, blood pressure and loss of sleep. It also causes loss of life at workshop floors and labour sites due to lack of concentration and a burden of heavy work commitments. Many of the students in the age group of 15–20 years commit suicides because they are unable to handle their mental stress (it may be exam stress, workload, etc.). So, it is important to assess the implications of workload of the human being [1]. It has a wide applicability such as apart from saving human lives, it can also help us in understanding the various causes of road accident. There are many reasons other than alcohol content, one of which is stress life. This assessment is useful in biomedical areas and also in detection of human brain activity. Hence, we have chosen this topic for our research work (Fig. 1).

Workload means person doing any mental activity for 7–8 h. The signal before activity and after activity of human being is taken and studied. The participant here is a human, and there are four prominent signals that are captured from the human body, namely electrocardiography (ECG), electroencephalography (EEG), pulse pressure (PP) and photoplethysmography (PPG).

In this, ECG and EEG signals are electrical, PP is pressure signal and PPG is optical signal. As shown in the diagram, there are two processing techniques that we will use, namely signal and image processing. The signal processing will be on the physical signals acquired from the human, and the software used is MATLAB. The image processing will be done by capturing the blinking of eye. When a person is stressed, he/she is tired and the blinking rate tends to increase which gives an indication of the amount of stress [2, 3].

Below shown (Fig. 2) is the flowchart of the entire system. It explains the exact flow of the entire process. The main blocks comprise the initial real-time signal

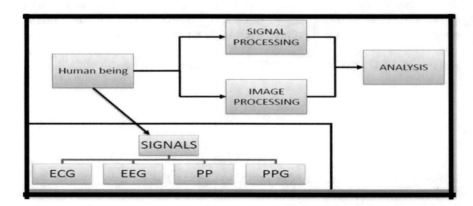

Fig. 1 Block diagram of the system

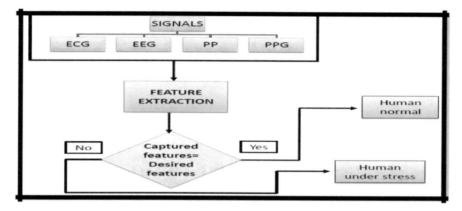

Fig. 2 Flowchart of the system

acquisition, conditional block and the final display of result based on the decision at the prior level.

The feature extraction is a common block. However, the features of each signal will be distinctly different from each other and the amounts of the parameters measured are also different. The exact parameters and their features are discussed in detail in the later sections of the paper.

For each feature extracted, there is a nominal value (corresponding to a normal person) which is checked with the real-time data and if the deviation observed is more than desired, the conditional block triggers the "human under stress" block else "human normal" is displayed.

2 Signal Acquisition and Analysis

2.1 ECG Signal Analysis

Electrocardiography (ECG or EKG) is the process of capturing the electrical activity of the heart for a period of time using electrical conductors placed over the skin. ECG waveform tells us about the electrical activity of the heart. If a person is suffering from a stressful life or has a lot of workloads, the heart conditions change and become abnormal. The ECG wave has a regular and repetitive pattern. It is popularly known as the PQRST wave. The P wave represents the depolarization of atria, QRS complex represents the depolarization of ventricles and repolarization of atria, and T wave represents the repolarization of ventricles. When a person is stressed, the person becomes sweaty and his/her heart beats at a faster pace [1–4]. The parameters that tend to show variations on heavy workload are mentioned in detail in the observation and result section.

The normal heart rate for adults is usually between 60 and 100 beats per minute, and frequency is 1.00–1.67 Hz. The optimum value for R amplitude is 0.5 mV (for Lead I configuration). The range for normal amplitude value for the T peak is from 0.1 to 0.5 mV [5]. Keeping the normal values in mind, the real-time data and the deviation between the two help in analysing the stress intensity.

2.2 EEG Signal Analysis

Electroencephalography (EEG) is the process of capturing electrical activity of the brain. EEG measures changes in voltage fluctuations resulting from ionic current present inside the neurons of the brain from the scalp. According to the frequency range, EEG wave consists of the following waves:

- Delta: below 4 Hz
- Theta: 4–8 Hz
- Alpha: 8–14 Hz
- Beta: 15–30 Hz
- Gamma: above 30 Hz.

When a person is in stress, rhythmic property of these waves is changed. Frequency refers to the rhythmic activity, and voltage refers to the average voltage or peak voltage of EEG activity. Analysis of these parameters when a person is in relax state and when a person is in stress helps us in analysing the stress intensity [6–9].

2.3 PP Signal Analysis

The difference between systolic blood pressure also known as highest arterial blood pressure and lowest arterial blood pressure that is diastolic blood pressure is termed as the pulse pressure (PP). Unit used to measure the pulse pressure is millimetres of Hg.

It is the force that the heart produces each time when it pumps. The systemic pulse pressure is approximately proportional to the amount of blood ejected from the left ventricle during systole (pump action) which is also known as stroke volume and inversely proportional to the elasticity property of the aorta.

The pulse pressure in young healthy, fully grown person without any stress is in the range of 30–40 mmHg. It increases with physical exercise or mental stress due to fast pumping and increased stroke volume.

For most of the individuals, during stressful period the blood pressure particularly the systolic pressure continuously increases while another part that is diastolic pressure remains the same.

In some cases when an individual has more physical stress, for example players and athletes, the lowest arterial blood pressure which is also known as diastolic pressure will fall as the highest arterial blood pressure that is systolic pressure increases. Due to this behaviour, we can observe a much greater increase in the volume of the blood pumped and cardiac output at a lower mean arterial pressure which enables much greater aerobic capacity and physical performance [10].

3 Observations and Results

3.1 ECG

When a person is stressed, various ECG parameters change. The main portions of the ECG wave are the P wave, QRS complex and the ST wave. The amplitudes of R and T waves show deflection from their optimum values when subjected to stress. Some of the changes are listed below:

- The R amplitude of the ECG waveform decreases.
- The heartbeat increases.
- The frequency of heartbeat decreases.
- T wave amplitude increases [11].

The figure below shows the graph of a sample. The initial sample captured has three sections. The raw ECG signal has predominant artefacts present which include baseline wander, power line interference (PLI) and motion artefacts (EM). Hence, the signal should be initially filtered and made free of all the noise-causing components (Fig. 3).

The final section contains the detected R peaks. These are the prominent peaks of the signal. On performing analysis of this sample, the following values are extracted from the signal.

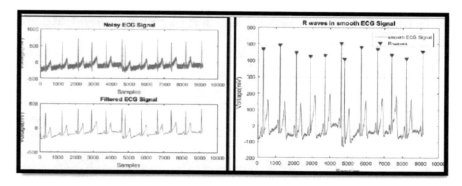

Fig. 3 The ECG analysis of 1 sample

On comparing it with the optimum values, it has been seen that the deviation is minimal and hence one may comment that the person is normal. Below is the table of the parameters and the extracted values in the sample (Table 1).

3.2 EEG

When a person is in stress, various EEG wave parameters change. Most important change is in beta rhythm. Beta rhythm increases when a person is in stress, and also alpha rhythm decreases due to stress. Changes in alpha and beta waves can be determined by various parameters. Some of the parameters are:

- Frequency, amplitude (voltage), entropy of beta wave
- Frequency, amplitude (voltage), entropy of alpha wave
- Phase and magnitude coherence of normal and stressed wave.

Figure 4a shows the initial sample of captured EEG wave which is noisy. Hence, the signal should be initially filtered and made free of all the noise-causing components. So, Chebyshev filter is used to remove noise from the wave. Figure 4b shows the filtered EEG waveform. It is required to extract the band frequency from EEG wave. Using wavelet analysis, it is possible to extract band frequency [12]. EEG wave is decomposed into several steps using wavelet toolbox. db6 1-D wavelet decomposition code is used here. It is required to determine how many levels are there to extract and achieve band frequency.

Initial captured EEG signal sampling frequency is 500 Hz. So, seven-level decomposition is needed. This frequency is decomposed up to 4 Hz that is up to seventh step decomposition. Decomposition is shown in Fig. 5. This decomposition is done in the following ways:

(a) 'S' is original filtered EEG wave.
(b) When it is decomposed, then level 1 produces two sets of coefficients, approximate coefficient cA1 and detailed coefficient cD1.
(c) Accordingly, it is decomposed into seven levels.
(d) To detect the frequency contained in each level, FFT is employed. The EEG signal contains all frequencies. Every frequency indicates a specific class.
(e) Using FFT, one can get frequency contained in each level and identify type of EEG signal. It is given in Table 2.

Table 1 ECG observations

Parameters	Sample 1
The average amplitude of R peak	0.4525 mV
The amplitude of T peak	0.1577 mV
Samples (N)	9160
Frequency	1.3100
Beats per minute (BPM)	78.6026

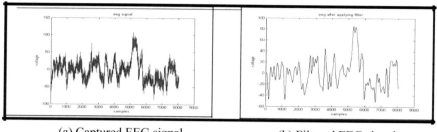

(a) Captured EEG signal   (b) Filtered EEG signal

Fig. 4 Raw and filtered EEG wave

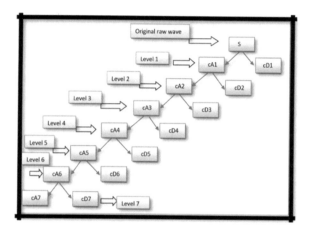

Fig. 5 EEG signal decomposition

Table 2 Observations of EEG wave

Wave	Frequency (Hz)	Voltage (micro-volts)
cD7	2	3.5
cD6	4.35	1.54
cD5	12.19	0.06787
cD4	27.79	0.001524
cD3	59	$2.527e^{-0.5}$

Below is the table of the parameters and the extracted values in the sample wave. From observations by comparing it with optimum values of EEG waves, the following are the types of wave at particular levels of decomposition (Fig. 6).

Fig. 6 Wavelet toolbox

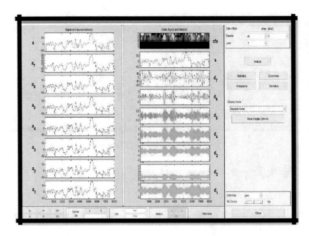

From above observations:

1. D7: Delta wave
2. D6: Theta wave
3. D5: Alpha wave
4. D4: Beta wave.

3.3 Pressure Pulse (PP)

Diastole is when the heart relaxes. When the heart relaxes, the chambers of the heart fill with blood, and a person's blood pressure decreases known as diastolic blood pressure (DBP). Systole is when the heart muscle contracts. Stroke volume (SV) is the amount of the blood pumped out from the left ventricle per beat during a cardiac cycle which is used to determine the cardiac output (CO), defined as the multiplication of stroke volume and heart rate. When a person is stressed, various pressure pulse wave parameters change.

- Systolic blood pressure (SBP) increases.
- Diastolic blood pressure (DBP) increases.
- Mean arterial pressure (MAP) increases.
- Cardiac output (CO) increases.
- Cardiac index.
- Stroke volume (SV) increases.
- Stroke index.

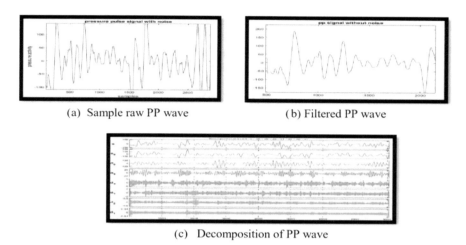

(a) Sample raw PP wave   (b) Filtered PP wave

(c) Decomposition of PP wave

Fig. 7 Sample, filtered and decomposition of PP wave

Table 3 PP wave observations

Parameters	Observed values	Normal values
Systolic pressure	142 mmHg	90–140 mmHg
Diastolic pressure	90 mmHg	60–90 mmHg
Heart rate	79.1 bpm	60–100 bpm
Stroke volume	78 ml	50–100 ml
Cardiac output	6.16 L/min	4–8 L/min

Figure 7 shows the graph of a sample with and without noise and its decomposition.

Similar to ECG and EEG signals, on comparing PP parameters with the optimum values, one can find that the deviation is minimal and hence one may comment that the participant is normal. Table 3 shows the parameters and the extracted values in the sample

4 Discussion

1. There are many signals that show variation when a person is under heavy workload. Some of these are ECG, EEG, PP, PPG, breathing rate, skin temperature and impedance.
2. Currently, work is done on three signals, namely ECG, EEG and PP. Individual signal analysis is performed by many people. However, it has not been collectively used for analysing workload stress. A system will be made in which data of all the four signals will be fed. Depending on real-time signal

acquisition, the results of each signal will be compared with their respective normal values. Depending on the deviation, the message will be displayed.

3. There is a lot of research works done for the ECG signal. However, a similar research work is not found for the EEG and PP signal.

4. All the three signals have distinct features, and the analysis is different for each. No two signals test the same parameters. Hence, the analysis is based on a wide database and includes study of various parameters.

5 Conclusion

1. Three main signals ECG, EEG and PP (two are electrical and one is pressure signal) have been studied. The signals are captured from the frontal region of brain, and hence they represent the brain activity. Signals are captured, and various parameters that get affected and show variation due to stress/workload have been studied. Depending on the results, it can be commented on whether or not the person is stressed.

2. For future work, fourth PPG sensor will be ordered and similar analysis will be done using this sensor. A wide database of signals will be formed, and the signal acquisition by digital image processing will be done. The digital image processing will be performed by capturing images of the eyes. The person will blink, and depending on the blinking of the person we will be able to comment on whether the person is stressed. Arduino programming will be used for programming and displaying the message.

3. The database of all the four signals will be collected and fed into a system. Depending on the real-time signals and the deviation from their normal values, the message will be displayed as to the mental condition of the person in test.

Acknowledgements The entire database made for the project purpose was recorded in Cummins College. The paper titled "Workload Assessment Based on Physiological Parameters" is an outcome of guidance and moral support bestowed on us throughout the project tenure, and for this we would like to acknowledge and express our profound sense of gratitude to our guide Dr. Revathi Shriram for her constant motivation.

We would also like to thank everybody in our Instrumentation and Control Department who have indirectly guided and helped us in completing this final-year project.

Last but not least, we would also like to thank all the people/subjects who volunteered and cooperated with us and helped us in collecting the database and results for our project.

All the subjects were willing volunteers who participated in the data collection, and they have given verbal consent for using the data for research purpose and for further publication (i.e. all the three signals ECG, EEG and PPG recorded and analysed). None of the ethical committee is involved in it.

References

1. Vanitha, V., and P. Krishnan. 2016. *Real Time Stress Detection System Based on EEG Signals*. Special Issue: S271S275, Biomedical Research.
2. Haak, M., S. Bos, S. Panic, and L.J.M. Rothkrantz. 2014. *Detecting Stress Using Eye Blinks And Brain Activity from EEG Signals*. Researchgate.
3. Haak, M., S. Bos, S. Panic, and L.J.M. Rothkrantz. 2009. Detecting Stress Using Eye Blinks During Game Playing. In *10th International Conference on Intelligent Games and Simulation*, 75–82, GAME-ON.
4. Ranganathan, G., R. Rangarajan, and V. Bindhu. 2011. Evaluation of ECG Signals for Mental Stress Assessment using Fussy Technique. *International Journal of soft Computing and Engineering (IJSCE)* 1 (4). ISSN: 22312307.
5. Bhide, A., R. Durgaprasad, L. Kasala, V. Velam, and Hulikal, N. 2016. Electrocardiographic Changes During Acute Mental Stress. *International Journal of Medical Science and Public Health (Online First)* 5, 835–838.
6. Jena, S.K. 2015. Examination Stress and Its Effect on EEG. Int J Med Sci Public Health 4:1493–1497.
7. Choong, W.Y., W. Khairunizam, M. I. Omar, M. Murugappan, A.H. Abdullah, H. Ali, and S. Z. Bong. EEG-Based Emotion Assessment Using Detrended Fluctuation Analysis (DFA). *Journal of Telecommunication, Electronic and Computer Engineering* 10 (1–13). ISSN: 2180–1843. e-ISSN: 2289–8131.
8. Jebelli, Houtan, Mohammad Mahdi Khalili, Sungjoo Hwang, and Sang Hyun Lee. 2018. Supervised Learning Based Construction Workers' Stress Recognition Using a Wearable Electroencephalography (EEG). Device, Construction Research Congress. https://doi.org/10.1061/9780784481288.005. Conference Paper March.
9. Kennedy, Lauren, and Sarah Henrickson Parker. Timing of Coping Instruction Presentation for Realtime Acute Stress Management. 2018. *Potential Implications for Improved Surgical Performance*. Springer Nature.
10. Schultz, Martin G., and James E. Sharman. 2013. Exercise Hypertension. *Pulse* 1: 161–176. https://doi.org/10.1159/000360-975, April (2014).
11. Malhotra, Vikas, and Mahendra Kumar Patil. 2013. Mental Stress Assessment of ECG Signal using Statistical Analysis of Bio-orthogonal Wavelet Coefficients. *International Journal of Science and Research (IJSR)* 2 (12): 430–434.
12. Al-shargie, Fares, and Tong Boon Tang, Nasreen Badruddin, Masashi Kiguchi. 2017. Mental Stress Quantification Using SVM with ECOC: An EEG Approach. *Medical & Biological Engineering & Computing* s11517017–1733-8.

Realistic Handwriting Generation Using Recurrent Neural Networks and Long Short-Term Networks

Suraj Bodapati, Sneha Reddy and Sugamya Katta

Abstract Generating human-like handwriting by machine from an input text given by the user may seem as an easy task but is very complex in reality. It might not be possible for every human being to write in perfect cursive handwriting because each letter in cursive gets shaped differently depending on what letters surround it, and everyone has a different style of writing. This makes it very difficult to mimic a person's cursive style handwriting with the help of a machine or even by hand for a matter of fact. This is why signing names in cursive is preferable on any legal documents. In this paper, we will try to use various deep learning methods to generate human-like handwriting. Algorithms using neural networks enable us to achieve this task, and hence, recurrent neural networks (RNN) have been utilized with the aim of generating human-like handwriting. We will discuss the generation of realistic handwriting from the IAM Handwriting Database and check the accuracy of our own implementation. This feat can be achieved by using a special kind of recurrent neural network (RNN), the Long Short-Term Memory networks (LSTM).

Keywords Handwriting generation · Recurrent neural networks (RNN) · Long Short-Term Memory networks (LSTM) · IAM handwriting database

S. Bodapati · S. Reddy (✉) · S. Katta
Department of Information Technology, Chaitanya Bharathi Institute
of Technology, Hyderabad, India
e-mail: snehareddycsr@gmail.com

S. Bodapati
e-mail: Surajbodapati97@gmail.com

S. Katta
e-mail: ksugamya_it@cbit.ac.in

© Springer Nature Singapore Pte Ltd. 2020
K. S. Raju et al. (eds.), *Proceedings of the Third International Conference on Computational Intelligence and Informatics*, Advances in Intelligent Systems and Computing 1090, https://doi.org/10.1007/978-981-15-1480-7_55

651

1   Introduction

The power of deep learning can be demonstrated by recurrent neural networks (RNNs) which can be viewed as a top class of dynamic models that is used to generate sequences in multiple domains such as machine translations, speech recognition, music, generating captions for input images and determining emotional tone behind a piece of text. RNNs are used to generate sequences by processing real data sequences and then predicting the next sequence. A large enough RNN is capable of generating sequences of subjective complexity. The Long Short-Term Memory networks (LSTMs) are a part of RNN architecture which is capable of storing and accessing information more efficiently than regular RNNs. In this paper, we use the LSTM memory to generate realistic and complex sequences [1].

The handwritten text in English used to train and test handwritten text recognizers and to perform identification and verification experiments of the writer is taken from the IAM Handwriting Database which contains forms of unconstrained handwritten text, which were saved as PNG images with 256 gray levels after being scanned at a resolution of 300 dpi [2]. This dataset is very small; about 50 Mb when parsed once. The dataset is a result of contribution from 657 writers, and each dataset has a unique handwritten style. The data used in this paper is a 3D time series with three coordinate axes. The (x, y) coordinates are the normal first two dimensions, whereas the third dimension is a binary 0/1 value. The 1 in this third coordinate has around 500 pen points and an annotation of ASCII characters which is used to signify the end of a stroke.

2   Concepts

2.1   Artificial Neural Networks

Artificial neural networks try to simulate the human brain by modeling its neural structure on a much smaller scale. When you consider a typical human brain, it consists of billions of neurons in contrast to ANN which has one by 1000th processor units of that of a human brain. A neural network consists of neurons, which are simple processing units, and there are directed, weighted connections between these neurons. A neural network has a number of neurons processing in parallel manner and arranged in layers. Layers consist of a number of interconnected nodes which contain an activation function. The patterns from the input layer are processed using the weighted connections, which help with the communication with the hidden layer. The output from one layer is fed to the next and so on. The final layer gives the output. For a neuron j, the propagation function receives the outputs of other neurons and transforms them in consideration of the weights into the network input that can be further processed by the activation function [3]. Every neuron operates in two modes: training and using mode.

2.2 Recurrent Neural Networks

The major difference between a traditional neural network and an RNN is that the traditional neural networks cannot use its reasoning about previous events in the process to inform the later neurons or events. Traditional neural networks start thinking process from scratch again. RNNs address this issue, by using loops in their network, thus making information persist. The RNN obtains input x_t and gives output value h_t [4]. The passing of information from one step of the network to the next is allowed by a loop. An RNN network can be taken as one of many copies where each one of the copy passes information to the previous. In RNNs, each word in an input sequence in RNN will be mapped with a particular time step. This results in the number of time steps equal to the maximum sequence length. A new component called a hidden-state vector h_t is associated with each time step, and this constitutes each iteration output. h_t seeks to encapsulate and summarize all of the information that was seen in the previous time steps when seen from a high level. In similar fashion, x_t is a vector that encapsulates all the information of a specific input word; the hidden-state vector at the previous time step is a function of both the hidden-state vector and current word vector. The two terms sum that will be passed through an activation function is denoted by sigma. Final hidden layer output can be given as $h_t = \sigma(W^{H*} h_{t-1} + W^{*X} x_t)$, where Wx is weight matrix to be multiplied with input x_t and is variable. W^H is recurrent weight matrix that is multiplied with hidden-state vector of earlier step. W^H is a constant weight matrix. These weight matrices are updated and adjusted via an optimizing process known as backpropagation through time. Sigma denotes the two terms sum is passed via an activation function, usually sigmoid/Tanh.

2.3 Long Short-Term Memory Networks (LSTMs)

Long Short-Term Memory networks (LSTMs) are capable of learning the long-term dependencies and are a special kind of RNN. LSTMs are modules that you can place inside an RNN which are explicitly used to avoid the long-term dependency problem. The long-term dependency problem can be described as a situation where the gap between the point where the network is needed and the relevant information becomes very large, and as a result, the RNNs become incapable to connect the previous information. The computation in LSTMs can be broken down into four components, an input gate which determines mount of emphasis to be put on each gate, a forget gate determines the unnecessary information, an output gate which will determine the final hidden-vector state based on intermediate states and a new memory container [5].

2.4 Backpropagation

A backpropagation neural network which is backward propagation of error uses the delta rule which is one of the most used rules of neural networks. A neural network assigns a random weight when it is provided with an input pattern, and later, based on its difference from output, it changes the weight accordingly. The delta rules use supervised method of learning that occurs with each cycle of epoch where error is calculated using backward error propagation of weight adjustments and forward flow of outputs. The network runs in forward propagation mode only once the neural network is trained. The new inputs are no longer used for training the network instead they are processed only in forward propagation mode to obtain the output. Backpropagation networks are usually slower to train when compared to some other types of networks and sometimes require a very large number of epochs.

2.5 One-Hot Encoding

One-hot encoding is used when categorical variables need to be converted into a form which can be given to the machine learning algorithms for regression or classification which helps in better prediction. For training the model, binarization of category is done using the one-hot encoder and including its features.

3   Model Implementation

The dataset used in this paper was originally used to as training dataset for hand-writing recognition models, which inputs a series of pen points and identifies the corresponding letter [6]. Our goal is to reverse this model using RNNs and train a model which takes letters as inputs and produces a series of points which can then be connected to form letters. In this paper, we use a model which can be viewed as a three sub-models stacked on top of one another. The models mentioned are trained using gradient backpropagation (Fig. 1).

The model used above is a combination of three different types of models. The sequences of pen points of the input handwriting are produced using the RNN cells. The one-hot encoding of the input text is given to the attention mechanism. The style and variation in the handwriting can be generated by MDM which achieves this by modeling the randomness in the handwriting. The MDN cap of the model predicts the x, y coordinates of the pen by drawing them from the Gaussian distribution, and the handwriting can be varied by modifying the distribution. We start with the inputs and work our way upwards through the computational graph. Step 1: The first step in implementing this model is to download and preprocess the data. The script was written for preprocessing searches for local directory for the files

Fig. 1 Model overview

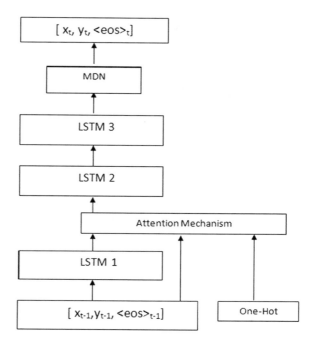

containing data we downloaded and performs preprocessing tasks such as normalizing the data, data cleaning, data transformations, splitting strokes in lines. Step 2: Build the model.

3.1 The Long Short-Term Memory Cells

This paper deals with three LSTM-RNN cells for developing a model. These three LSTM cells act as the backbone of the model. The above-mentioned three LSTM networks can keep the trail of independent patterns by using a differentiable memory. LSTMs used in this model use three different tensors to perform read, write and erase tasks on a memory tensor. We use a custom attention mechanism which digests a one-hot encoding of the sentence we want the model to write. We can add some natural randomness to our model with the help of the Mixture Density Network present on the top which chooses appropriate Gaussian distributions from which we are able to sample the next pen point. We start with the inputs and move upwards through the computational graph. Tensorflow seq2seq API is used to create the LSTM cells present in this model [7].

3.2 Attention Model

In this model, we use a differentiable attention mechanism to get the information about characters that make up a sentence. An attention mechanism can be described as a Gaussian convolution over a one-hot encoding of input texts using mixtures of Gaussians. As the model writes from character to character, it learns to shift the window. This is possible as the attention mechanism tells the model what is to be written and since all the parameters of the window are differentiable. The final output is a soft window into the one-hot encoding of the character the model thinks it is drawing. A heat map is obtained on stacking these soft windows vertically over time. This attention mechanism obtains inputs by looping with the output states of LSTM-1. Next, we concentrate the outputs of this attention mechanism to LSTM's state vector and concatenate the original pen stroke data for good measure (Figs. 2 and 3).

3.3 Mixed Density Networks (MDN)

Randomness and style in handwriting in this model are generated by using mixture density networks. MDNs can parameterize probability distributions hence are a very good method to capture randomness in the data. At the beginning of the strokes, a Gaussian with diffuse shapes is chosen, and another Gaussian is chosen at the middle of strokes with peaky shapes. They can be seen as neural networks which can measure their own uncertainty (Fig. 4).

Fig. 2 Heat map by stacking soft windows vertical over time

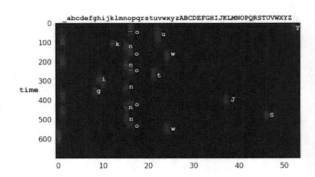

Fig. 3 Modified LSTM State vector

| LSTM 1 Output | Soft Window | Stroke data |

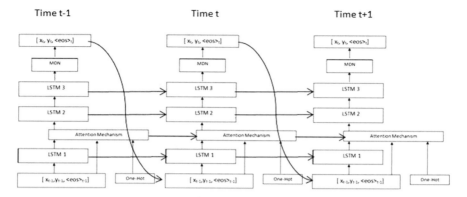

Fig. 4 Recurrent model to feedforward information from past iterations

Our main motive is to have a network that predicts an entire distribution. In this paper, we are predicting a mixture of Gaussian distributions by estimating their means and covariance with the output from a dense neural network.

This will result in the network being able to estimate its own uncertainty. When the target is noisy, it will predict diffuse distributions, and where the target is really likely, it will predict a peaky distribution.

3.4 Load Saved Data

After we complete building our model, we can start a session and load weights from a saved model. We can start generating handwriting after loading and training the weights (Figs. 5 and 6).

Fig. 5 Phis plot

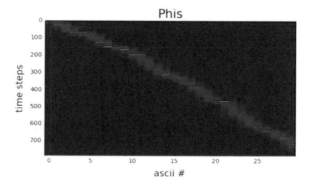

Fig. 6 Soft attention
Window plot

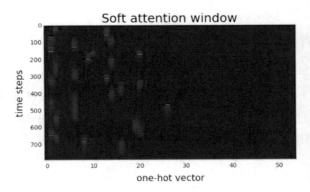

3.5 *Plots*

After we complete building our model, we can start a session and load weights from
a saved model. We can start generating handwriting after loading and training the
weights.

Phis: It is a time series plot of the window's position with vertical as time and
horizontal axis as sequence of ASCII characters that are being drawn by the model.
We complete building our model, and we can start a session and load weights from
a saved model. We can start generating handwriting after loading and training the
weights.

Soft Attention Window: This is a time series plot of one-hot encodings produced
by the attention mechanism with horizontal axis as one-hot vector and vertical axis
as time.

Step 2: On building the model given above, we can start a session and load
weights from a saved model. We can start generating handwriting after loading and
training the weights. We write script train.py to train our model. Running the script
will launch training with default settings (we can use argparse options to experiment). A summary directory with separate experiment directory is created by
default for each run. We need to provide a path to the experiment we would like to
continue if we wish to restore training.

$$python\ train.py - restore = summary \backslash experiment - 0$$

Losses can be visualized in command line or by using tensor board.

Using tensorflow 1.2 and GTX 1080 on default settings, the training time was
approximately five hours.

Step 3: Generate handwriting using the model:

A script generate.py is used to test the working of the model after training of the
model (Fig. 7).

Fig. 7 Loss plot

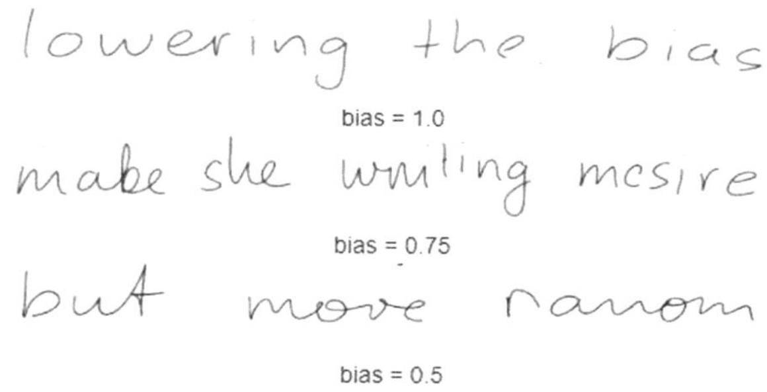

Fig. 8 Results for
Bias = 0.5, 0.75, 1.0

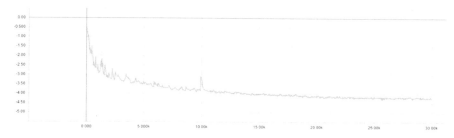

bias = 1.0

bias = 0.75

bias = 0.5

python generate.py −model = path_to_model

The text argument is used to input the text and test the working of the model. Different options for text generation:

- bias: We introduce a bias term which redefines parameters according to the bias term, and higher the bias, the clearer the handwriting is (Fig. 8).
- noinfo: Generated handwriting without attention window is plotted.
- animation: To animate the writing
- style: Specifies handwriting style.

4   Results and Discussions

In the final step of plotting, we will be able to generate some handwriting. Since the LSTMs' states start out as zeros, the model generally chooses a random style and then maintains that style for the rest of the sample. On repeatedly sampling for a couple of times, we will be able to see the change from everything messy, scrawling cursive to a neat print.

After a few hours, the model was able to generate readable letters and continuing to train for a day, and the model was able to write sentences with a few trivial and

Fig. 9 Results for bias 1, output of (1)

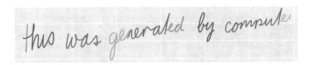

minor errors. After the model has generated few readable letters, as compared to most of the training sequences (as generated above) that were 256 points long in this model, we were able to sample sequences up to 750 points long. The handwriting can look cleaner or messier. This can be done by drawing the coordinates (x, y) from the Gaussian distribution network that has been predicted by the MDN cap of the model described. We introduce a bias term which redefines parameters according to the bias term.

Run command *Python generate.py –noinfo –text="this was generated by computer" –bias=1* (1), on running *python generate.py –noinfo –animation – text="example of animation" –bias=1* we can specify the animation style (Fig. 9).

5   Conclusion

Deep learning is a thriving field that is seeing lots of different advancements. With help of recurrent neural networks, Long Short-Term Memory networks, convolution networks and backpropagation to name a few, this field is quickly turning explosive and successful. The model used in this paper is a proof of the strength of deep learning in general. A text to speech generator can be generated by using the same model used in the paper; this can be achieved by using appropriate dataset and increasing the parameter number. The possibilities of deep learning are limitless.

References

1. Getting Started-DeepLearning 0.1 documentation. Deeplearning.net, 2017 [Online]. Available http://www.deeplearning.net/tutorial/gettingstarted.html. Accessed 19 Feb 2018.
2. Bridle, J.S. 1990. Probabilistic Interpretation of Feedforward Classification Network Outputs, with Relationships to Statistical Pattern Recognition. In *Neurocomputing: Algorithms, Architectures and Applications*, ed. F. Fogleman-Soulie, and J. Herault, 227–236. Springer.
3. Perwej, Y., and A. Chaturvedi. 2011. Machine Recognition of Hand Written Characters Using Neural Networks. *International Journal of Computer Applications* 14 (2), 6–9.
4. LeCun, Y., L. Jackel, P. Simard, L. Bottou, C. Cortes, V. Vapnik, J.S. Denker, H. Drucker, I. Guyon, U. Muller and E. Sackinger. 2018. Learning Algorithms for Classification: A Comparison on Handwritten Digit Recognition. *Neural Networks: The Statistical Mechanics Perspective* 261 (1), 276–309.

5. Plamondon, R., and S. Srihari. 2000. On-Line and Off-Line Handwriting Recognition: A Comprehensive Survey. *IEEE Transactions on Pattern Analysis and Machine Intelligence* 22 (1): 63–78.
6. http://www.fki.inf.unibe.ch/databases/iam-handwriting-database.
7. Leverington, D. Neural Network Basics, Webpages.ttu.edu, 2009. [Online]. Available http://www.webpages.ttu.edu/dleverin/neural_network/neural_networks.html. Accessed 19 Feb 2018.

Decentralized Framework for Record-Keeping System in Government Using Hyperledger Fabric

S. Devidas, N. Rukma Rekha and Y. V. Subba Rao

Abstract Blockchains are spreading their popularity in a wide range of applications due to their capability to ensure decentralization, immutability, transparency, verifiability, resilience and consensus-based record-keeping system. Hyperledger Fabric is a general-purpose permissioned blockchain featuring pluggable consensus mechanism and chain code that can be written in any general-purpose programming language, providing a consistent distributed ledger that is shared by a set of peers in the network. In this paper, we propose a novel decentralized framework for record-keeping system in government, which brings immutability, resilience and verifiability in the system that is based on permissioned blockchain Hyperledger Fabric.

Keywords Blockchain · Distributed ledger · Hyperledger fabric · Consensus · Chaincode · Immutability · Resilience · Verifiability

1 Introduction

For any government, well-managed records are the foundation to preserve transparency, verifiability, immutability, resilience and collaboration. To assess the performance of the government, it is important to have well-managed records which are transparent. Rights and interests of the people can be protected by these well-managed records, and officials can be held responsible and accountable for

S. Devidas (✉) · N. Rukma Rekha · Y. V. Subba Rao
School of Computer and Information Sciences, University of Hyderabad,
Hyderabad, Telangana 500046, India
e-mail: devidas13@uohyd.ac.in

N. Rukma Rekha
e-mail: rukmarekha@uohyd.ac.in

Y. V. Subba Rao
e-mail: yvsrcs@uohyd.ac.in

© Springer Nature Singapore Pte Ltd. 2020
K. S. Raju et al. (eds.), *Proceedings of the Third International Conference on Computational Intelligence and Informatics*, Advances in Intelligent Systems and Computing 1090, https://doi.org/10.1007/978-981-15-1480-7_56

their actions in the future. In this paper, a decentralized blockchain-based framework is proposed for record keeping in government.

Blockchain [1] is widely known to everyone as a bitcoin's underlying technology. Even though the bitcoin is controversial and is one of the popular applications of blockchain, the underlying technology behind the bitcoin is not controversial and is successfully spreading its popularity in other financial and non-financial domains [2–9]. Essentially, blockchain is a distributed ledger or database shared and replicated among all the un-trusted nodes in the peer-to-peer network [10]. It is an ordered data structure of blocks of transactions exactly like linked-list but in reverse order. In the blockchain, the transaction can be anything like transferring digital money, exchange of assets, data captured from different sources, and any form of communication among the peers in the network. Each block of blockchain is identified by its hash value which is generated using SHA256 cryptographic algorithm on the header of the block and linked with its previous block [10]. This property forms a cryptographic link between blocks, creating so-called blockchain. From the literature, the current blockchain implementations fall into permissionless and permissioned categories. Permissionless blockchain protocols, as the name says, allow anyone to participate in the network without requiring any permission and achieve consensus based on Proof of Work (PoW) [1]. Anyone in the network can act as a public node by running the code on their local device, validate the transactions in the network and act as miners [10]. All valid transactions on the ledger can be read by any peer of the network. Examples for permissionless blockchains are bitcoin, Ethereum [11], Monero [12], Litecoin [13], etc. Permissioned blockchains, on the other hand, use the same blockchain technology but set up peers who know each other, and these peers can verify transactions internally [14]. This entire setup process deals with specific permissions to peers. These blockchains deal with scalability and state compliance of data privacy rules and security issues that include database management, auditing, etc. This category includes Tendermint [15], Chain [16], Quorum [17], Hyperledger Fabric [14], and Multichain [24] etc.

2 Related Work

The first blockchain was implemented by Satoshi Nakamoto for cryptocurrency bitcoin in the year 2008 to enable payment between two participants without any trusted third party [1]. The system builds trust among participants using cryptographic protocols. The earlier blockchain models were accessed by users without any permission; industries have since then effectively implemented other instances in a permission context, limiting the participants' environment with suitable permissions. Initially, blockchains were used for maintaining financial ledgers but recently have found place in various other domains. Peterson et al. [18] proposed blockchain as an apt system for sharing healthcare data. The system aims at achieving data interoperability between institutions that deal with healthcare data.

Protection is required for highly sensitive healthcare data from unwanted access. Towards this end, Ekblaw, Ariel, et al. presented MedRec [19], a decentralized record management system for electronic health records (EHRs), using blockchains. Victoria L. Lemieux presented a synthesis of his original research for several documenting cases for land transactions, medical and financial record keeping using blockchains [20]. Mohamed Amine Ferrag, Makhlouf Derdour et al., gave an overview of the applications [21] of blockchains in various domains such as the Internet of Cloud, IoT, Internet of Energy, Fog Computing and Internet of Vehicle.

However, [19, 20] and [21] used permissionless blockchain where a lot of mining and expensive computations are required to achieve consensus. This leads to wastage of power, energy and system resources. Permissioned blockchains have the advantage of defined membership and access rights within the business network. Also, the transaction in permissioned blockchain can be made transparent to network peers with the help of cryptographic credentials. Most importantly, expensive computations and mining are avoided completely making it more practical for even small organizations that cannot afford system resources. In the proposed framework, simple mechanism like kafka [22] can be used which gives guarantee for fault tolerance and to achieve consensus.

In this paper, the idea of utilizing permissioned blockchain Hyperledger Fabric for record-keeping systems in government is introduced. The rest of the paper is organized as follows. First, in Sect. 3 an overview of Hyperledger Fabric is discussed. As the main contribution, in Sect. 4, main phases in the proposed framework are presented in detail. Main contributions and properties of proposed framework are summarized in Sect. 5. Finally, in Sect. 6, conclusion of the paper by some possible future work is discussed.

3 Hyperledger Fabric

3.1 Overview

In general, blockchains follow order-execute architecture [14], where all peers of the network are required to execute every transaction and output of all transactions is to be deterministic. Both permissionless and permissioned blockchains use order-execute architecture. Because of the order-execute architecture, permissioned blockchains suffer from many limitations such as scalability, requirement for sequential execution of transactions and so on.

Hyperledger Fabric [14] overcomes these limitations and follows execute-order architecture [14]. The Hyperledger Fabric is a permissioned blockchain where participants require some credentials, and these participants are called peers. A membership service provider (MSP) takes the responsibility of associating all peers of the network by providing cryptographic credentials such that permissioned nature of the Hyperledger Fabric is maintained. Any standard certificate authority

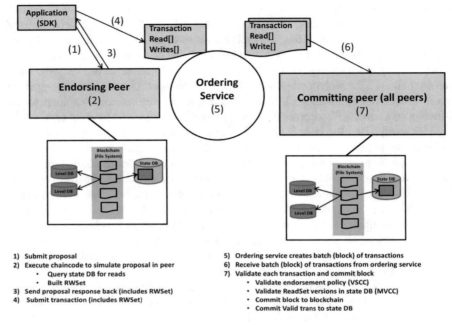

Fig. 1 Transaction lifecycle in v1.0 of Hyperledger Fabric [23]

(CA) such as X.509 [4] can be the MSP of a given system. Else, the system can come up with its own CA. The transaction life cycle in Hyperledger Fabric is as follows and also depicted in Fig. 1.

(1) The client or user prepares a transaction proposal and submits it to the endorsing peers to get endorsements.

(2) Endorsing peers verifies the authentication details of the client and proposal format, executes the chain code configured with them and prepares read/write set for the corresponding proposal.

(3) Endorsers send proposal response back to the client in the form of read/write set by signing it.

(4) Now the client submits the endorsed proposal along with read/write set to the ordering peers.

(5) The ordering peers then achieve the consensus among them by running consensus algorithm which is configured with them, forms a block of transactions and submits the block to the committing peers.

(6) Committing peers validates the endorsing policy and read/write set versions in state DB, marks the transactions as valid or invalid, commits all the valid transactions to state DB and broadcasts the block to all the peers to append their blockchain.

3.2 Merits

The major advantage of Hyperledger Fabric is that all transactions are irrefutable once they are committed. This, in turn, helps us to trace any modification if made in the transaction, helping us to expose corruption and fraudulent transactions.

This very property inspired us to embed Hyperledger Fabric in record keeping system in government offices such that all transactions are free of fraud and corruption, thereby increasing transparency in government records and its functionality.

4 Proposed Framework for Record-Keeping System with Hyperledger Fabric

In this section, a decentralized framework for record-keeping system in government offices using Hyperledger Fabric is proposed. Three main phases are proposed in our framework—endorsing phase, validation phase and ledger updating phase. The architecture of the proposed framework is shown in Fig. 2. Endorsing phase endorses the transaction proposal, checks the endorsing policy and sends the proposal response to the validator. The validating peer checks the proposal, sends the acknowledgement to the user, collects the transactions into a single block and broadcasts the block to all the peers to update their ledgers. Every communication in the system is happening through the TLS protocol to enforce the anonymity and security. The network has 'm' number of peers in the system and has 'l' endorsers and 'm' validators.

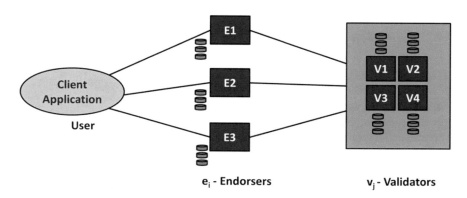

Fig. 2 Architecture of blockchain-based record-keeping system

4.1   Endorsing Phase

In this phase, user prepares a transaction proposal (record or document) T_x and signs the proposal using his/her secret key u_{sk}. Depending on the complexity of the network, user may submit the proposal to more than one endorser also. After receiving the proposal, the endorser checks the legitimacy of the user. If the user is legitimate, then the endorser executes their corresponding 'chain code' C, signs the proposal using his/her secrete key e_{sk}, sends it to the validator as a proposal response and sends the acknowledgement to the corresponding user.

Algorithm 1: Endorsement

Require: The user 'u' with key pair (u_{pk}, u_{sk}), transaction proposal T_x, Endorser e_i ($1<=i<=l$) with key pair (e_{pk}, e_{sk}) and corresponding chain code 'C'

1) 'u' signs T_x : $\rho=sign(T_x, u_{sk})$
2) 'u' sends ρ, T_x to e_i
3) e_i checks
    a.   cd1: verify(u_{pk}, T_x, ρ)
    b.   cd2: execute C and checks the format of T_x
4) if cd1 && cd2=true
    a. e_i signs ρ : $\rho^l = sign(\rho, e_{sk})$
    b.   sends ρ^l to validator;
    c.   returns acknowledgement to user;
    else
            return fail;

4.2   Validation Phase

In this phase, the endorsers $e_i (1 \leq i \leq l)$ submit endorsed proposal to the validator $v_j (1 \leq j \leq m)$ for validation and approval of the proposal to include in the distributed ledger. After receiving the proposal, validator verifies the signatures of both user and endorser and also verifies the endorsing policy specified. Endorsing policy is formulated by the administrator at the time of configuring the endorsing peer. Let us say if any organization has three endorsing peers, namely A, B and C, then to reach their endorsing policy the user has to get endorsed by all the peers A, B and C. If the user finds the endorsers legitimate and the proposal satisfies the endorsing policy, then validator marks the transaction proposal or record as valid and considers it for including in the blockchain, otherwise marks as invalid by specifying the reason why it is marked so and sends it back to the user along with acknowledgement. Now all the validators have to achieve the consensus to order the transactions into a block, which comprises of all the records marked as valid. The Kafka [22] mechanism has to be configured at all the validators to achieve the consensus.

Algorithm 2: Validation

Require: Endorser $e_i(1<=i<=l)$ with endorsed transaction proposal ρ^1, Validators v_j $(1<=j<=m)$ with key pair (v_{pk}, v_{sk}) and endorsing policy EP.

1) E sends ρ, ρ^1 to v_j
2) v_j checks
   a) cd1: verify $(e_{pk}, u_{pk}, \rho, \rho^1)$
   b) cd2: Endorsing policy EP
3) if cd1 & cd2 = true then
   a) marks the transaction as valid
   b) orders the valid transactions into a block 'b'

   else

   a) marks the transaction as invalid;
   b) sends the acknowledgement to client;

4.3 Ledger Updating Phase

Once the consensus is achieved by all the validators v_j $(1 \leq j \leq m)$ and block 'b' is created for all the valid transactions, then 'b' is broadcasted to all the peers P_k $(1 \leq k \leq n)$ of the particular channel to update their local copy of global data 'B'. Now all the peers p_k update their consistent distributed ledger B to B^1.

Algorithm: Ledger Update

Require: Validator v_j $(1<=j<=m)$, block 'b' and peers p_k $(1<=k<=n)$ and blockchain B

1) V_j sends b to p_k
2) P_k updates B $\rightarrow$ B^1

5 Main Contributions

The main contributions of our work in this paper are,

1. The proposed framework will validate the documents and creates the blocks only for valid documents. Invalid documents are sent back to the clients.
2. In government, there is no need to keep the ordering service and validating service separately as officers are assumed to be trusted and both the services can be done by the same authority. This will reduce the resource usage overhead on the system when compared to Hyperledger Fabric framework where ordering and validation are done by separate peers.
3. The documents/transactions are passed on in a hierarchical authority in government offices, and hence, the endorsed transaction is directly submitted for validation rather than sending it back to the client again. This reduces the communication overhead on the system, and the validated transaction is included in the ledger directly later.

More precisely, we list the properties that our framework satisfies.

Accountability: Each and every node participating in the network is accountable for every action on blockchain.

Immutability: The transactions once recorded in the blockchain can never be altered, thereby making the transactions irrefutable.

Verifiability: Each peer is able to verify records within the channel whenever required.

Auditability: All the transactions that are recorded in the blockchain are subject to audit any time in the future.

Resilience: The framework does not have any single point of failure (SPOF), and small group in the network cannot compromise the transactions.

The accountability and immutability properties ensure that all the transactions in the network are transparent, and no transaction can be altered making the system free from fraudulent transactions and corruption.

Verifiability and auditability properties help the peers/clients to check the sanctity of any transaction at any given point of time, thereby enhancing trust in government records and its functionality.

Resilience property ensures robustness of the framework where a SPOF will not affect the functionality of overall network. It means that even if an endorser or validator node gets failed, there are other endorsers or validators who can take up the job.

6 Conclusion

This paper proposed a novel decentralized framework for record-keeping system in government using Hyperledger Fabric. The user participating in the system has to prepare his transaction proposal of the document or record which is to be recorded in the distributed ledger. The user has to submit his/her proposal for endorsement to the endorsers; the endorser endorses it and submits to the validators to check the validity and then include it in the ledger. Every peer maintains the ledger; the newly suggested blocks from the validators are added to it.

The usage of permissioned blockchain Hyperledger Fabric helps the framework to avoid expensive computations as part of mining. The inbuilt property of Hyperledger Fabric makes all the transactions irrefutable, thereby avoids all sorts of fraudulent transactions and corruption. In addition to that, the ordering and validating services are combined together by taking advantage of government trusted officials. This helps in reducing resource usage overhead. Also, all the transactions flow in hierarchical pattern, except for invalid transactions. This reduces communication overhead compared to other systems where communication is bidirectional. Embedding invalid documents in the ledger and tracing the states update of the transaction proposed at each hierarchical level can be considered for future work.

Acknowledgements The first author acknowledges the financial support from Council of Scientific & Industrial Research (CSIR), Government of India in the form of a Senior Research Fellowship.

References

1. Nakamoto, Satoshi. 2008. Bitcoin: A Peer-to-Peer Electronic Cash System. Working Paper.
2. Vukolić Marko. 2017. Rethinking Permissioned Blockchains. In *Proceedings of the ACM Workshop on Blockchain, Cryptocurrencies and Contracts*, ser. BCC '17, 3–7. ACM, New York, NY, USA.
3. Li, Wenting, Sforzin Alessandro, Fedorov Sergey, and Ghassan O. Karame. 2017. Towards Scalable and Private Industrial Blockchains. In *Proceedings of the ACM Workshop on Blockchain, Cryptocurrencies and Contracts*, 9–14. ACM.
4. Adams, Carlisle, Farrell Stephen Tomi Kause, and Tero Mononen. 2005. Internet X. 509 Public Key Infrastructure Certificate Management Protocol (CMP). Tech. Rep.
5. How Consensus Algorithms Solve Issues with Bitcoin's Proof of Work. (n.d.). Retrieved from http://www.coindesk.com/stellar-ripple-hyperledger-rivals-bitcoinproof-work/.
6. Blockchain Adoption Moving Rapidly in Banking and Financial Markets: Some 65 Percent of Surveyed Banks Expect to be in Production in Three Years. (n.d.). Retrieved from https://www03.ibm.com/press/us/en/pressrelease/50617.wss.
7. Benhamouda, Fabrice, Shai Halevi, and Tzipora Halevi. 2018. Supporting private data on Hyperledger Fabric with Secure Multiparty Computation. In *Proceedings of IEEE International Conference on Cloud Engineering (IC2E)*, 357–363. IEEE.

8. Gao, Zhimin, Lei Xu, Glenn Turner, Brijesh Patel, Nour Diallo, Lin Chen and Shi Weidong. 2018. Blockchain-Based Identity Management with Mobile Device. In *Proceedings of the 1st Workshop on Cryptocurrencies and Blockchains for Distributed Systems*, 66–70. ACM.
9. Fernandez-Carames, Tiago M., and Paula Fraga-Lamas. 2018. A Review on the Use of Blockchain for the Internet of Things. *IEEE Access*.
10. Antonopoulos, A. M. (2017). *Mastering Bitcoin: Programming the Open Blockchain*. O'Reilly Media.
11. Ethereum. (n.d.). Retrieved from https://cryptocrawl.in/what-is-ethereum/Ethereum.
12. Menero. (n.d.). Retrieved from https://cryptonote.org/whitepaper.pdf.
13. Litecoin. (n.d.). Retrieved from https://litecoin.com/.
14. Androulaki, Elli, Artem Barger, Vita Bortnikov, Christian Cachin, Konstantinos Christidis, Angelo De Caro, David Enyeart, Christopher Ferris, Gennady Laventman, Yacov Manevich, et al. 2018. Hyperledger Fabric: A Distributed Operating System for Permissioned Blockchains. In *Proceedings of the Thirteenth EuroSys Conference*. ACM, 30.
15. Tendermint. (n.d.). Retrieved from http://tendermint.com.
16. Chain. (n.d.). Retrieved from https://chain.com/.
17. Quorum. (n.d.). Retrieved from http://www.jpmorgan.com/global/Quorum.
18. Peterson, Kevin, Rammohan Deeduvanu, Pradip Kanjamala, and Kelly Boles. 2016. A Blockchain-Based Approach to Health Information Exchange Networks. In *Proceedings of NIST Workshop Blockchain Healthcare*, 1–10.
19. Ekblaw, Ariel, Asaph Azaria, John D. Halamka, and Andrew Lippman. 2016. A Case Study for Blockchain in Healthcare: "MedRec" Prototype for Electronic Health Records and Medical Research Data. In *Proceedings of Big Data Conference*, 13. IEEE Open.
20. Lemieux, Victoria L. 2017. A Typology of Blockchain Recordkeeping Solutions and Some Reflections on Their Implications for the Future of Archival Preservation. In *Proceedings of IEEE International Conference on Big Data*, 2271–2278. IEEE.
21. Ferrag, Mohamed Amine, Makhlouf Derdour, Mithun Mukherjee, Abdelouahid Derhab, Leandros Maglaras, and Janicke, Helge. 2018. Blockchain Technologies for the Internet of Things: Research Issues and Challenges. arXiv preprint arXiv:1806.09099.
22. Apache Kafka. (n.d.). Retrieved from http://kafka.apache.org.
23. Hyperledger Fabirc Definition. (n.d.). Retrieved from https://www.investopedia.com/terms/h/hyperledger-fabric.asp.
24. Multichain. (n.d.). Retrieved from http://www.multichain.com/.

ECC-Based Secure Group Communication in Energy-Efficient Unequal Clustered WSN (EEUC-ECC)

G. Raja Vikram, Addepalli V. N. Krishna and K Shahu Chatrapati

Abstract With an advent of the Internet of things (IoT), wireless sensor networks (WSNs) are gaining popularity in application areas like smart cities, body area sensor networks, industrial process control, and habitat and environment monitoring. Since these networks are exposed to various attacks like node compromise attack, DoS attacks, etc., the need for secured communication is evident. We present an updated survey on various secure group communication (SGC) schemes and evaluate their performance in terms of space and computational complexity. We also propose a novel technique for secure and scalable group communication that performs better compared with existing approaches.

Keywords Wireless sensor network (WSN) · Elliptic curve cryptography (ECC) · Secure group communication (SGC) · Unequal clustering · WSN multicasting

1 Introduction

Wireless sensor network is a collection of sensor nodes deployed without any specific topology mainly in unattended or typical environments. These sensor node operations are limited by energy, storage, and computational capability. A sensor node needs to sense a specific physical phenomenon and send it to the base station (BS). The BS is supposed to be high powered and is available with sufficient

G. Raja Vikram (✉)
Department of CSE, Vignan Institute of Technology & Science, Hyderabad, Telangana, India
e-mail: grajavikram@gmail.com

A. V. N. Krishna
Department of CSE, Faculty of Engineering, CHRIST(Deemed to be University),
Bangalore, India
e-mail: hari_avn@rediffmail.com

K. S. Chatrapati
Department of CSE, J.N.T.U.H College of Engineering - Manthani, Karimnagar,
Telangana, India
e-mail: shahujntu@gmail.com

© Springer Nature Singapore Pte Ltd. 2020   673
K. S. Raju et al. (eds.), *Proceedings of the Third International Conference
on Computational Intelligence and Informatics*, Advances in Intelligent
Systems and Computing 1090, https://doi.org/10.1007/978-981-15-1480-7_58

resources. Since WSNs are used mainly in unattended and harsh environments without any physical protection, they are prone to security threats. Hence, security is a key aspect in WSN applications. For instance, in military applications a set of sensors will monitor the target movement and will exchange target position information. SGC provides location secrecy, which is required in homeland security, animal monitoring, and military supplications [1, 2]. Hence, a secured group communication is required to protect from threats like eavesdropping, impersonation, and node compromising.

Typically, a SGC for WSN needs to consider two issues: group management and secure group key distribution. Various clustering techniques were proposed to establish location-based group formation. In most of the techniques, every group will be headed by a cluster head responsible for data gathering, aggregation, and sending aggregated data to the base station. [3] presented a survey on energy-based clustering schemes for WSN. Regarding secured group key management, several surveys have been published in [4–7], and [8]. [9] presented a comprehensive survey on both group membership management and secure group key distribution.

The rest of the paper is organized as follows: Sect. 2 elaborates on security attacks and requirements of secure group key management. Section 3 explains existing schemes for clustering and SGC. In Sect. 4, we have explained the significance of elliptic curve cryptography. The proposed approach is explained in Sect. 5. The simulation results and comparative study are given in Sect. 6, and finally, Sect. 7 concludes the paper.

2   Requirements of Secure Group Communication

In this section, we first focus on the potential attacks on WSN group communication. Later, we list the requirements that SGC mechanism must satisfy to restrict these attacks.

2.1   Group Communication Potential Attacks

Group communication in WSN is prone to several attacks due to their features. Here, we enumerate the potential attacks on WSN group communication.

- **Eavesdropping**: An intruder can passively monitor the group communication and get to know all the messages being transmitted. For instance, in military applications the opponent wants to know the queries being sent. This attack does not leave any mark on the message being sent. Hence, it is very difficult to identify this attack. The solution to defy this attack is to encrypt messages.
- **Modification**: An attacker tries to modify the message being transmitted by replacing or deleting its content. This attack can be avoided by encrypting the message using secure group key.

- **Replay attack**: In this attack, the intruder may copy a previous message from channel and resend it later to disturb group communication. To restrict this attack, a nonce or random number can be added to each message to prove its relevance [10].
- **Masquerade attack**: An intruder will act as if he is a legal user and communicate or launch attacks using proxy identity. This attack can be avoided by providing secured authentication mechanism [7, 10].
- **Node compromise attack**: In this attack, the attacker may physically capture any node and get to know all the secret information like group key. Once this key is revealed, it becomes easy for him to break the entire group communication [7].
- **Denial of service (DoS) attack**: The main intention of intruder in launching this attack is to disrupt the services provided by group. DoS attacks can be launched either from a group member or outside node.

2.2 Secure Group Communication Requirements

Table 1 shows the list of security requirements to be followed to restrict the above-mentioned attacks on group communication.

3 Existing SGC Schemes

The following section elaborates on existing group communication schemes classified into three approaches.

3.1 Centralized Approach

In this approach, each group is monitored by a group head (GH) responsible for managing multicast activity. It manages both group establishment and group membership management. The robustness of private key encryption and secured key generation are major advantages of this approach.

In [11], a tree-based multicast approach named as logical key hierarchy (LKH) was proposed. In LKH, nodes are divided into groups and in turn into subgroups. The key encryption keys (KEK) are generated, which are used to send original keys. All the leaf nodes are associated with asymmetric keys, internal nodes are associated with KEK, and root node maintains the group key. The KEK helps us in rekeying process with a single multicast message. This results in the reduction of rekeying messages logarithmically.

Table 1 Secure group communication requirements

Attacks	Authentication	Integrity	Confidentiality	Backward and forward secrecy	Immediate re-keying	Robustness
Eavesdropping			Y	Y		
Modification		Y	Y		Y	Y
Replay	Y			Y	Y	Y
Masquerade	Y			Y	Y	Y
Node compromise	Y	Y	Y	Y	Y	Y
Denial of service			Y	Y		Y

The LKH scheme is further improved in [12]. An enhanced LKH was proposed by [12], where the group key was generated based on cryptographically strong one-way hash functions.

In [13], more powerful cluster head (CH) shares a secret key with all the cluster members. Session keys are distributed by using dedicated secret keys making this scheme robust. However, this approach requires more rekey messages to be sent at each join or leave.

3.2 Distributed Approach

In distributed approach, the group establishment and membership management tasks are distributed among all the members. The problem of a single point of failure is eliminated by collaborative work making it fault-tolerant.

In [13], a compromise-resilient group rekeying scheme (CRGR) is proposed based on polynomial distribution. Here, every group member sends a random number securely to the group head. Group head upon receiving random numbers will generate group key. A group key-based polynomial is sent to all members to extract the key. This approach restricts node compromise attacks.

In [14], a symmetric encryption and threshold-based clustering scheme is proposed. The group is constructed in a collaborative way. Cluster head generates group key based on the secret shares received from threshold t group members. This scheme also presents group membership management. Low storage and computational costs are major advantages of this approach. However, if an adversary can compromise t nodes, then the entire communication will be revealed.

3.3 Hybrid Approach

In this approach, group head is responsible for generating group key. The generated key will be distributed among the members in a decentralized way.

In [15], a ring-based secure group communication (RiSeG) was proposed. In this approach, role of the group controller is distributed among members by constructing a logical ring. A rekeying message is sent by GC in the ring until it reaches back to GC. Hence, the rekeying process takes $O(n)$ time. In this approach, all group membership operations like creation, join, and leave are defined. The major limitation here is that it is applicable in specific network model where a logical ring can be formed.

A distributed key forwarding scheme was proposed in [12], where each node stores two keys, namely local and global keys. Global keys are used for local key propagation. Local keys are calculated by subtree root and forwarded to reach all the leaf nodes. The major limitation of this scheme is that it supports only communication to/from the group controller.

4 Elliptic Curve Cryptography

ECC is a public key system based on the elliptic curves introduced by Neal Koblitz and Victor Miller. ECC has outperformed its public key counter parts RSA and ElGamal algorithms, especially in resource-constrained environments like sensor networks, mobile, and handheld devices. Shorter key lengths directly yield energy savings and require low computational complexity.

An elliptic curve is an algebraic curve defined by an equation in the form

$$y^2 = x^3 + ax + b \tag{1}$$

where $a, b \, \mathcal{E} \, K$ and elements of finite field K together with a point at infinity, and the condition $4a^3 + 27b^2 \neq 0$ holds. Figure 1 shows an elliptic curve.

Fig. 1 Sample elliptic curve

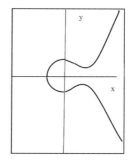

Table 2 Public key
cryptosystem key comparison

Security bits	Minimum size of public keys			Key size ratio
	DSA	RSA	ECC	ECC versus RSA/DSA
80	1024	1024	160	1:6
112	2048	2048	224	1:9
128	3072	3072	256	1:12
192	7680	7680	384	1:20
256	15,360	15,360	512	1:30

ECC is based on point arithmetic on elliptic curves. Considering an operation multiplication, it is easy to calculate the result point by multiplying any given point on the curve with a number. But it is very difficult to find out the number given source and result points.

Table 2 compares the key size ratio between prominent PKC algorithms: DSA, RSA, and ECC. As shown, ECC approach uses smaller key size compared with other contemporaries, which is suitable for constrained environments like WSN [16]. It consumes lesser memory and does faster computations [17].

5 ECC-Based Energy-Efficient Group Communication

In resource-constrained environments like WSN, to ensure security, lower computational cost approaches like ECC are favored. In our approach, symmetric cryptography along with ECC-based authentication is used to ensure secured group key distribution and communication.

Firstly, all the sensor nodes are divided into unequal clusters as explained in [1]. In this approach, based on the distance from BS clusters are formed with unequal size. Farther the group from BS, more its size. This allows GH near BS to handle less intracluster data than intercluster data. This scheme avoids hotspot problem by reducing overhead on cluster heads nearby base station. Group head is assumed to be a powerful member with better storage and computational capabilities. A logical tree is constructed rooted from group head connecting all the group members in a hierarchical manner. GH generates group key upon every member join or leave to ensure forward and backward secrecy.

Whenever a new node wants to join the group, it will send a join request to the base station. Base station will forward that request to GH. GH will authenticate new member, and verification will generate a new group key. This new key will be forwarded to all the members through the neighbor logical tree constructed.

The proposed approach is a contributory binary tree based one, with cluster head being the root. Each node is associated with an announced public key (Pb) and a secret private key (Pr).

The public key Pb_i is calculated as given below:

$$Pb_i^h = Pr_i^h \cdot G \tag{2}$$

where Pr_i^h and Pb_i^h are the private and public keys of a node i at a height h in binary tree. G is the base point on the given elliptic curve E. '•' is the scalar multiplication operation on given elliptic curve. Both E and G are predistributed to all the users.

The Pr_i^h of all leaf nodes is a random number generated secretly by the node itself. Pr_i^h and Pb_i^h keys of all intermediate nodes are calculated as shown below:

$$Pr_i^h = Pr_i^{2i} \cdot Pb_i^{2i+1}$$
$$\text{or} \tag{3}$$
$$= Pb_i^{2i} \cdot Pr_i^{2i+1}$$

$$Pb_i^h = Pr_i^h \cdot G \tag{4}$$

where for ith node left child will be at $2i$ position and right child will be at $2i + 1$ position in the constructed binary tree. Thus, the private key of the root node will be considered as a secret group key for communication.

$$G_k^h = Pr_i^h \tag{5}$$

where G_k^h is the secret group key for all the members rooted by node k.

The generated group key (G_k^h) is then distributed among all the members using the following process:

Algorithm G_K-Distribution

Input: Public keys of all nodes, binary tree, private keys of each node

Output: Distribution of G_K to all nodes rooted at K

1. The root node generates group key using $G_k^h = Pr_i^h$
2. For each child node 'i'

$$G_i^h = G_j^{h+1} (\text{xor}) \overset{h}{\underset{i}{Pr}} \tag{6}$$

where G_j^{h+1} is the group key of parent. This can be calculated by scalar multiplication of ith node private key and its siblings public key.
3. Repeat step 2 until group key is propagated till the leaf nodes.

ECC-based authentication scheme helps in reducing communication overhead by distributing group key in a cost-effective way. We utilize proactive procedure to speak with in the group and receptive methodology for among the bunch and furthermore utilize the ECC validations conspire.

6 Simulation Result and Analysis

To simulate our multicast communication, we have used ns-3 simulator. We have conducted simulation study using the parameters shown in Table 3. We compared our proposed approach EEUC-ECC with existing multicast algorithms like M-LEACH and MAODV. The simulation was carried out by varying node count to 500,700.

In Fig. 2, it may be observed that when the node count is 500, the proposed EEUC-ECC has approach clearly performed better compared with others. For node count of 700, our approach performed equally well with other approaches.

In Fig. 2, it may be observed that our approach has improved network lifetime significantly. When node count is 500, our approach has less mean first node die time. But when the node count increases to 700, the first node dies soon compared with M-LEACH and M-AODV.

Table 3 Simulation parameters

Network parameter	Value
Network size	25 m * 25 m
Number of sensor nodes	500, 700
Data packets length	2000 bits
Initial energy of sensor nodes	0.1 J
Data packet processing delay	0.1 ms
Transmission speed	100 bit/s
Bandwidth	5000 bit/s

Fig. 2 Network lifetime analysis for 500 nodes

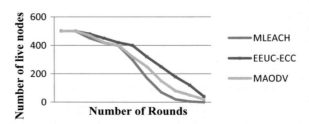

7 Conclusion

In this paper, we have exhibited that unequal size clustering with ECC-based secure group communication enables CHs close to the BS to effectively deal with more intergroup relay traffic, consequently maintaining a strategic distance from the hotspot issue. The simulation results demonstrated that the proposed approach significantly improved network lifetime.

Exploratory outcomes demonstrate that the proposed approach performs superior to M-LEACH and MAODV in terms of network lifetime and system correspondence overhead. In future work, we would direct reenactment strong to check the execution of other parameters like congestion effect and security level. We will plan to apply our approach in IoT-based environments to provide secured, scalable, and congestion-free multicast communication.

References

1. Abuzneid., A.S., T. Sobh, and M. Faezipour. 2015. An Enhanced Communication Protocol For Location Privacy In Wsn. *International Journal of Distributed Sensor Networks* 20, 297–317.
2. Chen., H., and W. Lou. 2015. On Protecting End-to-End Location Privacy Against Local Eavesdropper In Wireless Sensor Networks. *Pervasive and Mobile Computing* 16 (Part A), 36–50.
3. Raja Vikram, G., A.V.N. Krishna, and K. Shahu Chatrapati. 2014. Energy Aware Clustering-based Routing Schemes for Wireless Sensor Networks—A Survey. *International Conference on Innovations in Computer Science & Engineering (ICICSE-2014)* 405–409.
4. He. X., M. Niedermeier, and H. De Meer. Dynamic Key Management in Wireless Sensor Networks: A Survey. *The Journal of Network and Computer Applications* 36 (2), 611–622.
5. Naranjo., J., and L. Casado. 2007. An Updated View on Centralized Secure Group Communications. *Logic Journal of the IGPL* 36 (2), 611–622.
6. Klaoudatou., E., E. Konstantinou, G. Kambourakis, and S. Gritzalis. 2011. A Survey on Cluster-Based Group Key Agreement Protocols for WSNs. IEEE *Communications Surveys & Tutorials* 13 (3), 429–442.
7. Sakarind., P., and N. Ansari. 2007. Security Services in Group Communications Over Wireless Infrastructure, Mobile Adhoc and Wireless Sensor Networks. *Wireless Communications, IEEE Computer Society* 8–20.
8. Xiao., Y., V. K. Rayi., B. Sun, X. Du, F. Hu, and M. Galloway. 2007. A Survey of Key Management Schemes in Wireless Sensor Networks. *Computer Communications* 30 (11–12), 2314–2341.
9. Cheikhrouhou, O. 2015. Secure Group Communication in Wireless Sensor Networks: A Survey. *Journal of Network and Computer Applications* 41, 426–433.
10. Venkatraman., K., J. Vijay Daniel, and G. Murugaboopathi. 2013. Various Attacks in Wireless Sensor Network: Survey. *International Journal of Soft Computing and Engineering* 3, 107–120.
11. Cheikhrouhou., O., A. Kouba, G. Dini, H. Alzaid, and M. Abid. 2012. LNT: A Logical Neighbor Tree for Secure Group Management in Wireless Sensor Networks. *Procedia Computer Science* 5, 198–207.

12. Cheikhrouhou., O., A. Koubaa, O. Gaddour, G. Dini, and M. Abid. 2010. RiSeG: A Logical Ring Based Secure Group Communication Protocol For Wireless Sensor Networks. *The International Conference on Wireless and Ubiquitous Systems (ICWUS2010)*, Tunisia, 615–22.

13. Dini., G., and L.Lopriore. 2015. Key Propagation in Wireless Sensor Networks. *Computers & Electrical Engineering* 41, 426–33.

14. Bao, X., J. Liu, L. She, and S. Zhang. 2014. A Key Management Scheme Based on Grouping Within Cluster. In *11th World Congress on Intelligent Control and Automation* 126–33.

15. Cheikhrouho., O., A. Koubaa, G. Dini, and M. Abid. 2011. RiSeG: A Ring Based Secure Group Communication Protocol for Resource-Constrained Wireless Sensor Networks. *Personal and Ubiquitous Computing* 356–363.

16. Raja Vikram, G., A.V.N. Krishna, and K. Shahu Chatrapati. 2017. Variable Initial Energy and Unequal Clustering (VEUC) Based Multicasting in WSN. *IEEE WiSPNET 2017 International Conference* 82–86.

17. Kalra., S., and S.K. Sood. 2011. Elliptic Curve Cryptography: Survey and Its Security Applications. In *Proceedings of the International Conference on Advances in Computing and Artificial Intelligence (ACAI'11)*, USA, 102–106.

Performance Comparison of Random Forest Classifier and Convolution Neural Network in Predicting Heart Diseases

R. P. Ram Kumar and Sanjeeva Polepaka

Abstract The applications of machine learning (ML) in the digital era become inevitable. Few domains include virtual personal assistants, predictions while commuting, audio and video surveillance, filtering of email spam and malware(s), service and support in online and social media, refining the search engine performance, online fraud detection, product recommendations, healthcare, finance, travel, retail, media, and so on. Among the various functionalities, the applications of ML in the health domain play a momentous role. The objective of the paper is to focus the applications of ML in predicting the cardiac arrest/heart attack based on the earlier health records. Though there exists opulence of data on the history regarding the cardiac diseases, the inadequacy in analyzing and predicting the heart attack leads to sacrifice the human life. The research focuses on predicting the cardiac arrest/heart attack using the ML approaches based on the patient's historical data. Among the various ML techniques, the paper focuses on random forest classifier (RFC) and convolution neural network (CNN)-based prediction methods. The experimentation was conducted on the standard datasets available in the UCI repository. The results concluded that RFC had outperformed the other classifier regarding the classification accuracy.

Keywords Machine learning techniques · Healthcare · Cardiac arrest/heart attack prediction · Random forest classifier · Convolution neural network

1 Introduction

The machine learning (ML) is a subset of artificial intelligence which makes the system to learn automatically from the previous experience without the explicit programming knowledge. The objective of ML is to develop the solutions using the available data and learn themselves with that data. The learning process is initiated

R. P. Ram Kumar (✉) · S. Polepaka
Department of CSE, MREC(A), Secunderabad, India
e-mail: rprkvishnu@gmail.com

© Springer Nature Singapore Pte Ltd. 2020
K. S. Raju et al. (eds.), *Proceedings of the Third International Conference on Computational Intelligence and Informatics*, Advances in Intelligent Systems and Computing 1090, https://doi.org/10.1007/978-981-15-1480-7_59

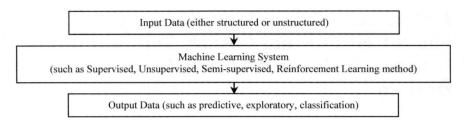

Fig. 1 ML technology

with the data collected from various sources such as sensors, applications, mobile devices, security systems, networks, and even from other appliances. Figure 1 shows the basics of ML technology summarized from Gartner [1]. After collecting the data, extraction of knowledge or determining the patterns from the observations takes place. Based on the observations, the ML is divided into four subdivisions, namely (1) supervised learning, (2) unsupervised learning, (3) semi-supervised learning, and (4) reinforcement learning [2].

The process of ML can be viewed in two phases, namely (1) learning and (2) prediction. When the training data is fed to the learning phase, functional processing (such as preprocessing, learning, and error-analysis) takes place. The predicted data results in the prediction phase for the new data concerning the trained model [2].

2 Motivation: Predicting the Cardiac Attack Using ML Approach

Even in the digital era, the cardiac disease becomes a predominant cause for the death of human beings. As per the World Health Organization (WHO) statement, approximately 17.9 million deaths occurs every year due to cardiovascular diseases (CVD), also called a heart attack or heart failure [3]. Further, the WHO estimated that the death rate might rise to 23.6 million during 2030 [4]. In this case, the prediction of cardiac arrest/attack becomes inevitable to reduce the risk levels. Few attributes of cardiac attack include age, sex, family history, chest pain type, ECG reading, blood sugar level, physical inactivity, tobacco usage, and cholesterol [5]. Even though the data mining techniques are used to explore and expand the relationship between the attributes, it is essential to predict the cardiac attacks in advance based on the symptoms, i.e., based on the historical data. The proposed method aims to predict the heart attack based on the historical information using machine learning algorithms.

2.1 Related Study

The following section deals with few existing techniques of heart attack/disease predictions using data mining and ML approaches.

Nikhar and Karandikar [6] proposed a system for predicting the heart disease using data mining and Naïve Bayes and decision tree classifiers. The present method tried to improve the performance of the Naïve Bayesian classifier by removing unnecessary and irrelevant attributes from the dataset. MAximal Frequent Itemset Algorithm (MAFIA) was applied for mining maximal frequent model in heart disease datasets. Experimental results showed that the decision tree-based predictions have better performance than the Naïve Bayesian classification.

Patel, Upadhyay, and Samir [7] proposed a heart disease prediction approach using ML and data mining techniques. The system aimed to extract hidden patterns through data mining approaches and predict heart disease in patients. Decision tree classification algorithm for heart disease diagnosis includes a J48 algorithm, decision stump, random forest, and LMT tree algorithm. Among the various classification methods, J48 tree technique has the highest accuracy of 56.76% with 0.04 s execution time.

Sen [8] proposed an automated system for predicting and diagnosing the heart disease using Naïve Bayesian classifier, support vector machine, decision tree, and K-nearest neighbor. Experimental results concluded that the Naïve Bayesian classifier has better performance than the other classifiers.

Howlader et al. [9] presented a heart disease classification approaches regarding severity prediction. Few existing methods illustrated in [10, 11] and [12] adopted decision tree, J 4.8 algorithm, K-NN, and ensemble neural network for effective diagnosis of heart diseases on publicly available research dataset from Cleveland dataset, University of California, with 34 attributes. The classifiers used were Bayes Net, Naïve Bayes, sequential minimal optimization, K-NN, K-Start, Heoffring, and J48 methods. The parameters used to evaluate the present method were precision, recall, and F1-score. When evaluated with the dataset of 116 patients collected from various hospitals in Bangladesh, the Naïve Bayes classifier outperformed the other classifiers.

Akhtar et al. [13] presented a study on data mining techniques to predict cardiac diseases. The dataset was collected from the Faisalabad Institute of Cardiology. The data cleaning, data selection, normalization, and attribute construction were done to stabilize the data. Later follows the outlier analysis, clustering, classification, and prediction operations. The classifiers used were J48, Naïve Bayes, and neural networks for pruned, pruned with selected attributes, unpruned, unpruned with selected attributes, respectively. The Naïve Bayes classifier outperformed the other classifiers regarding the classification accuracy.

Takci [14] proposed an enhanced cardiac attack prediction method based on feature selection method. Various machine learning algorithms examined in this study include C4.5, C-RT, SVM with linear, polynomial, radial and sigmoid kernels, ID3, K-NN, MLP, Naïve Bayes, and logistic regression models. Experimentation showed that the SVM with linear kernel and relief method were the best machine learning and feature selection methods with 84.81% of prediction accuracy.

Florence, Amma, Annapoorani, and Malathi [15] suggested a method for predicting the risk of heart attacks using decision tree (classification and regression tree (CART), iterative dichotomized 3 (ID3), and C 4.5) and neural networks. The data preprocessing stage includes cleaning, transformation, and reduction of data. Later follows the classification process. The knowledge about the classification leads to better prediction. Experimental results concluded that the heart attack could be predicted even with six (major) attributes in a dataset.

Vijiyarani and Sudha [16] proposed an efficient prediction method using classification tree algorithms such as decision stump, logistic model trees, and random forest. The Cleveland dataset [17] form UCI repository was used for evaluation with the parameters, namely true positive rate (TPR), F-measure, RoC area, and kappa statistics. The experimental study illustrates that the decision stump approach outperformed other methods regarding least error rate.

3 Materials and Methods

The proposed approach aims to predict the heart attack risk with the patient's data collected from Cleveland database, UCI Machine Learning Repository [17]. Though the database comprises of 76 attributes, researchers use 14 attributes, namely (1) age, (2) sex, (3) chest pain type (cp), (4) resting blood pressure (trestbps), (5) serum cholesterol (chol), (6) fasting blood sugar (fbs), (7) resting electrocardiographic results (restecg), (8) maximum heart rate achieved (thalach), (9) exercise-induced angina (exang), (10) induced ST depression (old peak), (11) peak exercise ST segment's slope (slope), (12) major vessels count (ca), (13) thal, and (14) diagnosed heart attack (num), for evaluation purpose.

Figure 2 shows the architecture of the proposed method. There are four stages in the prediction of heart attack in the present method. They are (1) data preprocessing, (2) feature selection and extraction (FSE), (3) classification, and (4) performance analysis. The preprocessing stage includes cleaning the unclear, missing, and duplicate values in the dataset. The FSE stage deals with the selection of relevant and dominant features from the dataset, thus enabling the dimensionality reduction of data being processed. The classification process deals with the data analysis to categorize the unknown data using a model. The performance analysis stage compares the classifier outcome and chooses the best classifier.

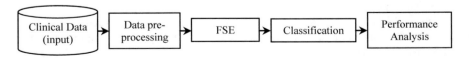

Fig. 2 Architecture of the proposed method

The preprocessing stage has four substages, namely cleaning, transformation, integration, and reduction. The process of identifying and removing the noise or missing values is called cleaning the data. The method of changing the data format through normalization, generalization, aggregation, and smoothing techniques is called transformation. The data which is not required for processing is wither from single or multiple sources which are integrated and eliminated. The reduction stage deals with the dimension reduction of complex data for easier handling. The rest of the paper deals with the evaluation of Cleveland database from UCI repository dataset with random forest and CNN classifiers.

3.1 Random Forest Classifier (RFC)

The RFC is an ensemble algorithm that combines either same or different types of algorithms for classification of objects. The objective of RFC is to form a set of decision trees from the subset of the randomly selected training set. The aggregation of votes from various decision trees and voting on the aggregated data determines the test object's class [18]. Figure 3 shows the architecture of the RFC method summarized from [19].

Working principle: RFC is a divide-and-conquer method of generating the decision trees (DT) based on the randomly split dataset. The group of DT classifiers is termed as forest. The attribute selection indicator generates an individual DT, and each such DT depends on a random sample. During the classification stage, the most popular among the DT votes is chosen as the prediction result. Table 1 shows the RFC procedure.

Fig. 3 Architecture of random forest classifier

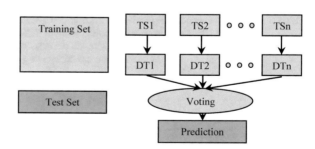

Table 1 RFC procedure

Steps	Procedure
Step 1	Select random training samples (TS) from the training set
Step 2	The DT is constructed from each TS and results are predicted
Step 3	Voting of predicted results
Step 4	The most popular vote is deliberated as the prediction result

Convolution neural network (CNN): The applications of CNN include object detection, recognition, image classification, and much more in the computer vision domain. Figure 4 shows the CNN stages. Table 2 shows the procedure of CNN. After the preprocessing and FSE stage, the random forest and CNN classifiers are applied to predict the heart attacks based on the past medical history of the concerned patient. Later follows the comparison of heart attack prediction accuracy between the two classifiers, namely RFC and CNN.

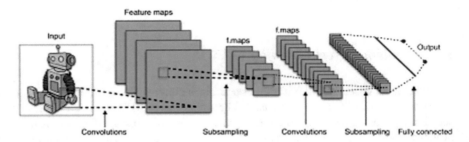

Fig. 4 Stages in CNN. *Source* [20]

Table 2 CNN procedure

Steps	Procedure
Step 1	Convolution layer estimates the output neurons that are locally connected in the input image, and feature maps of the input image are extracted during each computation
Step 2	Subsampling takes place through max pooling for dimension reduction. The process of convolution and subsampling is repeated subsequently
Step 3	Fully connected layers flatten the high-level features, and combining these features makes the CNN learn and predict the input label. In the present method, the CNN is implemented with the following credentials. They are activation = 'relu', hidden layer size = (14, 14, 14) with constant learning note and maximum iterations = 200

Table 3 K-fold cross-validations in the CNN and RFC methods

Accuracy of fold levels	CNN (accuracy level in %)	Random forest
Fold 1	95	90
Fold 2	90	80
Fold 3	85	75
Fold 4	75	70
Fold 5	90	85

4 Experimental Results

The proposed method is implemented using Python 3.6 installed on Intel Pentium P6200 processor with 2 GB DDR RAM. Out of 303 instances from the Cleveland database from UCI repository, 297 instances were selected after the preprocessing process discussed in Sect. 3. Among 297 instances, 236 and 61 instances were selected for training and testing processes, respectively. Table 3 shows the various K-fold cross-validations (with folds = 5) in the CNN and RFC methods. The performance measures of RFC and CNN are determined using the precision (Eq. 1), recall (Eq. 2), F1-score (Eq. 3), and accuracy (Eq. 4), respectively [21]. The Eqs. (1), (2), (3), and (4) depict the procedure for determining the performance measures.

$$Precision = \frac{TP}{TP + FP} \tag{1}$$

$$Recall = \frac{TP}{TP + FN} \tag{2}$$

$$F1 \text{ - } Score = 2 * \frac{Recall * Precision}{Recall + Precision} \tag{3}$$

$$Accuracy = \frac{TP + TN}{TP + TN + FP + FN} \tag{4}$$

where TP, TN, FP, and FN represents true positive (TP), true negative (TN), false positive (FP), and false negative (FN), respectively.

Table 4 shows the performance comparison of CNN and RFC regarding the precision, recall, F1-score, and classification accuracy, respectively.

The classification accuracy of RFC and CNN is 80.327 and 78.688% respectively. Even though there is no much difference between the classifications accuracy, DT has higher classification accuracy in predicting the cardiac attack risk.

Table 4 Performance comparison of CNN and RFC

Method	Precision	Recall	F1-score	Classification accuracy (in %)
RFC	82	80	80	80.327
CNN	80	79	78	78.688

5 Conclusion

The paper proposed a predictive analysis of risks regarding the cardiac attack on the standard UCI dataset. In spite of many classifiers, the study on existing methods made to work on RFC and CNN classifiers. In a few existing techniques, the authors claimed the higher accuracies (above 95% and sometimes even 99.5%) with various attributes of the classifiers. The experimental results disclosed that the characteristics of the chosen classifier influenced the performance of the classifier. In future, the authors are willing to work on tuning the performance of RFC and CNN to improve the prediction performance. Further, the present approach might be evaluated with other standard databases.

Acknowledgements I want to represent my gratitude to Ms. Rubina Tabassum (Hall Ticket No.: 15J41A0547) and Ms. Anjali Reddy Bhumanapally (Hall Ticket No.: 15J41A0505), IV Year B. Tech., CSE students (MR15) for their cooperation in preparing the manuscript.

References

1. Carlton, E. Sapp. 2017. Gartner, Preparing and Architecting for Machine Learning, Technical Professional Advice, Analyst(s). https://www.gartner.com/binaries/content/assets/events/keywords/catalyst/catus8/preparing_and_architecting_for_machine_learning.pdf.
2. Machine Learning: What it is and Why it Matters. https://www.simplilearn.com/what-is-machine-learning-and-why-it-matters-article.
3. Cardiovascular disease. http://www.who.int/cardiovascular_diseases/en/.
4. Marikani, T., and K. Shyamala. 2017. Prediction of Heart Disease using Supervised Learning Algorithms. *International Journal of Computers and Applications* 165 (5): 41–44.
5. Kaur, B., and W. Singh. 2014. Review on Heart Disease Prediction system using Data Mining Techniques. *International Journal on Recent and Innovation Trends Computing and Communication* 2 (10), 3003–3008.
6. Nikhar, S., and A.M. Karandikar. 2016. Prediction of Heart Disease Using Machine Learning Algorithms. *International Journal of Advanced Engineering, Management and Science* 2 (6): 617–621.
7. Patel, J., T. Upadhyay, and S. Patel. 2016. Heart Disease Prediction Using Machine learning and Data Mining Technique. *International Journal of Computer Science & Communication* 7 (1): 129–137.
8. Sen, S.K. 2017. Predicting and Diagnosing of Heart Disease Using Machine Learning Algorithms. *International Journal of Engineering and Computer Science* 6 (6): 21623–21631.
9. Howlader, K.C., S. Satu, and A. Mazumder. 2017. Performance Analysis of Different Classification Algorithms that Predict Heart Disease Severity in Bangladesh. *International Journal of Computer Science and Information Security (IJCSIS)* 15 (5): 332–340.
10. Tu, M.C., D. Shin, and D. Shin. 2009. Effective Diagnosis of Heart Disease Through Bagging Approach. In: *2nd International Conference on Biomedical Engineering and Informatics*, 1–4, IEEE Press.
11. Polat, K., S. Sahan, and S. Gunes. 2007. Automatic Detection of Heart Disease Using an Artificial Immune Recognition System (Airs) with Fuzzy Resource Allocation Mechanism and k-NN Based Weighting Preprocessing. *Expert Systems with Applications* 32 (2), 625–631.

12. Das, R., I. Turkoglu, and A. Sengur. 2009. Effective Diagnosis of Heart Disease Through Neural Networks Ensembles. *Expert Systems with Applications* 36 (4): 7675–7680.
13. Akhtar, N., M.R. Talib, and N. Kanwal. 2018. Data Mining Techniques to Construct a Model: Cardiac Diseases. *International Journal of Advanced Computer Science and Applications* 9 (1): 532–536.
14. Takci, H. 2018. Improvement of Heart Attack Prediction by the Feature Selection Methods. *Turkish Journal of Electrical Engineering & Computer Sciences* 26: 1–10.
15. Florence, S., N.G.B. Amma, G. Annapoorani, and K. Malathi. 2014. Predicting the Risk of Heart Attacks using Neural Network and Decision Tree. *International Journal of Innovative Research in Computer and Communication Engineering* 2 (11), 7025–7030.
16. Vijiyarani, S., and S. Sudha. 2013. An Efficient Classification Tree Technique for Heart Disease Prediction. In: *International Conferene on Research Trends in Computer Technologies*, 6–9.
17. UCI Machine Learning Repository, Cleveland Heart Disease Dataset. https://archive.ics.uci.edu/ml/datasets/heart+Disease.
18. Random Forest Classifier. https://medium.com/machine-learning-101/chapter-5-random-forest-classifier-56dc7425c3e1.
19. Understanding Random Forests Classifiers in Python. https://www.datacamp.com/community/tutorials/random-forests-classifier-python.
20. Convolutional Neural Networks in Python with Keras. www.datacamp.com/community/tutorials/convolutional-neural-networks-python.
21. Accuracy, Precision, Recall & F1 Score: Interpretation of Performance Measures. https://blog.exsilio.com/all/accuracy-precision-recall-f1-score-interpretation-of-performanc e-measures/.

Food Consumption Monitoring and Tracking in Household Using Smart Container

Y. Bevish Jinila, V. Rajalakshmi, L. Mary Gladence
and V. Maria Anu

Abstract Tracking food consumption in household and monitoring the usage of items in kitchen everyday is incredible. Manual intervention of this process consumes time and is annoying. This paper presents a smart container for monitoring and tracking the food items stored in household kitchen. The proposed solution serves to provide an automated response on the level of the items stored in the container and also generates an alert on its expiry. A prototype of the proposed method is developed and tested for level and expiry detection.

Keywords Smart container · Food consumption · Kitchen · Household

1 Introduction

In India, food waste is a growing area of concern. In household, much of the food waste comes from the kitchen due to inadequate planning of storage and lack of monitoring the level of food items, namely the pulses and rice. Those food items stored in containers are susceptible to be spoiled with pests if unnoticed for a long time, and it is difficult to maintain the level and expiry date for all the products stored in the containers. Also, the amount of items consumed daily, weekly, or monthly is not monitored.

Smart container is a storage container which is used to store the pulses, rice, etc., and intimate the level of the items in the container, to generate an alert when the item is nearing the expiry date, and to detect the presence of pests in the container. It is quite common in a household kitchen; grains are stored in containers to be used for cooking. In an urban area, people spend most of their time in work place, find difficult to manage, and monitor the grains stored in containers. The stored grains

Y. Bevish Jinila (✉) · V. Rajalakshmi · L. Mary Gladence · V. Maria Anu
Sathyabama Institute of Science and Technology, Chennai 600119, India
e-mail: bevishjinila.it@sathyabama.ac.in

© Springer Nature Singapore Pte Ltd. 2020
K. S. Raju et al. (eds.), *Proceedings of the Third International Conference on Computational Intelligence and Informatics*, Advances in Intelligent Systems and Computing 1090, https://doi.org/10.1007/978-981-15-1480-7_60

that remain unnoticed for a long time lead to the inculcation of pests which spoils the entire grain in the container. Further, it is difficult to maintain the expiry dates of all the items stored and trace the amount of grains consumed. This smart container is equipped with a level sensor and vibration sensor. Whenever initiated, the quantity of the items in the container is detected. Whenever the container is refilled, the expiry date of the item is stored in the storage unit. The collected information from the sensor network is forwarded by the IoT microcontroller to a Wi-fi router and sent to the storage. In the cloud server, the information about the containers and their levels with expiry is stored as records. The analysis is done on a 24-hour basis and is presented to the end user using an android application. In addition, a manual switch is installed to know the status of the containers anytime. This acts as a relay and enables the sensors in the sleep mode, collects the information, and delivers to the end user. This relay also enables a text-to-voice conversion. This facility also imparts the usage of the system for blind people.

2   Related Work

This section briefs on the various solutions proposed on food consumption monitoring [1]. Carla et al. [2] proposed the usage of smart shelf based on RFID [3, 4] for book identification. It employed microstrip transmission lines for proximity tag detection. The methodology ensured electromagnetic isolation barriers. The same is applied for varied applications.

Carla et al. [5] proposed a RFID cabinet which detects and recognizes the items stored in it. The method employed ensures the proper usage of electromagnetic fields, preventing the detection of items stored outside the container. It uses four non intrusive antennas for uniform electromagnetic coverage. Another, smart kitchen cabinet for home is proposed by Pal et al. [6], shows the usage of bluetooth [8, 9] to access the information from smart containers.

Rosarium et al. [7] have developed a smart grain container for tracking the food consumption and dietary habits of inhabitants. In this prototype, the ultrasonic sensor is used to detect the level of the container and transmits the information to the remote mobile or personal computer via Bluetooth. However, this prototype does not utilize any mechanisms for expiry detection of items stored in the container.

3   Proposed Framework

The architecture of the proposed system is shown in Fig. 1.

The containers are networked together to form a sensor network. The IoT-enabled microcontroller retrieves the level of the items in the containers, records the expiry date of the item during refill, generates an intimation when the

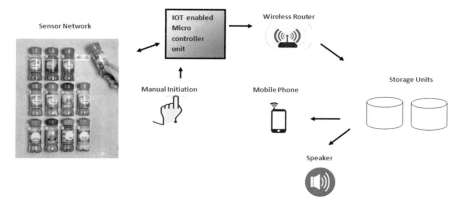

Fig. 1 Architecture of the smart container

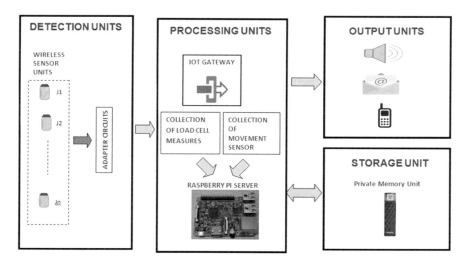

Fig. 2 Block diagram of smart container

item nears expiry and detects the presence of pests, and forwards it through the wireless router to the private memory unit. The data retrieved and processed can be visualized using a mobile application or in a speaker.

The block diagram of the smart container is shown in Fig. 2. It contains three process blocks and storage, namely the detection unit, processing unit, output unit, and storage unit. Load cells are used in the container to measure the level of the item in the container. Further, movement sensor is used to detect the pest in the item. The Raspberry pi server is used to process and send the data to the private memory unit and output units.

Figure 3 shows the flow diagram of data entry. Firstly, initialize the storage and connections. Whenever a container is moved, the sensor is activated and the weight of the container is calculated. If the weight computed is greater than the initial weight, the record is updated in the memory.

Figure 4 shows the flow diagram of the empty container detection. For every container, identify the item and activate the load cell, read the value, and calculate

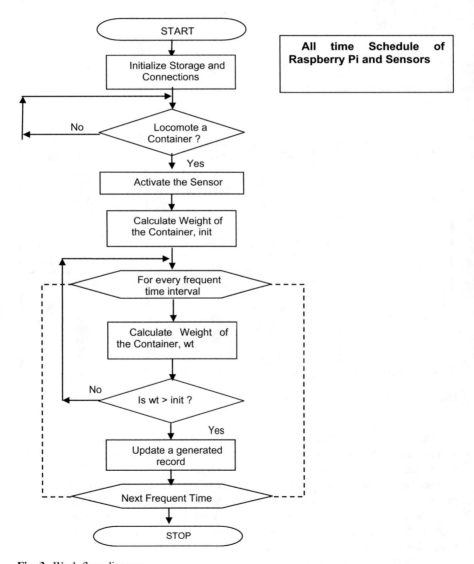

Fig. 3 Work flow diagram

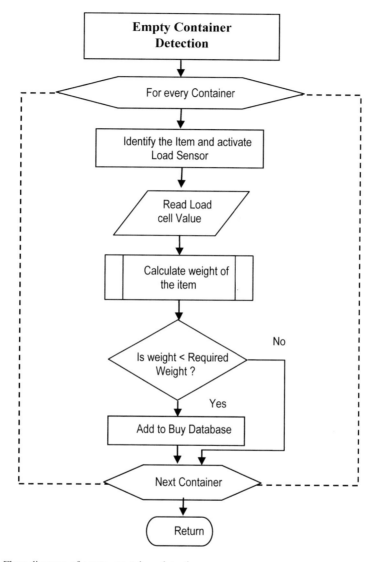

Fig. 4 Flow diagram of empty container detection

the weight of the item. If the calculated weight is less than the required weight, add the item to the buy database.

Figure 5 shows the flow diagram of expiry detection. For every container, identify the item and read the expiry date from the storage. Calculate the difference between the current system date and the expiry date. If it is less than or equal to two, add the item to the expiry database.

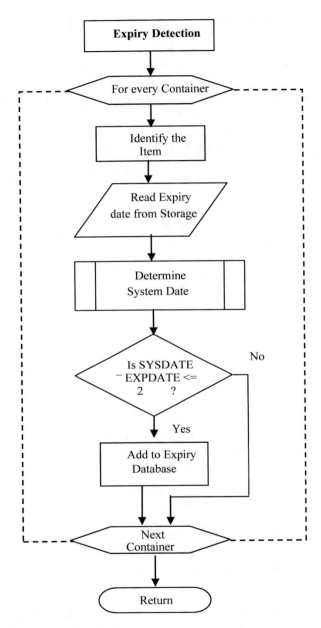

Fig. 5 Flow diagram of expiry detection

4 Experimental Results

The prototype for the proposed framework is developed to detect the level and the expiry of the items stored in the container. The system includes components, namely

(a) Load cells
(b) Vibration sensor
(c) Mobile application.

The level of the items in the containers is measured by the load cells. The load cells determine the weight of the items in the container. If the measured weight is below a fixed threshold, the concerned container is marked, and whenever there is a refill, the expiry date of the item is updated in private storage unit. The data gathered is processed by a Raspberry pi server, and through IoT gateway, it is stored in the storage. The processed data is updated on every 24-h basis automatically to the storage unit. For visualization, an android mobile application is developed, and using text-to-voice conversion, the output is given to a speaker. Figure 6 shows the mobile application that links to the prototype.

Table 1 shows the attributes listed for the items in the smart container. They include the container number, item name, and weight in kilograms, date of measurement, time of measurement, and date of expiry.

Fig. 6 Mobile application

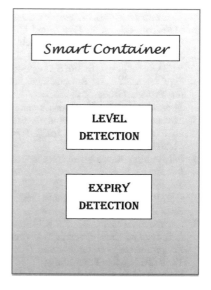

Table 1 Attributes of smart container

S. no.	Container number	Item name	Weight (kg)	Date of measurement	Time of measurement	Date of expiry
1	C1	Dhall	0.45	20/10/2018	11:50:24	1/11/2018
2	C2	Rice	0.5	20/10/2018	11:50:24	24/11/2018
3	C3	Green gram	1.0	20/10/2018	11:50:24	23/11/2018
4	C4	Black gram	0.65	20/10/2018	11:50:24	19/11/2018
5	C5	Pepper	0.85	20/10/2018	11:50:24	12/12/2018

5 Conclusion

This paper gives a clear idea on the development of the prototype smart container for household needs. The container detects the level of the items stored and generates an alert prior to the expiry of the items stored in the container.

References

1. Chi, P.Y., J.H. Chen, H.H. Chu, and B.Y. Chen. 2007. Enabling Nutrition-Aware Cooking in a Smart Kitchen. In *Conference on Human Factors in Computing Systems: CHI'07 Extended Abstracts on Human Factors in Computing Systems*, vol. 28, 2333–2338.
2. Medeiros, C.R., J.R. Costa, and C.A. Fernandes. 2008. RFID Smart Shelf with Confined Detection Volume at UHF. *IEEE Antennas and Wireless Propagation Letters* 7: 773–776.
3. Blessingson, Jeba, and Y. Bevish Jinila. 2010. Multi-Utility/Tracing Kit for Vehicles using RFID Technology. In *IEEE Conference on RSTS&CC*, 273–276.
4. Jaya Priyaa, Y., and Bevish Jinila. 2015. Secured Short Time Automated Toll Fee Collection for Private Group Transportation. In *IEEE International Conference on Innovations in Information, Embedded and Communication Systems*, 1–4.
5. Medeiros, Carla R., Cristina C. Serra, Carlos A. Fernandes, and Jorge R. Costa. 2011. UHF RFID Cabinet. In *IEEE International Symposium on Antennas and Propagation (APSURSI)*, 1429–1432.
6. Pal Amutha, K., Chidambaram Sethukkarasi, and Raja Pitchiah. 2012. Smart Kitchen Cabinet for Aware Home. In *The First International Conference on Smart Systems, Devices and Technologies*, SMART 2012, 9–14.
7. Pila, Rosarium, Saurabh Rawat, Indar Prakash Singhal. 2017. eZaar, The Smart Container. In *2017 2nd International Conference on Telecommunication and Networks (TEL-NET)*, 1–5.
8. Ramakrishna Murty, M., J.V.R Murthy, P.V.G.D. Prasad Reddy, and Suresh C. Sapathy. 2012. A Survey of Cross-Domain Text Categorization Techniques. In *International Conference on Recent Advances in Information Technology RAIT-2012*, ISM-Dhanabad, 978-1-4577-0697-4/ 12. IEEE Xplorer Proceedings.
9. Dolly, D.R.J., G.J. Bala, and J.D. Peter. 2017. A Hybrid Tactic Model Intended for Video Compression Using Global Affine Motion and Local Free-Form Transformation Parameters. *Arabian Journal for Science and Engineering*.

Accuracy Comparison of Classification Techniques for Mouse Dynamics-Based Biometric CaRP

Sushama Kulkarni and Hanmant Fadewar

Abstract Combining more than one authentication schemes enhances the robustness of a system against cyber-attacks. In this paper, we introduce a novel mouse dynamics-based biometric CaRP (CAPTCHA as gRaphical Password). It combines mouse dynamics-based biometric authentication scheme with knowledge-based authentication scheme. This study primarily focuses on the comparison of classification accuracy of binary decision tree, SVM, and ANN for proposed mouse dynamics-based authentication scheme of CaRP.

Keywords Completely Automated Public Turing Test to Tell Computers and Humans Apart (CAPTCHA) · Web security · Binary decision tree · Support Vector Machines (SVM) · Artificial Neural Network (ANN) · Mouse dynamics · CAPTCHA as gRaphical passwords (CaRP)

1 Introduction

CAPTCHA (Completely Automated Public Turing Test to Tell Computers and Humans Apart) is a reverse turing test to classify human and bot users. CAPTCHA as graphical passwords (CaRP) is a type of graphical password system built on top of CAPTCHA technology. CaRP is used for authenticating a user. Authentication schemes are broadly divided into following spectrums:

- Token-based authentication
- Knowledge-based authentication
- Biometric-based authentication [1].

Existing CaRP (CAPTCHA as gRaphical Passwords) has so far explored the knowledge-based authentication schemes. Knowledge-based authentication schemes require user to answer a secret question.

S. Kulkarni (✉) · H. Fadewar
School of Computational Sciences, S. R. T. M. University, Nanded, India
e-mail: sushama.s.kulkarni@gmail.com

© Springer Nature Singapore Pte Ltd. 2020
K. S. Raju et al. (eds.), *Proceedings of the Third International Conference on Computational Intelligence and Informatics*, Advances in Intelligent Systems and Computing 1090, https://doi.org/10.1007/978-981-15-1480-7_61

Biometric-based authentication techniques such as fingerprints, iris scan, or facial recognition, keyboard pattern recognition, and mouse dynamics pattern recognition are not widely adopted. Yet these techniques provide the highest security as compared to other group of techniques. Major obstacle in adoption of most of the biometric techniques is their expensiveness. But techniques like keyboard pattern recognition and mouse dynamics pattern recognition are not necessarily expensive.

Combination of authentication schemes enhances the security and ensures prevention of imposters and bots. Hence, we combined knowledge-based authentication scheme with biometric-based authentication technique.

2 Literature Review

2.1 Graphical Password Techniques

Graphical password techniques are classified as recognition-based techniques, recall-based techniques, and cued recall-based techniques [2].

Recognition based Techniques. Recognition-based technique requires user to memorize some set of previously selected images and recognize those images from a set of random images at the time of login. Click Text requires user to click a sequence of characters which are randomly arranged in set of 33 characters on a 2D space. It authorizes user if password characters are clicked in specified sequence [2]. ClickAnimal uses sequence of animal names as password. CAPTCHA is generated by arranging 2D animal images on a cluttered background [2]. Here, an alphabet consists of similar animals, e.g., dog, horse, pig, etc. It has smaller password space as compared to Click Text CaRP. Passfaces needs user to choose a set of human faces during registration phase and identify those preselected human faces from a random set of human faces at the time of authentication [3]. Déjà vu scheme uses similar approach like Passfaces, but it uses random art pictures instead of human faces [4]. These art images are difficult to remember for a user; thus, login phase takes longer time.

Recall-based Techniques. Recall-based technique requires user to reproduce or select something which he had produced or selected in registration phase. Draw-A-Secret technique (DAS) asks the user to draw something in a 2D grid canvas and reproduce the same drawing during authentication phase [1]. Similarly, Passdoodle allows user to use freehand drawing without a visible grid as a password [5].

Cued Recall-based Techniques. Cued recall-based techniques provide user an image or set of images from which user has to select click point as the password. It also provides some hints which help users to reproduce their passwords with high accuracy. Greg blonder proposed a method which presents the user with prestored images and asks to tap region by pointing location on image. It is more vulnerable as clicking region is small and simple [6]. Passpoints requires the user to set sequence of clicks on an image as his password and reproduce the same sequence of clicks during authentication phase [7]. Cued click point (CCP) method reduced hotspots and improvised usability of Passpoints. This method asks user to click on one point per image for a sequence of images [8]. It displays the next image on the basis of location of the previous click point.

2.2 Mouse Dynamics-Based Authentication

Mouse dynamics-based authentication systems can be classified as static and continuous authentication schemes. Static approach collects and verifies user's mouse dynamics data at particular times, whereas continuous approach collects and verifies user's mouse dynamics data continuously throughout the session [9]. A risk-based authentication scheme developed by combining keystroke and mouse dynamics biometrics achieved 8.21% of equal error rate [10]. It observed that mouse dynamics-based authentication accuracy degrades with decrease in session length for continuous authentication scheme. A study improved mouse dynamics biometric performance to meet European standard for commercial biometric technology [11]. It used variance reduction approach for performance improvement. It emphasized that the session length to record mouse dynamics data should be kept shorter so that attacker gets very less time window to interact the system. It highlighted that the short session length will produce less mouse dynamics data and result in reduction in performance [11].

3 Proposed Mouse Dynamics-Based Biometric CaRP

3.1 Overview of Experiment

Our experiment was performed by using static verification approach. We chose this approach because it requires short session length, ideally suitable for Web-based environment.

Environment. Major limitation in any mouse dynamics based authentication scheme is uncontrolled environmental variables related to software or hardware. Software dependencies like screen resolution, pointer speed, acceleration settings, mouse polling rate, and hardware variations like type of pointing device could interfere with the authentication results. But ideally we cannot mandate a user to use particular software and hardware to interact with a Web-based authentication scheme. Thus, we preferred to allow volunteers to use any available software settings and hardware devices for using the proposed scheme of mouse dynamics-based authentication.

Procedure and Experimental Task. We have implemented the proposed mouse dynamics-based biometric CaRP using PHP and requested 9 individual users to handle the system. The volunteers, 2 males and 7 females, were all teaching and non-teaching staff of the department of computer science. All 9 volunteers used the mouse dynamics-based biometric CaRP for 10 days and provided their mouse dynamics data for the classification study. On each day of interaction, each of the volunteer used the authentication system for 10 times. Thus, by the end of tenth day, we were able to collect at least 100 mouse data log sessions for each of the volunteer. Effectively, we have collected approximately 900 legitimate user mouse data log sessions for classification study. Volunteers have also acted as imposter for one of the other volunteer in this group. Each volunteer posing an imposter attacked other user account for 5 times on every day. Thus, we were able to collect 450 imposter user mouse data log sessions by the end of 10th day of experiment. All of the volunteers were informed about the purpose of experiment and given instructions for handling the system. Volunteers repeatedly accessed the system for 10 consecutive days. The experimental task was essentially drawing the same pattern on a 9-Dots panel for authentication that was drawn during the registration phase. Mouse dynamics data logs were automatically captured by the system and saved in separate Microsoft Excel file for each login session. Every volunteer was allowed to choose and register any possible desired pattern on 9-Dots panel.

3.2 Modules

The proposed mouse dynamics-based biometric CaRP technique combines two security methods to repel imposters and bots.

9-Dots panel CaRP. We have incorporated a 9-Dots panel which acts as mouse dynamics-based CaRP. It is similar to Draw-A-Secret (DAS) CaRP scheme, but instead of a visible grid, it provides a 9-Dots panel with a background image which is embedded dynamically. Thus, user does not get same static image at background of the 9-Dots panel. User has to draw the same pattern which was registered during sign up phase. Pattern is drawn using a mouse or touchpad by connecting the desired dots on the 9-Dots panel. As most of the mobile devices offer the same system of 9-Dots panel-based password security, this 9-Dots panel provides user-friendly interface and simplicity. Figure 1 shows a screenshot of the 9-Dots panel CaRP.

Mouse dynamics-based biometric authentication. This module records the mouse dynamics event by logging the pixel to pixel movement with the timestamp. During registration and login process, user is requested to draw a secret pattern on the 9-Dots panel and this module logs all mouse events for further biometric authentication process. We have extracted 80 features from this raw mouse dynamics data and further applied three classification techniques, namely

Fig. 1 Screenshot of 9-Dots panel CaRP

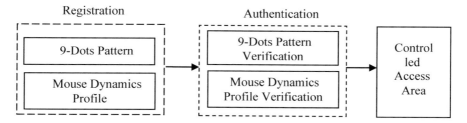

Fig. 2 Workflow of the proposed mouse dynamics-based biometric CaRP

binary decision tree, SVM, and ANN to authenticate a user. It allows maximum of 3 attempts to draw the registered pattern for a user in order to authenticate her.

Most of the mouse dynamics-based authentication schemes require user to allow mouse dynamics monitoring for longer duration [12]. But the proposed mouse dynamics-based biometric CaRP requires user to spend only few seconds to draw the secret pattern. Hence, the proposed scheme is less intrusive as compared to existing mouse dynamics-based authentication schemes. Figure 2 depicts the workflow of the proposed mouse dynamics-based biometric CaRP.

3.3   System Description

Methodology. The mouse dynamics-based user authentication is done by following methodologies on mouse dynamics dataset.

Mouse Dynamics Data Acquisition. The mouse dynamics data acquisition module allows the user to perform the mouse operation. The x-coordinates, y-coordinates, and elapsed times are recorded when user is performing the task.

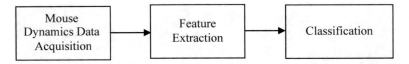

Fig. 3 Workflow of the mouse dynamics-based biometric authentication module

Feature Extraction. Features are extracted to classify the user's unique behavior and accomplish the classification task.

Classification. The classification consists of training phase and testing phase. In training phase, the legitimate user's profile and imposter's profile are built based on the mouse operation. In testing phase, the new mouse features are compared with legitimate user's profile and imposter's profile.

Figure 3 shows the workflow of the mouse dynamics-based biometric authentication module. Mouse dynamics data acquisition phase collects x-coordinate, y-coordinate, and elapsed time, while the user draws 9-Dots pattern. Then, acquired mouse dynamics data is analyzed and features are extracted. In classification submodule, three types of classifiers, namely binary decision tree, binary support vector machine, and artificial neural network, are tested for accuracy. The classifier classifies the given input data as legitimate user or imposter. The classifier performance is evaluated by false acceptance rate, false rejection rate, and identification rate.

3.4 Implementation

As depicted in Fig. 2, a user can gain access to controlled access area after successful completion of registration and authentication process. End user interface for performing registration and authentication process consists of a 9-Dots panel and mouse dynamics data acquisition modules. The end user interface is developed using PHP and JavaScript. As shown in Fig. 3, mouse dynamics-based biometric authentication module is divided into three submodules as follows:

Mouse Dynamics Data Acquisition. 9-Dots panel allows user to draw registered pattern, while mouse dynamics data acquisition module captures the mouse dynamics data generated for it. This data consists of x-coordinate, y-coordinate, and elapsed time of users when drawing secret pattern on 9-Dots panel. It includes both legitimate and imposter user's data.

Feature Extraction. The mouse dynamics data is further transformed into set of features called feature vector. The feature vector is used to represent the user's unique mouse dynamics behavior. Eighty individual features are extracted from the raw mouse dynamics data. Features extracted from raw mouse dynamics data include time, point-to-point distance, length, movement offset, point-to-point velocity, point-to-point acceleration, slope, slope angle, point-to-point direction angle change, distribution of cursor *X*-coordinate position, distribution of cursor *Y*-coordinate position, horizontal velocity, vertical velocity, tangential velocity, tangential acceleration, tangential jerk, curvature, curvature change rate, angular velocity, speed per direction, point-to-point distance per direction, number of inflection points, number of trajectory points, mouse operation frequency, etc. The extracted feature called feature vector is used as input to the classifier.

Classification. The output of the feature extraction is given as the input for the classifier. The classifier evaluates whether the given input is legitimate or imposter user sample. In an attempt to identify a user as an authenticated user or imposter, we have tested following three types of classifiers to distinguish the user as legitimate or imposter user:

Binary Decision Tree. A binary decision tree has two entities, namely decision nodes and leaves. The leaves are the decisions or the final outcomes of learning process. And the decision nodes are where the data is split. We have used ClassificationTree object in MATLAB for identifying the user as authenticated user or an imposter.

Binary Support Vector Machines. Binary support vector machines view a data point as a p-dimensional vector and attempt to identify whether the given point belongs to one of the two classes. We have used svmtrain and svmclassify functions from the Statistical toolbox of MATLAB for classifying a user as authenticated or imposter one.

Artificial Neural Network. Artificial neural network (ANN) is a set of connected input output network. Each connection is associated with a weight. ANN consists of one input layer, one or more intermediate layer, and one output layer. Weight of connection is adjusted to perform learning in ANN. Iterative updating of weight improves the performance of ANN. We have used feed-forward neural network from the Neural Network toolbox of MATLAB to perform classification of authenticated user and an imposter.

General procedure used by above mentioned classifiers to classify a user as legitimate user or imposter is as follows:

Step1: Train the classifier using training feature vector samples of legitimate user and imposter.
Step2: Check whether the feature vector samples from testing dataset match with the trained feature vector samples of legitimate user or imposter.
Step3: Classify the feature vector samples according to testing result as either legitimate user or imposter.

3.5   Evaluation Metrics

The performance of three classifiers used for classification task is evaluated using following metrics.

False Acceptance Rate. The false acceptance rate (FAR) describes the proportion of identification or verification transactions in which an impostor is incorrectly matched to a legitimate user profile stored within a biometric system. FAR reflects the ability of an unauthorized user to access a system.

$$FAR = \frac{Number\ of\ False\ Acceptance}{Number\ of\ imposter\ sample}$$

False Rejection Rate. The false rejection rate (FRR) describes the proportion of identification or verification transactions in which legitimate user is incorrectly rejected from a biometric system.

$$FRR = \frac{Number\ of\ False\ Rejection}{Number\ of\ legitimate\ samples}$$

Identification Rate (IR). The identification rate (IR) describes the proportion of identification transactions with the correct identifier returned compared to the total number of identification transactions [13].

$$Identification\ Rate\ (IR) = \frac{Identification\ transactions\ with\ correct\ identifier\ returned}{Total\ number\ of\ identification\ transactions}$$

3.6 Result of Experiment

Binary Decision Tree. Binary decision tree yielded 80.57% of average accuracy across all the users. Highest accuracy of binary decision tree was 94.73% for User7. Average false acceptance rate (FAR) across all of the users was 8.1%, and average false rejection rate (FRR) across all of the users was 11.93%.

Binary Support Vector Machines. Binary SVM achieved an average of 81.44% accuracy across all of the 9 individual users. Highest accuracy of binary SVM was 100% for User8. Average FAR across all of the users was 5.61%, and average FRR across all of the users was 12.69%.

Artificial Neural Network. ANN produced an average of 81.92% accuracy across all of the 9 individual users. Highest accuracy of ANN was 97.56% for User1. Average FAR across all of the users was 5.47%, and average FRR across all of the users was 15.56%.

As the results suggest the artificial neural network-based classification model is better than binary decision tree and binary SVM. We were not able to witness a great difference in reported accuracy. But combining mouse dynamics-based biometric authentication with knowledge-based authentication scheme definitely provides better security than the existing knowledge-based authentication scheme of

Table 1 Accuracy comparison of applied classification techniques

Classification technique	Binary decision tree			Binary SVM			ANN		
User	FAR (%)	FRR (%)	Accuracy (IR) (%)	FAR (%)	FRR (%)	Accuracy (IR) (%)	FAR (%)	FRR (%)	Accuracy (IR) (%)
User1	7.31	4.87	87.80	0	2.43	97.56	0	2.43	97.56
User2	0	9.37	90.62	0	6.25	93.75	0	3.12	96.87
User3	5.88	32.35	61.76	8.82	38.23	52.94	11.76	23.52	64.70
User4	11.32	15.09	79.24	11.32	7.54	81.13	11.32	26.41	62.26
User5	10.63	17.02	72.34	6.38	14.89	78.72	2.12	21.27	76.59
User6	8.51	2.12	89.36	4.25	14.89	78.72	4.25	4.25	91.48
User7	2.63	2.63	94.73	15.78	0	84.21	15.78	0	84.21
User8	8.69	0	91.30	0	0	100	0	4.34	95.65
User9	18	24	58	4	30	66	4	28	68
Average	8.10	11.93	80.57	5.61	12.69	81.44	5.47	15.56	81.92

CaRP. Intruding the proposed mouse dynamics-based biometric CaRP would require increased operating cost and efforts for an imposter. Table 1 depicts accuracy comparison of applied classification techniques.

4 Security Analysis

Mouse dynamics-based biometric CaRP does not use any characters, text, or audio as challenge; hence, any segmentation, OCR, and dictionary attacks are not possible on it. Mouse dynamics-based biometric CaRP does not fully depend upon any image or video as challenge for authentication; thus, no computer vision attack is applicable to it. As user has to match his previous biometric identity for accessing the proposed scheme, relay attacks are also not applicable to it. Biometric-based authentication makes proposed mouse dynamics-based biometric CaRP repellant to the phishing attack as well. Table 2 shows security analysis for mouse dynamics-based biometric CaRP.

Table 2 Security analysis for mouse dynamics-based biometric CaRP

Type of attack	Applicability on mouse dynamics based biometric CaRP
Segmentation	×
Computer vision techniques	×
OCR	×
Dictionary attack	×
Relay	×
Phishing	×

5 Conclusion

Continuous improvement in intrusion techniques has made the current authentication schemes more vulnerable to attacks. Combination of knowledge-based authentication with biometric-based authentication makes the proposed mouse dynamics-based biometric CaRP more robust. Accuracy comparison of the classification techniques indicates that the artificial neural network-based classification model offers better accuracy than binary decision tree and binary SVM. But we did not observe any big difference in accuracy offered by ANN, binary decision tree, and binary SVM. Any of the applied classification techniques do not produce satisfactory accuracy level. Therefore, we suggest that the mouse dynamics-based biometric authentication as a single means of authentication is not much reliable due to lower accuracy score. It should be noted that the mouse dynamics monitoring was performed for very short time span of few seconds for each user. Notably, we have conducted this experiment in uncontrolled environment so that realistic mouse dynamics dataset could be generated. Hence, we recommend the combination of multiple authentication schemes with mouse dynamics-based biometric authentication for designing future generation of CaRP.

Ethical approval. All procedures performed in studies involving human participants were in accordance with the ethical standards of the institutional research ethics committee and with the 1964 Helsinki declaration and its later amendments or comparable ethical standards.

Informed consent. Informed consent was obtained from all individual participants included in the study.

Acknowledgements We wish to acknowledge volunteer participants of this study who took time to handle the proposed system and provided their permission to collect mouse dynamics data.

References

1. Jermyn, I., A. Mayer, F. Monrose, M. Reiter, K. K., and A.D. Rubin. 1999. The Design and Analysis of Graphical Passwords. In: *Proceedings of the 8th USENIX Security Symposium*.
2. Zhu, B.B., J. Yan, G. Bao, M. Yang, and N. Xu. 2014. Captcha as Graphical Passwords—A New Security Primitive Based on Hard AI Problems. *IEEE Transactions on Information Forensics and Security* 9 (6): 891–904.
3. Real User Corporation. 2005. How the Passface System Works.
4. Dhamija, R., and A. Perrig. 2000. Déjà Vu: A User Study Using Images for Authentication. In *Proceedings of the 9th conference on USENIX Security Symposium*.
5. Varenhorst, C. 2004. *Passdoodles: A Lightweight Authentication Method*. MIT Research Science Institute.
6. Blonder, G. 1996. Graphical Passwords. U.S. Patent 5559961.

7. Wiedenbeck, S., J. Waters, J.C. Birget, A. Brodskiy, and N. Memon. 2005. PassPoints: Design and Longitudinal Evaluation of a Graphical Password System. *International Journal of Human-Computer Studies (Special Issue on HCI Research in Privacy and Security)* 63: 102–127.
8. Chiasson, S., A. Forget, R. Biddle, and P.C. van Oorschot. 2008. Influencing Users Towards Better Passwords: Persuasive Cued Clickpoints. In *Proceedings of HCI*, 121–130. Liverpool, UK: British Computer Society.
9. Jorgensen, Z., and T. Yu. 2011. On Mouse Dynamics As a Behavioral Biometric for Authentication. In *Proceedings of the 6th ACM Symposium on Information, Computer and Communications Security, ASIACCS 2011*, 476–482. New York: ACM.
10. Traore, I., I. Woungang, M.S. Obaidat, Y. Nakkabi, and I. Lai. 2012. Combining Mouse and Keystroke Dynamics Biometrics for Risk-Based Authentication in Web Environments. In *Fourth International Conference on Digital Home*, 138–145. Guangzhou, China: IEEE.
11. Nakkabi, Y., I. Traore, and A.A.E. Ahmed. 2010. Improving Mouse Dynamics Biometric Performance Using Variance Reduction via Extractors With Separate Features. *IEEE Transactions on Systems, Man, and Cybernetics—Part A: Systems and Humans* 40 (6): 1345–1353.
12. Mondal, S., and P. Bours. 2013. Continuous Authentication Using Mouse Dynamics. In *International Conference of the BIOSIG Special Interest Group (BIOSIG)* 1–12, Darmstadt.
13. Monroe, D. 2012. Biometrics Metrics Report v3.0. http://www.usma.edu/ietd/docs/BiometricsMetricsReport.pdf.

Implementation of Children Activity Tracking System Based on Internet of Things

M. Naga Sravani and Samit Kumar Ghosh

Abstract In this article, we discussed IoT-based children activity tracking system through which we can ensure the safety of the child. In this research, we implement the device which will be present with the child consisting of Global System for Mobile communications (GSM) and Global Positioning System (GPS) modules through which we know the text messages about the situation of the child and it also gives the exact location of the child in Google maps. The device consists of different types of sensors like temperature, light dependent resistor (LDR) and micro-electro-mechanical sensor (MEMS). If the temperature and sun light around the child increases more than the limit, then it automatically sends the text message to the parents and the buzzer will also be activated. Buzzer is used to indicate the bystanders if the child is in any danger conditions. MEMS accelerometer sensors are used to detect the tilt varies directions. The voice module is used to record the voice and sends the information automatically when the child falls down. Even it is also possible to track the child location at anytime, anywhere by sending one text message to the GSM module present with the child. The device is portable, and the child can wear it for secure purpose.

Keywords Microcontroller lpc2148 · Child location · Playback voice module · Monitoring of children activities · Mobile communications

M. Naga Sravani (✉) · S. K. Ghosh
Department of Electronics and Communication Engineering,
MLR Institute of Technology, Hyderabad 500043, India
e-mail: sravani2795@gmail.com

S. K. Ghosh
e-mail: samitnitrkl@gmail.com

713

1 Introduction

Nowadays human's life is changed with the help of the Internet and its connectivity with anyone, anytime, and anywhere due to the advancement of technology. Today's Internet is now expanding toward the Internet of Things (IoT) which is an advanced automation and analytics system which exploits networking, sensing, big data, and artificial intelligence technology to deliver complete systems for a product or service. At present, the sensors, processors, transmitters, receivers, etc., are now available in very cheap rate due to the rapid growth of technology. IoT systems have applications across industries due to the unique flexibility and ability which is to be suitable in any environment [1–3]. Jain et al. [4] have proposed a technique based on Global Positioning System (GPS) technology. In this application, it describes the location tacking and sharing using GPS by using Internet connectivity through web server. Anderson et al. [5] introduced a transportation information system by using GSM and GPS. In this work, it examines the framework and early testing and the advancement suggestions for a scope of urban and rustic conditions where transportation is rare or wasteful and where a focal specialist or foundation is not in a situation to give strong data assets to clients. Fleischer et al. [6] describes vehicle tracking and alert system based on GPS/GSM. This framework permits between city transport organizations to follow their vehicles continuously and gives security from furnished theft and mishap events. The proposed system is designed aim of safety to the child with the help of a tracking device and the parents can track or know the information of the child that where the child is present and what are the conditions around him so that they can guard their child from their present location. In present days, a large portion of the wearable available are focused on providing the location, activity, etc., in between the child and the parents for the source of data transfer via Wi-Fi and Bluetooth. Consequently, it is proposed to utilize SMS as the method of correspondence between the parent and child's wearable device, as this has less chances of failing compared to Wi-Fi and Bluetooth [7–9]. The work is based on the microcontroller-based platform. The functions of sending and receiving calls, SMS, and associating with the Internet are given by the Arduino GSM shield using the GSM network. Likewise, extra modules utilized which will give the present location of the child to the parents via SMS. In this work, it mainly uses the sensors like temperature, LDR, and MEMS accelerometer sensors which work according to the conditions given by the microcontroller. All the information from these sensors is given to the microcontroller from where the user gets the text messages through GSM module, and also it uses APR33A3 voice module to record the voices of the child when he falls down. Buzzer is also present in this device which will automatically activate when the child's temperature or sunlight around him increases so that the bystanders can help the child before parents arrive [10–12]. The overview of the project is that for a child safety by using this wearable device the process goes on in this way; that is, when the temperature and UV radiations around the child increases more than the given limit by the parents through instructions to the controller, then they immediately get a text message that their child is in sunlight or the temperature around him is high and also the buzzer that is with the child automatically activates so that the bystanders can help the child before parents arrive. Also when the child falls down, parents get the text message and also the location will be updated to the cloud so that if the parents want to know the location of the child, they can check it in Google maps by clicking the values or the link that they received as a text message; also at any time, if they want to track their child, they simply can send a text to the GPS module that is present with the child with a keyword like *T, then they can receive the location of the child. By using all these modules, safety of child is managed by the parents [13–15].

2 System Design and Architecture

In this section, we explain the architecture and the design methodologies for the device what we proposed in our work. The proposed block overview of the device is shown in Fig. 1.

2.1 System Overview

In the proposed design, we are utilizing a few sensors, like temperature sensor and LDR sensor, to measure the temperature and light intensity. Additionally, we are utilizing MEMS sensor to gather the information from the environment, and voice module is utilized to record the voice occurs surrounding places. These all sensors are utilized to view child condition. By utilizing proposed design, we are defeated the present issue in the existing system; in this design, if any of the sensors identify

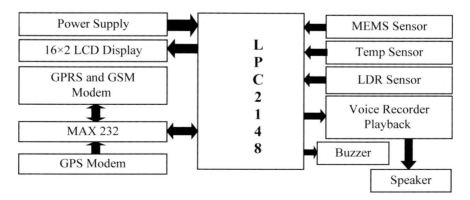

Fig. 1 Blocks overview of the device

irregular condition, then microcontroller will send instant message to the parents by utilizing GSM module likewise device will track the present area of the child by utilizing GPS and send to the parents. Here, we are utilizing buzzer to locate the unusual state of kid to the onlookers in order to help the kid in any danger. On the off chance that a parent needs to track their kid, they basically can send a content watchword like *T to the GSM module that is available with the kid, then they can get the present area of the child.

2.2 Wearable IoT device

The wearable loT device entrusted with gaining different information from all the diverse modules associated with it. It gets the information from its different physically associated modules, dissects this information, and refines the information in a more client reasonable organization to the different accessible user interfaces. The user, therefore, can conveniently view the information on their cell phone. In this system, we are using temperature sensor, LDR sensor, MEMS sensor, Voice IC, LCD, and buzzer; all are connected to the ports pins of LPC2148 microcontroller and GSM. GPS modules are connected to the serial ports of microcontroller, that is, UART0 and UART1 as shown in Fig. 2. The microcontroller will continuously read the sensors data and send to the LCD; also it will compare with predefined values; if the sensor data is greater than predefined values, then microcontroller will generate signal to the GSM module to send the alert message to the registered mobile number. If child falls down that detected by the MEMS sensor, then microcontroller will send alert message along with its current location of child; the location of child is read from GPS module. Also microcontroller will activate the buzzer and voice module. The voices of the child and voices of the surrounding people will be recorded when the child falls down, and later we can hear recorded voice to know what was happened at that moment. Here, we are using IoT technology so that microcontroller will update child location into IoT. Another feature of this system is if parents want to know the child current location, then simply they can send a text command to the GSM module with a keyword like *T, then that command will be received by the GSM module and given to the microcontroller, then microcontroller will read the location from GPS module and send to the particular mobile number.

In this system, we are using temperature sensor, LDR sensor, MEMS sensor, Voice IC, LCD, and buzzer; all are connected to the ports pins of LPC2148 microcontroller and GSM; GPS modules are connected to the serial ports of microcontroller that is UART0 and UART1 as shown in Fig. 3.

The microcontroller will continuously read the sensors data and send to the LCD display; also it will compare with predefined values; if the sensor data is greater than predefined values, then microcontroller will generate signal to the GSM module to send the alert message to the registered mobile number. If child falls down that detected by the MEMS sensor, then microcontroller will send alert

Fig. 2 Proposed wearable IoT device

message along with its current location of child; the location of child is read from GPS module. Also microcontroller will activate the buzzer and voice module. The voices of the child and voices of the surrounding people will be recorded when the child falls down, and later we can hear recorded voice to know what was happened at that moment. Here, we are using IoT technology so that microcontroller will update child location into IoT. Another feature of this system is if parents wants to know the child current location, then simply they can send a text command to the GSM module with a keyword like *T, then that command will be received by the GSM module and given to the microcontroller, then microcontroller will read the location from GPS module and send to the particular mobile number. The hardware components used in the proposed system are shown in Fig. 2. The system consists of the following components:

(a) **LPC2148 Microcontroller**: It is one of the widely used ICs which belongs to the Advanced RISC Machine (ARM 7) family. It requires power supply, crystal oscillator, reset circuits, RTC crystal oscillator, and UART to work properly.
(b) **GPS/GSM Module**: GPS sensor is used to determine the real-time location of the child. The location is received by the GPS module from the different satellites which is present in the NAVSTAR GPS system. The latitude and

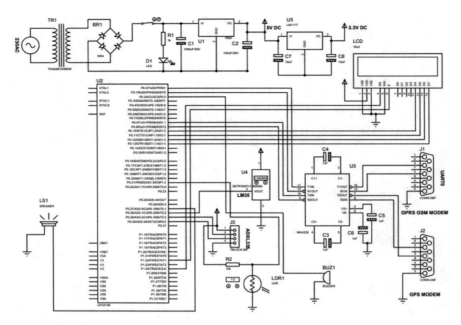

Fig. 3 Schematic diagram of IoT-based children activity tracking system

longitude coordinates are stored in the GPS system, and when the child falls down automatically through the GSM system, a text message goes to the parent as child has fallen at latitude and longitude, and also anytime, if the parent wants to know the location of the child where he is, then they can send one text message to the GSM module with a keyword as "*T," then they can get the reply with latitude and longitude of the child like 17.5952308, N, 78.441411, E. Then we will have a link provided to the GPS system as bosembedded.com/VTRACK1/read.asp, so by clicking on that we can see the location of the child in Google maps.

(c) **Buzzer**: In this device, we use buzzer which helps the child when he/she is in danger like if the temperature or the sunlight around the child increases; this buzzer automatically activates so that even the bystanders can approach the child and help him to get out of that danger. Also this can be used in another way like if the child separates from his/her parents, then they can locate their child by sounding the loud alarm on the device.

(d) **Voice Module**: In this wearable device, we use the apr33a3 voice module. The main use of this voice module is that if the child falls down, then this voice module automatically starts recording the voices of the child or the surrounding voices so that later on by hearing to those voices, we can know what was happened to the child at that moment. So this will be one added source to the safety of the child.

(e) **Temperature Sensor**: In order to measure the surrounding temperature of the child, this sensor called LM35 is used in this device. The output voltage of this sensor is linearly proportional to the centigrade temperature. This sensor mainly has an advantage over linear temperature sensors calibrated in Kelvin; it will be easy for the user as there is no need of subtracting a large constant voltage from the output in order to get the convenient Centigrade scaling. This sensor does not require any external calibration to provide accuracies of −55–150 °C room temperature. Single or plus/minus power supplies are used for this device. When the temperature around the child increases more than the given limit, then it automatically sends a text message through GSM module to the parent.

(f) **LDR Sensor**: In this design, light dependent resistor called LDR sensor is used to sense the child when he/she is in sunlight, as it is danger to the child to stay more time in the light we are using this sensor. In this, whenever the child is in sunlight, it automatically sends an alert text message to the parents and along with that the buzzer also activates, so that before the parents arrive, the bystanders will approach the child by hearing to that sound and may help the child.

Keil4 software is mainly used to activate ARM7 (lLPC2148) microcontroller according to the input received by it, and the code is written in "Embedded C." In this work, coding is written for GPS and GSM which is interfaced with ARM7 board by using flash magic software at the transmitter end. The interfaced modules generate proper output at the receiving side as per the code embedded in the controller. It automatically senses the current position of the child and sends its data to microcontroller when supply is provided to GPS board. The latitude and longitude values of the child's current position are received by the GSM module, and it sends to the receivers.

3 Results

In this section, we execute the experimental test and determine the various components of the proposed wearable device. The implementation of this work is basically focused on tracking of a child location and also its position. The location is tracked by the GPS module and sends location to parents using GPRS module.

A. **Temperature and LDR Sensor**: By testing the wearable device, we can get the reply as shown below. Whenever the temperature or the sunlight around the child increases, then automatically we get the text message to the provided mobile numbers as shown below.

Room Temp High

Your Child is in Sun Light

B. **GPS Location Sensor**: As described above, this GPS sensor gives the real-time location of the child. When the child falls down, then automatically parent gets the text message that child has fallen down along with the updated location of the child as latitude and longitude. Also when we give a keyword to the wearable device as "*T," we get the updated location of the child as shown below.

C. **Location on Google Maps**: We can see the location of the child in Google maps as shown below.

4 Conclusions

The device acts as a smart IoT device. It provides the information to the parents like real-time location, temperature surrounding the location, ultraviolet radiation index, SOS light neighboring that area along with a distress alarm buzzer, and also alert bystanders in acting to rescue or comfort the child. The main feature of this application is to track the child's location in a simple and cost-effective way with the help of GSM and SMS. It can be enhanced better way in future.

References

1. Dorsemaine, B., P. Gaulier, P. Wary, N. Kheir, and P. Urien. 2015. Internet of Things: A Definition and Taxonomy. In *2015 9th International Conference on Next Generation Mobile Applications, Services and Technologies*, 72–77, Cambridge.
2. Ghosh, Samit Kumar, Tapan Kumar Dey. Internet of Things: Challanges and its Future Perspective. *International Journal of Computer Engineering and Applications* XI(IX): 1–6. ISSN 2321-3469.
3. Zheng, Jun. D. Simplot-Ryl, C. Bisdikian, H.T. Mouftah. 2011. The Internet of Things. In *IEEE Communications Magazine* 49(ll): 30–31, Nov 2011. https://doi.org/10.1109/mcom.2011.606970.
4. Jain, S., A. Chandra, and M.A. Qadeer. 2011. GPS Locator: An Application for Location Tracking and Sharing Using GPS for Java Enabled Handhelds. In *International Conference on (CICN) Computational Intelligence and Communication Networks*, 406–410, Gwalior, India. https://doi.org/10.1109/cicn.2011.85.
5. Anderson, Ruth E., et al. 2009. Building a Transportation Information System Using Only GPS and Basic SMS Infrastructure. In *2009 International Conference on Information and Communication Technologies and Development (ICTD)*, IEEE.
6. Fleischer, P.B., A.Y. Nelson, R.A. Sowah, A. Bremang. 2012. Design and Development of GPS/GSM Based Vehicle Tracking and Alert System for Commercial Inter-City Buses. In *2012 IEEE 4th International Conference on Adaptive Science & Technology (ICAST)*, 1–6, 25–27 Oct 2012.
7. Moustafa, H., H. Kenn, K. Sayrafian, W. Scanlon, and Y. Zhang. 2015. Mobile Wearable Communications. *IEEE Wireless Communications* 22 (1): l0–ll.
8. Braam, K., Tsung-Ching Huang, Chin-Hui Chen, E. Montgomery, S. Vo, and R. Beausoleil. 2015. Wristband Vital: A wearable Multi-Sensor Microsystem for Real-Time Assistance Via Low-power Bluetooth Link. In *2015 IEEE 2nd World Forwn on Internet of Things (WF-IoT)*, 87–9l, Milan. 10.1109/WF-IoT.2015.7389032.
9. Digital Parenting: The Best Wearables and New Smart Baby Monitors. 2015. *The Latest Smart Baby Monitors and Connected Tech for Your Peace of Mind, 'Tech.* Rep., 2015.
10. Prince, N.N. 2013. Design and Implementation of Microcontroller Based Short Message Service Control System. In *2013 8th International Conference for Internet Technology and Secured Transactions (ICITST)*, 494–499, London.
11. Nasrin, S., and P. Radcliffe. 2014. Novel Protocol Enables DIY Home Automation. In *2014 Australasian Telecommunication Networks and Applications Conference (ATNAC)*, 212–216, Southbank, VIC.
12. Silva, F.A. 2014. Industrial Wireless Sensor Networks: Applications, Protocols, and Standards (Book News). *IEEE Industrial Electronics Magazine* 8 (4): 67–68.
13. WiFi, and WiMAX—Break Through in Wireless Access Technologies. In *lET International Conference on Wireless, Mobile and Multimedia Networks*, 141–145, Beijing.
14. Ahanathapillai, Vijayalakshmi, James D. Amor, Zoe Goodwin, and Christopher J. James. 2015. Preliminary Study on Activity Monitoring Using an Android Smart-Watch. *Healthcare Technology Letters* 2 (1): 34–39.
15. Kim, Ki Joon, and Dong-Hee Shin. 2015. An Acceptance Model for Smart Watches: Implications for the Adoption of Future Wearable Technology. *Internet Research* 25 (4): 527–541. https://doi.org/10.1108/IntR-05-2014-0126.

Data Mining: Min–Max Normalization Based Data Perturbation Technique for Privacy Preservation

Ajmeera Kiran and D. Vasumathi

Abstract Data mining system deals with huge volume of information which may include personal and sensitive information about the individuals such as bank credential details, financial records, health-related information, etc. Data mining process utilizes this information for analyzing purpose but privacy preservation of such sensitive data is very much crucial in data mining process in ordered to prevent the privacy about the individuals. In recent years, privacy preservation is an ongoing research topic because of the high availability of personal data which consist of private and sensitive information about the individuals. Data perturbation technique is a well-known data modification technique to preserve the privacy of sensitive values and achieves accurate data mining results. In data perturbation method, original data is perturbed (modified) before the data mining process begins. In the existing method, data modification is takes place by adding noise (Gaussian) to the original data. In this method, loss of data loss is little high, and to overcome such issue, proposed method is established. In this paper, min–max normalization-based data transformation method is used to protect the sensitive information in a dataset as well as to achieve good data mining results. The proposed method is applied on the adult dataset and the accuracy of the results is compared with Naïve Bayes classification algorithm and J48 decision tree algorithm with minimum information loss by having high data utilization. The performance of the proposed method is examined with two major considerations like maintaining the accuracy of the data mining application along with privacy preservation of original data.

Keywords Data mining · Data privacy · Data utility · Privacy preservation · Data transformation normalization · Min–max normalization · Naïve bayes classification · J48 decision tree algorithm

A. Kiran (✉) · D. Vasumathi
Department of CSE, JNTUHCEH, Hyderabad, Telangana 500085, India
e-mail: Kiranphd.jntuh@gmail.com

D. Vasumathi
e-mail: rochan44@gmail.com

© Springer Nature Singapore Pte Ltd. 2020
K. S. Raju et al. (eds.), *Proceedings of the Third International Conference on Computational Intelligence and Informatics*, Advances in Intelligent Systems and Computing 1090, https://doi.org/10.1007/978-981-15-1480-7_66

1 Introduction

In recent years, there are many developments in the field of information technology to store, collect, and analyze the terabytes of data in a limited time period [1]. Most of these data could be personalized sensitive data such as medical data records; criminal-based data, customers shopping data, credit card information or banking data, etc. [2]. Consequently, many business and public sectors shares such personal information which will benefit their business growths and also many companies disclose their confidential data for the purpose of data analysis activities [3]. But these data cannot be shared to everyone because individual data privacy may violate. Therefore, in order to preserve privacy without disclosing the private data and also obtaining good mining results is a challenging issue in the data mining process. To overcome such issue and also to obtain an accurate data mining results, privacy-preserving data mining comes into the existence.

Different privacy-preserving techniques have been invented and used in data mining [4]. In the existing method, data modification takes place by adding noise (Gaussian) to the original data. In this method, loss of data is little high, and to overcome this issue, proposed method is established [5] [6]. In this research work, a new data distortion technique is established using min–max normalization to preserving sensitive data. In this technique, numerical attributes are distorted in privacy preserving of sensitive individual data values. In this method, data values are distorted before the data mining application starts. Generally, Min-max normalization technique behaves like a preprocessing step in the field of data mining for the distortion of original data into a given range. The proposed Methodology used two well-known methods like Naïve Bayes classification and J48 decision tree methods for evaluating the privacy and accuracy levels of the data mining application. Original dataset and the perturbed dataset are compared with both the algorithms.

The remaining sections of this paper are organized as follows: Sect. 2 provides an overview of existing literature work which is carried out in the field of data transformation techniques with normalization concepts. Section 3 is the expansion of the proposed algorithm and also elaborates the min–max normalization technique. Section 4 explains the implementation procedure. Finally, Sect. 5 explains the conclusion of the research work.

2 Literature Study

There have been many literature studies are done in the field of privacy preserving data mining (PPDM) [7, 8]. These studies can be divided into two major categories. In the first type of category, methods primarily modify the different data mining algorithms so that all data mining operations are carried out without knowing the exact values of data. In the second type of category, methods are modifying the values of the data sets to preserve the privacy of data values. Several, research has been done in data perturbation and data distortion are as follows:

C. K. Liew, U. J. Choi, and C. J. (1985) Introduced first data distortion method by using probability distribution with three major steps: Initially, identification of underlying density function, secondly, generation of distorted series with the help of density function and finally, mapping of the original series onto distorted series.

Sweeney [9] has described a novel approach on the k-anonymity model with a consideration of a problem that data owner can share the individual information without violating the privacy of an individual. The main goal of the research is to protect individual sensitive information by using data suppression and data generalization methods.

Chen and Liu [10] have discovered a novel technique based on rotation data perturbation. The proposed technique has shown that many classifier techniques have zero loss of accuracy. The experimental results are proved that rotation-based technique gives good quality of privacy is improved without losing accuracy.

Wang et al. [11] have introduced a new technique for data mining called nonnegative matrix factorization(NNMF) which is work combined with distortion processing. Proposed method gives a better solution for the data mining issues by using discriminate functions.

R. Agrawal and R. Srikant (2007) elaborated an efficient technique for decision tree classifier by using the additive data perturbation technique. Random noise is added to every data element in the dataset. Random noise generated with the help of Gaussian distribution function.

Saif M. A. Kabir, Amr M. Youssef, and Ahmed K. Elhakeem (2007) have made an investigation to use nonnegative matrix factorization for data distortion to provide privacy in data mining. The experimental results shown that the proposed method is a very efficient tool in providing privacy preservation in data mining.

Liu et al. [12] have presented a data distortion method using wavelet transformation for privacy preservation of sensitive data in data mining. Wavelet-based privacy-preserving technique maintains good data mining and data privacy utilities. The experimental results proved that proposed method maintains the statistical properties of original data remain same by maximizing the data utility.

Zhenmin Lin, Jie Wang, Lian Liu, and Jun Zhang (2009) have illustrated a technique for data perturbation. In this method, the original data matrix is vertically partitioned into different submatrices by different owners of data. For each data holders can perturb their individual sensitive data by choosing a rotation matrix with independently or randomly. The experimental results have shown that the proposed method maintained the privacy without sacrificing the accuracy.

Peng et al. [13] have proposed four kinds of data distortion methods got preserving privacy in data mining. The main goal of the research is conducting distortion on submatrices of original data with several methods. The experimental results are proved that the proposed framework is very efficient in preserving privacy of individual sensitive information.

Jain and Bhandare [14] have made a research on data transformation technique with min–max normalization method with different real-life datasets and results shown that transformation-based approach is effective to protect individual sensitive information.

Manikandan et al. [15] have defined clustering-based data transformation technique for preserving the privacy of the sensitive data and achieved good results for accuracy.

Manikandan et al. [16] proposed a data transformation technique using normalization for data achieved good accuracy and also enhanced the performance of data mining algorithms.

3 Proposed Method

In this paper, the proposed framework used to protect the sensitive attributes in a dataset for privacy-preserving data mining using data transformation technique called min–mx normalization [17].

3.1 Min Max Normalization

Min–max normalization is a well-used data transformation technique to preserve the sensitive attributes in a dataset about the individuals [18], [19]. In this method, original dataset values are normalized by using min–max normalization function by considering minimum and maximum values in a dataset. Min–max normalization technique is performing a linear transformation on original dataset. It is mainly useful for classification purpose and it has utilized in many applications like artificial intelligence, clustering, nearest neighbor classification, neural networks, etc. [20].

The main objective of the proposed method is to normalize the original dataset **D** into a preserved form of dataset **D'** that fulfills the privacy requirements with minimum loss information with high data privacy. The proposed method mainly concentrates on data transformation using min–max normalization for modifying the original data values [21, 22].

Each attribute in a dataset is normalized by mounting its values so that they fall within the minimum user-specified range such as 0.0–1.0. For mapping a value, V_i of attribute A from the range of [min_A, max_A] to [new_min_A, new_max_A] is computed by following function

$$V_i' = \frac{V_i \min_A}{\max_A - \min_A} \left(\text{new}\_\max_A \_\text{new}\_\min_A \right) + \text{new}\_\min_A \qquad (1)$$

where V_i is the newly calculated value in the user-specified range. By using the min-max normalization method (Eq. 1), the relationships among the original values preserved.

3.2 Proposed Min–Max Normalization Algorithm

Procedure: Transformation of data using min–max normalization method.
Input: Original data set D, sensitive attributes S
Intermediate Output: Modified (Perturbed) data set D'
Output: Classification results R and R' of datasets D and D'
Step 1: Given input dataset D with tuple size n and extract sensitive attributes [S]
Step 2: Apply Min-Max Normalization method on Sensitive Attributes [S] $$V_i' = \frac{V_i \min_A}{\max_A - \min_A}(\text{new}\_\max_A \_\text{new\_min}_A) + \text{new\_min}_A$$
Step 3: Create perturbed dataset D' by replacing sensitive attributes in original dataset D.
Step 4: Apply Naïve Bayes classification algorithm on original dataset D having sensitive attributes S.
Step 5: Apply Naïve Bayes classification algorithm on perturbed dataset D' having perturbed sensitive attributes S.
Step 6: Compare and analyze the results of Steps 4 and Step 5 for analyzing accuracy of proposed method with Naïve Bayes classification algorithm
Step 7: Apply J48 decision tree algorithm on original dataset D having sensitive attributes S
Step 8: Apply J48 decision tree algorithm on perturbed dataset D' having perturbed sensitive attributes S.
Step 9: Compare and analyze the results of Steps 7 and Step 8 for analyzing accuracy of proposed method with J48 decision tree algorithm.

3.3 Proposed Min–Max Normalization Architecture

The primary goal of the proposed research work is to preserve the sensitive information available in the dataset. To accomplish such task, proposed method implemented data transformation technique using min–max normalization is to modify the original dataset D into perturbed dataset D' without neglecting privacy requirements.

In data mining, there are various Data modification techniques in privacy preservation are available for preserving sensitive attribute about the individuals which are mainly concentrates on two facts: primarily focus on privacy guarantee level and Secondly, high data utility level. The primary goal of each data modification technique is to obtain data privacy at higher level by providing privacy to the sensitive information along with higher data utilization (Fig. 1).

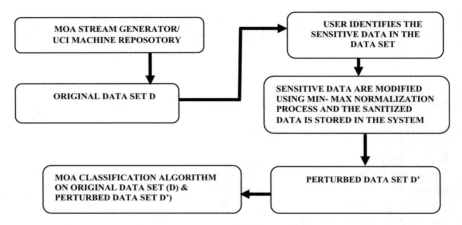

Fig. 1 Illustrates the proposed architecture which describes collection of data from, repository and identifying sensitive attributes applying min–max transformation ion sensitive data apply classification algorithms on perturbed and original data

4 Implementation

4.1 Datasets

In UCI machine repository [23], there are many number of datasets are available in data mining field to do research like bank dataset, cancer dataset, cover-type dataset, retail dataset, finance dataset, and medical dataset. But, in our proposed methodology Adult dataset is utilized for analysis purpose (Table 1).

4.2 Pre-processing of Data

In data mining, the initial step is pre-processing. The main objective of the pre-processing step is to eliminate irrelevant or null values data from the dataset. Some datasets may contain unnecessary data values in the dataset like unspecified attributes values and delimiters. By performing preprocessing, the performance of every data mining algorithm will improve.

Table 1 Adult dataset description	

Dataset	Description
Adult dataset	Total instances: 48,842
	Attributes: 6

4.3   *Experimental Results*

As part of implementation step, proposed method implemented data transformation technique using min–max normalization [19] to modify the original dataset D into perturbed dataset D' without neglecting privacy requirements and both the datasets D and D' are compared after implementing two well-known data mining algorithms like Naïve Bayes (Naïve Bayes classifier algorithm) and J48 (decision tree analysis algorithm.) on perturbed as well as original data sets.

Table 2 llustrates comparison values of both algorithms NB (Naïve Bayes classifier algorithm) and J48 (decision tree analysis algorithm.) for single column age and education attributes for original values and modified values and their efficiencies. In experimental step, the proposed algorithm is evaluated for an adult dataset. Proposed work has effectively overcome the limitations of existing work.

Table 2 shows the comparison results of Naïve Bayes, J48 algorithms on the original dataset and perturbed dataset on two numerical columns: Age and Education. Table 2 also displays different privacy measurement values like kappa statistics (KS), root means square error value (RMSE), mean absolute error (MAE), the time taken to complete the task, relative absolute error (RAE), root relative squared error (RRSE), etc.

From Table 3, accuracy of the proposed method is nearly same for Naïve Bayes algorithm after perturbation of original dataset into modified dataset for the both attributes age and education. That is Naïve Bayes algorithm on original dataset for correctly classified instances is 83.25% accuracy and for incorrectly classified instances it was 16.74%. The Naïve Bayes algorithm on perturbed dataset for correctly classified instances was 83.15% accuracy for age attribute and 83.04% for the education attribute as well as for incorrectly classified instances was 16.84% for age attribute and 16.95% for the education attribute. Finally, results from the both algorithms proven that data transformation technique using min–max normalizations can be a good alternative solution for the data owner where security of the data is very much important and accuracy of the proposed framework also maintained with little information loss. The privacy of the original data can be preserved by little accuracy loss.

Correspondingly, after applying the J48 algorithm on the original adult dataset.

From Table 4, accuracy of the proposed method is nearly same for J48 algorithm after perturbation of original dataset into modified dataset for the both attributes age and education. That is J48 algorithm on original dataset for correctly classified instances is 86.07% accuracy and for incorrectly classified instances it was 13.92%. The J48 algorithm on perturbed dataset for correctly classified instances was 85.99% accuracy for age attribute and 85.92% for the education attribute as well as for incorrectly classified instances was 14.00% for age attribute and 14.07% for the education attribute. Finally, results from the both algorithms proven that data transformation technique using min–max normalizations can be a good alternative solution for the data owner where security of the data is very much important and accuracy of the proposed framework also maintained with little information loss.

Table 2 Accuracy comparison results of NB and J48 using min–max normalization method for age and education attributes

Dataset 1: Adult dataset

	Age				Education			
	NB		J48		NB		J48	
	Original	Perturbed	Original	Perturbed	Original	Perturbed	Original	Perturbed
Correctly classified instances (%)	83.25	83.15	86.07	85.99	83.25	83.04	86.07	85.92
Incorrectly classified instances (%)	16.74	16.84	13.92	14.00	16.74	16.95	13.92	14.07
Kappa statistics	0.49	0.49	0.58	0.58	0.49	0.48	0.58	0.58
Mean absolute error	0.1746	0.176	0.1956	0.1945	0.1746	0.1772	0.1956	0.1957
Root mean squared error	0.3741	0.375	0.3201	0.3211	0.3741	0.3755	0.3201	0.3211
Relative absolute error	0.4795	0.4859	0.5371	0.5343	0.4795	0.4868	0.5371	0.5374
Root relative squared error	0.8768	0.8798	0.7502	0.7525	0.8768	0.8800	0.7502	0.7526
Time consumed (sec)	0.22	0.47	6.46	9.02	0.22	0.44	6.46	11.72

Table 3 Comparison of Naive Bayes algorithm for two numeric columns using min–max normalization

NB	Original (%)	AGE	Education
		Perturbed (%)	Perturbed (%)
Correctly classified instances	83.25	83.15	83.04
Incorrectly classified instances	16.74	16.84	16.95

Table 4 Comparison of J48 algorithm for two numeric columns using min–max normalization

J48	Original (%)	AGE	Education
		Perturbed (%)	Perturbed (%)
Correctly classified instances	86.07	85.99	85.92
Incorrectly classified instances	13.92	14.00	14.07

Table 5 Comparison of Naive Bayes algorithm on proposed method with existing method

Adult dataset	Existing method NB value using Gaussian noise (%)	Proposed NB value using min–max normalization (%)
Single attribute (age)	83.08	83.15
Single attribute (education)	82.47	83.04
Two attributes (age and education)	81.93	82.93

From Table 5, comparison of the accuracy with NB on Adult dataset for single column age is 83.08% in existing and for the proposed method is 83.15%. as well as for the education column the accuracy is 82.47% in existing method and for the proposed method is 83.04% and also for two columns (age and education) at a time the accuracy is 81.93% in existing method and for the proposed method is 82.93%. So privacy of the original dataset is preserved with minimum information loss.

From Fig. 2, comparison of the accuracy with NB on Adult dataset for single attribute age for existing method and proposed method is represented and also for the education attribute also represented.

From Table 6, comparison of the accuracy with J48 on Adult dataset for single column age is 85.90% in existing and for the proposed method is 85.99%. as well as for the education column the accuracy is 85.79% in existing method and for the proposed method is 85.92% and also for two columns (age and education) at a time the accuracy is 85.70% in existing method and for the proposed method is 85.82%. So privacy of the original dataset is preserved with minimum information loss.

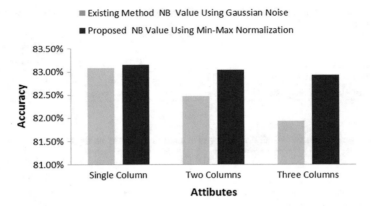

Fig. 2 Comparison of Naive Bayes algorithm on proposed method with existing method

Table 6 Comparison of J48 algorithm on proposed method with existing method

Adult dataset	Existing method J48 value using Gaussian noise (%)	Proposed J48 value using min–max normalization (%)
Single attribute (age)	85.90	85.99
Single attribute (education)	85.79	85.92
Two attributes (age and education)	85.70	85.82

From Fig. 3, comparison of the accuracy with J48 on Adult dataset for single attribute age for existing method and proposed method is represented and also for the education attribute also represented.

5 Conclusion and Future Work

In this paper, data is perturbed by using min–max normalization which is a data transformation technique for maximizing the data privacy. Proposed method modifies original data into min–max based perturbed data to preserve the sensitive personalized data. The experimental work is performed on the adult dataset with 48842 instances of six attributes. Proposed have succeeded in minimizing information loss while achieving both data accuracy and privacy. The proposed framework to protect sensitive privacy in data mining mainly concentrates on providing a higher level of privacy to the individual sensitive or private data;

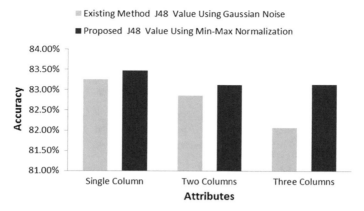

Fig. 3 Comparison of J48 algorithm on proposed method with existing method

proposed model takes input as sensitive attributes and produces output as min-max based perturbed dataset. The proposed method makes use of two well-known algorithms, namely naïve Bayes classification algorithm and J48 decision tree algorithm for analyzing the accuracy and privacy levels of the data mining model. Original dataset and min–max-based normalized data are compared with both the algorithms. The experimental results have shown that the accuracy of the modified dataset is minimum than the original dataset with minimum information loss but achieved higher level of data privacy to the individual sensitive data values. This research work can be extended further by applying more transformation technique like decimal scaling and Z-Score normalization to maximize the data accuracy with no data loss.

References

1. Chen, M., J. Han, and P. Yu. 1996. Data Mining: An Overview from a Database Prospective. *IEEE Transaction on Knowledge and Data Engineering* 8 (6): 866–883.
2. Chen, C.L.P., and C.Y. Zhang. 2014. Data Intensive applications, challenges, Techniques and technologies: A survey on Big Data. *Information Sciences* 275: 314–347.
3. Atallah, M., A. Elmagarmid, M. Ibrahim, E. Bertino, and V. Verykios. 1999. Disclosure Limitation Of Sensitive Rules. In *Workshop on Knowledge and Data Engineering Exchange*.
4. Agrawal, R., and R. Srikant. 2000. Privacy-preserving Data Mining. In *Proceeding of the ACM SIGMOD Conference on Management of Data*, 439–450, Dallas, Texas, U.S.A.
5. Mahalle, V.S., Pankaj Jogi, Shubham Purankar, Samiksha Pinge, and Urvashi Ingale. 2017. Data Privacy Preserving using Perturbation Technique. *Asian Journal of Convergence in Technology* III(III). ISSN No.:2350–1146, I.F-2.71.
6. Patel, Nikunj Kumar. 2015. *Data Mining: Privacy Preservation Using Perturbation Technique*.
7. Senosi, Aobakwe, and George Sibiya. 2017. *Classification and Evaluation of Privacy Preserving Data Mining: A Review*. IEEE.

8. Vaghashia, Hina, and Amit Ganatra. 2015. A Survey: Privacy Preservation Techniques in Data Mining. In *Proceedings of International Journal of Computer Applications (0975 – 8887)* 119(4).
9. Sweeney, L. 2012. k-anonymity: A model for protecting privacy. *International Journal of Uncertainty, Fuzziness and Knowledge-Based Systems* 10 (05): 557–570.
10. Chen, K., and Liu, L. 2005. A random rotation perturbation approach to privacy preserving data classification. In *Proceedings of International Conference on Data Mining (ICDM)*, IEEE, Houston, TX.
11. Wang, J., Zhong, W., and Zhang, J. 2006. NNMF-based factorization techniques for high-accuracy privacy protection on non-negative-valued datasets. In *Sixth IEEE International Conference on Data Mining - Workshops (ICDMW'06)*, 18–22 Dec. https://doi.org/10.1109/ICDMW.2006.123.
12. Liu, L., Wang, J., and Zhang, J. 2008. Wavelet-based data perturbation for simultaneous privacy-preserving and statistics-preserving. In *2008 IEEE International Conference on Data Mining Workshops*, 15–19 Dec. https://doi.org/10.1109/ICDMW.2008.77.
13. Peng, B., Geng, X., and Zhang, J. 2010. Combined data distortion strategies for privacy-preserving data mining. In *3rd International Conference on Advanced Computer Theory and Engineering (ICACTE)*, 20–22 Aug. https://doi.org/10.1109/ICACTE.2010.5578952.
14. Jain, Y.K., and Bhandare, S.K. 2011. A study on normalization techniques for privacy preserving data mining. In *Proceedings of International Journal of Computer & communication Technology*, 2 (VIII).
15. Manikandan, G., Sairam, N., Sudhan, R., and Vaishnavi, B. 2012. Shearing based data transformation approach for privacy preserving clustering. In *Third International Conference on Computing,Communication and Networking Technologies (ICCCNT-2012)*, July 26–28. SNS College of Engineering, Coimbatore.
16. Manikandan, G., Sairam, N., Sharmili S., and Venkatakrishnan S. 2013. Data masking – a few new techniques. In *International Conference on Research and Development Prospects on Engineering and Technology (ICRDPET-2013)*, March 29–30. E.G.S Pillay Engineering college, Nagapattinam.
17. Rajalaxmi, R.R., and A.M. Natarajan. 2008. An Effective Data Transformation Approach for Privacy Preserving Clustering. *Journal of Computer Science* 4 (4): 320–326.
18. Jain, Yogendra Kumar, and Santosh Kumar Bhandare. 2011. A Study on Normalization Techniques for Privacy Preserving Data Mining. In *Proceedings of International Journal of Computer & communication Technology* 2(VIII).
19. Saranya, C., and G. Manikandan. 2016. A Study on Normalization Techniques for Privacy Preserving Data Mining. *Proceedings of International Journal of Engineering and Technology (IJET)* 5 (3): 2701.
20. Mendes, Ricardo, and Joao P. Vilela. 2017. *Privacy-Preserving Data Mining: Methods, Metrics, and Applications*, 27 June 2017.
21. Kabir, Saif M.A., Amr M. Youssef, and Ahmed K. Elhakeem. 2007. On Data Distortion for Privacy Preserving Data Mining. In *Proceedings of IEEE Conference on Electrical and Computer Engineering (CCECE 2007)*, 308–311.
22. Xu, S., J. Zhang, D. Han, and J. Wang. 2005. Data Distortion for Privacy Protection in a Terrorist Analysis System. In *Proceeding of the IEEE International Conference on Intelligence and Security Informatics*, 459–464.
23. UCI Machine Learning Repository. http://archive.ics.uci.edu/ml/datasets.htm.

Role and Impact of Wearables in IoT Healthcare

Kamalpreet Singh, Keshav Kaushik, Ahatsham and Vivek Shahare

Abstract Introduction of Internet of things (IoT) in healthcare domain is affecting the life of people. Involvement of latest and advance wearable devices is making things simpler for patients who are suffering from some disease. These wearable devices are manufactured with some characteristics, which make them an acceptable part of human body. Latest technologies like edge computing are also assisting wearables to deliver the expected performance. This paper attempts to highlight the role of wearables in IoT healthcare, the reference architecture of latest technologies used in wearables and various communication protocols used in IoT. In addition, this paper also highlights the multiple wearables used in IoT healthcare with their characteristics.

Keywords IoT · Healthcare · Smart wearables

1 Wearables: Connecting Physical and Digital World

Wearables are electronic devices that are used to capture data from body with the help of sensors [1]. These wearables can either be worn or implanted on the body. Though these wearables help in monitoring and analyzing the health and well-being of a human, there are many challenges faced by them, such as location of the

K. Singh (✉) · K. Kaushik · Ahatsham · V. Shahare
School of Computer Science, UPES, Dehradun, India
e-mail: kpsingh@ddn.upes.ac.in

K. Kaushik
e-mail: keshavkaushik@ddn.upes.ac.in

Ahatsham
e-mail: ahatsham@ddn.upes.ac.in

V. Shahare
e-mail: vshahare@ddn.upes.ac.in

© Springer Nature Singapore Pte Ltd. 2020
K. S. Raju et al. (eds.), *Proceedings of the Third International Conference on Computational Intelligence and Informatics*, Advances in Intelligent Systems and Computing 1090, https://doi.org/10.1007/978-981-15-1480-7_67

sensor, distortion of signal and comfort of the person wearing the wearable [2]. However, the consumption of battery and usage of energy by the wearable have always been a constraint issue.

Most of the wearable devices either use Bluetooth to connect with the smart phones or Wi-Fi to communicate via Internet [3]. One such example of wearable device is shown in Fig. 1. An activity tracker has inbuilt sensors to gather data and electronics and software to connect with other devices like smart phones and exchange data without any human intervention. Other examples like fit bands and smart jewelry are discussed in the upcoming section.

Integrating IoT with medical devices improves the quality and efficiency of service. The data generated by IoT device (usually in real time) needs to be processed using data analytics such as descriptive analysis, predictive analysis and prescriptive analysis. Implantable wearables are the wearables that are surgically implanted in the body of the patient under their skin. These implantable wearables may be used to monitor the working of heart and insulin pumps. For addressing the issues related to healthcare, various IOT devices and its applications are emerging almost everyday [4]. These wearable devices help in monitoring and managing our day-to-day activities in perspective of our health and state of well-being [5]. Few of these devices that enable in revolutionize the delivery of healthcare globally are discussed below.

Fig. 1 Fitbit fitness tracker

- **Fit Bands and Smart Watches**:

Ranging from sophisticated companies like Apple and Samsung to bulk manufacturer of China, everyone has introduced their variant of smart watches and bands with different features ranging from connectivity with the smart phone to measuring the steps walked in a particular time period, calories burnt, monitoring heartbeat and sleeping hours. The companies are targeting in making these devices more sophisticated by using advanced sensors at certain points that help in recording data from human body. Most of these devices work independently, i.e., they do not depend upon the smart phone for Internet connectivity, enabling them to transmit the data captured directly to the cloud on their own. However, the data send by the sensors needs to be processed by software analysis to infer/deduce results regarding health of the user [6]. This processing is carried out on the cloud platform [7]. The result of this analysis helps the individual as well as their loved ones to be proactive against any progressively alarming health symptoms.

- **Smart Fabrics**:

Fabrics, being the most common assets nearest to the human skin, are considered the most favored platform for sensors in healthcare applications for monitoring physiological nature as a long-term signal. With help of ubiquitous computing merged with nanoelectronics, textiles are converted to smart textiles or e-textures using flexible sensors. These flexible sensors are reactive to environmental conditions and respond to changes as programmed. Based on the reaction to change in environmental conditions, e-textile can be categorized into following types: passive e-textiles: Sensors in such textiles are capable of only collecting the data from the surroundings and the human body and transmitting this captured data either to the smart phone or the cloud with help of Internet. Passive textiles need only sensors for their working.

- **Active e-textiles**:

Such textiles have more inbuilt intelligence as compared to passive textiles. Along with the collection of information, these textiles are also capable of re-acting and adapting to the external environment. This reaction and adaption, however, depend on the way they are programmed. Active textiles require sensors as well as actuators to respond and accommodate to the environmental conditions.

- **Smart textile**:

In addition to the properties of active and passive textiles, these textiles are also context-aware. In similar environmental conditions, these textiles act differently depending on the context in which the environment is. Very smart textiles along with sensors, actuators and processing units also have context-aware materials to recognize, respond and accommodate to the context. In comparison with the smart watch and wrist bands, smart textiles being closer to skin with large body coverage area are capable of collecting a large amount of data from human body as well as the environment surrounding the body [8]. This helps in better analysis of

(a) (b)

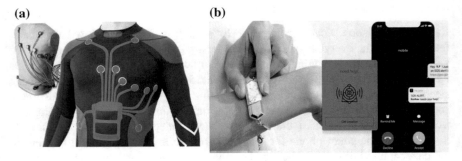

Fig. 2 **a** Smart textile used in health industry, **b** Smart jewelry

healthcare application aiding the well-being of a person as a whole. The smart textiles in the health industry are helpful for monitoring ECG, heart rate, EMG and EEG. Figure 2a shows a sample of smart textile used in medical science.

– **Smart Jewelry**:

With the advancement in wearable devices, the demand for making the IoT devices more stylish and fashionable for fashion-conscious generation is also increasing. This demand requires the IoT devices to be more trendy giving rise to beautiful smart jewelry incorporated with sensors which operate like smart wrist bands, having the capability to capture user's movement and other parameters on regular intervals required for health analysis. For example, the Fitbit, a pendant necklace released by Tory Burch in different trendy colors, is capable of performing same functions as a smart wrist band [9]. Being durable and water resistant, these devices are more easy and beneficial to use as these are capable of recording user's movements like sleep rate, heart rate and even menstrual cycle.

2 Technology Behind Wearables: Edge Computing

As a number of connected devices are connected day by day, there is a tremendous increase in the variety and volume of data. Data corresponding to wearables in IoT needs to be sensed, transferred, processed and stored in an efficient and hassle-free manner. Cloud computing provides the platform and support for storage and implementing IoT technology [10]. Certain factors like huge network bandwidth, high storage cost, irregular and high latency are drawbacks of using cloud computing in wearables for IoT. These drawbacks are positively handled by edge computing as it eliminates the problem of high latency in cloud computing by processing data at the source end. By doing so, devices with irregular or discontinuous connectivity can be used for analyzing and processing of data [11]. As data is available in different varieties and in different formats, it becomes difficult for IoT developers to analyze data. This problem is solved by edge computing by introducing the concept of Middleware, where data is converted into a common format and can be used in healthcare monitoring and diagnosis. Edge computing is defined as an additional processing and storage layer between IoT devices and cloud, which consists of fog or edge nodes. Edge computing is a practice of processing the data at the edge of the network, instead of processing it at the centralized node. Edge computing is much more than computing and processing data for IoT devices. Actually, edge computing also acts as an interface between IoT devices and cloud, i.e., physical world and computational world [12]. Applications based on edge computing use the flexibility and power of IoT devices to pre-process, filter, collect and store data in order to take actions and support decisions for physical world. There are various reasons for using edge computing with cloud computing in IoT ecosystem, and some of them are:

- **Robust connectivity-**Computational applications designed using edge computation not only reduce the latency of the network but also ensure the smooth functioning of application also with limited connectivity. This feature of edge computing is very useful for applications present at remote location with limited or poor network connectivity, as it gives an overhand over using expensive cellular technologies at remote locations.
- **Maintain Privacy-**Edge computing facilitates in maintaining the privacy, as sensitive data is pre-processed on-site, and by using the anonymity concept, privacy complaint data is sent to the cloud for analysis. Data captured by IoT devices can be sensitive in many senses like GPS location, audio and video by mic and camera, Internet-accessing habits, etc.
- **Latency-**With the advent of advance machine learning algorithms, edge computing is making IoT applications to run directly on devices, by capturing data with training models. This was not possible before with the default features of cloud computing, where the data was retrieved and processed with high cost of latency. But this is made possible by edge computing in support with latest machine learning algorithms.

Figure 3 shows the architecture of edge computing for wearable devices. Talking about this architecture, generally, devices are assigned with some roles, and they are categorized as edge devices, edge gateways, edge sensors and actuators. Sensors and actuators use low power radio technologies for their communication with devices or gateways. Edge devices are intelligent, battery powered and equipped with latest and updated operating system. These devices can interact with cloud directly or with the help of some gateway. Gateways actually act as an interface between edge devices and cloud. In the above figure, the cloud is treated as the root node, edge gateways as a middle tier, edge devices and actuators and sensors as the leaf nodes. Applications based on edge computing are divided into various modules, like analytics module for performing analysis on data coming from edge gateways and devices. A dashboard in cloud with complete global data view for proper understanding and analytics. Applications based on edge computing should demonstrate the transparent communication between various edge devices, visibility restrictions, security and privacy rules and data flow between various components.

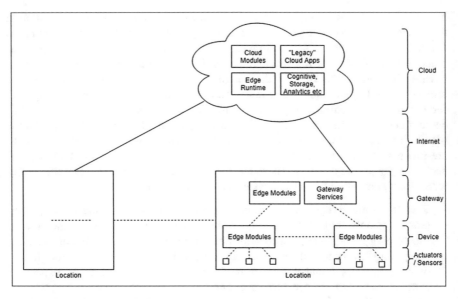

Fig. 3 Edge computing architecture

3 IoT Protocols for Wearable Devices

IoT devices communicate with each other with the help of certain predefined rules and regulations known as protocols. These protocols play crucial role in communication among multiples IoT devices. Availability of a wide range of protocols raises the question for applicability of these in various domains with their set of advantages and disadvantages. This section discusses application of these IoT protocols in the field of healthcare, especially wearables. Table 1 compares and contrasts the widely used protocols for communication between wearables and smart phones.

The above table lists various IoT protocols used in communication between wearables and smart phones. In addition, there are certain other protocols as well which are used in wireless body area network (WBAN). WBAN is a combination of various sensors, which can be implanted, in human body, which can monitor the physiological changes in human body and can transmit those changes to the coordinator devices placed in WBAN. As the data sensed and transmitted by WBAN sensors is highly private, so there are some M2 M protocols, which can take care of security and privacy of this data. Some protocols of such kind are:

- **CoAP**: CoAP stands for constrained application protocol, which works on UDP layer and is lightweight in nature. CoAP is designed for simple IoT devices, which uses low memory like Zigbee. CoAP can be easily mapped with HTTP because it works on request/response model. CoAP also provides the proxy

Table 1 IoT protocols for communication between wearables and smart phones

Sr. No.	Protocols	Power consumption	Max. data rate	Max. range	IEEE standard	Frequency
1	Zigbee	30 mW	250 kbps	75 m	802.15.4	0.9 2.4 GHz
2	Bluetooth	2.5–100 mW	3 MBps	100 m	802.15.1	2.4 GHz
3	Wi-Fi	1 W	150 MBps	250 m	802.11	2.4 5 GHz
4	3G LTE	1 2 W	100 MBps	100 km	X	0.7 3.6 GHz
5	WiMAX	10 W	50 MBps	50 km	802.16	2 10 GHz
6	Z-Wave	2.5 mA	100 kbps	30 m	802.15.4	915 MHz

support for CoAP-HTTP communication. All the features of CoAP make it best choice for open source and commercial applications.

– **MQTT**: It stands for Message Queuing Telemetry Transport protocol. It is a lightweight application protocol, which works on TCP layer. It works for communication on unreliable network, and under limited resources like processing power, memory etc. It is a messaging protocol, which works on publish and subscribe principle. It comprises three main components MQTT server/broker, MQTT publisher and MQTT subscriber. Broker is responsible for delivery of messages from publisher to subscriber.

– **REST**: REST stands for Representational State Transfer protocol, which uses HTTP as an application protocol and works over TCP and UDP layers to transfer data. It follows request-response principle, in which the client sends a request to the server and server gives back the response after processing the request. The verbose nature of REST and its text-based format makes it a suitable choice for many commercial applications.

– **AMQP**: It stands for Advanced Message Queuing Protocol, which is open standard in nature and works on TCP layer. There are various features provided by AMQP like routing, queuing, publish/subscribe, security and reliability, which makes it asynchronous compliment of HTTP. AQMP is used in some large-scale M2 M IoT message queue-based project.

4 Conclusion

The paper enables to explore the role of wearables in IoT healthcare. With the increase in use of IoT in healthcare domain, there is a need of revolution in wearable devices industry. As technology is changing day by day, more sophisticated and advance devices with better computational and processing power are coming in market. These wearable devices are influencing the IoT healthcare largely. Introduction of these wearable devices ensures the use of latest IoT protocols for communication. IoT is also adapting itself with more dynamic flow and architecture. Therefore, there is a need of continuous advancements in software as well as hardware parts in order to deliver the expected performance in near future.

References

1. Bansal, G., F.M. Zahedi, and D. Gefen. 2010. The Impact of Personal Dispositions on Information Sensitivity, Privacy Concern and Trust in Disclosing Health Information Online. *Decision Support Systems* 49 (2): 138.
2. Gao, Yiwen, He Li, and Yan Luo. 2015. An Empirical Study of Wearable Technology Acceptance in Healthcare. *Industrial Management and Data Systems* 115 (9): 1704.
3. Chan, M., D. Estve, J.Y. Fourniols, C. Escriba, and E. Campo. 2012. Smart Wearable Systems: Current Status and Future Challenges. *Artificial Intelligence in Medicine* 56 (3): 137–156.
4. Argus Insights. 2015. As Fitbit Prepares for IPO, New Consumer Research Reveals Areas of Wearables Market Vulnerability for Fitness Band Leader. (Online). Available http://www.argusinsights.com/fitbit-ipo-release/.
5. Angst, C.M., and R. Agarwal. 2009. Adoption of Electronic Health Records in the Presence of Privacy Concerns: The Elaboration Likelihood Model and Individual Persuasion. *MIS Quarterly* 33 (2): 339–370.
6. Atallah, L., G.G. Jones, R. Ali, J.J. Leong, B. Lo, and G.Z. Yang. 2011. Observing Recovery from Knee-Replacement Surgery by Using Wearable Sensors. In *Proceedings of the IEEE 2011 International Conference Body Sensor Networks*, 2934, May 2011.
7. Metcalf, D., R. Khron, and P. Salber (eds.). 2016. *Health-e Everything: Wearables and the Internet of Things for Health: Part Two: Wearables and IoT*. Orlando, FL: Moving Knowledge.
8. Cicero, M.X., B. Walsh, Y. Solad, T. Whitfill, G. Paesano, K. Kim, C.R. Baum, and D.C. Cone. 2015. Do You See What I See? *Insights from Using Google Glass for Disaster Telemedicine Triage, Prehospital and Disaster Medicine* 30 (01): 48.
9. Metcalf, D., R. Khron, and P. Salber (eds.). 2016. *Health-e Everything: Wearables and the Internet of Things for Health: Part One: Wearables for Healthcare*. Orlando, FL: Moving Knowledge.
10. Shi, W., J. Cao, Q. Zhang, Y. Li, and L. Xu. 2016. Edge Computing: Vision and Challenges. *IEEE Internet of Things Journal* 3 (5)
11. Ashton, K. 2009. That Internet of Things Thing. *RFiD Journal* 22 (7): 97114.
12. Dinev, T., and P. Hart. 2006. An Extended Privacy Calculus Model for e-commerce Transactions. *Information Systems Research* 17 (1): 61.

A Survey of IDS Techniques in MANETs Using Machine-Learning

Mohammed Shabaz Hussain and Khaleel Ur Rahman Khan

Abstract The mobile ad hoc networks (MANETs) had move towards the wireless networking technology. Due to its dynamic nature MANETs face the challenges towards critical attacks in OSI layers but research shows that in network layer the attacks are effectively done by intruders. In this survey, many of the attacks at network layer are identified in MANETs by most of the researchers which are outlined in this paper. Mostly, AODV routing protocols and other protocols are used for transferring packets to the destination. The communicated packets information is accumulated in the log files, to surveillance these routing of packets from these log files, the techniques used in MANETs are data mining, support vector machines (SVM), genetic algorithms (GA) and other machine-learning approaches. Further, the methodologies and techniques proposed for detecting and predicting these attacks from various kinds of intrusions within the MANETS are discussed.

Keywords Mobile ad hoc networks (MANETS) · Intrusion detection system (IDS) · Support vector machines (SVM) · Genetic algorithms (GA) · Ad hoc on-demand distance vector (AODV)

1 Introduction

Today universally, wireless networks are enormously spreading and are used by many people due to its immense necessity and usage of mobile devices. MANETs still need to be evolved for practical use in real-time applications which act like a secure medium, excessively used in communication and broadcasting of information resulting into complex problems which is of very important concern [1, 2].

M. S. Hussain (✉)
Department of Computer Science, Rayalaseema University, Kurnool, A.P, India
e-mail: mshabazh@gmail.com

K. U. R. Khan
Department of CSE, ACE Engineering College, Hyderabad, India
e-mail: khaleelrkhan@gmail.com

© Springer Nature Singapore Pte Ltd. 2020
K. S. Raju et al. (eds.), *Proceedings of the Third International Conference on Computational Intelligence and Informatics*, Advances in Intelligent Systems and Computing 1090, https://doi.org/10.1007/978-981-15-1480-7_68

MANET network is a group of various mobile nodes that make use of wireless interface from provisional networks. The significance of MANET is that it do not have any fixed infrastructure due to which there is more occurrence of intrusions in the network and MANETS do not have any centralized control.

Intrusion detection is the approach used to detect malicious activity through pattern recognition in enormous datasets involving methods like artificial intelligence and machine learning [2]. Intrusion detection is used for detecting any unauthorized access done for personal computer or any thinking machine. Due to this, providing security is becoming a critical task with the growing of Internet applications that make use of MANETs. The existing security can be enhanced and improved by using appropriate intrusion detection technique at specific network layer in which very little research is done. The patterns of end-user activities are examined and intrusions are identified using SVM and decision trees (DT).

2 Problems of Mobile Ad Hoc Networks in a Wireless Network Environments

The wireless networking technology with the developments in MANETs is subsequently evolving towards next-generation technology. The MANET characteristics are susceptible for critical attacks in most of the OSI layers, among which Network layer requires special attention, when these MANETs routing protocols are developed the developers think in their minds that there is no malevolent intruder node. But, intrusion detection system has different attacks in MANETs as mentioned below. Particular-based section, sign-based detection and spoofing.

3 Literature Review

In MANET, the idea of prevention is not enough from the security point, therefore intrusion detection system is the one more another concept of facilitating the security inside the networks; this intrusions detection system aids in spying these mischievous nodes inside the network. IDS are a unique strategy that is exclusively used for detection of malicious activities in MANETS. IDS need to be configured just like any software application to predict malicious nodes from network. Thus, to overcome these, IDS are being installed on these nodes to avoid them in the MANET due to its centralized behaviour [3, 1]. The main categories of MANETS are signature-based detection, specification and anomaly-based detection [2]. Similarly, the IDS have representative type of attacks in the context of MANETS as discussed in [4, 5]. In this paper, the author has evaluated the intrusion detection techniques based on intrusion datasets. The dataset used by the author is Aegean Wifi intrusion detection dataset. This dataset is tested by machine-learning techniques from which feature-based reduction techniques are used. The feature reduction technique gave a good gain, similarly the statistics of chi-squared is applied for the evaluation of the dataset accuracy with the help of feature reduction [6]. Results obtained are shown using the feature reduction which may lead to show little good results in accuracy point of view which had resulted from 110 to 41 is 2.4% [6]. Categorization of intrusion detection strategies are categorized as misuse-based detection, anomaly-based detection or combination of both [7]. These methods of intrusion detection had been misused; therefore, these mechanisms are needed to be checked using network packet flow guided by known miscellaneous threat patterns. These techniques give accuracy of higher grade, but it has easy implementation, whereas it cannot detect unknown intrusions [8]. Anomaly-based detection is capable of identifying unknown attacks but it has accuracy in lower state. The continuous evaluation of dataset is done to improve the performance.

The methodology of genetic algorithm to select optimum protocol with the context based on network is proposed [9]. The comprehensive performance of three protocols is used such as DSR, OLSR and AODV. To optimize the performance, the WPAN is linked with fuzzy logic techniques, and neuro fuzzy technique is used to improve the performance with each routing protocol. Similarly, in another study, genetic algorithm is used to improve the performance [9].

Another author has discussed in his paper about various detection techniques and the most effective is the anomaly-based detection in which the author has concentrated over machine-learning technique, which identifies the attack traffic in online, by rewriting the intrusion detection system rules for a fly-air experiments is conducted with the aid of Dockemu emulation tool assisted by Linux Containers, IPV6, and OLSR routing protocol. In this method, the detection engine is used as a mechanism for an intrusion detection system, that predicts attack online mode with the aid of machine-learning technique. These techniques are based on support vector machine and has been tested with a critical analysis of Dockemu emulation tool which is presumed to be a realistic case study. This proposal is succeeded in creating of rule patterns that can segregate the Denial-of-service attack accurately. Finally, this method of approach has given a support vector machine of good improved accuracy and performance [10]. The MANETSin an open medium as well as in a changing topology, the lack of appropriate resources mediocre current supply have chances of susceptible attacks based on the basic characteristics. In this paper, the author has discussed about detecting a malevolent node in MANETs by secure intrusion detection system which uses dynamic source routing protocol (DSRP). Thus, DSRP consistently has the ability to evade the fake acknowledgement [11]. In one of the proposed method, DSR protocol is taken into account for dynamic source routing purpose. This DSRP is mostly utilized in both the wired as well as wireless networks extensively. DSRP constitutes of two major steps as given below [12].

Route discovery and route protection. The method used in the route discovery is where a source will deliberately need to pass the packets to the target node.

The route path is monitored and maintained by watching if there is any loss of packet at the intended place from where the route discovery will start sending "ROUTE ERROR" directive to origin node.

In the proposed system, the packets are classified which includes 2 bit DSR protocol packet header. DSRP header had few pre-defined bits as different packet type and flags. The packet type and packet bits are like Data(00), Reply back(01), @CA(10) and DMNC(11).

There are other add-On techniques used in intrusion detection system, namely watchdog, 2ACK as well as AACK but these are suffering from few hidden problems. To overcome, another technique is proposed for the above three deficiencies of watchdog that has mediocre broadcast power, recipient collision and false mischievous nodes. In this, DSRP assists in skipping of the altered acknowledgements with the aid of cryptographic approaches, i.e., Rivert-Shamir Adleman RSA [13].

The system is integrated with cross-layer defence system. The intrusion detection takes the audit data and performs the reasoning of the data and tries to identify whether the server is about to be attacked. Intrusion detection is expected to be of either network-based or else host-based, it matters. Network-based IDS keep a hidden eye on the gateway of network through which packets are transferred into the network hardware through an interface. Host-based rely on OS that checks the data and examines the data [14].

4 Attacks in Manets

There are different types of intrusion attacks on network layer which are well versed in MANETs. In this section, we explore the categorization of attacks that are identified which occurs mostly in the network layer in the MANETs, and are sub-categorized into two The attacks are broadly classified as passive attack and the another on as active attack [15]. In which, further, There are many other attacks such as grey hole attack, black hole attack and the Sybil attack [16–18], Sybil Attack [18], point detection algorithms [15].

5 Secure Dropping of Data Packets

Extensive survey conveys that most of the data packets get dropped in-between due to various issues and problems. The research shows that 13% of the issues are related to weak communication channels and congestion problems, whereas 19% of issues are due to intruders that attack at specific network layer as discussed in various approaches [15]. In the protection scheme, failure of data packet is due to cooperative involvement of node behaviour. This scheme requires that node during the sending of packets should monitor the behaviour of other neighbouring nodes; On identifying such packet dropping, it will initiate the procedure for spying and then administers to investigate an attack, only after finding of a weak node that initiated for dropping of packets this scheme uses a function named as the trust collector and this is used to collect the values of trust within the nodes of neighbours under the suspicious nodes [17]. The performance of this technique is compared with watchdog algorithm [18] and the results obtained for improving the false alarm rates [14]. Further, few approaches based on neighbour watch system also discussed that were proposed earlier for identifying of the misbehaving nodes resulting in dropping of packets [19]. Packet forwarding and misbehaviour detection lie on flow conservation [20], where nodes continuously examine near by nodes, by maintaining a full list of such nodes they listen to and then decides behaviour of nodes from time to time in a time slice. Misbehaving of those nodes are found by comparison of the estimated proportion of data packets which gets fall or loss with the threshold of misbehaviour is pre-established. An enhanced version of policy of adaptation for this algorithm is proposed as [21].

Alterations of the nodes are achieved in two ways. Firstly, by using a method that calculates the number of nodes with that of misbehaviour detection threshold value. Secondly, the adaptability of the protection mechanism policies that accumulates the evolving of network environment alongside with that of the pre-defined objectives [22]. Similar to that, another such technique is proposed named as SCAN, which detects mischievous nodes at network layer which can secure the packet delivery in MANETs [23]. In this, the nodes can over hear the packets which are captured by corresponding neighbour and keeps a duplicate neighbour copy in table of routing. Therefore, whenever neighbouring nodes encounters to know the loss of packets, then these nodes move ahead for next hop using predefined routing table. Then, designate those packets as dropped in a scenario where spying node does ignore those packets which had been forwarded from its consecutive neighbouring node [24]. To lower this adverse effect of missing packets in MANETs, another mechanism is proposed that has two parts that work together, namely watchdog and path rater [14].

6 IDS Categorization

IDS are fundamentally segregated into three main categories due to the type of methodology used for detection which are employed, namely:

- Anomaly-based intrusion detection (ABID).
- Knowledge-based intrusion detection (KBID).
- Specification-based intrusion detection (SBID) [15].

We categorize as per the intrusion detection methodologies that are used as either signature-based intrusion detection, knowledge-based intrusion detection, anomaly-based intrusion detection, or its hybrid of these, or some other mechanism. Similarly, many of the IDS techniques used are depicted in Table 1 is which is shown.

Table 1 Various types of attacks, detection technique used, and routing protocols

Title of paper	Name of Algorithm	Architecture	Type of attack	Detection technique	Routing protocol	Source of data	Contribution
Zhang et al. [17]	Not used	Distributed	Black hole	Checking RREP	AODV	Sequence number of RREP's from intermediate nodes	Black hole detection mechanism
Sen.et al. [19]	Not used	Hierarchical	Grey hole attack	Monitoring behaviour in terms of RREP	AODV	Neighbour data collection module transmitted	Grey hole detection for AODV
Yang et al. [24]	Scan	Distributed	Data packet dropping	Information cross validation	Isolate attackers	Collabarative monitoring	Provide secure packet delivery in MANET's
Zhang and Lee [28]	Not used	Distributed peer to peer	Various network layer	ABID	AODV, DSR	System events	Agent-based IDS architecture
Yi et al. [23]	Not used	Hierarchical distributed	DOS attacks, routing loop	Other IDS	DSR	DSR routing specification	FSM to detect attacks
Nadeem and Howarth [15]	AIDP	Hierarchical, clustered	DoS attacks	ABID	AODV	Routing information	ABID for detecting DoS attacks in MANETs

6.1 Anomaly-Based Intrusion Detection

Anomaly-based intrusion detection (ABID) system is another well-known approach which detects anomalous activities that drastically change the normal behaviour. ABID systems are also known as behaviour-based intrusion detection, in that the model normal behaviour is tested and then compared with the present behaviour of the system of these existing network for predicting of intrusions among the network. One of the detection techniques such as anomaly-based technique consists of two distinct phases, namely training and testing [25].

6.2 Knowledge-Based Intrusion Detection (KBID)

Knowledge based intrusion detection (KBID) is a well-versed technique that stores nodes information along with the information of patterns which are also referred as one of the category of attack, similarly such patterns are matched that assist in detecting of misbehaving nodes. In other words, knowledge-based intrusion detection system has much information about precise attacks and tries an attempt to use them. This knowledge-based intrusion detection system initiates an alert in the form of an alarm when an attack is identified [15]. Ultimately, this knowledge-based intrusion detection system depends on knowledge of previous attacks to surveillance new attacks.

6.3 Other Intrusion Detection Proposals

In another study, the clustered approach looks out for single node and then selects the same as monitoring node where it detects the mischievous intruders [26].

Similarly, in another survey, there are many more methods of intrusion detection which are used widely for different range of attack as with the use of type ID in which a brief study of nodes are analysed by the mobile agents for MANET for the purpose of intrusion attacks and then concluded that many of these mobile agent's tries to capture the features which satisfy the requirements for MANETs using IDS. These mobile agents build the trust and start to expel their programs without being disturbing other nodes by the originating node status which will mis-interpret the nodes with the relevance to the other nodes similar to those inconsistencies Present in MANETS which reduces the network load by using a high bandwidth by creating the congestion problems and creating robust fault-tolerant behaviour [27]. Further, tries to pin-point hidden mobile agents which have security issues, which might be one of that reasons they are still not yet used extensively for IDS [15].

7 Conclusion and Future Scope

The key feature of MANETs is that it has less infrastructure and hence has minor control over the centralized behaviour and falls prey to various types of attacks. Specifically, an intrusion detection attacks which has become bit difficult task and thus, makes MANETS more vulnerable which can be easily harmed at the network layer.

This surveillance system for overcoming the intruders can be avoided by the use of effective machine-learning techniques, at the standard network layer and also by using appropriate intrusion detection techniques that detect and predict the hidden mischievous malfunctioning of those nodes which are altered by intruders which is my future work.

References

1. Elboukhari, Mohamed, Mostafa Azizi, and Abdelmalek Azizi. 2014. Intrusion Detection Systems in Mobile Ad Hoc Networks: A Survey. In *5th Workshop on Codes, Cryptography and Communication Systems (WCCCS)*, IEEE, 136–141, Morocco, Nov 2014.
2. Chadli, Sara, Mohamed Emharraf, Mohammed Saber, and Abdelhak Ziyyat. 2014. Combination of Hierarchical and Cooperative Models of an IDS for MANETs. In *Tenth International Conference on Signal-Image Technology and Internet-Based Systems*, IEEE, 230–236, Morocco, Nov 2014.
3. Saxena, Aumreesh Kumar, Sitesh Sinha, and Piyush Shukla. 2017. A Review on Intrusions Detection System in Mobile Ad-Hoc Network. In *Proceeding of International conference on Recent Innovations is Signal Processing and Embedded Systems (RISE-2017)*, IEEE, 27–29 Oct 2017.
4. Banerjee, Sayan, Roshni Nandi, Rohan Dey, and Himadri Nath Saha. 2015. A Review on Different Intrusion Detection Systems for MANET and itsvulnerabilities. *International Conference and Workshop on Computing and Communication (IEMCON)*, IEEE, 1–7, India, Oct 2015.
5. Poongothai, T., and K. Duraiswamy. 2014. Intrusion Detection in Mobile AdHoc Networks Using Machine Learning Approach. In *International Conference on Information Communication and Embedded Systems (ICICES2014)*, IEEE, 1–5, India, Feb 2014.
6. Pavani, K., and A. Damodaram. 2013. Intrusion Detection Using MLP for MANETs. In *Third International Conference on Computational Intelligence and Information Technology (CIIT 2013)*, 440–444, IEEE, India, Oct 2013.
7. Thanthrige, Udaya Sampath K. Perera Miriya, Jagath Samarabandu, and Xianbin Wang. 2016. A Machine Learning Techniques for Intrusion Detection on Public Dataset. In *Proceedings of IEEE Canadian Conference on Electrical and Computer Engineering (CCECE)*, IEEE.
8. Cannady, J., and J. Harrell. 1996. A Comparative Analysis of Current Intrusion Detection Technologies. In *Proceedings of the Fourth Technology for Information Security Conference*, Citeseer, vol. 96.
9. Salour, M., and X. Su. 2007. Dynamic Two-Layer Signature-Based IDS with Unequal Databases. In *ITNG '07. Fourth International Conference on Information Technology*, 77–82, Apr 2007.
10. Saeed, N.H., M.F. Abbod, and H.S. Al-Raweshidy. 2008. *An Intelligent MANET Routing System*. IEEE.

11. Peterson, Erick, and Marco Antonio To. 2017. A Novel Online CEP Learning engine for MANET IDS. In *IEEE 9th Latin-American Conference on Communications (LATINCOM)*.
12. Kazi, Samreen Banu, and Mohammed Azharuddin Adhoni. 2016. Secure IDS to Detect Malevolent Node in MANETS. In *International Conference on Electrical, Electronics and Optimization Techniques (ICEEOT)*.
13. Johnson, David B., David A. Maltz, Josh Broch. *2001. DSR: The Dynamic Source Routing Protocol for Multi-Hop Wireless Ad Hoc Networks*. ACM.
14. Marti, S., T.J. Giuli, K. Lai, and M. Baker. 2000. Mitigating Routing Misbehaviour in Mobile Ad Hoc Networks. In *Proceedings of the International Conference on Mobile Computing and Networking*, 255–265.
15. Nadeem, Adnan, Michael P. Howarth. 2013. A Survey of MANET Intrusion Detection & Prevention Approaches for Network LayerAttacks. *IEEE Communications Surveys & Tutorials*, vol. 15.
16. Kurosawa, S., and A. Jamalipour. 2007. Detecting Blackhole Attack on AODV-Based Mobile AdHoc Networks by Dynamic Learning Method. *International Journal of Network Security* 5 (3): 338–345.
17. Sen, J., M. Chandra, S.G. Harihara, H. Reddy, and P. Balamuralidhar. 2007. A Mechanism for Detection of Gray Hole Attacks in Mobile Ad Hoc Networks. In *Proceedings of the IEEE International Conference on Information Communication and Signal Processing ICICS*, Singapore.
18. Piro, C., C. Shields, and B. Levine. 2006. Detecting the Sybil Attack in Mobile Ad hoc Networks. In *Proceedings of the IEEE International Conference on Security and Privacy in Communication Networks*, IEEE.
19. Lee, S.B., and Y.H. Choi. 2006. A Resilient Packet Forwarding Scheme against Maliciously Packet Dropping Nodes in Sensor Networks. In *Proceedings of the ACM Workshop on Security of Ad Hoc and Sensor Networks (SANS2006)*, 59–70, USA.
20. Gonzalez-Duque, O.F., M. Howarth, and G. Pavlou. 2013. Detection of Packet Forwarding Misbehaviour in Mobile Ad hoc Networks. In *Proceedings of the International Conference on Wired/Wireless Internet Communications (WWIC 2007)*, 302–314, Portugal, June 2007.
21. Gonzalez-Duque, O.F., G. Ansa, M. Howarth, and G. Pavlou. 2008. Detection and Accusation of Packet Forwarding Misbehaviour in Mobile Ad hoc Networks. *Journal of Internet Engineering* 2 (8): 181–192.
22. Pavani, K., and A. Damodaram. 2013. Intrusion Detection Using MLP for MANETs. In *Third International Conference on Computational Intelligence and Information Technology (CIIT 2013)*, 440–444, India.
23. Yi, P., Y. Jiang, Y. Zhong, and S. Zhang. 2005. Distributed Intrusion Detection, for Mobile Ad Hoc Networks. In *Workshop on Proceedings of the IEEE Application and Internet*.
24. Yang, H., J. Shu, X. Meng, and S. Lu. 2006. SCAN: Self-Organized Network- Layer Security in Mobile Ad Hoc Networks. *IEEE Journal on Selected Areas in Communications* 24 (2): 261–273.
25. IEEE Std 802.11-2007, IEEE Standard for Information Technology-Telecommunication and Information Exchange Between Systems-Local and Metropolitan Area Network-Specific requirement, Part 11 Wireless LAN Medium Access Control and Physical Layer Specifications.
26. Yi, P., Y. Jiang, Y. Zhong, and S. Zhang. 2005. Distributed Intrusion Detection for Mobile Ad Hoc Networks. In *Proceedings of the IEEE Application and Internet Workshop*.
27. Hijazi, A., and N. Nasser. 2013. Using Mobile Agent for Intrusion Detection in Wireless Ad-Hoc Networks. In *Proceedings of the IEEE Wireless Communication and Networking Conference WCNC, March 2005*; Pavani, K., and A. Damodaram. 2013. Intrusion Detection Using MLP for MANETs. In *Third International Conference on Computational Intelligence and Information Technology (CIIT 2013)*, 440–444, India.
28. Zhang, Y., and W. Lee. 2000. Intrusion Detection in Wireless Ad-Hoc Networks. In *Proceedings of the ACM International Conference on Mobile Computing and Networking (MobiCom)*, 275–283, Boston, US.

Hand Gesture Recognition Based on Saliency and Foveation Features Using Convolutional Neural Network

Earnest Paul Ijjina

Abstract The rapid growth in multimedia technology opened new possibilities for effective human–computer interaction (HCI) using multimedia devices. Gesture recognizing is an important task in the development of an effective human–computer interaction (HCI) system and this work is an attempt in this direction. This work presents an approach for recognizing hand gestures using deep convolutional neural network architecture from features derived from saliency and foveation information. In contrast to the classical approach of using raw video stream as input, low-level features driven by saliency and foveation information are extracted from depth video frames are considered for gesture recognition. The temporal variation of these features is given as input to a convolutional neural network for recognition. Due to the discriminative feature learning capability of deep learning models and the invariance of depth information to visual appearance, high recognition accuracy was achieved by this model. The proposed approach achieved an accuracy of 98.27% on SKIG hand gesture recognition dataset.

Keywords Hand gesture recognition · Human computer interaction (HCI) · Convolutional neural network (CNN) · Depth information

1 Introduction

Hand gesture recognition is an active field of research due to its significance in the design of interfaces for human–computer interaction (HCI), which has wide applications in the areas of virtual reality, gaming, medicine, robotics and entertainment [16]. A human hand gesture is a non-verbal communication used for short-range visual communication that involves the usage of state and movement of fingers and/or hands to communicate a message. Some examples include sign

E. P. Ijjina (✉)
Department of Computer Science and Engineering, National Institute of Technology Warangal, Warangal, Telangana 506004, India
e-mail: iep@nitw.ac.in

© Springer Nature Singapore Pte Ltd. 2020
K. S. Raju et al. (eds.), *Proceedings of the Third International Conference on Computational Intelligence and Informatics*, Advances in Intelligent Systems and Computing 1090, https://doi.org/10.1007/978-981-15-1480-7_69

language used among people with hearing disabilities, during military combat, fight deck signals used in aviation and among deep-sea divers. The recent introduction of Kinect, a low-cost gaming peripheral that can capture depth information has opened new opportunities in hand gesture recognition and computer vision [3]. The availability of Microsoft SDK and OpenNI Libraries for human skeletal prediction and tracking led to their wide usage in full-body human gesture recognition, but hand gesture recognition is a problem yet to be fully addressed. This is the motivation, to develop an approach for hand gesture recognition using depth information only.

Over the years, several approaches for hand gesture recognition were proposed in the literature [13]. In contrast to the conventional approaches, the recent approaches focus on processing RGB-D video captured using low-cost depth sensors like Kinect for hand gesture recognition [14]. Ohn-Bar et al. [9] proposed an approach for hand gesture recognition, that uses RGB-D descriptors for object detection and activity recognition. Ren et al. [11] proposed a part-based hand gesture recognition system using Finger-Earth mover's distance metric [12] that matches finger parts to measure the distance between hand shapes. Kurakin et al. [5] used action graphs to recognize hand gestures by sharing states among different gestures and normalizing hand orientation. Jaemin et al. [4] used HMM to recognize hand gestures from features capturing hand shape and arm movement. Dominio et al. [2] segmented the RGB-D hand information into palm, finger regions and employed feature descriptors to recognize hand gestures using a support vector machine.

In the last few years, the effectiveness of deep convolutional neural networks for pattern recognition has been explored in various visual recognition challenges and tasks. This paper presents an approach for human hand gesture recognition using convolutional neural network utilizing the features derived from depth in-formation. In the remainder of the paper, Sect. 2 describes the proposed feature extraction for hand gesture recognition using convolutional neural networks and the experimental results. Finally, Sect. 3 concludes this work.

2 Proposed Approach

The recognition of hand gestures in videos is a challenging task due to its complex articulations, segmentation in complex backgrounds and illumination conditions. This work considers the hand gestures from SKIG dataset [6], which are captured by a Kinect sensor at a resolution of 320×240 pixels under varying hand-pose, background and illumination conditions as shown in Fig. 1.

The proposed approach uses a deep convolutional neural network architecture to recognize hand gestures from the range and nature of variation of features extracted from depth information. The feature extraction procedure and the CNN architecture used for hand-gesture recognition are elaborated in the following sub-sections.

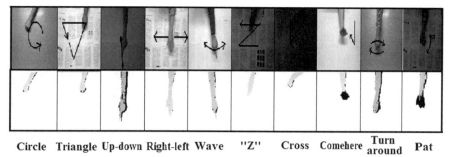

Circle Triangle Up-down Right-left Wave "Z" Cross Comehere Turn around Pat

Fig. 1 SKIG hand gestures. (Figure 4 from Liu et al. [6])

2.1 Feature Extraction

Since depth information is invariant to illumination changes and background clutter, depth information captured by Kinect sensor is considered in this approach to recognize human hand gestures. From the given depth information, the region of the hand (region of interest, ROI) is obtained by applying a mask generated through connected component analysis, as shown in Fig. 2.

Once the region of interest has been selected, the following attributes (driven by saliency and foveation) are computed from each video frame: (1) the x and y co-ordinates of the bottom left corner of the hand and (2) the average gray value of the ROI. These attributes are normalized and scaled to compute the five features using the procedure given in Algorithm 1. The optimal values of threshold used in the algorithm are determined empirically.

(a)   **(b)**   **(c)**

Fig. 2 Region of interest selection from depth information: (**a**) Depth frame from Kinect (**b**) ROI mask for selection of hand and (**c**) the ROI in depth frame

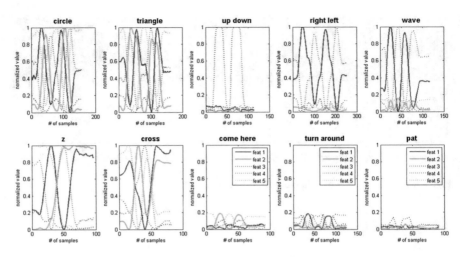

Fig. 3 Temporal variation of features for each SKIG hand gesture

The temporal variation of these features for the ten hand gestures is shown in Fig. 3. It can be observed that, the first seven hand-gestures namely: *circle, triangle, up-down, right-left, wave, z*, and *cross* have better discriminative pattern than the remaining three hand-gestures *come here, turn around* and *pat* due to the presence of significant hand displacement from the initial position. A convolutional neural network (CNN) is trained to recognize these hand gestures even when some of them lack a discriminate pattern. A 2D representation of dimension 22×50, representing the variation of features is obtained by down-sampling the temporal information to 50 samples, the features duplicated vertically with a two-element (pixel) margin from borders and between the features. Our experiments suggest that this representation of features ensure better recognition accuracy. The following subsection elaborates the deep convolutional neural network architecture used for gesture recognition.

ALGORITHM 1: Computation of features from attributes

Input: attributes $attr$ ($n \times 3$ matrix)
Output: $feat$ ($n \times 5$ matrix) feature matrix
$x_{threshold}$=0.3;
$y_{threshold}$=0.32;
$h_{threshold}$=0.5;
$attr[1..n, 1] = attr[1..n, 1]/320$;
$attr[1..n, 2] = attr[1..n, 2]/240$;
$attr[1..n, 3] = attr[1..n, 3]/255$;
$x_{range} = \mathbf{max}(attr[1..n, 1])\text{-}\mathbf{min}(attr[1..n, 1])$;
$y_{range} = \mathbf{max}(attr[1..n, 2])\text{-}\mathbf{min}(attr[1..n, 2])$;
$h_{range} = \mathbf{max}(attr[1..n, 3])\text{-}\mathbf{min}(attr[1..n, 3])$;
$feat[1..n, 1]=attr[1..n, 1]$;
$feat[1..n, 2]=attr[1..n, 2]$;
if $h_{range} > h_{threshold}$ **then**
 | $feat[1..n, 3]=1 - attr[1..n, 3]$;
else
 | $feat[1..n, 3]=h_{range} - attr[1..n, 3]$;
end
if $x_{range} > x_{threshold}$ **then**
 | $feat[1..n, 4]=1 - attr[1..n, 1]$;
else
 | $feat[1..n, 4]=x_{range} - attr[1..n, 1]$;
end
if $y_{range} > y_{threshold}$ **then**
 | $feat[1..n, 5]=1 - attr[1..n, 2]$;
else
 | $feat[1..n, 5]=y_{range} - attr[1..n, 2]$;
end

2.2 Gesture Recognition Using CNN

The ability of a convolutional neural network (CNN) [1] to learn the discriminative local patterns from input data is exploited to recognize hand gestures from the extracted features. A classical CNN classifier [10] with alternating convolution and subsampling layers with a neural network classifier shown in Fig. 4 with the configuration given in Table 1 is considered in this work. The back-propagation algorithm is used to train this model in batch mode.

The temporal variation of features is scaled-down to generate a 2D representation of 22×50 dimensions, which is given as input to the recognition model. Thus, the CNN classifier recognizes the hand gestures from the temporal variation of the proposed features. The experimental study on SKIG dataset is discussed in the following subsection.

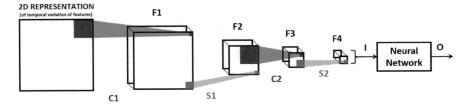

Fig. 4 The CNN classifier used for gesture recognition

Table 1 CNN configuration considered for SKIG hand gesture recognition

Layer: Configuration	Feature map: #, configuration
C1: 3×3 filters	F1: 4, 20×48 feature maps
S1: 2×2 filters	F2: 4, 10×24 feature maps
C2: 3×3 filters	F3: 8, 8×22 feature maps
S2: 2×2 filters	F4: 8, 4×11 feature maps
I: 352 vector	O: 10 outputs

2.3 Experimental Results

The experimental study was conducted on SKIG dataset [6], consisting of ten hand-gestures: *circle, triangle, up-down, right-left, wave, z, cross, come here, turn around* and *pat*. The performance of the proposed approach using 3-fold cross-validation when trained for 1500 epochs is 98.27%. The confusion matrix of the proposed approach is shown in Fig. 5. In this matrix, the non-diagonal elements represent the misclassified observations. The plot of test error versus training iteration for the three-folds is plotted in Fig. 6. The performance of the proposed approach against existing approaches is given in Table 2. From the table, it can be observed that the proposed approach using only depth information is comparable with the state-of-the art results using color, depth, and optical-flow information.

3 Conclusion

In this work, a deep learning framework for hand gesture recognition is proposed. In contrast to existing approaches that use raw video data as input, the salience and foveation information are used to generate an input representation capturing the

	circle	triangle	up down	right left	wave	z	cross	come here	turn around	pat
circle	**97.64**	0	0	0	**2.35**	0	0	0	0	0
triangle	**0.92**	**99.07**	0	0	0	0	0	0	0	0
up down	0	0	**99.07**	0	0	0	0	**0.92**	0	0
right left	0	0	0	**96.71**	**3.28**	0	0	0	0	0
wave	0	0	0	**2.35**	**97.64**	0	0	0	0	0
z	0	0	0	0	0	**100**	0	0	0	0
cross	0	0	0	0	0	0	**100**	0	0	0
come here	0	0	0	0	0	0	0	**96.29**	0	**3.70**
turn around	0	0	0	**0.92**	0	0	0	0	**99.07**	0
pat	0	0	0	0	0	0	0	**2.77**	0	**97.22**

Fig. 5 Confusion matrix of proposed approach for SKIG dataset

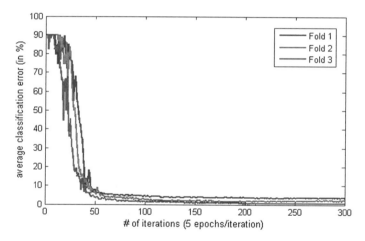

Fig. 6 Test error versus training iteration of the proposed approach for the 3-folds of SKIG dataset

change in spatial location of the hand. This representation is used by a CNN for a robust hand gesture recognition. Experimental study on SKIG hand gesture dataset demonstrates the effectiveness of this approach. The future work will extend this approach to other parts of the human body like limbs and face.

Table 2 Performance of the proposed approach on SKIG dataset

The approach	Modality	Accuracy
Liu et al. [6]	Depth + color	88.7
Tung et al. [15]	Depth + color	96.5
Pavlo et al. [7]	Depth + color	97.7
Nishida et al. [8]	Depth + color + optical flow	97.8
Pavlo et al. [7]	Depth + color + optical flow	98.6
Proposed approach	**Depth**	**98.27**

References

1. Bengio, Y. 2009. Learning Deep Architectures for AI. *Foundation and Trends in Machine Learning* 2 (1): 1–127.
2. Dominio, F., M. Donadeo, G. Marin, P. Zanuttigh, and G.M. Cortelazzo. 2013. Hand Gesture Recognition with Depth Data. In *4th ACM/IEEE International Workshop on Analysis and Retrieval of Tracked Events and Motion in Imagery Stream.* 9–16. ARTEMIS '13, ACM, New York, NY, USA. https://doi.org/10.1145/2510650.2510651.
3. Han, J., L. Shao, D. Xu, and J. Shotton. 2013. Enhanced Computer Vision with Microsoft Kinect Sensor: A Review. *IEEE Transactions on Cybernetics* 43(5): 1318–1334. https://doi.org/10.1109/TCYB.2013.2265378.
4. Jaemin, L., H. Takimoto, H. Yamauchi, A. Kanazawa, and Y. Mitsukura. 2013. A Ro-bust Gesture Recognition Based On Depth Data. In: *19th Korea-Japan Joint Workshop on Frontiers of Computer Vision (FCV).* 127–132. https://doi.org/10.1109/FCV.2013.6485474.
5. Kurakin, A., and Z.Z. Liu, Z. 2012. A Real Time System For Dynamic Hand Gesture Recognition with a Depth Sensor. In *20th European Signal Processing Conference (EUSIPCO),* 1975–1979.
6. Liu, L., and L. Shao. 2013. Learning Discriminative Representations From RGB-D Video Data. In *International Joint Conference on Arti cial Intelligence (IJCAI),* Aug 2013.
7. Molchanov, P., X. Yang, S. Gupta, K. Kim, S. Tyree, and J. Kautz. 2016. Online Detection and Classification of Dynamic Hand Gestures With Recurrent 3D Convolutional Neural Networks. In *IEEE Conference on Computer Vision and Pattern Recognition (CVPR).* 4207–4215.
8. Nishida, N., and H. Nakayama. 2016. Multimodal Gesture Recognition Using Multi-Stream Recurrent Neural Network. In *Revised Selected Papers of the 7th Pacific-Rim Symposium on Image and Video Technology,* PSIVT 2015, vol. 9431, 682–694, New York, NY, USA.
9. Ohn-Bar, E., and M. Trivedi. 2013. The Power is in Your Hands: 3D Analysis of Hand Gestures in Naturalistic Video. In *IEEE Conference on Computer Vision and Pattern Recognition Workshops (CVPRW),* 912–917. https://doi.org/10.1109/CVPRW.2013.134.
10. Palm, R.B. 2012. *Prediction as a Candidate for Learning Deep Hierarchical Models of Data.* Master's thesis, Technical University of Denmark, Asmussens Alle, Denmark.
11. Ren, Z., J. Yuan, J. Meng, and Z. Zhang. 2013. Robust Part-Based Hand Gesture Recognition Using Kinect Sensor. *IEEE Transactions on Multimedia* 15(5): 1110–1120. https://doi.org/10.1109/TMM.2013.2246148.
12. Ren, Z., J. Yuan, and Z. Zhang. 2011. Robust Hand Gesture Recognition Based On Finger-Earth Mover's Distance With a Commodity Depth Camera. In *Proceedings of the 19th ACM International Conference on Multimedia,* MM '11, ACM, 1093–1096, New York, NY, USA. https://doi.org/10.1145/2072298.2071946, http://doi.acm.org/10.1145/2072298.2071946.

13. Sarkar, A.R., G. Sanyal, and S. Majumder. 2013. Article: Hand Gesture Recognition Systems: A Survey. *International Journal of Computer Applications* 71(15): 25–37, Foundation of Computer Science, New York, USA.

14. Suarez, J., and R. Murphy. 2012. Hand Gesture Recognition With Depth Images: A Review. In *21st IEEE International Symposium on Robot and Human Interactive Communication (RO-MAN)*, 411–417. https://doi.org/10.1109/ROMAN.2012.6343787.

15. Tung, P.T., and L.Q. Ngoc. 2014. Elliptical Density Shape Model for Hand Gesture Recognition. In: *Fifth Symposium on Information and Communication Technology*, SoICT 14, ACM, 186–191, New York, NY, USA.

16. Wachs, J.P., M. Kolsch, H. Stern, and Y. Edan. 2011. Vision-Based Hand-Gesture Applications. *Communications of the ACM* 54(2): 60–71. https://doi.org/10.1145/1897816. 1897838, http://doi.acm.org/10.1145/1897816.1897838.

Human Fall Detection Using Temporal Templates and Convolutional Neural Networks

Earnest Paul Ijjina(ORCID)

Abstract The emerging areas of smart homes and smart cities need efficient approaches to monitor human behavior in order to assist them in their activities of daily living. The rapid growth in technology led to the use of depth information for privacy-preserving video surveillance, thereby making depth-based automatic video surveillance, a major area of research in academic and industrial communities. In this work, a temporal template representation of depth video is used for human action recognition using convolutional neural networks. A new temporal template representation capturing the spatial occupancy of the subject for a given time-period is proposed to recognize human actions. The ConvNet features extracted from these temporal templates that capture the local discriminative features of these actions are used for action detection. The efficacy of the proposed approach is demonstrated on SDUFall dataset.

Keywords Human action · Recognition · Temporal template · Convolution · Neural network

1 Introduction

Human behaviour analysis is a major area of research, due to its ability to recognize subject's intent, which has applications in various research/industrial domains including human–computer interaction, autonomous video surveillance, video categorization/tagging, robotics, gaming and entertainment. The analysis of human behaviour is sub-categorized into hand gesture recognition, facial expression recognition, full-body action detection, the interaction of a subject with another subject/object, a group activity or a crowd behaviour depending on the regions of interest and number of individuals considered in the study. It can also be classified

E. P. Ijjina (✉)
Department of Computer Science and Engineering, National Institute of Technology
Warangal, Warangal, Telangana 506004, India
e-mail: iep@nitw.ac.in

© Springer Nature Singapore Pte Ltd. 2020
K. S. Raju et al. (eds.), *Proceedings of the Third International Conference
on Computational Intelligence and Informatics*, Advances in Intelligent
Systems and Computing 1090, https://doi.org/10.1007/978-981-15-1480-7_70

into an activity, action or a motion depending on the temporal length of the analysis. In this work, we focus on recognizing human actions and interactions with an object/environment. Due to the rapid technological advancements, the need for automatic recognition of human actions in smart homes and cities is an essential requirement for providing assistance in activities of daily living. This work contributes to these efforts by designing a new temporal template for effective representation of human actions in depth videos. The features extracted by the convolutional neural network (CNN) on the temporal template representation of videos are used for action recognition.

A review of various deep learning approaches to activity recognition using different sensory equipment is given by [1]. An automated event detection system using audio features like spectrogram, Mel-frequency cepstral coefficients (MFCCs), linear predictive coding and matching pursuit computed on audio observation recorded by a smart-phone is proposed in [2]. The sensitivity to noise and the subject-specific nature (i.e. subject needs to carry the phone to ensure clear audio recording) limits its practical use. A smartphone-based approach using the reading from the accelerometer and the electronic compass for detecting the fall event of subject was proposed in [3]. An approach for recognizing daily activities using surface electromyography and an accelerometer is presented in [4], which uses the angle and amplitude information generated along the three axes to detect fall events. An accelerometer-based fall detection system using signal magnitude vector and signal magnitude area features computed from the temporal variation of acceleration is discussed in [5]. The variation in effectiveness of accelerometer information for action detection with its position and its subject-specific nature makes these accelerometer-based approaches unsuitable for reliable action detection. An embedded telehealth system using radar system for fall detection was proposed in [6]. The need for specialized sensory environmental set-up for recognition restricts its use for the general public. A multi-sensor fusion based fall detection system [7] using heterogeneous data fusion from multiple sensors and an evidential network for recognition was proposed for efficient recognition of human actions. The need for a specialized multi-sensory environment and its sensitivity to noise in recording conditions affects its suitability for practical use. A wireless body sensor network for monitoring 3D body motions is used with cloud computing to recognize fall events [8]. The subject-specific nature and specialized sensory set-up to be worn by the subject limit its suitability for prolonged monitoring of common subjects. A wireless body area network approach using air pressure sensors placed on the subject's body to recognize human falls is proposed in [9]. The need for specialized sensors and the necessity for wearing them during fall makes it impractical for regular use. A multi-camera video-based system using orthogonal views for calculating surface of the person in contact with the ground to detect a fall pose is proposed in [10]. This can misclassify a person lying on the floor as a fall. A support vector machine (SVM) based event detection system utilizing the tracking information of key human joints predicted from depth video is discussed in [11]. A two-stage event detection system using Microsoft Kinect is presented in [12], with the first stage segmenting the depth video frames to retrieve subject's vertical state followed by tracking and recognition in the second stage. A bag of curvature-scale-space words model, computed from human silhouettes obtained from depth video is used for fall detection in [13]. These depth-based approaches use computationally expensive methods like tracking, foreground computation and skeletal joint prediction, thereby unsuitable for real-time processing. In [14], shape and location features extracted from posture silhouettes computed from background subtraction of RGB video, are given as input to the support vector machine (SVM) to recognize human actions. The computation cost involved in background subtraction and shape/location feature extraction affects the suitability of this approach for practical use.

From the above discussion, some of the major limitations of existing approaches for fall detection are: (1) the use of sensors that needs to be worn/carried by the subject to be monitored, (2) the need for specialized sensory-environment/sensors for monitoring the subject, (3) the use of complex features that are computationally expensive, which might hinder their use for real-time monitoring and (4) the utilization of subject's pose after fall (that may not contain the most discriminative information) for recognizing fall events. To overcome these limitations of existing approaches, we use visual information in depth videos to compute a temporal template representation for human action detection. The effectiveness is further improved due to the robustness of ConvNet features used in action recognition. The novelty of this work is the new temporal template for action representation, computed on depth video with low computation cost and the efficient action recognition framework based on convolutional neural networks. The paper is organized such that: Sect. 2 presents the CNN based human action detection framework and the steps involved in computing the new temporal template. Section 3 describes the experimental study on SDUFall video dataset. Finally, Sect. 4 presents the concluding remarks and future work.

2 Proposed Method

In this work, we present a depth-based action recognition approach using a new temporal template representation and deep learning (CNN) based feature extraction. The ConvNet features [15] extracted from this temporal template representation are used by a support vector machine (SVM) [16] for human action detection. The block diagram depicting the major steps in this approach is given in Fig. 1.

In this work, the temporal template representation of videos is considered due to its ability to represent the motion in a sequence of frames as a single image. The ability of convolutional neural networks to extract robust (ConvNet) features from input representation is exploited to design an efficient action recognition framework. The next subsection explains the motivation, need and computation of the new temporal template proposed in this work.

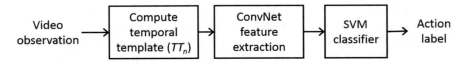

Fig. 1 Block diagram of the proposed approach

2.1 Temporal Template Representation of Depth Video

The traditional temporal templates like Motion Energy Image (MEI) and Motion History Image (MHI) are computed from the frame difference. To recognize motionless actions like sleeping, the temporal templates computed from frame difference will be ineffective. To overcome this limitation, we consider foreground information obtained by eliminating the background in depth video frames in the computation of the proposed temporal template. If $F(t)$ represents the tth video frame of observation with a background image B and (x, y) denotes the spatial coordinates. Then, the foreground image $FG(t)$ corresponding to the tth video frame is computed by removing the background (B) from the video frame $F(t)$, using Eq. 1.

$$FG(x, y, t) = |F(x, y, t) - B(x, y)| \tag{1}$$

This foreground information (FG) is binarized using a threshold (α) to obtain the binarized foreground information (BFG), using Eq. 2.

$$BFG(x, y, t) = FG(x, y, t) > \alpha \tag{2}$$

In the computation of a traditional motion history image (MHI), the use of max function leads to the suppression of overlapping foreground of former frames by the later frames, due to their higher gray value. As a result, the past foreground information gets overwritten at the overlapping regions, as shown in Fig. 2a. In a motion energy image (MEI), the overlapping and non-overlapping regions will have the same gray value, thereby losing the duration of overlap as shown in Fig. 2b. To overcome these limitations of traditional temporal templates, we propose a new temporal template (TT_n), that aggregating the scaled-by-n $\left(\frac{1}{n}\right)$ foreground information across n frames, shown in Fig. 2c. As a result, this temporal template captures the duration of spatial occupancy of the subject. This representation of observation with n frames computed by summing the scaled binarized foreground information (BFG) with $\frac{1}{n}$ is given in Eq. 3. A visual illustration of this computation is shown in Fig. 3.

$$TT_n(x, y) = \frac{1}{n} \sum_{t=1}^{n} BFG(x, y, t) \tag{3}$$

(a) MHI   (b) MEI   (c) TT_n

Fig. 2 The temporal templates (**a**) traditional Motion History Image (MHI), (**b**) Motion Energy Image (MEI), (**c**) New temporal template, TT_n computed from the foreground information for Bending action

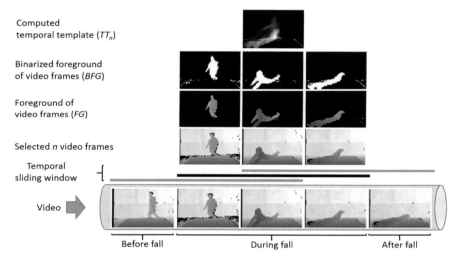

Fig. 3 Temporal template TT_n computation on a video stream

As video pre-processing using background subtraction removes other non-subject elements like chair and mattress on the floor, only subjects appearance in the observation is used for computing the temporal templates, which is, in turn, used for human action detection as shown in Fig. 3.

As a result, for observation with no subject motion, this temporal template representation captures the subject's pose. From Fig. 2c, it can be observed that scaling the foreground information reduces the random noise, that is common in low-cost depth video cameras. The following section presents the experimental study on SDUFall video dataset.

3   Experimental Study

This section presents the results of the proposed approach on SDUFall dataset [13]. Since a majority of time in activities of daily living is spent on motion-less actions, this dataset with motion-less actions is considered to evaluate the effectiveness of the approach for recognizing motionless actions. From the block diagram shown in Fig. 1, it can be observed that ConvNet features extracted by a convolutional neural network (CNN) [17] from temporal template representation is given as input to a support vector machine (SVM) classifier [18] for action recognition. This work considers a pre-trained CNN architecture, AlexNet to extract ConvNet features for action recognition.

The proposed approach is evaluated on SDUFall dataset,[1] which consists of six human actions, namely: *bending, falling, lying, sitting, squatting* and *walking*.

These actions were performed by 20 subjects among which 10 are male and 10 are female. Each subject performs the action 10 times by changing either (1) lighting conditions, (2) the object carried by the subject during the execution of the action, or (3) the relative directions and position of the subject with respect to the camera. As a result, 60 observations/subjects are captured, resulting in 1200 observations in total. The observations are recorded in three channels, namely, RGB video, depth video and the tracking information of 20 skeletal joints. The captured video is of 320×240 resolution at 30 fps and the average duration of each observation is 5.7 s. The leave one subject out (LOSO) evaluation scheme mentioned in [13] is used for evaluating the proposed approach.

The typical variation of frame difference for actions in this dataset is shown in Fig. 4. From the figure, it can be observed that there is no significant motion in the beginning and end of observation, due to the delay in subject's entry into the capturing region and the absence of subject's motion in the end, respectively. For an optimum representation, the temporal template should be computed on the frames with maximum motion information.

To demonstrate the variation in motion overtime for these actions, we consider a temporal sliding window of 20 frames and a shift width of 10 frames, to compute the temporal templates for observations of various actions. The temporal templates for the six actions are shown in Fig. 5.

From Fig. 5, it can be observed that (except for *walking* action), the computed temporal templates remain the same at the end of observations. The templates computed from the last video frames of *falling* and *lying* actions are identical, as the subject will be lying on the floor on completing either of these actions. As a result, the template corresponding to the last temporal window of the observation is not suitable for discriminating these actions. For effective discriminative representation of videos, we consider the template nearest to the end that is different for the last template, based on a threshold on the number of pixels they differ in. This template computed for each observation in the dataset is used as the input representation for

---

[1]http://www.sucro.org/homepage/wanghaibo/SDUFall.html.

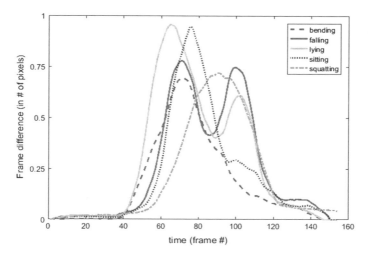

Fig. 4 Typical variation in frame difference for actions in SDUFall dataset

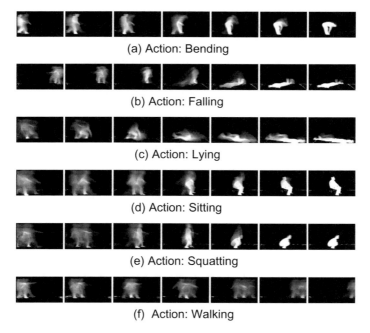

(a) Action: Bending

(b) Action: Falling

(c) Action: Lying

(d) Action: Sitting

(e) Action: Squatting

(f) Action: Walking

Fig. 5 The typical temporal templates for actions in SDUFall dataset, with a temporal window length of 20 frames and a shift width of 10 frames

Table 1 Performance comparison of the proposed approach on SDUFall dataset

Approach	Accuracy
Shape feat. + SVM class [13]	32.4
Shape feat. + ELM class [13]	51.31
Shape feat. + PSO-ELM class [13]	51.94
Shape feat. + VPSO-ELM class [13]	54.16
FV-SVM [19]	64.67
Fisher Vector-DVM [20]	64.67
Silhouette orientation Vol. [21]	89.63
Proposed approach	**91.58**

the video. The ConvNet features extracted from the template representation of observation are given as input to a support vector machine (SVM) for action recognition. An accuracy of 91.58% is obtained by the proposed approach for recognizing the six actions. The performance comparison of the proposed approach with existing approaches is given in Table 1. It can be observed that the proposed approach outperforms existing approaches, which could be a result of, the use of temporal template representation of observations and ConvNet feature for action detection.

In this dataset, multiple realistic variations of the same actions were considered making action recognition a challenging task. In spite of these variations in input representation of the same action, the proposed action recognition frame-work was able to efficiently discriminate these actions, which could be due to the robustness of ConvNet features. The next section presents the conclusions of this work.

4 Conclusions and Future Work

In this work, we propose a fall detection framework, using a new temporal template representation capturing the spatial occupancy of the subject over a given time period. A CNN is utilized to extract discriminative features, from the template representation of depth videos, to recognize the human actions. The ability of this template representation to capture the subject even in the absence of motion, makes this approach suitable for recognizing activities of daily living and human-fall. The noise tolerance and robustness of ConvNet features are also exploited for efficient recognition of human actions. The computation time of this approach can be significantly reduced for real-time applications by using general-purpose GPUs (GPGPUs) and hardware implementation of ConvNet feature extraction. The future work will extend this approach to other modalities and to other recognition tasks like hand gestures and group activities.

References

1. Wang, J., Y. Chen, S. Hao, X. Peng, and L. Hu. 2018. Deep Learning for Sensor-Based Activity Recognition: A Survey. *Pattern Recognition Letters.*
2. Cheena, M. 2016. Fall Detection Using Smartphone Audio Features. *IEEE Journal of Biomedical and Health Informatics* 20(4): 1073–1080.
3. Kau, L.J., and C.S. Chen. 2015. A Smart Phone-Based Pocket Fall Accident Detection, Positioning, and Rescue System. *IEEE Journal of Biomedical and Health Informatics* 19(1): 44–56.
4. Cheng, J., X. Chen, and M. Shen. 2013. A Framework For Daily Activity Monitoring and Fall Detection Based on Surface Electromyography and Accelerometer Signals. *IEEE Journal of Biomedical and Health Informatics* 17(1): 38–45.
5. Cheng, W.C., and D.M. Jhan. 2013. Triaxial Accelerometer-Based Fall Detection Method Using A Self-Constructing Cascade-AdaBoost-SVM Classifier. *IEEE Journal of Biomedical and Health Informatics* 17(2): 411–419.
6. Garripoli, C., M. Mercuri, P. Karsmakers, P.J. Soh, G. Crupi, G.A.E. Vandenbosch, C. Pace, P. Leroux, and D. Schreurs. 2015. Embedded DSP-Based Telehealth Radar System for Remote in-Door Fall Detection. *IEEE Journal of Biomedical and Health Informatics* 19(1): 92–101.
7. Aguilar, P.A.C., J. Boudy, D. Istrate, B. Dorizzi, and J.C.M. Mota. 2014. A Dynamic Evidential Network for Fall Detection. *IEEE Journal of Biomedical and Health Informatics* 18(4): 1103–1113.
8. Lai, C.F., M. Chen, J.S. Pan, C.H. Youn, and H.C. Chao. 2014. A Collaborative Computing Framework of Cloud Network and WBSN Applied to Fall Detection and 3-D Motion Reconstruction. *IEEE Journal of Biomedical and Health Informatics* 18(2): 457–466.
9. Lo, G., S. Gonzlez-Valenzuela, and V.C.M. Leung. 2013. Wireless Body Area Network Node Localization Using Small-Scale Spatial Information. *IEEE Journal of Biomedical and Health Informatics* 17(3): 715–726.
10. Mousse, Mikael A., C. Motamed, and E.C Ezin. 2017. Percentage of Human-Occupied Areas for Fall Detection from Two Views. *The Visual Computer* 33(12): 1529–1540.
11. Bian, Z.P., J. Hou, L.P. Chau, and N. Magnenat-Thalmann. 2015. Fall Detection Based on Body Part Tracking Using a Depth Camera. *IEEE Journal of Biomedical and Health Informatics* 19(2): 430–439.
12. Stone, E.E., and M. Skubic. 2015. Fall Detection in Homes of Older Adults Using the Microsoft Kinect. *IEEE Journal of Biomedical and Health Informatics* 19(1): 290–301.
13. Ma, X., H. Wang, B. Xue, M. Zhou, B. Ji, and Y. Li. 2014. Depth-Based Human Fall Detection Via Shape Features and Improved Extreme Learning Machine. *IEEE Journal of Biomedical and Health Informatics* 18(6): 1915–1922.
14. Yu, M., Y. Yu, A. Rhuma, S.M.R. Naqvi, L. Wang, and J.A. Chambers. 2013. An Online One Class Support Vector Machine-Based Person-Specific Fall Detection System for Monitoring an Elderly Individual in A Room Environment. *IEEE Journal of Biomedical and Health Informatics* 17 (6): 1002–1014.
15. Krizhevsky, A., I. Sutskever, and G.E. Hinton. 2012. Imagenet Classification With Deep Convolutional Neural Networks. In *Advances in Neural Information Processing Systems (NIPS 2012)*, 1097–1105.
16. Cortes, C., and V. Vapnik. 1995. Support Vector Networks. *Machine Learning* 20 (3): 273–297.
17. Lecun, Y., L. Bottou, Y. Bengio, and P. Haner. 1998. Gradient-Based Learning Applied to Document Recognition. In *Proceedings of IEEE* 86(11): 2278–2324.
18. Huang, G. B., Q. Zhu, and C. Siew. 2006. Extreme Learning Machine: Theory and Applications. *Neurocomputing* 70 (13): 489–501.

19. Aslan, M., A. Sengur, Y. Xiao, H. Wang, M.C. Ince, and X. Ma. 2015. Shape Feature Encoding Via Fisher Vector for Efficient Fall Detection in Depth-Videos. *Applied Soft Computing* 37: 1023–1028.
20. Aslan, M., O.F. Alin, A. Sengr, and M.C. Ince. 2015. Fall Detection With Depth-Videos. In *Proceedings of IEEE Signal Processing and Communications Applications Conference (SIU)*, 443–446.
21. Akagunduz, E., M. Aslan, A. Sengur, H. Wang, and M. Ince. 2016. Silhouette Orientation Volumes for Efficient Fall Detection in Depth Videos. *IEEE Journal of Biomedical and Health Informatics* 99: 1–8.

Action Recognition Using Motion History Information and Convolutional Neural Networks

Earnest Paul Ijjina

Abstract Human action recognition is a key step in video analytics, with applications in various domains. In this work, a deep learning approach to action recognition using motion history information is presented. Temporal templates capable of capturing motion history information in a video as a single image, are used as input representation. In contrast to assigning brighter values to recent motion information, we use fuzzy membership functions to assign brightness (significance) values to the motion history information. New temporal templates highlighting motion in various temporal regions are proposed for better discrimination among actions. The features extracted by a convolutional neural network from the RGB and depth temporal templates are used by an extreme learning machine (ELM) classifier for action recognition. Evidences across classifiers using different temporal templates (i.e. membership function) are combined to optimize the performance. The effectiveness of this approach is demonstrated on MIVIA video action dataset.

Keywords Temporal templates · Motion history information ·
Human action recognition · Convolutional neural networks(CNN) ·
Extreme learning machines(ELM)

1 Introduction

The study of human behaviour using visual analytics is an active area of research that focuses on analyzing subjects motion using computer vision algorithms to determine the nature of the behaviour. Human behaviour analysis can be classified into motion detection, gesture recognition, action recognition, event detection and activity detection depending on the duration for which the subject is analyzed. This

E. P. Ijjina (✉)
Department of Computer Science and Engineering, National Institute
of Technology Warangal, Warangal, Telangana 506004, India
e-mail: iep@nitw.ac.in

© Springer Nature Singapore Pte Ltd. 2020
K. S. Raju et al. (eds.), *Proceedings of the Third International Conference
on Computational Intelligence and Informatics*, Advances in Intelligent
Systems and Computing 1090, https://doi.org/10.1007/978-981-15-1480-7_71

can further be divided into single person behaviour, person to person interaction, interaction of person with object, group behaviour and crowd behaviour depending on the entities considered in motion analysis. Single person behaviour can be sub-classified into facial expressions, hand gestures, upper body actions and full-body actions based on the region of interest. Even with these wide variations in human behaviour analysis, visual analytics is the most widely used approach for detection. The introduction of Microsoft Kinect, a less expensive depth and colour camera, for gaming platform, led to the rapid growth in multi-modal computer vision research [1].

Over the years, various vision-based features for human action recognition were proposed. Some of these low-level features include space–time interest points (STIP) [2], motion boundary histograms (MBH) [3], histograms of optical flow (HOF) [4] and histogram-oriented gradients (HOG) [5]. These low-level local features are used to define the benchmark performance for datasets. Some of these features extend to spatio-temporal domain resulted in 3D SIFT [6], HOG3D [7] and 3D covariance descriptor [6]. In addition, temporal templates [8] are also used for human action recognition. In recent years, discriminative feature learning approaches like dictionary learning and deep learning have evolved, leading to the design of models that are adaptable to new domains. Some of these features learned from spatio-temporal data are EXMOVES [9] and C3D [10] features. In contrast to these approaches that use the raw data for feature learning, some approaches used deep learning model for recognizing patterns in extracted features. In [11], convolutional neural network (CNN) classifier is used to recognize local patterns in action bank features [12] for action recognition. A hierarchical extreme learning machine utilizing extreme learning machine autoencoders to learn local receptive fields (filtres) at multiple layers is proposed for visual recognition in [13]. A sparse canonical temporal alignment (SCTA) approach to action recognition with deep tensor decomposition is proposed in [14] by using SCTA for selecting and alignment of keyframes between observations. A tensor factorization method is used to find tensor subspace for extracting discriminative features from the selected keyframes.

In this paper, new temporal templates for action recognition in RGB + depth videos are proposed. In the computation of the temporal templates, the motion information between video frames is assigned weight (significance) by a fuzzy membership function. The novelty of this approach is the design of temporal templates that highlight the discriminative information using fuzzy membership functions and the fusion of evidence across models using temporal templates generated by different membership functions. The straight forward computation of temporal templates and the low computation cost of ConvNet features make this a suitable approach for real-time applications. In this paper, Sect. 2 presents the approach and the steps involved in computing the ConvNet features on the generated temporal templates. Section 3 gives the experimental results on MIVIA action dataset. Finally, Sect. 4 concludes this work.

2 Proposed Approach

A deep learning approach for human action recognition in videos using new temporal templates is presented in this work. The major components of the proposed approach are shown in Fig. 1. For a given RGB-D video, temporal templates are computed for the RGB video and depth video streams separately using a fuzzy membership distribution function (μ). We use a fuzzy membership function to determine the grayscale intensity (significance) assigned to temporal information in various frames, in the computed templates. The features extracted by a Convolutional neural network from RGB and depth templates are used by an ELM classifier for detecting actions. By utilizing the ConvNet features extracted on both modalities for action recognition, we aim to address the limitations of individual modalities, like the sensitive to illumination conditions or noise. Evidences across models using different temporal templates are combined to optimize the performance. The various steps involved in this approach are discussed in detail in the following subsections.

2.1 Representation of Videos Using Temporal Templates

We consider temporal template representation of videos for human action detection due to their capability of aggregating motion in a sequence of images into a single image. One of the classical temporal templates is motion history image (MHI), that can be computed as the sum of scaled motion information between video frames. In this MHI, recent motion is given a higher grayscale value (significance). Similarly, equal significance (gray value of 255) is given to motion information in all the frames, in the computation of motion energy image (MEI).

For a video with n frames, the motion information of the frame t is represented by $\psi(t)$. The formulation for computing the new temporal template TT is given in Eq. 1. Here, $\psi(t)$ is the grayscale image containing the motion information (frame difference) and $\mu(t)$ is the value of the fuzzy membership function for frame t. The traditional MHI attributes higher significance to past motion information and a motion energy image (MEI) assigns equal weight. To overcome these limitations

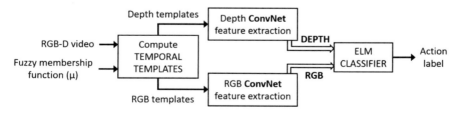

Fig. 1 The multi-modal recognition system

of traditional temporal templates, this new formulation for computing temporal templates that can assign higher weight to motion in any temporal region with a fuzzy membership function is presented.

$$\text{TT}(n) = \max{}_{t=1}^{n} \{\psi(t) \times \mu(t)\} \tag{1}$$

In this work, we consider six fuzzy membership functions μ1–μ6 whose distributions are given in Fig. 2. These functions highlight motion in the three major temporal regions (i.e. beginning, middle and ending of observation) and are considered in this work to demonstrate the use of function's distribution for weight assessment in Eq. 1. From the plots, it can be observed that motion energy image was computed by μ1, motion history image (MHI) was computed by μ2. In the remaining functions, μ3 assigns brighter value to motion in the beginning frames, {μ5, μ6} give more weightage to motion in middle frames, μ4 gives more weight to motion in beginning, ending frames. Thus, by defining temporal templates that highlight motion in various temporal regions i.e. beginning, middle, ending, we aim to identify the templates that have most discriminative information for recognizing the actions. The details of the proposed action recognition model using temporal templates are discussed in the following section.

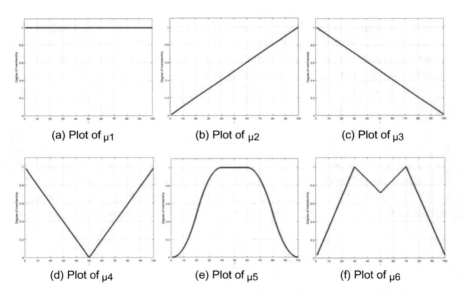

Fig. 2 Graphical representation of membership functions μ1–μ6 assuming # of frames in the observation (n) to be 100

2.2 Human Action Recognition Using CNN

The temporal template representations of depth and RGB video are used by a CNN [15] for ConvNet feature extraction. The extracted features are used by an extreme learning machine (ELM) classifier [16, 17] for human action recognition. In this work, the Matlab implementation of CNN given in [18] is used, optimizing the number of convolution kernels and their size empirically. The CNN classifier architecture consists of a convolution layer and an average pooling layer, followed by an ELM classifier. The filters in the convolution layer are set with orthogonal kernels and the output weights of the pooling layer are initialized randomly and the input weights of the output nodes in ELM classifier are learnt using least squares regression. The CNN classifier architecture consists of a convolution layer and an average pooling layer, followed by an ELM classifier.

This CNN + ELM classifier model is trained to predict the action associated for a given input, be generating the binary coded output. The predicted class label is determined by applying *argmax* on the (binary coded) outputs of the output layer. The following section presents the results on MIVIA dataset.

3 Experimental Study

The proposed action recognition model using temporal templates is evaluated on MIVIA [19, 20] action dataset, with RGB-D videos of seven actions, namely: *drinking, interacting with a table, opening a jar, random motion, sitting, sleeping* and *stopping*. The 7 actions were performed by 14 subjects with an average duration of 5 s. The RGB and depth videos of these observations are provided with this dataset and leave-one-subject-out (LOSubO) evaluation scheme is used for evaluating an approach on this dataset.

Since some of the actions in this dataset like *sleeping* and *sitting* barely have any motion, the RGB temporal templates computed from frame difference cannot discriminate these actions. To overcome this limitation, the depth temporal templates are computed from binarized depth data. The templates computed from both modalities are down-sampled to a 63×63 representation. The optimum number of nodes in the hidden layer of ELM classifier, the number and size of convolution kernels are empirically determined to be 5000, 28 and 7×7. The comparative study of this approach for the proposed membership functions using LOSubO scheme is in Table 1. It can be observed that μ_6 achieved better discrimination compared to other functions. Considering the binarized depth data in computing depth temporal template, assigning higher significance to motion in the middle frames, and the extraction of ConvNet features from both modalities are the potential contributors to the achievement of better discrimination with μ_6. Since the new templates assign higher significance to motion in one temporal region only, we combine evidence across multiple temporal templates for maximizing the

Table 1 Performance of the membership functions for depth, RGB and RGB and depth ConvNet features on MIVIA action dataset

Membership function	ConvNet features + ELM classifier		
	Depth data	RGB data	RGB and depth data
μ1	84.52	49.77	88.87
μ2	77.79	35.30	89.56
μ3	84.43	35.30	87.77
μ4	81.78	28.62	86.68
μ5	84.98	28.85	89.47
μ6	83.88	32.05	92.22

Fig. 3 Confusion matrix for MIVIA dataset with Average fusion across {μ2, μ3, μ5}

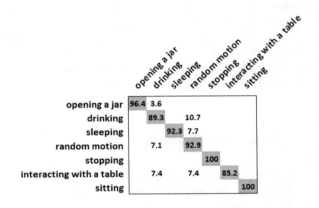

effectiveness of the recognition model. When evidences from {μ2, μ3, μ5}, emphasizing motion in {ending, beginning, middle} temporal regions are combined using average fusion-rule, 93.92% accuracy is obtained, whose confusion matrix is shown in Fig. 3. The performance comparison is given in Table 2. The proposed approach achieved better results than the existing approaches, which could be due to the various temporal templates and the use of multi-modal ConvNet features with ELM classifier for recognition.

The key contributors to the effectiveness of this work are: (1) the use of new templates capable of emphasizing motion in any temporal region of the observation, (2) the binarization of depth data when computing depth temporal template, (3) the

Table 2 Performance comparison on MIVIA dataset

Approach	Accuracy (in %)
Reject mechanism in [19]	79.8
HaCK in [21]	80.1
BoW in [20]	84.1
Deep Learning in [22]	84.7
Edit distance in [23]	85.2
Proposed approach	**93.92**

use of features extracted by CNN from both RGB and depth steams of video for action recognition (early fusion) and (4) combining evidences across models using different membership functions (late fusion). The next section concludes this work.

4 Conclusion

In this work, new temporal templates highlighting various temporal regions were proposed to emphasize the discriminative information necessary for human action recognition. A convolutional neural network coupled with an ELM classifier are used for human action recognition from temporal templates. Experiment study on MIVIA action dataset suggests the effectiveness of the proposed approach. For optimal performance, early fusion of ConvNet features extracted from both RGB and depth temporal templates are used for action detection, and late fusion of evidences across classifier using different temporal templates, are employed. The future work will extend this to other modalities and tasks like upper body action and hand gesture recognition.

References

1. Han, J., L. Shao, D. Xu, and J. Shotton. 2013. Enhanced Computer Vision With Microsoft Kinect Sensor: A Review. *IEEE Transactions on Cybernetics* 43(5): 1318–1334.
2. Laptev, I., and T. Lindeberg. 2003. Space-Time Interest Points. In *International Conference on Computer Vision (ICCV)*, 432–439.
3. Dalal, N., B. Triggs, and C. Schmid. 2006. Human Detection Using Oriented Histograms of flow and Appearance. In *European Conference on Computer Vision*, ECCV 06, II, 428–441.
4. Laptev, I., M. Marszalek, C. Schmid, and B. Rozenfeld. 2008. Learning Realistic Human Actions from Movies. In *Conference on Computer Vision and Pattern Recognition*, June 2008.
5. Dalal, N., and Triggs, B.: Histograms of oriented gradients for human detection. In Computer Vision and Pattern Recognition (CVPR), 886–893.
6. Yuan, C., X. Li, W. Hu, H. Ling, and S.J. Maybank. 2014. Modeling Geometric-Temporal Context With Directional Pyramid Co-Occurrence For Action Recognition. *IEEE Trans-actions on Image Processing* 23(2): 658–672.
7. Klaser, A., M. Marszalek, and C. Schmid. 2008. A Spatio-Temporal Descriptor Based on 3D-Gradients. In *British Machine Vision Conference (BMVC)*, 995–1004, Sept 2008.
8. Bobick, A.F., and J.W. Davis. 2001. The Recognition Of Human Movement Using Temporal Templates. *IEEE Transactions on Pattern Analysis and Machine Intelligence (PAMI)* 23 (3): 257–267.
9. Tran, D., and L. Torresani. 2013. EXMOVES: Classifier-Based Features for Scalable Action Recognition, CoRR abs/1312.5785.
10. Tran, D., L.D. Bourdev, R. Fergus, L. Torresani, and M. Paluri. 2014. C3D: Generic Features for Video Analysis. CoRR abs/1412.0767.

11. Ijjina, E.P., and C.K. Mohan. 2014. Human Action Recognition Based on Recognition of Linear Patterns in Action Bank Features Using Convolutional Neural Networks. In *International Conference on Machine Learning and Applications (ICMLA)*, 178–182, Dec 2014.
12. Sadanand, S., and J.J. Corso. 2012. Action Bank: A High-Level Representation of Activity in Video. In *IEEE Conference on Computer Vision and Pattern Recognition (CVPR)*, 1234–1241, June 2012.
13. Zhu, W., J. Miao, L. Qing, and G.B. Huang. 2015. Hierarchical Extreme Learning Machine for Unsupervised Representation Learning. In *International Joint Conference on Neural Networks (IJCNN)*, 1–8, July 2015.
14. Jia, C., M. Shao, and Y. Fu. 2017. Sparse Canonical Temporal Alignment With Deep Tensor Decomposition for Action Recognition. *IEEE Transactions on Image Processing* 26(2): 738–750.
15. McDonnell, M.D., and T. Vladusich. 2015. Enhanced Image Classification With A Fast-Learning Shallow Convolutional Neural Network. CoRR abs/1503.04596 2015.
16. Huang, G.B., H. Zhou, X. Ding, and R. Zhang. 2012. Extreme Learning Machine for Regression and Multiclass Classification. *IEEE Transactions on Systems, Man, and Cybernetics, Part B (Cybernetics)* 42(2); 513–529, Apr 2012. https://doi.org/10.1109/TSMCB.2011.2168604.
17. Huang, G.B., Q.Y. Zhu, and C.K. Siew. 2006. Extreme Learning Machine: Theory and Applications. *Neurocomputing* 70 (1): 489–501.
18. Zhu, W., J. Miao, and L. Qing. 2015. Constrained Extreme Learning Machines: A Study on Classification Cases. CoRR abs/1501.06115.
19. Carletti, V., P. Foggia, G. Percannella, A. Saggese, and M. Vento. 2013. Recognition of Human Actions from RGB-D Videos Using a Reject Option, 436–445, Springer: Berlin, Heidelberg.
20. Foggia, P., G. Percannella, G., A. Saggese, and M. Vento. 2013. Recognizing Human Actions by a Bag of Visual Words. In *IEEE International Conference on Systems, Man, and Cybernetics*, 2910–2915, Oct 2013.
21. Brun, L., G. Percannella, A. Saggese, M. Vento, and A. Hack. 2014. A System for the Recognition of Human Actions by Kernels of Visual Strings. In *IEEE International Conference on Advanced Video and Signal Based Surveillance (AVSS)*, 142–147, Aug 2014.
22. Foggia, P., A. Saggese, N. Strisciuglio, and M. Vento. 2014. Exploiting the Deep Learning Paradigm for Recognizing Human Actions. In *IEEE International Conference on Advanced Video and Signal Based Surveillance (AVSS)*, 93–98, Aug 2014.
23. Brun, L., P. Foggia, A. Saggese, and M. Vento. 2015. Recognition of Human Actions Using Edit Distance on Aclet Strings. In *International Conference on Computer Vision Theory and Applications (VISAPP)*, 97–103.

A Review of Various Mechanisms for Botnets Detection

Rishikesh Sharma and Abha Thakral

Abstract In recent years, the threat of botnets has increased exponentially. Escaping the barriers formed by firewall security, these bots attack a large number of hosts and infiltrate with the malicious content. To evade the botnets an efficient tracking is performed. In this paper, we have discussed the various attacking protocols adapted by botnets. It also includes the different ways in which the bots can be detected and stopped from harming the computer. The tracking is done by classifying the IP addresses in separate lists according to their respective features. The particular content embedded in images and spam e-mails are mentioned. The botmaster uses Command and Control server (C&C) to monitor the bots. The several ways preferred for the interaction include Internet Relay Chat (IRC), HyperText Transfer Protocol (HTTP) and Peer to Peer (P2P). Observing the threats and problems caused to a common person, important steps are being taken for consideration. Recently, Google has taken a step and removed 300 apps from play store which were designed to perform DDoS attacks on android devices. The detection of botnets helps us by taking the necessary steps and making us alert before any malicious activity takes place. The attackers have also occupied several social networking platforms with the network of bots that are further being implemented in our day to day lives to manipulate and influence the users. The platforms are henceforth misused to spread a message by posting it socially time and again with the unreal profile of bots [1].

Keywords Command and control (C&C) · Internet relay chat (IRC) · HyperText transfer protocol (HTTP) · Peer to peer (P2P)

R. Sharma · A. Thakral (✉)
Amity University, Noida, India
e-mail: athakral@amity.edu

R. Sharma
e-mail: rishikeshsharma501@gmail.com

© Springer Nature Singapore Pte Ltd. 2020
K. S. Raju et al. (eds.), *Proceedings of the Third International Conference on Computational Intelligence and Informatics*, Advances in Intelligent Systems and Computing 1090, https://doi.org/10.1007/978-981-15-1480-7_72

1 Introduction

A botnet is a network comprising of computers infected with malware which responds to the attackers to perform any activity they want. These infected computers are also known as bots or zombies and their master is called the botmaster. The illegal authorisation to the users' systems via bots enables the cybercriminal to control these infected computers remotely. These botnets are usually spread with the help of installed trojan, rootkits and worms which may have launched with some malicious purposes. These platforms are easily available on social networks and all over the web. With these networks, the botmaster can lead the cyber deeds such as financial frauds, malware distribution, identity theft, mass mailing of spam and storing of illegal content. With a single network, cybercriminals can commit several crimes within no time and capture sensitive data.

Meanwhile, many a times the computers don't even know that they are infected and the contaminated system can be used as a bot to infect another user. As proposed by The Spamhaus Project in 2017, there has been an enormous increase in the number of bots i.e. 32% of C&C botnets and the number of IoT botnets is increased by twice in 2017 compared to 2016 [1]. With the recent growth in botnets we must keep a track that despite the increase in botnets, there is a substantial decrease in the number of bots. This has been acclaimed because the bots today are more efficient and well trained to perform the required activities. The increase in botnets has led to the enhancement of a huge amount of traffic on online web services. But the generated traffic can be further bifurcated and studied to detect the bots according to the similarities. The DoS (Denial of service) attacks are prevailing in a huge number, but as claimed in [2], such attacks can be reduced with a few steps i.e. controlling them remotely and coordinating their included activities performed. Nowadays most of the population has easy access to the internet via their gadget. It has been observed that the network is not protected and the system is not programmed in a way to successfully detect the bot. The mobile phones are very favourable to be attacked via MMS, download links and Bluetooth sharing but can be avoided with the evaluation process. Even the laptops are not properly secured. Passwords are easy to decipher and the installed antivirus is often outdated. The attacker takes the benefit of such circumstances and connects to the system with one or the other authentication.

Once authorised, attackers are able to infect the system with their own server. The botnet also functions with the spread of spam emails that contain unsolicited messages. Recent survey has shown that there has been a drastic growth in spam malware and phishing attacks [3]. The phishing spam emails sent by the botmaster can steal the data by learning the credentials from the user. These messages are just the dummy, sent to the users to convince that it has been sent by a legit business firm. The user often believes to trust the vague financial institutions and land in trouble by disclosing personal information. The detection of such threats can be done by filtering the subject consisting of different keywords and contents based on the 'bad words' commonly used. Separate black and white lists are formed differentiating the sender's address and blocking it beforehand if it lies in the bad list. The recent such spam emails contain links hidden in the content i.e. texts and images which redirect to the false webpages that further redirect the victim to different domains. These included links need to be detected and made sure that they disallow any harmful redirection.

2 Bots and Botnets

The size of the botnets may further have a wide range and vary according to the aim and functions required to perform as mentioned in [4]. The recent actions to be taken are communicated to the bots from the botmaster. Usually, the bots are rested and do not perform any activities. The bots get activated only under specific circumstances. They contact the botmaster periodically and are invoked if they receive any messages.

The bots can be controlled by sending commands from school, home, park, government organisation or any other place in the world having the connectivity. It is vital to protect the systems from getting infected. The bots must be detected at a very early stage so that it can be avoided before it could perform any function. It has been stated that the detection based on DNS traffic and data mining are the most efficient instead of depending upon the basic style used for detection. It has also been proved in the research by applying different algorithms, that data mining techniques provide the best results [5]. A few researches [6] have been conducted to differentiate the phases in initialisation, processing and the final attacks; launched to successfully identify the botnets. The various methods which have been introduced for the detection of botnets and for securing our system from their serious threat are discussed in this paper. The attacks from botnet are not just limited to certain boundaries, but reach to every aspect possible. The attacks from botnet are not just limited to certain boundaries but reach to every aspect possible. Recently, the Russian president, Vladimir Putin had released a statement talking about nearly 25 million cyberattacks which were prevented during the FIFA world cup 2018 [7]. DETECTION BASED ON PROTOCOLS—There has been a constant approach for the detection of the contagious networks. The most widely and efficient methods adapted are as follows.

2.1 Command and Control (C&C)

The most common way of communication among the bots prevailing today is Command and Control. The attacker sends the command to the bots and controls them whenever needed as per the requirement. The messages are sent periodically guiding them to perform certain jobs. Depending upon the purpose and the action the same message can be sent to a single bot or to all the bots in the network

Table 1 Command and Control (C&C)

Author(s)	Discussions	Features	Benefits
Gu et al. [8]	Network based anomaly detection	– Spatial temporal correlation – Similarity analysis of bots	– Detect both host and server – Very efficient – Can even detect small botnets
Cho et al. [9]	Inference of protocol	– Finiteness and completeness of state machines – Reverse engineering	– Less number of queries – Time is saved – Operated by both human and tools
Gu et al. [10]	Distinguish between Communications	– Hypothesis testing – False results	– More reliable than passive anomaly techniques

simultaneously. The bots respond to the message immediately and provide feedback to the botmaster. Attackers are able to attack with distributed denial of service attacks (DDoS attacks) instructing the bots to attack a specific system. C&C server provides a platform for low-latency communication and is a very dependent source for botnets. A well-structured server comprising of interconnected bots favour redundancy and keep the server functioning. If it experiences any failure the bots still keep working (Table 1).

It is easy to detect a bot as the way of interaction is very uniform compared to humans; and once the C&C server is revealed, the whole botnet can be destructed. The communication among the bots and conversations held among humans can be separated efficiently and the detection can be done providing the false-positive results [10]. By applying fuzzy logic and the concepts of reverse engineering to enhance the botnet protocol according to state machines, the devices can be secured from the identified bot-attacks [9]. The bots of the same botnet respond in the same manner within the same pattern and time which can be noticed very easily. The traffic generated is very less and is identical to the general traffic. This makes it very difficult to distinguish between the necessary traffic generated and the excess ones which have no use. The detection of this traffic is very important, to make the network free from any unwanted creation of platform which may lead to infectious actions. The BotSniffer has applied the anomaly detection technique on a similar kind of behaviour among the detection of channel is made, all the bots and servers can be stopped. After the reveal of botnet, the whole server is shut down and the other vulnerable systems can be protected. The infected ones are to be notified so that they can remove such destructive functioning present in the system.

2.2 Internet Relay Chat (IRC)

IRC is an application layer protocol used quite extensively in the field of botnet. Internet Relay Chat is basically text messaging over the internet. However, instead of communicating from one person to another, we have the option to do group communication with as many people as we want, in segregated rooms or channels as commonly known. Not only chat, IRC also provides us the facility to share the files from one place to another and transfer the data among many bots. The usage of IRC is declining evenly but still, there is a large number of users opting for it.

Since a botnet consists of many bots at an instance and IRC provides a real-time channel to instruct the command to all, it is very helpful for the attackers saving the time and cost. On the contrary, if the bots are unable to connect the IRC server at centre, the botmaster faces difficulty to connect with the group. IRC can be traced easily as each bot needs a server for communication. Every bot has to respond and it needs to take place within a certain span of time. By noticing this synchronisation, the botnet is detected. This helps in reading the messages and breaking the flow of any further communication in the channel. The IRC bots do not perform any activity if no command is given to them. The IRC connections can also be detected by noticing the initial bytes as every IRC period includes PASS, NICK and USER keywords uniquely (Table 2).

Table 2 Internet Relay Chat (IRC)

Author(s)	Discussions	Features	Benefits
Gu [11]	Clustering of traffic	– Tracing of real networks – Free of command and control protocol – Intrinsic botnet communication and characteristic	– No prior knowledge is required – Efficiently grouping of traffic
Binkley and Singh [12]	Detection based on TCP	– IP channel names – Collection of IRC statistics	– Client botnets can easily be detected
Goebel and Holtz [13]	Scoring based on n-gram analysis	– Maintaining network traffic – IRC nicknames and IRC servers	– Detection of rare communication channels – Determine IP address – Warning e-mail is generated
Ma [14]	Characterisation of packet size sequence of TCP conversation	– Unknown algorithm measuring quasi-periodicity	– High accuracy with low positive rates – Detection can be done at a very early stage

2.3 Hypertext Transfer Protocol (HTTP)

HTTP is the most commonly used protocol in the world providing a convenient way to quickly and reliably move any data on the web. It is an application layer protocol that allows web-based applications to communicate and exchange data. It acts as a messenger on the web that can be used to deliver content such as images, audio files, video files, documents, etc. If two computers want to communicate and exchange data, they act like a client-server relation and take part in a request-respond cycle. The computers that communicate via the HTTP must understand and follow the HTTP protocol (Table 3).

Often, HTTP is preferred by the botmasters over IRC because HTTP is a connectionless protocol, i.e. after making the request the two computers disconnect from each other and when the response is ready, the connection re-establishes to deliver the response and then it gets closed. This feature makes it difficult for tracking but if a bot is tracked once, the bots need to be connected again. The bots need to provide information to each other anew and the connection is handled as the very first time. The performed activities and repetitive connections to destination nodes are monitored. The scoring approach has been proposed as well for the detection of HTTP: if the score increases more than the threshold, the domain is

Table 3 HyperText Transfer Protocol (HTTP)

Author(s)	Discussions	Features	Benefits
Balupari et al. [15]	Scoring based on repetitive attempts for the Connection with the Help of Heuristics	– Connection between specific domain and resources with distinction node – Botnet activities are monitored	– A very efficient way working with heuristics and provide an active result
Garasia et al. [16]	Application of data mining algorithms in combination with timestamp	– Operations being held on network traffic – Filtration and segregation of traffic	– A huge amount of data can be operated – More efficient than heuristics and signature based methods
Koo et al. [17]	Detection through filtering on repeatability P2P technique is adapted	-A comparing approach is adapted -Standard deviation is applied to record the repeatability	– Cost of implementation is lowered – More resilience effect
Lee et al. [18]	Periodic repeatability	– Activities by bots are to be monitored – The degree calculations are done by standard deviation	– Better than methods regarding DNS queries – More organised and intelligence

said to be malicious [15]. The HTTP protocol can also be solved by applying algorithms to detect the botnets with the help of timestamp. The application of such algorithms helps in detection of botnets without any known information about the network [16]. The HTTP botnets are also to be detected by recording the span of time taken during the connection as they request the commands regularly [17]. It has been proposed that DDoS attacks can be tracked with the help of recent testbeds designed according to the HTTP network. This also helps to point out the malicious HTTP bots by linking the relations made between clients and servers [18].

2.4 Peer to Peer (P2P)

Compared to C&C, peer to peer network differed in a very particular manner. Wherein C&C the communication is centralised, in P2P it takes place between several peers and provides the same harvesting, operating and distribution of data [19]. One of the most favourable features of the network is its resilient coefficient which helps it to recover quickly and perform with more efficiency [20]. P2P network consists of a group of bots with same features and attributes. The P2P network follows the architecture of dividing the activities among the bots which has to be processed. It is not a centralised network and each node is considered to be a client as well as a server. The peers within the network arises a problem for the hosts as their discovery may lead to a failure of network. The data is shared via internet from one peer to another without the involvement of any central server. It allows the sharing of files within all the connected peers and leaks the personal information from any bot. The P2P networks are much more complex compared to the others but it provides with one huge advantage. Even if a single infected system is detected, the whole network still survives as there is no way to track other bots as well through that bot. The survey shows that there has been a growth in peer to peer networks being distinctive from C&C servers. The linking of behavioural tasks performed by P2P network helps to generate an algorithm that further helps in enhancing the detection of bots. The algorithm based on the activities is performed within a specific interval of time and provide adaptable results [21]. A very high accuracy is observed while detecting the P2P botnets based on nodes. This is achieved by monitoring the flow, classifying the bots with the help of different algorithms and then evaluating them [22]. In research, it has been shown that P2P attacks can be effectively stopped even before the launching of threats by applying machine learning algorithms. The P2P network can also be differentiated based on the different components like attacking, propagating, controlling, updating, etc. This way of categorisation provides better results and is able to detect new botnets. It is also possible to detect the botnets at a working stage by studying the traffic and clustering accordingly. This leads to specifying the flow of traffic as well as the dependence upon different states of network while functioning [23], (Table 4).

Table 4 Peer to Peer (P2P)

Author(s)	Discussions	Features	Benefits
Grizzard et al. [19]	An architectural overview	– Harvesting dispersion and processing of information	– No centralised point
Davis et al. [20]	Simulation is used over the network	– Different strategies as random, tree-like and global have been adapted	– Overnet is less robust but has more resilience
Al-Hammadi and Aickelin [21]	Application of algorithms to correlate activities	– Analysing function and signal log file	– Pre defined bots signatures is not required – False alarms are reduced
Yin [22]	Node based detection	– Focuses network characteristics – Monitoring of network flow is done	– Has been provided very efficient up to 99–100% – Comprises of very low false rate of 0–2%
Huy Hang et al. [23]	The Social Behaviour of Botnets is taken into consideration	– Tracing traffic to analyse graphically – Clustering and graph mining	– Botnets can be stopped at waiting stage

3 Conclusion

In this paper, we have discussed the different methods that are adapted to detect malicious botnets. From these ways, our system can be protected from the severe threat prevailing in the cyber world. The botmasters practice several techniques to perform nefarious activities, but they can be tackled defensively. We have surveyed the mechanisms thoroughly and compared the efficiency of botnet detecting ways according to the attributes and environment of the attack. The agenda of attackers is to spread the network purposely performing the cyber-crimes. They come up with new concepts for gathering private data and information by adapting different protocols. Hence, every user must be prepared and be ready to take required actions to protect their data and evade the system from such attacks.

References

1. https://www.spamhaus.org/news/article/772/spamhaus-botnet-threat-report-2017.
2. Freiling, F.C., T. Holz, and G. Wicherski. 2005. Botnet Tracking: Exploring A Root-Cause Methodology to Prevent Distributed Denial-of-Service Attacks. In *European Symposium on Research in Computer Security*, 319–335, Springer, Berlin, Heidelberg.

3. https://hackercombat.com/phishing-attacks-increased-59-percent-2017-kaspersky-report/.
4. Fabian, M.A.R.J.Z., and M.A. Terzis. 2007. My Botnet is Bigger Than Yours (Maybe, Better Than Yours): Why Size Estimates Remain Challenging. In *Proceedings of the 1st USENIX Workshop on Hot Topics in Understanding Botnets*, Cambridge, USA.
5. Liao, W.H., and C.C. Chang. 2010. Peer to Peer Botnet Detection Using Data Mining Scheme. In *2010 International Conference on Internet Technology and Applications*, IEEE, 1–4.
6. Liu, L., S. Chen, G. Yan, and Z. Zhang. 2008. Bottracer: Execution-based Bot-Like Malware Detection. In *International Conference on Information Security*, 97–113, Springer, Berlin, Heidelberg.
7. https://www.infosecurity-magazine.com/news/russia-fends-off-25-million-world/.
8. Gu, G., J. Zhang, W., and W. Lee. 2008. *BotSniffer: Detecting Botnet Command and Control Channels in Network Traffic*.
9. Cho, C.Y., E.C.R. Shin, and D. Song. 2010. Inference and Analysis of Formal Models of Botnet Command and Control Protocols. In *Proceedings of the 17th ACM conference on Computer and communications security*, ACM, 426–439.
10. Gu, G., V. Yegneswaran, P. Porras, J. Stoll and W. Lee. 2009. Active Botnet Probing to Identify Obscure Command and Control Channels. In *2009 Annual Computer Security Applications Conference, ACSAC'09*, 241–253, IEEE.
11. Gu, G., R. Perdisci, J. Zhang, W. Lee. 2008. Botminer: Clustering Analysis of Network Traffic for Protocol-and Structure-Independent Botnet Detection.
12. Binkley, J.R., and S. Singh. 2006. An Algorithm for Anomaly-based Botnet Detection. *SRUTI* 6: 7–7.
13. Goebel, J., and T. Holz. 2007. Rishi: Identify Bot Contaminated Hosts by IRC Nickname Evaluation. *HotBots* 7: 8–8.
14. Ma, X., X. Guan, J. Tao, Q. Zheng, Y. Guo, L. Liu, and S. Zhao. 2010. A Novel IRC Botnet Detection Method Based on Packet Size Sequence. In *2010 IEEE International Conference on Communications (ICC)*, IEEE, 1–5
15. Balupari, R., V. Mahadik, B. Madhusudan, C.H. Shah. 2014. U.S. Patent No. 8,677,487, Washington, DC, U.S. Patent and Trademark Office.
16. Garasia, S.S., D.P. Rana, and R.G. Mehta. 2012. HTTP Botnet Detection Using Frequent Patternset Mining. In *Proceedings of [Ijesat] International Journal of Engineering Science and Advanced Technology 2*: 619–624.
17. Koo, T.M., H.C. Chang, and G.Q. Wei. 2011. Construction P2P Firewall HTTP-Botnet Defense Mechanism. In *2011 IEEE International Conference on Computer Science and Automation Engineering (CSAE)*, IEEE, 1, 33–39.
18. Lee, J.S., H. Jeong, J.H. Park, M. Kim, and B.N. Noh. 2008. The Activity Analysis Of Malicious http-Based Botnets Using Degree of Periodic Repeatability. In *International Conference on Security Technology, 2008, SECTECH'08*, IEEE, 83–86.
19. Grizzard, J.B., V. Sharma, C. Nunnery, B.B. Kang, and D. Dagon. 2007. Peer-to-Peer Botnets: Overview and Case Study. *HotBots* 7: 1–1.
20. Davis, C.R., S. Neville, J.M. Fernandez, J.M. Robert, and J. Mchugh. 2008. Structured Peer-To-Peer Overlay Networks: Ideal Botnets Command and Control Infrastructures? In *European Symposium on Research in Computer Security*, 461–480, Springer, Berlin, Heidelberg.
21. Al-Hammadi, Y., and U. Aickelin. 2010. Behavioural Correlation for Detecting P2P Bots. In *Second International Conference on Future Networks, ICFN'10*, IEEE, 323–327.
22. Yin, C. 2014. Towards Accurate Node-Based Detection of P2P Botnets. *The Scientific World Journal*.
23. Hang, H., X. Wei, M. Faloutsos, and T. Eliassi-Rad. 2013. Entelecheia: Detecting p2p Botnets In Their Waiting Stage. In *IFIP Networking Conference*, 1–9.

Malicious URL Classification Using Machine Learning Algorithms and Comparative Analysis

Anshuman Sharma and Abha Thakral

Abstract Exponential expansion in the application of the internet in each and every field has resulted in the escalation of data traffic over the internet. In the vastness of this data it has become important for engineers to classify the data as malicious and non-malicious so that different traffic can be treated differently. Rule-based and port-based classification exhibited a number of limitations which ultimately led to the steep decline in their usage to classify the internet traffic and gave rise to the machine learning techniques which are more promising and efficient. In this paper four popularly known machine learning classifiers: KNN, Naive Bayes, Decision Trees and Random forest have been implemented to classify the internet traffic based on whether the traffic is malicious or not and then compare their results on the basis of their accuracy score.

Keywords Classification · Machine learning · KNN · Naïve Bayes · Decision tree · Random forest

1 Introduction

In today's technologically advanced world, the internet plays an important role and is considered to be the backbone of the future. every other thing in this world is connected with the internet be it the television, cars, traffic lights, etc. To one's surprise even a trash can is now connected with the internet to inform the sanitation crew when it is full. With such extensive use of internet, the traffic generated by the internet has also increased and to improve the performance, quality of services, user experience and network security it is very crucial to classify the traffic.

A. Sharma (✉) · A. Thakral
Amity University, Noida, India
e-mail: sharma.anshuman97@gmail.com

A. Thakral
e-mail: athakral@amity.edu

© Springer Nature Singapore Pte Ltd. 2020
K. S. Raju et al. (eds.), *Proceedings of the Third International Conference on Computational Intelligence and Informatics*, Advances in Intelligent Systems and Computing 1090, https://doi.org/10.1007/978-981-15-1480-7_73

Internet service vendors should know what is going through the traffic to take various actions promptly if required. The ISP's are also responsible for providing the information accessed by an individual at any particular time to the government if asked for. As 'lawful Interception' helps keep track of any suspicious activity that may give rise to crime or terrorism. Malicious cyber activities across the internet make the system vulnerable to threats that might infiltrate and damage the computer without the owner's consent. Malicious URL's are extensively being used in cyber-attacks. Out of every 14 users, 1 user is tricked to click on a link containing malicious content [1, 2], which then gives the data access to the hacker and demand a handsome ransom or misuse the data, therefore, evaluation of risk associated with each type of URL must be done before clicking on to some random URL. The most common technique that the web browsers follow is to "blacklist" some type of websites that have been labeled after taking various feedbacks but this eventually is not a good technique as data is ever-growing as a repercussion of which the blacklist is very difficult to be updated before a malicious URL is clicked. Virus attacks and unauthorized use of systems lead to great financial losses and computer crimes [3]. The evolution of technology day by day makes it difficult to secure web systems. For example, Wanna cry attack that took place in 2017 has been recorded as one of the worst cyber-attacks in history. A type of malicious software that bars access to the files until the compensation is paid in exchange for the decryption key [4]. Therefore, it is of extreme importance to classify the malicious and non-malicious URL's beforehand. Also, traffic classification is critical for improving the IP network performance and user experience, as there's only limited bandwidth available and there is a lot of traffic that needs to get through. For example, traffic like online transactions, video conferencing, etc. need to get through on immediate basis, whereas other traffic like internet browsing, over the top streaming video, etc. may be less sensitive to delay but more sensitive to delay variation i.e. they can wait for the immediate traffic to pass and then they will pass. And in the end, there is all other that needs to get through but can still wait a bit. Therefore, when we classify the traffic before putting it on network we make the best use of bandwidth available and also bypass the pitfall to the computer system by classifying the malicious and non-malicious traffic beforehand which thereby enhances the user's experience and improves the quality of internet services and security [5, 6].

Machine Learning is a new and promising technique that learns from the patterns and statistics and classifies accordingly. Series of steps are followed in the application of Machine Learning. The arch step is to gather the data and split the data into test and train dataset followed by the training, modeling and predicting (Fig. 1).

Fig. 1 Machine learning model

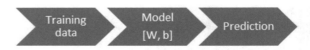

In this paper internet traffic classification techniques are discussed with the comparative analysis of four supervised machine learning classifiers.

The experimental result shows that random forest classifier came up with the highest accuracy as compared to other machine learning classifiers which was 84.98%.

2 Literature Review

Numerous studies have been conducted in the field of Internet traffic classification, researchers are pondering over techniques which yield a higher accuracy with a minimum false-negative rate to avoid the threat to systems from malicious URL's and to improve the network performance.

Port-based and Payload based techniques are the traditional techniques in the domain of Internet Traffic Classification. Port-based classification uses port numbers accredited by IANA (Internet Assigned Number Authority) to classify internet traffic [7]. For example, the port number 23 is reserved for Telnet whereas port number 20 is reserved for web applications. But eventually due to dynamic port number assignment this method is now obsolete (Table 1).

Payload is the part of the private user text which may perform the encryption of data. In this technique, the contents of the package are examined looking for characteristic signatures that are further used for the classification. But like every other technique, it too has its limitation, it cannot be used for encrypted networks because of the cryptographic technique, second, payload-based classification requires extravagant priced software which is not at all economical [8].

Ma et al. [9, 10] emphasized on online learning techniques/algorithms for analysing malevolent sites, utilising host-based and lexical based features for classification:

- Lexical features: It allows to capture the property of malicious URL's which might try to look nonidentical to benign URLs. The malicious URLs tend to cheat by using the flag keywords where they attempt to "spoof" the page. For example, www.amazon.com the presence of token ".com" is normal whereas it is not normal if it appeared in www.amazon.com/phishy.biz and for the implementation of the same URL may comprise of a bunch of tokens where '/', '.', '?', '-' etc are delimiters. The distinction is made by taking the token that appears in Top-level Domains and last token of path.

Table 1 Some IANA assigned port numbers

Assigned port	Application
23	Telnet
41	Graphics
165	Xerox
385	IBM application

- Host-Based Features: Uses features related to the host to collect information regarding the same. For example, it allows us to know where the 'malicious' site is hosted and who owns them, etc. using Whois Information, connection speed, location, etc.

In their experiments [9, 10], they concluded that a newly developed algorithm like CW (Confidence weighted) can be highly accurate with accuracies up to 99%.

Whereas in Ahmed [11] have used k means algorithm to classify the traffic using large and diverse dataset so that they have enough data for groupings and then filtering the traces to remove the noise which is crucial in order to form a reliable baseline for the training set. 3 out of 4 applications were correctly matched with a moderate rate of accuracy.

A hybrid approach was suggested by Erman et al. [9, 12] in early 2007. The main motivation behind him proposing this technique was to overcome the limitation of scarce labeled data. In this semi-supervised algorithm the machine is fed with the fusion of labeled as well as unlabelled data into a clustering model where the labeled data maps the clusters to already known classes. A probabilistic method is used to map the clusters which thereby allows some clusters to remain. This method was successful in overcoming the limitation that many new applications may come up with time about which has not been fed into the machine earlier and the machine was able to map the unknown instances without even detecting the type of flow, to the nearest cluster.

3 Machine Learning Techniques

Machine learning is an application of artificial intelligence that uses statistical techniques to give computers the ability to "learn" and improve with time from experience, without being explicitly programmed. The classification techniques involve a machine learning from pre-classified examples from which the model is prepared to classify the unseen data [10, 13].

The importance of classifying the ever-growing traffic over the internet led the machine learning techniques to efficiently classify the traffic when once trained.

Following supervised classifiers have been used to classify the traffic as malicious and non-malicious:

- K Nearest neighbor
- Naïve Bayes
- Decision tree
- Random forest.

Algorithm used in this study are easy to implement and interpret. They are powerful classifiers with creditable accuracies.

3.1 K Nearest Neighbor

K nearest neighbor classification model is the most basic and fundamental model in the supervised machine learning techniques. Here the function is approximated locally and all other calculations are conceded until classification [14, 15].

Steps:

(a) Choose the number of K of Neighbors
(b) Take the K nearest neighbors of the new data point, according to Euclidean distance.
(c) Among these K neighbors, count the number of datapoints in each category.
(d) Assign the new datapoint to the category where you counted the most neighbors.

3.2 Naïve Bayes

It is a type of probabilistic classifier based on strong independence assumptions between the features. A conditional probability is derived by analysing the relationship between each attribute and their class [16, 17]. It assumes that there is no dependency between different attributes (also called class conditional probability).

This classifier is based on Bayesian theorem which gives the conditional probability of an event D given B.

$$P(D|B) = \frac{P(B|D)P(D)}{P(B)}$$

3.3 Decision Tree

It uses a tree-like model of decisions. Tree-like structure of the algorithm helps to arrive at a solution by following any branch of the tree.

They work well with categorical (malicious, non-malicious, left, right) as well as numerical (1, 0) data and can be used for classification as well as regression problems [18]. In general, a decision tree asks a question and on the basis of the answer, it classifies accordingly. It starts at the top and works the way down.

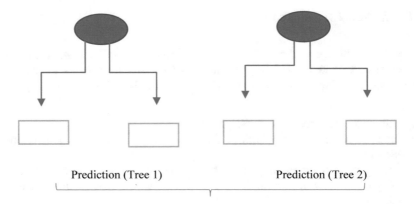

Fig. 2 Random forest classifier

3.4 Random Forest

It is one of the most powerful and popular supervised learning algorithms, capable of performing both classification and regression tasks (CART) (Fig. 2).

$$\text{Random Forest} = \sum \text{Prediction}(\text{Tree 1}) + \text{Prediction}(\text{Tree 2}) + \cdots$$

It is a collection of decision trees that tend to have a higher accuracy than decision trees as it uses a collection of decision trees and averages there result to predict an answer. It is the combination of tree predictors where each tree depends on the value of random vector that has been examined autonomously [19]. They combine the simplicity of decision trees with flexibility resulting in high accuracy.

4 Traffic Classification Using Machine Learning

Traffic Classification plays a notable role in the field of Internet science and Network security hence a classifier with good accuracy can enhance the performance of the network. The overall effectiveness of each algorithm is evaluated and then compared. It is evident from the results (Table 3) that K nearest neighbor is the weakest to classify the traffic as it takes comparatively more time with the lowest accuracy. Decision tree comes up with good accuracy and less time of execution than all other algorithms.

Table 2 Confusion matrix

Classified as	Y	$\bar{Y}$
Y	True positive	False positive
$\bar{Y}$	False negative	True negative

Table 3 Results

Classifiers	Accuracy (%)	Time (in sec)
K nearest neighbor	83.38	152.8
Naïve Bayes	83.30	5.24
Decision tree	84.55	2.36
Random forest	84.98	33.38

Confusion Matrix is the most common way to evaluate the performance of a classifier Table 2.

Suppose there is a traffic class Y and a trained traffic classifier is being used to classify the unseen packets.

(a) True Positive: % of members of class Y correctly identified as belonging to Y
(b) False Positive: % of members of other classes wrongly identified as class Y
(c) False Negative: % of members of class Y wrongly identified as not belonging to class Y
(d) True Negative: % of members of other class rightly identified as not belonging to class Y

$$\text{Accuracy} = \text{True Positive} + \text{True Negative}$$

In this paper, accuracy score has been used for the comparative analysis.

4.1 Comparative Analysis

A decision tree is built on an entire dataset accepting all features whereas Random forest arbitrarily selects the rows and features to build a set of decision trees and then takes the average of the results. Decision tree stands out with the second-best accuracy and turns out to be the best when time is taken into consideration after random forest. Random forest being the subset of decision tree has nearly the same hyperparameters as decision tree. The main difference is that random forest is a fusion of decision trees. K Nearest neighbor classifier is a lazy learner having local

Fig. 3 Comparative analysis

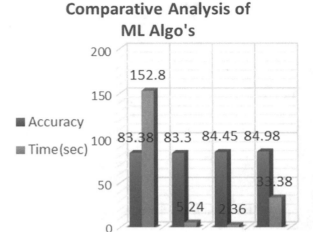

heuristics. KNN is a bit difficult to use in real-time prediction as it is a lazy learner whereas in contrast Naïve Bayes is an eager learner and is much faster than KNN as it is evident from the comparative analysis (Fig. 3). It takes a probabilistic estimation way and generates probabilities for each class figure.

5 Conclusion and Future Scope

In this paper four classifiers have been evaluated, our analysis is based on how accurately the classifier classifies the malicious and non-malicious traffic and how efficient it is. High accuracy and speed can be achieved by removing the redundant features and by expanding the size of the dataset, i.e. training the machine with more examples Technology is ever-growing factor and is improving with each fraction of second hence a lot of work can be done in this domain to improve the quality of internet services. Spam Detection, Internet traffic classifiers and Web Page classifiers all use almost same features to solve their problems, these can be used in compliment to enhance the classifying power by modifying them appropriately. Machine learning techniques work efficiently with good precision and overcomes the limitations exhibited by the traditional internet traffic classification techniques, therefore more work can be done in this field in the future using or modifying the decision tree algorithms as it comes up with good accuracy in lesser time.

References

1. Elovici, Y., A. Shabtai, R. Moskovitch, G. Tahan, and C. Glezer. 2007. Applying Machine Learning Techniques for Detection of Malicious Code in Network Traffic. In *Annual Conference on Artificial Intelligence*, 44–50. Berlin, Heidelberg: Springer.
2. Goseva-Popstojanova, K., G. Anastasovski, A. Dimitrijevikj, R. Pantev, and B. Miller. 2014. Characterization and Classification of Malicious Web traffic. *Computers & Security* 42: 92–115.
3. Gordon, L.A., M.P. Loeb, W. Lucyshyn, and R. Richardson. 2005. 2005 CSI/FBI Computer Crime and Security Survey. *Computer Security Journal* 21 (3): 1.
4. Mohurle, S., and M. Patil, M. 2017. A Brief Study of Wannacry Threat: Ransomware Attack. *International Journal of Advanced Research in Computer Science*, 8(5).
5. Bernaille, L., R. Teixeira, I. Akodkenou, A. Soule, and K. Salamatian. 2006. Traffic Classification on the Fly. *ACM SIGCOMM Computer Communication Review* 36 (2): 23–26.
6. Lim, Y.S., H.C. Kim, J. Jeong, C.K. Kim, T.T. Kwon, and Y. Choi, Y. 2010. Internet Traffic Classification Demystified: on the Sources of the Discriminative Power. In *Proceedings of the 6th International Conference*, 9. ACM.
7. IANA. Internet Assigned Numbers Authority (IANA). http://www.iana.org/assignments/port-numbers.
8. Karagiannis, T., A. Broido, A., and M. Faloutsos. 2004. Transport Layer Identification of P2P Traffic. In *Proceedings of the 4th ACM SIGCOMM Conference on Internet Measurement*, 121–134. ACM.
9. Erman, J., A. Mahanti, M. Arlitt, I. Cohen, I., and C. Williamson. 2007. Semi-Supervised Network Traffic Classification. In *ACM SIGMETRICS Performance Evaluation Review*, vol. 35, No. 1, 369–370. ACM.
10. Ma, J., Saul, L. K., Savage, S., Voelker, G. M. (2009, June). Identifying suspicious URLs: an application of large-scale online learning. In *Proceedings of the 26th annual international conference on machine learning*, 681–688. ACM.
11. Ahmed, T. 2017. Internet Traffic Classification Using Machine Learning.
12. Qi, X., and B.D. Davison. 2009. Web Page Classification: Features and Algorithms. *ACM Computing Surveys (CSUR)* 41 (2): 12.
13. Shafiq, M., X. Yu, A.A. Laghari, L. Yao, N.K. Karn, and Abdessamia, F. 2016. Network Traffic Classification Techniques and Comparative Analysis Using Machine Learning Algorithms. In *2016 2nd IEEE International Conference on Computer and Communications (ICCC)*, 2451–2455. IEEE.
14. Wikipedia. k-Nearest Neighbor Algorithm. http://en.wikipedia.org/wiki/K-nearest_neighbor_algorithm.
15. Ashari, A., I. Paryudi, and A.M. Tjoa. 2013. Performance Comparison Between Naïve Bayes, Decision Tree And k-Nearest Neighbor in Searching Alternative Design in an Energy Simulation Tool. *International Journal of Advanced Computer Science and Applications (IJACSA)*, 4(11).
16. Williams, N., S. Zander, and G. Armitage. 2006. A Preliminary Performance Comparison of Five Machine Learning Algorithms for Practical IP Traffic Flow Classification. *ACM SIGCOMM Computer Communication Review* 36 (5): 5–16.
17. Moore, A.W., and D. Zuev. 2005. Internet Traffic Classification Using Bayesian Analysis Techniques. In *ACM SIGMETRICS Performance Evaluation Review*, vol. 33, No. 1, 50–60. ACM.
18. Quinlan, J.R. 1996. Improved Use of Continuous Attributes in c4.5. *Journal of Artificial Intelligence Research* 4: 77–90.
19. Jun, L., Z. Shunyi, L. Yanqing, Z. Zailong. 2007. Internet Traffic Classification Using Machine Learning. In *Second International Conference on Communications and Networking in China, 2007. CHINACOM'07*, 239–243. IEEE.

Optimization in Wireless Sensor Network Using Soft Computing

Shruti Gupta, Ajay Rana and Vineet Kansal

Abstract Wireless Sensor Network (WSN) plays a significant role in today's era. It supports various applications in terms of monitoring, tracking, communication, sensing, preventing. By Systematic Literature Study, the objective is to analyze the various issues and limitations in WSN has facing and their solution to improve their performance in networking. For good performance of WSN first we identified optimization technique that is modelled to solve WSN issues. Afterwards various optimization algorithms for optimized the result at local search space for a global survival. Through this study main objective is to identify various optimization techniques they help in WSN to solve for their issues. Besides that this study helps in identifying the loopholes in existing techniques and their future scope and limitations.

Keywords Genetic algorithm · Firefly algorithm · Soft computing · Wireless sensor network

1 Introduction

Wireless Sensor Network consists of a sensor field that composes of sensor nodes, base stations, radio receivers for the various multipurpose functions and applications. Each node comprises of memory, sensors, analog to digital converter, microcontroller transmitter all this enable to work in one unit as single node when power is supply to them. The purpose of this is for tracking, detecting, monitoring

S. Gupta (✉) · A. Rana
Amity Univeristy, Noida, Uttar Pradesh, India
e-mail: shrutigupta.it@gmail.com

A. Rana
e-mail: ajay_rana@amity.edu

V. Kansal
IET Lucknow, Lucknow, Uttar Pradesh, India
e-mail: Vineetkansal@yahoo.com

© Springer Nature Singapore Pte Ltd. 2020
K. S. Raju et al. (eds.), *Proceedings of the Third International Conference on Computational Intelligence and Informatics*, Advances in Intelligent Systems and Computing 1090, https://doi.org/10.1007/978-981-15-1480-7_74

of various real-life application [1]. Although they are having many issues and limitations regarding their limited battery life because whole architecture is working using power only. If power supply is limited it's affected the performance of network that results into increase in the number of dead nodes. As the dead nodes increase, the performance of the network is affected and degraded the lifetime of network. To serve the same objective many researchers provided efficient models for better performance of Wireless Sensor Network. The applications of WSN are available in different fields such as in military, health sector, aviation sector, telecommunication sector [2, 3]. So, it is required to analyze and study the performance of WSN on various optimization techniques. This study helps to determine how the performance will be affected by the various optimization techniques. Marie et al. [4] explained wireless sensor network as a network of small and low-cost devices called sensor nodes, which are randomly placed and work together to transfer information from the monitored field through wireless links. Sensors are smaller, cheaper and smarter, therefore, WSNs are used in various applications. WSNs are used in different fields like military, healthcare, environment and security. Soft computing methodology is used to optimize results. It is helpful to solve problems including NP problems i.e. not solvable in polynomial time. Optimization techniques are used for optimized results when we have more than one solution for a single problem [5]. This technology is useful to identify maximum solutions in a given search space. In this, WSN identified the paper that already makes use of soft computing technologies for their better results. Many models and solutions are provided by researchers in the field of WSN for the given specific or general problem [6]. The purpose is to conduct the study is to know about various possibilities limitations its scope in the field of WSN. Through this study helps a beginner to identify which algorithm must have to choose for a particular time and situation [3, 7]. This study helps in identifying the way or procedures to apply soft computing methods With WSN and different approaches.

This paper explores about systematic literature review process according to Keele [8] study. In this, Keele and staff team explain the procedure to conduct the systematic literature review with its importance for conducting the study. In this paper, Sect. 2.1 research questions are designed to meet with objective of the paper. Section 2.2 explores various online resources and databases available for the selection of the paper. Section 2.3 shortlisted the paper on the basis of defined included and excluded criteria. Section 2.3 explains about various information obtained from selected research papers RP's. In last Sect. 2.4 checks the quality information of selected sources on the basis of quality check matrix.

2 Systematic Literature Review

SLR is a review process that provides a researcher to summarization of all information and facts for the purpose to get the systematic knowledge in their research field. This study also helps in identifying research goals and objectives [9]. It also provides a narrow view to conduct the entire study. In SLR we can identify convenient database and suitable data sources in respect of our research field. SLR is conducting in different steps.

2.1 Research Question

Before conducting a literature review research questions are generated. The survey is one of the method to obtain the solutions of generating questions. This step helps in the narrow down the scope of the literature and provides the first step to reach the objective. So in the title "Performance optimization in WSN using soft computing technique" we identify some key research questions (RQ) so that by proposing the solutions.

RQ1. What are the various recent research issues and areas in WSN?
RQ2. Empirically in literature survey, what are the various soft computing techniques used in WSN.
RQ3. Can all those techniques be able to increase the performance in WSN, in factual study?
RQ4. Can any classification is possible on these techniques on the basis of the empiric study? How?
RQ5. How we can show superiority among these techniques in WSN?

2.2 Search Strategy

Based on the title, we approach 13,200 papers from different sources of library from Google scholar including various sources of information from different digital libraries and databases. There is list of various data sources that explored individually (Table 1).

Table 1 Different data sources

Data source name	Source type	Source link
Ieee Xplore	Digital library	http://ieeexplore.ieee.org/Xplore/home.jsp
ACM Digital Library	Digital library	http://dl.acm.org/dl.cfm
Science Direct	Digital library	http://www.sciencedirect.com
Compendex	Digital library	http://adat.crl.edu/databases/about/compendex
Springer Link	Digital library	http://link.springer.com/
ISI Web of Science	Digital library	http://www.webofknowledge.com
Sensors	Digital library	http://www.mdpi.com/journal/sensors
Google Scholar	Search engine	http://scholar.google.com
Scopus	Digital library	http://www.scopus.com/

2.3   *Specify Information*

Information specification is done to achieve the information from the Research Paper that is documented in Sect. 2.3. These papers help us to identify many information for further study or their relevant work. The specification the information as follows:

A. *Years of publication*: The papers that we selected put in the range of 2005–2015. The 10 years range will be considered for the entire study. There is maximum publication in year 2011 in comparison with all other years.
B. *Source of RP's*: In this table, identification of the papers from various journals and conferences with their name has been identified from the selected paper. This study is helpful in identifying a kind of research and their suitable journal (Table 2).

Table 2 Discussion on research papers

Citation	Technique used	Author	Brief discussion
[10]	Fuzzy logic, Multi-agent system, Q learning, i.e. reinforcement learning	Samira Kalantari et al.	• In this approach, positive reward will be given to the agent if knowledge is increases and negative reward is given if knowledge is decreased on the basis of formulae • On behalf of its agent will be trained. Now using fuzzy logic input and output select member function then rules • Next state can be identified by Next state = power/distance Decide the efficiency of nodes by making fuzzy rules
[11]	Comprehensive list of issues in WSN with active research area	Gowrishankar S, Gowrishankar S, T. G. Basavaraju, Manjaiah D. H, Subir Kumar Sarkar	In this paper, authors discuss in details regarding issues in WSN in various areas of Sensor network including • Hardware and operating system for WSN • Medium access schemes • Deployment and localization • Synchronization and calibration

(continued)

Table 2 (continued)

Citation	Technique used	Author	Brief discussion
[12]	Evolutionary algorithm, genetic algorithm Nonconvex problem	Massimo Vecchioa, Roberto López Valcarcea, Francesco Marcelloni	According to paper, till now, to solve this problem nonconvex method is used. The problem is until the connectivity is not high, its problem to identify unique location of node in the network The two purposes or objective solved by is • Location Accuracy • Certain topological constraint to solve connectivity issues
[13]	Dynamic and scalable tree aware of spatial Correlation (YEAST)	Leandro A. Villas et al.	• In this paper researchers work on data aggregation and collection. In huge sensors node are distributed for the applications of monitoring i.e. for accuracy or monitoring • So the problem that occurs spatially correlated information and redundant and duplicate data can be detected by several nodes • On the basis of these tree parameters, we calculate network lifetime. By the spatial correlation they achieved 97% high accuracy and residual energy of the node network is 75%
[14]	Dense deployment and power assignment problem (d-DPAP), algorithm based on decomposition (MOEA/D)	Andreas Konstantinidisa, Kun Yang	• In this paper, researchers focused on the optimal localization and power assignment • Researcher proposed two algorithms MOEA/D AND GSH algorithm and show the results on the basis of simulation that specific MOEA algorithm works better than GSH

(continued)

Table 2 (continued)

Citation	Technique used	Author	Brief discussion
[15]	(1) Distributed constraint Satisfaction problems (discsps) having two algorithms i.e. Distributed breakout algorithm Dba (Dsa)	Weixiong Zhang, Guandong Wang et al.	• In this paper, researcher works on distributed environment on WSN where resources are scare when we deal with real-time data • So by using distributed constraint satisfaction problems (DisCSPs) algorithm they analysis the different behaviour and transitions by analyzing DSA and DBA algorithm bases on degree of parallel executions which tracked by colouring graphs by which they can make an optimal solution for distributed environment • This paper having a limitations of tracking of each mobile node is necessary
[16]	Computational geometry Virtual forces Mobility Disjoints sets Sleep scheduling Computational sets	Chuan Zhu a, b, Chunlin Zheng a, LeiShu c, n, Guangjie Han	• Researchers focus on the category of coverage deployment, strategy, sleep scheduling adjustable coverage radius for that they apply simulations on various tools • Results on the tools by analyzing there feature of WSN deployment strategy (random, deterministic) location (known or unknown) either its distributed or centralized nodes are distributed on the basis of these results are simulated on various tools so that they can improve connection and coverage problem
[17]	*Genetic fuzzy* Neuro-fuzzy	Seyed Mohammad Nekooei M. T. Manzuri-Shalmani	Using range-free localization, researcher estimated received signal

(continued)

Table 2 (continued)

Citation	Technique used	Author	Brief discussion
			strength (RSS) from the anchor nodes by using genetic fuzzy and neuro fuzzy membership functions. In this centroid, localization measures is targeted and results are compared with the view of neuro fuzzy and genetic fuzzy. In which neuro fuzzy shows results of this target more optimized results
[18]	"Network role-based routing algorithm, network role-based routing intelligent algorithm" ZigBee connected	Antonio M. et al.	• Researcher show their novelty by approaching new algorithm NORIA which is role-based assigning intelligent algorithms during routing. This algorithm is one iterative step of algorithm NORA using fuzzy logic by fuzzy functions • Through these algorithms that show the comparison of energy-efficient, fast, and effective manner with two more algorithms CDS and ZigBee. Fuzzy rules are executed for low-resourced nodes that make Wireless sensor networks
[19]	Localization using GA	Guo-Fang Nan et al.	• In this paper, single GA localization problem is solved Using fitness function and other different operators

2.4 Quality-Based Evaluation Criteria

Retrieving the various information from defined RP's, the last step of SLR has been approached [20, 21]. As the quality words occur, the expectations in terms of trust and reliability of the work will increase. Designing of the questions totally rely on quality input and output of all research papers that are listed in Table 3. There is list of designed questions that are mentioned below:

Table 3 Quality-based matrix

Design questions	R1	R2	R3	R4	R5	R6	R7	R8	R9	R10	R11	R12	R13	R14	R15	R16	R17	R18	R19	R20
A	Y	Y	Y	Y	Y	Y	Y	Y	Y	Y	Y	Y	Y	Y	Y	Y	Y	Y	Y	Y
B	Y	Y	Y	Y	Y	Y	Y	Y	Y	Y	Y	Y	Y	Y	Y	Y	Y	Y	Y	Y
C	Y	Y	Y	Y	Y	Y	Y	Y	Y	Y	Y	Y	N	Y	Y	Y	Y	Y	Y	Y
D	Y	Y	Y	Y	Y	Y	Y	Y	Y	Y	Y	Y	Y	Y	Y	Y	Y	Y	Y	Y
E	Y	Y	Y	Y	Y	Y	Y	Y	Y	Y	Y	Y	Y	Y	Y	Y	Y	Y	Y	Y
F	N	Y	Y	N	N	N	N	Y	Y	Y	Y	Y	N	Y	Y	N	Y	N	N	N
G	N	Y	Y	Y	Y	Y	Y	Y	Y	Y	Y	Y	N	Y	Y	Y	Y	Y	Y	Y
H	Y	Y	Y	Y	Y	Y	Y	Y	Y	Y	Y	Y	Y	Y	Y	Y	Y	Y	Y	Y

A. Objectives are clearly explained
B. Paper contains full methodology
C. Any novelty
D. Reference citations
E. Clearly descript about WSN
F. Used of any soft Computing technology
G. Development of any new model and design
H. Measures used in studied are properly defined.

3 Conclusion

The purpose of writing this paper is to get the knowledge about the literature related to optimization in WSN using soft computing techniques. For that, the systematic literature review method is opted for conducting the literature review. The reason for conducting the study is to know about the latest technology used in the field of WSN to optimize the performance. The optimization helps in of soft computing area to identify their potential area for research and their upcoming solution for mankind. This study helps to narrow down the scope of the study and provide directions and steps for future work.

References

1. Losilla, F., A.J. Garcia-Sanchez, F. Garcia-Sanchez, J. Garcia-Haro, and Z.J. Haas. 2011. A Comprehensive Approach to WSN-Based ITS Applications: A Survey. *Sensors* 11 (11): 10220–10265.
2. Huircan, Juan, Carlos Muñoz, Hector Young, Ludwig Von Dossow, Jaime M. Bustos, et al. 2010. ZigBee-Based Wireless Sensor Network Localization for Cattle Monitoring In Grazing Fields. *Computers and Electronics in Agriculture* 74 (2): 258–264.
3. Martins, Flávio V.C., et al. 2011. A Hybrid Multiobjective Evolutionary Approach for Improving the Performance of Wireless Sensor Networks. *Sensors Journal, IEEE, 11* (3): 545–554.
4. Rawat, Priyanka, Kamal Deep Singh, Hakima Chaouchi, and Jean-Marie Bonnin. 2014. Wireless sensor networks: a survey on recent developments and potential synergies. *The Journl of Supercomputing* 68 (1): 1–48.
5. Jang, J. Roger, C. Sun, and E. Mizutani. 1997. Neuro-Fuzzy and Soft Computing; A Computational Approach to Learning and Machine Intelligence.
6. Zhang, W., G. Wang, Z. Xing, and L. Wittenburg. 2005. Distributed Stochastic Search and Distributed Breakout: Properties, Comparison and Applications to Constraint Optimization Problems in Sensor Networks. *Artificial Intelligence* 161 (1): 55–87.
7. Nan, Guo-Fang, et al. 2007. Estimation of Node Localization with a Real-Coded Genetic Algorithm in WSNs. In *2007 International Conference on Machine Learning and Cybernetics,* vol. 2. IEEE.
8. Bara'a, A., et al. 2012. A New Evolutionary Based Routing Protocol for Clustered Heterogeneous Wireless Sensor Networks. *Applied Soft Computing, 12* (7): 1950–1957.

9. Ozdemir, O., et al. 2009. Channel Aware Target Localization with Quantized Data in Wireless Sensor Networks. *IEEE Transactions on Signal Processing* 57 (3): 1190–1202.
10. Ng, L.S., et al. 2011. Routing in Wireless Sensor Network Based on Soft Computing Technique. *Scientific Research and Essays* 6 (21): 4432–4441.
11. Shankar, G. 2008. Issues in Wireless Sensor Networks. In *Proceedings of the World Congress on Engineering*, vol. 1.
12. Massimo, Vecchio, Roberto López, et al. 2012. A Two-Objective Evolutionary Approach Based on Topological Constraints for Node Localization in Wireless Sensor Networks. *Applied Soft Computing* 12 (7): 1891–1901.
13. Villasab, Leandro A., Azzedine Boukerchea, Horacio A.B.F. de Oliveirac Regina, B.de Araujod Antonio, and A.F. Loureiro. 2011. Multi-objective Energy-Efficient Dense Deployment in Wireless Sensor Networks Using a Hybrid Problem-Specific MOEA/D. *Applied Soft Computing*, *11* (6): 4117–4134.
14. Villas, L.A., A. Boukerche, H.A. De Oliveira, R.B. De Araujo, and A.A. Loureiro. 2014. A spatial correlation aware algorithm to perform efficient data collection in wireless sensor networks. *Ad Hoc Networks*, 12: 69–85.
15. Ramakrishna Murty, M., J.V.R. Murthy, and P.V.G.D. Prasad Reddy. 2011. Text Document Classification Based on a Least Square Support Vector Machines with Singular Value Decomposition. *International Journal of Computer Application (IJCA)* 27 (7): 21–26.
16. Yang, K., et al. 2011. Multi-objective Energy-Efficient Dense Deployment in Wireless Sensor Networks Using a Hybrid Problem-Specific MOEA/D. *Applied Soft Computing* 11 (6): 4117–4134.
17. Zhu, Chuan, et al. 2012. A Survey on Coverage and Connectivity Issues in Wireless Sensor Networks. *Journal of Network and Computer Applications* 35 (2): 619–632.
18. Nicoli, Monica, et al. 2011. Localization in Mobile Wireless and Sensor Networks. *EURASIP Journal on Wireless Communications and Networking* 2011 (1): 1–3.
19. Ma, Di, et al. 2012. Range-Free Wireless Sensor Networks Localization Based on Hop-Count Quantization. *Telecommunication Systems* 50 (3): 199–213.
20. Nekooei, S.M., et.al. 2011. Location Finding in Wireless Sensor Network Based On Soft Computing Methods. In *2011 International Conference on Control, Automation and Systems Engineering (CASE*. IEEE.
21. Ortiz, Antonio M., et al. 2013. Fuzzy-Logic Based Routing for Dense Wireless Sensor Networks. *Telecommunication Systems* 52 (4): 2687–2697.

Analyzing Effect of Political Situations on Sensex-Nifty-Rupee—A Study Based on Election Results

N. Deepika and P. Victer Paul

Abstract Stock market is considered as a key driver of modern market. Stock market is based on the economy; it is a major source of raising resources for Indian corporate. It is known that the stock market fluctuates time to time, day to day depending on socio-political factors. There exists a relationship between the stock market indices. When even the governments change from one political party to other political parties, the stock market changes. The stock market also changes depending on the changes that take place in the world. The two stock market indices of India are Sensex and Nifty. The value of rupee, gold and crude oil depends on the socio-political situations. The prediction of stock market is an important issue because it has a challenging characteristic. There exists a relationship between them. The relationship between two variables can be found by rank correlation, which is a statistical tool. There are many mathematical software to find it. In this paper it is attempted to find the correlation between the Sensex–nifty, Sensex–rupee and nifty–rupee. On a politically important day of India on 15th May 2018. On this day, the counting of general elections of the Karnataka assembly took place. It is found that there is strong relationship between Sensex and nifty.

Keywords Sensex · Nifty · Rupee · Correlation · Karnataka elections 2018

N. Deepika (✉)
Research Scholar, Department of Computer Science and Engineering,
Vignan Foundation for Science, Technology and Research, Guntur,
Andhra Pradesh, India
e-mail: deepika.kitss@gmail.com

P. Victer Paul
Associate Professor, Research Scholar, Department of Computer Science
and Engineering, Vignan Foundation for Science, Technology and Research,
Guntur, Andhra Pradesh, India
e-mail: victerpaul@gmail.com

© Springer Nature Singapore Pte Ltd. 2020
K. S. Raju et al. (eds.), *Proceedings of the Third International Conference on Computational Intelligence and Informatics*, Advances in Intelligent Systems and Computing 1090, https://doi.org/10.1007/978-981-15-1480-7_75

1 Introduction

The stock markets are affected by a set of macro factors, such as interest rates, inflation, economic outlook, changes in policies, wars, and also by politics. Politicians and the decisions done by them under their governance can directly or indirectly influence business and consequently the stock prices. The economic strength of the country depends on its socio-political situations. The politically strong and stable political situation gives the strength to economy. To know how the political situation influences the Sensex and nifty the present study is made. On 15th May 2018, the counting of votes of Karnataka assembly general elections took place; in this context Sensex fluctuations are studied. National Stock Exchange (NSE) is the leading stock exchange of India. Full form of NIFTY is "National Stock Exchange Fifty". Full Form of "Sensex" is Sensitive Index. People are happy when the Sensex goes up and upset when it goes down [1].

From Fig. 1 shows that the Sensex on 15th may 2018 started low from the previous day. As the assembly election results are being declared, the early indications for stable government is possible, the Sensex improved its value from 10 to 10.30 am. From 10.30 am to till the end of the day it trends to decrease. Similarly we have taken the nifty and rupee value in the day at an interval of 30 min. The trends of assembly elections were known from 8.30 am, the leadings of various participants in the elections were given. These trends were changing from time to time. This also shown effect on the Sensex value.

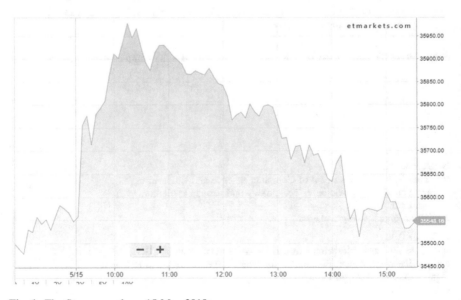

Fig. 1 The Sensex graph on 15 May 2018

2   Literature Review

In recent days, many researchers, economic analysts trying to study the nature of stock market and they try to predict its nature. The possibility of the increase in rates on December 16 is decreased with election of new president in US [2]. This story of the Indian economy casts light on the gradual process of economic change in a democratic polity. Democratic participation in India could be playing a silent role in making the growth story more inclusive [3]. A positive relation exists between FII and Sensex, exchange rate and Sensex have a negative relation for the time period from April 2005–March 2012 [4, 5]. Indian stock market index is represented by Sensex and Nifty. Increasing crude oil prices can increase the production cost which is able to have an effect on income and can decrease stock costs. The Gold price gets influenced by stock market index [6, 7].

Stock market is considered an extremely momentous factor of the financial status of a country [8]. A statistical measure named correlation is used to identify how the pair of variables are strongly related to each other. A correlation coefficient(r), $r < 0$ indicates negative relationship, $r > 0$ indicates positive relationship while $r = 0$ indicates no relationship (or that the variables are independent and not related). If $r = +1.0$ a perfect positive correlation and $r = -1.0$ a perfect negative correlation exists [9]. The markets suffered further on concerns about President Donald Trump's administration's ability to implement its pro-growth agenda and reports of a terrorist attack in Barcelona [10]. The behavior of Indian market between 1990 and 2001 concluded that Indian stock market is becoming informational efficient and efficiency has increased over time [11, 12].

3   Data Collection

The indices namely Sensex, Nifty and Rupee values per day are considered from the economic times of India website namely https://economictimes.indiatimes.com [13] which is the form of primary data. The data of election results is obtained from various television news channels and online news web sites and maintained as a dataset in a timely manner which is also in the form of bar graphs and pie graphs. The comments given about the formation of stable government can also be considered as an impact factor for the stock market. It represents the form of a graph with time in the X-axis and the Sensex value in the Y-axis. There is a provision of obtaining the data of one year, 6 six months, 3 months, 1 month, 5 days and 1 day. The Web page is provided with the historical data of Nifty, rupee, gold. The data with respect to various companies were also provided with numerical figures and graphs along with percentages under today's performance and performance over a period of month section. The sample collection of dataset is presented as follows (Table 1).

Table 1 Sample data

Time	bjp	Congress	Jds	Independent	Sensex	Nifty	inrusd	usdinr
8.45	57	57	25	2	35,514	10,786	0.015	67.522
8.50	57	57	25	2	35,514	10,786	0.015	67.522
8.55	79	73	25	2	35,514	10,786	0.015	67.522
9.00	77	75	25	2	35,514	10,786	0.015	67.889
9.05	77	75	25	2	35,514	10,786	0.015	67.522
9.10	86	78	25	2	35,514	10,786	0.015	67.522
9.15	85	78	26	2	35,514	10,786	0.015	67.522
9.20	86	76	30	2	35,514	10,796	0.015	67.843
9.25	93	79	39	2	35,755.41	10,856	0.015	67.522
9.30	93	80	40	2	35,767.42	10,864	0.015	67.843

4   Method of Study

The data collected in the data collection section is maintained in the form of CSV format, where the analysis of it considered using the statistical method namely Regression analysis. The regression analysis is considered to be the most powerful statistical method that allows to investigate the relationship among two or more variables of interest. It explains the impact of changes in an independent variable on the dependent variable. The several forms of regression models are presented as shown in Fig. 2.

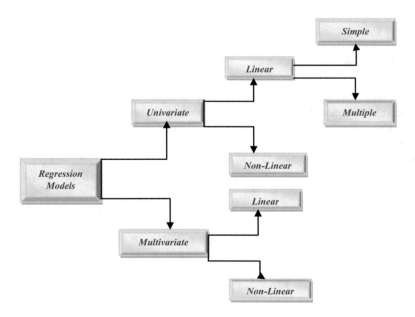

Fig. 2 Types of regression model

A Simple Linear Regression model is considered for the analysis of Sensex and nifty as response variables with respect to some political parties polling percentages. In the Simple Linear Regression, we consider two variables x and y, where y is dependent on x or influenced by x. Therefore, y is referred as dependent, or criterion variable and x is independent variable or predictor variable. The regression line estimation of y on x is depicted using the equation with regression parameters a, b.

$$y = a + bx,$$

where a is a constant, b is a regression coefficient.

The correlation is used to identify the strength of association among the variables as either positive or negatively correlated. A correlation coefficient measures the degree to which two variables be liable to change together. Thus coefficient describes the strength and the direction of the relationship.

5 Results Analysis

5.1 Metrics

For analyzing the results, we have considered the metrics namely mean, standard deviation, skewness and kurtosis where mean and standard deviation are the basic metrics used to understand the data. Skewness is the symmetry measure to estimate the similarity in the distribution with respect to the center point in the histogram plot. Kurtosis is the measure used to identify the outliers in the distribution. The histogram plot is an efficient graphical representation for the skewness and kurtosis of data set. From the results, we can observe rupee value is more uniform. Sensex value is more symmetric and all the factors data skewed right only. Rupee and independent having the positive kurtosis values, where independent is the heavy-tailed distributions than Rupee. Sensex, Nifty are having negative kurtosis values, which are light-tailed distributions. Relationship between Sensex, Nifty and Rupee is presented as (Fig. 3; Table 2)

(i)   sensex $= -0.00027280802896711975 + 3.2918684584186186 * \text{Nifty}$
(ii) sensex $= -0.003941933122405317 + 2,431,061.1934943437 * \text{Rupee}$
(iii) Nifty $= -0.0012076703933416866 + 738,504.9691678312 * \text{Rupee}$

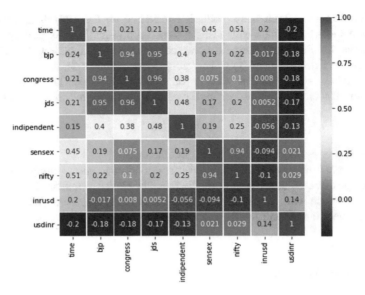

Fig. 3 Correlation matrix

Table 2 Metric calculations

Variable	Minimum	Maximum	Mean	Std. deviation	Skewness	Kurtosis
Sensex	35,514	35,974.600	35,772.28	86.94	0.6775	−1.133
Nifty	10,786	10,923.000	10,861.62	27.77	0.5353	−1.327
Usdinr	67.521	68.352700	68.05	0.17	0.6125	−1.415
Inrusd	0.0146	0.014810	0.01	0.00	1.6366	1.583
Jds	6.000	37	31.26	5.14	1.5318	0.715
Bjp	7.000	104	82.78	15.84	1.4458	0.346
Congress	17.000	78	63.81	8.54	1.3853	0.077
Independent	2.0000	3	2.38	0.22	5.5726	33.697

5.2 Graphs

The generated graphs describe the normal distribution which illustrates the mean and standard deviation. These graphs represent the impact of political situations on stock indices data distribution.

5.3   Discussion of the Results

Sometimes, political shocks can lead to significant changes in stock prices. The stock market analysis is influenced by political factors and thus we have considered counting date of the election for the Karnataka state and observed the variations in a periodical manner in the form of timestamp. From the observations it can be stated that the majority of political parties had a direct impact on the stock market.

The study of the stock market on a politically interested day 15 May 2018 gave important things to note. On this day the Karnataka assembly elections' counting was done. As the election results are being known, it trends as a political party gaining sufficient strength the Sensex, nifty strengthened. From the opening of market of the day the index moved up. At 9.30 am it was 35,767 at 10.25 AM it was 35,963 points. So, it was increased by 196 points. Then decreased gradually with ups and downs. Finally it ends up with 35,554 points. The thing happened between this time interval was one of the political party seems to get absolute majority. Then the results of election have shown that no political party is going to get the magic number of majority 113. So it may be a reason for decline in Sensex index. A similar trend has happened with Nifty. But the change in the rupee value did not show the relation its value increased from 67.56757 to 68.25939, it may be concluded that the rupee value is not dependent on this election results. The study of correlation between the Sensex and nifty show that strong relation exists with a correlation coefficient of 0.94, whereas Sensex and rupee is -0.094 and Nifty and rupee is -0.01. The negative sign indicates that there exists negative correlation as the values are very less, we can consider as no relation between them (Fig. 4).

6   Conclusion

It is proved many times that the stock market is independent of the socio-political situations. When situation is good the markets become strong and helps the economy. The instable political situation weakens the economic strength of the country. The market changes from time to time in a day depending on the situation, if the situation going to lead for stability, the stock market trends to move for strengthening. In case of the Karnataka assembly elections result day on 15th May 2018, when the results were leading to form a stable government the Sensex improved. As it is noted that, the stability attainment not possible, the Sensex again tends to decrease. When it is confirmed that stable government information not possible, the Sensex settled at the least point. The correlation between the Sensex and nifty is perfect coinciding, which means, the nifty also depends on the stable socio-political situation. Whereas the rupee value depends on many other factors, so it does not show a positive relation with Sensex and nifty. The brokers felt that the Sensex, Nifty, etc. were fallen down due to hung assembly in Karnataka. On the same day North Korea cancelled the high-level talks with Seoul so the other world

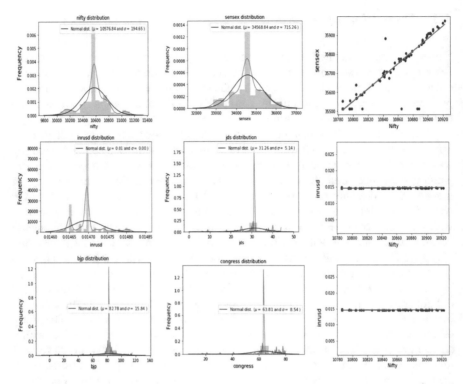

Fig. 4 Normal distribution and correlation plots

indicators also come down. Various limits of the study can be stated as: the political situations may not be the sole cause the fluctuations. The values of the Sensex, Nifty and rupees are taken from graphs, which may be slightly differing from actual values.

Acknowledgements We thank the Economic Times of India and online newspaper for providing the information.

References

1. https://www.quora.com/What-is-the-difference-between-Nifty-and-Sensex.
2. Rebeco, Canara. Twin events impacting Indian markets. Twin Events impacting the Indian Markets_Nov2016.pdf.
3. Mukherji, Rahul. 2009. The Political Economy of Development in India. In Conference Trade and Industry in the Asia Pacific Australian National University, Nov 20, 2009.

4. Makan, Chandni, Avneet Kaur Ahuja, and Saakshi Chauhan. 2012. A Study of the Effect of Macroeconomic Variables on Stock Market: Indian Perspective. In *Munich Personal Archive*, Nov 12.
5. Victer Paul, P., K. Monica, M. Trishanka. 2017. A Survey on Big Data Analytics using Social Media Data. In International IEEE Conference on Innovations in Power and Advanced Computing Technologies [i-PACT2017], Chennai, 1–5.
6. Nirmala Devi, K., and V. Murali Bhaskaran. 2015. Impact of Social Media Sentiments and Economic Indicators in Stock Market Prediction. In *International Journal of Computer Science & Engineering Technology (IJCSET)* 6 (04). ISSN: 2229–3345.
7. Naveen Balaji, S., P. Victer Paul, and R. Saravanan. 2017. Survey on Sentiment Analysis based Stock Prediction using Big data Analytics. In International IEEE Conference on Innovations in Power and Advanced Computing Technologies [i-PACT-2017], Chennai, 1–5.
8. Nataraja, N.S., L. Ganesh, and Sunil Kumar. Dynamic Linkages Between Cnxbank Nifty and Exchange Rates: Evidence From Indian Market.*International Journal of Business and Management Invention*. ISSN (Online) 2319-8028, ISSN (Print): 2319-801X.
9. Polisetty, Aruna, D. Prasanna Kumar, Jikku Susan Kurian. 2016. Influence of Exchange Rate on BSE Sensex & NSE Nifty. *IOSR Journal of Business and Management (IOSR-JBM)* 18 (9): 10–15. e-ISSN 2278-487X, p-ISSN 2319-7668.
10. MONTHLY NEWSLETTER—AUGUST 2017. Global Equity Markets Overview UNITED STATES file:///C:/Users/WIPRO/ Downloads/bajaj-finserv-monthly-August-2017.pdf.
11. Pradhan, H.K., and Lakshmi S. Narasimhan. 2002. Stock Price Behaviour In India Since Liberalization. In *Asia-Pacific Development Journal* 9 (2).
12. Victer Paul, P., S. Yogaraj, H. Bharath Ram, A. Mohamed Irshath. 2017. Automated Video Object Recognition System. In International IEEE Conference on Innovations in Power and Advanced Computing Technologies [i-PACT2017], Chennai, 1–5.
13. https://economictimes.indiatimes.com/indices/sensex_30_companies.

Study and Analysis of Modified Mean Shift Method and Kalman Filter for Moving Object Detection and Tracking

S. Pallavi, K. Ramya Laxmi, N. Ramya and Rohit Raja

Abstract In the scenario of Vehicle Tracking is the process of locating a moving object or multiple entities over time using a camera. It can be a time-consuming process due to the amount of data that is contained in video. Capture objects in video sequence of surveillance camera is nowadays a demanding application to improve tracking performance. The advanced technology makes video acquisition devices better and less costly, thereby increasing the number of applications that can effectively utilize digital video. Compared to still images, video sequences provide more information about how objects change over time. Thus, in this review, we discussed the several techniques attempted to site the available technique and various merits and demerits of it. In the present work we are detecting the moving object done by using simple background subtraction. For tracking of single moving object has been done using modified mean shift method and Kalman filter. Further result of both algorithms is compared on basis on time and accuracy.

Keywords Object tracking · Kalman filter · Mean shift method

S. Pallavi · K. Ramya Laxmi · N. Ramya · R. Raja (✉)
Department of CSE, Sreyas Institute of Engineering and Technology,
Nagole, Hyderabad, India
e-mail: drrohitraja1982@gmail.com

S. Pallavi
e-mail: pallavi.s@sreyas.ac.in

K. Ramya Laxmi
e-mail: kunta.ramya@gmail.com

N. Ramya
e-mail: ramya.n@sreyas.ac.in

© Springer Nature Singapore Pte Ltd. 2020
K. S. Raju et al. (eds.), *Proceedings of the Third International Conference on Computational Intelligence and Informatics*, Advances in Intelligent Systems and Computing 1090, https://doi.org/10.1007/978-981-15-1480-7_76

821

1 Introduction

For object tracking in this paper, the novel approach for automatic surveillance system detection of the moving object and tracking of single and multiple moving objects has been done using Kalman filter [1]. In the present work, author has used Kalman filter and optical flow algorithms for tracking multiple objects from standard databases for video. They tried to modify exciting algorithms for better performance. The proposed method is used to track the target through sequences of images for tracking low moving per with the help of Kalman filter algorithms. This present work is most important for real-time surveillance system [2].

Collins [3] used mean-shift algorithm and adapted feature scale selection for tracking multiple moving objects. Crucial parameter is used for mean-shift kernel method for which is suitable in changing in size. Zivkovic and Krose [4] has developed and proposed DOF of 5 degrees for finding histogram Kernel-based estimate for finding density with the help of function for a natural extension of the 'mean-shift' procedure. Comaniciu et al. developed [5] has developed non-rigid object tracking through the sequence of images with is similar to Kernel-based tracking using similarity. With reference to the size a robust and efficient tracking approach has been developed for detection motion according to the size.

Bradski [6] has developed a modified Mean Shift Algorithm Continuously Adaptive Mean Shift (CAMSHIFT) algorithm developed for tracking the moving person and face. The present work is based on a robust non-parametric technique for climbing density gradients to find the mode (peak) of probability distributions called the mean shift algorithm. For segmentation novel approached is used by author Comaniciu and Meer [7] successfully applied mean shift algorithms for segmentation. Shantaiya, Comaniciu [8, 9] have developed nonparametric technique for the analysis of complex multimodal feature space and to delineate arbitrarily shaped clusters.

2 Problem Formulation and Solution Methodology

In the present work, we have used two key steps for tracking moving objects. In first step is moving objects are detected and in second step tracking of object is done frame by frame. In the present work major challenge is tracking of object if it moves beyond the region and second challenge is how to eliminate the lighting and occlusion for moving object tracking Rawat & Raja [10–13]. We can attempt to solve the particular problem by predicting the location of the target and searching in a region centered on that location. Measurements are likely not accurate. This work will discuss how these problems can be addressed with the Kalman filter, and how well they are solved.

Algorithm is applied to two different datasets:

First dataset used here is taken from laboratory of intelligent and safe automobiles (LISA), which contains a video of moving vehicles captured at different months and at different locations. The database can be found at following link http://cvrr.ucsd.edu/LISA/vehicledetection.html

The Tracking algorithm has been successfully applied to standard surveillance video of CAVIAR [24] and PETS [25] databases also. Which can be found at the following link?

http://groups.inf.ed.ac.uk/vision/caviar/caviardata1/

http://www.hitech-projects.com/euprojects/cantata/datasets_cantata/dataset.htm

Where $\hat{q}$ is the target model, $\hat{q}_u$ is the probability of the uth element of $\hat{q}$, δ is the Kronecker delta function, $b\{X_i^*\}$ associates the pixel X_i^* to the histogram bin, and k (x) is an isotropic kernel profile.

Initialize the iteration number $k \leftarrow 0$.

Calculate the target candidate model $\hat{p}(y_0)$ in the current frame. The probability of the feature u in the target candidate model from the candidate region centered at position y is given by

$$
\begin{cases}
\hat{p}(y) = \{\hat{p}_u(y)\}_{u=1...m} \\
\hat{p}_u(y) = C_h \sum_{i=1}^{n_h} k \left\| \frac{y-x_i}{h} \right\|^2 \delta[b(x_i) - u]
\end{cases}
\tag{1}
$$

where C is constant, Calculate the weight vector $\{w_i\}_{i=1...n}$

$$
w_i = \sum_{u=1}^{m} \frac{\hat{q}_u}{\hat{p}_u(y_o)} \delta[b(x_i) - u]
\tag{2}
$$

Calculate the new position y_1 of the target candidate model In the mean shift iteration, the estimated target moves from y to a new position y_1, which is defined as

$$
y_1 = \frac{\sum_{i=1}^{n_h} x_i w_i g \left[\left\| \frac{y-x_i}{h} \right\|^2 \right]}{\sum_{i=1}^{n_h} w_i g \left[\left\| \frac{y-x_i}{h} \right\|^2 \right]}
\tag{3}
$$

Let $d \leftarrow \| y_1 - y_0 \|$, $y_0 \leftarrow y_1$. Set the error threshold ε (default 0.1) and the maximum Iteration number N (default 15).

$$
\text{If} (d < \varepsilon \text{ or } k \geq N)
$$

Estimate the width, height and orientation from the target candidate model (Cov) Coefficient can be used to adjust M_{00} in estimating the target area, denoted by A

Fig. 1 Workflow diagram

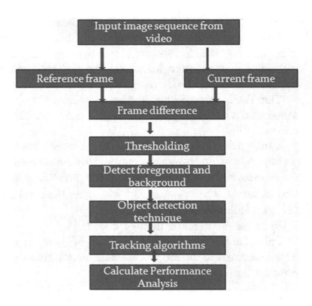

$$A = c(\rho)M_{00} \tag{4}$$

where $c(\rho)$ is a monotonically increasing function with respect to the Bhattacharyya coefficient $\rho(0 \leq \rho \leq 1)$, M_{00} is estimated frame If λ_1 and λ_2 height of the target that (Fig. 1).

$$k = \sqrt{A/(\pi\lambda_1\lambda_2)} \tag{5}$$

$$a = \sqrt{\lambda_1 A/(\pi\lambda_2)} \tag{6}$$

3 Results on Applying Both the Algorithms on Moving Vehicles

On applying MMST Algorithm for Vehicle Tracking (When background also changes with Vehicle) (Fig. 2).

Above results show the output when objects are tracked by using modified mean shift algorithm from a video where background also changes with vehicle very rapidly, in such video, if we apply modified mean shift algorithm then it can track selected object very efficiently. Here user needs to select the object to be tracked (vehicle) first, then tracking of selected object (vehicle) starts, such tracking have no effect of rapid change in background because it just tracks the selected region by the user hence not affected by background changes.

Fig. 2 a–c Tracking results on applying MMST algorithms on vehicle video from LISA database. The frames 20, 40 and 80 are displayed

Fig. 3 a–c outputs of Kalman filter for detection and tracking moving multiple cars when frame size is varied

On applying Kalman Filter Method for Vehicle Tracking (When background also Changes With Vehicle) (Fig. 3).

Above results show the output when objects are tracked by using Kalman filter method from a video where background also changes with vehicle very rapidly, in such video if we apply Kalman filter method then it will automatically select the moving object so user need not select any object for tracking. Here tracking of moving object (vehicle) automatically starts, such tracking affected by rapid change in background because it just compares the rapid change in region form previous frame, so all the objects which are changing their position are selected for tracking.

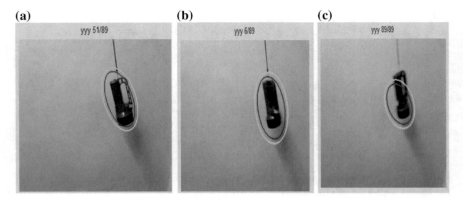

Fig. 4 a–c Outputs of MMST algorithms for tracking single objects when the camera is fixed. Three different sizes of frames with different intervals are displayed

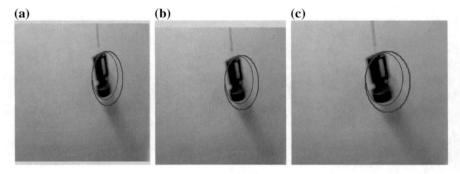

Fig. 5 a–c Outputs of Kalman Filter method for tracking single object when camera is fixed. Three different sizes of frames with different intervals are displayed

On Applying MMST Algorithm for Object Tracking (When background is Fixed) (Fig. 4).

On Applying Kalman Filter Method for Object Tracking (When background is Fixed) (Figs. 5 and 6; Table 1).

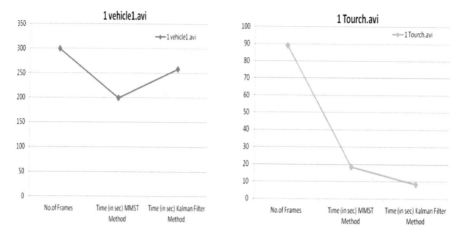

Fig. 6 Graphical representation of comparison of two methods when object and camera both are moving and both are still

Table 1 Comparison of MMST and Kalman filer object tracking methods

S. No	Video	No of frames	MMST (in s)	Kalman filter (in s)
1	Car	300	324	244
2	Camera	720	100	73

4 Conclusion and Future Scope

Form Result and its comparison it is clear that both the algorithm can track the object clearly, Table 1 show the time required by both the algorithms for tracking objects and it is very clear from the table that Kalman filter generates a good output with respect to time.

References

1. Patel, Hitesh A., and Darshak G. Thakore. 2013. Moving Object Tracking Using Kalman Filter. *International Journal of Computer Science and Mobile Computing (IJCSMC)* 2 (4): 326–332. ISSN 2320–088X.
2. Prasad, Kalane. 2012. Target Tracking Using Kalman Filter. *International Journal of Science and Technology* 2 (2).
3. Collins, R. 2003. Mean-Shift Blob Tracking through Scale Space. In *IEEE Proceedings Conference Computer Vision and Pattern Recognition*, Wisconsin, USA, 234–240.
4. Zivkovic, Z., and B. Krose. 2004. An EM-Like Algorithm for Color-Histogram-Based Object Tracking. In *IEEE Proceedings Conference Computer Vision and Pattern Recognition*, Washington, DC, USA, Vol. 1, 798–803.

5. Comaniciu, D., V. Ramesh, and P. Meer. 2003. Kernel-Based Object Tracking. *IEEE Transactions on Pattern Analysis and Machine Intelligenc* 25 (2): 564–577.
6. Bradski, G. 1998. Computer Vision Face Tracking for Use in a Perceptual User Interface. *Intel Technology Journal* 2 (Q2): 1–15.
7. Comaniciu, D., and P. Meer. 2002. Mean Shift: A Robust Approach Toward Feature Space Analysis. *IEEE Transactions on Pattern Analysis and Machine Intelligence* 24 (5): 603–619.
8. Shantaiya, Sanjivani, Kesari Verma, and Kamal Mehta. 2015. Multiple Object Tracking Using Kalman Filter and Optical Flow. *European Journal of Advances in Engineering and Technology* 2 (2): 34–39.
9. Comaniciu, D., V. Ramesh, and P. Meer. 2000. Real-Time Tracking of Non-Rigid Objects Using Mean Shift. In IEEE *Proceedings IEEE Conference on Computer Vision and Pattern Recognition* 2: 142–149, Hilton Head, SC.
10. Rawat, Nikita, and Raja Rohit. 2016. A Survey on Vehicle Tracking with Various Techniques. *International Journal of Advanced Research in Computer Engineering and Technology (IJARCET).* 5 (2): 374–377, ISSN 2278-1323.
11. Rawat, Nikita, and Raja Rohit. 2016. Moving Vehicle Detection and Tracking using Modified Mean Shift Method and Kalman Filter and Research. *International Journal of New Technology and Research (IJNTR),* 2 (5): 96–100, ISSN 2454-4116.
12. Raja, Rohit, Tilendra Shishir Sinha, Raj Kumar Patra, and Shrikant Tiwari. 2018. Physiological Trait Based Biometrical Authentication of Human-Face Using LGXP and ANN Techniques. *International Journal of Information and Computer Security, Issue on: Multimedia Information Security Solutions on Social Networks.* 10 (2–3): 303–320.
13. Raja, Rohit, Tilendra Shishir Sinha, Ravi Prakash Dubey. 2015. Recognition of Human-Face from Side-View Using Progressive Switching Pattern and Soft-Computing Technique. *Association for the Advancement of Modelling and Simulation Techniques in Enterprises, Advance B.* 58 (1): 14–34. ISSN 1240-4543.

Image Registration and Rectification Using Background Subtraction Method for Information Security to Justify Cloning Mechanism Using High-End Computing Techniques

Raj Kumar Patra, Rohit Raja, Tilendra Shishir Sinha and Md Rashid Mahmood

Abstract The present research paper deals with the registration and rectification of images for study of multimodal biometric systems through image cloning mechanisms using high-end techniques. Image cloning is a key research problem in information security applications. Most of the researcher addresses the above-said problem using template-based method which failed to deals with the soft computing and high-end computing techniques. Here, we propose knowledge-based or model-based technique for cloning. In this method, one can achieve more reliable information without any password or keys. Here the algorithm that has been proposed is going to be used for image cloning and at the same time securing reliable information through a double-authenticated based generation of secure codes after employing image cloning mechanism using high-end computing techniques. The proposed algorithm is tested and the results found are satisfactory.

R. K. Patra
Department of CSE, CMR Technical Campus, Kandlakoya (v) Medchal Road,
Hyderabad, India
e-mail: patra.rajkumar@gmail.com

R. Raja (✉)
Department of CSE, Sreyas Institute of Engineering and Technology,
Nagole, Hyderabad, India
e-mail: drrohitraja1982@gmail.com

T. S. Sinha
Department of Information Technology, Institute of Engineering and Management,
Salt Lake, Kolkata, India
e-mail: tssinha1968@gmail.com

M. R. Mahmood
Department of ECE, Guru Nanak Institutions Technical Campus, Hyderabad, India
e-mail: er.mrashid@gmail.com

© Springer Nature Singapore Pte Ltd. 2020
K. S. Raju et al. (eds.), *Proceedings of the Third International Conference on Computational Intelligence and Informatics*, Advances in Intelligent Systems and Computing 1090, https://doi.org/10.1007/978-981-15-1480-7_77

Keywords Artificial neural network (ANN) · Facial animation parameters (FAP) · Means value coordinates (MVC) · Fuzzy set rules (FSR) · Genetic algorithms (GA) · Region of interest (ROI)

1 Introduction

Many works on image cloning have been carried out by many researcher, in the year 2003, Hsu et al. [1] carried out the work in cloning facial expressions by applying expressive ratio image and FAP to texture conversion. They developed a 3-D facial model and by using ERI mechanism, transfer a human's subtle change to another one. Hence it modeled the MPEG-4 FAP to ratio image conversion.

After a gap of six years, in the year 2009, Farbman et al. [2] carried out the work in coordinates for instant image cloning. They introduced an alternative, coordinate-based approach commonly known as mean value coordinate (MVC). This mean value coordinate having advantages like high speed, easy to implement, etc.

In 2011, Alex et al. [3] carried out the work for combinatorial Laplacian image cloning in which they detected efficient and flawless cleaning methods, but those techniques beard several limitations too such as concavities and dynamic-shape-deformation. In their work, they presented a novel technique for image cloning using a combination of fast-Cholesky factorization and Laplacian for ensuring concavities, interactive image manipulation, handling holes, and dynamic. Further in the year 2011, Fu et al. [4] carried out the work for arbitrary image cloning, in which an image cloning method called arbitrary cloning was presented. To overcome the problem of image-matting, Poisson image editing techniques were applied. They compared the results with the existing techniques and it has been concluded that their work has given more realistic results. Raja et al. [6–8, 10] proposes hybrid computing mechanism. This involves soft computing (Neural Networks, Fuzzy logic and Genetic Algorithm).

These variations have been studied and analyzed further for the improvement of the system performance. Thus the mechanism of image cloning has been adopted to justify the study and analysis of multimodal biometric security systems Jai et al. [9] & Siddhesh et al. [10]. Only few researchers have been reported using high-end computing. In this paper, ANN and FSR have been employed as soft-computing tools. The present research carried out using modeling and simulation. In modeling, images at input side are enhanced and compressed to remove the distortion with more accuracy. Next, the object of interest is selected for cropping and trimming and hence image warping has been applied, later on the trimmed image has been segmented for edge detection then the region of interest detected Brand [11], Neha [12], & Morkel [13]. Afterward, the relevant features have been extracted and with the help of neural networks, features vectors have been utilized for framing a knowledge-based model called multimodal biometric system model. In the simulation, the training and test pattern is being done.

2   Problem Formulation and Solution Methodology

In this paper, the problem has been formulated in the following stages that can be associated with multimodal biometrics system for information security using image cloning technique with the help of soft-computing method. First one is the finding of noises and lossless information in the image. Noise may degrade the performance of the system. Hence for obtaining better results, noise must be removed and it should be compressed while doing simulation for information security. Next is related to object selection and it's localization for proper cropping and trimming. The third problem is the employment of image warping technique over trimmed object. In this proper localization has to be done with proper scaling and hence rectification of an object. Afterwards, reproduce the image by applying cloning mechanism of image processing. The next stage is related to computation of the boundaries of the original and cloned image. In the next stage, selection and extraction have been done of the relevant features from the enhanced, cropped, trimmed, segmented and cloned image. Next, framing of knowledge has been done by using model as corpus for analysis of multimodal biometric information security systems. The last one is authentication and authorization has been done in the simulation for information security of trained and test patterns. Here the main objective is to investigate and develop a systematic approach for analysis of biometric security systems by using artificial neural networks and fuzzy logic within a single framework of biometrical study and analysis.

Mathematical Analysis

In a noisy image, approximate wavelet shape can be computed by using simple structural analysis and template matching technique. Estimation of time occurrence can also be done by this approach. To calculate the wavelet shape for detection in noisy image, consider a wavelet class having $Si(t)$, $I = 0, \ldots, N - 1$, all having some common structure. Considering the corrupted image can be modeled by the equation considering, noise is additive then Based on this assumption that noise is additive, then the corrupted image has been modeled by the equation:

$$X(m, n) = i(m, n) + G\, d(m, n) \tag{1}$$

where, $i(m, n)$ is the clean image, $d(m, n)$ is the noise and G is the term for signal-to-noise ratio control. To de-noise this image, wavelet transform has been applied. Let the mother wavelet or basic wavelet be $\psi(t)$, which yields to,

$$\psi(t) = \exp(j2\pi ft - t2/2) \tag{2}$$

Further as per the definition of continuous wavelet transform CWT (a, τ), the relation yields to,

$$\text{CWT}(a, \tau) = \left(1/\sqrt{a}\right) \int x(t)\psi\{(t - \tau)/a\}dt \tag{3}$$

The parameters obtained in Eq. (3) has been discretized, using discrete parameter wavelet transform, DPWT (m, n), by substituting $a = $ aom, $\tau = n\tau_0$aom. Thus Eq. (3) in discrete form results to Eq. (4),

$$\text{DPWT}(m, n) = 2 - m/2\Sigma k\Sigma l \times (k, l)\psi(2 - mk - n) \tag{4}$$

where 'm' and 'n' are the integers, $a0$ and $\tau0$ are the sampling intervals for 'a' and 'τ', $x(k, l)$ is the enhanced image. The wavelet coefficient has been computed from Eq. (4) by substituting $a_0 = 2$ and $\tau_0 = 1$. Further the enhanced image has been sampled at regular time interval 'T' to produce a sample sequence $\{i\,(mT, nT)\}$, for $m = 0,1,2,\ M - 1$ and $n = 0, 1, 2, \ldots, N - 1$ of size $M \times N$ image. After employing discrete Fourier transformation (DFT) method, it yields to the equation of the form,

$$I(u, v) = \sum_{m=0}^{m-1} \sum_{n=0}^{n-1} i(m, n) \exp(-j2\pi(um/M + vn/N)) \tag{5}$$

For $u = 0,1,2, \ldots, M - 1$ and $v = 0, 1, 2, \ldots, N - 1$.

2.1 Solution Methodology for Proposed Algorithm

The objective of this paper is to investigate and develop a systematic approach for analysis of biometric security systems using Artificial Neural Network (ANN) and Fuzzy Logic within a single framework of biometrical study and analysis. The steps of accomplishing this objective are:

A. **Modeling of original and cloned image has been done using the following steps**:

    I. Enhanced image (trained & test image) has been obtained.

    II. Cropping and trimming of the obtained object has been done.

    III. Warping of the obtained trimmed image has been done as follows:

        (a) Image registration of trimmed object.

        (b) Image rectification of trimmed object.

    IV. Cloned image reproduction has been done.
    V. Segmentation of original and cloned image has been done.
    VI. Features of original and cloned images have been extracted.
    VII. Knowledge-based model of the original and cloned image has been obtained.

B. **Simulation for information security has to be done using the following steps**:

    I. Simulation of trained and test patterns has been done.
    II. Final conclusion for trained and test patterns has been obtained.

The methodology adopted in the present paper has been depicted below:

Algorithm 1 Modeling of Original and Clone Image

Step 1. Read a test image.
Step 2. Convert into grayscale image, say R.
(Enhancement stage)
Step 3. Use DCT to filter the image.
(Cropping and trimming stage)
Step 4. Enhance the image using median filtering and convolution theorem.
Step 5. Perform image equalization.
Step 6. Select the object of interest for image.
Step 7. Crop the selected object and rectify it using image warping technique.
Step 8. Store the image ready to be cloned in a template for the further simulation phase.

3 The Results and Discussions

The experimental setup of the present work has been carried out. To start with, first an unknown image has been read as input and converted into grayscale image. This shows the grayscale image of the original image, and the discrete cosine transformation of the grayscale image. The transformation has been done for enhancing the image with lossless information.

Figure 1 shows the original image and after enhancement phase when it gets noise-free. The enhanced noise-free object image has further processed for obtaining sharpened image for appropriate perception and for proper calculation of object image features.

Figure 2 shows matching of common points on various features for reconstruction of image for cloning and it shows the computation of connected component over the selected objected of interest on the image. The computation of the connected component has been done after proper enhancement and equalization of the image. Once this is ready, next the image has been obtained for the analysis for image cloning.

Original Image   Enhanced / Noise-free Image

Fig. 1 Output figure showing original image and enhanced noise free image

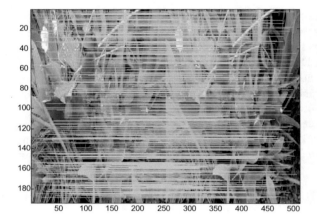

Fig. 2 Matching of common points on various features for reconstruction of image for cloning

(a)   Reconstructed   **(b)**   Image

Fig. 3 **a** Reconstructed Image and **b** Image ready to be cloned

Figure 3 shows the median filtration applied on the original image and the segmentation using convolution theorem and hence its distribution.

Further analysis has to be carried out with image warping over the cropped and trimmed object. Hence segmentation and extraction of features for the formation of the knowledge-based model have to be done. Further analysis has to be carried out for the simulation of the trained and test patterns for the justification of the multimodal biometric security system through image cloning using soft-computing techniques of soft-computing techniques.

Figure 4a boundary has been formed as shown in figure for the detection of object code using unidirectional temporary memory technique of artificial neural network. Histogram equalization of the image after application of iterative root power and iterative logarithmic root power. From the above observations, it has been found that there are lots of variations in the color of the image after employment of iterative-root power and its logarithmic value. When histogram equalization has been employed these color variations have been observed very clearly. It has been observed from Fig. 4b that the performance of the developed algorithms with optimal number of features selected is between 90 and 96% accuracy.

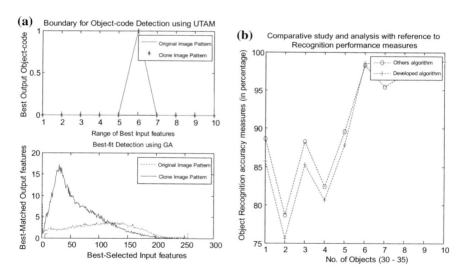

Fig. 4 **a** Boundary for object code detection using uni-directional temporary associative memory and best-fit detection using genetic algorithms for any object image. **b** Comparative study and analysis of the developed algorithms with other algorithms using any object image pattern for detection of behavioral traits of original image and cloned image

4 Conclusions and Further Scope of the Work

As stated earlier, with the help of high-end computing techniques for image cloning mechanism there are eight problems associated with multimodal biometrics system for information security. In the present work the features are extracted from original and cloned image and forming knowledge-based model of the original and cloned image. The simulation for trained and test patterns has been done and result has been found satisfactory.

References

1. Hsu, Pei-Wei, Yao-Jen Chang, Chao-Kuei Hsieh, and Yung-Chang Chen. 2003. Facial Expression Cloning: Using Expressive Ratio Image and FAP to Texture Conversion. Department of Electrical Engineering National Tsing Hua University, Hsinchu, Taiwan 300, R. O. C.
2. Farbman, Zeev, Gil Hoffer, Yaron Lipman, Daniel Cohen-Or, and Dani Lischinski. 2009. *Coordinates for Instant Image Cloning*. Hebrew University.
3. Vargas, Alex Cuadros, Luis Gustavo Nonato, and Valerio Pascucci. 2011. *Combinatorial Laplacian Image Cloning*.
4. Xinyuan, Fu, He Guo, Yuxin Wang, Tianyang Liu, and Han Li. Arbitrary image cloning. 2011. In Department of Computer Science and Technology, Dalian University of Technology Dalian, China, 116024.
5. Raja, Rohit, Tilendra Shishir Sinha, Ravi Prakash Dubey. 2015. Recognition of Human-Face from Side-View Using Progressive Switching Pattern and Soft-Computing Technique. *Association for the Advancement of Modelling and Simulation Techniques in Enterprises, AMSE Journals* 58 (1): 14–34.
6. Raja, Rohit, Tilendra Shishir Sinha, Ravi Prakash Dubey. 2016. Soft Computing and LGXP Techniques for Ear Authentication using Progressive Switching Pattern. *International Journal of Engineering and Future Technology*, 2 (2).
7. Rohit, Raja, Tilendra Shishir Sinha, and Ravi Prakash Dubey. 2016. Orientation Calculation of Human Face Using Symbolic techniques and ANFIS. *International Journal of Engineering and Future Technology* 7 (7): 37–50.
8. Raja, Rohit, Tilendra Shishir Sinha, Raj Kumar Patra, and Shrikant Tiwari. 2018. Physiological Trait Based Biometrical Authentication of Human-Face Using LGXP and ANN Techniques. *International Journal of Information and Computer Security Special Issue on: Multimedia Information Security Solutions on Social Networks* 10 (2–3): 303–320.
9. Jia, Jia, Shen Zhang, and Lianhong Cai. 2010. Facial Expression Synthesis Based on Motion Patterns Learned from Face Database. In 2010 IEEE 17th International Conference on Image Processing, Hong Kong, Sept 26–29, 2010.
10. Siddhesh, Angle, Reema Bhagtani, and Hemali Chheda. Biometrics: A further echelon of security. In *Department of Biomedical Engineering*, Thadomal Shahani Engineering College, T. P. S III, Bandra, Mumbai-50.
11. Brand, J.D., and J.S.D. Mason. Visual speech: A Physiological or Behavioral Biometrics. Colombia Sylvain Department of Electrical Engineering, University of Wales Swansea *SA2 8PP*, UK.
12. Neha, Sharma, J.S. Bhatia, and Neena Gupta. *An Encrypto-stego Technique Based Secure Data Transmission System*. PEC Chandigarh.

13. Morkel T., J.H.P. Eloff, and M.S. Olivier. An overview of image stegnography. In Information and Computer Security Architecture (ICSA) Research Group, Department of Computer Science, University of Pretoria, 0002, Pretoria, South Africa.
14. Bhensle, A.C., and Rohit Raja. 2014. An Efficient Face Recognition Using PCA and Euclidean Distance Classification. *International Journal of Computer Science and Mobile Computing*, 3 (6): 407–413. ISSN 2320–088X.

Analyzing Heterogeneous Satellite Images for Detecting Flood Affected Area of Kerala

R. Jeberson Retna Raj and Senduru Srinivasulu

Abstract Kerala flood is one of the most disastrous in recent years which affects millions of people's lives into standstill and thousands of people lost their houses and properties. Landslides and water inundation really hit the normal life of the people. The effects of climate change influences the environment by changing landscape, incessant rainfall, raise of temperature, failure of monsoon, etc. in this paper, the change detection of Kerala flood is analyzed and compared. Two different satellite images of before and after flood are considered and the changes in the flood-affected area are detected. The satellite image is co-registered, calibrated and geometric correction made for processing. The pre-processing algorithms are used to filter the speckle noise and making the image as noise-free. The image is analyzed and classified with supervised and unsupervised algorithms. The unsupervised K means algorithm and supervised algorithm such as Random forest, K-Nearest Neighborhood (KNN), KDTree-KNN, Maximum Likelihood (ML) and Minimum Distance (MD) classifiers are applied and the performance of the algorithms are compared. Finally, the changes in the image are demarcated and analyzed.

Keywords Change detection · Satellite images · Classification · Floods

1 Introduction

Climate change is globally important due to its impact influences the environment. The effects of climate change which include failure of monsoon, loss of precipitation, drought, flood, downpour, rising of temperature, raising of sea level and changes in coastal landscape. These factors affect the state's economy into standstill and weaken the agriculture growth. Furthermore, the level of groundwater is dipped

R. Jeberson Retna Raj (✉) · S. Srinivasulu
Department of Information Technology, Sathyabama Institute
of Science and Technology, Chennai, India
e-mail: jebersonretnarajr@gmail.com

© Springer Nature Singapore Pte Ltd. 2020
K. S. Raju et al. (eds.), *Proceedings of the Third International Conference on Computational Intelligence and Informatics*, Advances in Intelligent Systems and Computing 1090, https://doi.org/10.1007/978-981-15-1480-7_78

which aggravates the crisis further. From June to August 2018, Kerala receives a recorded high of 2226 mm rainfall as against a normal of 1620 mm. This is a huge of 41% higher than compared with the normal rainfall. The incessant rain has totally devastated the state of totally 483 causalities reported and 15 people went missing. The total loss of property estimated as over 40 thousand crores of rupees (5.6 billion dollars). Idukki is one of the worst-affected districts in the state recorded highest of 1419 in history. India Meteorological Department (IMD) data shows the decreasing trend of rainfall from 1875 to 2017. Therefore, the state didn't expect and prepare for the downpour. Severe landslides reported and more than 1 lakh peoples stay in relief camps.

1.1 Literature Survey

Detecting changes in two heterogeneous images are interesting and features of two images are compared and identified. Generally, the clustering and classification algorithm helps to differentiate the physical boundaries of a satellite image. We have to show the clear difference between two images of their vegetation cover, bare soil, water cover, buildings, etc. Numerous works have been discussed in literature regarding the change detection in satellite images. In the recent past, wei zhao et al. presented a paper for change detection using unsupervised neural network model [1]. The reference images are supposed to be compared and is considered for pixel by pixel comparison and the difference image is generated. In [2], land cover changes in Lampedusa Island of Italy have been discussed. The Landsat 5 and Landsat 8 OLI multispectral images from 1984 to 2014 are taken for detecting the spatial and temporal changes in the multispectral images. Initially, the images are co-registered and atmospherically correction takes place. The image is classified by classification algorithms and the vegetation distribution analysis is made. MLC (Maximum likelihood classifier) is used to determine the lack of land cover and vegetation examination and distribution is identified through NDVI (Normalized Difference Vegetation Index). The change detection can be analysed through MLC and NDVI. Geomorphic Flood Area (GFA) tool introduced by Università for identifying various flood-affected areas located in Europe and United States. It works based on the binary classification of Geomorphic Flood Index (GFI). Rejaur et al. proposed a method for analyzing SAR image for flood propagation in Kendrapara district of Orissa [3]. Initially, the SAR image is calibrated, geometrically corrected and filtered. Threshold method is used to detect the open water in the image. The permanent water bodies like pond and river are delineated from the image to detect the actual extension of flood in the image. Feedback-based genetic programming for satellite image classification is proposed in [4]. It is based on learning user preferences and region descriptors to encode spectral and textual properties. Nurwanda et al. presented their work in [5] regarding land cover change in Batanghari Regency, Jambi Province. Satellite image of the area has been taken for supervised classification using maximum likelihood method. Largest Path Index

(LPI) is used to assess the fragmentation of the land cover region. Gounaridis et al. introduced random forest-based classification algorithm to determine the land cover of Greece. The urban area of land cover is classified into five classes [6]. Darius Phir et al. presented different pre-processing methods used to improve classification accuracy. The Dark Object Subtraction (DOS) and Moderate Resolution Atmospheric Transmission (MODTRAN) correction methods integrated to improve the classification accuracy [7]. Rao Zahid et al. presented a land cover classification method using maximum likelihood classifier to classify the Okara district of Pakistan [8]. Caterina et al. introduced GFA tool used to detect flood-prone areas in Romania. The tool in QGIS used to perform binary classification of the input image which used to identify the flood areas [9]. In [10], a bee swarm optimization based classification of kidney lesions is presented. Renal masses of kidney images are classified and extraction of texture features is also discussed and analyzed [11].

2 Methods and Tools Used

Two Landsat images of Kerala with different date are considered for this work. The dataset of these images is openly available and provided by NASA [12]. The first satellite image kerala_oli_201837_lrg.jpg of Kerala is taken by the LANDSAT8 OLI sensors on February 6, 2018. The second satellite image kerala_msi_2018234_lrg.jpg is taken by LANDSAT8 MSI sensor on August 24, 2018. The image consists of three bands Red, Green and Blue. Initially, the images are co-registered and calibrated for processing. Furthermore, the LANDSAT images are geometrically corrected for pre-processing. Generally, the LANDSAT image itself having noise due to the factors include atmosphere and time of the image taken. Speckle noise is one of the default noise present in the image and it should be removed. Filtering methods can be used to pre-process the images.

2.1 Image Analysis and Classification

The preprocessed images of before and after the flood are then applying with image analysis and clustering algorithms. The image analysis can be done by Principal Component Analysis (PCA) method with five components [13]. PCA applied with the three bands Red, Green and Yellow of the image. The supervised and unsupervised classification algorithm applied with both of the images. For unsupervised classification, the K means cluster analysis approach can be used [14]. For supervised classification, this work considers Random Forest (RF) [15], K-Nearest Neighborhood (KNN) [16], KDTreeKNN, Maximum Likelihood Classifier (MLC) [17] and Minimum Distance Classifier (MDC) is used. The image analysis and classification algorithms are applied to both before flood and after flood images.

Finally, the images can be compared and the changes are detected. The changes are demarcated and it shows that the impact of flood in Kerala and areas which are inundated.

3 Experimental Results

The Sentinal Application Platform (SNAP) 5.0 is an open-source platform for spatial and temporal data processing. The Landsat image of Kerala before and after flood is loaded in the platform for geometric correction and calibration. The image is further pre-processed by various filtering methods. The speckle noises are removed so that the image is ready for further classification. The importance of preprocessing is used for increasing the accuracy of the classification. QGIS 2.18 is another tool used for image analysis. The grey level co-occurrence matrix is one of the methods for differentiating images. OTB toolbox is open source software for image analysis used for applying unsupervised classification with the images. The images of before and after flood satellite images of Kerala are considered for experiment. The images are co-registered, calibrated and geometric correction made for processing. In general, the satellite image itself contains speckle noise which is to be removed. Therefore, the images preprocessed by removing speckle noise for making the image noise-free. Now, the classification algorithms can be applied with the two images. The class label is defined with different classes like water, vegetation, bare lands, built-in areas and clouds. The classification algorithms such as K means, KNN, KDTreeKNN, Random Forest (RF) and Maximum Likelihood are applied with both before flood and after flood images. Figure 1 shows the outcome of the above classification algorithms. The flood-affected areas are clearly differentiated from the images. Figure 2a, b shows the resultant images of K means clustering algorithm. The flood-affected areas are labeled with light green color. The classified image of after flood shows that Idukki district of Kerala is completely inundated. Furthermore, many places the water stagnation is shown. Figure 2c, d shows the outcome of Principal Component Analysis (PCA) algorithm. The two images show the clear difference of flood in that region. Figure 2e, f shows the result of Random Forest Algorithm. The flood-affected areas are labeled as dark green which clearly differentiating with before flood image. Figure 2g, h shows the resultant image of KNN classier algorithm. The water in the image is labeled as green color which correctly classifies the pixels. The classified image KDTreeKNN classifier is shown in Fig. 2i, j. Figure 2k, l show the result of Maximum likelihood classifier.

Figure 3 shows the subset of the image with a window size of 2125 × 3450 is taken. This area is highly affected flood zone. The PCA algorithm is applied and the resultant image clearly shows the boundary of inundation.

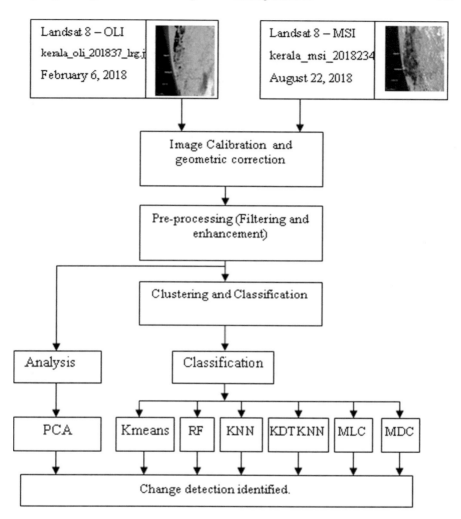

Fig. 1 Architecture diagram of change detection

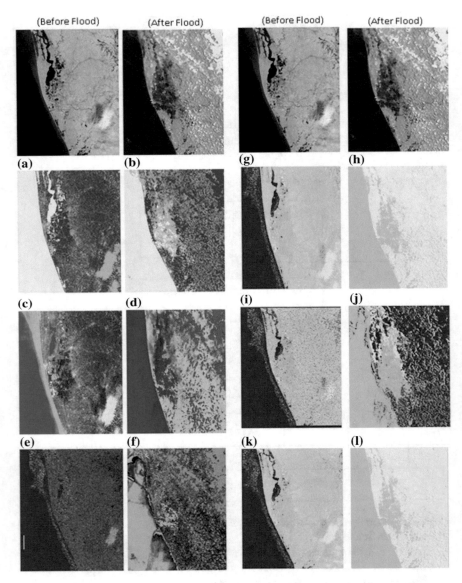

Fig. 2 **a** and **b** K means clustering algorithm **c** and **d** PCA algorithm **e** and **f** Random forest algorithm **g** and **h** KNN algorithm **i** and **j** KDT-KNN algorithm **k** and **l** Maximum likelihood clustering algorithm

Fig. 3 Subset of the image
and the classified image

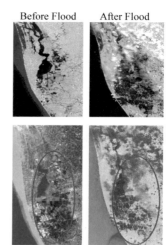

Before Flood   After Flood

3.1 Performance Analysis

The classification algorithms are tested with both images and the performance is compared with different metrics. Mean Absolute Error (MAE) and Signal to Noise Ratio (PSNR) is computed and the results of six algorithms are shown in Table 1. The performance characteristics of MAE is defined as lower the value is better the performance. The metric Peak Signal-to-Noise Ratio (PSNR) is defined as higher the value is better.

In the above algorithms, Random Forest (RF) algorithm represents better performance comparing with other algorithms for MAE calculations. The pixel count difference is also calculated. In that case, RF provides better performance. For PSNR comparison, RF algorithm is higher comparing with other algorithms. This result may vary according to other factors like noise in preprocessing stages. To visualize the difference between images, K means algorithm provides better results.

Figure 4 shows the comparison graph of MAE values of six classification algorithms. Each algorithm has its distinct features. For this dataset, the Random

Table 1 Comparison of different classification algorithms

Algorithms	MAE	PSNR
K Means	54.82	7.9
Random forest (RF)	34.00	12.5
K-Nearest neighbor (KNN)	46.1	10.4
KDTree KNN (KDTKNN)	85.17	8.3
Maximum likelihood (ML)	37.91	11.7
Minimum distance (MD)	55.3	9.1

Fig. 4 Comparison of MAE values of different algorithms

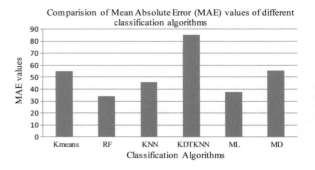

Fig. 5 Comparison of PSNR values of different algorithms

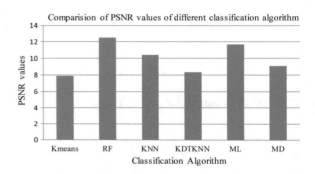

Forest provides better MAE values comparing with other algorithms. According to the metric MAE value, lower is better for good performance. Therefore, the classified result is more accurate comparing with other algorithms.

Figure 5 shows the graph compared with the PSNR values of different classification algorithms. According to the metric PSNR values higher the PSNR is better the result. The Random Forest algorithm poses the highest PSNR value which shows that it is the best among other algorithms.

4 Conclusion

Processing of satellite images is very important for critical analysis, flood preparedness and management. The relief work can be managed with the help of the right information. In this work, we applied with six classification algorithms on two heterogeneous satellite images. The algorithms work well against the images and the results show its strength. The algorithms correctly classified the images and we can interpret it. Random Forest algorithm is prominent among other algorithms in terms of accuracy. The results are compared and analyzed. The algorithm can work well against other kind of satellite images also.

References

1. Wei, Zhao, Zhirui Wang, and Maoguo Gong. 2017. Discriminative Feature Learning for Unsupervised Change Detection in Heterogeneous Images Based on a Coupled Neural Network. *IEEE Transactions on Geoscience and Remote Sensing* 55 (12).
2. Mei, A., C. Manzo, G. Fontinovo, C. Bassani, A. Allegrini, and F. Petracchini. 2015. Assessment of Land Cover Changes in Lampedusa Island (Italy) Using Landsat TM and OLI Data. *African Earth Sciences*. https://doi.org/10.1016/j.jafrearsci.2015.05.014.
3. Rejaur, Md Rahman, and Praveen K. Thakur. 2018. Detecting, Mapping and Analysing of Flood Water Propagation using Synthetic Aperture Radar (SAR) Satellite Data and GIS: A Case Study from the Kendrapara District of Orissa State of India. *The Egyptian Journal of Remote Sensing and Space Sciences*, 21 (1): S37–S41.
4. dos Santos, J.A., C.D. Ferreira, R.D.S. Torres, M.A. Gonçalves, and R.A.C. Lamparelli. 2011. A Relevance Feedback Method Based on Genetic Programming for Classification of Remote Sensing Images. *Information Sci*ences 181: 2671–2684.
5. Nurwandaa, Atik, Alinda Fitriany Malik Zainb, and Ernan Rustiadic. 2016. Analysis of Land Cover Changes and Landscape Fragmentation in Batanghari Regency, Jambi Province. *Procedia-Social and Behavioral Sciences* 227: 87–94.
6. Gounaridis, Dimitrios, and Sotirios Koukoulas. 2016. Urban Land Cover Thematic Disaggregation, Employing Datasets from Multiple Sources and Random Forests Modeling. *International Journal of Applied Earth Observation and Geoinformation* 51: 1–10.
7. Phiria, Darius, Justin Morgenrotha, Cong Xua, and Txomin Hermosilla. 2018. Effects of Pre-processing Methods on Landsat OLI-8 Land Cover Classification using OBIA and Random Forests Classifier. *International Journal Application Earth Obs Geoinformation* 73: 170–178.
8. Rao Zahid, Khalil, and Saad-ul-Haque. InSAR Coherence-Based Land Cover Classification of Okara, Pakistan. *The Egyptian Journal of Remote Sensing and Space Science*, http://dx.doi.org/10.1016/j.ejrs.2017.08.005.
9. Caterina, Samela, Raffaele Albano, Aurelia Sole, and Salvatore Manfreda. A GIS Tool for Cost-Effective Delineation of Flood-Prone Areas. *Computers, Environment and Urban Systems*, https://doi.org/10.1016/j.compenvurbsys.2018.01.013.
10. Himabindu, G., and M. Ramakrishna Murty et al. 2018. Classification of Kidney Lesions Using Bee Swarm Optimization. *International Journal of Engineering &Technology* 7 (2.33): 1046–1052.
11. Himabindu, G., and M. Ramakrishna Murty et al. 2018. Extraction of Texture Features and Classification of Renal Masses from Kidney Images. *International Journal of Engineering &Technology* 7(2.33): 1057–1063.
12. https://earthobservatory.nasa.gov/images/92669/before-and-after-the-kerala-floods.
13. Ran, He, Bao-Gang Hu, Wei-Shi Zheng, and Xiang-Wei Kong. 2011. Robust Principal Component Analysis Based on Maximum Correntropy Criterion. *IEEE Transactions on Image Processing* 20 (6): 1485–1494.
14. Yaoguo, Zheng, Xiangrong Zhang, Biao Hou, and Ganchao Liu. 2014. Using Combined Difference Image and k-Means Clustering for SAR Image Change Detection. *IEEE Geoscience and Remote Sensing Letters* 11 (3): 691–695.
15. Ham, J., Yangchi Chen, M.M. Crawford, and J. Ghosh. 2005. Investigation of the Random Forest Framework for Classification of Hyperspectral data. *IEEE Transactions on Geoscience and Remote Sensing* 43 (3): 492–501.
16. Yu, Zhiwen, Hantao Chen, Jiming Liu, Jane You, Hareton Leung, and Guoqiang Han. 2016. Hybrid k-Nearest Neighbor Classifier. *IEEE Transactions on Cybernetics* 46(6): 1263–1275.
17. Bruzzone, L., and D.F. Prieto. 2001. Unsupervised Retraining of a Maximum Likelihood Classifier for the Analysis of Multitemporal Remote Sensing Images. *IEEE Transactions on Geoscience and Remote Sensing* 39 (2): 456–460.

Action Recognition in Sports Videos Using Stacked Auto Encoder and HOG3D Features

Earnest Paul Ijjina⊙

Abstract Sports analytics is an emerging area of research with applications to personalized training and entertainment. In this work, an approach for sports action recognition using stacked autoencoder and HOG3D features is presented. We demonstrate that actions in sports videos can be recognized by 2D interpretation of HOG3D features, extracted from the bounding box of the player as input to a deep learning model. The ability of a stacked autoencoder to learn the underlying global patterns associated with each action is used to recognize human actions. We demonstrate the efficacy of the proposed classification system for action recognition on ACASVA dataset.

Keywords Stacked autoencoder (SAE) · Sports action recognition · HOG3D features

1 Introduction

Among the various fields in computer vision research, human behaviour analysis is a major area due to its applications in various domains ranging from medical to entertainment sector. Some of the variations in human actions include gesture recognition, abnormal activity detection, event detection, violent scene detection, etc., which has applications in various industrial and research domains. One among these application areas is the sports video analytics, where the intricate aspects of real-time events are analyzed to understanding the performance of players and the overall dynamics of the game. The identification of various events and their attributes through automatic video content analysis is a challenging task, which is gaining attention due to its ability to enhance viewer experience. The training of sports athletes is now moving towards the use of computer vision algorithms with

E. P. Ijjina (✉)
Department of Computer Science and Engineering, National Institute of Technology Warangal, Warangal, Telangana 506004, India
e-mail: iep@nitw.ac.in

© Springer Nature Singapore Pte Ltd. 2020
K. S. Raju et al. (eds.), *Proceedings of the Third International Conference on Computational Intelligence and Informatics*, Advances in Intelligent Systems and Computing 1090, https://doi.org/10.1007/978-981-15-1480-7_79

motion data captures in motion capture (MOCAP) environment to identify their strengths and weaknesses to design better training schedules.

Over the years, various approaches were proposed in the literature for action recognition. Some of these relied on single modalities like audio, visual (RGB and depth), Motion Capture (MOCAP) information, whereas others relied on multi-modal information. The frequency representation of audio observations is used with a bag of visual words model for audio event detection [1]. The action bank representation of videos is used to train a convolutional neural network (CNN) [2] with an evolutionary algorithm for action recognition in [3]. Low-level features extracted from video are used by a 3D convolutional neural network for action recognition in RGB videos. The discriminative representation computed from MOCAP information [4] is used as input to an SAE for action recognition. The temporal templates computed from both RGB and depth video [5] are used as input to CNN for multi-modal action recognition. We consider HOG3D features for recognizing sports actions in this paper.

In this work, we propose a deep learning model for sports action recognition in videos using HOG3D features. The novelty of this work lies in: (1) the 2D inter-pretation of HOG3D descriptor for representing video observations and (2) the use of deep learning model to learn the underlying features associated with each action. The remaining sections are organized as follows: Sect. 2 describes the approach and Sect. 3 covers the experimental setup and results. Finally, Sect. 4 gives the con-clusions and future work.

2 Proposed Approach

In this work, we propose a deep learning approach for human action recognition using the stacked autoencoder and HOG3D descriptor. We use HOG3D descriptor on videos for human action recognition due to their ability to capture the spatio-temporal motion of visual elements. The block diagram of the proposed approach is shown in Fig. 1. For a given input video, the bounding box of each subject in the video is identified. The HOG3D descriptor extracted from the bounding box is used as input to a stacked autoencoder (SAE) for sports action recognition. The individual components of this approach are explained in the fol-lowing subsections.

2.1 HOG3D Representation

The HOG3D descriptor proposed by Klser et al. [6], is considered an extension of SIFT descriptor [7] capturing both shape and information at the same time. Given a 3D patch, the histograms of 3D gradient orientations are used to compute this one-dimensional feature descriptor. The 960×1 one dimension HOG3D vector is

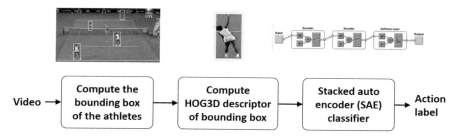

Fig. 1 Overview of the proposed approach (best viewed in color)

resized to a 20×48 2D representation, to be given input to the deep learning recognition model. The next subsection explains the use of this 2D representation for sports action recognition using stacked autoencoder.

2.2 Stacked Autoencoder (SAE) Classifier

In this work, we use a stacked autoencoder to recognize actions from HOG3D descriptor of videos. The block diagram of the SAE classifier used in this work is shown in Fig. 2. The architecture consists of two layers of encoding with n hidden nodes in the first layer and $n/2$ nodes in the second layer. The output layer is a soft-max layer with an output vector of size c, where c is the number of actions to be discriminated.

The first two layers are trained to minimize the reconstruction error during unsupervised training and the soft-max layer is trained to minimize the error between the predicted and actual action label. The weights in the SAE architecture are fine-tuned in the end using Back-Propagation Algorithm (BPA). The details of the experimental study of the proposed approach are discussed in the following section.

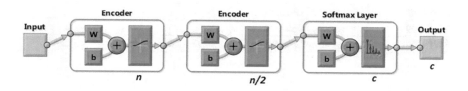

Fig. 2 Architecture of stacked autoencoder (SAE) classifier

3 Experimental Study

The proposed approach is evaluated on Adaptive Cognition for Automated Sports Video Annotation ACASVA dataset [8–10], which consists of footage of TV broadcasts of tennis/badminton games. The dataset consists of HOG3D features extracted from the bounding boxes of players. The features are of 960×1 dimensions as the HOG3D features are extracted using a 4×4 grid in space with 3 splits in time for each sub-block using a histogram of edge orientations is quantized using an icosahedron (20 faces regular polyhedron). The dataset consists of three actions: *Non-Hit*, *Hit* and *Serve*, performed by near and far player, whose details are given in Table 1. An illustration depicting the players and their actions in the dataset is given in Fig. 3. Since deep learning models require large number of training samples to learn the discriminative features, we partition the dataset into two splits with all observations of single's games in one split and all observations of double's games in second split and use two-fold cross-validation.

We consider an SAE classifier with 100 (n) hidden nodes in first layer and 50 hidden nodes in the second layer to classify the 3 (c) sports actions. The SAE model is trained for 500 epochs with a sparsity value of 0.15. The features learnt by the SAE for the two splits during twofold validation are shown in Fig. 4.

This two-dimensional HOG3D representation is also given as input to a CNN for comparative study. The performance of the proposed approach with SAE against CNN is shown in Table 2. From the table, it can be observed that SAE is able to perform better than CNN in both cases (splits). The performance is better when Single's is used for training, which could be due to a large number of training samples providing more discriminative information of recognition compared to Doubles.

Table 1 Description of observations in ACASVA dataset

Game	Gender-Type-Sport	Server	Hit	Non-hit
TWSA03	Women-Single's-Tennis	72	214	944
TMSA03	Men-Single's-Tennis	123	469	1881
TWSJ09	Men-Single's-Tennis	59	224	859
BMSB08	Men-Single's-Badminton	8	458	706
TWDA09	Women-Double's-Tennis	36	135	1064
TWDU06	Women-Double's-Tennis	77	405	3269

Fig. 3 Annotation of players in single's and double's tennis and the three human actions (Figure from [12])

The confusion matrix matrices corresponding to the two splits for CNN and SAE are given in Fig. 5a–d. From these tables, it can be observed that SAE was able to recognize Hit and Serve actions more effiectively than CNN.

The next section presents the conclusions of this work.

4   Conclusion and Future Work

This work proposed a deep learning approach for action recognition using 2D representation of HOG3D features. The 2D representation of the HOG3D descriptor is given as input to a stacked autoencoder to discriminate sports actions. Experimental studies on ACASVA dataset suggest that the proposed approach is efficient in recognizing sports actions. The future work will consider other spatio-temporal features like exmoves [11] as input to this recognition model.

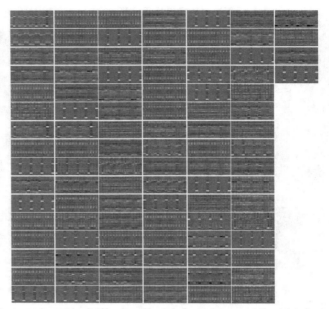

(a) Features learnt by SAE using Single's for training and Double's for testing

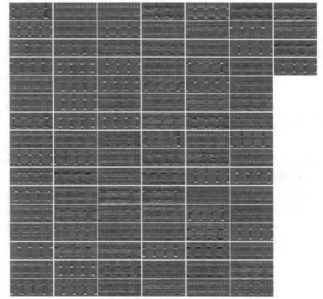

(b) Features learnt by SAE using Double's for training and Single's for testing

Fig. 4 The features learnt in the first layer of SAE, using Single's for training and Double's for testing in (**a**) and vice versa in (**b**)

Table 2 Performance comparison on ACASVA dataset

Approach	Train	Test	Accuracy (in %)
HOG3D with CNN	Single's	Double's	95.04
Proposed approach with SAE	**Single's**	**Double's**	**96.37**
HOG3D with CNN	Double's	Single's	90.43
Proposed approach with SAE	**Double's**	**Single's**	**93.78**

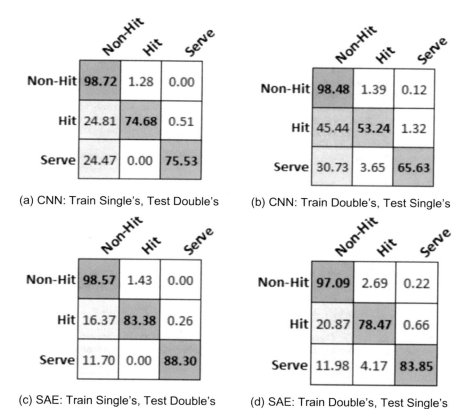

(a) CNN: Train Single's, Test Double's   (b) CNN: Train Double's, Test Single's

(c) SAE: Train Single's, Test Double's   (d) SAE: Train Double's, Test Single's

Fig. 5 The confusion matrix for CNN when (**a**) trained on single's and tested on double's, (**b**) trained on double's and tested on single's; and for SAE when (**c**) trained on single's and tested on double's, (**d**) trained on double's and tested on single's

References

1. Foggia, P., N. Petkov, A. Saggese, N. Strisciuglio, and M. Vento. 2015. Reliable Detection of Audio Events in Highly Noisy Environments. *Pattern Recognition Letters* 65: 22–28.
2. Simard, P.Y., D. Steinkraus, J.C. Platt. 2003. Best practices for convolutional neural networks applied to visual document analysis. In *ICDAR'03 Proceedings of the Seventh International Conference on Document Analysis and Recognition*, vol. 2, 958, Washington, DC, USA: IEEE Computer Society. http://dl.acm.org/citation.cfm?id=938980.939477.
3. Ijjina, Paul E., and Mohan, Krishna C. 2017. Human Behavioral Analysis Using Evolutionary Algorithms and Deep Learning. Chap. 7, 165–186. Wiley. https://doi.org/10.1002/9781119242963.ch7.
4. Ljjjina, Earnest Paul. 2016. Classification of Human Actions Using Pose-Based Features and Stacked Auto Encoder. *Pattern Recognition Letters* 83: 268–277.
5. Ijjina, E.P., and K.M. Chalavadi. 2017. Human Action Recognition in RGB-D Videos Using Motion Sequence Information and Deep Learning. *Pattern Recognition* 72: 504–516. https://doi.org/10.1016/j.patcog.2017.07.013.
6. Klaser, A., M. Marszalek, and C. Schmid. 2008. A Spatio-Temporal Descriptor Based on D-Gradients. In *BMVC 2008—19th British Machine Vision Conference*, vol. 275, eds. M. Everingham, C. Needham, and R. Fraile, 1–10. UK: British Machine Vision Association, Leeds. https://hal.inria.fr/inria-00514853.
7. Lowe, D.G. 2004. Distinctive Image Features from Scale-Invariant Keypoints. *International Journal of Computer Vision* 60 (2): 91–110.
8. De Campos, T., M. Barnard, Mikolajczyk, K., Kittler, J., Yan, F., Christmas, W., Windridge, D. 2011. An Evaluation of Bags-of-Words and Spatio-Temporal Shapes for Action Recognition. In *IEEE Workshop on Applications of Computer Vision (WACV)*, 344–351. https://doi.org/10.1109/WACV.2011.5711524.
9. FarajiDavar, N., T. De Campos, J. Kittler, F. Yan. 2011. Transductive Transfer Learning for Action Recognition in Tennis Games. In *IEEE International Conference on Computer Vision Workshops (ICCV Workshops)*, 1548–1553.
10. Farajidavar, N., T. De Campos, D. Windridge, J. Kittler, W. Christmas. 2011. Domain Adaptation in the Context of Sport Video Action Recognition. In *Domain Adaptation Workshop, in Conjunction with NIPS*, 1–6.
11. Tran, D., L. Torresani. 2013. EXMOVES: Classifier-Based Features for Scalable Action Recognition. CoRR abs/1312.5785.
12. FarajiDavar, N., De Campos, T. 2018. Adaptive Cognition for Automated Sports Video Annotation (ACASVA). Accessed 5 Oct 2018. http://www.cvssp.org/acasva/Downloads.

An Analytical Study on Gesture Recognition Technology

Poorvika Singh Negi and Praveen Kumar

Abstract Over the years, mankind has adapted and evolved, opening up new horizons in terms technology. The intellect and inventiveness of human beings have led to the development of many tools, gesture recognition technology being one of them, to help extend the capabilities of our sense, combining natural gestures to operate technology and making the optimum use of our body gestures, decreasing human effort and going beyond human abilities. Devices these days usually have one type of sensor installed in them; however, this paper will be covering the effect of multiple sensors on recognition accuracy. Furthermore, this paper proposes a new tool for communication made by combining the British sign language with gesture-based technology, which can quickly translate sign language into text, with one mobile phone. Principles of generating EOG, methods of sampling EOG signals, basic eye modes, blinking and fixation modes.

Keywords Accelerometer · British sign language (BSL) · Human–computer interaction (HCI) · Non-hearing-impaired (NHI) · Sensors

1 Introduction

Gesture Recognition Technology (GRT) is an emerging technology that allows the user to interact and control devices using simple and natural body gestures. This technology uses mathematical algorithms that are created using a variety of coding programs to communicate with devices and allows the user to control the device without physically touching it.

However, it is important to know the difference between body movements and gestures. Gestures communicate significant meaning to the observer, whereas the

P. S. Negi · P. Kumar (✉)
Amity University Uttar Pradesh, Noida, India
e-mail: Pkumar3@amity.edu

P. S. Negi
e-mail: negikpoorvika@gmail.com

© Springer Nature Singapore Pte Ltd. 2020
K. S. Raju et al. (eds.), *Proceedings of the Third International Conference on Computational Intelligence and Informatics*, Advances in Intelligent Systems and Computing 1090, https://doi.org/10.1007/978-981-15-1480-7_80

movements or motions that do not communicate a significant meaning to the observer are not a gesture. So, waving hello to a person standing at a distance will be an example of a gesture as the hand being waved is sending a greeting whereas pressing a key on the keyboard will be a simple movement.

Gestures usually originate from the face or hands, but can also originate from any motion or state of the body.

2 Problem Statement

When playing most video games, apart from eye-hand coordination and precision, speed is the essence to win [1]. Operating the game using a mouse, a joystick or any other input device usually slows down the reaction time of the player [2]. That is why players now prefer to operate the games using movements or gestures [3].

Apart from this, people who are physically disabled are either unable to or have trouble to provide the strength or precision required to operate traditional computer input devices also benefit from gesture recognition technology [4]. It helps them control devices and enter information via simple eye blinks, finger movements, hand gestures head motions, etc. [5].

Gesture Recognition Technology not only reduces human effort but may also be seen as a major solution to problems that we have been facing since the beginning of the technological era [6].

Systems using this technology identify human gestures, check what type of task is to be performed and then give outputs to the user.

3 Applications

Researchers continue to improve Gesture Recognition Technology by working on present day technology, making it more accurate and robust. This upcoming technology has a wide range of applications. Some of them are given as follows:

- It enables very young children to interact with computers when they are not old enough to use conventional input devices [7].
- It helps in communication between the deaf and disabled with the non-hearing impaired [4].
- It recognizes sign language and improves communication [2].
- It can be used in navigation or manipulating a virtual environment.
- It can be of aid in terms of distance learning or can be used for tele-teaching assistance [8].
- It can be used to improve already existing technology, making it easy to use.

4 Literature Survey

Communication is the essence of existence for all human beings. We communicate and interact with not just each other but with our surroundings, our environment and machines as well [6]. The interaction between two human beings can be either in the form of verbal communication or in terms of non-verbal communication [2]. Similarly, we can draw a comparison with the interaction between a user and a device.

Verbal communication is a form of communication, which involves words—written, spoken or signed. Therefore, when a user is interacting with a device using codes and written algorithms, it can be referred to as verbal communication. Non-verbal communication is communication which takes place with the help of body language, such as gestures, facial expressions, eye and posture [9]. The hosts or hostess in a plane explain the safety rules and guides to the passengers using gestures before the airplane takes off to avoid the possibility that the passengers may be unable to understand their languages [10]. Also this way, the aged or hear-impaired passengers can also understand the instructions.

Gestures are used for communication between people and machines as well. The communication between the user and the computer requires accuracy in terms of gesture recognition [6]. This interaction between the computer and user is known as the Human–computer interaction [11].

Interactions between human and computer are based on a precise gesture recognition system which can be used as Humanized Computer Interaction [12]. This can be used as a means of communication of meaningful information.

4.1 The Beginning

It all began with the Theremin in 1919, which was the first gesture recognition device invented [13]. It consisted of two antennas that emitted electro-static waves. One hand was hovered around each antenna resulting in the production of different sounds at different pitches and volume. The first pen-input device was the RAND tablet created in 1963 [2]. It allowed users to create freehand drawings on their computers. There existed an electric grid under the surface of the tablet and the stylus would emit pulses for the grid to detect, making virtual drawing a real thing. Later, in the late 1970s, Myron Krueger developed Videoplace, an artificial reality laboratory that created an environment for the users where the users' movements and actions without the use of gloves or goggles [14]. It has a very good feature recognition technique which can easily distinguish between hands and fingers, whether fingers are extended or closed and can even identify the difference between the fingers being used for the gesture.In 1989 Kramer developed Cyber Glove which could easily detect sideways movement of the fingers [7].

4.2   The Present

We know that keyboards are the most used input devices. The introduction of virtual keyboards was the next new breakthrough [13]. These use infrared sensors. For the projection of an image any flat surface can be used. Typing can be done on the virtual keyboards as they would do in normal keyboards. An infrared light beam thrown by a device onto the projected keyboard detects the user's fingers and the device monitor calculates the time taken by a pulse of infrared light to reflect back to the sensor from the user's fingertips [14].Another application called Navigaze, allows the user to control windows features and applications using slight head movements to perform functions like controlling the mouse pointer, clicking the mouse, etc., without using hands. It requires a webcam and infrared sensor for detection. Selected facial features perform particular tasks [6]. For example, closing an eye one, two, four seconds performs single, double or right-click. Navigaze also recognizes the difference between closed and open eyes and thus can respond to eye blinks.

4.3   The Future

Dhirubhai Ambani Institute of Information and Communication Technology, Gandhinagar, India, developed CePal, an infrared gesture-based remote control and gizmo that can be worn like a watch [10]. It is capable of helping motor-impaired and other limb-related disordered to perform day-to-day tasks like operating a TV, switching on lights, fans, air conditioners, etc. IIT-Madras developed ADITI which helps debilitating diseases like serious muscular-skeletal disorders and cerebral palsy to communicate with the help of gestures [5]. It is a USB device that perceives movement within a radius of six-inch. It is screen-based and provides a list of visual options like alphabets or pictures or words to express them [14]. With the help of ADITI patients can easily communicate through gestures like nodding of head or moving feet to generate a mouse click.

The user makes a gesture by static pose of the body or by physical motion of the body. Gesture recognition technology operates on mainly two different types of systems: 2D recognition system and 3D recognition system [6]. The software takes gesture as input or a command from the user. This input is translated into simple letters or words and then the computer gives output according to the command given by the user.

Many devices contain small sensors in them for gesture recognition purposes. Most of the devices and mobile phones have an accelerometer installed in them [1]. It helps in making operations easy and highly intuitive in devices like video game controllers for PS3 or Wii, android devices, iOS devices, etc. In present-day devices, for the purpose of saving battery energy, the number of accelerometers is restricted to one [15]. Accelerometers also use gyroscope with the help of which slightest of dimensions of the information are taken into account by the tracking and rotations.

In practice, gyroscope fills the information about the resolution of tilt or the orientation with accuracy [1]. This accuracy is unattainable without the accelerometer alone; hence the usage of the combination of accelerometer and gyroscope is very much in practice [8]. This decreases the chance of failure in recognition of similar movements or gestures [2]. Failure in recognizing gestures degrades the usability of the interface (Table 1).

An experiment was conducted by Kazuya Murao and Tsutomu Terada (2011), from the Graduate School of Engineering, Japan, where they explored and conducted an investigation by taking into account the effects of gesture recognition accuracy by changing the positions and numbers of sensors placed in a device. This experiment was held taking into account 27 gestures, nine gyroscopes and nice accelerometers [1].

Nine sensors were placed on board and 2160 samples were collected for the experiment to take place [16]. The board is shown in Fig. 1.

The equipment used was WAA-006 sensor which was made by Wireless-T Inc, a wireless 3 axis gyroscope and 3 axis accelerometer. The experiment board model was W117 X H 115 X D16 (mm) [1]. The numbers of subjects involved in the experiment were 8 and they were all right-handed. The subjects performed gestures

Table 1 List of gestures

ID	Gesture	Figure	ID	Gesture	Figure	ID	Gesture	Figure
1	Tilt to the near side		10	Tap lateral as though sifting		19	Shift up	
2	Tilt to the far side		11	Scoop		20	Shift down	
3	Tilt to the left side		12	Lay cards		21	Shift left	
4	Tilt to the right side		13	Gather cards		22	Shift right	
5	Tap upper side twice		14	Rap table with the longer lateral		23	Shift diagonally up	
6	Tap left side twice		15	Rap table with the face of the board		24	Shift diagonally down	
7	Swing twice to the left side quickly		16	Knock the board twice		25	Draw a circle	
8	Swing twice to the right side quickly		17	Turn the board over		26	Draw a triangle	
9	Shuffle cards		18	Rotate clockwise on the table		27	Draw a square	

Fig. 1 Sensor board

while holding the lower right side of the board [1]. To minimize the chances of error due to discrete explication by the subjects, one of the authors demonstrated the subjects all the time during the observation. The starting and endpoints were indicated by pausing and staying still. The sensor value was observed and its mean and variance were calculated [8]. Two basic results were drawn from the experiment.

The first one was that the use of multiple sensors set at specific positions affects the accuracy such that accuracy slightly increases, being followed by a downfall with the increase in the number of sensors, and the second one was that gesture performed are independent and some gestures that are specific in nature improve recognition accuracy more than others [8]. The number and kind of sensors and gestures do inputs and commands o the devices, however, this may not be the case when the subjects have hearing disability [4]. In order to achieve efficient communication between the deaf and the non-hearing impaired (NH1), we need something more easy to use in our day to day life then sign language.

Loss of hearing not only makes it difficult for the person to enjoy technology but also has a tremendous effect on their working and social life. British Sign Language (BSL) is one of the tools that are used by people having hearing impairments or who are deaf to communicate [13]. BSL was declared as an official minority language by the British government in the year 2003 [16]. However, the NHIs are not familiar with BSL, creating a huge rift between the deaf and the NHI in terms of communication. Moreover, BSL is neither fast nor easy to be understood by the non-hearing-impaired. This creates inefficient grounds for communication between deaf and non-hearing impaired [4] (Fig. 2).

Fig. 2 This depicts some gestures that can be used and remembered by deaf people by adding them with BSL

British Sign Language	Mobile Sign Language	Description
		You: click the device outside 1 time
		Me: click the device inside 1 time
		Questions like why, what, who, how, when: rotate the device in a circle
		Yes: bend the device two times

The input method will use touch screen technology instead of using gestures because fingerspelling (spelling a word or sentence using hand gesture) is time-consuming and tedious.

This model comprises of two parts-to design the basic gestures such that they recognize what their meanings are and then to organize these set of gestures so that they can be converted or can be used to have a conversation [4]. By this model of a gesture-based mobile phone, the conversation time drastically decreases, reducing the real-time discussion to three major steps: (1) giving input to the mobile using the touch screen technology (2) relating the word with limited words from the location (3) creating a sentence.

Another type of upcoming gesture recognition technology is based on the detection of eye movements or gestures [9]. The eye is a host to a continuous potential field. Detection of this electric field can be done with the eye open or close in darkness. It is a fixed dipole with the positive pole at the cornea and negative pole at the retina [9, 14].

This technology establishes a direct channel for communication between electrical devices and the user [4]. Instead of using motion gestures (as mentioned above), this technology uses bio-electric signals which help disabled people to communicate with their surroundings.

Our body gives out millions of electric signals every day. Researchers mainly focus on electroencephalogram, electrooculogram and electromyogram [10]. Electroencephalography (EEG) is used to record electrical activity of the brain whereas Electrooculography (EOG) is used to measure the potential between the retina and the cornea [6]. The signals received are called the electrooculograms [4].

Electromyogram signals (EMG) are used to design a Human–computer Interaction (HCI) interface [15]. Electromyography (EMG) is used for observing and recording the electrical signals that help muscles in the body to relax and contract [14]. The signals recorded are translated into graphs which are known as electromyogram. When studied carefully, they can detect when the cells were active and when the potential was generated [16].

Out of EEG, EMG and EOG, EOG is better understood. It is simpler to extract features and analyse EOG signals [6]. However, for precise recognition, successfully recognizing the eye–computer interaction interface is important [16].

According to a Min Lin and Guoming Mo (2011) of Shanghai University, Shanghai Medical Instrumentation College, 101 Yingkou Road, Shanghai, China, there are mainly two modes of eye gesture recognition [14].

They are:

1. Basic Eye Moving Mode of Recognition
2. Recognition of Blink Mode.

4.4 Basic Eye Moving Mode of Recognition

The basic moving modes that the eyes are capable of performing consist of moving the eye to right, left, up and down [14]. Figure 4 depicts the waves that are seen when these modes take place. Taking centre to be the initial position for all the modes, figure (a) shows a negative graph when the eye moves left, figure (b) shows a positive graph when the eye moves right, figure (c) shows negative graph when the eye moves up and figure (b) shows a positive graph when the eyes move down, all having different amplitude ranges [10].

The two graphs, when the eye moves in the left and eye, belong to horizontal EOG signals. They are regular in nature and have sharp ends.

The two graphs, when the eyes move up and down, belong to vertical EOG signals. They are irregular in nature unlike the horizontal EOG [14].

For diagonal eye moments, we break the movement into its basic horizontal and vertical components, and therefore analyzing them by their combined basic eye moving directions [9].

For diagonal eye moments, we break the movement into its basic horizontal and vertical components, and therefore analyzing them by their combined basic eye moving directions [9].

4.5 Recognition of Blink Mode

As the name suggests, this mode consists of blinks, i.e. the sudden or involuntary moment the eyelid to cover the eyes. When we blink, the eyeballs shoot upward, resulting in the generation of a potential difference across the vertical electrodes [14]. Blinking is a quicker action than the normal action of the eye to look upwards as seen in the previous mode. The difference between the two can be seen by comparing their relative pulse widths [5]. Comparing Fig. 3 with Fig. 4c, the pulse width of the blink is much narrower.

Fig. 3 Potential field in the eye

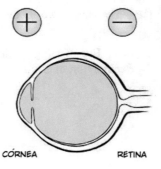

CÓRNEA   RETINA

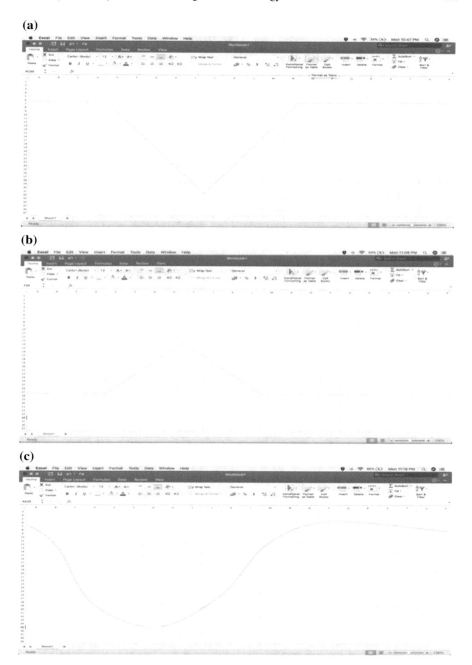

Fig. 4 **a** Negative graph when the eye moves left. **b** Positive graph when the eye moves right. **c** Negative graph when the eye moves up

Another problem that arises is differentiating the reflexive blink and the reflective blink from the voluntary wink signal [8]. For this, we use the difference in their amplitudes. The amplitude of the reflexive wink is higher as compared to the other two, making it easy to be detected by the use of internal algorithms [14].

5 Conclusion

In this paper, the effects of position and number of sensors and gestures on gesture recognition technology were discussed. Twenty-seven kinds of gestures were observed using nine accelerometers, nine gyroscopes and a board. However, more effective positions of the accelerometers and gyroscope can yet be investigated.

It also covered an upcoming system design for efficient and fast communication between the deaf and the NHI. In the future, further technologies will be created by combining touch screen technology and accelerometer techniques.

The paper also discusses the importance in the design of eye–computer interface, Recognition of Basic Eye Moving Mode and Recognition of Blink Mode.

References

1. Murao, Kazuya, Tsutomu Terada, Ai Yano, and Ryuichi Matsukura. 2011. Evaluating Gesture Recognition by Multiple-Sensor-Containing Mobile Devices. In *15th Annual International Symposium on Wearable Computers*, San Francisco, USA.
2. Lai, Ching-Hao. 2011. A Fast Gesture Recognition Scheme for Real-Time Human-Machine Interaction Systems. In *Conference on Technologies and Applications of Artificial Intelligence*.
3. Zabidi, Nur Syabila, Noris Mohd Norowi, and Rahmita Wirza O.K. Rahmat. 2016. A Review on Gesture Recognition Technology in Children's Interactive Storybook. In *4th International Conference on User Science and Engineering (i-USEr)*.
4. Xue, Haoyun, and Shengfeng Qin. 2011. Mobile Motion Gesture Design for Deaf People. In *Proceedings of the 17th International Conference on Automation & Computing*. Huddersfield, 10, Sept 2011, UK: University of Huddersfield.
5. Rawat, Seema, Parv Gupta, and Praveen Kumar. 2014. Digital life assistant using automated speech recognition. In *2014 Innovative Applications of Computational Intelligence on Power, Energy and Controls with their impact on Humanity (CIPECH)*, 43–47, IEEE.
6. Geer, D. 2004. Will Gesture Recognition Technology Point the Way. *Computer* 37 (10): 20–23. Los Alamitos, CA, USA.
7. Sonkusare, J.S., Nilkanth. B. Chopade, Ravindra Sor, Sunil L. Tade. 2015. A Review on Hand Gesture Recognition System. In *International Conference on Computing Communication Control and Automation*.
8. Murao, Kazuya, Ai Yano, Tsutomu Terada, Ryuichi Matsukura. 2012. Evaluation Study on Sensor Placement and Gesture Selection for Mobile Devices. In *Proceedings of the 11th International Conference on Mobile and Ubiquitous Multimedia*, New York, USA.
9. Konwar, P., H. Bordoloi. 2015. An EOG Signal Based Framework to Control a Wheel Chair. USA: IGI Global.

10. Wenhui, Wang, Chen Xiang, Wang Kongqiao, Zhang Xu, and Yang Jihai. 2009. Dynamic Gesture Recognition based on Multiple Sensors Fusion Technology. In *31st Annual International Conference of the IEEE EMBS*, 2–6, Sept, Minneapolis, Minnesota, USA.
11. Kumar, Praveen, Bhawna Dhruv, Seema Rawat, and Vijay S. Rathore. 2014. Present and future access methodologies of big data. *International Journal of Advance Research In Science and Engineering* 3: 541–547. 8354.
12. Kumar, Manish, Mohd Rayyan, Praveen Kumar, and Seema Rawat. 2016. Design and development of a cloud based news sharing mobile application. In *2016 Second International Innovative Applications of Computational Intelligence on Power, Energy and Controls with their Impact on Humanity (CIPECH)*, 217–221, IEEE.
13. Saeed, Atif, Muhamamd Shahid Bhatti, Muhammad Ajmal, Adil Waseem, Arsalan Akbar, and Adnan Mahmood. 2013. Android, GIS and Web Base Project, Emergency Management System (EMS) Which Overcomes Quick Emergency Response Challenges. In *Advances in Intelligent Systems and Computing*, vol 206, Springer: Berlin.
14. Chaturvedi, Anshika, Praveen Kumar, and Seema Rawat. 2016. Proposed noval security system based on passive infrared sensor. In International Conference on Information Technology (InCITe)-The Next Generation IT Summit on the Theme-Internet of Things: Connect your Worlds, 44–47, IEEE.
15. Lin, Min, Guoming Mo. 2011. Eye gestures recognition technology in Human-computer Interaction. In *4th International Conference on Biomedical Engineering and Informatics (BMEI)*, Shanghai, China.
16. Phade, G.M., P.D. Uddharwar, P.A. Dhulekar, and S.T. Gandhe. 2014. "Motion Estimation For Human–Machine Interaction. *International Symposium on Signal Processing and Information Technology (ISSPIT)*.

Artificial Intelligence Approach to Legal Reasoning Evolving 3D Morphology and Behavior by Computational Artificial Intelligence

Deepakshi Gupta and Shilpi Sharma

Abstract The motive of this paper is to provide a general introduction to the area or field of Artificial Intelligence and Legal Reasoning. Legal reasoning is an initiative that makes a new set of demands on artificial intelligence methods. Artificial life has now begun to be a mature inter-discipline. In this contribution, its origin is discovered by investigation, its key questions are elevated to a higher position, its main methodological tools are examined, and finally, its applications are discussed. As part of the flourishing knowledge at the junction between life science and computing, artificial life will continue to thrive and benefit from further scientific and technical procedures on both sides, the biological and the computational. The development of the framework draws remarkably on the philosophy of law, in which the understanding of legal reasoning is an important topic.

Keywords Artificial intelligence · Computational legal science · Computational science · Machine learning · Network-based inference

1 Introduction

Artificial Intelligence is also known as machine intelligence, i.e. machine itself knows the functions they have to perform. Artificial Intelligence is the research of artificial systems that disclose behavior characteristic of natural living systems. This involves computer simulations, biological and chemical experiments, and absolutely theoretical endeavors. Processes materialized on the molecular, cellular, neutral, social, and evolutionary plates are subject to investigation [1].

D. Gupta (✉)
Amity School of Engineering and Technology, Noida, India
e-mail: deepakshi730@gmail.com

S. Sharma
Amity University, Noida, India
e-mail: ssharma22@amity.edu

© Springer Nature Singapore Pte Ltd. 2020
K. S. Raju et al. (eds.), *Proceedings of the Third International Conference on Computational Intelligence and Informatics*, Advances in Intelligent Systems and Computing 1090, https://doi.org/10.1007/978-981-15-1480-7_81

There are four possible goals to follow in AI which are given as follows:

1. Systems that think like humans.
2. Systems that think rationally.
3. Systems that act like humans.
4. Systems that act rationally.

Artificial Intelligence and legal reasoning introduce a thought—precipitate outlook to both computational models of legal reasoning and the application of evolutionary thinking about the law. Artificial Intelligence including legal reasoning recommends the sight of law as a comprehensive organization on the edge of developing into a systematized computer system of legal resources [2].

The field of AI and Law consistently has two clear-cut motivations: viable and hypothetical. On the viable side, intelligent legal information systems, systems that can accommodate both lawyers and non-lawyers in their collaboration with both legal and non-legal standards. On the hypothetical side, the process of legal thinking and legal argumentation, utilizing computational models and systems. To strengthen a legal research system with contemporary accomplished systems technology, it is mandatory to convey a very penniless goal. Instead, we would like to permit a user who previously knows something about his province, to tell his own story using an introductory natural language interface conceivably and then have the system accomplish a partial interpretation before it inquiry for advance information [3].

In our existing surrounding, the title artificial intelligence can possibly foremost considered not as determined by parallelism, from the conscientious invention of theoretician, clinician and semantic technologists, besides description accustomed to introduce what it looks that specific PC frameworks hold to a certain extinct. Such framework, haven hence delineate and establish to implement those errands and tackle those issues that concurrently in the event that accomplished by people are withdrawing us have been representative of brainpower, used to said reveal Artificial Intelligence [4].

Our potential to progress logical reasoning is emphatically studied by the ability to formulate and grow new gadgets, better approaches to investigate the world; apparatus permit science and, abruptly, strengthen in science conduct to the formation of new gadgets in a category of constant revolution. The role bounce by systematic policies here described as non-segregated equipment and programming infrastructure represented to direct data and assist the scientific investigation. The occurrence and consequence of a research forecast prospecting unconventional and cross-methodological calculational perspective to the educational and inspecting research of offense [5].

The branch of artificial intelligence focuses on clone lifelike etiquette in computer simulations. The objective of this research extend from constructing a novel artificial life—forms to building imitations which will help researchers better apprehend biological life. Many artificial intelligence researchers have cautiously shifted focus from huge inflexible artificial intelligence environments to examine intelligence as an infant behavior of simple interactions [6].

The field of artificial intelligence is predicated on the presumption that a necessary characteristic of the human being: intelligence is an attribute corresponding only with the family known as Homo sapiens. This property can be so exactly depicted to the point where the same attribute of reasoning can be synthetic by the mechanism. The assumption of this outlook since the pivot of the century has lifted some analytical, experiential and existing affairs about the sort of mentality as it associates with the human being who formerly now, was examining the only beings chartered luminary of assigning the attribute of intelligence [7].

The possibility of using computers to legal reasoning processes is likely to produce a typically lawyer-like response; so what if we understand the legal reasoning or legal dispute formation better? The attempt to create programs to electrifying legal reasoning processes should have at least two conveniences; it should provoke more systematic study of the legal problem-solving, and it should proceed knowledge of the problem-solving potential of the computer [8].

There has been much argument over the nature of the legal reasoning process. Oliver Wendall Holmes, one of the American legal realists, stated: 'the life of the law has not been logic; it has been experiencing'. They would insist that lawyers make the prophecy of forensic and official bearing and judges prognosticate whether or not laws work for society and it is difficult to concern with such augury. Susskind would argue 'it is inordinately restricted to think that constructing expert systems in law is simply about computerizing legal reasoning: legal knowledge engineering extend into the very essence of jurisprudence and philosophy' [9].

2 Literature Review

A review of the literature was completed to understand the nature of artificial intelligence and legal reasoning.

Reference [1] highlights the artificial life which would eventuate from artificial intelligence. It clearly pointed out that artificial life has inherited its origin from artificial intelligence. It also pointed out the merits and demerits of artificial life. There is some hypothesis of artificial life that is described. There is also an enormous number of applications outline which can be examined to make use of artificial life techniques in the broad sense [1].

Pamela N. Gray pointed out the legal approach of artificial intelligence rather than the technological approach. He describes the legal issues rules and social policies concerning artificial intelligence pioneering in our life. He highlights the contradictory perspectives to study artificial intelligence and legal reasoning [2].

L. Thorne McCarty highlights the two main scopes of investigation—the practical work on penetrating legal information systems, and the conceptual work on computational models of legal reasoning. In each case, he pinpoints the knowledge characterization problem as the most serious problem covering AI and law [3].

Reference [4] is concerned with a characteristic of computer science. The general motive of this is to provide the institution with the branch of artificial intelligence and legal reasoning. More particularly, the issue of Expert Systems in Law is conveyed, and one perspective to the establishment of these networks is recommended [4].

Reference [5] highlights the interaction joining the issues and the research areas accommodate how computational instruments and strategies can esteem the exposure of new approaches in legal analysis and application [5].

Reference [6] describes a 3D environment to artificial life. It includes an illuminate object-oriented language, an OpenGL dispose engine, collision detection, as well as experimental, reinforce for articulated body physical imitation and collision resolution with static and dynamic friction [6].

Reference [7] pointed out that artificial intelligence furnishes the devastating and intense development in technology and commendatory tools used every day to strengthen socialize on earth. It also pointed the thoroughgoing legal implementation and appropriation of the 23 newly accustomed artificial intelligence proposition as one of the appropriate estimates for saving humanity from the forthcoming ultimatum posed by reorganizations hopped-up by advances in A I [7].

Reference [8] suggests that computer software may serve attorneys in both the work and accomplishment of their reasoning processes. It's claimed that the time has come for consequential interdisciplinary work between lawyers and computer scientists to investigate the computer's potential in law [8].

Reference [9] focuses on the enlargement of AI and Law and looks at the process of technical reorganization with regard to the association for the management of technology. One particular branch of AI concerns Expert Systems (ES) which will be investigated with particular reference to Computer-Assisted Document Drafting (CADD) applications [9].

3 Research Methodology

A mix of 20 journals, research papers and articles were thoroughly examined and studied to acquire knowledge about artificial intelligence approach to legal reasoning evolving 3D morphology and behavior by computational artificial intelligence. The qualitative approach was adopted to understand the legal and non-legal issues of Artificial Intelligence through computational AI. This topic became more popular in recent years. Hence, the domain of AI and legal reasoning has extended critical mass and manifests all the signs of being a source of penetrative research that should be of interest to both regulations.

Since Artificial Intelligence and Big Data, moreover Natural Language Processing, are becoming favored in legal science. There is a significant variation in research methodologies adopted by scholars. This not only highlights the legal issues of Artificial Intelligence but also highlights Computational Artificial Intelligence. The concepts of Artificial Intelligence in robotics are also highlights in this research. It hence implements that the field of Artificial Intelligence is progressing day by day.

4   Result

Law and legal reasoning are natural earmark for artificial intelligence systems. Like medical diagnosis and other tasks for Oracle systems, legal analysis is a matter of transliterating data in terms of higher-level notions. But in law the documentation is more like those for a system aimed at perception natural language: they tell a story about the human eventuality that may lead to the proceedings.

Statements of the law, too, are scribbled in natural language, and legal arguments are ordinarily arguments about what that language means or ought to mean. This study is one of the few research attempts in this fertile area. It is unique in advancing a computational model for examining legal problems in a way that brings these perspectives of AI research together and makes consciousness from a jurisprudential perspective as well.

Gardner first analyzes various positions in Anglo-American jurisprudence and their materiality for work in artificial intelligence. It distinguishes aspects of legal reasoning that any precisely expert system in law must make a place for and recommends a way to deteriorate the process of legal analysis that takes these characteristics into account (Figs. 1 and 2).

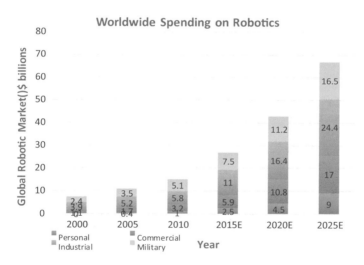

Fig. 1 Worldwide spending on robotics

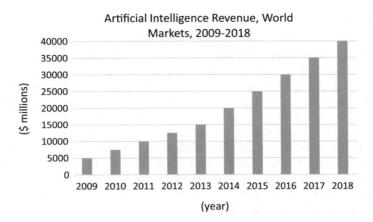

Fig. 2 Artificial intelligence revenue, World markets, 2009–2018

5 Future Work

Looking back over 25 years of AI and Law permit us to see a great deal of enlargement both in technologies and understanding and in the technology and the part played by AI and Law. The expansion of the World Wide Web (Internet), the immeasurable depletion in the cost of information stockpiling and the immense expansion in computational ability have to unite to change the idea of AI and Law applications totally, both in accessibility and range.

On the theoretical side, various relationships between cases and regulations and rules, between legal knowledge and judgment knowledge, and between formal and informal perspectives have provided a compatible source of inspiration and explicit progress has been made in interpreting these relationships better.

References

1. Steels, Luc. 1993. The Artificial Life Roots of Artificial Intelligence. *Artificial Life* 1(1_2).
2. Gray, Pamela N., V.T. Brookfield, and Stephen M. McJohn. 1998. *Artificial Legal Intelligence*. Harvard Journal of Law & Technology 12(1).
3. McCarthy, L.T. 1990. AI and Law: How to Get There from Here. *Notes of Workshop on Artificial Intelligence and Legal Reasoning*.
4. Susskind, Richard E. 1986. Expert Systems in Law: A Jurisprudential Approach to Artificial Intelligence and Legal Reasoning. *The Modern Law Review* 49(2).
5. Wing, Jeannette M. 2006. Ex Machina: Analytical Platforms, Law & The Challenges Of Computational Legal Science. *Communications of the ACM* 49(3).
6. Barella, Antonio, Carlos Carrascosa, and Vicente Botti. 2007. Breve: A 3D Environment for the Stimulation of Decentralized System & Artificial Life. In *Proceedings of the 2007 IEEE/WIC/ACM International Conference on Intelligent Agent Technology*. IEEE Computer Society.

7. Wogu, I.A.P., F.E. Olu-Owolabi, P.A. Assibong, B.C. Agoha, M. Sholarin, A. Elegbeleye, D. Igbokwe, and H.A. Apeh. 2017. Artificial Intelligence, Alienation and Ontological Problems of Other Minds: A Critical Investigation Into the Future of Man and Machines. In *Computing Networking and Informatics (ICCNI)*.
8. Buchanan, Bruce G., and Thomas E. Headrick. 1970. Some Speculation About Artificial Intelligence And Legal Reasoning. *Stan. L. Rey.* vol. 23.
9. Harris, Vicky. 1992. Artificial Intelligence and the Law—Innovation in a Laggard Market. *JL & Inf. Sci.*, vol. 3.

Data Mining in Healthcare and Predicting Obesity

Anant Joshi, Tanupriya Choudhury, A. Sai Sabitha and K. Srujan Raju

Abstract Humanity has been in the pursuit of lengthening its lifespan for centuries. Healthcare provides many solutions to illnesses, injuries, and data mining, which provides useful patterns in data, is being actively used in healthcare sectors for the betterment of services provided by healthcare institutions. Obesity is a major issue in the current world and is a known cause of a multitude of other health problems. Data mining is a rapidly growing field with many applications in the field of healthcare. One of its many uses is the prediction of diseases. In this paper, we use classification, a data mining technique, to predict obesity in patients. We compare several classification techniques, such as random forest generation, decision trees, and neural net, to find the most accurate technique.

Keywords Healthcare · Data mining · Obesity · Classification · Prediction · Time-use surveys · Random forest · Health informatics

1 Introduction

Healthcare has been a part of human life since the dawn of time. Even the early ages, every village had a doctor, a medic, a shaman, someone who could cure illnesses and rejuvenate people. As time progressed, so did science. Man's curiosity

A. Joshi · A. Sai Sabitha
Amity University, Noida, UttarPradesh, India
e-mail: anant.joshi@live.com

A. Sai Sabitha
e-mail: saisabitha@gmail.com

T. Choudhury (✉)
University of Petroleum and Energy Studies, Dehradun, India
e-mail: tanupriya1986@gmail.com

K. Srujan Raju
CMR Technical Campus, Hyderabad, India
e-mail: drksrujanraju@gmail.com

© Springer Nature Singapore Pte Ltd. 2020
K. S. Raju et al. (eds.), *Proceedings of the Third International Conference on Computational Intelligence and Informatics*, Advances in Intelligent Systems and Computing 1090, https://doi.org/10.1007/978-981-15-1480-7_82

led him to explore the wonders of the human body, understanding it one piece at a time. Healthcare became an important part of society, with doctors and medics being one of the most respected and highly paid jobs in the world and it is a field that is constantly being improved.

Data mining is the process of extracting or finding useful information from data repositories. It is part of knowledge discovery in databases or KDD. Data mining is rapidly becoming a popular field of research as the gathering of data has increased exponentially over the course of the past century and vast amounts of data are collected every second and stored in warehouses. Analysis of this collected data is top priority for many companies and organizations. Application of data mining ranges from bioinformatics and cheminformatics to climate prediction and healthcare.

Healthcare is also one such field in which techniques of data mining are widely used. These techniques of data mining are used in healthcare to classify medical insurance fraud, predict diseases, analyze effectiveness of treatment, correlate symptoms and diseases, etc. Data mining is a powerful tool that allows healthcare institutions to analyze the massive amounts of data generated in medical transactions. In the healthcare sector as a whole, data mining plays a crucial role and is rapidly becoming an essential part of this field.

Obesity is a major problem in many developed countries of the world. Obesity is classified as a medical condition in which a person has gained an excessive amount of body weight. This leads to many health problems and is generally considered to have an ill effect on the human body. It is linked to a plethora of diseases from heart attacks, diabetes to mental health problems such as depression. It is caused by combination of eating a surplus amount of food, very little exercise and in some cases genetic susceptibility. Obesity can be controlled by moderation in food intake and large amount of exercise. Obesity is also often ignored because a lot of people believe they are not overweight. This is a major problem in the world. The need to predict obesity is becoming clearer. The world is moving in a healthier direction and prediction of obesity will play a crucial role in the future.

This is where data mining comes in. It allows researchers to use classification techniques to predict a disease from a variety of other factors. We use classification to predict obesity in sample of American population.

2 Literature Review

Providing medical care to people is the core of healthcare and it also encompasses many activities which aim to improve human health. Practitioners of healthcare are doctors, dentists, chiropractors, psychiatrists, surgeons, etc. Physical well-being of an individual is the main focus of healthcare and mental well-being is becoming increasingly important in healthcare institutions in many parts of the world. Healthcare is given importance to reduce suicide rates among different age groups.

Healthcare institutions fund a variety of activities from researching cures, treatments, improving current treatments to judging their effectiveness [1]. This has led to rapid development in the field of medicine and continues to play an integral role in modern medicine.

From an economic perspective, healthcare is often a debatable issue even in many well-developed countries such as the USA. Universal healthcare was one of the most debated topics during the US Presidential elections. Statistically, healthcare is a major part of the GDP of many countries and contributes to more than 10% of the GDP in countries such as USA, the Netherlands, France, Germany, Canada, and Switzerland. Many countries around the world struggle to provide their citizens with proper healthcare.

Data mining is the process of finding novel and useful information from basic data. It also helps in visualization of data which has become an integral part of the modern business landscape. Healthcare institutions use data mining as it has many relevant applications in the field.

One of the applications of data mining in healthcare is healthcare management. Data mining is used to track frequent patients, patients with high-risk ailments, chronic disease identification, etc. This allows doctors to monitor a patient's condition over a period of time. Novel or unknown patterns help in detecting rarer diseases which sometimes get misdiagnosed [2]. Data mining is also used to compare the effectiveness of drugs for a certain disease. We can comparatively study the effectiveness of several treatments and use that information to identify the appropriate treatment. The cost-to-effectiveness ratio is also an important factor that can be identified and analyzed. Using data mining techniques, the world of modern pharmaceuticals is constantly battling to provide the cheaper and more effective medicine. A more common use of is to mine customer data to find trends and information that can be useful to the providers. A large amount of billing data, ambulance data, and financial data, among others, is generated in the modern healthcare world. Prediction based modeling is used to find out which products are more likely to be purchased by the customer. Mining such data has also led healthcare indices which highlight the kinds of treatments more preferred by customers as well as which services they are more likely to use.

Obesity is a medical condition that is very common in the modern world especially in developed countries such as the USA and UK [3]. It is becoming so common that some people may not even be aware that they are obese and tend to falsely believe they have normal weight. As many as 53% of the American population does not believe it has a weight problem. There has been a trend in lower recognition of weight problems since the 90's and this trend has unfortunately been matched by an increasing rate of obesity [4]. Even children are affected by this. Research has shown that as many as 26% of parents of obese kids did not think that their kids were overweight or obese [5]. Obesity has also been linked around 100,000–400,000 deaths per year [6]. From a more economic perspective, obesity takes up to 6% of USA's yearly health budget and is on par with other major chronic diseases [7]. It is also regarded to be a more serious problem than smoking and drinking [8]. It is becoming clearer that people are not aware of their own weight problems. The need for predicting obesity is increasing as the world attempts to move toward a healthier future and data mining is proving to be a key tool in this endeavor.

3 Data Mining

The results obtained from data mining help in understanding the data and gain new knowledge from it. Data mining is also known as knowledge discovery in databases (or KDD). It is a quickly developing field of computer science in which we discover new patterns and find new information in datasets. Data mining involves a number of techniques which allow us to manipulate and transform data in many ways allowing us to find useful information [9].

Data mining involves six main types of tasks. These are:

1. Anomaly/Outlier Detection: This technique is used to try and find observation which does not match a predefined pattern. Observations which do not match the pattern are often considered to be actionable or critical. Anomaly detection has a wide variety of uses such as detecting fraud, safety systems fault detection, and even medical problems [10].
2. Association Rule Mining: Association rule mining is technique that is used to find relations between attributes and variables in large datasets. It is used to find recurring rules which define the relationship of variables. It is often used in marketing to find relationships between items and use that information to optimize sale [11].
3. Clustering: Clustering is used to group similar items or variables together. The groups are similar by some criterion of similarity [12]. Many fields use clustering technique. For example, it is commonly used in bioinformatics, computer graphics, pattern recognition, and information retrieval.
4. Regression Analysis: This technique is borrowed from statistical mathematics and it is used to find relationship among variables. It helps us in finding variance in one variable with respect to variance in another.
5. Data Visualization/Summarization: This technique is used to summarize or visualize a dataset using the most important factors or attributes of the dataset. It is used to help visualize the dataset and make it easier to understand.
6. Classification: This technique is used find a model that helps categorize the data and distinguish classes. It is used in predictive analysis to predict a class label which is essentially an attribute in the dataset. We use other attributes to try and predict this class label. Classification is the technique that is used in this paper so we will go into a bit more detail.

4   Classification

Classification is used to build models and can be used to make predictions in data. These models help in identifying the attributes or variables that most impact the prediction of the class label. It is one of the major techniques that are being actively used in predictive analysis especially in the healthcare research field [13, 14]. The models we have used are:

1. Logistic Regression Model: This model is a regression model in which a dependent variable can take two values (usually 0 or 1) which are used as a basis for predicting the result. We can use several variables as independent variables which will be used to predict the data.
2. Decision Tree Model: This model creates a tree structure by dividing the dataset in smaller subsets. After a certain number of divisions, only end nodes (also called leaf nodes) are obtained and a complete tree forms. This is also used in predictive analysis and helps in identifying the most common factors affecting a certain class label [15].
3. Random Forest Model: This technique of modeling constructs a large amount of decision trees and uses a random variable to split the nodes. Random forest creates a "forest" of decision trees and finds the mean class or decision tree for regression [16].
4. Neural Net Model: This model creates, in a way, an artificial brain and just like a brain, it contains many connected neurons or nodes. The data is passed between nodes and uses a transfer function. This artificial brain has variable power which depends on the number of nodes or neurons chosen to pass the data through [17].

5   Methodology

Initially, the dataset was collected from the American Time Use Survey(ATUS) website. This dataset is a part of the Eating and Health module. The survey was conducted during the periods of 2006–2008 and 2014–2016 by the Bureau of Labor Statistics which is part of the US Department of Labor. The Rattle interface for the R programming language is used for classification and analysis. Rattle provides an easier way to mine data and obtain results by using approved R packages. The following steps illustrate the methodology followed:

1. Data Collection: As previously mentioned, the data was collected from the ATUS EH website which is part of the US Department of Labor. Two datasets of different time periods were considered. The more recent 2014–2016 dataset was selected because of the dynamic nature of obesity as a growing health concern.

2. Preprocessing: The dataset contained an attribute for body mass index (BMI) "erbmi" which was used to create our class label "obesity." A BMI of ≥ 30 is the international standard for obesity. This relation was used to create our categorical class label. It has a value of 1 for participants with BMI ≥ 0. We also remove the "erbmi" and fields related to weight and height as these factors are used to calculate BMI.

3. Classification: To perform classification, the dataset was divided into training and testing parts. We use Rattle automatic data partition algorithm to evenly sample our dataset. Different classification techniques were used to model the dataset.

4. Modeling: We create the decision tree, random forest, neural net, and logistic regression models for our dataset. We set "obesity", the class label, as the target for our modeling techniques.

5. Evaluation: The models were evaluated using an error matrix on the whole dataset. The output of the error matrix gives us the total error in prediction of the class label and the percentage accuracy of the model in predicting obesity.

6 Experimental Setup and Analysis

6.1 Dataset

The dataset used in this study was retrieved from the Bureau of Labor Statistics of United States Department of Labor. The American Time Use Survey (ATUS) Eating and Health module dataset was used. In this dataset, data is taken from surveys done from 2014 to 2016 in which the participants were asked various questions related to their health, eating habits, exercise habits, etc. The compilation of this data was used to create the dataset (Fig. 1).

	tucaseid	tulineno	eeincome:	erbmi	erhhch	erincome	erspemch	ertpreat	ertseat	ethgt	etwgt	eudietsoda
1	2.01E+13	1	-2	33.2	1	-1	-1	30	2	0	0	-1
2	2.01E+13	1	1	22.7	3	1	-1	45	14	0	0	-1
3	2.01E+13	1	2	49.4	3	5	-1	60	0	0	0	-1
4	2.01E+13	1	-2	-1	3	-1	-1	0	0	0	-1	2
5	2.01E+13	1	2	31	3	5	-1	65	0	0	0	-1
6	2.01E+13	1	1	30.7	3	1	1	20	10	0	0	1
7	2.01E+13	1	1	33.3	1	1	5	30	5	0	0	-1
8	2.01E+13	1	1	27.5	3	1	-1	30	5	0	0	-1
9	2.01E+13	1	1	25.8	3	1	-1	117	10	0	0	-1
10	2.01E+13	1	1	28.3	3	1	5	80	0	0	0	2
11	2.01E+13	1	1	40.5	3	1	-1	35	20	0	0	1
12	2.01E+13	1	2	28	1	5	-1	0	5	2	0	-1
13	2.01E+13	1	1	27.9	3	1	5	25	10	1	0	-1
14	2.01E+13	1	2	30.4	3	5	5	150	5	0	0	1
15	2.01E+13	1	2	26.8	3	5	-1	0	300	0	0	-1
16	2.01E+13	1	1	32.9	3	1	5	80	0	0	0	-1
17	2.01E+13	1	3	25.8	3	2	-1	105	2	0	0	-1
18	2.01E+13	1	1	26.5	3	1	-1	60	5	0	0	-1
19	2.01E+13	1	-2	35.2	3	-1	-1	30	1	0	0	2

Fig. 1 ATUS EH dataset

6.2 Attribute Selection

Attributes were selected based on their relevance for classification of the data. The obesity attribute was set as the target as this is the attribute we are trying predict. Attributes related to the height and weight of the participant were removed because the "erbmi" attribute which contained the body mass index (BMI) of the participant and was used in creating the class label.

6.3 Tools Used

We used the R programming language to obtain the various results. We used Rattle, a data mining interface as well as R scripts.

7 Results

The dataset contains many fields which correspond to various questions answered by the participants. To find out which fields are most relevant to the prediction of obesity, we perform logistic regression analysis (LGA) on the dataset. We used LGA because the response value (or target value) is categorical, i.e. it holds a true/ false or 0/1 value (Fig. 2).

Logistic regression analysis shows us the factors that most affect the value of obesity are eeincome1, eusnap, eugenhth, erincome, erspemch, ertpreat, euexfreq, eufastfdfrq, and eutherm (Table 1).

We also used random forest generation method to classify and predict the dataset (Fig. 3).

We also find the variable importance of the model made using random forest classification. Evaluation is done using an error matric to find the percentage error in predicting the target.

The result of evaluation of these models is:

1. Logistic regression model (Fig. 4)
2. Random forest model (Fig. 5)
3. Neural net model (Fig. 6)
4. Decision tree model (Fig. 7).

The random forest model only has 9% error which means it can predict obesity with 91% accuracy. Random forest is one of the most accurate predictive algorithms. LGA model gives us 28% error and can predict obesity with 72% accuracy, however, because logistic regression models tend to suffer from overfitting and give

```
Coefficients:
              Estimate Std. Error z value Pr(>|z|)
(Intercept) -1.1780172  0.5435154  -2.167 0.030204 *
eeincome1    0.1168160  0.0447756   2.609 0.009083 **
erhhch      -0.0787709  0.0573391  -1.374 0.169512
erincome    -0.1200231  0.0575096  -2.087 0.036887 *
erspemch     0.0314523  0.0101985   3.084 0.002042 **
ertpreat    -0.0017432  0.0005819  -2.996 0.002737 **
ertseat      0.0005156  0.0005110   1.009 0.312925
eudietsoda   0.0314280  0.0279607   1.124 0.261011
eudrink     -0.1854903  0.1506371  -1.231 0.218184
eueat        0.0813562  0.0489491   1.662 0.096502 .
euexercise  -0.0443438  0.0771850  -0.575 0.565620
euexfreq    -0.0390584  0.0128551  -3.038 0.002379 **
eufastfd    -0.0827838  0.1128833  -0.733 0.463341
eufastfdfrq  0.0477122  0.0139888   3.411 0.000648 ***
euffyday     0.0256282  0.0423269   0.605 0.544858
eufdsit      0.0256345  0.0761138   0.337 0.736274
eusnap      -0.1566323  0.0666239  -2.351 0.018723 *
eugenhth     0.5474742  0.0271365  20.175 < 2e-16 ***
eugroshp    -0.0390406  0.0494141  -0.790 0.429487
euinclvl    -0.0200316  0.0693830  -0.289 0.772803
euincome2    0.1186912  0.0615269   1.929 0.053719 .
eumeat       0.1418788  0.1408744   1.007 0.313872
eumilk      -0.2322607  0.1360036  -1.708 0.087682 .
euprpmel     0.0515636  0.0517247   0.997 0.318819
eusoda      -0.0922919  0.0534125  -1.728 0.084004 .
eustores     0.0047323  0.0279125   0.170 0.865372
eustreason   0.0139361  0.0189566   0.735 0.462244
eutherm      0.1393691  0.0511935   2.722 0.006481 **
euwic       -0.0307961  0.0191671  -1.607 0.108117
exincome1   -0.0015287  0.0017930  -0.853 0.393885
---
Signif. codes:  0 '***' 0.001 '**' 0.01 '*' 0.05 '.' 0.1 ' ' 1
```

Fig. 2 Coefficients and their effect on the prediction

results of predicting more than they actually can, this model is not considered to be the best for analysis. The neural net model with ten hidden layer nodes has a minor improvement in accuracy at 73% accuracy. This accuracy was improved by increasing the number of hidden nodes but this model is infeasible because of processing power and time complexity for improving its accuracy. Decision tree model also gives us 72% accuracy but decision trees tend to not be very accurate [18], (Table 2; Fig. 8).

Table 1 Most important attributes using logistic regression

Attribute	Meaning
eeincome1	Total monthly household income (before taxes)
eusnap	Received SNAP (food stamps) benefits over the last month
eugenhth	General health of the participant
erincome	Relationship between income and poverty threshold
erspemch	Change in spouse/unmarried partner's employment status
ertpreat	Total amount of time spent in primary eating and drinking
euexecfreq	Times exercise activities were undertaken by participant in the last week
eufastfdfrq	Times participant purchased: deli food, carry-out, delivery food, or fast food
eutherm	Food thermometer used when preparing meals

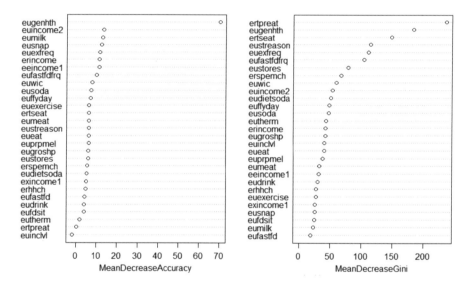

Fig. 3 Most important attributes based on random forest model

8 Conclusions

From these results, the conclusion is that the most important factor in predicting a person's obesity is their own perception of their health. The participants were asked to rate their own health for the attribute 'eugenhth' and we can see that multiple models support this attribute as the most important factor. Other important factors are frequency of exercise and frequency of eating fast food or eating outside. We can use these factors to accurately report obesity and refine medical surveys.

```
Error matrix for the Linear model on obesityprediction.csv (counts):

        Predicted
Actual    0    1
     0 7667  417
     1 2691  437

Error matrix for the Linear model on obesityprediction.csv (proportions):

        Predicted
Actual    0    1 Error
     0 0.68 0.04  0.05
     1 0.24 0.04  0.86

Overall error: 28%, Averaged class error: 46%

Rattle timestamp: 2017-04-04 18:36:00 Anant Joshi
======================================================================
```

Fig. 4 Error matrix for logistic regression

```
Error matrix for the Random Forest model on obesityprediction.csv (counts):

        Predicted
Actual    0    1
     0 7913  171
     1  805 2323

Error matrix for the Random Forest model on obesityprediction.csv (proportions):

        Predicted
Actual    0    1 Error
     0 0.71 0.02  0.02
     1 0.07 0.21  0.26

Overall error: 9%, Averaged class error: 14%

Rattle timestamp: 2017-04-16 16:32:58 Anant Joshi
```

Fig. 5 Error matrix for random forest

```
Error matrix for the Neural Net model on obesityprediction.csv (counts):

        Predicted
Actual    0    1
     0 7468  616
     1 2427  701

Error matrix for the Neural Net model on obesityprediction.csv (proportions):

        Predicted
Actual    0    1 Error
     0 0.67 0.05  0.08
     1 0.22 0.06  0.78

Overall error: 27%, Averaged class error: 43%

Rattle timestamp: 2017-04-16 16:33:04 Anant Joshi
```

Fig. 6 Error matrix for neural net

```
Error matrix for the Decision Tree model on obesityprediction.csv (counts):

      Predicted
Actual    0    1
    0  8084    0
    1  3128    0

Error matrix for the Decision Tree model on obesityprediction.csv (proportions):

      Predicted
Actual    0  1  Error
    0  0.72  0     0
    1  0.28  0     1

Overall error: 28%, Averaged class error: 50%

Rattle timestamp: 2017-04-16 16:32:56 Anant Joshi
```

Fig. 7 Error matrix for decision tree

Table 2 Accuracy of prediction with different models

Model	Accuracy (%)
Logistic regression	72
Random forest	91
Neural net	70–75
Decision tree	72

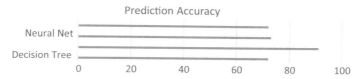

Fig. 8 Accuracy of prediction

We can also conclude that the random forest model allows us to predict obesity with the highest accuracy. Logistic regression and other models, however, can only allow us to predict accuracy at 70–75% rate. This is not sufficient as these models are prone several problems but their accuracy may increase with addition of variables to the dataset. These conclusions clarify that the awareness of obesity and its acceptance play a major role in the health of a person.

References

1. Indicators, O.E.C.D. 2015. Health at a Glance 2011. *OECD Indicators,* Paris: OECD Publishing. https://doi.org/10.1787/health_glance-2015-en. Accessed 15 Feb 2016.
2. Obenshain, M.K. 2004. Application of Data Mining Techniques to Healthcare Data. *Infection Control and Hospital Epidemiology* 25 (08): 690–695.

3. Cullen, K. 2011. A Review of Some of the Existing Literature on Obesity in Children and Young People and a Commentary on the Psychological Issues Identified. *Obesity in the UK: A Psychological Perspective*, 56.

4. Flegal, K.M., M.D. Carroll, C.L. Ogden, and C.L. Johnson. 2002. Prevalence and Trends in Obesity Among US Adults, 1999-2000. *JAMA* 288 (14): 1723–1727. https://doi.org/10.1001/jama.288.14.1723.

5. He, M., and A. Evans. 2007. Are Parents Aware that their Children are Overweight or Obese? Do they Care? *Canadian Family Physician* 53 (9): 1493–1499.

6. Flegal, K.M., B.I. Graubard, D.F. Williamson, and M.H. Gail. 2005. Excess Deaths Associated With Underweight, Overweight, and Obesity. *JAMA* 293 (15): 1861–1867.

7. Wolf, A.M. 1998. What Is the Economic Case for Treating Obesity? *Obesity Research* 6: 2S–7S. https://doi.org/10.1002/j.1550-8528.1998.tb00682.x.

8. Sturm, R. 2002. The Effects of Obesity, Smoking, and Drinking on Medical Problems and Costs. *Health Affairs* 21 (2): 245–253.

9. Han, J., J. Pei, and M. Kamber. 2011. *Data Mining: Concepts and Techniques*, Elsevier.

10. Chandola, V., A. Banerjee, and V. Kumar. 2009. Anomaly Detection: A Survey. *ACM Computing Surveys (CSUR)* 41 (3): 15.

11. Agrawal, R., T. Imieliński, and A. Swami. 1993. Mining Association Rules Between Sets of Items in Large Databases. In *Acm Sigmod Record*, ACM, 22(2): 207–216.

12. Estivill-Castro, V. 2002. Why So Many Clustering Algorithms: A Position Paper. *ACM SIGKDD Explorations Newsletter* 4 (1): 65–75.

13. Kumari, M., and S. Godara. 2011. *Comparative Study of Data Mining Classification Methods in Cardiovascular Disease Prediction 1.*

14. Antonie, M.L., O.R. Zaiane, and A. and A. Coman. 2001. Application of Data Mining Techniques for Medical Image Classification. In *Proceedings of the Second International Conference on Multimedia Data Mining*, 94–101, Springer-Verlag.

15. Dobra, A. 2009. Decision Tree Classification. In *Encyclopedia of Database Systems*, 765–769, Springer, US.

16. Breiman, L. 2001. Random Forests. *Machine Learning* 45 (1): 5–32.

17. Zhang, G.P. 2009. Neural Networks for Data Mining. In *Data Mining and Knowledge Discovery Handbook*, 419–444, Springer, US.

18. James, G., D. Witten, T. Hastie, and R. Tibshirani. 2013. *An Introduction to Statistical Learning*, Vol. 6. New York: Springer.

An Android-Based Mobile Application to Help Alzheimer's Patients

Sarita, Saurabh Mukherjee and Tanupriya Choudhury

Abstract Alzheimer's is an incessant neurodegenerative ailment that to a degree influences around 44 million people around the world. Because of its infection results, Alzheimer's patients confront a few memory troubles, which is firmly associated with their day-by-day lives. So as to assist Alzheimer's patient with coping up with the memory misfortunes, specialists propose the incitement of the memory of patient by utilizing an application which will incorporate sustenance data, exercise data, mind diversions, and so on. In order to take care of this developing issue, in this research paper we build up a simple-to-utilize versatile use cases that provide assistance to Alzheimer's Patients mainly the people associated with them like their family members or guardian by assisting the patient and their guide to recall the life occasions. Thus, it essentially includes three key highlights: Workout, Games, and Food. Games can be a method used for photo validation and so forth. Exercise segment incorporates data about different exercise for Alzheimer's patient. Food section incorporates data in regard to the eating routine graph and diet plan of the patient. Reminder segment includes the expansion of cautions highlight. The principal highlight of this work is constructing an application that could energize the customary step-by-step nearness of a person influenced by Alzheimer's infection.

Keywords Alzheimer's · Dementia · Family · Games · Food · Reminder

Sarita (✉) · S. Mukherjee
Banasthali Univeristy, Banasthali Vidyapith, Tonk, Rajasthan 304022, India
e-mail: sarita10103@gmail.com

S. Mukherjee
e-mail: mukherjee.saurabh@rediffmail.com

T. Choudhury
UPES, Dehradun, India
e-mail: tanupriya1986@gmail.com

© Springer Nature Singapore Pte Ltd. 2020
K. S. Raju et al. (eds.), *Proceedings of the Third International Conference on Computational Intelligence and Informatics*, Advances in Intelligent Systems and Computing 1090, https://doi.org/10.1007/978-981-15-1480-7_83

1 Introduction

Alzheimer's disease maybe defined as a dementia of a kind that creates issues identified with memory, thinking limit, and conduct changes. Its manifestations as a rule create at a moderate pace and deteriorate with time, getting to be basic enough to meddle with everyday schedule undertakings. It is a standout among the most commonly occurring types of dementia, rather a general term for memory misfortune and other such subjective capacities. The illness represents around a range of sixty to eighty percent of all dementia cases. In contrast to various interminable sicknesses, by 2050 Alzheimer's infection has the pervasiveness on the ascent, whereby 160 million individuals are thought to be all inclusive anticipated to experience the ill effects of this malady. Not exclusively does the ailment influence the patient, anyway conjointly likewise the parental figures, and in addition family, companions and care experts.

During the underlying progressions, the memory misfortune is black out, yet at a later progression, it is unable to convey any discussion further and answer to nature. In the USA, the disease has been noted to account as the sixth primary reason for fatal cases.

The earliest time sign of Alzheimer's is trouble in looking into recollecting the newly attained information because Alzheimer's movements regularly begin in that area of the mind which impacts learning. As Alzheimer's taking through the brain, it prompts an outrageous addition in the symptoms, which consolidate direct changes, demeanor, and disarray; broadening conflicts about the place, time, and events; continuously real memory setback and actions changes; and inconvenience walking, swallowing, and talking.

2 Stages/Phases of Alzheimer's Disease

The disease presents with three major phases, namely mild, moderate, and severe.

2.1 Phase One—Mild Alzheimer's Disease/Early Stage

Minor memory hardships could join dismissing all around respected assertions. Various indications of this stage are:-

- Issue finding the right words to finish the formation of sentences.
- Difficulties with remembering or checking on the name upon preface to new individuals.
- Losing or wrongly placing items in the home without consciously knowing.

2.2 Phase Two—Moderate Alzheimer's Disease/Middle Stage

This dimension goes on for an all-encompassing time span than different stages. In this fragment, while the patient may retain a mess of memory left, certain assignments together with paying bills or composing may likewise rise as definitely extreme. At this stage, social issues which incorporate nervousness, perplexity, tumult, and disappointment that is phenomenal for the individual may also develop as imperative. Loss of rest, bother recollecting their own name, gloom or state of mind adjustments are a portion of alternate signs that parental figures can likewise note.

2.3 Severe Alzheimer's Disease (Late Stage)

On the extreme levels, the individual totally transforms into reliant on parental figures for completing simple ordinary obligations. Generous or add up to memory misfortune is obvious, close by an absence of ability to do any of the common substantial undertakings including brushing teeth or hair or taking strolls.

3 Literature Review

In literature, couple of versatile-based framework is proposed [9] for Alzheimer's patients [8]. The smart phone gadget application will be utilized to direct the Alzheimer's patients to help them in their everyday exercises and additionally focus on the closest relative of patients utilizing uncommon android-based application to control them. The application gives different diversions and test to support understanding mind capacities and show advance report. The primary point of these kinds of use is to make a workplace for patient at home and decrease wellbeing costs and furthermore put less weights on human services experts.

The motivation behind this work was to deliberately seek and depict the writing on versatile applications utilized in intercessions on AD/dementia and to build up another potential application for AD/dementia patients.

Pirani [1] have created Android application that could facilitate the regular day-to-day existence of a man influenced by Alzheimer's sickness and give different functionalities, for example, following developments of the patient through GPS, giving medication and nourishment timing warnings, day-by-day schedule tracker, test/games to increment psychological working of the patient, and timely alarm to provide a reminder of the app.

Yamagata [2] concentrated on creative gadgets, for example, iPads and tablets, which are standard and simple to utilize, can't just help decide phase of dementia, yet additionally give incitement to enhance subjective working. It is trusted that this exploration will break down that exceptionally made applications and existing assistive programming can be utilized to diminish the side effects and enhance perception of more seasoned grown-ups experiencing AD- or other dementia-related illnesses.

Sposaro [3] portrayed a device which enhances the nature of treatment for dementia patients utilizing versatile applications. iWander, keeps running on a few Android-based gadgets with GPS and correspondence abilities. This considers guardians to cost adequately screen their patients remotely. The information uncollected from the gadget is assessed utilizing Bayesian system procedures which gauge the likelihood of meandering conduct. Upon assessment, a few game plans can be taken dependent on the circumstance's seriousness, dynamic settings, and likelihood.

Habash [4] actualized Android-based portable innovation to help specialists to oversee and screen their Alzheimer's patient. The application additionally fills in as a help apparatus for Alzheimer's patient by reminding them about the medicines.

Coppola [5] tended to psychological working and personal satisfaction for individuals determined to have dementia by means of innovation. Furthermore, centered around inventive gadgets, for example, iPads and tablets, which are standard and simple to utilize, cannot just help decide phase of dementia, yet additionally give incitement to enhance intellectual working.

Tabaki [6] planned a total framework to screen and record patient's areas, pulse, and rest which sorts out the manner in which that the parental figures fathom and fulfill patient's and displayed an observing framework for individuals living with Alzheimer's sickness and comprises a wearable and an Android application.

4 Algorithm

Steps for the proposed algorithms are:-

Step 1: Create application in Android Studio long with the New Project wizard.
Step 2: Settings, Permissions, and API Key for the Project.
Step 3: Add a few perspectives and exercises to interface with the code.
Step 4: If the application is opened for the first time then Fill all the details of the user and go to step 5.
Step 5: Home Screen consists of following sections:
(I) My Info $\rightarrow$ Where client can get all the data which he filled in Step 5.
(ii) Games $\rightarrow$ Mind Games for the client.
(iii) Workout $\rightarrow$ Exercise and exercise plan for the client.
(iv) Diet Plan $\rightarrow$ Food Recipes and Diet Plan for the client.
(v) Reminder $\rightarrow$ Reminder for the client for any reasons (e.g., taking medication).
(vi) My exercises $\rightarrow$ User can include a rundown of exercises he needs to do and tail them appropriately.

(vii) My Home → a guide/map view is opened with a mark on the user's home location which he filled in Step 5 and he can easily navigate to that location which ultimately gives open the way to home in the Google Maps app.

Step 6: Test the code on the bases of input and finish the Project.

Components illustrated in Fig. 1 are the interrelated parts of proposed app.

5 Our Proposed Approach

5.1 Screenshots of App:-

This app fundamentally includes four key highlights: games, exercise, food, and reminder also (Figs. 2, 3, 4, 5, 6, 7, 8, 9, 10, 11, 12 and 13).

Games:- Games are of a sort-like picture acknowledgment of different flying creatures, animals, vegetables, and so forth and score will be produced on the bases of right answers.

Workout:- Exercise segment incorporates data about different exercise for Alzheimer's patient.

Food:- Food area holds data in regards to the eating regimen diagram and diet plan of the patient.

Reminder:- Update area includes the addition of alarm for food, exercise activities, etc.

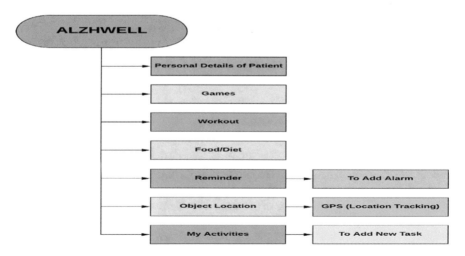

Fig. 1 Diagrammatic illustration of proposed system

Fig. 2 Splash screen activity

These functionalities prove useful to the most vital requirements of the patients and furthermore keep them associated with their loved ones. This application is as of now being produced completely for an android telephone.

6 Hardware/Software Requirements

This application requires Microsoft Windows with proficient memory stockpiling and for testing or execution, it requires android working framework. Android Studio is the official software marked by Google used for the development of Android applications comprising of integrated emulating and coding abilities [10]. This application utilizes the idea of machine learning. Machine learning is a field of computer science which enables the software to learn features based on already known data and then execute the required task on new data after going through

Fig. 3 Registration details

Fig. 4 Home page

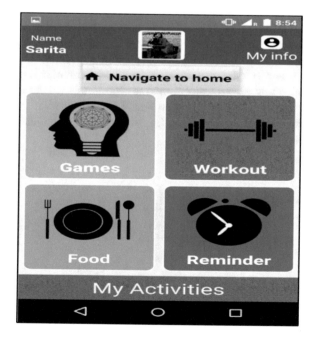

Fig. 5 Self info activity

suitable period of learning through a good amount of data. In future, this work can be connected by using machine learning counts [7] which is confined into two groupings: Supervised learning and unsupervised learning. In supervised learning, system is offered point of reference information sources and their typical yields, given by an "educator," and the objective is to take in a general standard that maps responsibilities from commitment to yield, while in unsupervised acknowledging, there is no tags given in learning figuring, relinquishing it in solitude to discover structure in its data.

Fig. 6 About Alzheimer's disease

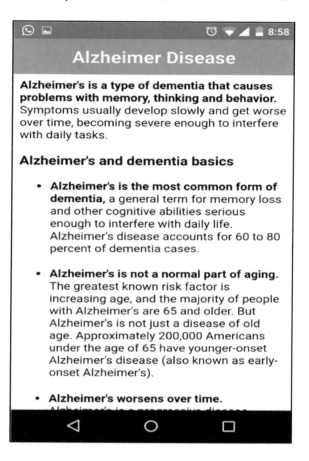

7 Merits/Use Cases

1. This application could have an effect in the battle against the disease.
2. This application would incite better attention and improved memory. Research in addition confirmed that as the capacity to perform various tasks enhanced, senior cerebrum action changed and subjective advantages moved into regular day-to-day existence. All the more essentially, the study affirmed that psychological capacity can be enhanced with appropriate preparing strategies and training methods.

Fig. 7 My Home navigation activity

8 Conclusion

There is a neglected need in geriatric offices for invigorating dementia patients, and in addition giving no-nonsense information to demonstrating expanded subjective capacities with innovation. Research has demonstrated that innovation like a versatile application animates those with dementia. It is trusted that this research will analyze extraordinarily made applications, and existing assistive programming can be utilized to diminish the manifestations and enhance discernment of more

Fig. 8 Game activity

established grown-ups experiencing an AD or other dementia-related sicknesses. Such versatile applications could enable dementia people to end up less unsettled and remain in their homes longer, while additionally giving mindfulness and constructive difference in frame of mind by those of another age toward the elderly. This application can be changed by utilizing machine learning approaches. We can fuse new highlights as and when we require. Reusability is conceivable as and when required in this application.

Fig. 9 Workout related activity

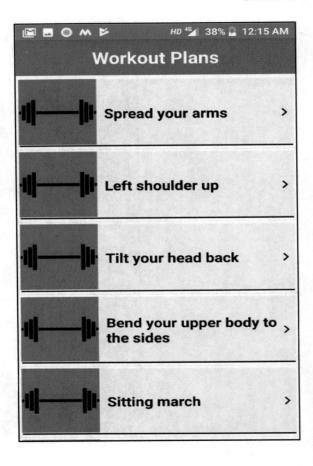

Fig. 10 Food activity

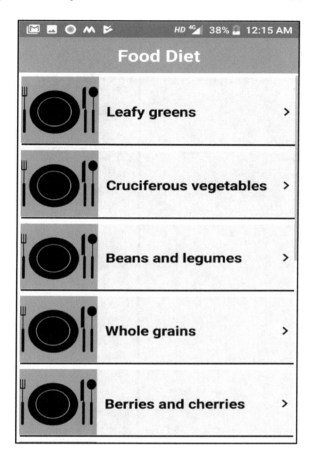

Fig. 11 Reminder activity

Fig. 12 Addition of new task

Fig. 13 Activity added

References

1. Pirani, E.Z., F. Bulakiwala, M. Kagalwala, M. Kalolwala, and S. Raina. 2016. Android Based Assistive Toolbox for Alzheimer. *Procedia Computer Science* 79: 143–151.
2. Yamagata, C., J.F. Coppola, M. Kowtko, and S. Joyce. 2013. Versatile Application Advancement and Ease of Use Research to Encourage Dementia and Alzheimer Patients. In *Systems, applications and innovation meeting (LISAT)*, 1–6.
3. Sposaro, F., J. Danielson, and G. Tyson. 2010. iWander: An Android Application for Dementia Patients. In *Engineering in Medicine and Biology Society (EMBC), 2010 yearly worldwide meeting of the IEEE*, 3875–3878. IEEE.
4. Habash, Z.A., W. Hussain, W. Ishak, and M.H. Omar. 2013. Android-Based Application to Assit specialist with Alzheimers persistent. In *International Conference on Computing and Informatics*, vol. 28, pp. 511–516.
5. Coppola, J.F., M.A. Kowtko, C. Yamagata, and S. Joyce. 2013. *Applying Portable Application Advancement to Encourage Dementia and Alzheimer Patients*, 16. Paper: Wilson Center for Social Entrepreneurship.
6. Tabakis, I.M., M. Dasygenis, and M. Tsolaki, 2017. An Observing Framework for Individuals Living with Alzheimer's Ailment. In *2017 Panhellenic Conference on Electronics and Telecommunications (PACET)*, 1–4. IEEE.
7. Kumar, P., T. Choudhury, S. Rawat, and S. Jayaraman. 2016. Analysis of Various Machine Learning Algorithms for Enhanced Opinion Mining Using Twitter Data Streams. In *2016 International Conference on Micro-Electronics and Telecommunication Engineering (ICMETE)*, 265–270. IEEE.
8. Acharya, M.H., T.B. Gokani, K.N. Chauhan, and B.P. Pandya. 2016. Android Application for Dementia Understanding. In *2016 International Conference on Inventive Computation Technologies (ICICT)*, vol. 1, 1–4. IEEE.
9. Gulia, S., S. Mukherjee, and T. Choudhury. 2016. An Extensive Literature Survey on Medical Image Steganography. *CSI transactions on ICT* 4 (2–4): 293–298.
10. https://developer.android.com/.

Author Index

A

Ahatsham, 735
Ali, Mohammed Mahmood, 443
Anjani Varma, C., 337
Ansari, Sakil, 85

B

Balamurugan, K., 97
Berin Jones, C., 65
Bevish Jinila, Y., 693
Bhagyalaxmi, D., 39
Bhat, Mrugali, 619, 639
Bhavani, Y., 481
Biksham, V., 417
Bodapati, Suraj, 651
Bremiga, G.G., 65

C

Chandrasekaran, K., 557
Chandra Sekhar Reddy, P., 427
Chatrapati, K. Shahu, 673
Chavva, Subba Reddy, 259
Choudhury, Tanupriya, 877, 889

D

Dandu, Sujatha, 361
Deepika, N., 811
Dendage, Tejaswini, 639
Deoskar, Vaidehi, 639
Devidas, S., 663
Dhavale, N.P., 49
Dilip, G., 97

E

Eedi, Hemalatha, 289
Eshack, Ansiya, 281
Eswara Reddy, B., 403
Ezhilarasi, T.P., 97

F

Fadewar, Hanmant, 701

G

Gagnani, Lokesh, 567
George, Neenu, 619
Ghosh, Samit Kumar, 713
Govardhan, A., 311, 627
Guguloth, Ravi, 301
Gummalla, Sridhar, 601
Gupta, Deepa, 349
Gupta, Deepakshi, 869
Gupta, Shruti, 801

H

Habib, Faheem, 49
Hanisha Durga, G., 337
Harish, Vemula, 249
Hussain, Mohammed Shabaz, 743

I

Ijjina, Earnest Paul, 753, 763, 773, 849

J

Jain, Rishabh, 579
James, Jinu, 619

© Springer Nature Singapore Pte Ltd. 2020
K. S. Raju et al. (eds.), *Proceedings of the Third International Conference on Computational Intelligence and Informatics*, Advances in Intelligent Systems and Computing 1090, https://doi.org/10.1007/978-981-15-1480-7

Janaki, V., 481, 491
Janet, B., 49
Jayasree, H., 17, 29
Jeberson Retna Raj, R., 839
Joshi, Anant, 877
Jyotishi, Amalendu, 349

K
Kamakshi Prasad, V., 85
Kansal, Vineet, 801
Katta, Sugamya, 651
Kaushik, Keshav, 735
Khaing, Kyi Kyi, 213, 219
Khan, Khaleel Ur Rahman, 743
Khare, Sangita, 349
Khin, Thuzar, 213
Kiran, Ajmeera, 723
Kolan, Anusha, 17, 29
Koppula, Neeraja, 153
Korani, Ravinder, 427
Krishna, Addepalli V.N., 673
Krishnakumar, S., 281
Kulkarni, Pooja, 639
Kulkarni, Shrinidhi, 619
Kulkarni, Sushama, 701
Kumara Swamy, M., 145
Kumar, Praveen, 579, 857
Kumar, Sheo, 145
Kyi, Tin Mar, 213

L
Lakshmaiah, K., 403
Lakshman Naik, R., 163
Latchoumi, T.P., 97
Latha, Y.L. Malathi, 387
Leburu, Rangaiah, 123
Lifna, C.S., 325

M
Mahender, U., 145
Mahmood, Md Rashid, 829
Majeti, Srinadh Swamy, 49
Managathayaru, N., 337
Mangathayaru, N., 271
Manju, D., 501
Manjula, B., 163
Manjula, R., 39
Maria Anu, V., 693
Mary Gladence, L., 693
Mathura Bai, B., 337
Mittal, Sanjeev K., 85
Mohammmed, Sheena, 237
Moukthika, Dasika, 29
Mukherjee, Saurabh, 889

Muppala, Chaitanya, 301, 361
Murali Krishna, S., 403
Murali Mohan, V., 467

N
Nagaraju, G., 271
Naga Sravani, M., 713
Narender, Kethavath, 547
Negi, Poorvika Singh, 857

P
Padigela, Praveen Kumar, 535
Padmaja Rani, B., 153, 271
Pallavi, S., 821
Parsewar, Sneha, 619
Patra, Raj Kumar, 829
Perumalla, Sai Reetika, 289
Polepaka, Sanjeeva, 683
Potluri, Anusha, 301, 361
Pradeep, S., 1
Pradeepthi, K.V., 513
Prakash, Ramesh, 557
Prasadu, Reddi, 523
Prashanth Reddy, Gadila, 123
Prathima, T., 627
Puttamadappa, C., 547

Q
Qaseem, Mohd S., 443

R
Rahman, Md Ateeq Ur, 443
Rahul, Gampa, 387
Rajalakshmi, V., 693
Raja, Rohit, 821, 829
Raja Vikram, G., 673
Rama Devi, Boddu, 109
Ramadevi, Y., 627
Ramanababu, V., 39
Ram Kumar, R.P., 683
Ramudu, Kama, 373
Ramya, N., 821
Ramya Laxmi, K., 821
Rana, Ajay, 801
Ranga Babu, Tummala, 373
Rao, Koppula Srinivas, 153
Rathnamma, M.V., 557
Reddy, Sneha, 651
Rukma Rekha, N., 663

S
Sai Sabitha, A., 877
Sai Sandeep, J., 337
Sai Sarath, V., 337

Sameera, Nerella, 187
Sammulal, P., 501
Sandeep, Polamuri, 387
Sangam, Ravi Sankar, 259
Sanjeev, K.V., 557
Sarita, 889
Sasank, V.V.S., 523
Satish Babu, B.V., 201
Satyanarayana, K.V.V., 467
Saxena, Ashutosh, 513
Seetha, M., 453, 501
Shahare, Vivek, 735
Shaik, Hafeezuddin, 145
Sharma, Anshuman, 791
Sharma, Rishikesh, 781
Sharma, Shilpi, 869
Sharma, Yogesh Kumar, 1
Sharmila, K., 491
Shashi, M., 187
Shiva Prakash, B., 557
Shriram, Revati, 619, 639
Shyam Sundar, K., 349
Sikander, Shaik Raza, 591
Singh, Arjeeta, 579
Singh, Kamalpreet, 735
Sinha, G.R., 213, 219
Sinha, Tilendra Shishir, 829
Sreedhar, K.C., 133
Sreevani, K.S.S., 17, 29
Sridevi, R., 237, 249, 481, 591
Srinivasulu, Senduru, 839

Srujan Raju, K., 109, 213, 219, 877
Subba Rao, Y.V., 663
Suguna, R., 535
Sunil, G., 337
Surekha, A., 523
Suresh Babu, K., 201
Swami Das, M., 311
Swamy, G.V., 601
Swathi, M., 133
Swe, Wit Yee, 219

T
Talware, Rajendra, 227
Thakral, Abha, 781, 791
Tiwari, Vikas, 513

V
Vakkalanka, Sairam, 523
Vasumathi, D., 417, 723
Vasundhara, D.N., 453
Velapure, Akshay, 227
Venkata Ramana, V., 557
Venkateswara Rao, G., 601
Victer Paul, P., 811
Vijaya Lakshmi, D., 311
Vijayalakshmi, M., 325
Vishnuvardhan, B., 163

W
Wandra, Kalpesh, 567

Printed in the United States
By Bookmasters